Green and Sustainable Approaches Using Wastes for the Production of Multifunctional Nanomaterials

Green and Sustainable Approaches Using Wastes for the Production of Multifunctional Nanomaterials

Edited by

Abhishek Kumar Bhardwaj
Department of Environmental Science, Amity School of Life Sciences, Amity University, Gwalior, Madhya Pradesh, India

Arun Lal Srivastav
Chitkara University School of Engineering and Technology, Chitkara University, Solan, Himachal Pradesh, India

Kuldip Dwivedi
Department of Environmental Science, Amity University Madhya Pradesh (AUMP), Gwalior, Madhya Pradesh, India

Mika Sillanpää
Department of Biological and Chemical Engineering, Aarhus University, Aarhus, Denmark

ELSEVIER

Elsevier
Radarweg 29, PO Box 211, 1000 AE Amsterdam, Netherlands
125 London Wall, London EC2Y 5AS, United Kingdom
50 Hampshire Street, 5th Floor, Cambridge, MA 02139, United States

ISBN: 978-0-443-19183-1

For Information on all Elsevier publications
visit our website at https://www.elsevier.com/books-and-journals

Publisher: Candice Janco
Acquisitions Editor: Jessica Mack
Editorial Project Manager: Aleksandra Packowska
Production Project Manager: Bharatwaj Varatharajan
Cover Designer: Miles Hitchen

Typeset by MPS Limited, Chennai, India

Contents

List of contributors

Naveed Qasim Abro National Centre of Excellence in Analytical Chemistry, University of Sindh, Jamshoro, Pakistan

Brahim Achiou Laboratory of Membranes, Materials and Environment, Faculty of Sciences and Technologies of Mohammedia, Hassan II University of Casablanca, Mohammedia, Morocco

Priyanka Adhikari National Institute of Pharmaceutical Education and Research, Guwahati, Assam, India

Saad Alami Younssi Laboratory of Membranes, Materials and Environment, Faculty of Sciences and Technologies of Mohammedia, Hassan II University of Casablanca, Mohammedia, Morocco

Banafsheh Haji Ali School of Environment, College of Engineering, University of Tehran, Tehran, Iran

Zafar Ali Chemistry Department, University of Turbat, Balochistan, Pakistan

Majid Baghdadi School of Environment, College of Engineering, University of Tehran, Tehran, Iran

Anand Kumar Bajpayee Department of Zoology, MLK [PG] College, Balrampur, Uttar Pradesh, India

Aamna Balouch National Centre of Excellence in Analytical Chemistry, University of Sindh, Jamshoro, Pakistan

Chhavi Baran Centre for Environmental Science, IIDS, University of Allahabad, Allahabad, Madhya Pradesh, India

A.P. Bayuseno Centre for Waste Management, Department of Mechanical Engineering, Diponegoro University, Semarang, Indonesia

Abhishek Kumar Bhardwaj Department of Environmental Science, Amity School of Life Sciences, Amity University, Gwalior, Madhya Pradesh, India

Laxmi Kant Bhardwaj Amity Institute of Environmental Toxicology, Safety and Management (AIETSM), Amity University, Noida, Uttar Pradesh, India

Abhi Sarika Bharti Faculty of Science and Technology, Dr. Shakuntala Misra National Rehabilitation University, Lucknow, Uttar Pradesh, India

Rajashree Bhuyan Kalinga Institute of Industrial Technology, Bhubaneswar, Odisha, India

Palakshi Bordoloi CSIR-North East Institute of Science and Technology, Jorhat, Assam, India

Majda Breida Laboratory of Membranes, Materials and Environment, Faculty of Sciences and Technologies of Mohammedia, Hassan II University of Casablanca, Mohammedia, Morocco

Minh Quy Bui TNU-University of Sciences, Thai Nguyen City, Thai Nguyen, Vietnam

Zakariya Chafiq Elidrissi Laboratory of Membranes, Materials and Environment, Faculty of Sciences and Technologies of Mohammedia, Hassan II University of Casablanca, Mohammedia, Morocco

Dinesh Chandola Govind Ballabh Pant National Institute of Himalayan Environment, Almora, Uttarakhand, India

Zhen Hong Chang Department of Chemical and Petroleum Engineering, Faculty of Engineering, Technology and Built Environment, UCSI University, Kuala Lumpur, Malaysia

Shikha Baghel Chauhan Amity Institute of Pharmacy, Amity University, Noida, Uttar Pradesh, India

Kristi Priya Choudhury Department of Chemistry, Shahjalal University of Science and Technology, Sylhet, Bangladesh

Gopala Krishna Darbha Environmental Nanoscience Laboratory, Department of Earth Sciences & Centre for Climate and Environmental Studies, Indian Institute of Science Education and Research Kolkata, Mohanpur, West Bengal, India

Parna Das Microbial Engineering & Algal Biotechnology Laboratory, Department of Biosciences, JIS University Agarpara, Kolkata, India

Ranjita S. Das Department of Chemistry, Visvesvaraya National Institute of Technology, Nagpur, Maharashtra, India

Shrestha Debnath Microbial Engineering & Algal Biotechnology Laboratory, Department of Biosciences, JIS University Agarpara, Kolkata, India

Y. Devika Microbial Engineering & Algal Biotechnology Laboratory, Department of Biosciences, JIS University Agarpara, Kolkata, India; Department of Biotechnology, Maulana Abul Kalam Azad University of Technology, Kolkata, India

Vibhash Dhyani Govind Ballabh Pant National Institute of Himalayan Environment, Almora, Uttarakhand, India

Anupam Dikshit Biological Product Laboratory, Department of Botany, University of Allahabad, Prayagraj, Uttar Pradesh, India

Doha El Machtani Idrissi Laboratory of Membranes, Materials and Environment, Faculty of Sciences and Technologies of Mohammedia, Hassan II University of Casablanca, Mohammedia, Morocco

Ahlam Essate Laboratory of Membranes, Materials and Environment, Faculty of Sciences and Technologies of Mohammedia, Hassan II University of Casablanca, Mohammedia, Morocco

Manish Gaur Centre of Biotechnology, Institute of Inter-Disciplinary Sciences, University of Allahabad, Prayagraj, Uttar Pradesh, India

Sachin Rameshrao Geed CSIR-North East Institute of Science and Technology, Jorhat, Assam, India

Partha Sarathi Ghosal School of Water Resources, Indian Institute of Technology Kharagpur, Kharagpur, West Bengal, India

Dipankar Ghosh Microbial Engineering & Algal Biotechnology Laboratory, Department of Biosciences, JIS University Agarpara, Kolkata, India

Chintya Gunarto Department of Chemical Engineering, Widya Mandala Surabaya Catholic University, Surabaya, East Java, Indonesia; Collaborative Research Center for Zero Waste and Sustainability, Widya Mandala Surabaya Catholic University, Surabaya, East Java, Indonesia

Ena Gupta Department of Family and Community Sciences, University of Allahabad, Prayagraj, Uttar Pradesh, India

R. Ismail Centre for Waste Management, Department of Mechanical Engineering, Diponegoro University, Semarang, Indonesia

Muhammad Saqaf Jagirani National Centre of Excellence in Analytical Chemistry, University of Sindh, Jamshoro, Pakistan; Institute of Green Chemistry and Chemical Technology, School of Chemistry & Chemical Engineering, Jiangsu University, Zhenjiang, P.R. China; School of Materials Science & Engineering, Jiangsu University, Zhenjiang, P.R. China

J. Jamari Centre for Waste Management, Department of Mechanical Engineering, Diponegoro University, Semarang, Indonesia

Saket Jha Department of Surgery, College of Medicine, University of Illinois, Chicago, IL, United States

Tanu Jindal Amity Institute of Environmental Toxicology, Safety and Management (AIETSM), Amity University, Noida, Uttar Pradesh, India

Mian Adnan Kakakhel College of Hydraulic & Environmental Engineering, Three Gorges University, Yichang, Hubei, P.R. China

Muhammad Siddique Kalhoro Institute of Physics, University of Sindh, Jamshoro, Pakistan

Navish Kataria Department of Environmental Sciences, J.C. Bose University of Science and Technology, YMCA, Faridabad, Haryana, India

Deeksha Katyal University School of Environment Management, Guru Gobind Singh Indraprastha University, Dwarka, New Delhi, India

Nitin Khandelwal Environmental Nanoscience Laboratory, Department of Earth Sciences & Centre for Climate and Environmental Studies, Indian Institute of Science Education and Research Kolkata, Mohanpur, West Bengal, India

Aftab Hussain Khuhawar Chemical Engineering and Technology, Ocean University of China, Qingdao, P.R. China

Abdul Hameed Kori National Centre of Excellence in Analytical Chemistry, University of Sindh, Jamshoro, Pakistan

Youness Kouzi Laboratory of Membranes, Materials and Environment, Faculty of Sciences and Technologies of Mohammedia, Hassan II University of Casablanca, Mohammedia, Morocco

Anupama Kumar Department of Chemistry, Visvesvaraya National Institute of Technology, Nagpur, Maharashtra, India

Naresh Kumar Amity Institute of Environmental Sciences (AIES), Amity University, Noida, Uttar Pradesh, India

Sakib Hussain Laghari National Centre of Excellence in Analytical Chemistry, University of Sindh, Jamshoro, Pakistan

Leow Hui Ting Lyly Research Centre for Sustainable Process Technology (CESPRO), Faculty of Engineering and Built Environment, Universiti Kebangsaan Malaysia, Bangi, Selangor, Malaysia

Sarfaraz Ahmed Mahesar National Centre of Excellence in Analytical Chemistry, University of Sindh, Jamshoro, Pakistan

Abhradeep Majumder School of Environmental Science and Engineering, Indian Institute of Technology Kharagpur, Kharagpur, West Bengal, India

Aastha Malik University School of Environment Management, Guru Gobind Singh Indraprastha University, Dwarka, New Delhi, India

Pubali Mandal Department of Civil Engineering, Birla Institute of Technology and Science Pilani, Pilani, Rajasthan, India

Najma Memon National Centre of Excellence in Analytical Chemistry, University of Sindh, Jamshoro, Pakistan

Charu Misra Department of Pharmacy, School of Chemical Sciences and Pharmacy, Central University of Rajasthan, Ajmer, Rajasthan, India

S. Muryanto Centre for Waste Management, Department of Mechanical Engineering, Diponegoro University, Semarang, Indonesia

Nishita Narwal University School of Environment Management, Guru Gobind Singh Indraprastha University, Dwarka, New Delhi, India

Newton Neogi Department of Chemistry, Shahjalal University of Science and Technology, Sylhet, Bangladesh

Uroosa Noor Department of Family and Community Sciences, University of Allahabad, Prayagraj, Uttar Pradesh, India

Rudra Prakash Ojha Department of Zoology, Nehru Gram Bharati (Deemed to be University), Prayagraj, Uttar Pradesh, India

Mohamed Ouammou Laboratory of Membranes, Materials and Environment, Faculty of Sciences and Technologies of Mohammedia, Hassan II University of Casablanca, Mohammedia, Morocco

Rahul Pandey Department of Biotechnology, S. M. M. Town P. G. College, Ballia, Uttar Pradesh, India

Ashok Kumar Pathak Department of Physics, Ewing Christian College, University of Allahabad, Prayagraj, Uttar Pradesh, India

Shalini Purwar Banda University of Agriculture and Technology, Banda, Uttar Pradesh, India

Nathania Puspitasari Department of Chemical Engineering, Widya Mandala Surabaya Catholic University, Surabaya, East Java, Indonesia; Collaborative Research Center for Zero Waste and Sustainability, Widya Mandala Surabaya Catholic University, Surabaya, East Java, Indonesia

Md. Refan Jahan Rakib Department of Fisheries and Marine Science, Faculty of Science, Noakhali Science and Technology University, Noakhali, Bangladesh

Ery Susiany Retnoningtyas Department of Chemical Engineering, Widya Mandala Surabaya Catholic University, Surabaya, East Java, Indonesia

Kulbhushan Samal CSIR- Central Mechanical Engineering Research Institute, Durgapur, West Bengal, India

Soumita Sarkar Microbial Engineering & Algal Biotechnology Laboratory, Department of Biosciences, JIS University Agarpara, Kolkata, India

Shikha Saxena Amity Institute of Pharmacy, Amity University, Noida, Uttar Pradesh, India

Devi Selvaraj PG Department of Chemistry, Cauvery College for Women (Autonomous), Tiruchirappalli, Tamil Nadu, India

Syed Tufail Hussain Sherazi National Centre of Excellence in Analytical Chemistry, University of Sindh, Jamshoro, Pakistan

Mohee Shukla Biological Product Laboratory, Department of Botany, University of Allahabad, Prayagraj, Uttar Pradesh, India

Rohit Shukla Biological Product Laboratory, Department of Botany, University of Allahabad, Prayagraj, Uttar Pradesh, India

Abhimanyu Kumar Singh Department of Physics, Shyama Prasad Mukherjee Govt. Degree College, University of Allahabad, Prayagraj, Uttar Pradesh, India

Ravikant Singh Department of Biotechnology, Swami Vivekanand University, Sagar, Madhya Pradesh, India

Felycia Edi Soetaredjo Department of Chemical Engineering, Widya Mandala Surabaya Catholic University, Surabaya, East Java, Indonesia; Collaborative Research Center for Zero Waste and Sustainability, Widya Mandala Surabaya Catholic University, Surabaya, East Java, Indonesia

Shashi Soni Department of Family and Community Sciences, University of Allahabad, Prayagraj, Uttar Pradesh, India

Md Abdus Subhan Department of Chemistry, Shahjalal University of Science and Technology, Sylhet, Bangladesh; Division of Nephrology, School of Medicine and Dentistry, Medical Center, University of Rochester, NY, United States

Yeit Haan Teow Research Centre for Sustainable Process Technology (CESPRO), Faculty of Engineering and Built Environment, Universiti Kebangsaan Malaysia, Bangi, Selangor, Malaysia; Department of Chemical and Process Engineering, Faculty of Engineering and Built Environment, Universiti Kebangsaan Malaysia, Bangi, Selangor, Malaysia

Pooja Thathola Govind Ballabh Pant National Institute of Himalayan Environment, Almora, Uttarakhand, India

Ajay Kumar Tiwari Department of Physics, Nehru Gram Bharati (Deemed to be University), Prayagraj, Uttar Pradesh, India

Rajat Tokas Amity Institute of Environmental Toxicology, Safety and Management (AIETSM), Amity University, Noida, Uttar Pradesh, India

Swati Rose Toppo Department of Microbiology and Bioinformatics, Atal Bihari Vajpayee Vishwavidyalaya, Bilaspur, Chhattisgarh, India

Shipra Tripathi Department of Physics, Faculty of Science and Technology, Dr. Shakuntala Misra National Rehabilitation University, Lucknow, Uttar Pradesh, India

Thi Thao Truong TNU-University of Sciences, Thai Nguyen City, Thai Nguyen, Vietnam

Kailash Narayan Uttam Saha's Spectroscopy Laboratory, Department of Physics, University of Allahabad, Prayagraj, Uttar Pradesh, India

Tharmaraj Vairaperumal Institute of Clinical Medicine, College of Medicine, National Cheng Kung University, Tainan, Taiwan ROC; Environmental Science and Technology Laboratory, Department of Chemical Engineering, SRM Institute of Science and Technology, Kattankulathur, Tamil Nadu, India

Jitendra Singh Verma CSIR-North East Institute of Science and Technology, Jorhat, Assam, India

Manisha G. Verma Department of Chemistry, Visvesvaraya National Institute of Technology, Nagpur, Maharashtra, India

Manoj Kumar Yadav Department of Civil and Environmental Engineering, Indian Institute of Technology Patna, Patna, Bihar, India

Preface

The unprecedented development and industrialization of the world generates enormous amounts of waste materials which creates severe environmental problems. Waste generated by diverse human activities (both industrial and household) can also cause human health risks. Hence, efficient approaches to waste management are the need of the hour.

In developing countries, garbage generation has increased in tandem due to exponential population growth. Many biodegradable wastes are currently disposed of in malicious ways, such as by burning, unscientific dumping, or direct discharge into the water bodies. Abundant biodegradable waste in the ecosystem can contaminate the environment as they promote the growth of many pathogenic microbial communities in the vicinity of wastes and these microbes can cause a variety of infectious diseases.

However, biodegradable waste or biomass can be used as raw material for nanoparticle production via green synthesis. This is because plant- and animal-related wastes have a treasure of biochemicals for the reduction of metal and nonmetal ions. Natural biological systems are used to produce nanomaterials through green material synthesis processes. NMs recycled from different types of nonbiogenic waste could be a pioneering approach to not only avoid hazardous effects on the environment but also to implement circular economy practices, which are crucial to attaining sustainable growth. Moreover, recycled NMs can be utilized as a safe and revolutionary alternative with outstanding potential for many biomedical applications.

The book discusses the current status and perspectives of biogenic and nonbiogenic waste generation rates throughout the globe along with holistic and sustainable approaches for the production of multifunctional nanomaterials using domestic waste, food waste, agriculture, and fruit wastes. Moreover, the book chapters have been discussed, to examine the characteristics of nonbiogenic synthesized nanomaterials, their applications, and limitations with the biogenic synthesized nanomaterials.

Further, the incorporation of the chapter on the application of nanomaterials, synthesized from agricultural wastes for wastewater treatment, provides an environment-friendly, toxic-free, and sustainable approach. The synthesis of nanoparticles from biowaste offers potential benefits over the chemical-based synthesis approach as it is eco-friendly, cost-effective, and easy. Moreover, the precursor of natural sources can be reused, recycled, and reduced.

The major challenge to scale up the synthesis of nanoparticles for industrial production from biowaste has been attributed to the monodispersity, size, and shape of the NPs, which have also been addressed in the chapters keeping in mind the recent progress and future prospects.

This book will be a pioneering compilation of the different strategies to be adopted for the green synthesis of multifunctional NPs and also for the effective management of the enormous amount of biogenic and nonbiogenic wastes. Thus the present book will be an asset to the students and researchers working on nanomaterial developments in multidisciplinary domains.

Chapter 1

Global status of biogenic and nonbiogenic waste production and their employability in nanomaterial production

Manisha G. Verma[1], Ranjita S. Das[1], Abhishek Kumar Bhardwaj[2] and Anupama Kumar[1]

[1]*Department of Chemistry, Visvesvaraya National Institute of Technology, Nagpur, Maharashtra, India,* [2]*Department of Environmental Science, Amity School of Life Sciences, Amity University, Gwalior, Madhya Pradesh, India*

1.1 Introduction

Waste is a natural by-product resulting from various activities by humans and nature. A by-product is a secondary product that is produced during a chemical reaction, manufacturing process, or production process, but does not have any use in the same process; however, it can be used for another application, for example, molass is a by-product of refining sugarcane, wheat germ is a by-product of wheat milling, etc. Generally, waste is nonusable material, which should be destroyed after its primary use. It can also be converted into another form so one can again use it (Elwan et al., 2014). Different communities define waste in a different way, such as engineers define waste as something that has been released from domestic and commercial sources or something that has no longer value to the owner, ecologist holds the view that nothing is waste in the atmosphere, while industrial ecologists define trash as "a right thing in a wrong place." Whereas anthropologists hold the view that trash is a forceful, illuminating, and honest testimony to a civilization. Waste can represent several things depending on how it is used: for example, those who dump their trash in a landfill may see it as useless, while those who recover objects from trash may see it as ore. Waste production is influenced by population expansion, urbanization, industrialization, economic progress, etc., (Pandey et al., 2016). Waste comes in many different forms, including wastewater, hazardous substances, municipal and commercial waste. Paper and paper board, food waste, yard trimming plastic, metals, wood, glass, rubber, leather, textiles waste, etc., are the components of municipal solid waste (MSW) (Sharma and Jain, 2020). The volume of waste is continuously increasing throughout the globe. Fig. 1.1 shows MSW generation in the year 2017 in various countries (statista.com).

The nature of waste differs greatly from one place to the other place, and it also changes significantly with the season and along the years (Rominiyi and Adaramola, 2020). Due to COVID-19 pandemic, around 3.4 billion mask and face shields were discarded daily throughout the planet (Benson et al., 2021). In 2019, the global plastic manufacturing industry was anticipated to reach around 370 million MT. According to recent predictions, over 25,000 million MT of plastic waste will be generated by 2050 (Vieira et al., 2022). Improper management of it can affect the environment at different scales such as, air composition, and water composition and also results in climate change (Lee et al., 2020). Fig. 1.2 represents composition of MSW produced in North America in the year 2016 (Kearns, 2019).

Open dumping endangers the health of the public living near. (Vergara and Tchobanoglous, 2012).

Due to its complexity and heterogeneity, MSW is difficult to dispose of sustainably, incurs significant financial losses, and adversely impacts the environment and public health such as plastic and electronic consumer products are scattered everywhere, some of which leads to serious environmental pollution and degradation (Farooq et al., 2021). Fig. 1.3 shows the expected e-waste generation up to 2030 (sciencenews.org), as a consequence of which, the esthetics of living area spoils and endangers the health and life of all species. Therefore, effective waste management should be an important factor for the growth of any country and welfare of its citizens.

In historical times, waste was that material which was of no longer use and was discarded, but nowadays, the SWACHCHH BHARAT MISSION has proved that appropriate waste management has become a top national priority

Green and Sustainable Approaches Using Wastes for the Production of Multifunctional Nanomaterials. DOI: https://doi.org/10.1016/B978-0-443-19183-1.00015-5

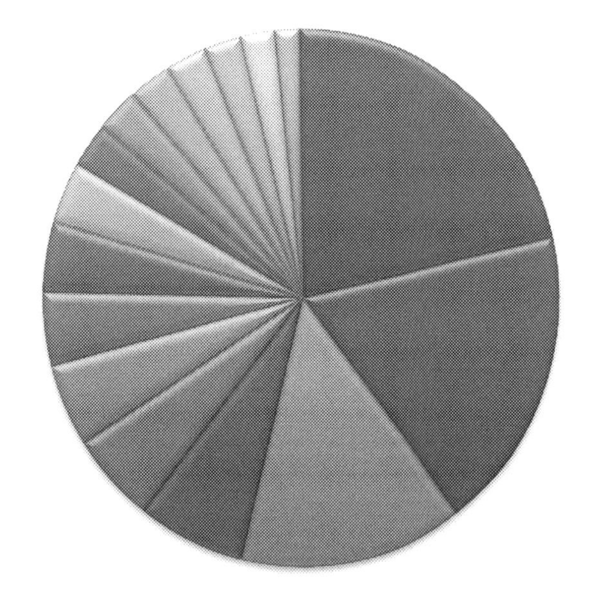

FIGURE 1.1 Municipal solid waste generation on a global scale in 2017.

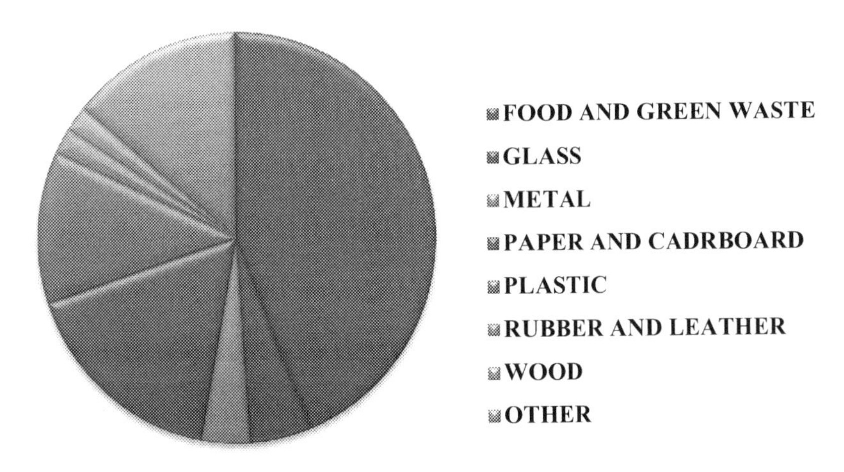

FIGURE 1.2 Composition of municipal solid waste.

in developing countries. Fig. 1.4 represents the estimated waste generation till 2050 (statista.com). The composition of waste is basically important in order to select the strategy and technology for its management. Biogenic waste can be composted into humus, which can then be used directly on agricultural fields, as well as anaerobically digested into biogas, liquid fuel, and humus (Dell'Abate et al., 2000; Braga Nan et al., 2020), while nonrecyclable dry debris with the high calorie content can be shredded into refuse-derived fuel (RDF) and is used for waste to energy facilities. All these processes decrease the amount of waste and also produce power in the form of green fuel (Ahring, 2003). The transformation of trash into fuels, value-added products, and energy solves two problems simultaneously, that is, reduces the amount of nonrecyclable waste and also produces bioproducts, so the increasing demands can be fulfilled in a sustainable way (Das et al., 2022, 2023).

MSW is often divided between biodegradable and nonbiodegradable waste. Biodegradable wastes are termed as biogenic and include food, agricultural, garden wastes, etc. However, nonbiodegradable wastes are termed as nonbiogenic (takes many years to degrade) and include plastic, glass, metal, ceramic, as well as the construction and demolition waste.

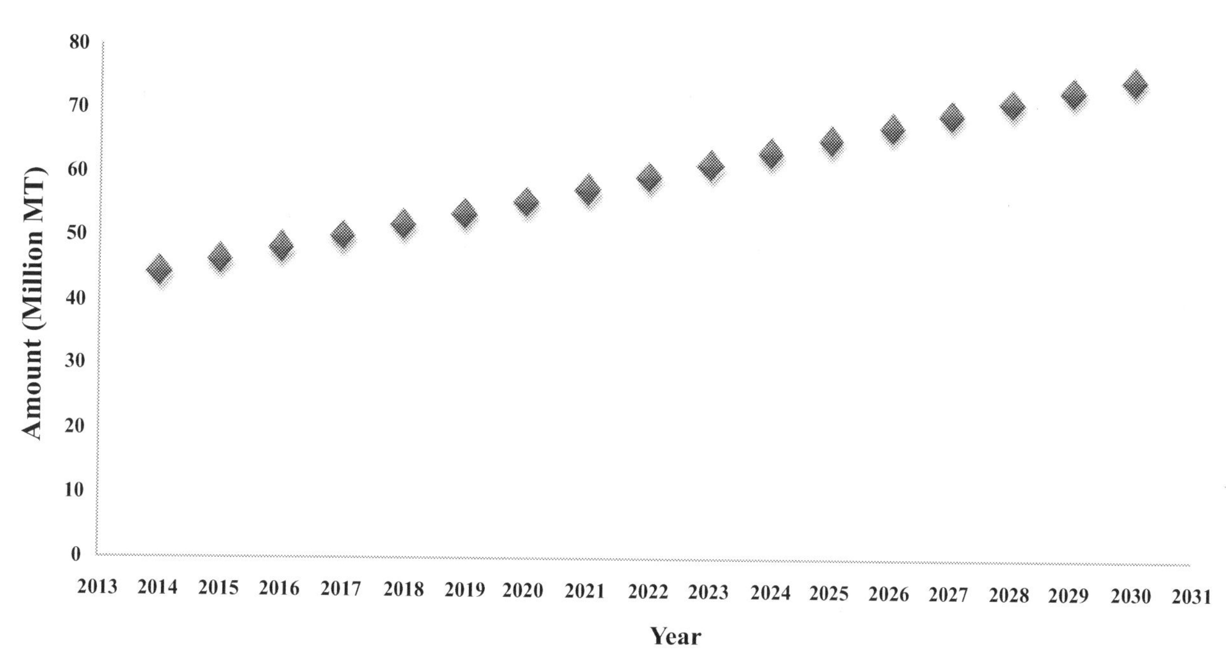

FIGURE 1.3 Global E-waste generation forecasted till 2030 (Million metric tons).

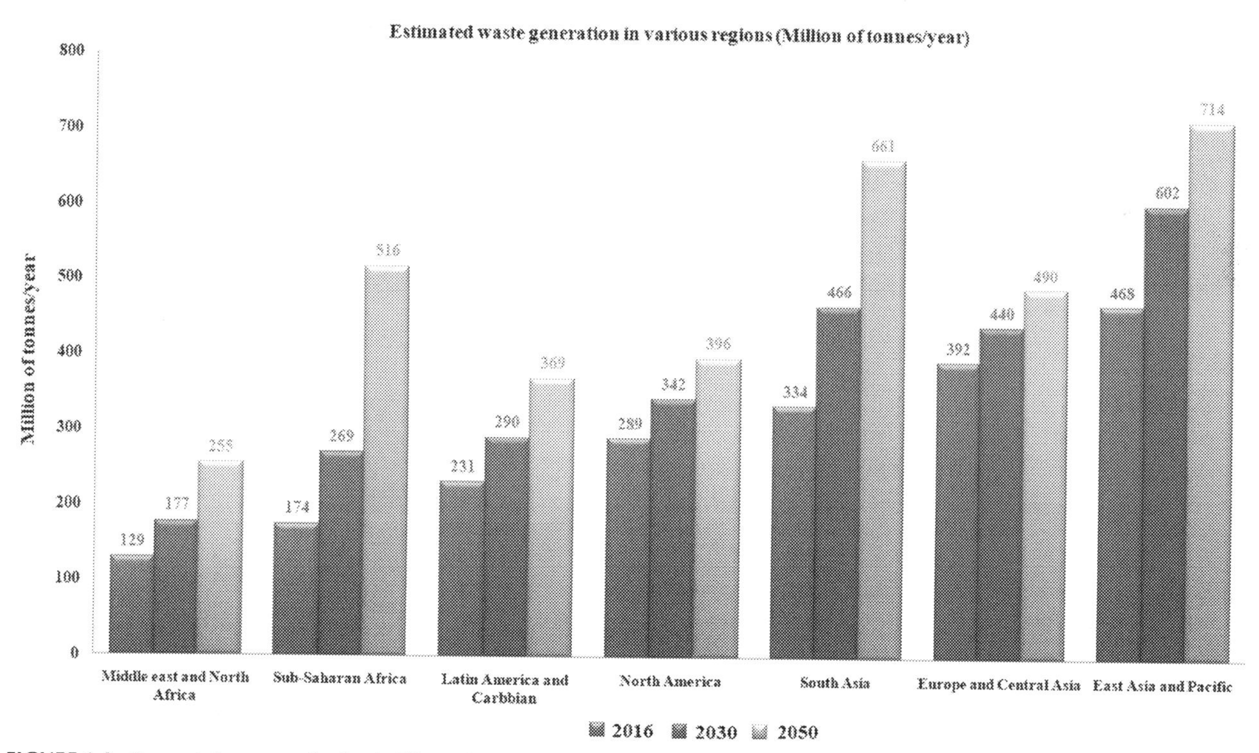

FIGURE 1.4 Forecasted waste production (millions of tonnes annually).

1.2 Biogenic waste

The category of trash that is biodegradable such as grass, clippings leaves, food waste, cardboard, leather products, wood paper, and leather products is termed as biogenic waste. These includes the waste resulting from living organisms, plants, and animals when their life cycle ends. It is usually considered that biogenic waste does not harm the environment. However, a close examination of our surrounding reveals that if trash is not disposed properly, it can result into health hazard.

1.2.1 Sources

The origins of biogenic waste are agricultural waste resulted due to harvesting residue, solid manure, rotten products, horticulture waste resulting from park and garden, plant and animal waste, peels and seeds of various fruits and vegetables. Food processing waste includes fruit peels, pulps, spent grain, pressed seeds, pomace, waste bread (Gaharwar et al., 2023; Weiland, 2000). Paper industry generates pulp as by-product. Pharmaceutical industrial waste includes bio-waste market waste, etc.

1.2.2 Impact on environment and human health

Waste production has no beneficial effect on the environment. Treatment and disposal of waste can have both favorable and adverse impacts on the ecosystem (Skillicorn et al., 2013). Rotting waste produces chemicals such as ammonia as well as greenhouse gases (GHGs) such as methane and carbon dioxide (Lacy and Rutqvist, 2015; Lamb et al., 2021). When there is an excessive amount of organic waste in water, oxygen levels of water drop, which impacts the marine life. Globally, every solid waste sector, even agricultural waste also makes a significant contribution to GHG emissions (Wood and Cowie, 2004). Large quantity of GHG emission is responsible for ozone depletion and global warming. Not only when waste is handled (such as during transport), but also when it is discarded, GHGs are emitted (Bader and Bleischwitz, 2009). Methane and carbon dioxide both GHGs promote global warming and climate change by their presence (Bozak, 2021; Meinshausen et al., 2009; Gavurova et al., 2021). Methane is a highly potent GHG with a global warming potential 25 times that of CO_2 when measured over a 100-year time horizon (Skillicorn et al., 2013). The intergovernmental panel on climate change estimated that waste management emits less than 5% of global GHG emission (and 9% of global methane emissions), although this estimate is uncertain and unpredictable because waste management can act as either a net source or a sink of GHG. Landfills are the main producers of GHGs, as methane is produced during the decomposition of organic waste (Olivier and Peters, 2020; Ahmadi et al., 2020; Hénault-Ethier et al., 2017). Decentralized composting and biomethanation can help to minimize the expenses of trash transportation over long distances, as well as the leachate and GHG emission caused by dumping mixed waste at the dumpsites, once biodegradable material has been sorted. Methane emissions from dump are hardly monitored in practice, instead being assumed for reporting purposes. In order to estimate methane emission from landfills, the first-order decay model is widely used. The rate of methane production is predicted by this model. Thus, minimization of GHG emissions has become an increasingly important concern of waste management initiatives.

1.3 Nonbiogenic waste

Nonbiogenic waste is defined as compounds that do not decompose or decay naturally, that is, plastic, rubber, rexene, pesticide, heavy metals, etc., that cannot be dissolved by natural forces or biological processes or eliminated by bacteria and abiotic elements. These materials are generally not produced via natural processes. Commonly, such material synthesizes via chemical reaction or industrial process or synthetic process.

1.3.1 Sources

This includes the majority of medical waste (biomedical waste), electric equipment construction and demolition debris, etc. Inorganic trash is mostly nonbiodegradable, which means it may remain longer in the environment (Bhardwaj et al. 2022). Nonbiogenic artifacts discovered during that time are crucial to our understanding of past civilizations.

Used cotton, dry fabric products, earphones, sanitary pads, syringes, shaving blades, fingernails, bandages, and other objects are examples of biomedical waste. Animal bones and body parts, as well as surgical waste, are also visible. Computer displays, motherboards, cathode ray tubes, televisions, phone chargers, CDs, headphones, LCD/plasma televisions, air conditioners, and refrigerators are a few examples of electrical equipment. These items are referred to as "e-waste" because they are no longer functional. Computer equipment makes up a large portion of e-waste.

Construction and demolition waste includes unused, damaged, or unwanted construction materials. Construction waste may contain lead, asbestos, paint, and other dangerous materials. According to estimates, around 10%–15% of the materials used in building construction are discarded (Lacy and Rutqvist, 2015).

TABLE 1.1 Global emission of greenhouses gases by various sectors in 2016.

Source	Quantity (%)
Construction & manufacture	24.3
Industrial & commercial	16.8
Transportation	16.5
Agriculture	11.9
Residential	11
Land use and forestry	6.6
Waste	3.2
Other	9.5

1.3.1.1 *Impact on environment and human health*

Organic matter decomposition in dumpsites leads to global warming through GHG emissions (Kennedy et al., 2009). Table 1.1 shows Global Emission of Greenhouses Gases by various sectors in 2016 (Peel and Schmidt, 2020). Chlorinated plastics are commonly used for the production of gloves, blood bags, and other medical applications. These chlorinated materials after their use are burnt for safe disposal. However, their burning releases harmful gases such as dioxins and furans. Vinyl chloride has been identified by International Agency for Research on Cancer as human carcinogen. Due to all these reasons, a number of nations, notably those who are in Europe, have banned soft PVC with phthalates meant for human interaction (for instance, toys).

1.3.1.2 *Applications of waste*

Animal and plant waste is frequently the source of waste that can degrade through the action of living creatures. A large portion of MSW consists of biogenic waste, such as agricultural waste and food waste. When aerobic decomposition is carried out, waste is broken down into simple compound with the liberation of CO_2 and water. While during anaerobic decomposition, methane and liquid slurry are obtained (Zsigraiová et al., 2009; Angelidaki et al., 2003). Methane gas produced in landfills during biomass decomposition are utilized to generate energy (Suberu et al., 2013). In contrast, it is possible to recover energy by burning nonbiogenic waste that is not toxic but has calorific value. Waste is crushed, dried, and processed into pellets or briquettes to maximize the energy output of RDF. Instead of coal, they can be used to create heat in a range of industries.

1.3.1.2.1 Composting

The process of composting involves transformation of organic waste into vital ingredients of humus-rich soil. Vegetable peel, dairy, cereals, bread, unbleached paper napkins, eggshells, meats, etc., can be composted. Microorganisms break down the biogenic waste material, which also results in a volume decrease of up to 50%. Compost or humus is useful as a mulch or soil enhancer. Composting is a useful way to clean and use waste and sewage sludge (Vehlow et al., 2007). The use of compost is expected to increase as landfill and solid-waste incineration choices are hampered by more strict environmental rules and geographical restrictions (Smidt et al., 2011). The entire process of composting involves pummeling, categorizing, size reduction, and digestion.

1.3.1.2.2 Concrete Production

Due to rising environmental concerns worldwide and its expensive cost, Portland cement is no longer the only binding material used in concrete. However, a substance that can be used as comparatively priced, ecologically friendly binders for concrete construction can be produced when enough regular Portland cement and ashes from agricultural or other biogenic wastes are combined. Experiments have shown that the properties of peanut shell, palm leaf, and rice husk ashes are sufficient for pozzolans. Additional experimental studies have concentrated on the fundamental properties of fresh and hardened concretes made of ordinary Portland cement (OPC) with coconut husk ash (CHA) and also with rice husk ash (RHA) (Arum et al., 2013). To make new concrete more workable than OPC alone, RHA or CHA can be used along with OPC. OPC with CHA and OPC with RHA concrete have low porosity than OPC alone. CHA and RHA can be used along with Portland cement for concrete production, as their strength activity index is better than 100% for up to 15% replacement, indicating that they are good pozzolans.

1.3.1.3 Utilization of plastic waste

Plastics are not terrible in and of themselves, but the rubbish they generate is not biodegradable. According to a recent estimate by the Indian government, just 20% of the waste plastic materials were recycled globally in the year 2019, 25% was burned, and 55% of global plastic trash was thrown in landfill, polluting streams, and groundwater resources in 2019 (Ahluwalia and Patel, 2018). According to the Central Pollution Control Board, plastic waste amounts to 8%–9% of total MSW in India, with 60% of that being recycled in the formal sector. According to a survey by the NCL Pune, PET recycled in India is predicted to reach 90%, compared with 72% in Japan, 48% in Europe, and 31% in the United States in the year 2017 (Ahluwalia and Patel, 2018). Despite India's plastic recycling rate being substantially better than the global average of around 15%, due to the mixing of various waste streams, there is still a substantial amount of unrecyclable plastic trash that is either landfilled or causes groundwater resources to become contaminated.

1.3.1.3.1 Reuse of polyethene

In order to prevent plastic from entering our ecosystem, reuse of polyethene is another method, therefore lowering the danger. The polyloom, a plastic weaving handloom invented by the Center for environment education, promotes the recycling and reuse of used plastic bags. The polyloom is used to make plastic woven fabric in a variety of designs, which is then customized and turned into complete products such as bags, pouches, and bottle holders.

1.3.1.3.2 Road construction by plastic waste

Plastic road construction is common in India especially in states such as Tamil Nadu and Himachal Pradesh. It is a novel idea to build motorways using shredded thin plastics. These thin polymers can double the longevity and quality of tar roads when used in hot-mix facilities to generate ready asphalt mixtures that are disseminated and compacted for road construction. Finely shredded thin films of plastic are put on the heated stones after they have been heat softened to form a baked-on polymer coating. Bitumen adheres to the covered stones effectively making the roads more durable, especially during rain storms. Plastic was used on 1400 km rural tar roads in Tamil Nadu in 2003–04, resulting in roughly 1 tonne of waste every km of single-lane road. The Central Pollution Control Board has released the results of comparative test as well as road-building guidelines. The Central Road Research Institute mandated plastic roads on all national highway up to 50 km from cities with population of over half a million people in 2015. Various steps involved in plastic road construction are shown in Fig. 1.5 (Santos et al., 2021).

1.3.1.4 Production of valuable nanoparticles using different wastes

Pd nanoparticles (Pd NPs) can be obtained by a green and environment-friendly method from waste papaya peel extract without the use of reducing chemicals, high-temperature calcinations, or reduction processes, in a one-pot green and sustainable technique (Dewan et al., 2018). Biomolecule found in discarded papaya peel extract converts Pd (II) to nano-size Pd(O). The innovative palladium nanoparticles work as a catalyst to provide a simple and effective technique for direct Suzuki, Miyura, and Sonogashira coupling with high yield under mild reaction conditions. Papaya peels contain flavin mononucleotide and flavin adenine dinucleotide, which may operate as a $Pd(OAc)_2$ reductant.

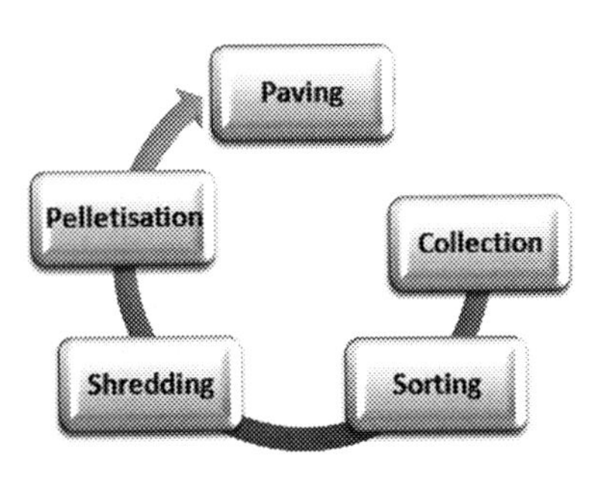

FIGURE 1.5 Process of road construction by plastic.

Silver nanoparticles (AgNPs) show a broad spectrum of antimicrobial and antifungal properties. This NP has been developed by a greener economic and ecofriendly approach using oyster mushroom Pleurotussajor-caju extract (Naraian and Abhishek, 2020).

Numerous metal nanoparticles have been developed by employing a green and efficient route of synthesis. These NPs have unique properties and excellent applications (Bhardwaj and Naraian, 2021). Pariel et al. have performed the Cyanobacteria-assisted intracellular synthesis of gold nanoparticle. Husain et al. have synthesized silver NPs with average size 30 nm using Nostoc muscorum NCCU 442, which shows antibacterial and antioxidant activity against Staphylococcus aureus MTCC 902.

Copper nanoparticles possess a broad range of multiple applications such as antimicrobial activity, as catalyst in organic synthesis, as nanofluid and lubricants, etc. (Bhardwaj et al., 2019). A rod-shaped cuprous oxide nanoparticle has been developed by Bhardwaj et al. using laser ablation technique. This NP shows bactericidal and inhibitory action against Staphylococcus aureus (Gram-positive bacteria) with minimum concentration of 140 and 120 ug/L.

In biogenic waste, coenzymes, enzymes, proteins, terpenoids, and other important components are present. These compounds are recognized to have an important role in the production of many 2D materials; hence, they have an impact on green synthesis methods for 2D materials (Patil and Chandrasekaran, 2020). The various substances from which graphene can be synthesized are as follows:

An interweaved few-layer vertical graphene has been developed in a single-step synthesis using inductively coupled plasma-enhanced chemical vapor deposition technique, from waste lard oil. Lard is a swine fat made from the animals' fatty portions. Waste lard oil is an abundant, nontoxic, environmentally benign, and low-cost carbon source (Singh et al., 2021).

Mosquito repellents, widely used throughout the world, are available in a range of formulations (plug-in vaporizer, spare, cream, oil, coil, and so on). The repelling vaporizer comprises an empty plastic bottle (refill) and a rod. Recycling of these waste bottles and rods is mandatory in terms of waste management. A few-layered graphene has been prepared from these empty mosquito repellent refills. These graphite rods were employed as a source of graphene during the manufacturing process (Singh et al., 2021).

For high-valued items, dead camphor leaves are used as a starting point. In an N_2 atmosphere, graphene was made from camphor leaves using a single-step pyrolysis technique.

The tea tree (*Melaleuca alternifolia*) is a Native Australian plant that is widely distributed over Australia, Europe, and North America. This is often used to make essential oil for traditional medicine and as an active element in a range of products. The tea tree leaves are used to extract oil using a steam distillation process. A high-quality graphene can be obtained by using tea tree extract oil and a radiofrequency capacitively connected plasma-enhanced CVD method at around 800°C. The graphene that resulted has a three-dimensional structure.

Rise husk is the rice grain's outer coating, which is usually separated as waste during large-scale rice milling. The majority of rice husk is made up of cellulose 50%, lignin (25–30)%, and silica (15–20)% (Ikram et al., 2020). Graphene can be prepared from rice husk (Singh et al., 2021). The abundance of carbon-rich minerals on the planet has sparked the scientific community's curiosity. Wheat straw can also be turned into a high-quality graphene sheet using a combined hydrothermal and graphitization technique. This is a low cost, environmentally friendly, catalyst-free method, providing a way for the production of green and energy-efficient carbon nanomaterials from renewable biomass waste (Chen et al., 2016).

An impermeable few-layered graphene can be developed, with a cost-effective and environment-friendly method, from plastic solid waste, in a two-step pyrolysis process, first at 400°C and then at 750°C (Pandey et al., 2019). Waste tyres are good source of carbon and sulfur; they are used to create a heteroatom-doped carbon nanosheet (Veksha et al., 2020).

GO and rGO can also be produced from a variety of natural wastes, as garbage, animal waste, and industrial trash such as newspaper and soot powders produced in exhaust of diesel vehicles (Akhavan et al., 2014).

Graphitic Carbon Nitride can be prepared from aloe vera gel using self-combustion process. H-BN is also known as white graphene. It has numerous good and unique features, including a large band gap of 5.97 eV, which make it a top type of insulator. H-BN nanosheets can be produced from bulk H-BN using extracts of 17 plant leaves/roots/stems cells/foliage/needles/cocklebur/seaweed, followed by ultrasound, H-BN nanosheet can be used as a dye remover, antioxidant, and nanofiller. Experimentally it has proved that biogenically synthesized H-BN nanosheet has better adsorption performance than the pristine H-BN and H-BN synthesized in isopropanol, and it can also be used in the removal of both cationic and anionic dyes. Biogenically produced H-BN nanosheets showed scavenging activity, whereas pristine H-BN and materials did not.

TABLE 1.2 Number of incineration plants in various countries.

Country	No. of WtE plants	WtE share in waste management (%)
Austria	11	30
France	123	34
Germany	65	38
Netherlands	13	39
Sweden	24	49
Japan	310	58
China	77	16
USA	86	7

1.3.1.5 Transformation of waste to energy

In order to provide a waste-to-energy (WtE) technology selection criterion, numerous studies have been done. The ability to convert trash to energy efficiently is mostly determined on the garbage quality (Rahman et al., 2017), so depending upon the type of trash, various methods of WtE transformation are as follows:

1.3.1.5.1 Incineration

Municipal solid waste is widely used to generate power in waste to energy plants (Ryu and Shin, 2013). Table 1.2 shows total number of incineration plants in various countries till the year 2013. In these plants, trash is burned to generate steam, which is then used to generate electricity. The WtE conversion also lowers the trash volume (Lariviere, 2011). WtE plants are used in several countries, notably Japan and portions of Europe, to create electricity from MSW (where the open space for landfills in very less). The mass-burn system is a widely used WtE technology in the USA. In the United States, WtE plants burned around 12% of the 292 million tonnes of MSW produced in the year 2018. Sixty-five power plants in the United States produce about 13.5 billion kilowatt hours of energy in 2020, using around 25 million tonnes of MSW.

1.3.1.6 Biofuel generation

1.3.1.6.1 Biomethanation

Anaerobic decomposition or biomethanation is a more advanced method for biochemical conversion of organic waste (Dabe et al., 2019; Chen et al., 2008). Microorganisms break down organic waste in an oxygen-free environment, releasing methane-containing biogas (O'Callaghan, 2016; Parkin and Owen, 1986). Biomethanation, like composting, can be carried out on a local scale or in huge centralized operations. LPG or CNG can be substituted for the gas. Furthermore, this can be packed as compressed biogas, which can be transformed into energy using generators at 30% conversion efficiency. India's largest biomethanation plant built in Solapur (Maharashtra) is capable of processing 400 tons of trash per day. In tatol, 60–80 tonnes per day of organic manure and 3 MW of power are generated from this plant. The Koyambedu wholesale vegetable market in Chennai, which has a 30-tonne-per-day biomethanation unit built, is a prime example of waste digestion and energy conversion at the source of waste (Ahluwalia and Patel, 2018). In cities all throughout India, such as Nashik, Pune, Bangalore, Coimbatore, Delhi, Vadodara, and Mumbai, small-scale biomethanation units have been installed. Streetlights are operated by biofuel produced from these units. With 121 TDP of biogenic waste digested by 25 plants of 5 TDP capacity and 2 with a capacity of three TPD, Pune tops the list.

1.3.1.6.2 Syngas production

Syngas is a combination of Hydrogen, Methane, Carbon dioxide, Hydrocarbons, and Carbon monoxide, with a few diluents and impurities, such as hydrogen sulfide and nitrogen (Foster et al., 2021). By treating dry waste with a high caloric content, pyrolysis, gasification, and plasma gasification can produce valuable by-products as syngas, ethanol, and biochar (Pham et al., 2015). Syngas has been identified as a viable fuel source for manufacturing commercial chemicals such as plastics, paints, textiles, and construction ingredients (Ramos et al., 2018). It can also supply energy for the combustion-based generation of electricity. The inorganic part of syngas can be used well by autotrophic acetogenic

TABLE 1.3 Average per capita CO_2 gas mission (tons) from various countries in the year 2020 (Emission Gap Report 2022-United Nations Environment Programme).

Country	Amount (tons)
United States	14
Russian Federation	13
China	9.7
Brazil & Indonesia	7.5
European Union	7.2
India	2.4
Other	2.3

bacteria to create a variety of commodity chemicals by using the Ljungdahl and wood pathways. Gasification is the process of turning a sizable amount of carbon present in any material into a gaseous form when heated in the presence of oxygen under controlled conditions. Synthesis gas or syngas is the result of all these processes and has a variety of uses, including energy generation. Syngas has the advantages of being able to be stored and transferred for later use.

1.3.1.6.3 Biotransformation of carbon dioxide

The concentration of CO_2 in air is rising at an alarming rate. Table 1.3 shows Average Per Capita CO_2 Gas Emissions from various countries in the year 2020 (Barzak, 2011). CO_2 is now getting popular as a possible bioeconomy feedstock. The scientific community is mainly concentrating on biological mechanism for CO_2 adsorption and conversion. Various compounds can be produced as a result of CO_2 adsorption (Thakur et al., 2018). Microalgae have the ability to absorb carbon dioxide and convert it to Fuels, Biodiesel, Commodity Chemicals, Polymers, Biomass, high-value specialties, etc., (Kircher, 2021). Algae farming can concentrate on biopolymers and high-value specialty molecules rather than low-value commodity chemicals.

A variety of commercial and bioelectricity can be produced using the bioelectrochemical system (BES) to valorize CO_2, and also, artificial photosynthesis may create biochemicals using CO_2 in daylight.

1.3.1.7 Agro waste as the tool in development of biocatalyst and its application

1.3.1.7.1 Agro waste as a source of Phytochemicals

Agricultural waste is very rich in cellulose and has high carbon content with multifunctional groups. It offers a variety of uses because of its high strength, eco-friendliness, affordability, availability, and reusability (Ghosh et al., 2017). Nowadays, agro waste is used for biodiesel production, environmental remediation, curative medicine, and catalyst creation (Protection and April, 2015; Capanoglu et al., 2021). New pharmacological leads for chief phytochemicals with a variety of therapeutic effects have been discovered from a variety of foods and agricultural waste (Gaharwar et al., 2022). Enzyme immobilization is a cost-cutting method that involves repurposing one enzyme in many processes. This makes creating a continuous process with minimal enzyme waste much easier. After the proper processing, agricultural waste can be used for immobilizing enzymes. Mango leaves have been used for thousands of years for their antioxidant, anticancer, and antiviral properties (Rao and Rathod, 2019).

Andrographolide, a phytoconstituent found in *Andrographis paniculata's* roots, seeds, and most significantly in leaves, is widely used around the world to treat the common cold, diarrhea, fever from various viral causes, jaundice, liver and cardiovascular tonic, and as an antioxidant. Among other pharmacological effects, Andrographolide is thought to have antiinflammatory, antiallergic, and antidiabetic hepatoprotective and anti-HIV properties.

1.3.1.7.2 Utilizing agro waste as a wastewater treatment tool

Heavy metals are one of the most common contaminants in wastewater from pesticides and chemicals that are often found in environment. These heavy metals can be separated from water with the help of agro waste. Peanut husk is one of the finest precursors for copper removal from wastewater and is inexpensive and readily available (Rao and Rathod, 2019).

1.3.1.7.3 Biocatalyst development and application using agro waste

Coconut shells and peanuts fruit shell produce a lot of agro waste. Both of these fruits have a lot of shells left behind. Shells from nuts and coconuts share several chemical properties. A good supply of very porous activated carbon is claimed to be found in coconut shells. Due to their larger surface area per unit volume, very porous carbon materials have been employed in heterogeneous catalysis. This can also be used to immobilize enzymes so that other enzymatic processes can be carried out.

1.3.1.7.4 Bioenergy

Pollution occurs when several times used vegetable oil for frying is dumped. As a result, it is preferable to use the extra oil to generate biodiesel. Biodiesel is a green, renewable, nontoxic fuel that can be used effectively, making it a viable alternative to traditional fuels. Long-chain mono alkyl esters of lipids generated from renewable sources such as virgin oil, cooking oil, algae, and animal fat make up biodiesel (Capanoglu et al., 2021).

1.3.1.8 Animal waste to value

1.3.1.8.1 Hydroxyapatite production from eggshell

Avian eggshell is a poultry industry by-product that has got a lot of interest and could be used in medical and dental treatment. Eggshell is a sort of agricultural waste that is normally deemed useless and abandoned, hence causes pollution. Hydroxyapatite is a good substance for bone and tissue regeneration. Eggshell is used to make hydroxyapatite, which causes the lowest waste pollution and allows the waste to be turned into a valuable product (Abdulrahman et al., 2014). Converting eggshell to hydroxyapatite and nanohydroxyapatite is an ecologically benign technique. The hydroxyapatite and nanohydroxyapatite derived from eggshells have a good probability of lowering the price of replacing or repairing a bone while having no adverse effects on the environment.

1.3.1.9 Utilization in bioeconomy

As per the Organization for Economic Co-operation and Development, green economy needs three elements: (1) renewable biomass, (2) biotechnological knowledge, and (3) integration of application (Venkata Mohan et al., 2018). Building a waste-based bioeconomy improves wealth development, employment opportunities environmental protection, waste management, and many more through the manufacture of a variety of bio-based items from biogenic waste.

The bioeconomy's major contributors include healthcare, biofuels, and biocommodity chemicals. The production of biofuels and materials from waste adds a new and long-term dimension by addressing the important concerns of pollution and energy in a comprehensive way (Schneider and Ragossnig, 2013). Biofuels generation is currently a key priority in the transition to a bio-based economy (OECD, 2013), and it significantly reduces GHG emissions when compared with fossil fuels. The SAHYOG project produced a strategic research agenda for bioeconomy in India and the EU. However, more research is needed to employ biowaste as a feedstock.

1.4 National and International effort in waste management

- "It is the duty of every citizen to conserve and enhance the natural environment, including forests, lakes, rivers, and wildlife," states Article 51 A.g of the Indian Constitution. Punishment is also meted out for noncompliance.
- The solid waste management rules 2016 establish a practical framework for dealing with India's MSW management concern. It represents the substantial advancement against India's original set of guidelines, the Municipal Solid Waste Management Laws of 2000.
- The Indian government's strategic direction and funding of national initiatives such as JNNURM, AMRUT, Smart City, and the Swachh Bharat Mission have also contributed to an atmosphere where the problem is receiving more, but insufficient attention.
- According to the Indian Government's 2016 Biomedical Waste Management Rules, which describe the duties of a health facility, the phase-out of chlorine-containing bags, gloves, and other products must be finished within 2 years of announcement, that is, March 28th, 2016.
- The United Nation Environment Programme (UNEP) is the United Nation's system principal environmental authority. UNEP is already working on a variety of projects and initiatives to help its member countries improve their waste management. Sustainable consumption and production are also part of this program's activities, as well as integrated solid waste management (ISWM), which is based on the 3R idea (reduce, reuse, and recycle), employing

waste plastic and biomass to handle hazardous garbage, obtain resources, and generate usable energy, managing e-waste and recycling trash for agriculture. By examining the climatological implications of all waste operations, from waste prevention to waste disposal, ISWM seeks to build the foundation for an extensive global plan. UNEP has also been supporting the development of a "Global Platform for Waste Management" (GPWM) to make it easier to coordinate international assistance for waste management. GPWM systems need to be explicitly linked to a global waste and climate change plan.

- Since 1990, numerous cities, including Pune, Rajkot, Pammal, and Mysore, have attempted to organize trash pickers. Additionally, they have collaborated with NGOs to improve the working conditions for sorting dry waste for recyclables, assigned them specific trash to collect, and connected their waste management efforts with the official solid waste management system.
- Pune was one of the first cities to include its citizens in source-side garbage collection. In 1990, SNDT Women's University organized waste pickers as a labor union, then as a cooperative association known as SWaCH, in order to collaborate with the Pune Municipal Corporation on waste collection and recycling.
- SWaCH garbage collectors are provided with uniforms, tools, identification cards, health insurance, and sheds for separating recyclable waste from nonbiodegradable garbage by the Pune Municipal Corporation. Waste collectors are permitted to sell recyclables and keep the revenues in addition to the pay they receive from their work at home. In 2017, eight SWaCH members held 53% of the real estate in Pune (Lacy and Rutqvist, 2015). They then make sure that the collected wet waste is periodically composted or sent to a nearby biomethanation facility, supporting the company's efforts to convert the collected wet waste into manure, biogas, and ultimately electricity.
- According to the Central Pollution Control Board (2013), plastic garbage makes about 8%–9% of MSW in India, of which more than 60% is covertly recycled. PET recycling in India is anticipated to reach 90%, compared with 72% in Japan, 48% in Europe, and 31% in the United States, according to a study by the National Chemical Laboratory in Pune (2017). India recycles plastic at a rate that is much greater than the average rate for the world, which is only about 15%.
- In Japan, according to the automobile recycling act of 2002, anyone purchasing an automobile must pay for recycling fee to the Japan automobile recycling promotion center's fund management Corpus and return the car to the dealer when it is no longer in use.

1.4.1 Business opportunity

By 2025, India's bioeconomy is predicted to grow significantly, generating several opportunities for both the public and private sectors (Venkata Mohan et al., 2018). Biogas production, energy generation, biochemical valorization, and other waste-derived bio-based business prospects all have market potential in India. A strategy to create electricity from renewable sources has been put in place. About 100 MT of bioethanol might be made using half of the total agricultural waste generated, with the possibility of more. In a project just launched in Kashipur, Uttarakhand, maize stover, cotton silk, wheat straw, wood chips, bagasse, bamboo, rice straw, and other agricultural leftovers are being transformed to bioethanol. A successful bioeconomy relies on the development of long-term business models that are intertwined with government policies and subsidies early on in the technology transfer process. Municipalities are the most important stakeholders in the solid waste management, and they must consider public private partnership as a source of revenue.

Municipally owned MSW businesses in Sweden combine three distinct types of operations: (1) public service operations involving the collection of solid waste from residences, places of business, and industries; (2) processing operations involving the transformation of waste; and (3) marketing operations assisting the re-entry of goods and recycled materials into the economy. The Husk Power System business model is based on creating mini-grids powered by bioenergy for rural communities that are not connected to the main grid but are within 10 km of agricultural waste.

India's waste-related industries are now dispersed throughout the entire country. In order to undertake a thorough examination, businesspeople are aggressively lobbying the government to industrialize and franchise the "fragmented" waste management sector. The majority of waste management companies are still start-ups because so few large organizations and international corporations have moved here from other nations. Franchise India is investigating waste-related economic potential in conjuction with SingEx, a Singaporean Investment organization. There would be potential for enterprises to enter the waste management area if it is given industrial status (REF). Government aid, such as low-interest loans or land allotments, will be more beneficial to this industry. India's energy strategy promotes renewable energy expansion by providing federal and state-level incentives. Business opportunities may arise as a result of "Swachh Bharat" initiative.

1.5 Disposal

Once the principles of waste reduction, recycling, and resource recovery have been implemented, the residual waste must be safely deposited in order to isolate nonrecoverable and other such waste from negatively impacting the environment (Treat, 1978). Unrecoverable, stabilized trash should be buried in a sanitary landfill, which has a protected bottom and side-liners (Vergara and Tchobanoglous, 2012).

1.6 Future perspective

At present, the world is moving toward sustainable bioeconomy. Biogenic waste generated in enormous quantity can be consider as a potential feedstock for making bio-based economy (Ward et al., 2008; Ferdous et al., 2021). Petroleum feedstock can be replaced with waste by pursuing proper technique. A step in this direction opens scope, opportunities for business and economic realm. Conversion of waste into value-added products such as fuel, chemicals, energy leads to environment protection and bioeconomy (Holm-Nielsen et al., 2009).

The use of biogenic and nonbiogenic waste in the synthesis of nanomaterials has been proved to be a novel and intriguing area of nanotechnology with the potential to revolutionize the future nanoscience and technology, but this field of study is still in its infancy; thus more opportunities for green synthesis to create nanomaterials from biogenic and nonbiogenic waste are needed. Finally, this thorough study is expected to stimulate scientists in this relatively new and uncharted field.

1.7 Concluding remarks

Waste-derived bioeconomy offers tremendous scope and opportunity, industrial-scale innovation, provided the appropriate expertise is built by integrating applicable science and technology with large financial incentives. Adopting a waste-derived bioeconomy can better address the majority of the sustainable development goals. The government, industry, and scientific community all have a vested interest in realizing this topic's enormous potential.

Waste should be pushed across the world as a high-value resource and future feedstock with a long-term business plan and models. It will be beneficial to develop long-term strategy and goals for a more uniform garbage collection and management system among local governments. Providing appropriate funding for talent and technology development, as well as demonstration, should be a top priority. Support for the waste derived bioeconomy will be beneficial if it incorporates a long-term incentive structure, regulatory stability, and less market distortions, among the factors.

References

Abdulrahman, I., et al., 2014. From garbage to biomaterials: an overview on egg shell based hydroxyapatite. Journal of Materials 2014, 1–6. Available from: https://doi.org/10.1155/2014/802467.

Ahluwalia, I.J., Patel, U., 2018. Working Paper No. 356 solid waste management in India an assessment of resource recovery and environmental impact isher judge ahluwalia, *Indian Council for Research on International Economic Relations*, (356), pp. 1–48.

Ahmadi, A., et al., 2020. Benefits and limitations of waste-to-energy conversion in Iran. Journal of Renewable Energy Research and Applications (RERA) 1 (1), 27–45. Available from: https://doi.org/10.22044/RERA.2019.8666.1007.

Ahring, B.K., 2003. Perspectives for anaerobic digestion. Advances in Biochemical Engineering/Biotechnology 81, 1–30. Available from: https://doi.org/10.1007/3-540-45839-5_1.

Akhavan, O., Bijanzad, K., Mirsepah, A., 2014. Synthesis of graphene from natural and industrial carbonaceous wastes. RSC Advances 4 (39), 20441–20448. Available from: https://doi.org/10.1039/c4ra01550a.

Angelidaki, I., Ellegaard, L., Ahring, B.K., 2003. Applications of the anaerobic digestion process. Advances in Biochemical Engineering/Biotechnology 82, 1–33. Available from: https://doi.org/10.1007/3-540-45838-7_1.

Arum, C., Ikumapayi, C.M., Aralepo, G.O., 2013. Ashes of biogenic wastes—pozzolanicity, prospects for use, and effects on some engineering properties of concrete. Materials Sciences and Applications 04 (09), 521–527. Available from: https://doi.org/10.4236/msa.2013.49064.

Bader, N., Bleischwitz, R., 2009. Measuring urban greenhouse gas emissions: the challenge of comparability. Cities and Climate Change 2 (3), 1–15.

Barzak, C., 2011. *The closing window, new labor forum.* Sage Publications Inc. Available from: http://10.0.16.83/NLF.202.0000015%0Ahttp://search.ebscohost.com/login.aspx?direct = true&db = a9h&AN = 61439533&site = ehost-live&scope = site.

Benson, N.U., Bassey, D.E., Palanisami, T., 2021. COVID pollution: impact of COVID-19 pandemic on global plastic waste footprint. Heliyon 7 (2), e06343. Available from: https://doi.org/10.1016/j.heliyon.2021.e06343.

Bhardwaj, A.K., et al., 2019. Bacterial killing efficacy of synthesized rod shaped cuprous oxide nanoparticles using laser ablation technique. SN Applied Sciences 1 (11), 1–8. Available from: https://doi.org/10.1007/s42452-019-1283-9.

Bhardwaj, A.K., Naraian, R., 2021. Cyanobacteria as biochemical energy source for the synthesis of inorganic nanoparticles, mechanism and potential applications: a review. Biotech 11 (10), 1–16. Available from: https://doi.org/10.1007/s13205-021-02992-5.

Bhardwaj, A.K., Naraian, R., Sundaram, S., Kaur, R., 2022. Biogenic and non-biogenic waste for the synthesis of nanoparticles and their applications. Bioremediation. CRC Press, pp. 207–218.

Bozak, N., 2021. Chapter 5. Waste, *The Cinematic Footprint*, pp. 155–188. Available from: https://doi.org/10.36019/9780813551968-007.

Braga Nan, L., et al., 2020. Biomethanation processes: new insights on the effect of a high H2partial pressure on microbial communities. Biotechnology for Biofuels 13 (1), 1–17. Available from: https://doi.org/10.1186/s13068-020-01776-y.

Capanoglu, E., Nemli, E., Tomas-Barberan, F., 2021. Novel approaches in the valorization of agricultural wastes and their applications. Journal of Agricultural and Food Chemistry [Preprint]. Available from: https://doi.org/10.1021/acs.jafc.1c07104.

Chen, F., et al., 2016. Facile synthesis of few-layer graphene from biomass waste and its application in lithium ion batteries. Journal of Electroanalytical Chemistry 768, 18–26. Available from: https://doi.org/10.1016/j.jelechem.2016.02.035.

Chen, Y., Cheng, J.J., Creamer, K.S., 2008. Inhibition of anaerobic digestion process: a review. Bioresource Technology 99 (10), 4044–4064. Available from: https://doi.org/10.1016/j.biortech.2007.01.057.

Dabe, S.J., et al., 2019. Technological pathways for bioenergy generation from municipal solid waste: renewable energy option. Environmental Progress and Sustainable Energy 38 (2), 654–671. Available from: https://doi.org/10.1002/ep.12981.

Das, R.S., Kumar, A., Wankhade, A.V., Mandavgane, S.A., et al., 2022. Antioxidant analysis of ultra-fast selectively recovered 4-hydroxy benzoic acid from fruits and vegetable peel waste using graphene oxide based molecularly imprinted composite. Food Chemistry 376, 131926–131939. Available from: https://doi.org/10.1016/j.foodchem.2021.131926. In this issue.

Das, R.S., Mohakar, V.N., Kumar, A., 2023. Valorization of potato peel waste: Recovery of p-hydroxy benzoic acid (antioxidant) through molecularly imprinted solid-phase extraction. Environmental Science and Pollution Research 30, 19860–19872. Available from: https://doi.org/10.1007/s11356-022-23547-y. In this issue.

Dell'Abate, M.T., Benedetti, A., Sequi, P., 2000. Thermal methods of organic matter maturation monitoring during a composting process. Journal of Thermal Analysis and Calorimetry 61 (2), 389–396. Available from: https://doi.org/10.1023/A:1010157115211.

Dewan, A., et al., 2018. Greener biogenic approach for the synthesis of palladium nanoparticles using papaya peel: an eco-friendly catalyst for C-C coupling reaction. ACS Omega 3 (5), 5327–5335. Available from: https://doi.org/10.1021/acsomega.8b00039.

Elwan, A., et al., 2014. Solid waste as a renewable feedstock: a review. ARPN Journal of Engineering and Applied Sciences 9 (8), 1297–1310.

Farooq, A., et al., 2021. A framework for the selection of suitable waste to energy technologies for a sustainable municipal solid waste management system. Frontiers in Sustainability 2, 1–17. Available from: https://doi.org/10.3389/frsus.2021.681690. May.

Ferdous, W., et al., 2021. Recycling of landfill wastes (tyres, plastics and glass) in construction – a review on global waste generation, performance, application and future opportunities. Resources, Conservation and Recycling 173, 105745. Available from: https://doi.org/10.1016/j.resconrec.2021.105745. December 2020.

Foster, W., et al., 2021. Waste-to-energy conversion technologies in the UK: processes and barriers – a review. Renewable and Sustainable Energy Reviews 135, 110226. Available from: https://doi.org/10.1016/j.rser.2020.110226. January 2020.

Gaharwar, S.S., Kumar, A., Mandavgane, S.A., Rahagude, R., Gokhale, S., Yadav, K., Borua, A.P., et al., 2022. Valorization of Punica granatum (pomegranate) peels: a case study of circular bioeconomy. Biomass Conversion and Biorefinery. Available from: https://doi.org/10.1007/s13399-022-02744-2. In this issue.

Gaharwar, S.S., Mohakar, V.N., Kumar, A., et al., 2023. Chapter 13 value addition of fruit and vegetable waste: a nutraceutical perspective. Elsevier 253–268. Available from: https://doi.org/10.1016/B978-0-323-91743-8.00002-2. In this issue.

Gavurova, B., Rigelsky, M., Ivankova, V., 2021. Greenhouse gas emissions and health in the countries of the European Union. Frontiers in Public Health 9, 1–13. Available from: https://doi.org/10.3389/fpubh.2021.756652. December.

Ghosh, P.R., et al., 2017. Production of high-value nanoparticles via biogenic processes using aquacultural and horticultural food waste. Materials 10 (8). Available from: https://doi.org/10.3390/ma10080852.

Hénault-Ethier, L., Martin, J.P., Housset, J., 2017. A dynamic model for organic waste management in Quebec (D-MOWIQ) as a tool to review environmental, societal and economic perspectives of a waste management policy. Waste Management 66, 196–209. Available from: https://doi.org/10.1016/j.wasman.2017.04.021.

Holm-Nielsen, J.B., Al Seadi, T., Oleskowicz-Popiel, P., 2009. The future of anaerobic digestion and biogas utilization. Bioresource Technology 100 (22), 5478–5484. Available from: https://doi.org/10.1016/j.biortech.2008.12.046.

Ikram, R., Jan, B.M., Ahmad, W., 2020. Advances in synthesis of graphene derivatives using industrial wastes precursors; prospects and challenges. Journal of Materials Research and Technology 9 (6), 15924–15951. Available from: https://doi.org/10.1016/j.jmrt.2020.11.043.

Kearns, D.T., 2019. Waste-to-energy with CCS: a pathway to carbon-negative power generation. Global CCS Institute 1–11.

Kennedy, C., et al., 2009. Greenhouse gas emissions from global cities. Environmental Science and Technology 43 (19), 7297–7302. Available from: https://doi.org/10.1021/es900213p.

Kircher, M., 2021. Bioeconomy – present status and future needs of industrial value chains. New Biotechnology 60, 96–104. Available from: https://doi.org/10.1016/j.nbt.2020.09.005. March 2020.

Lacy, P., Rutqvist, J., 2015. Waste to wealth. Waste to Wealth. Available from: https://doi.org/10.1057/9781137530707 [Preprint].

Lamb, W.F., et al., 2021. A review of trends and drivers of greenhouse gas emissions by sector from 1990 to 2018. Environmental Research Letters 16 (7). Available from: https://doi.org/10.1088/1748-9326/abee4e.

Lariviere, M., 2011. Methodology for allocating municipal solid waste to biogenic and non-biogenic energy, *Using Municipal Solid Waste for Fuel*, (May), pp. 141–161.

Lee, R.P., et al., 2020. Sustainable waste management for zero waste cities in China: potential, challenges and opportunities. Clean Energy 4 (3), 169–201. Available from: https://doi.org/10.1093/ce/zkaa013.

Meinshausen, M., et al., 2009. Greenhouse-gas emission targets for limiting global warming to 2°C. Nature 458 (7242), 1158–1162. Available from: https://doi.org/10.1038/nature08017.

Naraian, R., Abhishek, A.K.B., 2020. Green synthesis and characterization of silver NPs using oyster mushroom extract for antibacterial efficacy. Journal of Chemistry, Environmental Sciences and its Applications 7 (1), 13–18. Available from: https://doi.org/10.15415/jce.2020.71003.

O'Callaghan, K., 2016. Technologies for the utilisation of biogenic waste in the bioeconomy. Food Chemistry 198, 2–11. Available from: https://doi.org/10.1016/j.foodchem.2015.11.030.

Olivier, J.G.J., Peters, J.A.H.W., 2020. Trends in global CO_2 and total greenhouse gas emissions: 2019 Report, *PBL Netherlands Environmental Assessment Agency*, 4068(May), pp. 1–70. http://www.pbl.nl/en.

Pandey, B.K., et al., 2016. Municipal solid waste to energy conversion methodology as physical, thermal, and biological methods. Current Science Perspectives 2 (2), 39–44.

Pandey, S., et al., 2019. Bulk synthesis of graphene nanosheets from plastic waste: an invincible method of solid waste management for better tomorrow. Waste Management 88, 48–55. Available from: https://doi.org/10.1016/j.wasman.2019.03.023.

Parkin, G.F., Owen, W.F., 1986. Fundamentals of anaerobic digestion of wastewater sludges. Journal of Environmental Engineering 112 (5), 867–920. Available from: https://doi.org/10.1061/(asce)0733-9372(1986)112:5(867).

Patil, S., Chandrasekaran, R., 2020. Biogenic nanoparticles: a comprehensive perspective in synthesis, characterization, application and its challenges. Journal of Genetic Engineering and Biotechnology 18 (1). Available from: https://doi.org/10.1186/s43141-020-00081-3.

Peel, W.P., Schmidt, C.E., 2020. Emissions sources, *Managing hazardous air pollutants*, pp. 65–160. Available from: https://doi.org/10.1201/9780367812003-2.

Pham, T.P.T., et al., 2015. Food waste-to-energy conversion technologies: current status and future directions. Waste Management 38 (1), 399–408. Available from: https://doi.org/10.1016/j.wasman.2014.12.004.

Protection, U.S.E., April, A., 2015. Bio-based products and chemicals, waste-to-energy scoping analysis U. S. Environmental Protection Agency Office of resource conservation and recovery April 2015, 1995(April), pp. 1–14.

Rahman, S.M.S., Azeem, A., Ahammed, F., 2017. Selection of an appropriate waste-to-energy conversion technology for Dhaka City, Bangladesh. International Journal of Sustainable Engineering 10 (2), 99–104. Available from: https://doi.org/10.1080/19397038.2016.1270368.

Ramos, A., et al., 2018. Co-gasification and recent developments on waste-to-energy conversion: a review. Renewable and Sustainable Energy Reviews 81, 380–398. Available from: https://doi.org/10.1016/j.rser.2017.07.025. June 2016.

Rao, P., Rathod, V., 2019. Valorization of food and agricultural waste: a step towards greener future. Chemical Record 19 (9), 1858–1871. Available from: https://doi.org/10.1002/tcr.201800094.

Rominiyi, O.L., Adaramola, B.A., 2020. Proximate and ultimate analysis of municipal solid waste for energy generation. ABUAD Journal of Engineering Research and Development (AJERD) 3 (1), 103–111.

Ryu, C., Shin, D., 2013. Combined heat and power from municipal solid waste: current status and issues in South Korea. Energies 6 (1), 45–57. Available from: https://doi.org/10.3390/en6010045.

Santos, J., et al., 2021. Recycling waste plastics in roads: a life-cycle assessment study using primary data. Science of the Total Environment 751, 141842. Available from: https://doi.org/10.1016/j.scitotenv.2020.141842.

Schneider, D.R., Ragossnig, A.M., 2013. Biofuels from waste. Waste Management and Research 31 (4), 339–340. Available from: https://doi.org/10.1177/0734242X13482228.

Sharma, K.D., Jain, S., 2020. Municipal solid waste generation, composition, and management: the global scenario. Social Responsibility Journal 16 (6), 917–948. Available from: https://doi.org/10.1108/SRJ-06-2019-0210.

Singh, M.P., et al., 2021. Biogenic and non-biogenic waste utilization in the synthesis of 2D materials (graphene, h-BN, g-C2N) and their applications. Frontiers in Nanotechnology 3, 1–25. Available from: https://doi.org/10.3389/fnano.2021.685427. September.

Skillicorn, D.B., Zheng, Q., Morselli, C., 2013. Spectral embedding for dynamic social networks. Proceedings of the 2013 IEEE/ACM International Conference on Advances in Social Networks Analysis and Mining, ASONAM 2013, 316–323. Available from: https://doi.org/10.1145/2492517.2492522.

Smidt, E., et al., 2011. Transformation of Biogenic Waste Materials through Anaerobic Digestion and subsequent composting of the residues. A case study. Dyn. Soil Dyn. Plant 5, 63–69.

Suberu, M.Y., Bashir, N., Mustafa, M.W., 2013. Biogenic waste methane emissions and methane optimization for bioelectricity in Nigeria. Renewable and Sustainable Energy Reviews 25, 643–654. Available from: https://doi.org/10.1016/j.rser.2013.05.017.

Thakur, I.S., et al., 2018. Sequestration and utilization of carbon dioxide by chemical and biological methods for biofuels and biomaterials by chemoautotrophs: opportunities and challenges. Bioresource Technology 256, 478–490. Available from: https://doi.org/10.1016/j.biortech.2018.02.039. February.

Treat, N.L., 1978. Net energy from municipal solid waste. v(9), pp. 4255–4276.

Vehlow, J., et al., 2007. European Union waste management strategy and the importance of biogenic waste. Journal of Material Cycles and Waste Management 9 (2), 130–139. Available from: https://doi.org/10.1007/s10163-007-0178-9.

Veksha, A., et al., 2020. Heteroatom doped carbon nanosheets from waste tires as electrode materials for electrocatalytic oxygen reduction reaction: effect of synthesis techniques on properties and activity. Carbon 167, 104–113. Available from: https://doi.org/10.1016/j.carbon.2020.05.075.

Venkata Mohan, S., et al., 2018. Waste derived bioeconomy in India: a perspective. New Biotechnology 40, 60–69. Available from: https://doi.org/10.1016/j.nbt.2017.06.006.

Vergara, S.E., Tchobanoglous, G., 2012. Municipal solid waste and the environment: a global perspective. Annual Review of Environment and Resources. Available from: https://doi.org/10.1146/annurev-environ-050511-122532.

Vieira, O., et al., 2022. A systematic literature review on the conversion of plastic wastes into valuable 2D graphene-based materials. Chemical Engineering Journal 428. Available from: https://doi.org/10.1016/j.cej.2021.131399.

Ward, A.J., et al., 2008. Optimisation of the anaerobic digestion of agricultural resources. Bioresource Technology 99 (17), 7928–7940. Available from: https://doi.org/10.1016/j.biortech.2008.02.044.

Weiland, P., 2000. Anaerobic waste digestion in Germany - Status and recent developments. Biodegradation 11 (6), 415–421. Available from: https://doi.org/10.1023/A:1011621520390.

Wood, S., Cowie, A., 2004. For fertiliser production, *Development* [Preprint], (June).

Zsigraiová, Z., et al., 2009. Integrated waste-to-energy conversion and waste transportation within island communities. Energy 34 (5), 623–635. Available from: https://doi.org/10.1016/j.energy.2008.10.\015.

Chapter 2

Sustainable advances in the synthesis of waste-derived value-added metal nanoparticles and their applications

Nishita Narwal[1], Deeksha Katyal[1], Aastha Malik[1], Navish Kataria[2], Abhishek Kumar Bhardwaj[3], Md. Refan Jahan Rakib[4] and Mian Adnan Kakakhel[5]

[1]University School of Environment Management, Guru Gobind Singh Indraprastha University, Dwarka, New Delhi, India, [2]Department of Environmental Sciences, J.C. Bose University of Science and Technology, YMCA, Faridabad, Haryana, India, [3]Department of Environmental Science, Amity School of Life Sciences, Amity University, Gwalior, Madhya Pradesh, India, [4]Department of Fisheries and Marine Science, Faculty of Science, Noakhali Science and Technology University, Noakhali, Bangladesh, [5]College of Hydraulic & Environmental Engineering, Three Gorges University, Yichang, Hubei, P.R. China

2.1 Introduction

Nanotechnology involves the synthesis, manipulation, and application of materials at a nanosized scale less than 100 nm. At small scales, materials often exhibit latent features that diverge significantly from their larger counterparts, yielding a diverse range of useful properties (Salvador-Morales and Grodzinski, 2022). When the particle size is reduced, nanotechnology can develop remarkable characteristics including biological properties, electromagnetic properties, ferromagnetic properties, thermal properties, metal sensing, laser sensing, and enzymatic attributes, making it one of the most important scientific fields (Zhu et al., 2022). Nanoparticle synthesis has become one of the leading advanced technologies in the progress of science. The synthesis of nanoparticles requires stable conditions such as constant temperature, pressure, sophisticated instruments, extra safety, energy, and voltage. Due to their outstanding properties, the synthesis of nanoparticles becomes an important aspect. Nanoparticles can be created using chemical, physical, and biological methods. Chemical synthesis for nanomaterials requires substantial chemical quantities and generates toxic waste byproducts (Mahmud et al., 2022). Physical synthesis methods require extensive use of heavy machinery, specialized equipment, and standardized procedures to produce nanoparticles. Biological synthesis involves utilizing plant waste, agricultural waste, and animal or bacterial products as substrates for nanoparticle synthesis (Guan et al., 2022). Biological synthesis, utilizing green and sustainable approaches, is preferred over chemical and physical methods due to its safety, cost-effectiveness, and environmental friendliness. These biologically synthesized nanomaterials have versatile applications, including addressing various environmental challenges through the combination of green synthesis and chemical approaches. Fig. 2.1 illustrates the steps of green synthesis, using waste materials as raw materials to prepare the extract.

Waste-derived nanomaterials synthesized through biological methods have shown great potential in various industries such as food, medicine, drug delivery, paints, additives, and environmental cleanup (Ghosh et al., 2022). The integration of bioengineering and nanotechnology with green synthesis techniques and materials aims to reduce environmental impact and health risks associated with nanomaterial manufacturing, use, and disposal. These waste-based nanoparticles play a crucial role in pollution mitigation and waste reduction in the environment. Utilizing discarded natural resources and low-cost waste substrates like plants, seeds, bacterial extracts, industrial waste, crop residues, etc. has long been a focus of research. These materials offer advantages such as low toxicity, ease of replication, and economic viability for green synthesis. Metal nanoparticles synthesized from waste substrates, known as green nanoparticles, are gaining popularity. The bioreduction technique is one method used for green synthesis, where waste materials serve as reducing and stabilizing agents when combined with metal salts to create metallic nanoparticles. Various waste materials and environment-friendly resources can be employed to generate nanoparticles, including gold, iron, iron oxide, manganese, nickel, cobalt, copper,

Green and Sustainable Approaches Using Wastes for the Production of Multifunctional Nanomaterials. DOI: https://doi.org/10.1016/B978-0-443-19183-1.00013-1

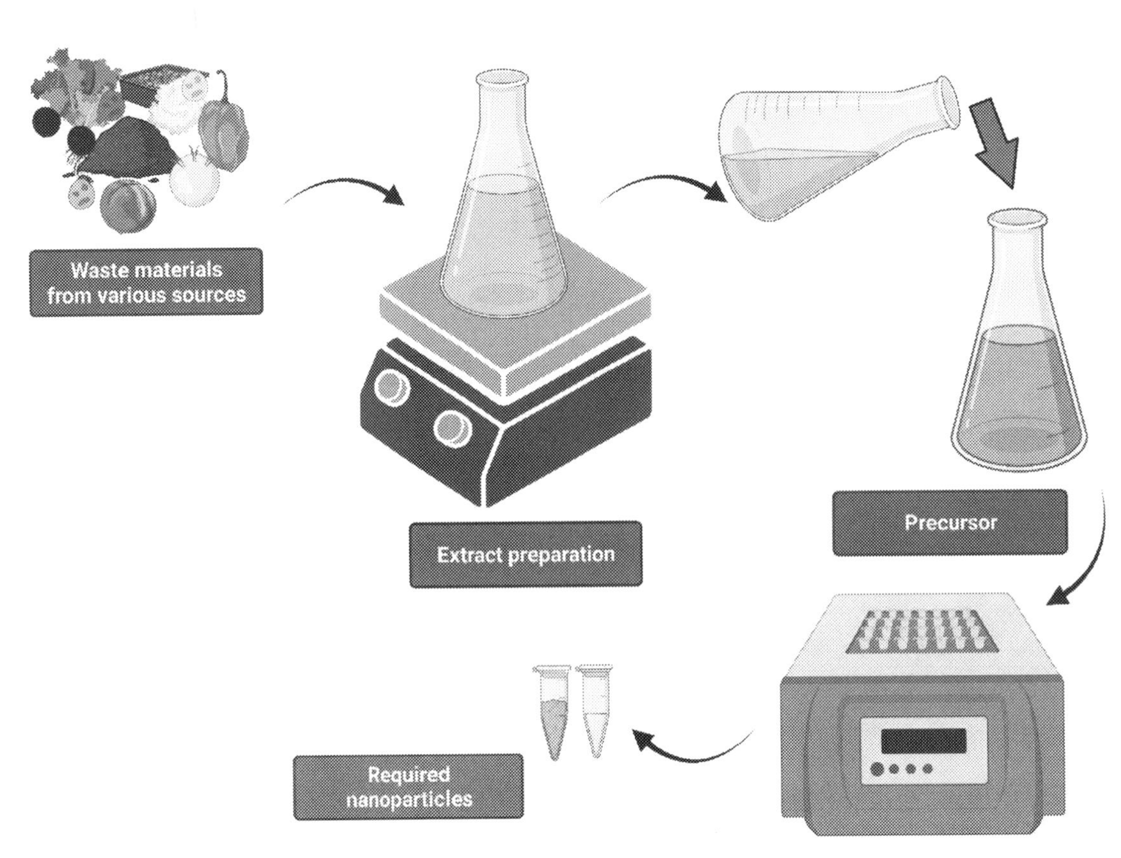

FIGURE 2.1 Synthesis of waste-derived nanoparticles.

silver, palladium, platinum, zinc, zinc oxide, and magnetite. Metallic nanoparticles have diverse applications in biochemistry, circuitry, optics, and nanodevice development (Chen et al., 2022), with greener synthesis techniques enabling the usage in the food and medical industries.

Sustainability can be defined as the capacity to continuously support or uphold a process over time. Because of its small size, nanotechnology keeps its sights as a leading technology. The application of nanotechnology is also made resonantly burdensome by a number of potential benefits, including improved food security, use in agrarian aspects, retention of nutrients in the soil, treatment of wastewater, and use in health care. Fossil fuels, farming, and food all face challenges related to vulnerability, sustainability, public health, and a good quality of life. In order to maintain a clean environment, nanoparticle synthesis must meet several requirements at various stages of production and processing using a sustainable strategy that includes recycling and wise use of generated waste couples with green synthesis (Brar et al., 2022). According to the most recent 20 years of data (2003–2023) obtained from ScienceDirect, a total of 356,079 articles on nanoparticle synthesis were reported, with 139,675 articles focusing on green nanoparticle synthesis. Fig. 2.2 represents the number of studies reported on ScienceDirect on green synthesis of (iron NP, zinc NP, copper NP, silver NP, gold NP, platinum NP, titanium NP, cerium NP, and thallium NP) nanoparticles. Nanomaterials' goal as a sustainable strategy is to lessen the amount of chemicals that are dispersed. According to this viewpoint, nanoparticles are cutting-edge nanotools that can be used to cleanup soils, water, agriculture, farming, and waste management. The health of agricultural plants can be preserved by using nanoparticles as sensors to monitor the quality of the soil in a field. Nanoparticles can be used as vectors to treat a variety of diseases. Nanoparticles have benefits in cosmetics, paints, and adhesives in addition to health and well-being. This chapter examines how nanoparticle synthesis is currently progressing throughout the world. Moreover, the development of metal nanoparticles using effective, viable, biological, green, and clean methodologies, that have been investigated by researchers in the field of nanotechnology, is also discussed (Bhardwaj and Naraian, 2020; Bhardwaj et al., 2018). Additionally, the production of eco-friendly and simple nanoparticles from plant materials and waste, specifically agro waste, plant waste, animal waste, forest residues, industrial waste, electronic waste, and mining waste, is also discussed. Furthermore, more emphasis is being placed on the uses of metal nanomaterials derived from waste in various fields, the production of nanomaterials and products that do not harm the environment or human health, and the production of nanoproducts that address environmental issues.

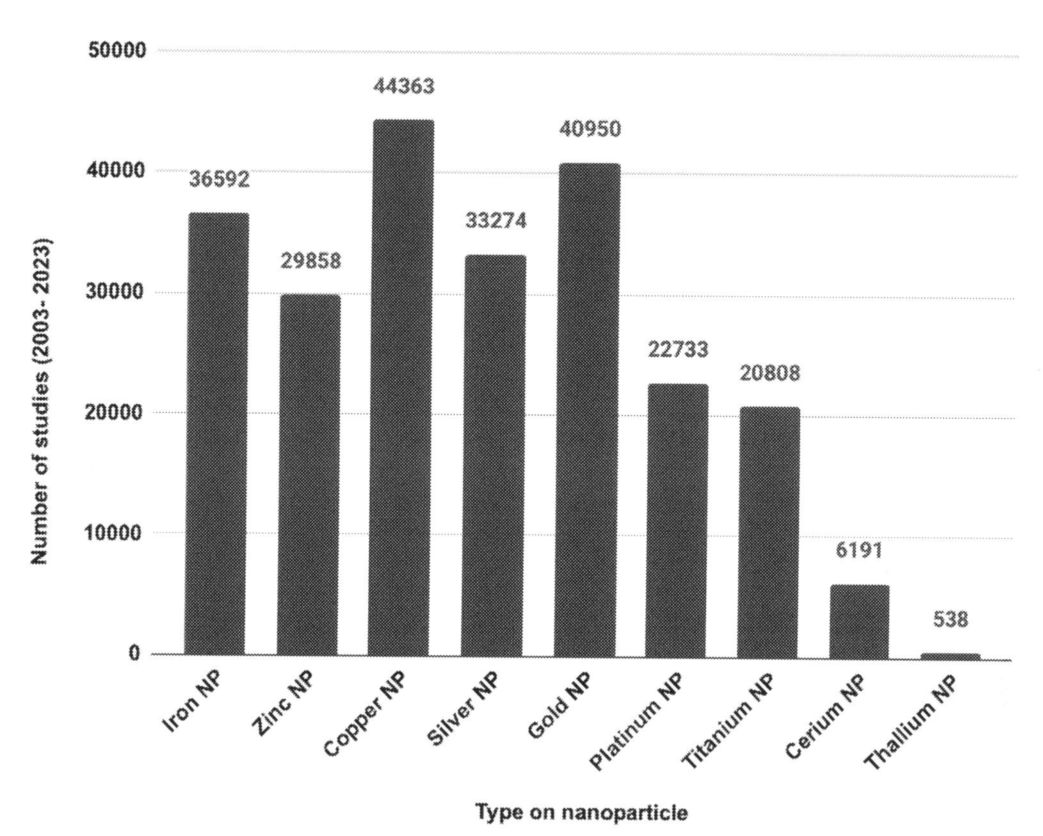

FIGURE 2.2 Number of studies reported on ScienceDirect on green synthesis of various nanoparticles from 2003 to 2023 (Keyword used-Green synthesis of each type of metal NP).

2.2 Nanoparticles: types and synthesis approaches

Nanoparticles can be classified based on dimensions, chemical structure, and properties, as shown in Fig. 2.3 and discussed in the following section. Nanoparticles can be categorized on the basis of dimensions: zero-dimensional (0D), one dimensional (1D), two dimensional (2D), and three dimensional (3D). Zero-dimensional nanoparticles have fixed dimensions in length, width, and height, with examples like graphene quantum dots and clusters. One-dimensional nanoparticles, such as nanowires and carbon nanotubes, possess length and breadth as their two axes. Two-dimensional nanoparticles, like nanofilms and nanolayers, have all three axes: length, breadth, and height. Materials like graphite, aerogels, and polycrystals that are not restricted to the nanoscale in any dimension are referred to as three-dimensional nanomaterials.

Nanoparticles can be classified based on their shapes, encompassing various geometries such as spheres, tubes, spirals, and cylinders. Another classification criterion is the chemical composition of nanoparticles, leading to three main categories: organic nanoparticles, inorganic nanoparticles, and carbon-based nanoparticles (Kango et al., 2013). Additionally, there exist composite nanoparticles, which combine organic and inorganic components. Organic nanoparticles are nanostructures composed of polymers and liposomal components, including micelles, dendrimers, solid-lipid nanoparticles, liposomes, and ferritin. These organic nanoparticles are biodegradable and find wide-ranging applications. On the other hand, inorganic nanoparticles, categorized as noncarbon-based, encompass metal and metal oxide nanoparticles, quantum dots, and ceramics. Inorganic nanoparticles exhibit enhanced chemical and mechanical stability and demonstrate high resistance to biological degradation and breakdown in biological matrices. Examples of inorganic nanoparticles include carbon nanotubes, gold nanoparticles, mesoporous silica, iron oxide nanoparticles, and quantum dots (Xu et al., 2006). In the upcoming section, the focus will be on metal nanoparticles, which fall under the category of inorganic nanoparticles. Composite nanoparticles, formed by combining organic and inorganic nanoparticles, create a distinct category with altered properties due to the presence of multiple materials. These composite nanoparticles have a robust shell that contains the necessary components, resulting in a material with combined properties (Lin et al., 2006). Another category in the classification based on chemical composition is carbon-based nanoparticles, consisting of carbon and carbon-derived compounds. Examples of carbon-based nanoparticles include carbon black, fullerenes, graphene, carbon nanotubes, and carbon

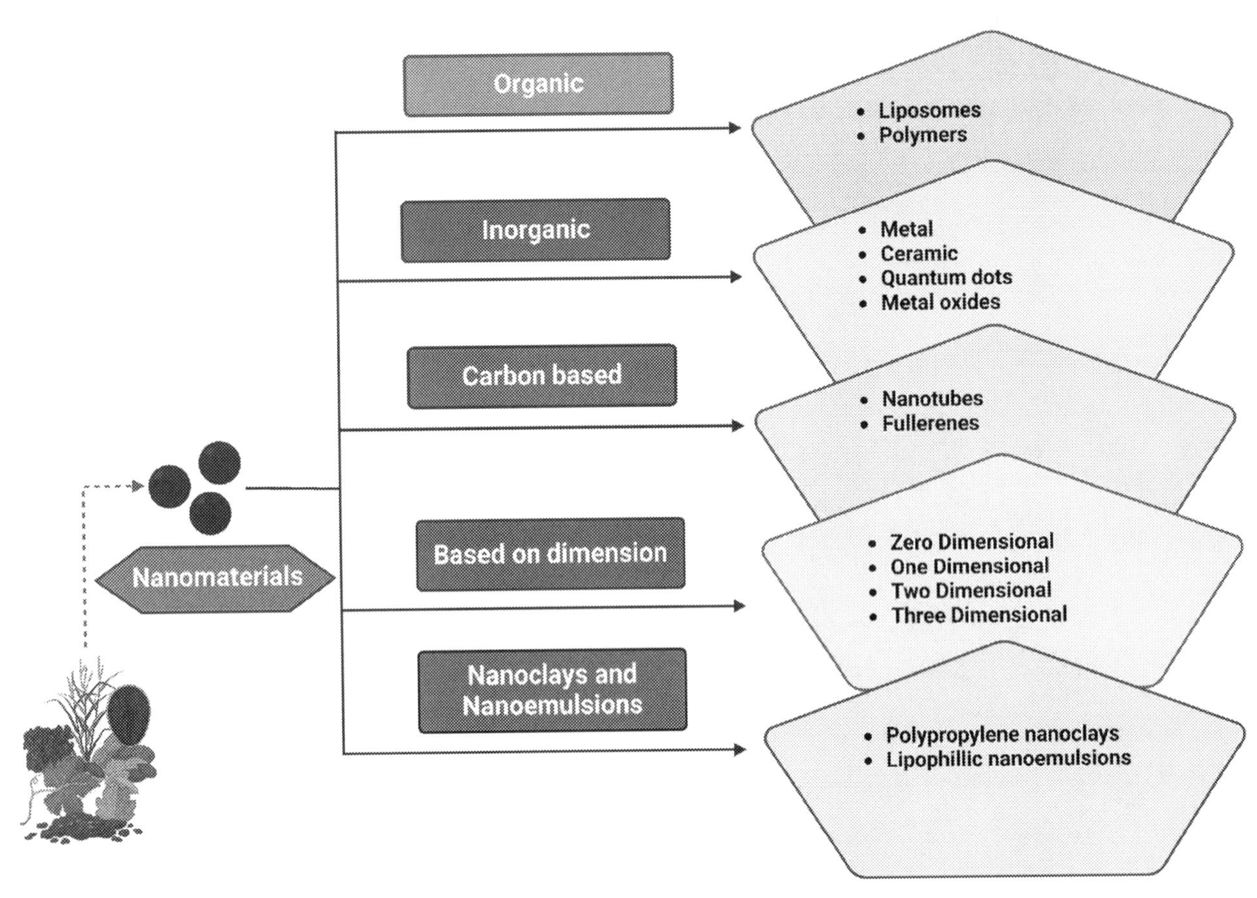

FIGURE 2.3 Classification of nanomaterials.

nanofibers (Scida et al., 2011). In addition, two noteworthy categories include nanoclays, which are stratified crystalline silicate minerals in the size range of 50 to 100 nm, such as bentonite, kaolinite, montmorillonite, hectorite, and halloysite. Another category is nanoemulsions, which are systems consisting of small droplets typically ranging from 10 to 1000 nm, with lipophilic nanoemulsions being a specific type.

Nanomaterials can be synthesized using two approaches: bottom-up and top-down. The bottom-up approach involves using small atoms to construct larger clusters, making it a constructive method of nanoparticle synthesis. Methods such as vapor deposition, condensation, sol-gel, spray pyrolysis, electrochemical deposition, and aerosol processes fall under this approach. On the other hand, the top-down approach involves breaking down bulk materials into powder form, making it a destructive method. Sputtering, thermal ablation, and ball milling are common top-down synthesis methods (Iqbal et al., 2012). Each type of nanomaterial has distinct properties that influence its applications and suitability. Waste-derived metal nanoparticles have diverse uses in various fields such as ecology, environmental cleanup, pharmacy, engineering, medicine, immunology, cell biology, and drug delivery. In this context, green synthesis offers a simple process to obtain nanoparticles with several advantages.

2.3 Metallic nanoparticles and their classification

Metallic nanoparticles are nanoscale entities containing metals or metal-derived compartments which have attracted scientists for a long time due to their use in biomedical applications, antimicrobial agents, adhesive industries, environmental cleanup, green engineering, public health, and antibacterial paints. Due to their enormous prospects in nanotechnology, they are indeed the subject of attention. The ability to synthesize and modify these materials with a variety of elements like chitosan, fibers, carbon, gel, etc., now enables the creation of composites that are extremely beneficial in the fields of biotechnology, microbiology, and immunology (Ahmad et al., 2010). Due to their specialized characteristics like electromagnetic properties, conductivity, antimicrobial properties, thermal conductivity, mechanical properties, tensile strength, flexibility, elasticity, and high reactivity, different metallic nanoparticles and their constituents have recently received more attention (Jamkhande et al., 2019). In general, metallic nanoparticles can be classified as metal nanoparticles, metal oxide nanoparticles, metal sulfide nanoparticles, doped metal nanoparticles, doped metal oxide nanoparticles, and metal organic frameworks (MOFs).

2.3.1 Metal nanoparticles

Metal-based nanoparticles are particles synthesized from metals with nanometric sizes ranging from 10 to 100 nm. Nanoparticles may be created from almost any scrap containing traces of metal. Aluminum (Al), cobalt (Co), copper (Cu), cadmium (Cd), gold (Au), iron (Fe), lead (Pb), silver (Ag), cerium (Ce), palladium (Pd), nickel (Ni), thallium (TI), molybdenum (Mo), selenium (Se), platinum (Pt), titanium (Ti), and zinc (Zn) are the most commonly used metals for nanoparticle synthesis (Bhattacharya and Mukherjee, 2008). Fig. 2.4 shows the most commonly used nanoparticles globally. Aluminum nanoparticles have a high reactivity and a large surface area, iron-based nanoparticles are air sensitive, silver nanoparticles have antibacterial properties, gold nanoparticles are also very reactive, cobalt nanoparticles have good magnetic properties, cadmium nanoparticles have good conductivity, copper-based nanoparticles have good electrical conductivity, and zinc-based nanoparticles are good antimicrobial agents. All of these metal nanoparticles can be created by utilizing environment-friendly ingredients, such as various waste products, animal-derived compounds, plants, and their derivatives.

2.3.2 Metal oxide nanoparticles

Metal oxide nanoparticles contain an oxide anion and metal cation. Due to their remarkable characteristics, they have been used for a wide range of applications. Metal oxides including Ag_2O, FeO, MnO_2, CuO, Bi_2O_3, ZnO, MgO, TiO_2, and Al_2O_3 are frequently employed (Durán and Seabra, 2012). The reactivity of different metal oxides is influenced by the differences in their characteristics. Iron oxides are very reactive, titanium oxides are frequently used to treat water, silicon oxides are relatively stable and aid in the functionalization of various other molecules, zinc oxide nanoparticles

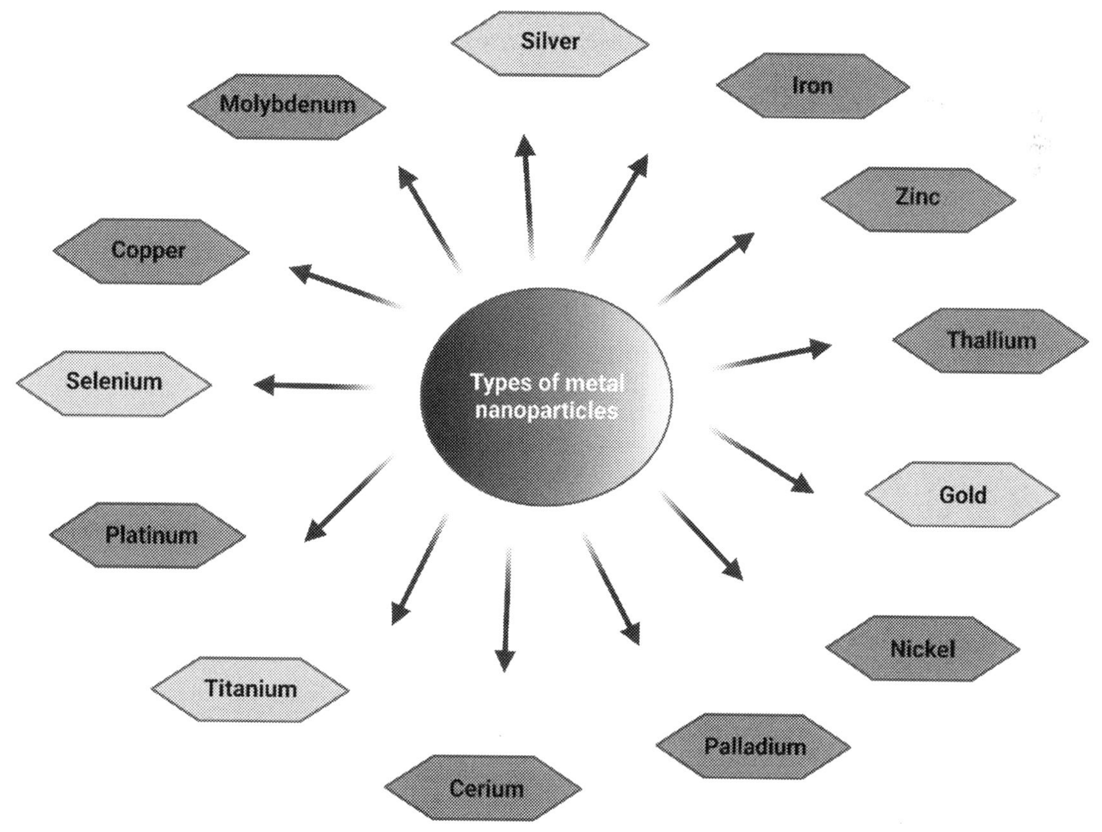

FIGURE 2.4 Various types of metal nanoparticles.

have good antimicrobial and antifungal properties and are safe to use in the food industry, cerium oxide nanoparticles act as good antioxidants, and aluminum oxides have a very high adsorption potential because they have a large surface area (Mansoor et al., 2022).

2.3.3 Metal sulfide nanoparticles

Metal ions and sulfur molecules combine to form the new class of metal containing nanostructures known as metal sulfide nanomaterials. The most promising materials for a wide range of technologically relevant applications, from agriculture to the environment, and beyond, are metal sulfides (Pham et al., 2022). As a result, taking advantage of the unusual qualities of metal sulfide nanoparticles, they open up appealing new applications. They are utilized in supercapacitors, which are energy storage devices, because they have good electrochemical and conductive properties. They work well in solar cells, batteries, LEDs, and catalysis. Copper sulfide nanoparticles are used in biomedical applications, zinc sulfide for antimicrobial activity, iron sulfides for photothermal properties, silver sulfides for biosensing properties, and gold sulfides for antimicrobial properties.

2.3.4 Doped metal and metal oxide nanoparticles

Doped metal and metal oxide nanoparticles are created to manipulate the characteristics of nanoparticles and achieve desired effects (Rajput et al., 2022). Nanoparticles are modified via doping, which improves their electrical, magnetic, thermal, optical, and photocatalytic properties. Doping might also enhance the antimicrobial effects. Conductivity is impacted by the use of a dopant because the quantity of valence electrons is changed, either raised or decreased. Doped metal and metal oxide nanoparticles are therefore capable of a wider range of wavelengths than the pure ones. Doping can be done in many different ways, including flame spray pyrolysis, precipitation technique, and sol/gel impregnation, to mention a few. Under solar light, TiO_2 and ZnO are already exploited for their photocatalytic functions; but, because of their poor absorption under visible light, they cannot be utilized in that range (Karthik et al., 2022). However, when these metal oxides are doped with transition metal ions like Fe, their photocatalytic system is altered to allow for their use in multimodal applications that can be performed both under visible light and solar light.

2.3.5 Metal organic framework

MOFs are biodynamic hybrid crystalline porous materials that have linker molecules of organic origin surrounding a positively charged center metal ion (Ettlinger et al., 2022). In these materials, organic or inorganic ligands act as linkers and metal ions as the major coordination center. As a result, a repetitive hollow structure with high porosity and an incredibly enormous internal surface area is created by the metal ions, which act as nodes to bind the arms of the linkers together (Jiang et al., 2022). MOFs are frequently utilized because of their ability to hold drugs and their biodegradability. The usage of Zn-based MOFs for drug loading and antidiabetic action is widespread. Mg MOF materials are employed as carriers for antioxidants. Ag-based MOFs are effective antibacterial substances.

2.4 Waste-derived metal nanoparticles

Waste products are a possible source of raw materials for the synthesis of nanoparticles. Nanoparticles made from waste can improve the security and purity of water. However, there may be certain difficulties, such as waste toxicity-related issues. For instance, the extraction of plastic and metal can involve some dangerous compounds. However, after being thoroughly examined for safety, the materials may have good economic, commercial, and environmental value. Therefore, it is imperative to create clean and green metal nanoparticles from waste that have additional value (Bhardwaj and Naraian 2021). Different kinds of metal nanoparticles can be obtained depending on the sources of waste produced as shown in Fig. 2.5. Waste from farms, businesses, and homes can all be used. Batteries, rubber, plastics, wires, electrical equipment, and other industrial trash are notable sources of carbon, nickel, iron, lead, zinc, and copper. Metal nanoparticles can be extracted from electronic waste. Additionally, plastic waste can be recycled to create nanocomposites (Akbayrak et al., 2020). The numerous precious metal nanoparticles created from diverse waste streams are included in this section.

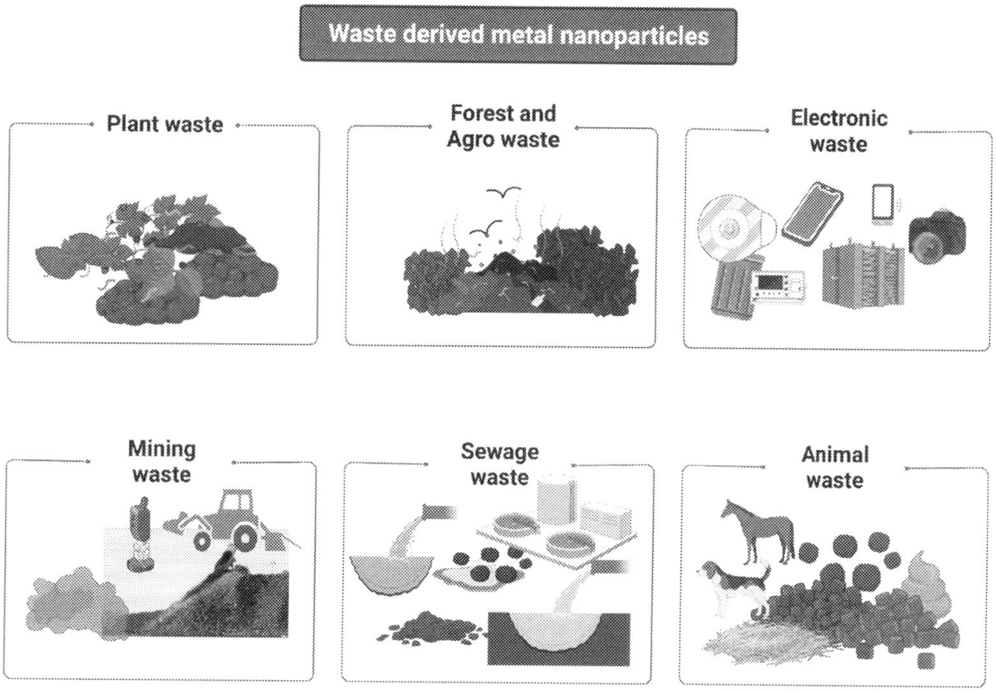

FIGURE 2.5 Different types of waste materials for the synthesis of metal nanoparticles.

2.4.1 Agro waste, forest residues, and plant waste

Agriculture waste, plant extracts, and other environment-friendly methods and materials are being extensively investigated for the manufacturing of nanoparticles. Fields and agricultural leftovers and plant waste like leaves, roots, bark, stems, and branches are relatively affordable substrates for extract preparation for synthesis of metal nanoparticles as mentioned in Table 2.1. Chemical-based reducing agents are not needed because plant-based abstracts already contain stabilizers and reducing agents. Numerous studies have already been done, and more are being performed (Sangeetha et al., 2017). Some substances, including coffee grounds and bilberry wastes, contain phenolic groups that have been effectively exploited as a starting point for the environment-friendly synthesis of silver nanoparticles (Baiocco et al., 2016). It is possible to create environment-friendly titanium dioxide nanoparticles by extracting fruit peel waste from kiwi, plum, and peach (Ajmal et al., 2019). For the manufacture of silver nanoparticles, *Moringa oleifera*, sugarcane bagasse, and *Citrus aurantifolia* peel extract can also be employed (Baiocco et al., 2016). Gold nanoparticles have been successfully created using grape and mango waste. Copper nanoparticles can be created using *Lantana camara*, and copper oxide nanoparticles can be created using *Gloriosa superba*. Platinum nanoparticles can be created using the fungus *Neurospora crassa*, and cadmium sulfide nanoparticles can be created using the fungus *Fusarium oxysporum*.

2.4.2 Animal waste

Animal leftovers, microorganisms, and animal wastes have been used in the environment-friendly creation of various metallic nanoparticles due to their capacity for reduction. Additionally gaining popularity are nanomaterials created from microbial extracts. Alginate and chitosan, two nanomaterials that can be used to adsorb a range of micropollutants from water, can also be linked to different metal nanoparticles (Gokila et al., 2017). *Klebsiella pneumoniae* was capable of producing cadmium sulfide nanoparticles on cell surface in the presence of cadmium ions. In the presence of the necessary reaction media containing gold salt, *F. oxysporum* releases reducing agents into the solution to produce gold nanoparticles (Ghazwani, 2015). In the presence of aqueous anionic complexes of Si and Ti, the fungus can also produce nanoparticles of silica and titania. The manufacture of monodispersed silver nanoparticles with an average particle size was demonstrated by *Aspergillus flavus* (Devipriya and Roopan, 2017). In the presence of aqueous anionic combinations of Si and Ti, respectively, this fungus can also be utilized to synthesize silica and titanium nanoparticles. Silver nanoparticles can also be created using *A. flavus*. Sericin and silk fibroin are biopolymeric substances that come from insects and spiders (Harisha et al., 2021). The silk produced by *Bombyx mori*, *Trichoptera*, and *Hymenoptera* is used to

TABLE 2.1 Synthesis of nanoparticles using different waste materials as substrates.

Category of waste	Nanoparticle synthesis	Size (nm)	Morphology	Synthesis process	Characterization	References
Waste printed circuit boards	Copper oxide nanoparticles	5	Polycrystalline	Biosynthesis	TGA, FESEM-EDX, FTIR, XRD	Ravi et al. (2023)
Grape pomace waste from *Vitis vinifera*.	Gold Nanoparticles	30	Polyhedral-	One-pot green approach	UV-Vis, XPS, SEM, ATR-FTIR, XRD	Gubitosa et al. (2022)
Watermelon peel waste	Titanium dioxide quantum dots	7	Polycrystalline	Biosynthesis	UV–visible, FTIR, XRD, SEM, EDX, mapping, TEM, and TGA	Ali et al. (2022)
Seed extract of *Lepidium sativum*	ZnO nanoparticles	25.6	Hexagonal	Bio-assisted synthesis	XRD, FTIR, HPLC, SEM	Meer et al. (2022)
Chicken feather waste	Carbon-based molybdenum oxide nanocomposite	–	–	Hydrothermal and copyrolysis process	HR-SEM, XRD, XPS	Ganesan et al. (2022)
Coconut husk ash	Zn NP	–	Spherical	Coprecipitation reaction	XRD, BET, SEM–EDX, UV–Vis, FTIR, photoluminescence spectroscopy	Wary et al. (2022)
Peel waste of jackfruit	Iron NP	–	–	Coprecipitation method	XRD, FT-IR	Praveena et al. (2022)
Barberry leaf	Iron NP	20–40	Spherical	Precipitation reaction	UV spectroscopy, XRD, SEM, TEM, FTIR	Samadi et al. (2021)
Agricultural waste	Titanium dioxide nanoparticles	5–10 nm	Spherical	Sol–gel method	FTIR, TGA, SEM, EDX, TEM	Farag et al. (2021)
Leaves of *Verbascum thapsus*	Zero-Valent Iron Nanoparticles	–	Irregular	Biosynthesis	FT-IR, SEM, TEM	Saleh et al. (2021)
Onion peel	Silver NP	–	Spherical	Precipitation reaction	UV–Vis, XRD	Santhosh et al. (2021)
Pineapple peels	Zinc oxide nanoparticles	10–60	–	Biologically synthesis	XRD, UV-Vis, TEM, X-ray spectroscopy	Mirgane et al. (2021)
Cuminum cyminum seed extract	Titanium dioxide nanoparticles	15.17	Spherical	Biosynthesis	UV–Vis, XRD, FTIR, SEM, TEM	Mathew et al. (2021)
Mangifera indica seed	Silver NP	26.85	Round	Precipitation reaction	UV-Vis, Zeta potential, FT-IR, XRD, TEM, SAED, EDX	Donga and Chanda (2021)
Rosa damascene, Urtica dioica, and Thymus vulgaris leaf extracts	Iron NP	33	Spherical	Biosynthesis	FTIR, TEM, XRD, SEM, EDX	Jain et al. (2021)

Safflower	Silver NP	8.67 ± 4.7	Spherical	Biosynthesis	TEM, SEM, electron diffraction, FTIR	Rodríguez-Félix et al. (2021)
Red peanut skin	Zero-Valent Iron Nanoparticles	10.6	Spherical	Biosynthesis	SEM, TEM, XRD, UV-Vis, FTIR	Pan et al. (2020)
Industrial waste	Ag nanoparticles	20	Spherical	Coprecipitation synthesis	UV-Vis, dynamic light scattering, zeta potential, FT-IR, proton nuclear magnetic resonance spectroscopy, TEM	Pandey et al. (2020)
Sugarcane waste ash	Silica nanoparticles	20	Spherical	Biosynthesis	FTIR, TGA	Rovani et al. (2020)
Tea waste	Activated carbon	50–60	C- shaped	Coagulation process	Scanning transmission electron microscopy (STEM), Dynamic mechanical thermal analysis (DMTA)	Akbayrak et al. (2020)
Ginkgo leaves	Pd NPs	7.61–13.13	Spherical	Coprecipitation method	UV-Vis, FTIR, TEM, XPS	Cui et al. (2020)
Printed circuit boards	Copper	5–32	Spherical	Reduction	SEM, TEM, XRD	El-Nasr et al. (2020)
Watermelon rind	Fe_3O_4 magnetic nanoparticles	–	Spherical	Bioreduction mechanism	FT-IR, XRD, SEM	Adebowale et al. (2020)
Pulp fibers	Cellulose nanocrystals	15–40	Spherical	Composite enzymolysis	SEM, XRD, DLS, TGA	Xu and Chen (2019)
Corn cob	Silicon carbide (SiC) nanoparticles	40–100	Spherical	Carbothermal reduction	FT-IR, Raman spectra, XRD, SEM, TEM, BET	An et al. (2019)
Silicon cutting waste	β-SiC nanoparticles	40 to 50	Spherical	Carbothermal reduction	XRD, FTIR, SEM, TEM, Raman, PL and DSC	Jiang et al. (2019)
Mango peel	Zero-Valent Iron Nanoparticles	1–10	–	Biosynthesis	XPS, FTIR	Desalegn et al. (2019)
Jatropha waste	Gold nanoparticles	14	Triangle, hexagonal and spherical	Reduction	UV–Vis, TEM, Selected area electron diffraction, Powder X-ray diffraction, FTIR	Kanchi et al. (2018)
Citrus peels	AgNPs	3–12	Spherical	Bioreduction	DLS, EDX, FTIR, FESEM, HRTEM	Omran et al. (2018)
Fresh plant leaves of olive, guava, fig, and lemon	Zinc oxide nanoparticles	7.1–28	Hexagonal	Biosynthesis	XRD, SEM, TEM, FTIR, TGA, PSA	El-Arab (2018)
DVD player	Iron NP	3–5	Hexagonal	Reduction	SEM, TEM, TGA, FTIR	Angamuthu et al. (2017)
Tire rubber	Carbon nanoparticles	30–40	Spherical	Chemical vapor deposition	XRD, Raman spectroscopy, XPS	Maroufi et al. (2017)

(Continued)

TABLE 2.1 (Continued)

Category of waste	Nanoparticle synthesis	Size (nm)	Morphology	Synthesis process	Characterization	References
Egg shell membrane	AgNPs	7.8–10.4	Spherical	Photochemical synthesis	Ultraviolet-visible diffuse reflectance spectroscopy, FE-SEM, EDX, XRD, TEM	Huang et al. (2016)
Pomelo peels	Carbon nanoparticles	–	Spherical	Carbonization	SEM, XRD, FTIR, XPS, and BET	Li et al. (2016)
Macadamia shell waste	Silicon carbide (SiC) and silicon nitride (Si3N4) nanopowders	20–80 and 40–100	–	Carbothermal reduction and nitridation	SEM, XRD, Raman spectroscopy, XPS, FTIR	Rajarao and Sahajwalla (2016)
Watermelon rind	AgNPs	109.97	Spherical	Biosynthesis	SEM, EDX, FT-IR, TG/DTG	Patra et al. (2016)
Punica granatum peel extract	Platinum nanoparticles	16–23	Spherical	Bioreduction	UV, DLS, TEM, XRD, EDX, FTIR	Dauthal and Mukhopadhyay (2015)
Cotton fibers	Cellulose nanocrystals	76.29–94.26	Rod-like shape particles	Sulfuric acid hydrolysis followed by ultrasonication	TEM, SEM, AFM, XRD, FTIR	Ibrahim et al. (2015)
Moringa Oleifera flower	Au NP	3–5	Hexagonal and triangular	Reduction	UV–Vis, XRD, TEM, DLS, FTIR, H-NMR 294 spectroscopy	Anand et al. (2015)
Watermelon rind	Gold nanoparticles	20–140	Spherical	Coprecipitation	UV–Vis, SEM, XRD, FT-IR, TGA	Patra and Baek (2015)

create nanoparticles like fibroin-TiO_2 and silk fibroin, which have numerous medical uses. The field of medicine a lso uses nanomaterials made from components extracted from sponges (Khan et al., 2019). Under carefully controlled reaction conditions, the extract of the earthworm *Eisenia andrei* could successfully manufacture nanosized red gold particles, and the marine worm *Polychaeta* is utilized to create silver nanoparticles. Additionally, ZnS/chitosan nanophotocatalysts made from the chitin found in the exoskeletons of crabs and starfish can be utilized to purify water.

2.4.3 Mining waste

High quantities of heavy metals and metalloids are produced during mining operations and are often deposited in deposits (Shirmehenji et al., 2021). Metals including copper, iron, aluminum, nickel, and gold can all be extracted from low-grade ores, tailings, slag, and electroplating waste (Iranmanesh and Hulliger, 2017). Using plum and tomato extract as a reducing agent and stabilizer for the green synthesis, research on mining waste has demonstrated that selenium nanoparticles can be produced. In mine tailings, heavy metals like arsenic, copper, lead, and zinc are present (Shirmehenji et al., 2021). Numerous studies on the biosynthesis of nanoparticles from mining waste have shown that *Pseudomonas stutzeri* bacterial biomass can be used to create copper nanoparticles from mine debris. Additionally, it is possible to create magnetic manganese ferrite nanoparticles and iron oxide nanoparticles from residual sludge and mud from mining.

2.4.4 Sewage and industrial waste

Waste produced by commercial or industrial activities is referred to as industrial waste. Sewage may also include pretreated industrial effluent as well as wastewater from households and workplaces. Hydroxyapatite nanoparticles can be made using leftover eggshells from industries (Horiguchi et al., 2022). Industrial waste water and sewage effluents are a rich source of metal elements. As a result, there has been a lot of interest in turning these wastes into value-added products. The fabrication of silica-based nanoparticles can be done from industrial waste (Oviedo et al., 2022). By using the coprecipitation approach, it has also been demonstrated that zinc ferrite nanopowders were made from steel industry wastes. Industrial wastes of citron juice, peel waste of lemon, orange, and mandarin can be used to make zero-valent iron nanoparticles (Amin et al., 2022). Fly ash of iron and steel industries has also been used to create nanometric TiO_2-based nanocatalysts. Using sewage sludge, TiO_2/Fe-based nanomeric catalysts can also be created (Khan et al., 2022). Dry sewage sludge can be used to create nanomaterials containing iron, iron oxides, and carbon-based nanoparticles (Jeyakumar and Vincent, 2022). Waste biomass can be used to create metallic nickel nanoparticles on biochar as catalysts. Copper-based nanomeric photocatalysts have also been derived from sewage sludge.

2.4.5 Electronic waste

All the discarded and unusable electronic and electric gadgets are considered as electronic garbage. Metallic nanoparticles can be recovered from these wastes and used again. Electronic devices have metallic parts and that metal portion can be recovered out of these devices (Brar et al., 2022). Adsorbent nanoparticles have been created from a range of electronic wastes; one such study on printed circuit boards resulted in the creation of nanoparticles that can be used to remove contaminants from a number of ecological matrices (Brar et al., 2022). Using bacterial biomass, electronic scrap leachate was converted into gold nanoparticles (Li et al., 2022). Additionally, e-waste can be used to recover gold and silver nanoparticles. Waste-printed circuit boards could be utilized to create copper nanoparticles (Rajkumar et al., 2022). One of the best strategies to decrease toxic electronic waste and synthesize metal nanoparticles of economic significance is to recycle e-waste.

2.5 Applications of metal nanoparticles

Metallic nanoparticles have applications in medicine and environmental cleanup (Olga et al., 2022). Metallic nanoparticles have antimicrobial properties. Silver nanoparticles are used in wound dressings. The mechanical strength and anticancer properties of gold nanoparticles are highly useful in drug delivery (Ko et al., 2022). Metal-based nanoparticles are used for their antibacterial properties, diagnostic assays, antifungal properties, gene delivery systems, soil pollutant removal, water treatment, packaging materials, food industry, industrial additives, electronics, paints and cosmetics, automobiles, petroleum refining, biosensors, catalysts, anticancer, radiotherapy enhancement, physics, mechanics, pharmaceuticals, optics, and fluorescent materials (Shnoudeh et al., 2019). Due to the strong antibacterial capabilities of silver nanoparticles, any living cell that comes into touch with them becomes poisonous and can aid in inhibiting protein synthesis. Gold nanoparticles are also employed in the production of window coatings and to reduce cellular activity (Dai and Compton, 2006). One of the best adsorbent materials is iron and iron oxide nanoparticles. They are used to remove heavy metals from aqueous solutions, including lead. Iron oxide nanoparticles are utilized in catalytic and photocatalytic reactions in addition to the removal of pollutants. Iron oxides are also used in propellant formulations to speed up combustion (Rahman et al., 2011). In combination with other nanomaterials, iron oxide nanoparticles aid in the retention of soil nutrients and improve water holding capacity by enhancing cation exchange capacity. Due to their magnetic and electrical characteristics, copper oxide nanoparticles have a wide range of applications. They are employed in the production of sensors, semiconductors, solar cells, and circuit boards (Eshkeiti et al., 2015). Cancer treatment involves the use of zinc oxide nanoparticles. They selectively kill cancerous cells while leaving healthy cells alone. Nanoparticles based on zinc are used to treat skin infections and in antidandruff shampoos (Abendrot and Kalinowska-Lis, 2018). The antibacterial properties of titanium dioxide nanoparticles have been exploited in catalytic processes. Titanium nanoparticles are used in water treatment techniques to remove contaminants (Saha et al., 2013). They also have a number of additional uses, such as the creation of nanopaints and antimicrobial paints. Large-scale uses have been employed for nickel nanoparticles. They have magnetic and catalytic activity and are chemically stable. They aid in the efficient operation of circuit boards, digital gadgets, and catalytic converters (Kuzminska et al., 2015). Nickel-based nanoparticles are also used in solar cells and supercapacitors. They can also be employed to reduce pollution. Platinum nanoparticles are also used in catalysis, photolysis, and biological applications. They are used as catalysts in semiconductors, solar cells, and catalytic converters (Senthilvelan et al., 2013). Platinum nanoparticles are used in biosensors as well. Platinum nanoparticles have been used because of their ability to fight cancer and diabetes. Nanoparticles are very valuable engineered agents as they have scope in large-scale applications starting from agriculture, water treatment, to food industry. Table 2.2 shows other studies reported by various researchers on waste-derived nanoparticles and their applications.

TABLE 2.2 Waste-derived nanoparticles and their applications.

Nanoparticle	Process used	Source	Sector	Application	References
Biochars	Adsorption	Corn stover, organic peel, and pistachio shell	Wastewater treatment	Removal of lead	Mireles et al. (2019)
Biochars	Adsorption	Cotton stalk	Wastewater treatment	Removal of lead	Gao et al. (2021)
Biochars	Adsorption	Peanut shell, chonta pulp, and corn cob	Wastewater treatment	Removal of cadmium and lead	Puglla et al. (2020)
Biochars	Adsorption	Banana stem and leaf	Wastewater treatment	Removal of cadmium and lead	Lie et al. (2022)
$CaCO_3$	Adsorption	Eggshell	Wastewater remediation	Lead adsorption	Wang et al. (2019)
Copper/Cuprous oxide nanoparticles decorated graphene oxide (Cu/Cu2O/rGO)	Electrocatalysis and biosensing	–	Wastewater treatment	Removal of dopamine	Devnani et al. (2019)
CuO	Photocatalysis	Printed circuit boards	Wastewater remediation	Dye removal	Nayak et al. (2019)
Graphene	Adsorption	Polyethylene terephthalate	Wastewater treatment	Dye removal	El Essawy et al. (2017)
Iron oxide nanoparticles	Adsorption	Mill scale	Wastewater treatment	Dye removal	Abdul Rahim Arifin et al. (2017)
Iron oxide nanoparticles	Adsorption, flocculation, coagulation	–	Wastewater treatment	Removal of oil droplets	Jabbar et al. (2022)

(Continued)

TABLE 2.2 (Continued)

Nanoparticle	Process used	Source	Sector	Application	References
Iron oxide nanoparticles	Adsorption	Steel waste	Wastewater treatment	Removal of congo red dye	Borth et al. (2021)
Magnetite nanoparticles	Adsorption	Mill scale waste	Wastewater treatment	Removal of Copper ions	Sulaiman et al. (2021)
Metal-doped ZnO	Photocatalysis	Fabric filter dust	Wastewater treatment	Dye removal	Wu et al. (2014)
Nano cuprous oxide	Complexation	E-waste	Pollutant monitoring in water	Detection of heavy metal mercury and dopamine in water	Abdelbasir et al. (2018)
$NiFe_2O_4$/ZnCuCr-LDH composite	Adsorption	Saccharin wastewater	Wastewater treatment	Dye removal	Zhang et al. (2020)
Porous aerogels	Adsorption	Paper, cotton textiles, plastic bottles	Wastewater treatment	Oil adsorption	Thai et al. (2019)
Porous silica nanoparticles	Complexation	Rice husks	Capturing of air pollutants	Capture of carbon dioxide gas	Zeng and Bai (2016)
Silica gel powder nanoparticles	Adsorption	Ceramic fiber honeycombs	Capturing of air pollutants	Capture of carbon dioxide gas	Wu et al. (2021)
Silica nanoparticles	Adsorption	Sugarcane waste ash	Wastewater remediation	Dye removal	Rovani et al. (2018)
SiO_2	Adsorption	Corn husk waste	Wastewater treatment	Dye removal	Peres et al. (2018)
ZnO	Photocatalysis	Printed circuit boards	Wastewater remediation	Dye removal	Nayak et al. (2019)

2.6 Conclusion and future paradigms

This book chapter explores the utilization of waste products in the production of nanoparticles through sustainable approaches. While the concept of recycling waste into environmentally beneficial materials is promising, there are still significant knowledge gaps concerning nanoscale materials that need to be addressed before implementing these ideas in practical applications. Key areas of concern include energy consumption, residual waste generation, behavior in different environmental settings, exposure routes, and potential toxic effects. It is crucial to conduct thorough cost-benefit analyses to ensure that nanoparticle processes are economically viable beyond the experimental stage.

Recent studies have expanded the exploration of nanotechnology applications by utilizing new materials, prompting the need for further research and development to identify safer alternatives with low toxicity and better properties. To address the challenges posed by existing waste management systems, experts suggest implementing life cycle evaluations and risk assessments earlier in the development of new methods, considering the potential environmental impacts of nanoparticles and their reaction products. Green chemistry principles provide a suitable framework for nanoparticle synthesis, but it does not guarantee their safety or environmental impact. Understanding the complex factors involved in waste-derived nanoparticles is crucial for making cautious and ethical choices to achieve sustainability. Standardizing the use of nanoparticles with proper protocols and prioritizing greener substitutes with minimal negative effects is important. By working together, science, industry, and the economy can enhance their effectiveness.

References

Abdelbasir, S.M., El-Sheikh, S.M., Morgan, V.L., Schmidt, H., Casso-Hartmann, L.M., Vanegas, D.C., et al., 2018. Graphene-anchored cuprous oxide nanoparticles from waste electric cables for electrochemical sensing. ACS Sustainable Chemistry & Engineering 6 (9), 12176–12186.

Abdul Rahim Arifin, A.S., Ismail, I., Abdullah, A.H., Shafiee, F.N., Nazlan, R., Ibrahim, I.R., 2017. Iron oxide nanoparticles derived from mill scale waste as potential scavenging agent in dye wastewater treatment for batik industry, Solid State Phenomena, Vol. 268. Trans Tech Publications Ltd., pp. 393–398.

Abendrot, M., Kalinowska-Lis, U., 2018. Zinc-containing compounds for personal care applications. International Journal of Cosmetic Science 40 (4), 319–327.

Adebowale, K., Egbedina, A., Shonde, B., 2020. Adsorption of lead ions on magnetically separable Fe_3O_4 watermelon composite. Applied Water. Science (New York, N.Y.) 10 (10), 1–8.

Ahmad, M.Z., Akhter, S., Jain, G.K., Rahman, M., Pathan, S.A., Ahmad, F.J., et al., 2010. Metallic nanoparticles: technology overview & drug delivery applications in oncology. Expert Opinion on Drug Delivery 7 (8), 927–942.

Ajmal, N., Saraswat, K., Bakht, M.A., Riadi, Y., Ahsan, M.J., Noushad, M., 2019. Cost-effective and eco-friendly synthesis of titanium dioxide (TiO_2) nanoparticles using fruit's peel agro-waste extracts: characterization, in vitro antibacterial, antioxidant activities. Green Chemistry Letters and Reviews 12 (3), 244–254.

Akbayrak, S., Özçifçi, Z., Tabak, A., 2020. Activated carbon derived from tea waste: a promising supporting material for metal nanoparticles used as catalysts in hydrolysis of ammonia borane. Biomass and Bioenergy 138, 105589.

Ali, O.M., Hasanin, M.S., Suleiman, W.B., Helal, E.E.H., Hashem, A.H., 2022. Green biosynthesis of titanium dioxide quantum dots using watermelon peel waste: antimicrobial, antioxidant, and anticancer activities. Biomass Conversion and Biorefinery 1–12.

Amin, A.M., Rayan, D.A., Ahmed, Y.M., El-Shall, M.S., Abdelbasir, S.M., 2022. Zinc ferrite nanoparticles from industrial waste for Se (IV) elimination from wastewater. Journal of Environmental Management 312, 114956.

An, Z., Xue, J., Cao, H., Zhu, C., Wang, H., 2019. A facile synthesis of silicon carbide nanoparticles with high specific surface area by using corn cob. Advanced Powder Technology 30 (1), 164–169.

Anand, K., Gengan, R.M., Phulukdaree, A., Chuturgoon, A., 2015. Agroforestry waste Moringa oleifera petals mediated green synthesis of gold nanoparticles and their anti-cancer and catalytic activity. Journal of Industrial and Engineering Chemistry 21, 1105–1111.

Angamuthu, M., Satishkumar, G., Landau, M.V., 2017. Precisely controlled encapsulation of Fe_3O_4 nanoparticles in mesoporous carbon nanodisk using iron based MOF precursor for effective dye removal. Microporous and Mesoporous Materials 251, 58–68.

Baiocco, D., Lavecchia, R., Natali, S., Zuorro, A., 2016. Production of metal nanoparticles by agro-industrial wastes: a green opportunity for nanotechnology. Chemical Engineering Transactions 47, 67–72.

Bhardwaj, A.K., Naraian, R., 2020. Green synthesis and characterization of silver NPs using oyster mushroom extract for antibacterial efficacy. Journal of Chemistry, Environmental Sciences and its Applications 7 (1), 13–18.

Bhardwaj, A.K., Naraian, R., 2021. Cyanobacteria as biochemical energy source for the synthesis of inorganic nanoparticles, mechanism and potential applications: a review. 3 Biotech 11 (10), 445.

Bhardwaj, A.K., Shukla, A., Maurya, S., Singh, S.C., Uttam, K.N., Sundaram, S., et al., 2018. Direct sunlight enabled photo-biochemical synthesis of silver nanoparticles and their bactericidal efficacy: photon energy as key for size and distribution control. Journal of Photochemistry and Photobiology B: Biology 188, 42–49.

Bhattacharya, R., Mukherjee, P., 2008. Biological properties of "naked" metal nanoparticles. Advanced Drug Delivery Reviews 60 (11), 1289–1306.

Borth, K.W., Galdino, C.W., de Carvalho Teixeira, V., Anaissi, F.J., 2021. Iron oxide nanoparticles obtained from steel waste recycling as a green alternative for Congo red dye fast adsorption. Applied Surface Science 546, 149126.

Brar, K.K., Magdouli, S., Othmani, A., Ghanei, J., Narisetty, V., Sindhu, R., et al., 2022. Green route for recycling of low-cost waste resources for the biosynthesis of nanoparticles (NPs) and nanomaterials (NMs)-A review. Environmental Research 207, 112202.

Chen, Z., Wei, W., Chen, H., Ni, B., 2022. Recent advances in waste-derived functional materials for wastewater remediation. Eco-Environment & Health.

Cui, Y., Lai, X., Liu, K., Liang, B., Ma, G., Wang, L., 2020. Ginkgo biloba leaf polysaccharide stabilized palladium nanoparticles with enhanced peroxidase-like property for the colorimetric detection of glucose. RSC Advances 10 (12), 7012–7018.

Dai, X., Compton, R.G., 2006. Direct electrodeposition of gold nanoparticles onto indium tin oxide film coated glass: application to the detection of arsenic (III). Analytical Sciences 22 (4), 567–570.

Dauthal, P., Mukhopadhyay, M., 2015. Biofabrication, characterization, and possible bio-reduction mechanism of platinum nanoparticles mediated by agro-industrial waste and their catalytic activity. Journal of Industrial and Engineering Chemistry 22, 185–191.

Desalegn, B., Megharaj, M., Chen, Z., Naidu, R., 2019. Green synthesis of zero valent iron nanoparticle using mango peel extract and surface characterization using XPS and GC-MS. Heliyon 5 (5), e01750.

Devipriya, D., Roopan, S.M., 2017. Cissus quadrangularis mediated ecofriendly synthesis of copper oxide nanoparticles and its antifungal studies against Aspergillus niger. Aspergillus flavus. Materials Science and Engineering: C 80, 38–44.

Devnani, H., Rashid, N., Ingole, P.P., 2019. Copper/Cuprous oxide nanoparticles decorated reduced graphene oxide sheets based platform for bio-electrochemical sensing of dopamine. ChemistrySelect 4 (2), 633–643.

Donga, S., Chanda, S., 2021. Facile green synthesis of silver nanoparticles using Mangifera indica seed aqueous extract and its antimicrobial, antioxidant and cytotoxic potential (3-in-1 system). Artificial Cells. Nanomedicine, and Biotechnology 49 (1), 292–302.

Durán, N., Seabra, A.B., 2012. Metallic oxide nanoparticles: state of the art in biogenic syntheses and their mechanisms. Applied Microbiology and Biotechnology 95 (2), 275–288.

El Essawy, N.A., Ali, S.M., Farag, H.A., Konsowa, A.H., Elnouby, M., Hamad, H.A., 2017. Green synthesis of graphene from recycled PET bottle wastes for use in the adsorption of dyes in aqueous solution. Ecotoxicology and Environmental Safety 145, 57–68.

El-Arab, N.B., 2018. Synthesis and characterization of zinc oxide nanoparticles using green and chemical synthesis techniques for phenol decontamination. International Journal of Nanoelectronics and Materials 11 (2), 179–194.

El-Nasr, R.S., Abdelbasir, S.M., Kamel, A.H., Hassan, S.S., 2020. Environmentally friendly synthesis of copper nanoparticles from waste printed circuit boards. Separation and Purification Technology 230, 115860.

Eshkeiti, A., Reddy, A.S., Emamian, S., Narakathu, B.B., Joyce, M., Joyce, M., et al., 2015. Screen printing of multilayered hybrid printed circuit boards on different substrates. IEEE transactions on components, packaging and manufacturing technology 5 (3), 415–421.

Ettlinger, R., Lächelt, U., Gref, R., Horcajada, P., Lammers, T., Serre, C., et al., 2022. Toxicity of metal–organic framework nanoparticles: from essential analyses to potential applications. Chemical Society Reviews.

Farag, S., Amr, A., El-Shafei, A., Asker, M.S., Ibrahim, H.M., 2021. Green synthesis of titanium dioxide nanoparticles via bacterial cellulose (BC) produced from agricultural wastes. Cellulose 28 (12), 7619–7632.

Ganesan, S., Sivam, S., Elancheziyan, M., Senthilkumar, S., Ramakrishan, S.G., Soundappan, T., et al., 2022. Novel delipidated chicken feather waste-derived carbon-based molybdenum oxide nanocomposite as efficient electrocatalyst for rapid detection of hydroquinone and catechol in environmental waters. Environmental Pollution 293, 118556.

Gao, L., Li, Z., Yi, W., Li, Y., Zhang, P., Zhang, A., et al., 2021. Impacts of pyrolysis temperature on lead adsorption by cotton stalk-derived biochar and related mechanisms. Journal of Environmental Chemical Engineering 9 (4), 105602.

Ghazwani, A.A., 2015. Biosynthesis of silver nanoparticles by Aspergillus niger, Fusarium oxysporum and Alternaria solani. African Journal of Biotechnology 14 (26), 2170–2174.

Ghosh, S., Sarkar, B., Kaushik, A., Mostafavi, E., 2022. Nanobiotechnological prospects of probiotic microflora: synthesis, mechanism, and applications. Science of The Total Environment 838, 156212.

Gokila, S., Gomathi, T., Sudha, P.N., Anil, S., 2017. Removal of the heavy metal ion chromiuim (VI) using Chitosan and Alginate nanocomposites. International Journal of Biological Macromolecules 104, 1459–1468.

Guan, Z., Ying, S., Ofoegbu, P.C., Clubb, P., Rico, C., He, F., et al., 2022. Green synthesis of nanoparticles: current developments and limitations. Environmental Technology & Innovation 102336.

Gubitosa, J., Rizzi, V., Laurenzana, A., Scavone, F., Frediani, E., Fibbi, G., et al., 2022. The "end life" of the grape pomace waste become the new beginning: the development of a virtuous cycle for the green synthesis of gold nanoparticles and removal of emerging contaminants from water. Antioxidants 11 (5), 994.

Harisha, K.S., Parushuram, N., Asha, S., Suma, S.B., Narayana, B., Sangappa, Y., 2021. Eco-synthesis of gold nanoparticles by Sericin derived from Bombyx mori silk and catalytic study on degradation of methylene blue. Particulate Science and Technology 39 (2), 131–140.

Horiguchi, G., Ito, M., Ito, A., Kamiya, H., Okada, Y., 2022. Controlling particle adhesion of synthetic and sewage sludge ashes in high temperature combustion using metal oxide nanoparticles. Fuel 321, 124110.

Huang, M., Du, L., Feng, J.X., 2016. Photochemical synthesis of silver nanoparticles/eggshell membrane composite, its characterization and antibacterial activity. Science of Advanced Materials 8 (8), 1641–1647.

Ibrahim, I.K., Hussin, S.M., Al-Obaidi, Y., 2015. Extraction of cellulose nano crystalline from cotton by ultrasonic and its morphological and structural characterization. International Journal of Materials Chemistry and Physics 1, 99–109.

Iqbal, P., Preece, J.A., Mendes, P.M., 2012. Nanotechnology: the "top-down" and "bottom-up" approaches. Supramolecular chemistry: from molecules to nanomaterials.

Iranmanesh, M., Hulliger, J., 2017. Magnetic separation: its application in mining, waste purification, medicine, biochemistry and chemistry. Chemical Society Reviews 46 (19), 5925–5934.

Jabbar, K.Q., Barzinjy, A.A., Hamad, S.M., 2022. Iron oxide nanoparticles: preparation methods, functions, adsorption and coagulation/flocculation in wastewater treatment. Environmental. Nanotechnology, Monitoring & Management 17, 100661.

Jain, R., Mendiratta, S., Kumar, L., Srivastava, A., 2021. Green synthesis of iron nanoparticles using Artocarpus heterophyllus peel extract and their application as a heterogeneous Fenton-like catalyst for the degradation of Fuchsin Basic dye. Current research in green and sustainable. chemistry (Weinheim an der Bergstrasse, Germany) 4, 100086.

Jamkhande, P.G., Ghule, N.W., Bamer, A.H., Kalaskar, M.G., 2019. Metal nanoparticles synthesis: an overview on methods of preparation, advantages and disadvantages, and applications. Journal of Drug Delivery Science and Technology 53, 101174.

Jeyakumar, R.B., Vincent, G.S., 2022. Recent advances and perspectives of nanotechnology in anaerobic digestion: a new paradigm towards sludge biodegradability. Sustainability 14 (12), 7191.

Jiang, S., Gao, S., Kong, J., Jin, X., Wei, D., Li, D., et al., 2019. Study on the synthesis of β-SiC nanoparticles from diamond-wire silicon cutting waste. RSC Advances 9 (41), 23785–23790.

Jiang, Y., Deng, Y.P., Liang, R., Chen, N., King, G., Yu, A., et al., 2022. Linker-compensated metal–organic framework with electron delocalized metal sites for bifunctional oxygen electrocatalysis. Journal of the American Chemical Society 144 (11), 4783–4791.

Kanchi, S., Kumar, G., Lo, A.Y., Tseng, C.M., Chen, S.K., Lin, C.Y., et al., 2018. Exploitation of de-oiled jatropha waste for gold nanoparticles synthesis: a green approach. Arabian Journal of Chemistry 11 (2), 247–255.

Kango, S., Kalia, S., Celli, A., Njuguna, J., Habibi, Y., Kumar, R., 2013. Surface modification of inorganic nanoparticles for development of organic–inorganic nanocomposites—a review. Progress in Polymer Science 38 (8), 1232–1261.

Karthik, P.E., Rajan, H., Jothi, V.R., Sang, B.I., Yi, S.C., 2022. Electronic wastes: a near inexhaustible and an unimaginably wealthy resource for water splitting electrocatalysts. Journal of Hazardous Materials 421, 126687.

Khan, M.S.J., Kamal, T., Ali, F., Asiri, A.M., Khan, S.B., 2019. Chitosan-coated polyurethane sponge supported metal nanoparticles for catalytic reduction of organic pollutants. International Journal of Biological Macromolecules 132, 772–783.

Khan, F., Kesarwani, H., Kataria, G., Mittal, G., Sharma, S., 2022. Application of titanium dioxide nanoparticles for the design of oil well cement slurry—a study based on compressive strength, setting time and rheology. Journal of Adhesion Science and Technology 1—17.

Ko, W.C., Wang, S.J., Hsiao, C.Y., Hung, C.T., Hsu, Y.J., Chang, D.C., et al., 2022. Pharmacological role of functionalized gold nanoparticles in disease applications. Molecules (Basel, Switzerland) 27 (5), 1551.

Kuzminska, M., Carlier, N., Backov, R., Gaigneaux, E.M., 2015. Magnetic nanoparticles: improving chemical stability via silica coating and organic grafting with silanes for acidic media catalytic reactions. Applied Catalysis A: General 505, 200—212.

Li, H., Sun, Z., Zhang, L., Tian, Y., Cui, G., Yan, S., 2016. A cost-effective porous carbon derived from pomelo peel for the removal of methyl orange from aqueous solution. Colloids and Surfaces A: Physicochemical and Engineering Aspects 489, 191—199.

Li, H., Ye, M., Fu, Z., Zhang, H., Wang, G., Zhang, Y., 2022. A freestanding, hierarchically porous poly (imine dioxime) membrane enabling selective gold recovery from e-waste with unprecedented capacity. EcoMat e12248.

Mahmud, J., Sarmast, E., Shankar, S., Lacroix, M., 2022. Advantages of nanotechnology developments in active food packaging. Food Research International 111023.

Mansoor, A., Khan, M.T., Mehmood, M., Khurshid, Z., Ali, M.I., Jamal, A., 2022. Synthesis and characterization of titanium oxide nanoparticles with a novel biogenic process for dental application. Nanomaterials 12 (7), 1078.

Maroufi, S., Mayyas, M., Sahajwalla, V., 2017. Nano-carbons from waste tyre rubber: an insight into structure and morphology. Waste Management 69, 110—116.

Mathew, S.S., Sunny, N.E., Shanmugam, V., 2021. Green synthesis of anatase titanium dioxide nanoparticles using Cuminum cyminum seed extract; effect on Mung bean (Vigna radiata) seed germination. Inorganic Chemistry Communications 126, 108485.

Meer, B., Andleeb, A., Iqbal, J., Ashraf, H., Meer, K., Ali, J.S., et al., 2022. Bio-assisted synthesis and characterization of zinc oxide nanoparticles from lepidium sativum and their potent antioxidant, antibacterial and anticancer activities. Biomolecules 12 (6), 855.

Mireles, S., Parsons, J., Trad, T., Cheng, C.L., Kang, J., 2019. Lead removal from aqueous solutions using biochars derived from corn stover, orange peel, and pistachio shell. International Journal of Environmental Science and Technology 16 (10), 5817—5826.

Mirgane, N.A., Shivankar, V.S., Kotwal, S.B., Wadhawa, G.C., Sonawale, M.C., 2021. Waste pericarp of ananas comosus in green synthesis zinc oxide nanoparticles and their application in waste water treatment. Materials Today: Proceedings 37, 886—889.

Nayak, P., Kumar, S., Sinha, I., Singh, K.K., 2019. ZnO/CuO nanocomposites from recycled printed circuit board: preparation and photocatalytic properties. Environmental Science and Pollution Research 26 (16), 16279—16288.

Olga, M., Jana, M., Anna, M., Irena, K., Jan, M., Alena, Č., 2022. Antimicrobial properties and applications of metal nanoparticles biosynthesized by green methods. Biotechnology Advances 107905.

Omran, B.A., Nassar, H.N., Fatthallah, N.A., Hamdy, A., El-Shatoury, E.H., El-Gendy, N.S., 2018. Waste upcycling of Citrus sinensis peels as a green route for the synthesis of silver nanoparticles. Energy Sources, Part A: Recovery, Utilization, and Environmental Effects 40 (2), 227—236.

Oviedo, L.R., Muraro, P.C.L., Pavoski, G., Espinosa, D.C.R., Ruiz, Y.P.M., Galembeck, A., et al., 2022. Synthesis and characterization of nanozeolite from (agro) industrial waste for application in heterogeneous photocatalysis. Environmental Science and Pollution Research 29 (3), 3794—3807.

Pan, Z., Lin, Y., Sarkar, B., Owens, G., Chen, Z., 2020. Green synthesis of iron nanoparticles using red peanut skin extract: synthesis mechanism, characterization and effect of conditions on chromium removal. Journal of Colloid and Interface Science 558, 106—114.

Pandey, S., De Klerk, C., Kim, J., Kang, M., Fosso-Kankeu, E., 2020. Eco friendly approach for synthesis, characterization and biological activities of milk protein stabilized silver nanoparticles. Polymers 12 (6), 1418.

Patra, J.K., Baek, K.H., 2015. Novel green synthesis of gold nanoparticles using Citrullus lanatus rind and investigation of proteasome inhibitory activity, antibacterial, and antioxidant potential. International Journal of Nanomedicine 10, 7253.

Patra, J.K., Das, G., Baek, K.H., 2016. Phyto-mediated biosynthesis of silver nanoparticles using the rind extract of watermelon (Citrullus lanatus) under photo-catalyzed condition and investigation of its antibacterial, anticandidal and antioxidant efficacy. Journal of Photochemistry and Photobiology B: Biology 161, 200—210.

Pham, D.T., Quan, T., Mei, S., Lu, Y., 2022. Colloidal metal sulfide nanoparticles for high performance electrochemical energy storage systems. Current Opinion in Green and Sustainable Chemistry 100596.

Praveena, S.M., Xin-Yi, C.K., Liew, J.Y.C., Khan, M.F., 2022. Functionalized magnetite nanoparticle coagulants with tropical fruit waste extract: a potential for water turbidity removal. Arabian Journal for Science and Engineering 1—10.

Puglla, E.P., Guaya, D., Tituana, C., Osorio, F., García-Ruiz, M.J., 2020. Biochar from agricultural by-products for the removal of lead and cadmium from drinking water. Water 12 (10), 2933.

Rahman, M.M., Khan, S.B., Jamal, A., Faisal, M., Aisiri, A.M., 2011. Iron oxide nanoparticles. Nanomaterials 3, 43—67.

Rajarao, R., Sahajwalla, V., 2016. A cleaner, sustainable approach for synthesising high purity silicon carbide and silicon nitride nanopowders using macadamia shell waste. Journal of Cleaner Production 133, 1277—1282.

Rajkumar, S., Elanthamilan, E., Wang, S.F., Chryso, H., Balan, P.V.D., Merlin, J.P., 2022. One-pot green recovery of copper oxide nanoparticles from discarded printed circuit boards for electrode material in supercapacitor application. Resources, Conservation and Recycling 180, 106180.

Rajput, V., Gupta, R.K., Prakash, J., 2022. Engineering metal oxide semiconductor nanostructures for enhanced charge transfer: fundamentals and emerging SERS applications. Journal of Materials Chemistry C .

Ravi, B., Duraisamy, P., Marimuthu, T., 2023. A novel integrated circular economy approach in green synthesis of copper oxide nanoparticles from waste printed circuit boards and utilization of its residue for preparation of carbon engulfed nano polymer membrane. Journal of Cleaner Production 383, 135457.

Rodríguez-Félix, F., López-Cota, A.G., Moreno-Vásquez, M.J., Graciano-Verdugo, A.Z., Quintero-Reyes, I.E., Del-Toro-Sánchez, C.L., et al., 2021. Sustainable-green synthesis of silver nanoparticles using safflower (Carthamus tinctorius L.) waste extract and its antibacterial activity. Heliyon 7 (4), e06923.

Rovani, S., Santos, J.J., Corio, P., Fungaro, D.A., 2018. Highly pure silica nanoparticles with high adsorption capacity obtained from sugarcane waste ash. ACS Omega 3 (3), 2618–2627.

Rovani, S., Santos, J.J., Guilhen, S.N., Corio, P., Fungaro, D.A., 2020. Fast, efficient and clean adsorption of bisphenol-A using renewable mesoporous silica nanoparticles from sugarcane waste ash. RSC Advances 10 (46), 27706–27712.

Saha, I., Bhattacharya, S., Mukhopadhyay, A., Chattopadhyay, D., Ghosh, U., Chatterjee, D., 2013. Role of nanotechnology in water treatment and purification: potential applications and implications. International Journal of Chemical Science and Technology 3 (3), 59–64.

Saleh, M., Isik, Z., Aktas, Y., Arslan, H., Yalvac, M., Dizge, N., 2021. Green synthesis of zero valent iron nanoparticles using Verbascum thapsus and its Cr (VI) reduction activity. Bioresource Technology Reports 13, 100637.

Salvador-Morales, C., Grodzinski, P., 2022. Nanotechnology tools enabling biological discovery. ACS nano 16 (4), 5062–5084.

Sangeetha, J., Thangadurai, D., Hospet, R., Purushotham, P., Manowade, K.R., Mujeeb, M.A., et al., 2017. Production of bionanomaterials from agricultural wastes. Nanotechnology. Springer, Singapore, pp. 33–58.

Santhosh, A., Theertha, V., Prakash, P., Chandran, S.S., 2021. From waste to a value added product: green synthesis of silver nanoparticles from onion peels together with its diverse applications. Materials Today: Proceedings 46, 4460–4463.

Scida, K., Stege, P.W., Haby, G., Messina, G.A., García, C.D., 2011. Recent applications of carbon-based nanomaterials in analytical chemistry: critical review. Analytica Chimica Acta 691 (1–2), 6–17.

Senthilvelan, S., Chandraboss, V.L., Karthikeyan, B., Natanapatham, L., Murugavelu, M., 2013. TiO_2, ZnO and nanobimetallic silica catalyzed photodegradation of methyl green. Materials Science in Semiconductor Processing 16 (1), 185–192.

Shirmehenji, R., Javanshir, S., Honarmand, M., 2021. A green approach to the bio-based synthesis of selenium nanoparticles from mining waste. Journal of Cluster Science 32 (5), 1311–1323.

Shnoudeh, A.J., Hamad, I., Abdo, R.W., Qadumii, L., Jaber, A.Y., Surchi, H.S., et al., 2019. Synthesis, characterization, and applications of metal nanoparticles. Biomaterials and Bionanotechnology. Academic Press, pp. 527–612.

Sulaiman, S.Y.A.Z.A.N.A., Ismail, I., Man, H.C., Rosdi, N., 2021. Rapid adsorption of magnetite nanoparticles from recycled mill scale waste as potential adsorbent for removal of Cu (II) ions, Solid State Phenomena, Vol. 317. Trans Tech Publications Ltd., pp. 270–275.

Thai, Q.B., Le, D.K., Luu, T.P., Hoang, N., Nguyen, D., Duong, H.M., 2019. Aerogels from wastes and their applications aerogels from wastes and their applications. JOJ Material Science 5, 1–5.

Wary, R.R., Baglari, S., Brahma, D., Gautam, U.K., Kalita, P., Baruah, M.B., 2022. Synthesis, characterization, and photocatalytic activity of ZnO nanoparticles using water extract of waste coconut husk. Environmental Science and Pollution Research 29 (28), 42837–42848.

Wu, J., Zhu, X., Yang, F., Ge, T., Wang, R., 2021. Easily-synthesized and low-cost amine-functionalized silica sol-coated structured adsorbents for CO_2 capture. Chemical Engineering Journal 425, 131409.

Wu, Z.J., Huang, W., Cui, K.K., Gao, Z.F., Wang, P., 2014. Sustainable synthesis of metals-doped ZnO nanoparticles from zinc-bearing dust for photodegradation of phenol. Journal of Hazardous Materials 278, 91–99.

Xu, J.T., Chen, X.Q., 2019. Preparation and characterization of spherical cellulose nanocrystals with high purity by the composite enzymolysis of pulp fibers. Bioresource Technology 291, 121842.

Xu, Z.P., Zeng, Q.H., Lu, G.Q., Yu, A.B., 2006. Inorganic nanoparticles as carriers for efficient cellular delivery. Chemical Engineering Science 61 (3), 1027–1040.

Zeng, W., Bai, H., 2016. High-performance CO_2 capture on amine-functionalized hierarchically porous silica nanoparticles prepared by a simple template-free method. Adsorption 22 (2), 117–127.

Zhang, H., Xia, B., Wang, P., Wang, Y., Li, Z., Wang, Y., et al., 2020. From waste to waste treatment: mesoporous magnetic $NiFe_2O_4$/ZnCuCr-layered double hydroxide composite for wastewater treatment. Journal of Alloys and Compounds 819, 153053.

Zhu, Q., Chua, M.H., Ong, P.J., Lee, J.J.C., Chin, K.L.O., Wang, S., et al., 2022. Recent advances in nanotechnology-based functional coatings for the built environment. Materials Today Advances 15, 100270.

Chapter 3

Fundamental scope of nanomaterial synthesis from wastes

Pooja Thathola[1], Priyanka Adhikari[2], Vibhash Dhyani[1] and Dinesh Chandola[1]
[1]*Govind Ballabh Pant National Institute of Himalayan Environment, Almora, Uttarakhand, India,* [2]*National Institute of Pharmaceutical Education and Research, Guwahati, Assam, India*

3.1 Introduction

Over the past years, engineered nanomaterials have demonstrated promising advanced techniques in many areas, ranging from the transformation of energy and storage (Verma et al., 2016), to monitoring of pollution, packaging (de Abreu et al., 2012), food ingredients in agriculture and controlled delivery, technology related to the membrane (Goh et al., 2014); treatment of the wastewater, delivery of the drug and in diagnosis purpose, engineering related to bone and tissue (Shadjou and Hasanzadeh, 2016), etc. Nowadays, nanomaterials engineering and nanotechnology are gradually increasingly concerned with sustainable approaches, for the reduction of negative environmental impacts and associated health risks, by using the disposal of novel nanomaterials. The synthesis of nanomaterials is done through several methods, which mainly depend on the end-user demand (Bhardwaj et al., 2021). Two major approaches are followed for the synthesis of nanomaterials, (1) top-down, in this approach, molecules of larger size are reduced up to the nanoscale level, and (2) bottom-up, in this approach, molecules having the smaller size are combined for the production of the nanomaterials. There are several synthesis methods involved in these approaches; the four major categories of these approaches are chemical, physical, mechanical, and biological. Materials of sizes lesser the 100 nm are called nanomaterials. The particle size of the material is only the way of basic distinction between nanomaterial and nonnanomaterial. A particle size lesser than 100 nm is referred to as a nanomaterial, and above 100 nm particle size is considered nonnanomaterial. The nanomaterials have a higher surface/volume ratio, which shows good performance, so they can be applied to different areas that is, nanosorbents, nanosensors, and fuel cells. Furthermore, the performance of nanomaterials is much higher in the adsorption of gas and liquid in both phases (Xiang et al., 2015). Considering the recent situation, if the waste is efficaciously recycled for the synthesis of NMs, waste management is beneficial for the environment. There are several studies reported on nanomaterial production from waste. Agricultural waste for the production of NMs is beneficial in terms of cost. Notably, various parts of plants have been expansively used to produce beneficial goods, including pencils, paper, and wooden furniture (Martinez-Alier, 2009). Furthermore, the use of waste in the production of nanomaterials provides a solution to the environmental burdens and directly reduces the impact of that waste on the environment. In addition, industrial waste is also a major problem nowadays. Approximately industries generate 7.6 billion tons of waste directly disposed to the environment every year (Li et al., 2005). Waste produced from industries has various type of negative impacts on the environment, as this waste is toxic, corrosive, flammable, highly reactive, infectious, and radioactive. Consequently, it is high time to recycle these industrial wastes to produce value-added products for the sake of environmental safety. Two main approaches are widely used to manage various waste materials, that is, energy recovery and mechanical recycling. Wastes can be incinerated for the energy recovery process. In contrast, waste material can be reprocessed into secondary raw material by keeping its intact basic structure in the mechanical recycling process. In terms of environmental safety, mechanical recycling is more favorable than energy recovery as incineration produces hazardous chemicals. However, mechanical recycling is often costly and nonprofitable. Hence, it is inevitable to recycle and valorize waste materials into products that have high economic value and appreciable end use. In view of these facts, many new processing routes have been explored for minimizing the associated cost and environmental impact of the recycling process.

Green and Sustainable Approaches Using Wastes for the Production of Multifunctional Nanomaterials. DOI: https://doi.org/10.1016/B978-0-443-19183-1.00009-X

3.2 Waste as a synthesis of nanomaterial

The source of emission depends on high materials related to the waste that is produced in massive amounts that should be classified into two categories: firstly, waste generated by industries, and secondly, consumers-generated waste. Both categories of waste should be used as value-added products. In the industrial sector, the waste generated is mainly batteries, rubber tires, and wastewater, which are the prominent sources of contamination, that is, carbon, lead, zinc, copper, etc. The consumer products include overwhelming amounts of electronic waste and plastic waste. Electronic waste, that is, e-waste, is the main source of recoverable and semiprecious metals, which mainly depends on the polymer type and the purity degree, the plastic waste is normally or mainly recycled and further reused for material packing, or the plastic waste should be used as the raw material in many other applications, which mainly include construction materials, composites of fiber, paper, new polymers, and nanoparticles of carbon.

3.2.1 Different types of waste for nanomaterial synthesis

A wide range of materials generated from waste is generally recycled for the retrieval of valuable products; there are many nanomaterials reported earlier that are mainly generated from the waste (Fig. 3.1). Furthermore, the recovery and synthesis of nanomaterials or nanoparticles from waste products have been reported by many researchers (Table 3.1). Researchers established the synthesizing method of materials from waste such as batteries, rubber tires, and waste sludge. Those materials are also processed for the NMs recovery, many of them are mentioned in the subsections. The used and waste items used for the recovery and the synthesis of nanomaterials of their metals and oxides of the metal should be classified into different categories (i.e.,) batteries, used instruments and machines, and sludges and wastes.

3.2.1.1 *Industrial waste*

For sustainable development, the scientific community around the globe selects bioresources that are mainly safe, conferring "green" credentials. Biomass generated from different types of industries has been long recognized as waste, and mainly this waste is the source of environmental contamination. Food industries mainly discharge proteins, celluloses, and carbohydrates. Moreover, a large amount of pressed cake during oil extraction is a major waste in the oil industry. Nowadays, consumers need high value of nutritive in their meal protein, so, for the extraction of high nutritive meal protein, the huge amount of fibrous waste material ("oil-and-protein") is generally discarded from the industries. At present, researchers reported very few ways to use this kind of waste; normally, the residue of this waste is usually disposed of in landfills and, hence, is an "end-of-pipe" waste. As industries find an alternative use for the waste, the upgradation of this "end-of-pipe" waste for its secondary use is prudent from both ecological and economic points of view. In industries, solid waste such as batteries, tires, and waste sludge is generated in huge amounts. Nowadays, researchers are trying to reduce this waste in value-added products or recycle those products for remediation purposes by designing materials such as nanomaterials.

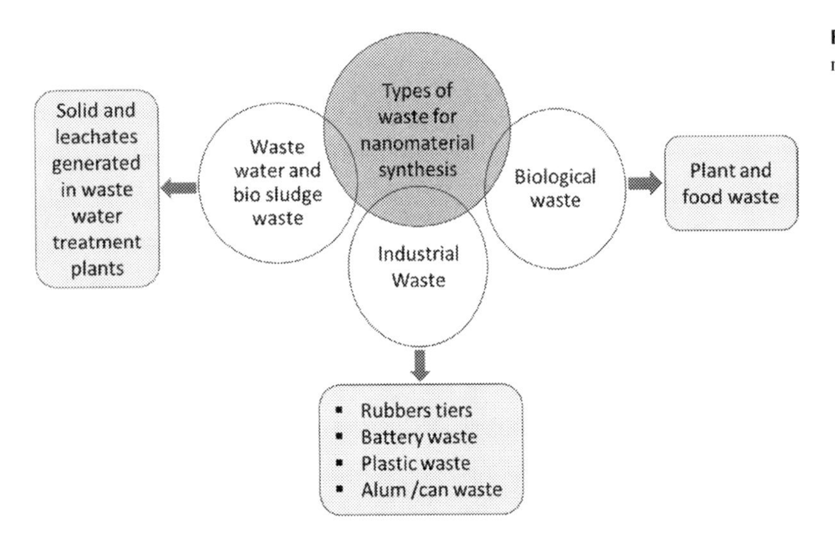

FIGURE 3.1 Different types of waste sources for nanomaterial synthesis.

TABLE 3.1 List of waste-derived nanomaterial.

S. no.	Nanomaterials	Waste material	Size (nm)	References
1	Nanoscale single crystalline g- and a-Al powder	Aluminum foil waste	20–300	El-Amir et al. (2016)
2	Activated Carbon	Waste tires	200–300	Ahmad et al. (2009)
3	Carbon nanosheets	Plant waste	410	Chen et al. (2016)
4	Zn nanoparticles	Spent zinc manganese battery waste	100–300	Xiang et al. (2015)
5	Pb nanoparticles	Cathode ray-tube funnel glass waste	4–34	Xing and Zhang (2011)
6	Nano Ni/Fe particles	Steel pickling waste liquor	20–50	Fang et al. (2011)
7	Biogenic silica nanoparticle	Rice husk waste	10–30	Alshatwi et al. (2015)
8	Au and Ag nanoparticles	Breynia rhamnoides biomass waste	37	Gangula et al. (2011)
9	Solid SiO_2e poly- (styrene sulfonic acid) nanocomposite	Polystyrene waste	67	Gangula et al. (2011)
10	Carbon nanoparticles	Plastic bag wastes	10–400	Hu et al. (2014)
11	Carbon nanoparticles	Pomelo peel wastes	–	Lu et al. (2012)
12	Carbon Nanosheets	Hemp waste	1–6	Wang et al. (2017)
13	Nanohydroxyapatite	Eggshell waste	5–20	Akter Jahan et al. (2017)
14	Graphene	cigarette filter waste	–	Chen et al. (2018)
15	Graphene sheets	Polypropylene, polyethylene, and polyethylene tetrathalate	–	Pandey et al. (2021)
16	Flakes of graphene sheets with silica nanoparticles	Rice husk ash	–	Singh et al. (2021)
17	Graphene film	Tea tree plant	–	Jacob et al. (2015)
18	Silver nanoparticles	Oyster mushroom	11–44	Bhardwaj and Naraian (2020)

3.2.1.2 Batteries

An economy of low carbon is mainly the target for the management of waste and recycling of waste. Moreover, assessing comprehensive energy and economic costs in every step of the recycling procedure is mandatory to standardize the lab-scale process to pilot-scale process in the operational mode. The recycling process primarily involves the following steps: the material is collected, then the nonrecyclable part of that material is discarded at the final step, the pretreatment a collection of waste material, discarding of nonrecyclable fractions, pretreatment, and recovery of valuable components. It is important that only a small fraction of the scrapped products is associated with valuable materials, and these fractions are also not homogeneous. Therefore, the selected category for worthy products recycling is viable designing and, more importantly, the effective technologies for recycling. The different types of spent batteries are discussed below in the subsections, another important factor is the costs involved in the recycling (i.e., energy, environmental, economic, and manpower outlays), NMs from the different waste categories.

3.2.1.2.1 Lead batteries

A million units of lead battery waste are produced annually around the globe; bearing the harmful effects and exposure of these lead batteries to the environment and the capacity of bioaccumulation and the recycling process of these batteries is a valuable and effective approach while for avoiding the problems related to ecological and public health (Zhou et al., 2017). The paste in the lead is made of sulfate, metal oxide, and dioxide; in this form, the batteries are highly difficult to reuse due to the insolubility of these compounds. Though, pyrometallurgical processes should be employed to recover the form of lead from the spent paste of the lead-acid battery. A synthesis technique is designed to make lead nanopowder from the lead paste spent by the pyrolytic route. This article reports the particle's submicron of lead oxide using poly (N-vinyl-2- pyrrolidone) by the thermal process, which mainly helps decrease the particle size (Zhou et al., 2017). Although the technique also overcomes the difficulties mainly associated with lead paste insolubility on the aqueous media, the processes named pyrometallurgical should have unfavorable consequences, which are mainly due to the requirements of high energy as well as the disagreeable emissions of pollutants in the air, which mainly includes dioxide of sulfur and lead dust.

3.2.1.2.2 Zinc-manganese (Zn–Mn) batteries

As similar to lead battery waste, a large number of Zn–Mn transportable batteries are also used in industrialized countries. A billion units of Zn–Mn batteries are produced every year (Xiang et al., 2015). Excessive zinc is toxic to plants and humans, the adverse effects of zinc are a hindrance in the uptake of calcium and deficiencies in the metabolism of humans and other living organisms (Valko et al., 2005). Therefore, getting Zn from recycled batteries of Zn–Mn is an appropriate approach for reusing the zinc metal in many other applications, whereas solutions for its uncontrolled distribution into the environment. Zn nanomaterials synthesis by use of recycled Zn–Mn batteries as preliminary material has been investigated in current years. Xiang et al. demonstrated a highly efficient and superficial method for separating zinc from used batteries. The recycled cathode zinc was put into the crucible and subjected to a vacuum pressure of 1 Pa, 800°C heating temperature, and 200°C condensing temperature. This technique can accomplish the separation of zinc efficiently up to 99.68%. By using this method for the separation in the presence of nitrogen gas inside the heating chamber, nanoparticles with different structures should be obtained through different morphologies, which mainly include fibers, hexagonal prisms, and sheets (Xiang et al., 2015).

3.2.1.3 Rubber tires

Every year 1000 million rubber tires are generated. And these tires are usually thrown or burned openly, which is the leading cause of environmental pollution (Moghaddasi et al., 2013). For this reason, the landfill disposal of rubber tires is restricted or banned in developed countries (Turgut and Yesilata, 2008). Therefore, this is necessary to find a suitable solution for managing the waste of rubber tires. Mostly, tires consist of natural rubber 45%–47%, black carbon 21%–22%, metals 12%–25%, textile 5–10%, additives 6%–7% and zinc 1%–2%, and sulfur ~1% (Evans, 2005; Samaddar et al., 2018). The Zn nanoparticle synthesis from the used tires mainly consists of an effective alternative through value-added potential. There are many researchers who found ways to systematize the nanoparticles from the tires. Moghaddasi et al. reported the recycled tire nanoparticles by applying milling of the ball for five consecutive hours. This method yielded rubber ash particles that were smaller than 500 nm; when they were attached to the silicon waste, the size of particle size was additionally reduced to 50 nm. These particles are a better source of zinc for the plant when it is compared with the commercial fertilizer $ZnSO_4$ (Moghaddasi et al., 2013).

The principal component of rubber tires is carbon, so these tires are also an efficient source of carbon for various potential applications. Maroufi et al. reported the synthesis of high-value carbon nanoparticles (CNPs) from rubber tire waste using the approach of high temperature, that is, 1550°C. The rubber tire transformation was carried out at various reaction times ranging from 5 seconds to 20 minutes. In the initial 5 minutes, the impurities, such as sulfur, zinc, and oxygen, have been removed (Maroufi et al., 2017). Another report by Gómez-Hernández et al. described the thermal transformation at 1000°C, and carbon nanoparticles were produced under self-pressurized conditions (~22 nm) with an 81% high yield under optimized conditions. These nanoparticles showed better thermal stability and conductivity, and the chemical analysis showed the partially oxidized that is, C, 84.9%; O, 4.9%; S, 10.21% (Gómez-Hernández et al., 2019).

3.2.1.4 Cathode ray tube

Cathode ray tube funnel glass is a hazardous waste due to the high lead content in it (20%–30%) (Xing and Zhang, 2011). In many certain cases, a fraction of this lead came out from the broken glass containing lead and entered landfills. And it is a tough task to extract this lead from landfills using conventional extraction methods. Therefore efforts are taken to prepare the nanoparticles from these leachates, Pb nanomaterials are prepared from these leachates using funnel glass following carbothermal vacuum reduction and consolidation of inert gas (Xing and Zhang, 2011).

Waste of funnel glasses containing lead waste mainly exists in the form of lead oxide, and its melting point and boiling point (MP 888°C, BP 1535°C) are much more complex with metallic lead (MP 328°C, BP 1740°C). Lead oxide evaporation is a tough task from the glass funnel at low temperatures. Therefore, carbon is added as a reducing agent in lead oxide to metallic lead for evaporation. After the reduction, the metallic lead should accumulate on the surface of the tube (Yot and Mear, 2009).

3.2.2 Wastewater and biosludge

The effluents from wastewater and sludges from industries are identified as the source of metals and their oxides; therefore the effluents are generally studied for the recovery of heavy and trace metals. Elements such as PbO, Mo, Ni, Co, Cu, Ag, and SiO_2, are rare elements. The concentration of elements varies according to the source of the waste and sludge and the association of the minerals with them (Smith and Hageman, 2004). These elements are used for the preparation of the nanomaterial in optimized conditions.

3.2.3 Electric and electronic wastes

The overwhelming generation of end-of-life products related to electronic and electrical instruments worldwide causes severe public health and environmental issues in terms of e-waste (Baldé et al., 2017). Over the last few years, there is a steady increase in recycling techniques development for e-waste. Moreover, the recycling of e-waste recycling is more challenging due to the complex mixtures of different types of metals, glass, plastics, and engineered nanomaterials (Abdelbasir et al., 2018). Moreover, in developing countries, the separation of the electronic components is continually being miniaturized, which diminishes the efficiency (Baldé et al., 2017). The approach to industrial recycling starts with energy-intensive mechanical operations such as crushing in bulk, grinding, and separation majorly based on the magnetic properties (Long et al., 2010); after that, the chemical treatments are applied for purification and the extraction of the metals after that these metals are used for the preparation of the nanomaterials (Quan et al., 2010).

3.2.4 Plastic waste

The extensive use of plastic, which is nondegradable in the natural environment, has brought problematic waste materials into the world. The polymers from the plastic are used intensely in fabricating several products. Synthetic plastic production has increased infinitely during the last few years (Gu and Ozbakkaloglu, 2016). Traditional plastics are designed for durability and external environmental conditions did not affect the plastic, this is the main reason for the persistence and accumulation for years in the environment (Pol and Thiyagarajan, 2010). Plastic is the major component of solid waste worldwide (Ritchie, 2018). In most countries, the separation process is done at the source, which means the materials are segregated at the disposal point, and initiated main collective effort at the household level is recycling, which is mainly aimed at the reduction need for the production of new plastics from waste materials. At the household level, the main items include many types of plastics and nylon. With the huge amount of solid waste generated worldwide, the solutions for plastic waste as starting material for the source of energy recovery or the values-added products are the solutions to reduce the solid waste of plastic from the environment (Ackerman, 2010). Kumar (2011) reported the thermodynamic perspective, and the recovery of energy is less favorable as the plastic energy content (42.6 MJ/L) at a minimum order of lower magnitude than conventional fuels, that is, heating oil (443.5 MJ/kg). Moreover, the use of types of plastic as an energy source should be seriously detrimental to the environment as well as to human health due to the release of harmful dioxins during the burning of plastics.

Other plastics, that is, polyvinyl alcohol and polyethylene terephthalate, are also used CNTs synthesis by integrating some other elements (Deng et al., 2016).

3.2.5 Graphene

Graphene is a packed carbon atom thick single sheet of sp2 hybridized containing structure like a lattice honeycomb. The main properties of this material are high surface area, good stability of chemicals, high electrical conductivity, and strong mechanical strength (Novoselov et al., 2012). Energy, health, and environment sectors are revolutionized by graphene (Yang et al., 2019). Its strength and properties (electrical and mechanical) (Marinho et al., 2012) make graphene application in many areas. There are two sources by which graphene is prepared, that is, graphite and organic molecules (Strudwick et al., 2015). Typical methods for graphene preparation include the bottom-up approach, the top-down approach, epitaxial growth on silicon carbide (Mishra et al., 2012), mechanical cleavage, and chemical reduction of graphene oxide (Abdolhosseinzadeh et al., 2015). The bottom-up approach uses chemical vapor deposition (CVD) on metallic films (Strudwick et al., 2015). The top-down approach uses liquid exfoliation of graphite crystal (Coleman, 2013).

3.2.6 Rice husk

The biomass of the rice husk is waste material from agricultural residues. The overall rive production was 497.7 billion metric tons in 2019–20 (Singh et al., 2021). Therefore the production of rice husk is also in tons. The rice husk is the external cover of rice that is separated during rice processing. The major composition in rice husk is cellulose ($\sim$38%), hemicellulose ($\sim$20%), lignin ($\sim$22%), and silica SiO_2 ($\sim$16%–20%) (Singh et al., 2021). Therefore, the generated husk is rich in carbon and a good source for preparing nanomaterials. There are several reports on the preparation of nanomaterials from rice husks. (Muramatsu et al., 2014) prepared the graphene from the rice husk using KOH as the activating agent. Similarly, (Sankar et al., 2017) also reported a crystalline graphene nanosheet with ultrathin multilayer prepared using brown rice husk.

3.3 Application of nanoparticles derived from waste

Worldwide enormous waste is produced by manufacturing industries, overpopulation, and expansion. This is measured as a noteworthy task globally that necessitates an imperative resolution (Lyu et al., 2021). In addition, outstanding developments in the arena of biomedicine have obstructed the whole gamut of healthcare and remedy. This has flagged the mode for additional purifying of the consequences of biomedical policies in the direction of early discovery and treatment of diverse ailments (Jeevanandam et al., 2016). Various nanomaterials (NMs) have been devoted to dissimilar biomedical applications counting the delivery of the drug, immunizations, imagery modes, and biosensors. Noxiousness is still the chief issue constraining their use (Lyu et al., 2021). A schematic diagram is shown in Fig. 3.2. NMs cast off from diverse types of wastes exist as a groundbreaking method to not solitary evade dangerous effects on the atmosphere but also implement spherical practices economically, which are vital to reaching sustainable development. Additionally, recycled NMs have been used as a harmless yet world-shattering substitute with outstanding possibilities for numerous biomedical applications (Aflori, 2021). This section focused on waste-recycled NM's potential applications. Their essential healing actions as antimicrobial, anticancer, antioxidant nanodrugs, and nanovaccine.

3.3.1 Antimicrobial activity

A major application of nanoparticles in biomedical science is their usage as advanced antimicrobial agents, which incapacitates the disadvantages of predictable antimicrobial agents (Khezerlou et al., 2018). NPs have assorted machinery of action for killing disease-causing microorganisms, that is, the release of metallic ions and oxidative pressure (Wang et al., 2017). It has been established that the lesser the particle size, the improved the activity of the antimicrobial agent (Murru et al., 2020). Carbon nanoparticles activated carbon and carbon nanotubes can cause membrane impairment in

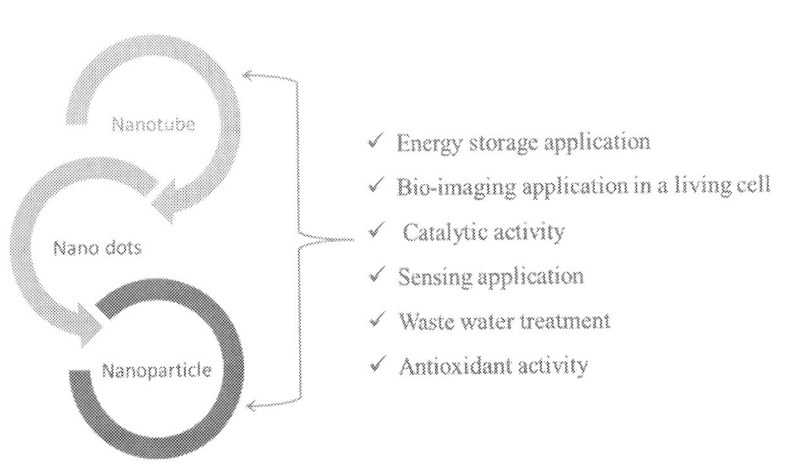

FIGURE 3.2 Application of nanomaterials derived from waste.

bacteria due to oxidative stress and additional methods. Carbon nanoparticles manufactured from an industrial waste spinoff (rapeseed oil cake) produced throughout the oil withdrawal procedure can stop the growth of numerous bacteria, for example, *Staphylococcus aureus, Pseudomonas diminuta, Yersinia enterocolitica, Salmonella enterica Typhimurium, Candida albicans*, and *Escherichia coli* (Das Purkayastha et al., 2014). Furthermore, graphene oxide, particularly synthesized from biomass bases, has shown larger antibacterial activity. Activated carbon nanoparticles produced from kitchen waste by the combustion method are also reported for antimicrobial activity. They kill bacteria by interacting with their cell walls, which directly leads to the inhibition of cell division. Gold nanoparticles from water extract of watermelon residue have shown antibacterial activity against *E. coli*, Bacillus cereus, and Listeria monocytogenes (Patra and Baek, 2015). Additional waste used for the biogenesis of gold nanoparticles was the peel extract of Citrus maxima; they have robust antibacterial activity for *S. aureus* and *E. coli* (Yuan et al., 2017).

3.3.2 Antioxidant activity

Antioxidants have a role in averting and avoiding particles inside a cell from oxidizing. Free radicals are generated mainly in the biological oxidation reaction. After that restraint, the response is started because of the reactivity of radicals, due to which the major threat is cells damaged, which may lead to cell death. Though, the competence of accepted and artificial antioxidants is imperfect due to unfortunate absorption, exertion crossing cell membranes, and dilapidation during distribution, all of which subsidize their near-to-the-ground bioavailability. They have been covalently linked to nanoparticles to deliver enhanced steadiness and continued the release, compatibility, and embattled distribution of antioxidants (Doghish et al., 2021). Carbon quantum dots were finished using a carbonization method that used leftover green tea shrubberies as a carbon source. Carbon quantum dots have a latent mechanism for DPPH reduction (Murru et al., 2020). Gold nanoparticles recycled as a waste product from leaves water extract of A. noeanum, bark extracts of *Abroma augusta* L., and peels of watermelon have good antioxidant properties. This might be because of the capability of gold nanoparticles to counteract and rummage DPPH free radicals by giving hydrogen ions or electrons (das Purkayastha et al., 2014; Mobeen Amanulla and Sundaram, 2019; Shahriari et al., 2019). Antioxidant activities have been too testified for titanium oxide nanoparticles, these nanoparticles finished from agro-waste such as peels of fruits, for example, *L. siceraria, P. dulce, P. guajava, Allium eriophyllum, A. haussknechtii*, and the foliage biomass of *M. citrifolia* (Ajmal et al., 2019; Alavi and Karimi, 2018).

3.3.3 Anticancer activity

Recently, cancer studies have inspired scholars and scientists to pursue more progressive actions apart from chemo and radiative treatment. High noxiousness and lacking specificity are the main anxieties of these traditional treatments (Doghish et al., 2021). The optimum cancer medicine should have an exact besieged delivery scheme, additionally to suitable drug release. Consequently, a diversity of nanoscale-based medicinal transporters has newly been industrialized and examined. The characteristic method of cancer handling with nanomaterials is grounded on the conveyance of the chemotherapeutical mediator, which has a narrow healing index, abridged cellular diffusion, and extremely toxic belongings (Negri et al., 2020). Biomass-based carbon nanomaterials have shown potential as they have outstanding biochemical and physical possessions, which makes them outstanding nanocarriers with large effectiveness and with low toxicity for embattled drug transport (Zhou et al., 2017). Activated carbon originating from biomass resources, for example, wood, coal, bamboo, and coconut shells, has shown outstanding photothermal activity against human lung cell line H-1299, activated carbon changes light energy into heat, which ultimately causes the death of tumorous cells (Thiyagarajulu and Arumugam, 2021). Inorganic quantum dots, specifically carbon quantum dots, have been established for their exceptional anticancer activity for malignant human umbilical vein endothelial cells and mesothelioma cell lines (H2452). The anticancer activity of metal oxide nanoparticles has been established, they bind to the cell surfaces, therefore, causing low stability. The development of reactive oxygen species by these nanoparticles is the chief mechanism by which they inhibit cancerous cells. Copper oxide nanoparticles also have shown anticancer activity, the main mechanism they follow is enhancing the strength of apoptosis in cancer cells (Jeevanandam et al., 2016). Additionally, they reduce the appearance of oncogenes by growing the appearance of tumor suppresser proteins (Singh, 2019). The synthesis of zinc nanoparticles from *Lauruso bilis* leaves has shown anticancer activity in A549 lung cancer (Vijayakumar et al., 2016). Additionally, the zinc nanoparticles synthesized from *Eclipta prostrate* and *Ziziphus nummularia* leaves have shown anticancer activity (Chung et al., 2015).

3.3.4 Nanodrugs and nanovaccine

Nanopharmaceuticals are the combination of nanotechnology with biomedical and pharmaceutical sciences. Nanomedicine is a novel and speedily rising field. Several drug-based nanoparticles have been registered in clinical trials, though drug-based nanoparticles face a ratio of encounters due to their toxic nature, which is essential for healthier classification, and the nonappearance of controlling strategies (Bobo et al., 2016). Iron oxide nanoparticles have been used in numerous medicines as iron replacement treatment. These include Ferrlecit, Venofer, Dexferrum, and Infant, which are entirely used in treating anemia related to chronic kidney illness (Bobo et al., 2016; Caster et al., 2017). Superparamagnetic iron oxide nanoparticles made from fruit peel biomass rejoin powerfully when bare to magnetic fields, cathartic energy that can be used in directing specific tumors (Bano et al., 2016). In the last decade, carbon nanoparticles recycled from agricultural waste have established more attention in research applications on the ground of drug distribution (Fathy et al., 2019). Gold nanoparticles are valuable for drug delivery, because of their large surface area, drug particles can easily bind to the gold nanoparticles by physical absorption, and one other mode of action is covalent or ionic bond formation. Previously leaf waste of M. foetida and peels of onion peel were used as a bioreductant in the synthesis of gold nanoparticles, which were then used professionally to treat cancer cells (Patra and Baek, 2015; Yallappa et al., 2015). Gold nanoparticles were also reported from Taxus baccata to extract for treating anticancer activity against human cell lines for breast (MCF-7), cervical (HeLa), and ovarian (Caov-4) cancer cells. Copper oxide nanoparticles recycled from Ficus religiosa and Camellia sinensis leaves also have strong anticancer activity against lung cancer and breast cancer respectively (Bobo et al., 2016; Emima Jeronsia et al., 2019). Apart from these, iron and titanium oxide nanoparticles are also derived from recycled waste biomass of Brown seaweed and Vigna radiata legume as a potential anticancer drug against breast, liver, and cervical cancer; osteosarcoma, and chondrosarcoma cancer cell lines respectively (Namvar et al., 2014; Sha et al., 2013).

3.3.5 Biosensors

There are biological indicators that prerequisite recurrent and unceasing measurement. Traditional approaches for biomarker finding are expensive, time-consuming, and want multifaceted strategies and specialists to work plans. The usage of biosensors with nanotechnology has been one of the finest biomedical submissions for the speedy and precise analysis of sickness as suitable treatment can be given at an initial phase for dropping the death rate (Nagraik et al., 2021). Biosensors are the amalgam of biological rudiments and physicochemical mechanisms, they are used to sense specific biological particles of a medium and they basically produce a quantifiable signal and after that examine the biological response (Singh, 2019; Townshend et al., 2021). Numerous nanoparticles recycled from diverse sources of waste have been applied as biosensors for initial exposure to different sicknesses. Copper oxide nanoparticles recycled from vegetable waste are reported as biosensors for the detection of glucose in the blood with the advantage of recyclability, selectivity, and low cost (Muthuchamy et al., 2018). Carbon nanotube nanoparticles from polymer biomass have the ability to detect early cancer and have an advantage due to their large surface area and fast electron transfer. Nanoparticles (nanographene oxide) recycled from agricultural waste are also used as DNA and RNA biosensors (Fang et al., 2011).

3.3.6 Wastewater treatment

In the expanse of water decontamination, nanotechnology proposed the option of an effectual exclusion of contaminants and microorganisms. Nowadays, nanoparticles, nanomembrane, and nanopowder are basically used for finding and eliminating chemical and biological materials comprising metals (e.g., cadmium, copper, lead, mercury, nickel, zinc), nutrients (e.g., phosphate, ammonia, nitrate, and nitrite), cyanide, organics, algae (e.g., cyanobacterial toxins) viruses, bacteria, parasites, and antibiotics (Tewari et al., 2013). Fundamentally, four modules of nanoscale resources are being assessed as functional resources for cleaning water for example, metal-containing nanoparticles, carbonaceous nanomaterials, zeolites, and dendrimers. Carbon nanotubes and nanofibers also demonstrate some optimistic outcomes. Nanomaterials disclose respectable outcome than other methods used for water treatment due to their high superficial extent. It is proposed that this nanomaterial can be used in upcoming times as large-scale water purification (Álvarez-Ayuso et al., 2003; Bhardwaj et al., 2021). It is also reported that the coliform and bacteria pickled with ultrasonic irradiation for a small span of time period earlier silver nanoparticle treatment at lower absorption, improved antibacterial results (Li et al., 2005).

3.4 Waste as a synthesis of nanomaterials

Various types of waste materials have been used in the synthesis of different types of nanomaterials after different types of treatment, that is, physical treatment, chemical treatment, and combinations of both treatments. Milling and grinding are the most common physical pretreatment methods. Heating and treatment with strong acids such as H_2SO_4, HNO_3, and HCl for the separation of contaminants from the waste come under chemical pretreatment (Samaddar et al., 2018). Sodium borohydride for metallic nanoparticle reduction is the most popular nanoparticle synthesis method. Generally, the sodium borohydride solution is freshly prepared and rapidly added to the solution of waste materials. Fang et al. implemented this method for the preparation of Fe nanoparticles from the waste material of steel pickling released from the plant of steel (Fang et al., 2011).

Another important method is the solvent thermal method, which mainly involves the chemical reactions autoclave reactor with solvent and heated for a particular time and constant critical temperature (Dubin et al., 2010). In this process, when the water is used in the place of solvent, it is known as the hydrothermal process. Tang et al. reported that hematite nanoparticle (α-Fe_2O_3) was synthesized using the thermal solvent method in this process, the solvent is ethanol, and the solution for the material mixture is $FeCl_3$ and NaOH at the temperature of 150°C for the time interval of 2 hours for getting the nanoparticle. Electrodeposition is another technique that uses voltage for the promotion of chemical reactions in the aqueous solutions for synthesizing nanomaterials (Tang et al., 2011). Jeevanandam et al. applied this method for the production of nanowires, nanoporous materials, and nanocylinders. These abovementioned techniques or methods are the green synthesis methods, which are mainly based on environmentally friendly and biocompatible materials such as plants and microorganisms that have recently arisen (Jeevanandam et al., 2016).

3.5 Characterization and synthesis of the nanomaterials

Various methods are reported for the synthesis and characterization of nanomaterials from various types of wastes, which may include vacuum vaporization and deposition, vapor deposition using physical and chemical methods, condensation of gas, mechanical attrition, combustion synthesis, physical vapor deposition, high-temperature synthesis using self-propagating, sol–gel methods or techniques, and electrodeposition unstable, etc. (Weintraub et al., 2010). In the nanomaterial synthesis using the vapor phase, the mixture vapor phase is thermodynamically unstable for the prepared solid material. This is also known as chemical supersaturation, where the condensed phase is produced, to provide the thermodynamics favoring for the vapor phase for chemical reactions. On desired levels of supersaturation, condensation allows the component of the particle to form nucleates (Swihart, 2003). The method vapor of deposition is further divided into chemical and physical vapor deposition, which have subcategories further. CVD may include chemical reaction, condensation, deposition, evaporation, coalescence, coagulation, etc., whereas physical vapor deposition consists of sublimation, condensation, coalescence, postprocessing, coagulation, deposition, etc. (Virji and Stefaniak, 2014). CVD includes plasma-enhanced electron cyclotron resonance CVD and plasma assistant CVD. While the physical vapor deposition is further subcategorized into arc vacuum, ion beam, pulsed laser, magnetron sputtering ion beam sputtering, electron beam deposition, and filtered arc deposition (Chen and Mao, 2007). Colloidal suspension formation and a sol take place in the process called sol–gel, through which the reaction of precursors was hydrolysis as well as polymerization, in which the formation of transition from liquid sol to solid gel phase, which may cause loss of solvent and the complete polymerization (Rahimi-Nasrabadi et al., 2016). The deposition of electrons is favorable for the synthesis and production of nanomaterials for catalysis conduction and reactions involved in electrochemical methods, where the deposition of nanomaterials occurs on the surface of electrode surfaces (Wang et al., 2017). In the combustion solution, the synthesis of nanomaterials was done in a homogeneous solution with different types of oxidizers and fuels for the nanomaterial's formation (Aruna and Mukasyan, 2008). The synthesized nanomaterials are characterized using field emission scanning electron microscopy, atomic force microscopy, cyclic voltammetry, X-ray diffraction, transmission electron microscopy, X-ray photoelectron spectroscopy, ultraviolet–visible, Fourier-transform infrared spectroscopy (FTIR), Raman spectroscopy, high-resolution TEM (Zhou and Wu, 2013). In the process of characterization, spectroscopy, especially Raman Spectroscopy is used as a nondestructive tool. This is used for the quality and structure determination of synthesized nanomaterials. In the working principle, incident light from a sample is inelastically scattered and by the energy of its characteristic molecular vibrations, shifted in frequency (Kneipp et al., 1999). Different versions of Raman spectroscopy are used for synthesizing several types of nanomaterials for the estimation of their shape, size, surface area, the surface of the material, and the properties of the materials. In nanomaterial of graphene, Raman spectroscopy helps to determine the number and layers quantity. Raman spectroscopy generates the peak at the carbon atom, in which bands (D and G) are disused more commonly, in which the graphene curved nanosheets

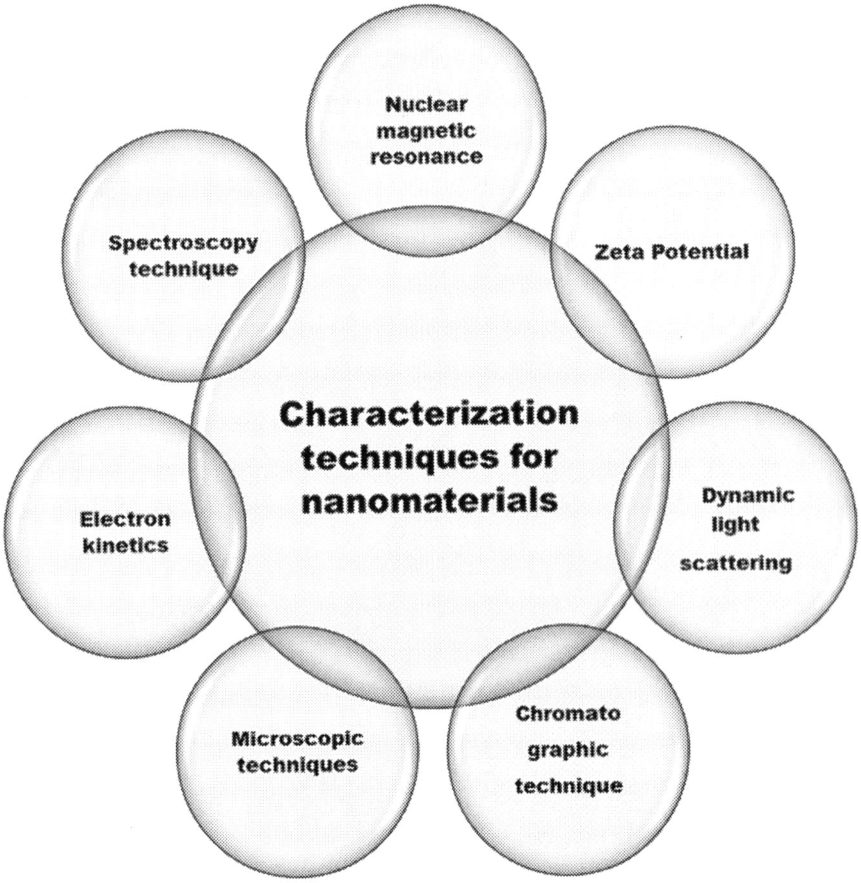

FIGURE 3.3 Techniques for nanomaterial synthesis.

defects are indicated first (Zhou and Wu, 2013). Electron microscopy is a popular tool for determining the morphological characteristics of materials. Equally, the applications of microscopy are available widely for nanomaterials characterization. The working principle of electron microscopy is generally based on the energetic electrons applications as the major source, which mainly focused on electromagnetic lenses (Bozzola, 2007). The electron microscopy 3D version is mainly used for qualitative and quantitative measurements. Material linkages and functionalities are confirmed using the FTIR analysis. Thermogravimetric analysis is used for the analysis of the thermal behavior of the materials. Various techniques for characterization are shown in Fig. 3.3. Apart from the abovementioned techniques and methods for the nanomaterial's synthesis, there are various other fabrication approaches also employed for the nanomaterial's synthesis. However, the most important thing in nanomaterial synthesis is the optimization of the optimum conditions for the enhancement of the nanomaterial's quantity and reproducibility.

3.6 Future directions and conclusions

Nanomaterials development from waste has been quickly employed in several areas, and it has many applications in every field, that is, in the field of microelectronics to industries related to bioengineering, nanomaterials is a multidisciplinary subject. But still, there are a few concerns that may addressed previously before these nanomaterials are accessible commercially in the future. Few are described below:

1. The nanomaterial toxicity should not be ignored. Thus, the study related to the toxicity of nanomaterials has been studied especially the use of the nonmaterial in living systems and in applications related to the environment.
2. The restrictions on the use of innovative nanomaterials because they are patented.
3. The development of nanomaterial that is environmentally friendly and economically important.
4. Production of nanomaterial-based products at the industrial level needs to be addressed.
5. Nanomaterial chemistry understanding, which is normally involved in the applications.
6. Equipment's high cost and limited scalability of materials synthesized from waste need further attention.

The uses of nanomaterials in different areas/fields mainly include sensing of nanoprobe, cancer biosensors, protein separation, water purification and remediation, microbial fuel cell, catalysis, etc. Although there are enormous, exciting potential, and feasible applications of nanomaterials, several concerns and challenges stay unsolved to date, and a few of the concerns are listed above. The wide-ranging efforts by researchers from various science sectors predicted that the various positive examples of waste-derived nanomaterial products would be commercially accessible in the future.

References

Abdelbasir, S.M., El-Sheikh, S.M., Morgan, V.L., Schmidt, H., Casso-Hartmann, L.M., Vanegas, D.C., et al., 2018. Graphene-anchored cuprous oxide nanoparticles from waste electric cables for electrochemical sensing. ACS Sustainable Chemistry & Engineering 6, 12176–12186. Available from: https://doi.org/10.1021/ACSSUSCHEMENG.8B02510/SUPPL_FILE/SC8B02510_SI_001.PDF.

Abdolhosseinzadeh, S., Asgharzadeh, H., Kim, H.S., 2015. Fast and fully-scalable synthesis of reduced graphene oxide. Scientific Reports 5. Available from: https://doi.org/10.1038/SREP10160.

Ackerman, F., 2010. Waste management and climate change. 5, 223–229. https://doi.org/10.1080/13549830050009373.

Aflori, M., 2021. Smart Nanomaterials for biomedical applications—a review. Nanomaterials 11, 396. Available from: https://doi.org/10.3390/NANO11020396.

Ahmad, A.A., Hameed, B.H., Ahmad, A.L., 2009. Removal of disperse dye from aqueous solution using waste-derived activated carbon: optimization study. Journal of Hazardous Materials 170 (2–3), 612–619. Available from: https://doi.org/10.1016/j.jhazmat.2009.05.021.

Ajmal, N., Saraswat, K., Bakht, M.A., Riadi, Y., Ahsan, M.J., Noushad, M., 2019. Cost-effective and eco-friendly synthesis of titanium dioxide (TiO_2) nanoparticles using fruit's peel agro-waste extracts: characterization, in vitro antibacterial, antioxidant activities. Green Chemistry Letters and Reviews 12, 244–254. Available from: https://doi.org/10.1080/17518253.2019.1629641/SUPPL_FILE/TGCL_A_1629641_SM5398.DOCX.

Akter Jahan, S., Mollah, M.Y.A., Ahmed, S., Susan, M.A.B.H., 2017. Nano-hydroxyapatite prepared from eggshell-derived calcium-precursor using reverse microemulsions as nanoreactor. Materials Today: Proceedings 4, 5497–5506.

Alavi, M., Karimi, N., 2018. Characterization, antibacterial, total antioxidant, scavenging, reducing power and ion chelating activities of green synthesized silver, copper and titanium dioxide nanoparticles using Artemisia haussknechtii leaf extract. Artificial Cells, Nanomedicine, and Biotechnology 46, 2066–2081. Available from: https://doi.org/10.1080/21691401.2017.1408121.

Alshatwi, A.A., Athinarayanan, J., Periasamy, V.S., 2015. Biocompatibility assessment of rice husk-derived biogenic silica nanoparticles for biomedical applications. Materials Science & Engineering. C 47, 8–16. Available from: https://doi.org/10.1016/j.msec.2014.11.005.

Álvarez-Ayuso, E., García-Sánchez, A., Querol, X., 2003. Purification of metal electroplating waste waters using zeolites. Water Research 37, 4855–4862. Available from: https://doi.org/10.1016/J.WATRES.2003.08.009.

Aruna, S.T., Mukasyan, A.S., 2008. Combustion synthesis and nanomaterials. Current Opinion in Solid State and Materials Science 12, 44–50. Available from: https://doi.org/10.1016/J.COSSMS.2008.12.002.

Baldé, C.P., Forti, V., Gray, V., Kuehr, R., Stegmann, P., 2017. Quantities, flows, and resources the global E-waste.

Bano, S., Nazir, S., Nazir, A., Munir, S., Mahmood, T., Afzal, M., et al., 2016. Microwave-assisted green synthesis of superparamagnetic nanoparticles using fruit peel extracts: surface engineering, T2 relaxometry, and photodynamic treatment potential. International Journal of Nanomedicine 11, 3833–3848. Available from: https://doi.org/10.2147/IJN.S106553.

Bhardwaj, A.K., Naraian, R., 2020. Green synthesis and characterization of silver NPs using oyster mushroom extract for antibacterial efficacy. Journal of Chemistry, Environmental Sciences and its Applications 7 (1), 13–18. Available from: https://doi.org/10.15415/jce.2020.71003.

Bhardwaj, A.K., Sundaram, S., Yadav, K.K., Srivastav, A.L., 2021. An overview of silver nano-particles as promising materials for water disinfection. Environmental Technology & Innovation 23, 101721.

Bobo, D., Robinson, K.J., Islam, J., Thurecht, K.J., Corrie, S.R., 2016. Nanoparticle-based medicines: a review of FDA-approved materials and clinical trials to date. Pharmaceutical Research 33, 2373–2387. Available from: https://doi.org/10.1007/S11095-016-1958-5.

Bozzola, J.J., 2007. Conventional specimen preparation techniques for transmission electron microscopy of cultured cells. Methods in Molecular Biology 1–18. Available from: https://doi.org/10.1007/978-1-59745-294-6_1.

Caster, J.M., Patel, A.N., Zhang, T., Wang, A., 2017. Investigational nanomedicines in 2016: a review of nanotherapeutics currently undergoing clinical trials. Wiley Interdisciplinary Reviews: Nanomedicine and Nanobiotechnology 9. Available from: https://doi.org/10.1002/WNAN.1416.

Chen, W., Lv, G., Hu, W., Li, D., Chen, S., Dai, Z., 2018. Synthesis and applications of graphene quantum dots: a review. Nanotechnology Reviews 7 (2), 157–185.

Chen, X., Mao, S.S., 2007. Titanium dioxide nanomaterials: synthesis, properties, modifications and applications. Chemical Reviews 107, 2891–2959. Available from: https://doi.org/10.1021/CR0500535/ASSET/CR0500535.FP.PNG_V03.

Chen, Y.M., Yu, X.Y., Li, Z., Paik, U., Lou, X.W., 2016. Hierarchical MoS2 tubular structures internally wired by carbon nanotubes as a highly stable anode material for lithium-ion batteries. Science Advances 2 (7), e1600021. Available from: https://doi.org/10.1126/sciadv.1600021.

Chung, I.M., Rahuman, A.A., Marimuthu, S., Kirthi, A.V., Anbarasan, K., Rajakumar, G., 2015. An investigation of the cytotoxicity and caspase-mediated apoptotic effect of green synthesized zinc oxide nanoparticles using eclipta prostrata on human liver carcinoma cells. Nanomaterials 5, 1317. Available from: https://doi.org/10.3390/NANO5031317.

Coleman, J.N., 2013. Liquid exfoliation of defect-free graphene. Accounts of Chemical Research 46, 14–22. Available from: https://doi.org/10.1021/AR300009F.

das Purkayastha, M., Manhar, A.K., Mandal, M., Mahanta, C.L., 2014. Industrial waste-derived nanoparticles and microspheres can be potent antimicrobial and functional ingredients. Journal of Applied Chemistry 2014, 1–12. Available from: https://doi.org/10.1155/2014/171427.

de Abreu, D.A.P., Cruz, J.M., Losada, P.P., 2012. Active and intelligent packaging for the food industry. 28, 146–187. Available from https://doi.org/10.1155/2014/171427.

Deng, J., You, Y., Sahajwalla, V., Joshi, R.K., 2016. Transforming waste into carbon-based nanomaterials. Carbon 96, 105–115. Available from: https://doi.org/10.1016/J.CARBON.2015.09.033.

Doghish, A.S., El-Sayyad, G.S., Sallam, A.A.M., Khalil, W.F., El Rouby, W.M.A., 2021. Graphene oxide and its nanocomposites with EDTA or chitosan induce apoptosis in MCF-7 human breast cancer. RSC Adv 11, 29052–29064. Available from: https://doi.org/10.1039/D1RA04345E.

Dubin, S., Gilje, S., Wang, K., Tung, V.C., Cha, K., Hall, A.S., et al., 2010. A one-step, solvothermal reduction method for producing reduced graphene oxide dispersions in organic solvents. ACS Nano 4, 3845–3852. Available from: https://doi.org/10.1021/NN100511A/ASSET/IMAGES/MEDIUM/NN-2010-00511A_0009.GIF.

El-Amir, A.A.M., Ewais, E.M.M., Abdel-Aziem, A.R., Ahmed, A., El-Anadouli, B.E.H., 2016. Nano-alumina powders/ceramics derived from aluminum foil waste at low temperature for various industrial applications. Journal of Environmental Management 183 (1), 121–125.

Emima Jeronsia, J., Allwin Joseph, L., Annie Vinosha, P., Jerline Mary, A., Jerome Das, S., 2019. Camellia sinensis leaf extract mediated synthesis of copper oxide nanostructures for potential biomedical applications. Materials Today: Proceedings 8, 214–222. Available from: https://doi.org/10.1016/J.MATPR.2019.02.103.

Evans, P., 2005. Scaling and assessment of data ISSN 0907-4449 quality.

Fang, Z., Qiu, X., Chen, J., Qiu, X., 2011. Degradation of the polybrominated diphenyl ethers by nanoscale zero-valent metallic particles prepared from steel pickling waste liquor. Desalination 267, 34–41. Available from: https://doi.org/10.1016/J.DESAL.2010.09.003.

Fathy, N.A., Basta, A.H., Lotfy, V.F., 2019. Novel trends for synthesis of carbon nanostructures from agricultural wastes. Carbon Nanomaterials for Agri-food and Environmental Applications 59–74. Available from: https://doi.org/10.1016/B978-0-12-819786-8.00004-9.

Gangula, A., Podila, R., M, Ramakrishna, Karanam, L., Janardhana, C., Rao, A.M., 2011. Catalytic reduction of 4-nitrophenol using biogenic gold and silver nanoparticles derived from Breynia rhamnoides. Langmuir 27 (24), 15268–15274. Available from: https://doi.org/10.1021/la2034559.

Goh, P.S., Ng, B.C., Lau, W.J., Ismail, A.F., 2014. Inorganic nanomaterials in polymeric ultrafiltration membranes for water treatment. Separation & Purification Reviews 44, 216–249. Available from: https://doi.org/10.1080/15422119.2014.926274.

Gómez-Hernández, R., Panecatl-Bernal, Y., Méndez-Rojas, M.Á., 2019. High yield and simple one-step production of carbon black nanoparticles from waste tires. Heliyon 5. Available from: https://doi.org/10.1016/J.HELIYON.2019.E02139.

Gu, L., Ozbakkaloglu, T., 2016. Use of recycled plastics in concrete: a critical review. Waste Management 51, 19–42. Available from: https://doi.org/10.1016/J.WASMAN.2016.03.005.

Hu, Y., Jensen, J.O., Zhang, W., Cleemann, L.N., Xing, W., Bjerrum, N.J., Li, Q., 2014. Hollow spheres of iron carbide nanoparticles encased in graphitic layers as oxygen reduction catalysts. Angewandte Chemie 53 (14), 3675–3679. Available from: https://doi.org/10.1002/anie.201400358.

Jacob, M.V., Rawat, R.S., Ouyang, B., Bazaka, K., Kumar, D.S., Taguchi, D., et al., 2015. Catalyst-free plasma enhanced growth of graphene from sustainable sources. Nano Letters 15 (9), 5702–5708. Available from: https://doi.org/10.1021/acs.nanolett.5b01363.

Jeevanandam, J., Chan, Y.S., Danquah, M.K., 2016. Biosynthesis of metal and metal oxide nanoparticles. ChemBioEng Reviews 3, 55–67. Available from: https://doi.org/10.1002/CBEN.201500018.

Khezerlou, A., Alizadeh-Sani, M., Azizi-Lalabadi, M., Ehsani, A., 2018. Nanoparticles and their antimicrobial properties against pathogens including bacteria, fungi, parasites and viruses. Microbial Pathogenesis 123, 505–526. Available from: https://doi.org/10.1016/J.MICPATH.2018.08.008.

Kneipp, K., Kneipp, H., Itzkan, I., Dasari, R., reviews, M.F.-C., 1999. Undefined ultrasensitive chemical analysis by Raman spectroscopy. Citeseer.

Kumar, K., 2011. Influence of ultrasonic treatment in sewage sludge. Journal of Waste Water Treatment & Analysis 02. Available from: https://doi.org/10.4172/2157-7587.1000115.

Li, P., Li, J., Wu, C., Wu, Q., Li, J., 2005. Synergistic antibacterial effects of β-lactam antibiotic combined with silver nanoparticles. Nanotechnology 16, 1912. Available from: https://doi.org/10.1088/0957-4484/16/9/082.

Long, L., Sun, S., Zhong, S., Dai, W., Liu, J., Song, W., 2010. Using vacuum pyrolysis and mechanical processing for recycling waste printed circuit boards. Journal of Hazardous Materials 177, 626–632. Available from: https://doi.org/10.1016/J.JHAZMAT.2009.12.078.

Lu, W., Qin, X., Liu, S., Chang, G., Zhang, Y., Luo, Y., Asiri, A.M., Al-Youbi, A.O., Sun, X., 2012. Economical, green synthesis of fluorescent carbon nanoparticles and their use as probes for sensitive and selective detection of mercury(II) ions. Analytical Chemistry 84 (12), 5351–5357. Available from: https://doi.org/10.1021/ac3007939.

Lyu, Q., Peng, L., Hong, X., Fan, T., Li, J., Cui, Y., et al., 2021. Smart nano-micro platforms for ophthalmological applications: the state-of-the-art and future perspectives. Biomaterials 270, 120682. Available from: https://doi.org/10.1016/J.BIOMATERIALS.2021.120682.

Marinho, B., Ghislandi, M., Tkalya, E., Koning, C.E., de With, G., 2012. Electrical conductivity of compacts of graphene, multi-wall carbon nanotubes, carbon black, and graphite powder. Powder Technology 221, 351–358. Available from: https://doi.org/10.1016/J.POWTEC.2012.01.024.

Maroufi, S., Mayyas, M., Sahajwalla, V., 2017. Nano-carbons from waste tyre rubber: an insight into structure and morphology. Waste Management 69, 110–116. Available from: https://doi.org/10.1016/J.WASMAN.2017.08.020.

Martinez-Alier, J., 2009. Social metabolism, ecological distribution conflicts, and languages of valuation. Capitalism Nature Socialism 20 (1), 58–87.

Mishra, N., Das, G., Ansaldo, A., Genovese, A., Malerba, M., Povia, M., et al., 2012. Pyrolysis of waste polypropylene for the synthesis of carbon nanotubes. Journal of Analytical and Applied Pyrolysis 94, 91–98. Available from: https://doi.org/10.1016/J.JAAP.2011.11.012.

Mobeen Amanulla, A., Sundaram, R., 2019. Green synthesis of TiO_2 nanoparticles using orange peel extract for antibacterial, cytotoxicity and humidity sensor applications. Materials Today: Proceedings 8, 323–331. Available from: https://doi.org/10.1016/J.MATPR.2019.02.118.

Moghaddasi, S., Khoshgoftarmanesh, A.H., Karimzadeh, F., Chaney, R.L., 2013. Preparation of nano-particles from waste tire rubber and evaluation of their effectiveness as zinc source for cucumber in nutrient solution culture. Scientia Horticulturae 160, 398–403. Available from: https://doi.org/10.1016/J.SCIENTA.2013.06.028.

Muramatsu, H., Kim, Y.A., Yang, K.S., Cruz-Silva, R., Toda, I., Yamada, T., et al., 2014. Rice husk-derived graphene with nano-sized domains and clean edges. Small (Weinheim an der Bergstrasse, Germany) 10, 2766–2770. Available from: https://doi.org/10.1002/SMLL.201400017.

Murru, C., Badía-Laíño, R., Díaz-García, M.E., 2020. Synthesis and characterization of green carbon dots for scavenging radical oxygen species in aqueous and oil samples, Antioxidants (Basel), 9. pp. 1–18. Available from: https://doi.org/10.3390/ANTIOX9111147.

Muthuchamy, N., Gopalan, A., Lee, K.P., 2018. Highly selective non-enzymatic electrochemical sensor based on a titanium dioxide nanowire–poly (3-aminophenyl boronic acid)–gold nanoparticle ternary nanocomposite. RSC Advances 8, 2138–2147. Available from: https://doi.org/10.1039/C7RA09097H.

Nagraik, R., Sharma, A., Kumar, D., Mukherjee, S., Sen, F., Kumar, A.P., 2021. Amalgamation of biosensors and nanotechnology in disease diagnosis: mini-review. Sensors International 2. Available from: https://doi.org/10.1016/J.SINTL.2021.100089.

Namvar, F., Rahman, H.S., Mohamad, R., Baharara, J., Mahdavi, M., Amini, E., et al., 2014. Cytotoxic effect of magnetic iron oxide nanoparticles synthesized via seaweed aqueous extract. International Journal of Nanomedicine 9, 2479–2488. Available from: https://doi.org/10.2147/IJN.S59661.

Negri, V., Pacheco-Torres, J., Calle, D., López-Larrubia, P., 2020. Carbon nanotubes in biomedicine. Topics in Current Chemistry (Cham) 378. Available from: https://doi.org/10.1007/S41061-019-0278-8.

Novoselov, K.S., Fal'Ko, V.I., Colombo, L., Gellert, P.R., Schwab, M.G., Kim, K., 2012. A roadmap for graphene. Nature 490, 192–200. Available from: https://doi.org/10.1038/NATURE11458.

Pandey, S., Karakoti, M., Surana, K., Dhapola, P.S., SanthiBhushan, B., Ganguly, S., et al., 2021. Graphene nanosheets derived from plastic waste for the application of DSSCs and supercapacitors. Scientific Reports 11, 3916. Available from: https://doi.org/10.1038/s41598-021-83483-8.

Patra, J.K., Baek, K.H., 2015. Novel green synthesis of gold nanoparticles using Citrullus lanatus rind and investigation of proteasome inhibitory activity, antibacterial, and antioxidant potential. International Journal of Nanomedicine 10, 7253. Available from: https://doi.org/10.2147/IJN.S95483.

Pol, V.G., Thiyagarajan, P., 2010. Remediating plastic waste into carbon nanotubes. Journal of Environmental Monitoring 12, 455–459. Available from: https://doi.org/10.1039/B914648B.

Quan, C., Li, A., Gao, N., 2010. Synthesis of carbon nanotubes and porous carbons from printed circuit board waste pyrolysis oil. Journal of Hazardous Materials 179, 911–917. Available from: https://doi.org/10.1016/J.JHAZMAT.2010.03.092.

Rahimi-Nasrabadi, M., Rostami, M., Ahmadi, F., Shojaie, A.F., Rafiee, M.D., 2016. Synthesis and characterization of $ZnFe_2 - xYbxO_4$–graphene nanocomposites by sol–gel method. Journal of Materials Science: Materials in Electronics 27, 11940–11945. Available from: https://doi.org/10.1007/S10854-016-5340-5.

Ritchie, H., 2018. Data MR-OW in, 2018 undefined plastic pollution. ourworldindata.org.

Samaddar, P., Ok, Y.S., Kim, K.H., Kwon, E.E., Tsang, D.C.W., 2018. Synthesis of nanomaterials from various wastes and their new age applications. Journal of Cleaner Production 197, 1190–1209. Available from: https://doi.org/10.1016/J.JCLEPRO.2018.06.262.

Sankar, S., Lee, H., Jung, H., Kim, A., Ahmed, A.T.A., Inamdar, A.I., et al., 2017. Ultrathin graphene nanosheets derived from rice husks for sustainable supercapacitor electrodes. New Journal of Chemistry 41, 13792–13797. Available from: https://doi.org/10.1039/C7NJ03136J.

Sha, B., Gao, W., Han, Y., Wang, S.Q., Wu, J., Xu, F., et al., 2013. Potential application of titanium dioxide nanoparticles in the prevention of osteosarcoma and chondrosarcoma recurrence. Journal of Nanoscience and Nanotechnology 13, 1208–1211. Available from: https://doi.org/10.1166/JNN.2013.6081.

Shadjou, N., Hasanzadeh, M., 2016. Graphene and its nanostructure derivatives for use in bone tissue engineering: recent advances. Journal of Biomedical Materials Research. Part A 104, 1250–1275. Available from: https://doi.org/10.1002/JBM.A.35645.

Shahriari, M., Hemmati, S., Zangeneh, A., Zangeneh, M.M., 2019. Biosynthesis of gold nanoparticles using Allium noeanum Reut. ex Regel leaves aqueous extract; characterization and analysis of their cytotoxicity, antioxidant, and antibacterial properties. Applied Organometallic Chemistry 33, e5189. Available from: https://doi.org/10.1002/AOC.5189.

Singh, R.P., 2019. Potential of biogenic plant-mediated copper and copper oxide nanostructured nanoparticles and their utility. Nanotechnology in the Life Sciences 115–176. Available from: https://doi.org/10.1007/978-3-030-16379-2_5.

Singh, M.P., Bhardwaj, A.K., Bharati, K., Singh, R.P., Chaurasia, S.K., Kumar, S., et al., 2021. Biogenic and non-biogenic waste utilization in the synthesis of 2D materials (graphene, h-BN, g-C2N) and their applications. Frontiers in. Nanotechnology 3, 53. Available from: https://doi.org/10.3389/FNANO.2021.685427/BIBTEX.

Smith, K., Hageman, P., … GP-UG, 2004. Undefined potential metal recovery from waste streams. researchgate.net.

Strudwick, A.J., Weber, N.E., Schwab, M.G., Kettner, M., Weitz, R.T., Wünsch, J.R., et al., 2015. Chemical vapor deposition of high quality graphene films from carbon dioxide atmospheres. ACS Nano 9, 31–42. Available from: https://doi.org/10.1021/NN504822M.

Swihart, M.T., 2003. Vapor-phase synthesis of nanoparticles. Current Opinion in Colloid & Interface Science 8, 127–133. Available from: https://doi.org/10.1016/S1359-0294(03)00007-4.

Tang, W., Li, Q., Gao, S., Shang, J.K., 2011. Arsenic (III,V) removal from aqueous solution by ultrafine α-Fe2O3 nanoparticles synthesized from solvent thermal method. Journal of Hazardous Materials 192, 131–138. Available from: https://doi.org/10.1016/J.JHAZMAT.2011.04.111.

Tewari, S., Jindal, R., Kho, Y.L., Eo, S., Choi, K., 2013. Major pharmaceutical residues in wastewater treatment plants and receiving waters in Bangkok, Thailand, and associated ecological risks. Chemosphere 91 (5), 697–704. Available from: https://doi.org/10.1016/j.chemosphere.2012.12.042.

Thiyagarajulu, N., Arumugam, S., 2021. Green synthesis of reduced graphene oxide nanosheets using leaf extract of Lantana camara and its in-vitro biological activities. Journal of Cluster Science 32, 559–568.

Townshend, B., Xiang, J.S., Manzanarez, G., Hayden, E.J., Smolke, C.D., 2021. A multiplexed, automated evolution pipeline enables scalable discovery and characterization of biosensors. Nature Communications 12. Available from: https://doi.org/10.1038/S41467-021-21716-0.

Turgut, P., Yesilata, B., 2008. Physico-mechanical and thermal performances of newly developed rubber-added bricks. Energy Build 40, 679–688. Available from: https://doi.org/10.1016/J.ENBUILD.2007.05.002.

Valko, M., Morris, H., Cronin, M., 2005. Metals, toxicity and oxidative stress. Current Medicinal Chemistry 12, 1161–1208. Available from: https://doi.org/10.2174/0929867053764635.

Verma, M.L., Puri, M., Barrow, C.J., 2016. Recent trends in nanomaterials immobilised enzymes for biofuel production. Critical Reviews in Biotechnology 36, 108–119. Available from: https://doi.org/10.3109/07388551.2014.928811.

Vijayakumar, S., Vaseeharan, B., Malaikozhundan, B., Shobiya, M., 2016. Laurus nobilis leaf extract mediated green synthesis of ZnO nanoparticles: characterization and biomedical applications. Biomedicine & Pharmacotherapy 84, 1213–1222. Available from: https://doi.org/10.1016/j.biopha.2016.10.038.

Virji, M.A., Stefaniak, A.B., 2014. A review of engineered nanomaterial manufacturing processes and associated exposures. Comprehensive Materials Processing 8, 103–125. Available from: https://doi.org/10.1016/B978-0-08-096532-1.00811-6.

Wang, L., Hu, C., Shao, L., 2017. The antimicrobial activity of nanoparticles: present situation and prospects for the future. International Journal of Nanomedicine 12, 1227–1249. Available from: https://doi.org/10.2147/IJN.S121956.

Weintraub, B., Zhou, Z., Li, Y., Deng, Y., 2010. Solution synthesis of one-dimensional ZnO nanomaterials and their applications. Nanoscale 2, 1573–1587. Available from: https://doi.org/10.1039/C0NR00047G.

Xiang, X., Xia, F., Zhan, L., Xie, B., 2015. Preparation of zinc nano structured particles from spent zinc manganese batteries by vacuum separation and inert gas condensation. Separation and Purification Technology 142, 227–233. Available from: https://doi.org/10.1016/J.SEPPUR.2015.01.014.

Xing, M., Zhang, F.S., 2011. Nano-lead particle synthesis from waste cathode ray-tube funnel glass. Journal of Hazardous Materials 194, 407–413. Available from: https://doi.org/10.1016/j.jhazmat.2011.08.003.

Yallappa, S., Manjanna, J., Dhananjaya, B.L., Vishwanatha, U., Ravishankar, B., Gururaj, H., 2015. Phytosynthesis of gold nanoparticles using Mappia foetida leaves extract and their conjugation with folic acid for delivery of doxorubicin to cancer cells. Journal of Materials Science. Materials in Medicine 26. Available from: https://doi.org/10.1007/S10856-015-5567-3.

Yang, Z., Tian, J., Yin, Z., Cui, C., Qian, W., Wei, F., 2019. Carbon nanotube- and graphene-based nanomaterials and applications in high-voltage supercapacitor: a review. Carbon 141, 467–480. Available from: https://doi.org/10.1016/J.CARBON.2018.10.010.

Yot, P.G., Mear, F.O., 2009. Lead extraction from waste funnel cathode-ray tubes 698 glasses by reaction with silicon carbide and titanium nitride. Journal of Hazardous Materials 172(1), 117–123. Available from https://doi.org/10.1016/j.jhazmat.2009.06.137.

Yuan, C.G., Huo, C., Gui, B., Cao, W.P., 2017. Green synthesis of gold nanoparticles using Citrus maxima peel extract and their catalytic/antibacterial activities. IET Nanobiotechnology/IET 11, 523–530. Available from: https://doi.org/10.1049/IET-NBT.2016.0183.

Zhou, H., Su, M., Lee, P.H., Shih, K., 2017. Synthesis of submicron lead oxide particles from the simulated spent lead paste for battery anodes. Journal of Alloys and Compounds 690, 101–107. Available from: https://doi.org/10.1016/J.JALLCOM.2016.08.094.

Zhou, Z., Wu, X.F., 2013. Graphene-beaded carbon nanofibers for use in supercapacitor electrodes: synthesis and electrochemical characterization. Journal of Power Sources 222, 410–416. Available from: https://doi.org/10.1016/J.JPOWSOUR.2012.09.004.

Chapter 4

Anticipated challenges in the synthesis of different nanomaterials using biogenic waste

Newton Neogi[1], Kristi Priya Choudhury[1] and Md Abdus Subhan[1,2]

[1]*Department of Chemistry, Shahjalal University of Science and Technology, Sylhet, Bangladesh,* [2]*Division of Nephrology, School of Medicine and Dentistry, Medical Center, University of Rochester, NY, United States*

4.1 Introduction

Nanoparticles (NPs) produced from plants are often manufactured by the use of physical and chemical processes for large-scale manufacturing, which may ultimately result in the NPs being released into the environment. NPs can pose a risk to human health and the health of other living organisms when they are released into the environment. This is due to the fact that NPs are frequently manufactured with the use of hazardous chemicals that serve as functional groups and set off dangerous metabolic reactions (Bhardwaj et al., 2021). Therefore, methods of biological synthesis have been developed as a means of mitigating the dangers associated with the dispersal of NPs based on chemical compounds into the surrounding environment (Barhoum et al., 2020). Biogenic waste products are associated with a broad variety of different biomasses in their component parts (Brosowski et al., 2019). The tea waste template was used to create cuboid/pyramid-shaped magnetic iron oxide (Fe_3O_4) NPs in the 5–25 nm size range. These NPs could be used for up to five adsorption cycles and were efficient at removing arsenic metal from water. In their cell walls, food waste has cellulose, hemicelluloses, pectins, lignins, proteins, and biodegradable polysaccharides. The waste products also contain phytochemicals such as polyphenols, carotenoids, flavonoids, dietary fibers, and essential oils. They control the shape and size of NPs as they are being synthesized by acting as templates. Metal salts are converted to metal or metal oxide NPs by the biomolecules. NPs produced through biosynthesis are used in photocatalysis, sensing, and biomedicine (Bhardwaj et al., 2018). The characteristics of these NPs are distinct from those made by other conventional and chemical processes due to the lack of capping agents and surfactants. Hence, NPs produced using trash and algae have a variety of unique applications (Bhardwaj and Naraian, 2021). These biomasses may be placed into one of five groups according to their characteristics. Categories are determined by the materials' origins, such as forest biomass, agricultural biomass, municipal waste biomass, industrial waste, and biomass derived from municipal waste as well as municipal waste derived from industrial waste, as well as biomass derived from roadside greenery and railway line side wood (Brosowski et al., 2019). The phrase "reusing" refers to the process of handling biogenic waste (BW). In recent years, scientists have discovered that many types of BW may be used in the synthesis of nanomaterials. Both aquaculture and horticulture result in the production of a significant amount of organic waste. These wastes include organic material, often known as biomass. These are used in the process of using nanomaterials. This is a fantastic chance to better manage these wastes (Ghosh et al., 2017). However, there are a few roadblocks to overcome when it comes to the production of a variety of nanomaterials from biogenic waste, even though there are a lot of obstacles to overcome (Fig. 4.1). There have been a lot of exciting new ways discovered for the synthesis of two-dimensional layered materials out of biogenic waste materials. Since then, no commercially available 2D material that is created from the biogenic waste has been produced. There is a significant potential for establishing a manufacturing method at an industrial scale for two-dimensional layered materials created from garbage. In light of this, the commercialization of the synthetic pathway using waste material is an essential component of waste management. Consequently, the significance of this synthetic pathway is beyond anyone's ability to fathom. In addition, the synthesis of two-dimensional layered materials via the

Green and Sustainable Approaches Using Wastes for the Production of Multifunctional Nanomaterials. DOI: https://doi.org/10.1016/B978-0-443-19183-1.00010-6

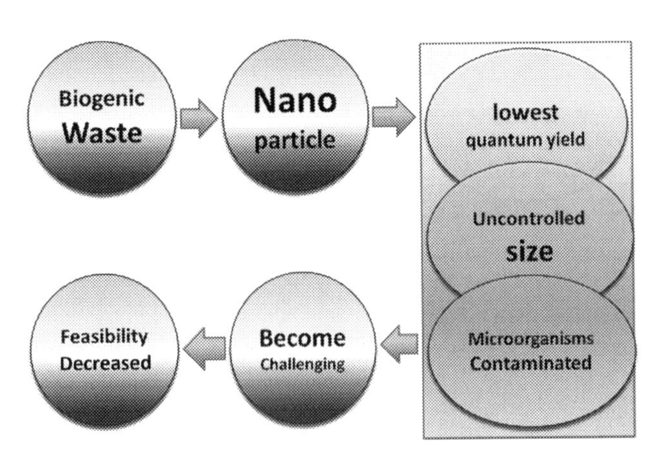

FIGURE 4.1 Disadvantages of synthetic route to nanoparticles from biogenic waste.

use of biogenic waste is a new and fascinating area of nanotechnology that has the potential to have a substantial influence on future developments in the fields of nanoscience and technology. Because of this, the difficulties associated with taking this path have become a significant cause for worry in this age of research (Singh et al., 2021).

4.2 Sustainable development of the green synthesis of nanoparticles

Green nanotechnology is a field that is growing quickly and producing a lot of useful things. It makes it possible to perform research, come up with new ideas, and find real, long-term solutions to problems around the world. The goal of green nanotechnology is to use whole living organisms or their molecules, cells, tissues, and organs as bioengineers or raw materials to make unique and useful NPs with long-term benefits. A lot of NPs are made today without using chemical reagents and organic solvents that could be bad for the environment. Because of their different properties, NPs made with eco-friendly nanotechnology can be used as functional materials (Raveendran et al., 2003). Regarding nanotoxicology, there is an urgent need for rules that control the safe use of nanomaterials, particularly MNPs created using green chemistry techniques. This requirement relates to the limitation of the environment's exposure to potentially dangerous nanosystems. Not only the size but also the surface charge and agglomeration/aggregation range of the metal nanoparticles (MNPs) provide some information regarding interactions between the MNPs and the biological medium. The surface charge and agglomeration/aggregation range of the MNPs can suggest conditions for the adsorption of metabolites. They can also reflect the interaction of a substance with ions and other dispersed solutes present in the organism or environment. The absorption peak associated with the surface plasmon resonance of the MNPs can rule out several tests that use the same wavelength range of absorption to measure the activity (Hossain et al., 2021; Iravani, 2011). Regulations are beginning to pay more attention to NPs, and the discussion compares and contrasts MNPs synthesized by traditional chemistry and green chemistry. The fact that green synthesis processes are less harmful, environment-friendly, cost-effective, quick, and simple to execute is a piece of crucial information about sustainability. Using plants, scientists need to study the mechanisms underpinning the bioreduction, nucleation, growth, and stabilization of MNPs. MNPs can be made from plants in a way that is predictable, repeatable, scalable, and safe for use in many different ways (Alshameri and Owais, 2022; Auffan et al., 2009).

4.3 Challenges of green approaches in nanoparticles synthesis

In the case of NP synthesis utilizing green procedures, there are several challenges including reproducibility, scale-up, and predictability.

4.3.1 Reproducibility

High reproducibility is a concerning challenge when it is the time of MNP synthesis with the green process. The raw materials for synthesis purposes which are generally plants as well as plant organs can vary with the solar year. In the case of the various environmental pressures of the solar year, there are also incidental and intentional anthropic actions. Plant organs' age, plant's genetic background, metabolite concentration, abiotic and biotic factors, metabolites production control, and genetically modification of organisms (GMOs) are the key factors for expected reproducibility of NPs

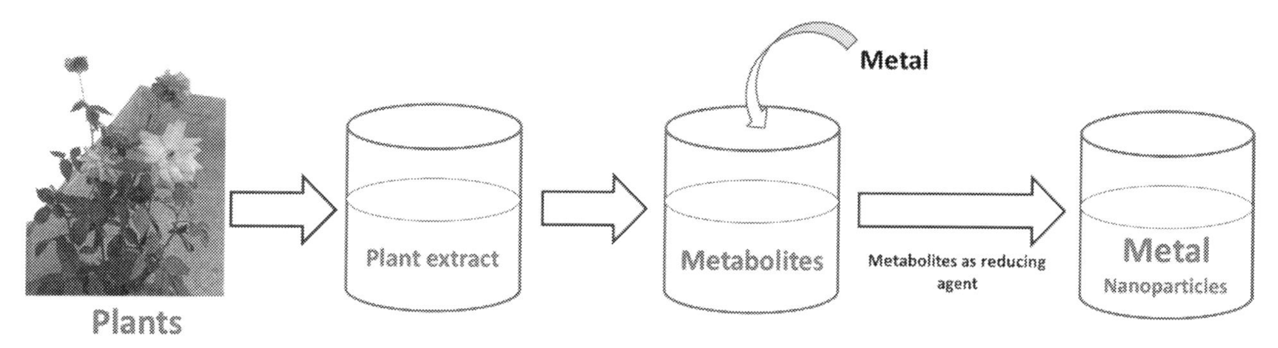

FIGURE 4.2 Green synthesis of metal nanoparticles from plant extract.

which is synthesized utilizing plants. A plant's organ age can influence the concentration of metabolites whereas a high concentration of metabolites is required for expected reproducibility. Mainly, metabolites can act as a stabilizer and a reducing agent during the synthesis of NPs. So, the variation of concentration of metabolites will affect the formation of NPs, and metabolites become structuring agents in the synthesis purpose (Fig. 4.2). As the metabolites' production system and the control of the production can influence the concentration of the metabolites, so it has become an important factor for this purpose. A plant's genetic background and abiotic and biotic factors are also important in the case of metabolite production. Thus, these are also concerning issues for green synthesis. To control the metabolite production, GMOs can play a vital role. Tissue culture can play a role in developing expected metabolites for NP synthesis via green processes. By using tissue culture, the metabolites can be secreted with a growth medium. Thus, the production of these can affect NP synthesis with a green process (Singh et al., 2018).

4.3.2 Scale-up

Scale-up is another challenge in the case of green synthesis of NPs. The low abundance of noncultivated plants, which are the raw materials and lack of proper equipment, is the factor here. For this, it becomes difficult to control the composition and heterogeneity of extracts. Extraction of metabolites from plants for a batch of plants should be done under the same condition in order to maintain homogeneity. Sometimes a plant species is selected as a potential green synthesis driver but later it was found inexpressive due to the presence of an inhibitor. Any modification to the extraction process cannot yield molecules which is best for green synthesis of MNPs. Again, extraction from noncultivated plants produces metabolites with different metabolic profiles. In order to avoid these, types of limitation extraction processes usually done under similar conditions for green synthesis of MNPs. Plant tissue culture is the best method for producing MNPs on a large scale because this method is independent and produces uniform biochemical products. This method is independent of the effect of biotic and abiotic elements including pathogens, temperature, humidity, etc. People choose plant tissue cultivation over plant crops because it is also emancipated from climate, soil, and most importantly regional characteristics. Extraction of metabolites is easy from plant tissue culture compared to whole plants. Automated synthesis reactors are very useful in the production of large-scale MNPs (Holkar et al., 2018; Singh et al., 2018).

4.3.3 Predictability

For the generation of NPs using green synthesis, predictability is the key factor. One option can be running a high-throughput screening. It is important to identify new substances that can enable the fabrication processes and the synthesized NPs physically, chemically, and biologically. Simulations are suitable for the predictability of the factors, which is impacting the process of green synthesis of MNPs by plants. Thus, it is essential to make use of computer-aided design and mathematical modeling (Singh et al., 2018; Samuel et al., 2022).

4.4 Challenges of synthesis of different nanomaterials using biogenic waste

4.4.1 Carbon dot and graphene

Carbon dots (CD) are a novel type of fluorescent nanomaterials that are composed of small carbon and have particle sizes of less than 10 nm. CDs were found inadvertently in 2004 while purifying single-wall nanotubes. These discrete,

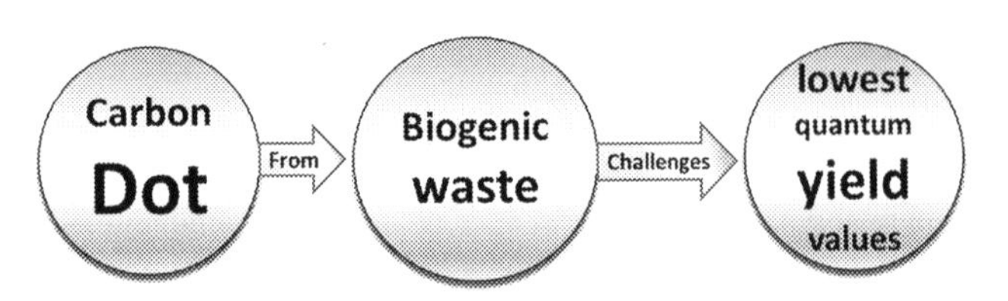

FIGURE 4.3 Obstacles in the synthesis of carbon dots from biogenic waste.

quasispherical particles have since been given the name "carbon dots" (Liu et al., 2020). CDs may be produced by the synthesis of biogenic waste. This waste includes waste from animals, waste from vegetables, and so on. The manufacture of CDs using animal feces resulted in the lowest quantum yield values (Fig. 4.3) (Wareing et al., 2021). It is not yet known what influence the reaction conditions had on the quantum yield of the CD when the hydrothermal technique was used. At these temperatures, the conversion process only takes a few minutes, but at lower temperatures, it may take many hours. This is a topic that requires more investigation and calls for a careful balance between time and temperature. The hydrothermal process is used most often for the production of CDs including animal byproducts. The waste was a result of its cheap production costs and positive effects on the surrounding environment. On the other hand, the size of the synthesized product cannot be adjusted in this manner when using the hydrothermal approach. Therefore, researchers will not be able to achieve the size they anticipated in their subsequent study (de Oliveira and da Silva Abreu, 2021). Graphene is the prototypical example of the first material with two-dimensional layers. With the assistance of Scotch tape, it was produced by the process of mechanical exfoliation of bulk graphite into a graphene nanosheet (Jian et al., 2009). After that, such stacked materials on a two-dimensional plane have gained a lot of interest from researchers for a variety of uses. As a result, the question of how to synthesize 2D-layered materials in a method that is both practical and practicable has emerged. These materials may be created by the use of waste products from biogenic processes (Wareing et al., 2021). 2011 was the first year when monolayer graphene was successfully produced from waste materials (Ruan et al., 2011). There is a significant amount of difficulty associated with these kinds of pathways using biogenic waste materials. Employing biogenic waste materials in a synthetic pathway of these materials presents a number of obstacles, including quality, repeatability, unpredictable characteristics of beginning waste materials, size and shape, and others. Researchers will not only be able to create 2D-layered materials with the assistance of the biogenic synthesis of 2D-layered materials utilizing waste materials, but they will also be able to enhance the knowledge of greener and safer issues associated with 2D-layered materials. The characteristics of critical evaluation for two-dimensional materials obtained from waste include a speedy synthesis that is also straightforward, a procedure that is both inexpensive and manageable, and environmental friendliness. As a consequence of this, the waste-derived two-dimensional layered materials that are produced should have a high quality and monolayer stability, together with exciting physicochemical features that may be suitable for upcoming useful applications (Singh et al., 2021).

4.4.2 Nanofluid

The primary difficulty at this time is the commercialization of vegetable oils, as well as the optimization of cutting fluids made from vegetable oils to be more environmentally friendly. More research has to be done on the effects that innovative vegetable oils have on tribology characteristics and other crucial process factors before they can be used commercially. Vegetable oils are in a position where they possess all of the potential traits necessary to seize these chances. Allow the use of these environment-friendly cutting fluids to make the industrial environment more sustainable, both economically and in terms of improved circumstances for the environment. The increased capacity for heat transmission and the improved lubricating performance of nanofluids are additional benefits that further support the use of vegetable oil in many applications. Because of the nanofluid form of vegetable oils, wear and the coefficient of friction may be reduced to a greater extent. This has a direct impact on the performance and consistency of machine tools. Oils such as coconut, canola, and sesame may be used in the production of nanofluids (Sankaranarayanan et al., 2021). Waste soybean oil and waste serpentine may also be used to produce nanofluid via another method of synthesis (Cui et al., 2022). These wastes of soybean oil processing have the potential to harbor microorganisms. Therefore, while working with these wastes, researchers run the risk of contacting microorganisms and will face a variety of physical challenges. In addition, soybean oil may be considered a kind of bio oil. The insufficient oxidative stability of raw bio-oils and the poor low-temperature properties of bio-oils would seem to be the most evident variables affecting the performance of bio-oils. Numerous investigations have shown that bio-oils are capable of cloudiness and precipitation at certain temperatures (Erhan et al., 2006). The viscosity of oxidation products has a significant impact on the viscosity of bio-oils. As a byproduct of secondary oxidation, the formation of molecules with a high molecular weight, such as

ketones, may occur. As bio-oils have poorer thermo-oxidation stability than petrol lubricating oils, they will deteriorate at a faster rate. This presents another obstacle. As a consequence of this, the deterioration produces abrasive particles, which in turn causes damage to the characteristics of the lubricant. Additionally, free fatty acids hasten the corrosive effect that worn surfaces have on the environment (Mannekote and Kailas, 2012). However, the recycling of bio-oil waste should be encouraged in order to safeguard the environment, as this will continue to encourage the production of goods that are more biodegradable, less poisonous, and more efficient in their energy conservation (Chowdary et al., 2021). Recent research has shown that it is feasible to synthesize a stable CuO nanofluid by using waste palm oil (Javed et al., 2018). In addition, the use of materials derived from oil palms as a capping, stabilizing, and reducing agent offers a significant amount of promise in environment-friendly and biosynthetic methods of producing metal and metal oxide NPs. Because palm oil matrices include polar ester bonds as well as long hydrocarbon chains, these matrices are excellent candidates for the role of stabilizers in the production of NPs. The cultivation of oil palms is presently under a serious challenge from a number of different pests and illnesses. When palm trees are plagued with pests and diseases, not only do they compete with the palm for nutrients but they also cause harm to the trees. This may result in a significant decrease in the crop production. Defoliators, which might include bagworms, moths, rhinoceros beetles, and other similar insects, are the most common types of pests that are found in oil palm plantations (Maluin et al., 2020). The additional challenges that the oil palm sector must contend with include a falling oil extraction rate and a massive increase in the output of residue and trash. When palm oil is produced in palm mills, the amount of oil extracted from the fruit bunch is calculated as a percentage of the total weight of the fruit bunches. The palm oil industry and palm oil mills are responsible for around 90% of the world's biowaste, despite the fact that oil only accounts for 10% of the biomass (Kahar et al., 2022). Because it may contribute to eutrophication, pollution, and any other sort of disturbance for both aquatic and terrestrial life, improper treatment of this residue and biowaste may lead to problems in the environment as well as the health of humans and animals. The oil palm sector is responsible for the production of enormous quantities of biomass waste, and the inappropriate management of this biowaste has the potential to pose significant risks to both human health and the natural environment. One of the options is to employ biowaste in the creation of nanocellulose, carbon-based nanomaterials, nanosilica, and other nano-oil-based palm derivatives. This would be one of the solutions for biomass waste management for the oil palm sector, and beneficial for nanosynthesis from biogenic waste (Maluin et al., 2020).

4.4.3 Metal-based nanoparticles

Biogenic synthesis technologies of metal-based NPs have been created in today's modern world. However, there are several obstacles to overcome when using these synthetic pathways. The production of metal-based NPs using a wide variety of biological organisms has been accomplished with some degree of success (Quester et al., 2013). On the other hand, not every biological creature can be used effectively. In addition to that, a few of them have chemicals that are poisonous. Further, since a wide range of biological entities may be used in the biogenic synthesis of NPs, the mechanism for biogenic synthesis can change based on the beginning biological material that is used. This is because the biogenic synthesis of NPs is an extremely versatile process. Therefore, there is a lack of a complete comprehension of it. There is a degree of variability in the biological entity's capacity for chemical redox reactions. As a result, the nanomaterial that is created may exhibit a variety of distinct chemical and physical properties. This is dependent on the circumstances under which the experiment is conducted. As a result, achieving precise control over the nanoparticle size distribution, shape, and chemical surface is one of the most difficult challenges associated with the biological production of nanomaterials. In this regard, the chemical and physical methods may provide a greater capacity to regulate the characteristics of the nanoparticle, in comparison to the biological routes. The comprehensive characterization of the biological agents that are responsible for the formation of NPs is an essential component of the biogenic synthesis of NPs that has to be further investigated. This is a crucial aspect of the biogenic synthesis of NPs (Tortella et al., 2021). An example of a green synthesis process is the production of metal-based NPs from biogenic waste. Producing a diverse array of silver and gold NPs may be accomplished in a straightforward and efficient manner using this method. At the atomic and molecular levels, there are some complexities involved. As a result, having insights into the reduction process at the molecular level would ultimately enable researchers to alleviate the deficiencies using more beautiful techniques. When considering some NPs, it is common to be worried about restrictions, such as rough surfaces, polycrystallinity, nonuniformity, poorly defined composition, and the difficulty of modifying the surfaces without affecting the parent NPs in any way. In addition, the creation of NPs with ultrafine structures remains a significant obstacle to overcome. The overall stability of the process to generate homogeneous and monodisperse nano-sized NPs is still a challenge, and it is difficult to manage the shape and size of the product after it has been manufactured. In a similar vein, the real structure, size distribution, and topographical qualities of the parent NPs are scarcely retained throughout the synthesis of multimetallic NPs from biogenic waste. This is because of the nature of the synthesis. In addition, achieving high reproducibility can be difficult due to the fact that fruit wastes are affected by a wide range of environmental factors, such as the amount of water present, the absence or presence of particular components, variations in pH, the presence or absence of parasites, the intensity of light exposure, and so on (Chandrasekaran and Bahkali, 2013). There is a need for deeper research into the particular planning for desired nanomaterials, the actual mechanism, and its practical usage. The functionality of the surface of silver and gold NPs is what defines their likely performance. For this reason, having a comprehensive understanding of the firm's status, size, and structural aspects, where NPs may be utilized, is of utmost importance. This may also give the most definitive findings yet about the functioning, structural characteristics, and segregation of NPs. Understanding the molecular-level interactions between the functional groups and the specific surface of the NPs is necessary in order to make further use of the reducing and capping agents from fruit wastes, both those that already exist and those that have not yet been discovered but which have the potential to regulate the properties of these NPs (Table 4.1).

In order to examine the effect of sunlight on the production of Ag NPs, a transparent glass container containing a solution of silver nitrate and a solution of mushroom extract was incubated for one hour in the presence of sunlight and in the absence of light. It was found that the solution placed in the dark does not alter color, but the creation of silver NPs produces a dark burgundy hue. In the presence of sunlight, photochemical kinetics were faster than in the absence of light. In addition to producing NPs with a smaller size, the reaction in sunlight also reduced the size of NPs. In sunlight, the output of Ag NPs was also quite high. When the reaction was exposed to sunlight using a blue polyethylene sheet, smaller and denser particles were produced than when red, green, and yellow light was used. Particles produced by monochromatic blue sunlight and the entire spectrum of sunshine (broadband light) had average particle diameters of 3.18 ± 0.72 nm and 7.08 ± 2.92 nm, respectively. A study used *P. citrinopileatus* mushroom extracts as a capping and reducing agent in the photobiochemical production of silver NPs. There is often no capping substance used in conventional or chemical techniques for producing NPs. This is crucial to the green approach. This method of synthesis produces silver NPs with a single distribution and is rapid, affordable, one-step, size-controllable, and efficient (Bhardwaj et al., 2018).

TABLE 4.1 Nanoparticle synthesis from waste materials.

NPs	Raw materials	Morphology	Size (nm)	References
Pd NPs	Watermelon rind	Spherical	96	Lakshmipathy et al. (2015)
Gold NPs	Grape skin, seeds and stalk	Quasi-spherical	20–25	Krishnaswamy et al. (2014)
	Rice bran	Spherical	50–100	Malhotra et al. (2014)
Iron NPs	Citrine juices	Cylindrical, Spherical, and irregular	3–300	Machado et al. (2014)
Cellulose	Waste cotton fibers	Spherical	40–90	Fattahi Meyabadi et al. (2014)
Silver NPs	*Satsuma mandarin* (Citrus unshiu) peel extract	Spherical	5–20	Basavegowda and Rok Lee (2013)
	Carob leaf extract	Spherical	5–40	Sengupta and Sarkar (2022)
	Eucalyptus hybrida peels	Spherical	50–150	Sengupta and Sarkar (2022)
	Aloe vera leaf extract	Spherical and triangular	50–350	Sengupta and Sarkar (2022)
	Orange and banana peels	Spherical	41.3 ± 5.2	Abdel-Khalek, Hamed and Hasheesh (2021)
	Neem leaf extract	Spherical, cubical, bean-like, and irregular shapes	31.9	Zafar and Iqbal (2022)
Cerium oxide NPs	*M. sapientum* peel	Irregular	4–13	Miri et al. (2021)
Zinc oxide NPs	Neem leaf extract	Spherical, cubical, bean-like, and irregular shapes	49	Zafar and Iqbal (2022)

4.4.4 Nanotubes

The production of multiwall carbon nanotubes (MWCNTs) by means of fluidized bed chemical vapor deposition was catalyzed with the use of red mud, which is a hazardous byproduct of the bauxite processing industry. The temperature of the reaction was kept constant at 650°C throughout. Plant extracts from garden grass (*Cynodon dactylon*), rose (*Rosa*), neem (*Azadirachta indica*), and wall nut (*Juglans regia*) have been used in the synthesis of catalysts for the development of MWCNTs utilizing the CVD technique. Because there is not enough published information on the quality and amount of the CNTs that are created, green synthesis of CNTs continues to be a difficult task, particularly for industrial manufacturing. Synthesis of CNTs of higher quality may be seen most often in green precursors. However, the more environment-friendly method to the synthesis of CNTs tackles the issues that are concerned about the environment. As a result, substantial optimization is required for the manufacture of CNTs at large scales using more environment-friendly methods (Makgabutlane et al., 2021). The production of MWCNT has been achieved using the pyrolysis of leftover rice straw. Within this context, the diameter of the nanotube was between 15 and 40 nm, and the surface area was 188 m^2/g (Makgabutlane et al., 2021). In recent years, there has been a shift toward the use of natural precursors in the manufacturing of CNTs, namely hydrocarbons derived from plants. Almost all components of plants (including the stem, leaves, seeds, and roots) have the potential to be used as CNT precursors (Zhu et al., 2012; Vanessa et al., 2013). Additionally, plant derivatives such as camphor, turpentine oil, palm oil, sesame oil, olive oil, and coconut oil have been investigated for their potential as CNT precursors. Pyrolysis is a straightforward and cost-effective method for the synthesis of CNTs, which may be carried out at low temperatures. The most significant obstacle that we face here is the lack of certainty around the acquisition of morphology (Hamid et al., 2017).

4.4.5 Metal-organic framework

Because of the wide variety of applications for metal-organic frameworks (MOFs), the synthesis of MOFs has become a topic of interest for both academic researchers and industry organizations. The synthesis of MOFs has already been transferred from the laboratory to the industrial level required for widespread use. A number of different firms are now working on the synthesis of these materials in order to further application development. MOFs have applications in a variety of economic areas, such as the energy industry, the environmental field, and the catalysis industry, amongst others. In addition to the design of materials, it is vital to take into consideration the difficulties associated with different synthesis methods. Concerns have been raised about the product yield, the reaction efficiency, and the corresponding effect on the environment (Julien et al., 2017). The convergence of environment-friendly chemistry and long-term sustainability carries with it the promise of great compatibility with the requirements of industrial production. The development of more environment-friendly synthesis methods has the potential to deliver fast growth in the real-world applications of MOFs, which is anticipated in the not-too-distant future. In addition to the development of more environment-friendly synthesis techniques, it is of utmost importance to evaluate the materials' suitability for use in actual field applications. Cytotoxicity is an additional concern in this context (Tamames-Tabar et al., 2014). For example, inorganic metal ions (such as iron, zinc, copper, nickel, and others) cannot be broken down by biological processes in their natural environments. Zinc(II) and iron(II) are the metal ions that are utilized for MOFs. These are used most often out of all the numerous metal ions that are mentioned for them. There have only been a few in vitro and in vivo investigations that have established the acute toxicity of certain MOFs (such as MIL-100, 88A, and MIL-88B) when administered in large concentrations (Kumar et al., 2019). Another obstacle to overcome is the deterioration of MOFs. In addition, the expulsion of digested excess metal ions might result in the formation of extremely hazardous metal oxides owing to exposure to the surrounding environment. On the other hand, the potential dangers to one's health posed by organic linkers are little known (Kumar et al., 2020). The use of green MOFs, which include biocompatible metal ions and bioligands (such as nucleobases, polypeptides, and amino acids), in combination with biogenic waste, may provide a solution to the problem. However, the manufacturing of these environment-friendly bioMOFs has a number of obstacles, including a complicated synthetic guiding methodology for the synthesis of bioMOFs, low thermal and chemical stability, and limited control over active sites and porosity (Cai et al., 2019). The microwave-assisted technique appears to be the best method for rapidly and efficiently preparing MOFs with high purity and yield while controlling of size when compared to the other synthesis methods of MOFs with biogenic waste. This is the conclusion reached after comparing the different synthesis methods of MOFs with biogenic waste (Sohrabi et al., 2021).

4.4.6 Nanocellulose

In the process of modifying the material characteristics of nanofiber cellulose, one of the obstacles that must be overcome is the statistical distribution of the internal stress as well as the nanofiber cellulose (NFC). The detection of numerous faults is made more difficult due to the reduced size of the material entities being examined. It is vital to have a solid grasp of the mechanical qualities related to NFC in order to have a thorough appreciation for the applicability of the technology. According to the findings of many different pieces of research, the ultimate tensile strength of NFC is 1 GPa, and its modulus is 30 GPa (Thomas et al., 2020). The fact that the majority of the polymers utilized in the production of composites are hydrophobic, while nanocellulose is hydrophilic, is one of the most significant challenges that arise when attempting to incorporate nanocellulose in these materials. This may be a challenge during the process of combining the components, as it increases the propensity of nanocellulose to form aggregates. These aggregates would not only result in nondispersed matrices but would also reduce the nanoscale characteristics of a reinforcing agent. As a means of increasing profitability, the manufacturing of nanocellulose may benefit from the use of the leftover biomass produced by these types of enterprises (Ribeiro et al., 2019). Cellulose nanofiber has been shown to have beneficial reinforcing properties on polymer nanocomposites. In most cases, it has been discovered that the incorporation of nanocellulose particles into a polymer matrix results in an increase in the matrix's modulus and strength while simultaneously resulting in a reduction in the material's permeability to gas. These materials' characteristics are highly reliant on the potential of robust interactions between the hydrogen bonds, which are bolstered even more by the fact that the particles' sizes play a role in the dynamic. Recent years have seen the emergence of a wide variety of prospects and possible uses for nanocellulose. New capabilities have been added to some of them, including acting as hydrogels or drug delivery systems, as antimicrobial agents, intelligent materials, and edible films for food packaging. In recent times, research has been done into a variety of biomasses, in addition to research into the extraction methods and manufacturing procedures for nanocellulose. Researchers produced biodegradable films consisting of hemicellulose and reinforced with CNC derived from wheat straw. They showed that nanocellulose reinforced the films, enhancing their mechanical and barrier qualities. Another work was presented, this time focusing on the development of chitosan (CS) biocomposites with nanocrystalline cellulose made from rice straw waste (Pires et al., 2019). The authors have provided data that demonstrates the better interfacial compatibility of CS/CNC biocomposites, which also have exceptional tensile strength (Pires et al., 2019). When the manufacturing of nanocellulose is seen as a concept for a biorefinery, the utilization of several lignocellulosic wastes from agricultural (nonprocessed) and industrial (processed) activities becomes the best proposal in terms of cost savings, energy savings, and economic growth. Agricultural biowaste is a source of cellulose that is readily accessible all over the globe, is relatively inexpensive, and has not been fully utilized. This source of cellulose has the potential to be employed in the manufacturing of nanocellulose products on a global scale. In this regard, the design of locally produced nanocellulose processes from industrial biowaste (food, beverage, pulp and paper, etc.), integrated within the process that generates them offers up an intriguing avenue for the use of these materials. This strategy should result in large cost savings for companies (due to the minimizing of residues), as well as significant profits and the creation of new materials based on nanocellulose (García et al., 2016). When it comes to the environment, the persistent use of biomass and the use of harsh chemicals throughout the process both have a negative influence on sustainability. As a result, the major objective is to concentrate on developing a process that is both economical and kind to the natural world, with the ultimate goal being the manufacture of high-quality microcrystalline cellulose (MCC) and nanocrystalline cellulose (NCC). In addition, it is desirable to establish a suitable approach for getting the intended end product of MCC and NCC from a specific source of biomass. This may be done by creating a suitable method for obtaining the desired end product of MCC and NCC. Therefore, the same process may be conducted on a variety of biomasses in order to identify an optimal reaction condition for the generation of MCC and NCC that is required while using a variety of source materials. This will allow for more flexibility in this field (Haldar and Purkait, 2020). Furthermore, the primary difficulties encountered are associated with the objective, which, in the context of the use of lignocellulosic residues resulting from agricultural and industrial activities, not only entails the obtaining of a high-value-added product such as nanocellulose from a renewable source but also the reduction on the environmental impact of these wastes and the economic benefit that the use of this kind of material could provide everywhere in the world (García et al., 2016). The unique technique has the potential to lead to certain high-value-added goods and applications. This potential is supported by the extraction of nanocellulose from a variety of biomass wastes for further development of functional materials. However, there are still a number of technological obstacles and financial concerns that need to be resolved before manufacturing on a big scale and widespread use. The collection, transportation, and storage of biomass waste, the purchase of pricey chemicals or enzymes, and the considerable amount of energy that must be used during the processing phase are the primary sources of the high costs that are incurred. It is required to discover effective ways to create a large number of cellulose nanofibers at a cheap cost and with outstanding characteristics and uniform size in order to accomplish production on a scale that is suitable for commercial use. The development of effective pretreatments as well as precise structural and functional modification is the primary focus of the problems presented by the technical side. The production of high-quality nanocellulose requires effective pretreatments that can remove various contaminants without disrupting the cellulose structure. This is a precondition for the process (Yu et al., 2021).

4.5 Conclusion

There are some obstacles in utilizing biogenic waste material to synthesize NPs. In the case of utilization of NPs, their size is the main feature. However the synthetic routes that include biogenic waste as an initial compound cannot control the size of NPs. Along with this, the lower quantum yield of the product is another challenge. Moreover, biogenic waste can be contaminated with microorganisms. Thus, before using as an initial reactant the waste material needs treatment of antimicrobial agents which can increase the cost of synthesis. Therefore, the conversion of biogenic waste material to a valuable asset through the production of innovative nanomaterials necessitates further intensive investigations.

References

Abdel-Khalek, A.A., Hamed, A., Hasheesh, W.S.F., 2021. Does the adsorbent capacity of orange and banana peels toward silver nanoparticles improve the biochemical status of Oreochromis niloticus? Environmental Science and Pollution Research 28 (25), 33445–33460. Available from: https://doi.org/10.1007/s11356-021-13145-9.

Alshameri, A.W., Owais, M., 2022. Antibacterial and cytotoxic potency of the plant-mediated synthesis of metallic nanoparticles Ag NPs and ZnO NPs: A review. OpenNano 8, 100077. Available from: https://doi.org/10.1016/j.onano.2022.100077. August.

Auffan, M., et al., 2009. Towards a definition of inorganic nanoparticles from an environmental, health and safety perspective. Nature Nanotechnology 4 (10), 634–641. Available from: https://doi.org/10.1038/nnano.2009.242.

Barhoum, A., et al., 2020. Plant celluloses, hemicelluloses, lignins, and volatile oils for the synthesis of nanoparticles and nanostructured materials. Nanoscale 12 (45), 22845–22890. Available from: https://doi.org/10.1039/D0NR04795C.

Basavegowda, N., Rok Lee, Y., 2013. Synthesis of silver nanoparticles using Satsuma mandarin (Citrus unshiu) peel extract: A novel approach towards waste utilization. Materials Letters 109, 31–33. Available from: https://doi.org/10.1016/J.MATLET.2013.07.039.

Bhardwaj, A.K., et al., 2018. Direct sunlight enabled photo-biochemical synthesis of silver nanoparticles and their bactericidal efficacy: Photon energy as key for size and distribution control. Journal of Photochemistry and Photobiology B: Biology 188, 42–49. Available from: https://doi.org/10.1016/j.jphotobiol.2018.08.019.

Bhardwaj, A.K., Naraian, R., 2021. Cyanobacteria as biochemical energy source for the synthesis of inorganic nanoparticles, mechanism and potential applications: A review. 3 Biotech 11 (10), 445.

Bhardwaj, A.K., Shukla, A., Singh, S.C., Uttam, K.N., Nath, G., Gopal, R., 2021. Green synthesis of Cu_2O hollow microspheres. Advanced Materials Proceedings 2 (2), 132–138.

Brosowski, A., et al., 2019. How to measure the impact of biogenic residues, wastes and by-products: Development of a national resource monitoring based on the example of Germany. Biomass and Bioenergy 127, 105275. Available from: https://doi.org/10.1016/j.biombioe.2019.105275. May.

Cai, H., Huang, Y.L., Li, D., 2019. Biological metal–organic frameworks: Structures, host–guest chemistry and bio-applications. Coordination Chemistry Reviews 378, 207–221. Available from: https://doi.org/10.1016/j.ccr.2017.12.003.

Chandrasekaran, M., Bahkali, A.H., 2013. Valorization of date palm (Phoenix dactylifera) fruit processing by-products and wastes using bioprocess technology – Review. Saudi Journal of Biological Sciences 20 (2), 105–120. Available from: https://doi.org/10.1016/J.SJBS.2012.12.004.

Chowdary, K., et al., 2021. A review of the tribological and thermophysical mechanisms of bio-lubricants based nanomaterials in automotive applications. Journal of Molecular Liquids 339, 116717. Available from: https://doi.org/10.1016/j.molliq.2021.116717.

Cui, X., et al., 2022. Preparation and application of sustainable nanofluid lubricant from waste soybean oil and waste serpentine for green intermittent machining process. Journal of Manufacturing Processes 77, 508–524. Available from: https://doi.org/10.1016/J.JMAPRO.2022.03.032.

de Oliveira, B.P., da Silva Abreu, F.O.M., 2021. Carbon quantum dots synthesis from waste and by-products: Perspectives and challenges. Materials Letters 282, 128764. Available from: https://doi.org/10.1016/j.matlet.2020.128764.

Erhan, S.Z., Sharma, B.K., Perez, J.M., 2006. Oxidation and low temperature stability of vegetable oil-based lubricants. Industrial Crops and Products 24 (3), 292–299. Available from: https://doi.org/10.1016/j.indcrop.2006.06.008.

Fattahi Meyabadi, T., et al., 2014. Spherical cellulose nanoparticles preparation from waste cotton using a green method. Powder Technology 261, 232–240. Available from: https://doi.org/10.1016/j.powtec.2014.04.039. July.

García, A., et al., 2016. Industrial and crop wastes: A new source for nanocellulose biorefinery. Industrial Crops and Products 93, 26–38. Available from: https://doi.org/10.1016/j.indcrop.2016.06.004.

Ghosh, P.R., et al., 2017. Production of high-value nanoparticles via biogenic processes using aquacultural and horticultural food waste. Materials 10 (8). Available from: https://doi.org/10.3390/ma10080852.

Haldar, D., Purkait, M.K., 2020. Micro and nanocrystalline cellulose derivatives of lignocellulosic biomass: A review on synthesis, applications and advancements. Carbohydrate Polymers 250, 116937. Available from: https://doi.org/10.1016/j.carbpol.2020.116937. August.

Hamid, Z.A., et al., 2017. Challenges on synthesis of carbon nanotubes from environmentally friendly green oil using pyrolysis technique. Journal of Analytical and Applied Pyrolysis 126, 218–229. Available from: https://doi.org/10.1016/j.jaap.2017.06.005. June.

Holkar, C.R., et al., 2018. Scale-up technologies for advanced nanomaterials for green energy: Feasibilities and challenges, nanomaterials for green energy. Elsevier Inc. Available from: 10.1016/B978-0-12-813731-4.00014-X.

Hossain, A., et al., 2021. Application of nanomaterials to ensure quality and nutritional safety of food. Journal of Nanomaterials 2021. Available from: https://doi.org/10.1155/2021/9336082.

Iravani, S., 2011. Green synthesis of metal nanoparticles using plants. Green Chemistry 13 (10), 2638–2650. Available from: https://doi.org/10.1039/c1gc15386b.

Javed, M., et al., 2018. Synthesis of stable waste palm oil based CuO nanofluid for heat transfer applications. Heat and Mass Transfer/Waerme- und Stoffuebertragung 54 (12), 3739–3745. Available from: https://doi.org/10.1007/s00231-018-2399-y.

Jian, Y.H., et al., 2009. In situ observation of graphene sublimation and multi-layer edge reconstructions. Proceedings of the National Academy of Sciences of the United States of America 106 (25), 10103–10108. Available from: https://doi.org/10.1073/pnas.0905193106.

Julien, P.A., Mottillo, C., Friščić, T., 2017. Metal-organic frameworks meet scalable and sustainable synthesis. Green Chemistry 19 (12), 2729–2747. Available from: https://doi.org/10.1039/c7gc01078h.

Kahar, P., et al., 2022. An integrated biorefinery strategy for the utilization of palm-oil wastes. Bioresource Technology 344, 126266. Available from: https://doi.org/10.1016/J.BIORTECH.2021.126266.

Krishnaswamy, K., Vali, H., Orsat, V., 2014. Value-adding to grape waste: Green synthesis of gold nanoparticles. Journal of Food Engineering 142, 210–220. Available from: https://doi.org/10.1016/j.jfoodeng.2014.06.014. June.

Kumar, P., et al., 2019. Regeneration, degradation, and toxicity effect of MOFs: Opportunities and challenges. Environmental Research 176, 108488. Available from: https://doi.org/10.1016/j.envres.2019.05.019. May.

Kumar, S., et al., 2020. Green synthesis of metal–organic frameworks: A state-of-the-art review of potential environmental and medical applications. Coordination Chemistry Reviews 420, 213407. Available from: https://doi.org/10.1016/j.ccr.2020.213407.

Lakshmipathy, R., et al., 2015. Watermelon rind-mediated green synthesis of noble palladium nanoparticles: Catalytic application. Applied Nanoscience (Switzerland) 5 (2), 223–228. Available from: https://doi.org/10.1007/s13204-014-0309-2.

Liu, J., Li, R., Yang, B., 2020. Carbon dots: A new type of carbon-based nanomaterial with wide applications. ACS Central Science 6 (12), 2179–2195. Available from: https://doi.org/10.1021/acscentsci.0c01306.

Machado, S., et al., 2014. Utilization of food industry wastes for the production of zero-valent iron nanoparticles. Science of the Total Environment 496, 233–240. Available from: https://doi.org/10.1016/j.scitotenv.2014.07.058.

Makgabutlane, B., et al., 2021. Green synthesis of carbon nanotubes to address the water-energy-food nexus: A critical review. Journal of Environmental Chemical Engineering 9 (1), 104736. Available from: https://doi.org/10.1016/J.JECE.2020.104736.

Malhotra, A., et al., 2014. Multi-analytical approach to understand biomineralization of gold using rice bran: A novel and economical route. RSC Advances 4 (74), 39484–39490. Available from: https://doi.org/10.1039/c4ra05404k.

Maluin, F.N., Hussein, M.Z., Idris, A.S., 2020. An overview of the oil palm industry: Challenges and some emerging opportunities for nanotechnology development. Agronomy 10 (3). Available from: https://doi.org/10.3390/agronomy10030356.

Mannekote, J.K., Kailas, S.V., 2012. The effect of oxidation on the tribological performance of few vegetable oils. Journal of Materials Research and Technology 1 (2), 91–95. Available from: https://doi.org/10.1016/S2238-7854(12)70017-0.

Miri, A., et al., 2021. Cerium oxide nanoparticles: Green synthesis using Banana peel, cytotoxic effect, UV protection and their photocatalytic activity. Bioprocess and Biosystems Engineering 44 (9), 1891–1899. Available from: https://doi.org/10.1007/s00449-021-02569-9.

Pires, J.R.A., de Souza, V.G.L., Fernando, A.L., 2019. Production of nanocellulose from lignocellulosic biomass wastes: Prospects and limitations. Lecture Notes in Electrical Engineering 505, 719–725. Available from: https://doi.org/10.1007/978-3-319-91334-6_98/COVER.

Quester, K., Avalos-Borja, M., Castro-Longoria, E., 2013. Biosynthesis and microscopic study of metallic nanoparticles. Micron (Oxford, England: 1993) 54–55, 1–27. Available from: https://doi.org/10.1016/j.micron.2013.07.003.

Raveendran, P., Fu, J., Wallen, S.L., 2003. Completely "green" synthesis and stabilization of metal nanoparticles. Journal of the American Chemical Society 125 (46), 13940–13941. Available from: https://doi.org/10.1021/ja029267j.

Ribeiro, R.S.A., et al., 2019. Production of nanocellulose by enzymatic hydrolysis: Trends and challenges. Engineering in Life Sciences 19 (4), 279–291. Available from: https://doi.org/10.1002/elsc.201800158.

Ruan, G., et al., 2011. Growth of graphene from food, insects, and waste. ACS Nano 5 (9), 7601–7607. Available from: https://doi.org/10.1021/nn202625c.

Samuel, M.S., et al., 2022. A review on green synthesis of nanoparticles and their diverse biomedical and environmental applications. Catalysts 12 (5). Available from: https://doi.org/10.3390/catal12050459.

Sankaranarayanan, R., et al., 2021. A comprehensive review on research developments of vegetable-oil based cutting fluids for sustainable machining challenges. Journal of Manufacturing Processes 67, 286–313. Available from: https://doi.org/10.1016/j.jmapro.2021.05.002. May.

Sengupta, A., Sarkar, A., 2022. Synthesis and characterization of nanoparticles from neem leaves and banana peels: A green prospect for dye degradation in wastewater. Ecotoxicology (London, England) 31 (4), 537–548. Available from: https://doi.org/10.1007/s10646-021-02414-5.

Singh, J., et al., 2018. "Green" synthesis of metals and their oxide nanoparticles: Applications for environmental remediation. Journal of Nanobiotechnology 16 (1), 1–24. Available from: https://doi.org/10.1186/s12951-018-0408-4.

Singh, M.P., et al., 2021. Biogenic and non-biogenic waste utilization in the synthesis of 2D materials (graphene, h-BN, g-C2N) and their applications. Frontiers in Nanotechnology 3, 1–25. Available from: https://doi.org/10.3389/fnano.2021.685427. September.

Sohrabi, H., et al., 2021. Nanoscale metal-organic frameworks: Recent DEVELOPMENTS in synthesis, modifications and bioimaging applications. Chemosphere 281, 130717. Available from: https://doi.org/10.1016/j.chemosphere.2021.130717. April.

Tamames-Tabar, C., et al., 2014. Cytotoxicity of nanoscaled metal-organic frameworks. Journal of Materials Chemistry B 2 (3), 262–271. Available from: https://doi.org/10.1039/c3tb20832j.

Thomas, P., et al., 2020. Comprehensive review on nanocellulose: Recent developments, challenges and future prospects. Journal of the Mechanical Behavior of Biomedical Materials 110, 103884. Available from: https://doi.org/10.1016/j.jmbbm.2020.103884. April.

Tortella, G., et al., 2021. Bactericidal and virucidal activities of biogenic metal-based nanoparticles: Advances and perspectives. Antibiotics 10 (7), 1–23. Available from: https://doi.org/10.3390/antibiotics10070783.

Vanessa, R., et al., 2013. Carbon nanotubes synthesized using sugar cane as precursor. International Journal of Energy and Power Engineering 7 (12), 1027–1030.

Wareing, T.C., Gentile, P., Phan, A.N., 2021. Biomass-based carbon dots: Current development and future perspectives. ACS Nano 15 (10), 15471–15501. Available from: https://doi.org/10.1021/ACSNANO.1C03886/ASSET/IMAGES/MEDIUM/NN1C03886_0012.GIF.

Yu, S., et al., 2021. Nanocellulose from various biomass wastes: Its preparation and potential usages towards the high value-added products. Environmental Science and Ecotechnology 5, 100077. Available from: https://doi.org/10.1016/j.ese.2020.100077.

Zafar, M., & Iqbal, T. (2022). Green synthesis of silver and zinc oxide nanoparticles for novel application to enhance shelf life of fruits. *Biomass Conversion and Biorefinery* [Preprint], (0123456789). Available from: https://doi.org/10.1007/s13399-022-02730-8.

Zhu, J., et al., 2012. Synthesis of multiwalled carbon nanotubes from bamboo charcoal and the roles of minerals on their growth. Biomass and Bioenergy 36, 12–19. Available from: https://doi.org/10.1016/j.biombioe.2011.08.023.

Chapter 5

Domestic waste utilization in the synthesis of functional nanomaterial

Abhi Sarika Bharti[1], Chhavi Baran[2], Abhishek Kumar Bhardwaj[3], Shipra Tripathi[4], Rahul Pandey[5] and Kailash Narayan Uttam[6]

[1]Faculty of Science and Technology, Dr. Shakuntala Misra National Rehabilitation University, Lucknow, Uttar Pradesh, India, [2]Centre for Environmental Science, IIDS, University of Allahabad, Allahabad, Madhya Pradesh, India, [3]Department of Environmental Science, Amity School of Life Sciences, Amity University, Gwalior, Madhya Pradesh, India, [4]Department of Physics, Faculty of Science and Technology, Dr. Shakuntala Misra National Rehabilitation University, Lucknow, Uttar Pradesh, India, [5]Department of Biotechnology, S. M. M. Town P. G. College, Ballia, Uttar Pradesh, India, [6]Saha's Spectroscopy Laboratory, Department of Physics, University of Allahabad, Prayagraj, Uttar Pradesh, India

5.1 Introduction

Humans' activity generates waste on daily basis, which is a major concern since the ancient period. As the population grows successionally with time, the amount and variety of waste also increase. Due to the increasing population and to meet their demands and needs, various facilities are growing, industries are expanding, and urbanization is occurring. As a result of these factors, a significant amount of daily waste is produced, which has now turned into a challenging issue on a global scale (Amasuomo and Baird, 2016). Waste could be of any type such as municipal, domestic, agricultural, and commercial (Kwon and Bard, 2012).

5.1.1 Categorization of the domestic waste

Domestic wastes mainly included solid and liquid waste excreted on a regular scale. It includes the dry as well as wet waste that is being generated day-by-day through domestic activity by humans. On the basis of the wastes accumulated from the household and municipal bins, they are been categorized as organic, hazardous, and recyclable waste as shown below in the classification of domestic waste.

Classification of domestic waste

1. *Organic waste*: this refers to the waste released from the kitchen; where plant waste includes the leftover of vegetables such as leaves, fruits, and flowers and animal waste as chicken and fish meat and bone waste.
2. *Hazardous waste*: this refers to the waste including medicines, paints, chemicals, bulbs, spray cans, fertilizer and pesticides container, batteries, shoe polish, etc.
3. *Recyclable waste*: this mainly consists of glass, metals, plastics, paper, cardboard products, etc.

Wastes that are solid and dry in nature and have 80% of the material are of the recyclable category. Therefore, this waste is been transferred to a specific sector analyzing its specific behavior and properties. Similarly, waste that is biodegradable and wet, is maximally (almost 100%) degraded biologically in the form that is being utilized mostly for livestock and in the agriculture field.

Globally, millions of tonnes of waste are generated regularly, which needs a proper minimization process for the complete removal of the waste without any harmful effect in return. The waste management and handling (WMH) team regularly gathers the waste from storage bins and door-to-door collection depending on the area. Proper storage and

Green and Sustainable Approaches Using Wastes for the Production of Multifunctional Nanomaterials. DOI: https://doi.org/10.1016/B978-0-443-19183-1.00002-7

handling procedure for the wastes makes it an easy-to-operate mechanism for diminishing the remains accordingly (Agarwal et al., 2015).

5.1.2 Domestic waste collection history

India was found to be generating waste per capita of 119.07 g/day (61 million tonnes per year) with a population of 1.38 billion (Annual Report on Solid Waste Management (2020–21), CPCB, Delhi). As per data recorded, the waste generation will be increasing by up to 300 million tonnes of waste annually with a population of 1.823 billion by 2051, showing a 1.33% increment per year in waste generation due to the growing population and changing lifestyle. India's waste generation counted off annually of about 61 million tonnes, predicted to generate 107 million tons by 2031, and which would be reaching 161 million tonnes in 2041 showing the five times increase in four decades (Pal and Bhatia, 2022).

The increase of waste in rural areas can greatly be due to the existing population in urban areas and due to the advances in their lifestyle. The data generated from municipal solid waste (MSW) management of the waste generated by different states have put Maharashtra state at the top in comparison to Delhi and followed by Kolkata, Chennai, and Bangalore (Loizia et al., 2021).

The global waste generation annually of municipal solid waste collection shows that nearly 50% of the food and green waste generated by low- to high-income cities can be seen in Fig. 5.1. Among that maximum percentage of waste, of about 33% is still settled as open dumping shown in the data collected (Fig. 5.2) for the global treatment and disposal of waste. India is not managing the volumes of waste generated regularly through the growing city population, and this influences the surrounding and public health.

5.1.3 Domestic waste material utilization in different sectors

There are varieties of waste being generated domestically such as discarded food, plastic, metal, and glass wastes that are being utilized accordingly to their properties. Both biodegradable and nonbiodegradable are the most fundamental and crucial categories of waste. The categories of waste segregated from various sectors are separated considering their condition. Separation of the waste from the preliminary condition as the dry and wet waste would maximally help from the proper separation and treatment procedure. Furthermore, the wastes collected are been transferred to the respective sectors or industries for utilizing or reusing the waste, either part of it or another form (Abdel-Shafy and Mansour, 2017). The problem originated due to improper waste management and has evolved several mitigation processes.

5.1.4 Nanomaterial synthesis using domestic waste

While using some stuff, people usually discard that thing after a single use. But if thinking deeply, whatever nature has given, from soil to plants to trees and forests, even their every part, is completely utilizable for the benefit of living

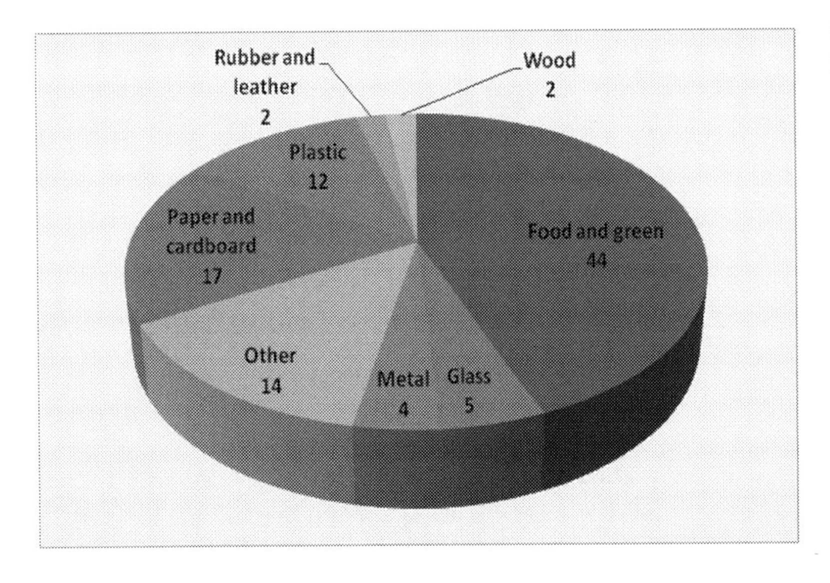

FIGURE 5.1 Global waste composition (in percent) (Kaza et al., 2018).

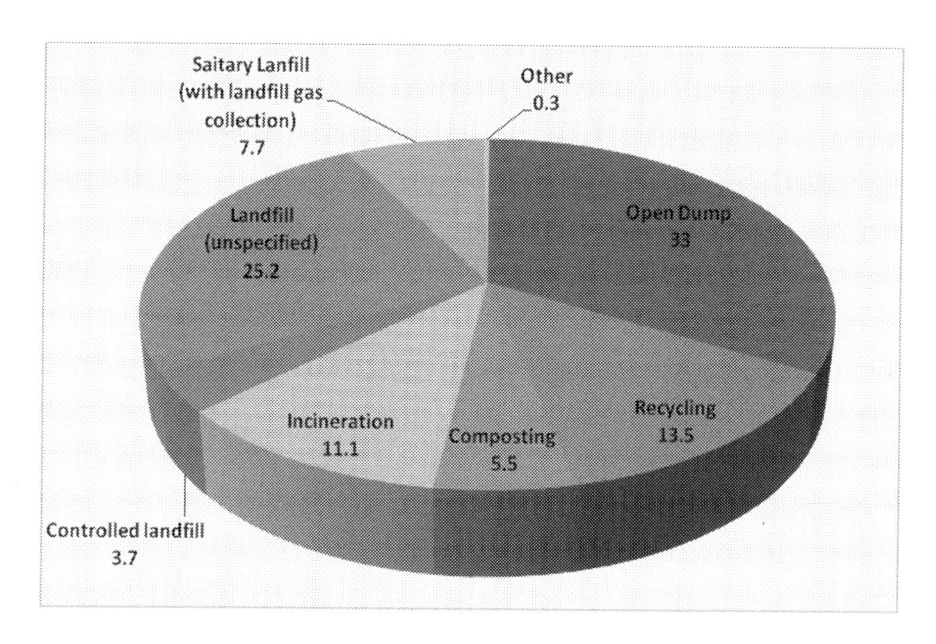

FIGURE 5.2 Global treatment and disposal of waste (in percent) (Kaza et al., 2018).

creatures with zero-waste production (Fu et al., 2013). The things that are thrown as waste are the things that can be utilized in another form.

In general, there is a large amount of production of waste cooking oil (WCO) in developed countries, a fact given by the United State Energy Information Administration (USEIA). The daily estimate of the waste cooking oil (WCO) is about 100 million gallons and its average per capita amount is found to be 9 pounds while the total production in Canada alone reaches about 135,000 tons annually (Chhetri et al., 2008). In the countries of the European Union, the amount of annual waste cooking oil (WCO) ranges from 750,000 to 1 million tons, worriedly in the United Kingdom, about 200,000 tons are produced annually. If this situation will continue, then the day is not far, and we will have to live among the garbage. Taking into account the abovementioned data, there is a need for an immediate solution that can deal with this present situation. It's high time to take some novel steps for the waste reclamation process, thereby protecting our environment. Along with the waste recycling and dumping procedure, there is a need for new innovation as well as a beneficial method for the waste minimization process by keeping in mind the sake of environmental safety.

Nature introduces various approaches and perceptions for the synthesis of the different nanoparticles. From the literature survey, it has been observed that a number of ways have been defined where biologically originated material is used for manufacturing and reducing the waste into metal and metal oxide nanoparticles by the biomimetic application (Sharma et al., 2019). The manipulation of the shape, size, structure, and dispersity is a critical element to be considered while manufacturing novel, facile, and economically feasible protocols. Analyzing all the other synthesis processes organically introduced and plant-mediated methods are specifically more appropriate as a reliable and eco-friendly approach.

The exploration of the unique properties of nanomaterials (NMs) and their probable applications has gained a lot of attention since the origination of the word "nanotechnology" (Bhardwaj et al., 2021). NMs are materials ranging in the 1–100 nm nanoscale dimensions and have vast applications in different industries and the surrounding fields such as scientific, pharmaceutical, business fields, and energy (Bhardwaj et al., 2017). Numerous types of NMs have been assembled using different techniques, and it has been thoroughly investigated how they might be utilized because of their ability to interact with the biological environment (Bhardwaj et al., 2018). Focusing on their accessibility and affordability, waste materials have the potential to be used as feedstock, and NMs synthesis can be utilized as an emerging cutting-edge technique for recycling waste and its treatment (Noviyantia et al., 2020). The production of NMs from regular household and industrial wastes is a revolutionary way of achieving waste-to-wealth and zero-waste goals. There have been various attempts made to generate NMs from the different biological and industrial wastes originating domestically.

On the other hand, firstly, agricultural waste was used as a feedstock for NMs, which may be a cost-effective method. But notably, many plant parts have been employed to make practical items such as pencils, paper, and wooden furniture, but these materials will eventually result in the formation of garbage after usage. Waste resources can be utilized to extract metal oxides, inorganic quantum dots, gold nanoparticles, and carbon nanoparticles. Utilizing biowaste also lessens the harm that the waste causes to the ecosystem and eases environmental restrictions (Noviyantia et al., 2020). Nanomaterials recovered from various types of waste offer a ground-breaking strategy. This would not only prevent harmful environmental consequences but also applies circular economy principles, which are essential to achieve sustainable growth. Also, the recovered nanoparticles have been utilized as a secure substitute in numerous biomedical applications (Purkayastha et al., 2014).

5.1.5 Techniques used for domestic waste–derived nanoparticles

There are several techniques that are being utilized for the synthesis and manufacturing of nanoparticles from the sensitive, capable, reactive, and most resourceful domestic waste (Bhardwaj et al., 2022).

5.1.5.1 *Mechanical milling/ball milling*

It is a novel method for creating metal-oxide nanoparticles, mechanochemical processing applied with or without a solid-state chemical reaction. Precursor materials such as metal, amalgam, or powdered inorganic compounds were utilized in accomplishing the solid-state reaction and producing the intended constituents of the nanoparticles using surface-modifying agents (e.g., carboxylic acid or other acids) and solutions (e.g., stearic acid or toluene) (Bhardwaj et al., 2019). The completely formed products are frequently used for bulk materials that are nano-grained or nanocomposites rather than specific applications. Since they typically have a wide size distribution, changeable form, and side by side also contain impurities and flaws (Virji and Stefaniak, 2014). Waste materials such as rubber tires have been considered the subject of numerous research, as disposing of them leads to pollution and environmental degradation problems. To tackle this situation, rubber tire waste can be managed sustainably by synthesizing ZnNPs using the mechanical/ball milling method. This technique produced rubber ash particles that were smaller than 500 nm; using silicon waste, the size of the particles was further reduced to 50 nm (Abdelbasir et al., 2020b).

5.1.5.2 *Hydrothermal method*

The method utilizes different chemicals that are directly hydrolyzed at high temperatures to generate nanomaterials. The products obtained at ambient temperature as a result of phase crystallization at high temperature and pressure can be used as intermediates in the synthesis. This technique offers the benefit of being able to produce massive, high-quality crystals as well as compounds that are unstable while being close to their melting points. Pomelo (*Citrus maximus*), a well-known fruit from Southeast Asia, has a flavor similar to grapefruit and a noticeably thicker peel. Some scientists go one step further by creating an optical sensor with recycled carbon nanoparticles from used pomelo peels using a hydrothermal process (Abdelbasir et al., 2020b). The cost of the equipment is the main limitation of its commercialization (Modan and Plaiasu, 2020).

5.1.5.3 *The sodium borohydride reduction method*

This is a common technique for producing metallic NPs in which waste materials are quickly mixed with a sodium borohydride solution to produce NPs through a reduction process. The NPs produced by the reaction were periodically washed with deoxygenated water and ethanol solution to help with the elimination of extra sodium borohydride. This technique was developed by scientists to produce Fe nanoparticles from steel pickling waste materials gathered from a steel mill (Samaddar et al., 2018).

5.1.5.4 *Sol–gel process*

For the fabrication of different nanoparticles, a sol–gel process is a bottom-up approach in which the molecules of the metal oxide are diffused in water followed by heating and mixing till the formation of a gel. Furthermore, it can be utilized as a molding substance in the production of ceramics and as a transitional layer between thin films of metal oxides. The preparation of high-purity products using this technique is simple and clear, it has a very high production efficiency and can result in a thicker coating that protects against corrosion. It has minimal startup cost and produces high-quality products, but it has drawbacks such as the contraction that takes place during processing and the prolonged processing time (Modan and Plaiasu, 2020). Eggshells waste released from homes, eateries, and bakeries is a type of solid waste that can be reduced by a prevalent practice of landfilling, but it may have negative environmental effects as such food wastes attract worms and insects and emit unpleasant odors when they biodegrade in landfills. Pure calcium carbonate with minimal porosity is the primary component of this solid waste and so as a $CaCO_3$ source, it can be recycled into useful goods. The waste eggshell might be viewed as a feasible calcium source with a wide range of applications. Using eggshells can be advantageous in terms of environmental issues as well as avoiding unnecessary landfill space. Studies have reported that CaO nanoparticles were synthesized using the sol–gel method, which has been utilized in the wastewater treatment system. The process involved raw eggshells in HCl to create $CaCl_2$ solution, stirring $Ca(OH)_2$ gel with dropwise additions of NaOH, and then drying the gel at a high temperature. This approach is viewed as affordable, simple, and environmentally beneficial (Habte et al., 2019).

5.1.5.5 *Spray pyrolysis*

Using this technique, nanostructures were formed when a precursor-containing solution dispersed onto a heated substrate in a furnace, leading to the breakdown of the precursor and the formation of suitable finalized material. Copper, titanium dioxide (TiO_2), cadmium sulfide (CdS), and cadmium selenide (CdSe) compounds are synthesized via spray pyrolysis. Like all other methods, it also has some drawbacks, including the difficulty of scaling up (yield is relatively low), the potential for sulfide oxidation when treated in air, and challenges estimating the growth in temperature (Sabnis et al., 2013). Furthermore, according to reported studies, an individual household in Indonesia consumes as much as of 5 L of palm oil per month, resulting in significant proportions of wasted cooking oil. Therefore, it is necessary to exploit the plentiful cooking oil waste as a source of raw materials. Using used cooking oil as carbon precursor, salacca peel-based activated carbon as substrate, and ferrocene as a metal catalyst, nanocarbons were synthesized by the nebulized spray pyrolysis process (Arie et al., 2017).

As mentioned above, nanoparticles can be produced in several ways, but producing nanoparticles by physical and chemical pathways could have a detrimental effect on the environment. The aforementioned physical and chemical methods used for nanoparticle synthesis are very advantageous and effective. But in addition, they have some drawbacks, such as the fact that physical methods take a long time to reach thermal stability and use a lot of energy while chemical methods only need a short time to achieve stability (Cele, 2020). As a result, the physical synthesis process is ineffective for producing nanoparticles. On the other hand, the main disadvantage of the chemical process of producing nanoparticles is the usage of chemicals and solvents that could be toxic to our fragile ecosystem. Because of these factors, biologically synthesized pathways were chosen for producing nanoparticles over physical and chemical synthesis techniques. A vital step in the production of eco-friendly nanoparticles is the bio-reduction process of metal ions into their reduced form to the size range of 1–100 nm. The green synthesis method is more effective, less complicated, more affordable, free of chemicals (i.e., nontoxic approach), environmentally friendly, and it is easy to scale up larger activities that have recently gained a lot of significance. Zinc oxide, silver, gold, palladium, and other metal oxides can now be easily produced using green synthesis method of the waste materials (Mohamed El Shafey, 2020). Similarly, one of the gifts of nature that is wasted is a chicken eggshell membrane (ESM), which also has been used to produce luminous gold nanoparticles (AuNPs) in a single step (Devi et al., 2012). Metal ions are biologically reduced by phytocompounds present in plant extracts, such as polyphenols, terpenoids, saponins, carbohydrates, flavonoids, and polyols, which synthesize the nanoparticles possessing strong antibacterial, antioxidant, and catalytic properties. These properties have led to a wide range of nanoparticles that are useful in the pharmaceutical, cosmetic, food, enzyme, and drug delivery industries (Pal et al., 2019).

5.1.6 Application of waste-derived nanomaterials (NMs)

Nanomaterials recovered from different wastes provide different methods for implementing circular economy principles, which are essential for achieving sustainable growth and avoiding harmful environmental consequences. Recycled NMs have been used as an innovative yet safe alternative with remarkable potential for numerous biomedical applications, wastewater treatments, food packaging, energy storage and for their pivotal therapeutic actions (antimicrobial, antioxidant, anticancer, antibacterial, antiviral activity, and nano drugs), etc. Therefore, nanoparticles manufactured by green synthesis are far superior to those produced by the physical and chemical methods. Green synthesis includes extracts of various plant parts, seeds, fruits, etc. (Salem and Fouda, 2021; Singh et al., 2018).

5.1.7 Antimicrobial application

Environmental waste, particularly food waste, serves as a "resource" that can be used to successfully synthesize NPs. Many scientific works recently describe the utilization of biodegradable food waste in the production of several NPs. In accordance, the peel of the mango fruit extract has been utilized for the synthesis of AuNPs (Yang and Hong, 2014). Grape waste, which is a source of plenty of organic chemicals, acts as a reducing agent and causes metals to be converted to NPs. Silver (Ag) ions were reduced into NPs in the presence of grape seed extract, resulting in spherical and polygonal silver nanoparticles (AgNPs) with a mean size of 25–35 nm. These NPs can effectively combat both Gram-positive and Gram-negative bacteria (Xu et al., 2015). The thrown-away peel of the edible tropical fruits, namely custard apples with medicinal, antibacterial, insecticidal, and anticancerous abilities, has been found to be useful in the synthesis of AgNPs and reduction of silver ions because its peel contains water-soluble hydroxyl and ketone groups (Dwivedi and Gopal, 2010; Madhumitha et al., 2012; MukhlesurRahman et al., 2005; Wagner et al., 1980; Joshi et al., 2018). Banana peels, one of the dietary wastes, exhibit both a reducing and capping agent that was used to synthesize hydroxyapatite and Mn_3O_4 nanoparticles (Chanakya et al., 2009; HappiEmaga et al., 2008;). Pomegranate and orange peel extracts were utilized for the synthesis of silver nanoparticles (AgNPs) using the low concentration of the silver nitrate ($AgNO_3$) solution. And, these AgNPs behave as an antifungal agent against the pathogen *Alternaria solani* early blight of tomato (Mostafa et al., 2021). Similarly, AgNPs prepared from the cover of *Arachis hypogaea* are founded to be effective in the control of *Aedes aegypti* (yellow fever) and *Aedes stephensi* (human malaria) (Velu et al., 2015). Additionally, pomegranate residual extract (*Punica granatum*), a significant solid waste used to synthesize silver nanoparticles by reducing silver ions, behaves as an antifungal agent against plant pathogens (Edison and Sethuraman, 2013).

5.1.8 Application in the wastewater treatment

Tea waste has been used to produce magnetic iron oxide ($Fe3O_4$) NPs of the size range 5–25 nm, exhibiting the cuboidal and pyramidal shape of the acquired nanoparticles. The Magnetic iron oxide (Fe_3O_4) NPs showed supermagnetic properties that were employed for up to five adsorption cycles and were extremely effective at removing arsenic metal from water (Lunge et al., 2014). Researchers have also employed defatted cashew nut shell (CNS) starch to produce silver nanoparticles and reduce silver ions (Velmurugan et al., 2015). Silver nanoparticles prepared from one of the biowastes generated lemon peel shows effective property as antidermatophytic activity, drug resistance, and broad-spectrum antibiotics (Nisha et al., 2014). In addition, lemon peel waste promotes the formation of water-soluble carbon quantum dots (wsCQDs) of 1–3 nm size using methylene blue dye. These wsCQDs equipped with a fluorescent probe have become a simple, fast, feasible technique for the identification of chromium (Cr^{6+}) VI in drinking water. Even titanium dioxide–doped soluble carbon quantum dots (TiO_2-wsCQDs) were observed to show higher (~2.5 times) photocatalytic activity than TiO_2 nanofibers (Tyagi et al., 2016). Similarly, green synthesis of the ZnONPs from the potato peel has also shown greater potential toward the wastewater treatment procedure through the photocatalytic degradation capability (Bhuvaneswari et al., 2017). *Cucumis sativus* peel silver nanoparticle (CP-AgNPs) synthesized from the *C. sativus* (Cucumber) peel extract shows the presence of some phytochemicals such as tannins, saponins, steroids, glycosides, terpenoids, and flavonoids, revealing the capping and reducing capability of the CP-AgNPs (Adamu et al., 2021). Hydrated aluminum nanoparticles formed from the tea waste and aluminum sulfate and NaOH have been coprecipitated with tea waste and anionic polyacrylamide to form the porous nanomaterial. The nanoparticles are prepared to act as a biosorbent and can be applied for the elimination of the excess fluoride from the drinking water (Huimei et al., 2015).

Iron nanoparticles formed through the biosynthesis method were used to remediate eutrophic wastewater with the help of Eucalyptus leaf extracts prepared from leaf litter (Wang et al., 2014). Many other leaf extracts have been used in the manufacture of iron nanoparticles for domestic wastewater treatment, including *Mangifera indica, Murrayakoenigii, Azadiracta indica*, and *Magnolia champaca* (Abdel-Shafy and Mansour, 2017). The nanoparticles produced were then tested for their ability to remove chemical oxygen demand (COD), total phosphates, and ammonia (Kanchi and Ahmed, 2018). The extract of *Averrhoa carambola* has been applied as a green and unique stabilizer in the manufacture of stabilized magnetite nanoparticles in another successful attempt in order to effectively remove Chlorazol Black E from wastewater (Ahmed et al., 2015). The achievement has also been attained in the biosynthesis of zinc nanomaterials (nanoparticles and nanobeads) from the aqueous extract of dried *Cuminum cyminum* fruit. The produced nanobeads were successfully used for the breakdown of Alizarin Red dye for its removal from the wastewater (Sirisha and Mary, 2016). Greenway of nanoparticle synthesis delivers a cost-effective, distinct, environmentally friendly, and conducive approach.

Currently, metal nanoparticles have shown various applications in medicine, pharmacy, and agriculture. Nanobiotechnology, in combination with green chemistry, has great potential for the development of novel products that are beneficial to human health, environment as well industries. In this study, silver nanoparticles (AgNPs) were synthesized using aqueous peel extracts of *C. sativus* by eco-friendly, inexpensive, and nontoxic biological methods. The silver precursor used was silver nitrate solution. A visual observation was used to confirm the formation of silver nanoparticles while the characterization of the synthesized silver nanoparticle was carried out using UV–Visible spectroscopy and Transmission Electron Microscopy (TEM). The phytochemical screening was also conducted to confirm the constituents of the synthesized *C. sativus* peel silver nanoparticle (CP-AgNPs). Visible color changes were observed after 1 hour from yellowish to reddish brown and became dark brown after 5 hours for 1 mM, 3 mM, and 5 mM $AgNO_3$ solution at 1:1 (v/v) to the extract, the color was more darker at 5 mM $AgNO_3$, and the UV-Vis spectroscopy at a fixed 5 mM $AgNO_3$ showed surface plasmon resonance with maximum absorbances of 434, 489, and 522 nm for the nanoparticles obtained at three (3) different concentration of 2%, 5%, and 20% (w/v) of the *C. sativus* peel extract, respectively. The TEM showed a spherical shape with no aggregate, and the visualization using Image J software was found to be about 45, 115, and 47 nm with polydispersity of 32.1%, 10.8%, and 46.0%, for CP-AgNPs at concentration of 2%, 5%, and 20% (w/v), respectively. The phytochemicals screening revealed the presence of tannins, saponins, flavonoids, glycosides, steroids, and terpenoids, which are responsible for the reducing and capping property of CP-AgNPs.

5.1.9 Applications in the biomedical field

Biowaste has been widely used as a renewable raw material for gasoline production, but nanomaterials prepared from biowaste have attracted more and more attention because of their strong catalytic activity and low cost. For example, nanocatalyst has been produced using the waste polymer. To produce silica NPs with a size range of 9–20 nm, stainless steel slag waste has been exposed to acid digestion using hydrothermal treatment of nanoparticle synthesis to serve as an oxidation catalyst (Bennett et al., 2012). Watermelon rinds extract was used to produce the gold nanoparticles exhibiting the application as the antioxidant activity, cancer treatment, biomedical, pharmaceutical, cosmetics, and food sectors (Patra and Baek, 2015). Potato peel powder is utilized for the production of the copper(II)-NPs that have been proven to work as an anticancerous agent in contrast to MCF-7 breast cancerous cells (Idhayadhulla et al., 2021). Significantly, alkaline protease from fungi *Beauveria* species has been successfully used to synthesize silver nanoribbons. These nanoribbons are elongated, anisotropic, nontoxic, and biocompatible silver nanostructures prepared from discarded X-ray films. This study describes a brand-new, environmentally benign technique for producing silver nanoribbons that could be successfully applied medically to treat burns, bacterial infections, and wounds (Shankar et al., 2015). Green synthesis of the highly pure zinc oxide nanoparticles from the garlic peel extract can further be useful for sensors, coatings, catalysis, water treatment, biotechnology, and the biomedical field (Modi and Fulekar, 2020). Furthermore, the waste lychee peels have been synthesized to produce carbon nanoparticles (CNPs). That showed the higher fluorescent property and greater water solubility, which could be useful in the nanotech field such as bioimaging and biolabeling application (Chaudhary and Bhowmick, 2015). Using biodegradable waste, green synthesis of silver nanoparticles (AgNP) is affordable, safe for the environment, and environmentally beneficial. To produce AgNPs, coconut shell extract (*Cocos nucifera*) has been employed. The AgNPs produced were possessed to have a potential antimicrobial property. Silver nanoparticles (AgNPs) have also been synthesized using the onion peel, and the nanoparticle developed is seen as resistant for both the Gram-negative and Gram-positive bacteria, showing the antibacterial activity of the AgNPs (Santhosh et al., 2021). That will be an appealing, economical, and environment-friendly path, which could possibly be used by the food, cosmetic, and medical fields (Das et al., 2021; Goswami and Mathur, 2022).

5.1.10 Applications in energy storage

Some studies have reported the production of green carbon nanodots from food waste. For every 100 kg of food waste, around 120 g of carbon nanodots of approximately 2–4 nm size particles can be produced. Nowadays, green wastes are being used to produce nanosized cellulose components for the enduring materials sectors. This was because nanocellulose that has numerous beneficial qualities and has also been known to emerge as a renewable biocide that will be a prominent study focusing on many fields. Moreover, high-quality single-layered graphene has been produced from raw wastes such as grass blades, grass, waste food, and waste polystyrene with the help of the green synthesis method. Researchers have utilized recycled sulfonated polystyrene to form a solid SiO_2-polystyrene sulfonic acid nanocomposite with the help of the sol–gel method using tetraethyl orthosilicate, to manufacture a soluble polystyrene sulfonic acid catalyst. These waste-derived sulfonic acid nanocomposite catalysts have been utilized to catalyze biomass valorization reactions, such as the conversion of xylose to furfural, the production of biodiesel, and the oxidation of furfural to maleic and succinic acids. Utilizing the different wastes, metals and metal oxide nanoparticles can be formed by using a simplified thermal breakdown process, which enhances its quality (Alonso-Fagundez et al., 2014). Furthermore, banana peel extract is used to form manganese oxide (Mn_3O_4) nanoparticles, which serves two key roles, i.e., by limiting the agglomeration of nanoparticles during processing and reducing the conversion of potassium permanganate ($KMnO_4$) to manganese oxide (Mn_3O_4). Pectin present in banana peel is essential for the surface modification of the NPs. The electrodes made from derived Mn_3O_4 NPs show higher electrochemical performance (Yan et al., 2014). The synthesis of high-quality and reduced graphene oxide sheets has been accomplished by scientists using a novel process from a variety of naturally occurring green wastes such as leaves, wood, and fruit waste, and carbon waste including pulp and paper industry wastes such as newspaper cardboard. It is an inexpensive approach of the high-pure-grade graphene sheet that is being utilized in different working areas such as drug delivery, cancer treatment, solar panels, anticorrosion coating and paints, and electrical and electronic industries (Akhavan et al., 2014).

5.1.11 Applications in the industrial and commercial sector

Some scientists published another fascinating study on the use of nanomaterials made from refuse by reverse micro-emulsions method (at room temperature), in which eggshell-derived calcium precursors effectively produce nanohydroxyapatite (nano-HAP). The produced nano-HAP is being applied for biomaterial utility such as bone implants (Samaddar et al., 2018). Some studies followed the process of thermo-oxidative degradation, carbonization, passivation, and polymerization. That shows the utilization of plastic waste bags for the first time as a source of carbon for the synthesis of fluorescent carbon nanoparticles. The derived carbon nanoparticle serves as the potential fluorescent probe for bioimaging (Hu et al., 2014). Advanced techniques, including autoclaves, crucibles, quartz tubes, and muffle furnaces, have been utilized to generate carbon nanotubes (CNTs) from waste plastic and rubber materials. The prepared carbon nanotubes have been employed in the body of aircraft and spacecraft, bulletproof jacket, which behave as semiconducting material and act as a catalyst in several reactions (Bazargan and McKay, 2012). The production of highly porous aerogels using recycled materials such as paper, plastic bottles, and cotton fabrics has been demonstrated by some researchers using a simplified technique. Similarly, cotton cloth scraps have been utilized to produce textile-derived aerogels. Aerogels strongly behave as a good absorbent material with greater absorption capacity and low thermal conductivity and can be utilized in the surgical and sanitary fields (Thai et al., 2019).

5.1.12 Applications of waste-derived nanoparticles in the environment

One of the major issues in the world is pollution, which puts people, animals, plants, and ecosystem at risk. Due to the requirements of experienced individuals and high-value instrumentation to operate, preliminary tests in the lab for identifying contaminants in water are prohibitively expensive in some areas, and to fix this issue, scientists are developing inexpensive sensors that use nanoparticles to assess contaminants. For instance, some researchers developed an optical sensor that synthesizes nanomaterials from hydrothermal processing that could be promising work in the near future. They have been also used to prepare highly sensitive and higher applicability sensors and other instruments to be utilized in different fields (Abdelbasir et al., 2020b). In the same way, onion peel waste was used for the manufacturing of the zinc oxide nanoparticle (ZnONP) that has been utilized as plant nutrients and for wastewater treatment. Synthesized ZnONPs from onion peel have been noticed to enhance the growth of mung bean and wheat seed than the control seed (Modi et al., 2022). Many other domestic waste–derived nanomaterials specifications with their potential applications are compiled in Table 5.1. In addition, Fig. 5.3 summarizes the applications of the synthesized nanomaterial from the domestic waste.

TABLE 5.1 Domestic waste-mediated synthesized nanoparticles.

Domestic waste	Used part	Synthesized nanoparticles	Size	Applications	References
Food waste					
Pisum sativum (Pea) and *Lagenaria siceraria* (bottle gourd)	Peels	AgNPs	20–70 nm	• Antibacterial potential	Deepa et al. (2022)
Onion	Peel	Zinc oxide (ZnONPs)	20–80 nm	• Nano-based nutrients for agricultural use	Modi et al. (2022)
Avian eggshell	Egg shell	$CaCO_3$ NPs	≤ 50 nm	• Nutritional supplements • Medicine	Ahmed et al. (2022)

(Continued)

TABLE 5.1 (Continued)

Domestic waste	Used part	Synthesized nanoparticles	Size	Applications	References
Chicken eggshell	Egg shell	Hydroxyapatite nanoparticle (HAP)	100 nm	• Adsorbent • Antibiotic, heavy metals, and organic dye removal from aqueous solutions	Alhasan et al. (2022)
		TiO_2 NPs	27.3 nm	• Antimicrobial efficacies • Anticancer • Photocatalytic applications	Ramya et al. (2022)
		CaO NPs	24.34 nm	• Removal of lead (II) from aqueous solutions	Jalu et al. (2021)
Cucumber (*Cucumis sativus*)	Peel	AgNPs	45–115 nm	Good medicinal properties and have exhibited good physiological activity. And possessed Strong antimicrobial effects • Antimicrobial property and good physiological activity	Adamu et al. (2021)
Potato	Peel	Cu(II)-NPs	~ 71 nm	• Anticancer medicament	Idhayadhulla et al. (2021)
Corn	Cob	AgNPs	55 nm	• Anti-*Trypanosoma cruzi* agent	Brito et al. (2020)
Corn	Cob	AgNPs	102 nm	• Antileishmanial and antifungal agents	Silva Viana et al. (2020)
Corn	Cob	AgNPs	11 nm	• Bioactive application • Catalysts for wastewater treatment.	Doan et al. (2020)
Corn	Cob	AuNPs	35 nm		
Garlic	Peel	ZnONPs	12.6 nm	And wat • Application in biotechnology, biomedical, catalysis, coatings, sensors, and water remediation.	Modi and Fulekar (2020)
Eggs	Outer shell	Hydroxyapatite (HA) nanorods and nanoplates	92.61 nm	• Biomaterial utility for bone implant	Noviyantia et al. (2020)
		Calcium Oxide nanoparticles (CaO NPs)	50 nm–198 nm	• Filler material in the automobile and paper industry	Habte et al. (2019)
Potato	Peel	ZnONPs	~ 71 nm	• Photocatalytic activity • Wastewater treatment	Bhuvaneswari et al. (2017)
Azadiracta indica	Leaf extracts	Iron nanoparticles (FeNPs)	96–110 nm	• Evaluated for the removal of chemical oxygen demand, total phosphates, ammonia, and nitrogen from waste water	Abdel-Shafy and Mansour (2017)
Murrayakoenigii			99–129 nm		
Magnolia champaca			120–200 nm		
Lagenaria siceraria, Luffa cylindrica, Solanum lycopersicum, Solanum melongena	Vegetable Peel	AgNPs	~ 20 nm	• Antibacterial properties • Anticancer properties	Sharma et al. (2016)

(Continued)

TABLE 5.1 (Continued)

Domestic waste	Used part	Synthesized nanoparticles	Size	Applications	References
Lemon	Peels	Carbon quantum dots (wsCQDs)	1–3 nm	• Identification of the chromium (Cr6 +) VI in the drinking water.	Tyagi et al. (2016)
Arachis hypogaea (peanut)	Peels	AgNPs	20–50 nm	• Significant larvicidal activity against the larvae of *Aedes aegypti* and *Aedes stephensi* mosquito	Velu et al. (2015)
Averrhoa carambola	Leaf extracts	Fe_3O_4NPs	2.38 nm	• Effectively remove Chlorazol Black E from wastewater	Ahmed et al. (2015)
Lemon	Peels	AgNPs	17–61 nm	• Antidermatophytic activity • Drug resistant • Antibiotics	Nisha et al. (2014)
Tea	Tea waste	Magnetite nanoparticles (Fe_3O_4NPs)	5–25 nm	• Eliminate arsenic metal from water	Lunge et al. (2014)
Eucalyptus leaf extracts	Leaf litter	FeNPs	20–80 nm	• Wastewater treatment	Wang et al. (2014)
Eggs	Outer shell	AuNPs	< 20 nm	• Biolabeling and bioimaging • Catalysis and sensor development	Devi et al. (2012)
Fruit waste					
Pomegranate peel and Orange peel	Peel extracts	Ag NPs	14 nm	• Antifungal agents against plant pathogen • Replacement for chemical fungicide	Mostafa et al. (2021)
Ananas comosus (pineapple)	Fruit peel	Au NPs	20.71 nm	• Biomedical applications	Pechyen et al. (2021)
Passiflora edulis (passion fruit)			18.68 nm		
Fruit latex of Achras sapota Linn	Fruit latex	Ag NPs and Cu alloy NPs	20–40 nm	• Biomedical applications • Biocompatible latex	Thakore et al. (2019)
Grape	Stalk waste	Ag NPs	~28 nm	• Suitable for sensing purposes	Bastos-Arrieta et al. (2018)
Custard apple peel	Peel extract	Ag NPs	20 nm	• Packaging material with increased shelf life of the packaged foods.	Joshi et al. (2018)
Grape	Seed extract	Silver nanoparticles (Ag NPs)	25–35 nm	• Resist gram-positive and gram-negative bacteria	Xu et al. (2015)
Lychee peels	Peel extracts	Carbon nanodots	40–70 nm	• Bioimaging and biolabelling	Chaudhary and Bhowmick (2015)
Watermelon rinds	Rinds extract	AuNPs	20–140 nm	• Antioxidant activity • Cancer treatment	Patra and Baek (2015)

(Continued)

TABLE 5.1 (Continued)

Domestic waste	Used part	Synthesized nanoparticles	Size	Applications	References
				• Biomedical, pharmaceutical, cosmetics, and food sectors	
Cashew nut shell (CNS)	Seed kernel	AgNPs	10–50 nm	• Cost effective than engineered nanoparticles • Pharmaceutical and biomedical applications	Velmurugan et al. (2015)
Banana peel	Peel extract	Manganese oxide (Mn_3O_4) nanoparticles	20–50 nm	• Higher electrochemical performance	Yan et al. (2014)
Mango peel	Peel extract	Gold nanoparticles (AuNPs)	6–18 nm	• No biological cytotoxicity • Applicative in drug delivery systems	Yang and Hong (2014)
Pomegranate peel	Peel extracts	Ag NPs	8 nm- 30 nm	• Reduce anthropogenic pollutant	Edison and Sethuraman (2013)
Corn	Cob	Corn-cob based nanoparticles	22 nm	• Drug delivery system	Kumar et al. (2010)
Dry waste					
Polysterene plastic	Plastic wastes	Polystyrene/ TiO_2/CuO Nanocomposites	14.94 nm	• Antibacterial • Antifungal • Antioxidant • Anticorrosion	Almarbd and Abbass (2022)
		Polystyrene/ Graphite Oxide/ CuO nanocomposites	16.93 nm		
Expanded polystyrene (EPS)	Plastic wastes	Carbon nanodots	2.8–5.2 nm	• Bioimaging, metallic ion detection • Optical imaging, and biological works • Electrocatalysts	Mustafa et al. (2021)
Aluminum foil	Foil	Aluminum and rice husk ash (Al-RHA composite)	Composite	• Enhances binding strength and fire retardant quality of the cement • Future fuel source	Balinee and Ranjith (2021)
Hairs	Human hair and green tea polyphenol	Human hair keratin & Epigallocatechin gallate (KE-EGCG) nanoparticles	50 nm	• Antioxidant and antiinflammation effects • Promote cell proliferation	Yi et al. (2021)
Newspapers	Deinked and untreated waste newspapers	Cellulose nanocrystals (CNC)	10–30 nm	• Enhances physical properties of paper sheets	Guan et al. (2021)
Cotton	Cotton wastes	Cellulose nanocrystals	1–100 nm	• Biomedical use • Composite films for packaging.	Rizal et al. (2021)

(Continued)

TABLE 5.1 (Continued)

Domestic waste	Used part	Synthesized nanoparticles	Size	Applications	References
Hairs	Human hair	Recombinant keratin nanoparticles (RKNPs)	306–308 nm	• Enhanced wound healing • Potential tissue-engineering • Applications • Development of keratin biomaterials	Gao et al. (2019)
Spent lithium-ion batteries	metals	Polymetallic nanoparticles	<50 nm	• Pollutants removal	Nascimento et al. (2018)
Nails	Human Fingernails	Carbon nanodots	3.5 nm	• High-quantum yield • Ultra-sensitive probe for the detection • Promote cell proliferation and tissue regeneration	Chatzimitakos et al. (2018)
Hairs	Human hair	Human hair keratin (HHK) nanoparticles	~75 nm	• Increase drug loading capacity	Liu et al. (2017)
Water bottles	Plastic wastes	Silicon carbide (SiC)	5–20 nm	• Reinforcement effect in epoxy resin to form nanocomposites	Meng et al. (2010)
Disposable boxes			30–70 nm		

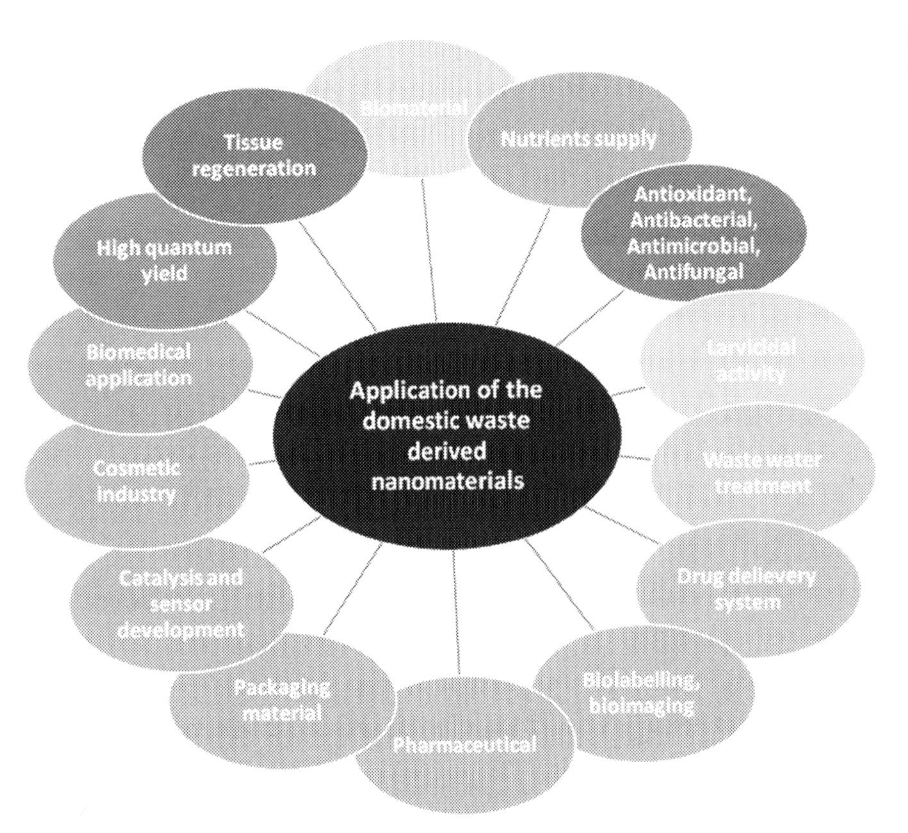

FIGURE 5.3 Application of the domestic waste–derived nanomaterials.

5.2 Conclusion

The chapter provides a thorough, insightful, and understandable summary of this groundbreaking and fascinating area that is to achieve the different approaches to waste-to-wealth, waste-to-energy, and zero-waste targets. It has been revealed that valuable nanoparticles could be extracted from waste sources originating from daily living, such as biomass wastes, cooking oil, and many other domestic wastes. Analyzing the synthesis and application of the nanoparticles using domestic waste was a sustainable way to minimize and utilize the waste load. Moreover, it has become a significant way in research in recent years to gain one of the sustainable goals by synthesizing nanomaterials derived from trash. Following that, various synthesis methods have been discussed, which have been utilized for the waste-derived nanoparticles preparation. Domestically produced biomass wastes are an intriguing and noteworthy source for the synthesis of different NMs. The prepared nanoparticles could offer several advantageous characteristics such as improved stability, decreased toxicity, cost-effectiveness, and environmentally friendly approach. The use of recycled nanoparticles from domestic waste has been proven to be substantially important in the biomedical field. Additionally, the use of biowaste alleviates environmental burdens on one hand, and on the other hand, side by side lessens the harmful effects of the waste on the ecosystem. The utilization of waste involves significant learning, which is the importance of public awareness, involvement of the local community and utilizing the waste as a source of revenue.

References

Abdelbasir, S.M., McCourt, K.M., Lee, C.M., Vanegas, D.C., 2020b. Waste-derived nanoparticles: synthesis approaches, environmental applications, and sustainability considerations. Frontiers in Chemistry 8, 782.

Abdel-Shafy, H.I., Mansour, M.S.M., 2017. Green Synthesis of Metallic Nanoparticles from Natural Resources and Food Waste and Their Environmental Application: Green Metal Nanoparticles. Wiley, pp. 321–386.

Adamu, A.U., Abdulmumin, Y., Mustapha, R.K., 2021. Green synthesis, characterization and phytochemicals analysis of silver nano-particles using aqueous peel extract of *Cucumis sativus*. Journal of Materials and Environmental Science 12, 1627–1636.

Agarwal, R., Chaudhary, M., Singh, J., 2015. Waste management initiatives in India for human well being. European Scientific Journal 1, 105–127.

Ahmed, J.K., Ahmaruzzaman, M., Bordoloi, M.H., 2015. Novel *Averrhoa carambola* extract stabilized magnetite nanoparticles: a green synthesis route for the removal of chlorazol black E from wastewater. RSC Advances 5, 74645–74655.

Ahmed, N.S., Kamil, F.H., Hasso, A.A., Abduljawaad, A.N., Saleh, T.F., Mahmood, S.K., 2022. Calcium carbonate nanoparticles of quail's egg shells: synthesis and characterization. Journal of the Mechanical Behavior of Materials 31, 1–7.

Akhavan, O., Bijanzada, K., Mirsepaha, A., 2014. Synthesis of graphene from natural and industrial carbonaceous wastes. Royal Society of Chemistry Advances 4, 20441–20448.

Alhasan, H.S., Alahmadi, N., Yasin, S.A., Khalaf, M.Y., Ali, G.A.M., 2022. Low-cost and eco-friendly hydroxyapatite nanoparticles derived from eggshell waste for cephalexin removal. Separations 9, 1–15.

Almarbd, Z.Z., Abbass, N.M., 2022. Recycling of plastic waste made of polystyrene and its transformation into nanocomposites by green methods. Chemical Methodologies 6, 940–952.

Alonso-Fagundez, N., Laserna, V., Alba-Rubio, A.C., Mengibar, M., Heras, A., Mariscal, R., et al., 2014. Poly-(styrene sulphonic acid): an acid catalyst from polystyrene waste for reactions of interest in biomass valorization. Catalysis Today 234, 285–294.

Amasuomo, E., Baird, J., 2016. The concept of waste and waste management. Journal of Management and Sustainability 6, 88–96.

Annual Report on Solid Waste Management, 2020–21. CPCB, Delhi. https://cpcb.nic.in/openpdffile.php?id = UmVwb3J0RmlsZXMvMTQwM-18xNjU1MzU0NzkxX21lZGlhcGhvdG8xNjQ3MS5wZGY = .

Arie, A.A., Hadisaputra, L., Susanti, R.F., Devianto, H., Halim, M., Enggar, R., et al., 2017. Synthesis of carbon nano materials originated from waste cooking oil using a nebulized spray pyrolysis. Journal of Physics: Conference Series 877, 1–7.

Balinee, B., Ranjith, P.G., 2021. Use of discarded aluminium foil in cementitious material as fire retardant – an experimental study. Journal of Building Engineering 44, 1–10.

Bastos-Arrieta, J., Florido, A., Perez-Rafols, C., Serrano, N., Fiol, N., Poch, J., et al., 2018. Green synthesis of Ag nanoparticles using grape stalk waste extract for the modification of screen-printed electrodes. Nanomaterials 8, 1–14.

Bazargan, A., McKay, G., 2012. A review—synthesis of carbon nanotubes from plastic wastes. Chemical Engineering Journal 195–196, 377–391.

Bennett, J.A., Wilsona, K., Lee, A.F., 2012. Catalytic applications of waste derived materials. Journal of Materials Chemistry A 00, 1–22.

Bhardwaj, A.K., Shukla, A., Mishra, R.K., Singh, S.C., Mishra, V., Uttam, K.N., et al., 2017. Power and time dependent microwave assisted fabrication of silver nanoparticles decorated cotton (SNDC) fibers for bacterial decontamination. Frontiers in Microbiology 8, 330.

Bhardwaj, A.K., Shukla, A., Maurya, S., Singh, S.C., Uttam, K.N., Sundaram, S., et al., 2018. Direct sunlight enabled photo-biochemical synthesis of silver nanoparticles and their Bactericidal Efficacy: photon energy as key for size and distribution control. Journal of Photochemistry and Photobiology B: Biology 188, 42–49.

Bhardwaj, A.K., Gupta, M.K., Naraian, R., 2019. Myco-nanotechnological approach for improved degradation of lignocellulosic waste: its future aspect. Mycodegradation of Lignocelluloses 227–245.

Bhardwaj, A.K., Sundaram, S., Yadav, K.K., Srivastav, A.L., 2021. An overview of silver nano-particles as promising materials for water disinfection. Environmental Technology & Innovation 23, 101721.

Bhardwaj, A.K., Naraian, R., Sundaram, S., Kaur, R., 2022. Biogenic and non-biogenic waste for the synthesis of nanoparticles and their applications. Bioremediation. CRC Press, pp. 207–218.

Bhuvaneswari, S., Subashini, G., Subramaniyam, S., 2017. Green synthesis of zinc oxide nanoparticles using potato peel and degradation of textile mill effluent by photocatalytic activity. World Journal of Pharmaceutical Research 774–785.

Brito, T.K., Silva Viana, R.L., Gonçalves Moreno, C.J., da Silva Barbosa, J., Lopes de Sousa Junior, F., Campos de Medeiros, M.J., et al., 2020. Synthesis of silver nanoparticle employing corn cob xylan as a reducing agent with anti-Trypanosoma cruzi activity. International Journal of Nanomedicine 15, 965–979.

Cele, T., 2020. Preparation of Nanoparticles. Engineered Nanomaterials—Health and Safety. Intechopen. Available from: https://doi.org/10.5772/intechopen.90771.

Chanakya, H.N., Sharma, I., Ramachandra, T.V., 2009. Micro-scaleanaerobic digestion of point source components of organic fraction of municipal solid waste. Waste Manage (New York, N.Y.) 29, 1306–1312.

Chatzimitakos, T.G., Kasouni, A., Troganis, A.N., Stalikas, C.D., 2018. Carbonization of human fingernails: towards the sustainable production of multifunctional nitrogen and sulfur co-doped carbon nanodots with highly luminescent probing and cell proliferative/migration properties. ACS Applied Materials & Interfaces 10, 16024–16032.

Chaudhary, V., Bhowmick, A.K., 2015. Green synthesis of fluorescent carbon nanoparticles from Lychee (*Litchi chinensis*) plant. Korean Journal of Chemical Engineering 32, 1707–1711.

Chhetri, A.B., Watts, K.C., Islam, M.R., 2008. Waste cooking oil as an alternate feedstock for biodiesel production. Energies 1, 3–18.

Das, G., Shin, H.S., Kumar, A., Vishnuprasad, C.N., Patra, J.K., 2021. Photo-mediated optimized synthesis of silver nanoparticles using the extracts of outer shell fibre of *Cocos nucifera* L. fruit and detection of its antioxidant, cytotoxicity and antibacterial potential. Saudi Journal of Biological Sciences 28, 980–987.

Deepa, Ameen, F., Amirul Islam, M., Dhanker, R., 2022. Green synthesis of silver nanoparticles from vegetable waste of pea *Pisum sativum* and bottle gourd *Lagenaria siceraria*: characterization and antibacterial properties. Frontiers in Environmental Science 10, 1–11.

Devi, P.S., Banerjee, S., Chowdhury, S.R., Kumar, G.S., 2012. Eggshell membrane: a natural biotemplate to synthesize fluorescent gold nanoparticles. RSC Advances 2, 11578–11585.

Doan, V.D., Luc, V.S., Nguyen, T.L.H., Nguyen, T.D., Nguyen, T.D., 2020. Utilizing waste corn-cob in biosynthesis of noble metallic nanoparticles for antibacterial effect and catalytic degradation of contaminants. Environmental Science and Pollution Research 27, 6148–6162.

Dwivedi, A.D., Gopal, K., 2010. Biosynthesis of silver and goldnanoparticles using *Chenopodium album* leaf extract. Colloids and Surfaces A: Physicochemical and Engineering Aspects 369, 27–33.

Edison, T.J.I., Sethuraman, M.G., 2013. Biogenic robust synthesis of silver nanoparticles using *Punica granatum* peel and its application as a green catalyst for the reduction of an anthropogenic pollutant 4-nitrophenol. Spectrochimica. Acta A 104, 262–264.

Fu, H., Yang, X., Jiang, X., Yu, A., 2013. Bimetallic Ag–Au nanowires: synthesis, growth mechanism, and catalytic properties. Langmuir: The ACS Journal of Surfaces and Colloids 29, 7134–7142.

Gao, F., Li, W., Deng, J., Kan, J., Guo, T., Wang, B., et al., 2019. Recombinant human hair keratin nanoparticles accelerate dermal wound healing. ACS Applied Materials and Interfaces 11, 18681–18690.

Goswami, P., Mathur, J., 2022. Application of agro-waste-mediated silicananoparticles to sustainable agriculture. Bioresources and Bioprocessing 9, 1–12.

Guan, Y., Li, W., Gao, H., Zhang, L., Zhou, L., Peng, F., 2021. Preparation of cellulose nanocrystals from deinked waste newspaper and their usage for papermaking. Carbohydrate Polymer Technologies and Applications 2, 1–8.

Habte, L., Shiferaw, N., Mulatu, D., Thenepalli, T., Chilakala, R., Ahn, J.W., 2019. Synthesis of nano-calcium oxide from waste eggshell by sol-gel method. Sustainability 11, 1–10.

HappiEmaga, T., Robert, C., Ronkart, S.N., Wathelet, B., Paquot, M., 2008. Dietary fibre components and pectin chemical features ofpeels during ripening in banana and plantain varieties. Bioresource Technology 99, 4346–4354.

Hu, Y., Yang, J., Tian, J., Jia, L., 2014. ChemInform abstract: green and size-controllable synthesis of photoluminescent carbon nanoparticles from waste plastic bags. RSC Advances 46, 47169–47176.

Huimei, C., Chen, G., Peng, C., Xu, L., Zhu, X., Zhang, Z., et al., 2015. Enhanced removal of fluoride by tea waste supported hydrous aluminium oxide nanoparticles: anionic polyacrylamide mediated aluminium assembly and adsorption mechanism. RSC Advances 5, 29266–29275.

Idhayadhulla, A., Manilal, A., Ahamed, A., Alarifi, S., Raman, G., 2021. Potato peels mediated synthesis of Cu(II)-nanoparticles from tyrosinase reacted with bis-(N-aminoethylethanolamine) (Tyr-Cu(II)-AEEA NPs) and their cytotoxicity against Michigan cancer foundation-7 breast cancer cell line. Molecules (Basel, Switzerland) 26, 1–10.

Jalu, R.G., Chamada, T.A., Kasirajan, R., 2021. Calcium oxide nanoparticles synthesis from hen eggshells for removal of lead (Pb(II)) from aqueous solution. Environmental Challenges 4, 1–20.

Joshi, S.S., Sahoo, A.K., Prasad, N.R., Udachan, I.S., 2018. Biosynthesis of silver nanoparticles (AgNPs) by extract of Annona Squamosa waste: characterization and their antimicrobial activity. International Journal of Research and Analytical Reviews 5, 907–914.

Kanchi, S., Ahmed, S., 2018. Green Metal Nanoparticles: Synthesis, Characterization and Their Applications. Wiley online library, Scrivener Publishing978-1-119-41886-3. Available from: https://doi.org/10.1002/9781119418900.

Kaza, S., Yao, L., Bhada-Tata, P., Van Woerden, F., 2018. Trends in Solid Waste Management, What a Waste 2.0: A Global Snapshot of Solid Waste Management to 2050. Urban Development. World Bank, Washington, DC. Available from: https://datatopics.worldbank.org/what-a-waste/trends_in_solid_waste_management.html.

Kumar, S., Upadhyaya, J.S., Negi, Y.S., 2010. Preparation of nanoparticles from corn cobs by chemical treatment methods. BioResources. 5, 1292–1300.

Kwon, S.J., Bard, A.J., 2012. DNA analysis by application of Pt nanoparticle electrochemical amplification with single label response. Journal of the American Chemical Society 134, 10777–10779.

Liu, J., Zheng, J., Rao, H., 2017. Preparation and drug-loading properties of human hair keratin nanoparticles. International Journal of Science 4, 93–96.

Loizia, P., Voukkali, I., Zorpas, A.A., Pedreno, J.N., Chatziparaskeva, G., Inglezakis, V.J., et al., 2021. Measuring the level of environmental performance in insular areas, through key performed indicators, in the framework of waste strategy development. The Science of the Total Environment 753, 141974.

Lunge, S., Singh, S., Sinha, A., 2014. Magnetic iron oxide (Fe_3O_4) nanoparticles from tea waste for arsenic removal. Journal of Magnetism and Magnetic Materials 356, 21–31.

Madhumitha, G., Rajakumar, G., Roopan, S.M., Rahuman, A.A., Priya, K.M., Saral, A.M., et al., 2012. Acaricidal, insecticidal, and larvicidal efficacy of fruit peel aqueous extract of *Annona squamosa* and its compounds against blood-feeding parasites. Parasitology Research 111, 2189–2199.

Meng, S., Wang, D.H., Jin, G.Q., Wang, Y.Y., Guo, X.Y., 2010. Preparation of SiC nanoparticles from plastic wastes. Materials Letters 64, 2731–2734.

Modan, E.M., Plaiasu, A.G., 2020. Advantages and disadvantages of chemical methods in the elaboration of nanomaterials. *The Annals of "Dunarea de Jos" University of Galati. Fascicle IX, Metallurgy and Materials Science* 43, 53–60.

Modi, S., Fulekar, M.H., 2020. Green synthesis of zinc oxide nanoparticles using garlic skin extract and its characterization. Journal of Nanostructure 10, 20–27.

Modi, S., Yadav, V.K., Choudhary, N., Alswieleh, A.M., Sharma, A.K., Bhardwaj, A.K., et al., 2022. Onion peel waste mediated-green synthesis of zinc oxide nanoparticles and their phytotoxicity on mung bean and wheat plant growth. Materials 15, 1–12.

Mohamed El Shafey, A., 2020. Green synthesis of metal and metal oxide nanoparticles from plant leaf extracts and their applications: a review. Green Processing and Synthesis 9, 304–339.

Mostafa, Y.S., Alamri, S.A., Alrumman, S.A., Hashem, M., Baka, Z.A., 2021. Green synthesis of silver nanoparticles using pomegranate and orange peel extracts and their antifungal activity against *Alternaria solani*, the causal agent of early blight disease of tomato. Plants (Basel) 10, 1–18.

MukhlesurRahman, M., Parvin, S., Ekramul Haque, M., EkramulIslam, M., Mosaddik, M.A., 2005. Antimicrobial and cytotoxicconstituents from the seeds of *Annona squamosa*. Fitoterapia 76, 484–489.

Mustafa, K., Kanwal, J., Musaddiq, S., 2021. Waste plastic-based nanomaterials and their applications. Topics in Mining, Metallurgy and Materials, Engineering 781–803. Available from: https://doi.org/10.1007/978-3-030-68031-2_27.

Nascimento, M.A., Cruz, J.C., Rodrigues, G.D., de Oliveira, A.F., Lopes, R.P., 2018. Synthesis of polymetallic nanoparticles from spent lithium-ion batteries and application in the removal of reactive blue 4 dye. Journal of Cleaner Production 202, 264–272.

Nisha, S.N., Aysha, O.S., Rahaman, J.S.N., Kumar, P.V., Valli, S., Nirmala, P., et al., 2014. Lemon peels mediated synthesis of silver nanoparticles and its antidermatophytic activity. Spectrochimica Acta—Part A: Molecular and Biomolecular Spectroscopy 124, 194–198.

Noviyantia, A.R., Akbara, N., Deawatia, Y., Ernawatia, E.E., Malika, Y.T., Fauziab, R.P., et al., 2020. A novel hydrothermal synthesis of nanohydroxyapatite fromeggshell-calcium-oxide precursors. Heliyon 6, 1–6.

Pal, M.S., Bhatia, M., 2022. Current status, topographical constraints, and implementation strategy of municipal solid waste in India: a review. Arabian Journal of Geosciences 15 (1176), 1–26.

Pal, G., Rai, P., Pandey, A., 2019. Green synthesis of nanoparticles: a greener approach for a cleaner future. Micro and Nano Technologies, Green Synthesis, Characterization and Applications of Nanoparticles. Elsevier, pp. 1–26. Available from: https://doi.org/10.1016/B978-0-08-102579-6.00001-0.

Patra, J.K., Baek, K., 2015. Novel green synthesis of gold nanoparticles using Citrullus lanatusrind and investigation of proteasome inhibitory activity, antibacterial, and antioxidant potential. International Journal of Nanomedicine 10, 7253–7260.

Pechyen, C., Ponsanti, K., Tangnorawich, B., Ngernyuang, N., 2021. Waste fruit peel—mediated green synthesis of biocompatible gold nanoparticles. Journal of Materials Research and Technology 14, 2982–2991.

Purkayastha, M.D., Manhar, A.K., Mandal, M., Mahanta, C.L., 2014. Industrial waste-derived nanoparticles and microspheres can be potent antimicrobial and functional ingredients. Journal of Applied Chemistry 2014, 1–12.

Ramya, S., Vijayakumar, S., Vidhya, E., Bukhari, N.A., Hatamleh, A.A., Nilavukkarasi, M., et al., 2022. TiO_2 nanoparticles derived from egg shell waste: eco synthesis, characterization, biological and photocatalytic applications. Environmental Research 214, 113829.

Rizal, S., Khalil, H.P.S., Oyekanmi, A.A., Gideon, O.N., Abdullah, C.K., Yahya, E.B., et al., 2021. Cotton wastes functionalized biomaterials from micro to nano: a cleaner approach for a sustainable environmental application. Polymers 13, 1–36.

Sabnis, S.M., Bhadane, P.A., Kulkarni, P.G., 2013. Process flow of spray pyrolysis technique. IOSR Journal of Applied Physics 4, 07–11.

Salem, S.S., Fouda, A., 2021. Green synthesis of metallic nanoparticles and their prospective biotechnological applications: an overview. Biological Trace Element Research 199, 344–370.

Samaddar, P., Ok, Y.S., Kim, K.H., Kwon, E.E., Tsang, D.C.W., 2018. Synthesis of nanomaterials from various wastes and their new age applications. Journal of Cleaner Production 197, 1190–1209.

Santhosh, A., Theertha, V., Prakash, P., Chandran, S.S., 2021. From waste to a value added product: green synthesis of silver nanoparticles from onion peels together with its diverse applications. Materials Today: Proceedings 46, 4460–4463.

Shankar, S., Prasad, R.G.S.V., Selvakannan, P.R., Jaiswal, L., Laxman, R.S., 2015. Green synthesis of silver nanoribbons from waste X-ray films using alkaline protease. Materials Research Express 5, 2158–5849.

Sharma, K., Kaushik, S., Jyoti, A., 2016. Green synthesis of silver nanoparticles by using waste vegetable peel and its antibacterial activities. Journal of Pharmaceutical Sciences and Research 8, 313–316.

Sharma, D., Kanchi, S., Bisetty, K., 2019. Biogenic synthesis of nanoparticles: a review. Arabian Journal of Chemistry 12, 3576–3600.

Silva Viana, R.L., Pereira Fidelis, G., Jane Campos Medeiros, M., Antonio Morgano, M., Gabriela Chagas Faustino Alves, M., DominguesPassero, L.F., et al., 2020. Green synthesis of antileishmanial and antifungal silver nanoparticles using corn cob xylan as a reducing and stabilizing agent. Biomolecules 10, 1–21.

Singh, J., Dutta, T., Kim, K.H., Rawat, M., Samaddar, P., Kumar, P., 2018. Green synthesis of metals and their oxide nanoparticles: applications for environmental remediation. Journal of Nanobiotechnology 16, 1–24.

Sirisha, S.A.D., Mary, A., 2016. Green synthesis of nanoparticle of zinc and treatment of nanobeads for waste water of alizarin red dye. International Journal of Environmental Research and Development 6, 11–17.

Thai, Q.B., Le, D.K., Luu, T.P., Hoang, N., Nguyen, D., Duong, H.M., 2019. Aerogels from wastes and their applications. Juniper Online Journal Material Science 5, 001–004.

Thakore, S.I., Nagar, P.S., Jadeja, R.N., Thounaojam, M., Devkar, R.V., Rathore, P.S., 2019. Sapota fruit latex mediated synthesis of Ag, Cu mono and bimetallic nanoparticles and their in vitro toxicity studies. Arabian Journal of Chemistry 12, 694–700.

Tyagi, A., Tripathi, K.M., Singh, N., Choudhary, S., Gupta, R.K., 2016. Green synthesis of carbon quantum dots from lemon peel waste: applications in sensing and photocatalysis. RSC Advances 6, 72423–72432.

Velmurugan, P., Park, J., Lee, S., Jang, J., Yi, Y., Han, S., et al., 2015. Reduction of silver (I) using defatted cashew nut shell starch and its structural comparison with commercial product. Carbohydrate Polymers 133, 39–45.

Velu, K., Elumalai, D., Hemalatha, P., Janaki, A., Babu, M., Hemavathi, M., et al., 2015. Evaluation of silver nanoparticles toxicity of *Arachis hypogaea* peel extracts and its larvicidal activity against malaria and dengue vectors. Environmental Science and Pollution Research 22, 17769–17779.

Virji, M.A., Stefaniak, A.B., 2014. A review of engineered nanomaterial manufacturing processes and associated exposures. Comprehensive Materials Processing 8, 103–125.

Wagner, H., Reiter, M., Ferstl, W., 1980. New drugs with cardiotonicactivity I. Chemistry and pharmacology of the cardiotonic activeprinciple of Annona squasmosa L. Planta Medica 40, 77–85.

Wang, T., Lin, J., Chen, Z., Megharaj, M., Naidu, R., 2014. Green synthesized iron nanoparticles by green tea and eucalyptus leaves extracts used for removal of nitrate in aqueous solution. Journal of Cleaner Production 83, 413–419.

Xu, H., Wang, L., Su, H., Gu, L., Han, T., Meng, F., et al., 2015. Making good use of food wastes: green synthesis of highlystabilized silver nanoparticles from grape seed extract and theirantimicrobial activity. Food Biophysics 10, 12–18.

Yan, D., Zhang, H., Chen, L., Zhu, G., Wang, Z., Xu, H., et al., 2014. Supercapacitive properties of Mn_3O_4 nanoparticles biosynthesized from banana peel extract. RSC Advances 4, 23649–23652.

Yang, N., Hong, L.W., 2014. Biosynthesis of Au nanoparticles using agricultural waste mango peel extract and its in vitro cytotoxic effect on two normal cells. Materials Letters 134, 67–70.

Yi, Z., Cui, X., Chen, G., Chen, X., Jiang, X., Li, X., 2021. Biocompatible, antioxidant nanoparticles prepared from natural renewable tea polyphenols and human hair keratins for cell protection and anti-inflammation. ACS Biomaterials Science & Engineering 7, 1046–1057.

Chapter 6

Panorama of microbial regimes toward nanomaterials' synthesis

Dipankar Ghosh[1], Soumita Sarkar[1], Shrestha Debnath[1], Parna Das[1] and Y. Devika[1,2]

[1]*Microbial Engineering & Algal Biotechnology Laboratory, Department of Biosciences, JIS University Agarpara, Kolkata, India,* [2]*Department of Biotechnology, Maulana Abul Kalam Azad University of Technology, Kolkata, India*

6.1 Introduction

The production of nanometer-size particles of various morphologies, diameters, and monodispersity is an important topic of nanoscience research (Iravani, 2014) due to the unique applications of metal nanoparticles (NP) in several industrial areas (Koul et al., 2021). Nanoparticles are synthesized using a variety of ways, including physical, chemical, and biological processes (Salunke et al., 2016). Due to the availability of numerous reductase enzymes that can reduce metal salts to metal particles, microorganisms are essential nanofactories that can accumulate and detoxify heavy metals (Singh et al., 2016) toward NP biogenesis compared to conventional approaches (Hasan, 2015; Shedbalkar et al., 2014). Understanding of metal-microbe interactions has recently focused on using microorganisms as "nanofactories" for the manufacture of metal nanoparticles (El-Shanshoury et al., 2011; Sharma et al., 2012) from metal ions into the element metal using enzymes produced by intracellular and extracellular activities. Moreover, metal ions have entered and accumulated within the microbial cells where these metal ions are sequestrated and nucleated via reducing microenvironment. It allows natural growth of microbial assemblages while metal ions' nucleation happens. It allows capture of metal ions on the cellular surfaces, port of entry within the cell, sequestration within the cytosolic regions, and reduction of metal ions to generate microbial-assisted nanoparticles using molecular catalysis of reducing enzymes. Extracellular creation of nanoparticles entails trapping metal ions on the cell surface and reducing ions in the presence of enzymes. While in case of extracellular biogenesis of NPs, extracellular secretion (microbial culture filtrates) of diverse microbial assemblages has been amassed and utilized for the in vitro synthesis of NPs. Microbial cells are typically parade resistant against the metal ions and continue to produce NPs in an effective manner (Li et al., 2011a,b; Bhardwaj and Naraian, 2021; Ovais et al., 2018). Green chemistry concepts are compatible with bioorganism synthesis because involvement of single-step nanoparticles made employing biological approaches or green technology to have various natures, improved stability, and acceptable dimensions (Parveen et al., 2016).

Nanomaterials are classified according to their form, size, characteristics, and constituents. The organic (biotic) and inorganic (abiotic) nanoparticles are involved (Chaudhary et al., 2020; Nogueira et al., 2012; Kolahalam et al., 2019; Abdelraof et al., 2019). The classification of nanoparticles into the biotic and abiotic forms is discussed in Fig. 6.1.

6.2 Application of nanomaterials in different fields

6.2.1 Medical uses

Nanomaterials' biomedical uses include regenerative pharmaceuticals in addition to the therapeutic uses of antibacterial and anticancer agents (Sharma and Mondal, 2020). Nanomedicine is a guided recovery strategy that uses manufactured biodegradable nanoplatforms such as poly L-lactic acid, polyglycolic acid, poly D,L-lactide-coglycolide, polycaprolactone, polyphosphazenes, and others to coordinate cell growth (Nance, 2019; Liu et al., 2020; Ulbrich et al., 2016; Weissig et al., 2014; Chien et al., 2012; Ahn et al., 2011). In CT scan, iodine-based liposomal nanotechnology has been

Green and Sustainable Approaches Using Wastes for the Production of Multifunctional Nanomaterials. DOI: https://doi.org/10.1016/B978-0-443-19183-1.00020-9

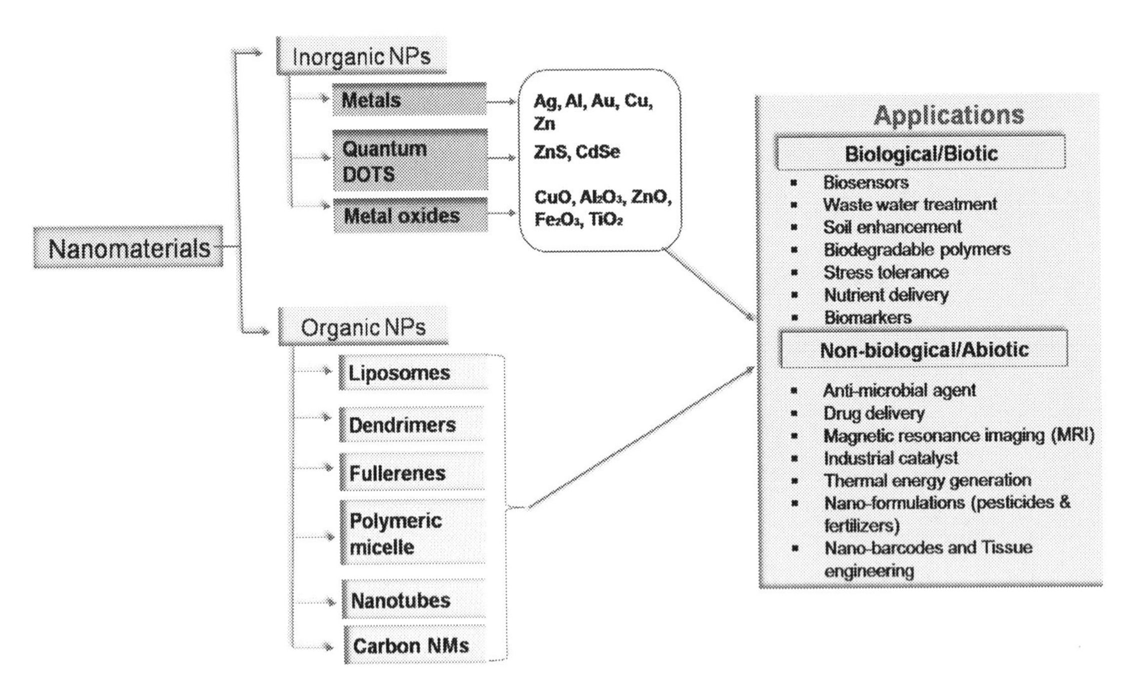

FIGURE 6.1 Classification of nanoparticles: organic and inorganic nanoparticles (Kumar and Lal, 2014; Berry, 2012; Al-Kayiem et al., 2013).

employed to reduce the danger of iodinated contrast agents (Han et al., 2019). Nanoparticles are substantially smaller than those currently used as ultrasonic contrast agents. As with other nanoparticles, the tiny size and extra surface labeling aid in lesion targeting, as proven by PFOB loaded with folate-PEG-chitosan (Hu et al., 2016; Polyak and Ross, 2018; Partlow et al., 2007).

6.2.2 Agricultural uses

Several new nanotechnology technologies have been demonstrated to have agricultural uses to boost yields for crop protection, toxicity evaluation, and disease detection of plants (Jose and Radhakrishnan, 2018) involving delayed release fertilizer (SRF) and controlled release fertilizer (CRF) (Guo et al., 2018). Even carbon-based nanoparticles, TiO_2, iron oxide, zinc oxide, and urea hydroxyapatite have been described as natural biofertilizers and plant growth-stimulating features (Mejias et al., 2021; Zahra et al., 2015; Raliya et al., 2017). Nano-GroTM, nano-Ag Answer, Biozar Nano-Fertilizer, Nano Max NPK Fertilizer, Master Nano Chitosan Organic Fertilizer, and TAG NANO (NPK, PhoS, Zinc, Cal, etc.) fertilizers are some commercial products of nanofertilizers (Prasad et al., 2017). Silver, carbon, silica, and aluminosilicate nanoparticles have all been studied for their potential as antifungal agents (Thomas et al., 2017; Malandrakis et al., 2019). As a novel antibacterial agent, copper nanoparticles serve as a significant function. Copper nanoparticles have been demonstrated to exhibit antifungal effect against plant pathogenic fungus such as *Phoma destuctiva*, *Curvularia lunata*, *Alternaria alternata*, and *Fusarium oxysporum* due to their surface ratio and huge volume (Kanhed et al., 2014; Chaud et al., 2021).

6.2.3 Industrial uses

Nanomaterials have a variety of industrial uses (He and Hwang, 2016). Antimicrobial nanoparticles, such as silver NPs, can be employed to improve food safety, including metal nanoparticles or metal oxides in polymeric nanomaterials used in food packaging, such as ZnO nanoparticles embedded in polystyrene films (Handford et al., 2014; Li et al., 2011a,b; Singh et al., 2017). Nanosensors aid in quality monitoring by detecting CO_2 levels released by food and food-contaminating chemicals, tracking shipments through various logistical phases, preventing contamination, and ensuring a quality item (Kumar et al., 2017). Nanomaterials can also improve membrane separation for water purification and desalination (Nnaji et al., 2018). When exposed to sunshine, zinc oxide (ZnO) or zinc tin oxide demonstrated photocatalytic breakdown efficiency of 50% and elimination of 77% of chemical oxygen demand from textile effluent (Danwittayakul et al., 2015). Because the inclusion of reduced graphene oxide (RGO) reduces the separation band of CuO nanoparticles in nanocomposites, the reduced copper oxide/graphene oxide (CuO/RGO) nanocomposites can function as a photocatalyst. At 120 minutes and 180 minutes, respectively, the breakdown of ortho- and para-nitrophenol was observed. Decomposition of CuO/RGO NC can remove nitrophenol from wastewater (Botsa and Basavaiah, 2019).

Nanoparticles have an essential function in the textile sector to enhance the textile properties. Fabric softness, durability, UV resistance, water repellency, fire resistance, and antibacterial qualities are among these features. Active nanoparticles such as TiO_2, chitosan, N-Halamine, Ag, Cu_2O, and metal/hemp fibers can be chemically or physically integrated into textiles to create an antibacterial textile (Li et al., 2007; Butola, 2019; Naikoo et al., 2021). UV protection materials are created by treating textiles with UV-blocking (UVB and UVA radiations) nanomaterials to increase the UV shielding. UV-sensitive nanomaterials like TiO_2 and ZnO may either scatter or absorb UV rays. These materials are stable and nontoxic and they are resilient to higher temperatures and many types of washing (Yang et al., 2004). ZnO nanorods have also been put on cotton textiles as an excellent UV diffusing layer. Furthermore, as an UV-absorbing layer, ZnO nanoparticles have been put to cotton and polyester textiles (Tania and Ali. 2021).

6.3 Pathway of biosynthesis of nanomaterials

Biological route for nanoparticle production processes differs from one regime to another (Fig. 6.2). Depending on where the reducing agent is placed, nanoparticles can be formed within or outside the cell, and other extracts such as flavonoids, phenolics, terpenoids, and cofactors aid to support the nanoparticles by serving as stabilizing or capping agents (Kharissova et al., 2013). It has now been published that biological systems may function as a "bio-laboratory" for the creation of metal oxide and pure metal particles at the nanoscale using a biomimetic technique. Bacteria (Shivaji et al., 2011), fungi (San Chan and Don, 2013), yeast (Kumar and Lal, 2014), plant extracts (Akhtar et al., 2013), and waste materials (Kanchi et al., 2018) have all been used as environmentally acceptable precursors to create a synthesis of NPs with potential applications (Sharma et al., 2019; Vijayaraghavan and Nalini, 2010). In the case of nanomaterials biosynthesis, three distinct response modes occur: brief incubation, growth phase, and termination phase (Lukman et al., 2011; Manti et al., 2008; Salunke et al., 2016).

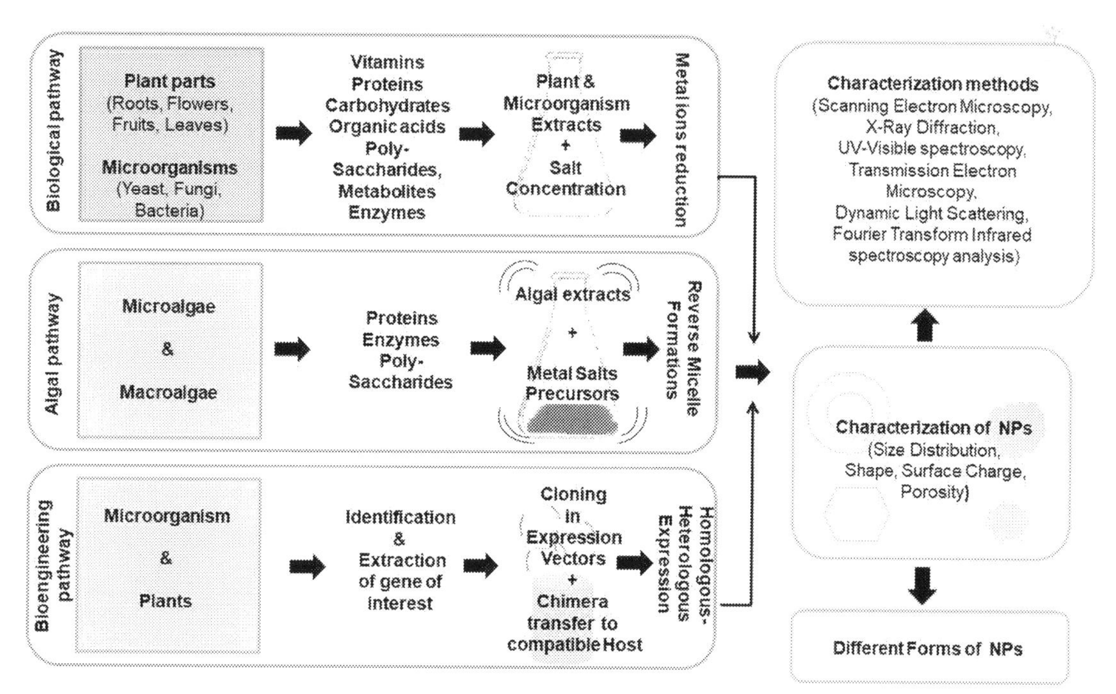

FIGURE 6.2 Generalized schematic representation of various pathways of nanoparticle biosynthesis (Gwala et al., 2021; Boedicker et al., 2021; Dhanker et al., 2021; Guan et al., 2022).

6.3.1 Physicochemical methods for nanomaterials' synthesis

There are a variety of physical and chemical preparation techniques for nanoparticle fabrication, such as radiation, chemical precipitation, photochemical, electrochemical, and Langmuir-Blodgett techniques (Keat et al., 2015). Thus, green nanoparticle production can be carried out by four methods as below:

6.3.1.1 *Polysaccharide method*

Metal nanoparticles were manufactured employing the polysaccharide approach, which employed water and polysaccharides as reducing, stabilizing, or both reducing and stabilizing agent. In a mild heating system, for example, silver nanoparticles can be produced by employing starch as a protectant and -d-glucose as a reducing agent. Consequently, the attraction between starch and particles is enhanced. Silver nanoparticles are weak and reversible at higher temperatures, allowing for the separation of the produced silver nanoparticles (Mochochoko et al., 2013).

6.3.1.2 *Tollens method*

The Tollens technique requires only one step. This method entails cutting down on Ag^+ ions by saccharides whenever there is ammonia, resulting in silver nanoparticles ranging in shape and size from 50 to 200 nm. Because ammonia has a great affinity for Ag^+ ions, $Ag(NH_3)^{2+}$ is a stable complex ion; thus, the concentration of ammonia and the type of the reducing agents plays a crucial part in the formulation of AgNPs (Dondi et al., 2012).

6.3.1.3 *Irradiation method*

Metal nanoparticles may be generated at room temperature using a variety of irradiation processes that do not need the use of reducing chemicals. As a result, temperature-dependent capping agents can be exploited in the irradiation approach as well. UV irradiation of *Coprinus comatus* extract and chloroauric acid solution, for example, can yield gold nanoparticles with characteristic form and size distribution (Naeem et al., 2021).

6.3.1.4 *Polyoxometalates method*

Polyoxometalates are a vast structurally diverse class of molecular metal oxide clusters. Meanwhile, their reduced forms have a larger ability to transmit and/or store electrons and protons and may therefore operate as an efficient donor or a multielectron acceptor without structural modification. As a result, soluble polyoxometalates can be applied to inertly produce noble nanoparticles via stepwise multielectron redox processes (Cauerhff and Castro, 2013; Wang and Weinstock, 2013).

6.3.2 Biological methods for nanomaterials synthesis

Diverse algal regimes (i.e. blue-green algae, brown algae, green algae, and red algae) are microbial factories because of their incredible ability to bioremediate toxic metals, and then change them into more malleable forms (Dhillon et al., 2012; Sahoo et al., 2014; Chaudhary et al., 2020) with the participation of NADPH-dependent reductase during nitrogen fixation, photosynthesis, and respiration (Sharma et al., 2016). Chloroauric acid was incubated with *Ulva intestinalis* and *Rhizoclonium fontinale* algae at 20°C for 72 hours to produce intracellular AuNPs and confirmed by TEM analysis (Sicard et al., 2010). Bioengineering technologies (i.e., coexpressing metallothioneins and phytochelatin synthase) can be used as simpler, safer, more ecologically friendly, low-cost, and less-hazardous alternatives to traditional treatments for nanomaterials synthesis (Iravani and Varma, 2019; Choi et al., 2018). Silver nanoparticles were created using scaffolds obtained from the host *Bacillus* sp. (Liu et al., 2012).

6.4 Synthesis of various nanomaterials by using microorganisms

Microbial biosynthetic pathways are potential framework for nanoparticle synthesis as shown in Table 6.1.

6.4.1 Metal nanomaterials

Microorganisms create inorganic compounds inside or outside the cell, frequently at nanoscale scale and with complex shape. *Stenotrophomonas maltophilia SELTE02*, a strain isolated from the rhizome soil of *Astragalus bisulcatus*, a selenium hyperaccumulator legume, shows promising conversion of selenium (SeO_3^{-2}) to elemental selenium (SeO) that

TABLE 6.1 Types of microorganisms and pathways used for biosynthesis of various nanomaterials.

Name of microorganisms	Name of produced nanomaterials	Name of the pathway	References
Aeromonas hydrophilia MTCC-1739	ZnO	Bacteria-derived pathway	Jayaseelan et al. (2012)
Bacillus cereus	Ag	Green pathway	Sunkar and Nachiyar (2012)
Bacillus subtilis ATCC11774	Ag	Bioengineering pathway	Soni and Prakash (2015)
Desertifilum sp. *EAZ03*	ZnO	Algal pathway	Ebadi et al. (2022)
E. coli DH5α	Au	Bacteria-derived pathway	Ghorbani and Rashidi (2013)
E. coli (NCTC-13216)	CdTe	Bacteria-derived pathway	Kominkova et al. (2017)
Erwinia herbicola	SnO_2	Green pathway	Srivastava and Mukhopadhyay (2014)
Pseudomonas aeruginosa JP-11	CdS	Bacteria-derived pathway	Raj et al. (2016)
Pseudomonas poae CO	Ag	Green pathway	Ibrahim et al. (2020)
Streptomyces griseus	Cu	Bacteria-derived pathway	Ponmurugan et al. (2016)
Thermus scotoductus SA-01	Au	Bacteria-derived pathway	Erasmus et al. (2014)
Arthroderma fulvum strain HT77	Ag	Fungi-derived pathway	Xue et al. (2016)
Aspergillus niger	Ag	Green pathway	Al-Zubaidi et al. (2019)
Aspergillus flaveus TFR7	Titanium dioxide	Fungi-derived pathway	Raliya et al. (2015)
Aspergillus terreus MTCC632	Silver	Fungi-derived pathway	Velhal et al. (2016)
Aureobasidium pullulans	Te, Se	Fungi-derived pathway	Liang et al. (2019)
Cephalosporium sp.	Ag	Fungi-derived pathway	Kaur et al. (2018)
Humicola sp.	CeO_2	Green pathway	Khan and Ahmad (2013)
Hypocrea lixii	NiO	Green pathway	Salvadori et al. (2015)
Phoma glomerata	Te, Se	Fungi-derived pathway	Liang et al. (2019)
Penicillium duclauxii	Ag	Green pathway	Almaary et al. (2020)
Penicillium oxalicum	Ag	Fungi-derived pathway	Du et al. (2015)
Saccharomyces cerevisiae	TiO_2	Green pathway	Peiris et al. (2018)
Sclerotinia sclerotiorum MTCC8785	Ag	Fungi-derived pathway	Saxena et al. (2016)
Anabaena cylindrical	CuO	Algal pathway	Bhattacharya et al. (2019)
Bifurcaria bifurcutta	CuO & Cu_2O	Algal pathway	Abboud et al. (2014)
Chlorella pyrenoidosa	Ag	Green pathway	Aziz et al. (2015)

(*Continued*)

TABLE 6.1 (Continued)

Name of microorganisms	Name of produced nanomaterials	Name of the pathway	References
Chlorella vulgaris	Ag	Algal pathway	El-Moslamy et al. (2016)
Chondrus crispus	Au & Ag	Green pathway	Castro et al. (2013)
Chroococcus minutus	Ag	Algal pathway	Sahoo et al. (2020)
Cylindrospermum stagnale	CuO	Algal pathway	Sonbol et al. (2021)
Gracilaria gracilius	ZnO	Green pathway	Francavilla et al. (2014)
Gilidiella acerosa	Ag	Algal pathway	Dahoumane et al. (2017)
Jania rubens	Fe_3O_4	Green pathway	Salem et al. (2020); Sudha et al. (2013)
Nostoc muscorum NCCU 442	Ag	Algal pathway	Husain et al. (2021)
Sargassum bovinum	Pd	Algal pathway	Momeni and Nabipour (2015)
Sargassum plagiophyllum	AgCl	Algal pathway	Dhas et al. (2014)
Sargassum muticum	Fe_3O_4	Algal pathway	Mahdavi et al. (2013)
Spirulina platensis	Au	Algal pathway	Suganya et al. (2015)
Synechocystis sp. *PCC 6803*	Au	Algal pathway	Liu and Choi (2021)
Turbinaria ornate/Brown sea weed	Au	Green pathway	Nahvi et al.(2021)
Ulva fasciata	ZnO	Algal pathway	Alsaggaf et al. (2021)

accumulates selenium granules either in the cytoplasm, or in the extracellular space (Di Gregorio et al., 2005). The activity of cysteine desulfhydrase (CS lyase) was found to be responsible for the formation of CdS nanocrystals and it was also found that bacterial cellulose isolated from strain *Gluconoacetobacter xylinus* was also used to synthesize 30 nm of CdS nanomaterials deposited on bacterial cellulose nanofibers(Narayanan and Sakthivel, 2010). Fungal MNP production can occur intracellularly (inside the fungal cell), extracellularly (outside the fungal cell), or at the cell surface, and can be accomplished by enzymatic reduction as well as cell wall-related mechanisms involving multiple biomolecules (Silva et al., 2016). The enzyme NADH-dependent nitrate reductase and anthraquinone from *F. oxysporum* are responsible for the reduction of silver ions (Devi and Joshi, 2015). According to one study, ZnO nanomaterial was synthesized by mixing dried *Sargassum muticum* algae powder with distilled water and heating until the mixture was completely, then adding zinc sodium acetate salt solution and stirring continuously for many hours until the mixture was completely dissolved until nanomaterial is generated. The synthesized ZnO have a hexagonal shape, range from 35 to 57 nm, and are limited by biologically active functional groups such as sulfate, amine, hydroxyl, and carbonyl (Azizi et al., 2014).

6.4.2 Gold nanomaterials (AuNPs)

The biosynthesis of gold nanomaterials is an environment-friendly method that requires only a short time to convert metal ions into nanomaterials. *Pithophora crispa* is one of the most exploited microalgae species involved in the biosynthesis of Au nanomaterials by desalting the precursor of chloroauric acid using intracellular and extracellular proteins and peptides (Xie et al., 2007). An endophytic fungus, *Colletotrichum* sp., isolated from the leaves of geranium (*Pelargonium tombolens*), and they rapidly reduce gold ions to nonvalent gold nanoparticles. It has a spherical shape and many slabs. These gold nanoparticles exhibit a mixture of flat disk and rod morphology (Gole et al., 2001). It has been observed that glutathione bound through a free amine group or a cysteine residue has a function in the evolution of gold nanoparticles. Experiments employing extracts from various regions of diverse plant species are being conducted out as evidence for the relevance of biogenesis of gold nanomaterials grows. The extracts are utilized to synthesize alkaloids, flavonoids, saponins, steroids, tannins, and other secondary biomolecules and metabolites capable of reducing gold salts to generate gold nanomaterials (Husen, 2017). When the dead tissue of *Avena sativa* was used to synthesize gold nanoparticles, the results showed that functional groups such as the carbonyl group and the amino and sulfhydryl groups present in the cell wall were also incorporated in the biological process of Au-(III) to gold nanoparticles. *Rhodopseudomonas capsulata*, a prokaryotic bacterium, has been shown to reduce Au^{3+} to Au^0 at room temperature (He et al., 2007).

6.4.3 Silver nanomaterials (AgNPs)

Extracellular AgNPs' production entails trapping metal ions on the cell's outer surface and reducing them in the existence of enzymes or biomolecules, whereas intracellular synthesis happens within the bacterial cells. Nanoparticle extracellular production has been proposed as a low-cost, large-scale manufacturing method that requires less processing (Mendrek et al., 2017). In a report, a silver-resistant strain of *Pseudomonas stutzeri AG259* accumulated silver nanoparticles intracellularly within size range of few nm to 200 nm using NADH-dependent reductase enzyme that supplies electrons and itself oxidizes to NAD^+. The transfer of electrons from NADH results in the bioreduction of silver ions into silver nanomaterials (Srivastava and Constanti, 2012). It is also showed that the use of *Aspergillus flavus* resulted in the silver nanoparticles proliferation on the surface of its cell wall when incubated with silver nitrate solution for 72 hours. The typical particle diameter was found to be 8.92 nm (Vigneshwaran et al., 2007). A study showed that protein present in the *Capsicum annuum* leaf extract first trapped the silver ions on their surface via electrostatic interaction force, so the silver ions are reduced by these proteins leading to the changes in the secondary structures of proteins and finally leads subsequently to the emergence of silver nuclei (Li et al., 2007). Almost all micro green algae species used extracellularly generate silver nanomaterials of varying size and morphology, such as *Pithophora oedogonia* (cubical and hexagonal, 24–55 nm), *Chlorococcum humicola* (spherical, 16 nm), *Chlorella vulgaris* (triangular, 28 nm), *Chlorella reinhardtii* (rectangular and rounded, 1–15 nm), and *Enteromorpha flexuosa* (circular, 15 nm) (Jena et al., 2014).

6.4.4 Metal oxide nanomaterials

Metal oxide nanoparticles such as Fe_3O, TiO_2, CuO, and ZnO are being investigated intensively as possible novel antibacterial agents. Because of their low toxicity to human cells, cheap cost, effective size-dependent inhibition of various bacteria, capacity to prevent biofilm formation, and even removal of spores, they are excellent for use as antibacterial agents in textiles, personal care products, biomedical, and secondary sectors (Stankic et al., 2016). The intracellular magnetosomes of some bacteria such as *Magneto spirillum* help encapsulate Fe_2O_3 nanomaterials in a bioavailable form with the help of certain multicellular proteins (eg, ferritin enzymes or iron reductase) (Barber-Zucker and Zarivach, 2017). CuO and SnO_2 metal oxide nanoparticles were previously made using the microorganisms *Morganella morganii* and *Erwinia herbicola*, respectively, using enzymes such as NADH involved in the redox reaction. Metabolites secreted by bacteria in culture media can induce reduction and stabilization of newly formed metal nanomaterials (Srivastava and Mukhopadhyay, 2014). Nanoparticles of spherical shape and sizes ranging from 100 to 150 nm were generated efficiently utilizing a variety of plant extracts, including titanium dioxide (TiO_2), *viz. Annona squamosa* peel (Roopan et al., 2012), *Cocos nucifera* coir, and *Psidium guajava* (Santhoshkumar et al., 2014). PbS nanocrystals can also be biosynthesized by controlling the concentration of poly-etheneglycol in the *Clostridiaceae* sp. where SO_4^{2-} is first reduced to S^{2-} by the sulfate-reducing bacteria, and then S^{2-} gradually combined with Pb^{2+} to precipitate as PbS-nanomaterial (Qi et al., 2016). The biosynthesized ZnO nanomaterials showed significant antibacterial efficacy against fish infections (Sanaeimehr et al., 2018).

6.4.5 Organic nanomaterials

Intracellular or extracellular production of nanomaterials is a component of a bacterial cell's detoxifying process via enzymatic reduction that results in confirmed alterations in metal ion reactivity to form NPs (Grasso et al., 2019) with the advent of extracellular polymeric substances (EPS) as an ascending bioreductants (Gahlawat et al., 2016) involving bacterial regimes to produce some organic nanostructures (i.e. cellulose nanofibers, nanocrystalline cellulose), especially *Gluconacetobacter* sp. (Golmohammadi et al., 2017).

6.5 Genomic approach of biosynthesis of nanomaterials

In prokaryotic regime, bacteria are foremost choice concerning convenient platform for genetic engineering compared to fungi (Zonooz and Salouti, 2011). Bacterial nitrate reductase has revealed to be responsible in order to reduce ions to nanoparticles under controlled temperature (Das et al. 2014; Lv et al., 2018). The bulk of these proteins in fungi are hydrolytic enzymes such as amylases, cellulases, and proteases. The majority of proteomic investigations have concentrated on proteins involved in metabolic processes (Jain et al., 2011). Two compounds of Zn, Mg, and Ti metals were subjected to fungal extracellular enzyme secrets based on research on the performance of precursor compounds and salt concentration (Raliya and Tarafdar, 2014). To further understand the nature of proteins, denaturation of proteins found in the fungal cell-free filtrate has been done. The creation of ZnO nanoparticles using both denaturized (heat treated) and native (untreated) proteins demonstrated that the natural form of proteins is not required for nanoparticle synthesis (Jain et al., 2013). Algae are used in the manufacturing of nanoparticles because their proteins lower the Ag^{+} ions, and these proteins diminish the nanoparticles and aid to preserve the stability of silver (Ag) nanoparticles. Hydroxyl groups in Tyr residues and carboxyl groups in Asp/Glu residues reduce Ag^{+} ions. This causes anisotropic development of silver nanoplates, resulting in rod-like particles with a mean length of 44 nm and a width of 16–24 nm. Marine algae metabolites such as *Chaetoceros calcitrans*, *Chlorella salina*, *Isochrysis galbana*, and *Tetraselmis gracilis* may also degrade silver ions and hence create silver nanoparticles (Roy et al., 2013).

6.6 Conclusion

Microbes are widely framed as nanobiotechnological biotic tool for microorganism-mediated nanoparticle production to eliminate the need for dangerous, poisonous, and costly chemical ingredients in the process of creation and stabilization processes. Though accelerated studies need to be required for ameliorating efficiency of microbial-assisted route and establish stable control over maintaining size, morphology, and stability of microbial NPs. Moreover, microbe-assisted NPs' biosynthesis is a very slow process (lasts from hours to days) in comparison to physical and/chemical approaches. Thus, reduction in biosynthesis tenure may make microbial-assisted route more attractive in near future. Even though particle size and mono-depressiveness of microbial NPs needs to be further accosted as an extensive investigation. Diverse ranges of scientific research have depicted that microbe-assisted NPs can easily been degraded after certain period of half life (Xiang et al., 2007; Hergt et al., 2005; Hergt and Dutz, 2007). Therefore, stability issues associated with microbe-assisted NPs need to be addressed to enhance the sustainability of microbial approach to biosynthesize NPs toward large-scale industrial application (i.e. health and medical sectors) in near future.

Acknowledgment

Authors would like to thank JIS University Kolkata and JIS Group of educational initiatives.

References

Abboud, Y., Saffaj, T., Chagraoui, A., El Bouari, A., Brouzi, K., Tanane, O., et al., 2014. Biosynthesis, characterization and antimicrobial activity of copper oxide nanoparticles (CONPs) produced using brown alga extract (*Bifurcaria bifurcata*). Applied Nanoscience 4 (5), 571–576.

Abdelraof, M., Hasanin, M.S., Farag, M.M., Ahmed, H.Y., 2019. Green synthesis of bacterial cellulose/bioactive glass nanocomposites: effect of glass nanoparticles on cellulose yield, biocompatibility and antimicrobial activity. International Journal of Biological Macromolecules 138, 975–985.

Ahn, S., Jung, S.Y., Seo, E., Lee, S.J., 2011. Gold nanoparticle-incorporated human red blood cells (RBCs) for X-ray dynamic imaging. Biomaterials 32 (29), 7191–7199.

Akhtar, M.S., Panwar, J., Yun, Y.S., 2013. Biogenic synthesis of metallic nanoparticles by plant extracts. ACS Sustainable Chemistry & Engineering 1 (6), 591–602.

Al-Kayiem, H.H., Lin, S.C., Lukmon, A., 2013. Review on nanomaterials for thermal energy storage technologies. Nanoscience & Nanotechnology-Asia 3 (1), 60–71.

Almaary, K.S., Sayed, S.R., Abd-Elkader, O.H., Dawoud, T.M., El Orabi, N.F., Elgorban, A.M., 2020. Complete green synthesis of silver-nanoparticles applying seed-borne *Penicillium duclauxii*. Saudi Journal of Biological Sciences 27 (5), 1333–1339.

Alsaggaf, M.S., Diab, A.M., ElSaied, B.E., Tayel, A.A., Moussa, S.H., 2021. Application of ZnO nanoparticles phycosynthesized with *Ulva fasciata* extract for preserving peeled shrimp quality. Nanomaterials 11 (2), 385.

Al-Zubaidi, S., Al-Ayafi, A., Abdelkader, H., 2019. Biosynthesis, characterization and antifungal activity of silver nanoparticles by *Aspergillus niger* isolate. Journal of Nanotechnology Research 1 (1), 23–36.

Azizi, S., Ahmad, M.B., Namvar, F., Mohamad, R., 2014. Green biosynthesis and characterization of zinc oxide nanoparticles using brown marine macroalga *Sargassum muticum* aqueous extract. Materials Letters 116, 275–277.

Aziz, N., Faraz, M., Pandey, R., Shakir, M., Fatma, T., Varma, A., et al., 2015. Facile algae-derived route to biogenic silver nanoparticles: synthesis, antibacterial, and photocatalytic properties. Langmuir: The ACS Journal of Surfaces and Colloids 31 (42), 11605–11612.

Barber-Zucker, S., Zarivach, R., 2017. A look into the biochemistry of magnetosome biosynthesis in magnetotactic bacteria. ACS Chemical Biology 12 (1), 13–22.

Berry, C.C., 2012. Applications of inorganic nanoparticles for biotechnology, Frontiers of Nanoscience, Vol. 4. Elsevier, pp. 159–180.

Bhardwaj, A.K., Naraian, R., 2021. Cyanobacteria as biochemical energy source for the synthesis of inorganic nanoparticles, mechanism and potential applications: a review. 3 Biotech 11 (10), 1–16.

Bhattacharya, P., Swarnakar, S., Ghosh, S., Majumdar, S., Banerjee, S., 2019. Disinfection of drinking water via algae mediated green synthesized copper oxide nanoparticles and its toxicity evaluation. Journal of Environmental Chemical Engineering 7 (1), 102867.

Boedicker, J.Q., Gangan, M., Naughton, K., Zhao, F., Gralnick, J.A., El-Naggar, M.Y., 2021. Engineering biological electron transfer and redox pathways for nanoparticle synthesis. Bioelectricity 3 (2), 126–135.

Botsa, S.M., Basavaiah, K., 2019. Removal of nitrophenols from wastewater by monoclinic CuO/RGO nanocomposite. Nanotechnology for Environmental Engineering 4 (1), 1–7.

Butola, B.S., 2019. Recent advances in chitosan polysaccharide and its derivatives in antimicrobial modification of textile materials. International Journal of Biological Macromolecules 121, 905–912.

Castro, L., Blázquez, M.L., Muñoz, J.A., González, F., Ballester, A., 2013. Biological synthesis of metallic nanoparticles using algae. IET Nanobiotechnology 7 (3), 109–116.

Cauerhff, A., Castro, G.R., 2013. Bionanoparticles, a green nanochemistry approach electronic journal of biotechnology, vol. 16, núm. 3, 2013, pp. 1–10 Pontificia Universidad Católica de Valparaíso Valparaíso, Chile. Electronic Journal of Biotechnology 16 (3), 1–10.

Chaud, M., Souto, E.B., Zielinska, A., Severino, P., Batain, F., Oliveira-Junior, J., et al., 2021. Nanopesticides in agriculture: benefits and challenge in agricultural productivity, toxicological risks to human health and environment. Toxics 9 (6), 131.

Chaudhary, R., Nawaz, K., Khan, A.K., Hano, C., Abbasi, B.H., Anjum, S., 2020. An overview of the algae-mediated biosynthesis of nanoparticles and their biomedical applications. Biomolecules 10 (11), 1498.

Chien, C.C., Chen, H.H., Lai, S.F., Hwu, Y., Petibois, C., Yang, C.S., et al., 2012. X-ray imaging of tumor growth in live mice by detecting gold-nanoparticle-loaded cells. Scientific Reports 2 (1), 1–6.

Choi, Y., Park, T.J., Lee, D.C., Lee, S.Y., 2018. Recombinant *Escherichia coli* as a biofactory for various single-and multi-element nanomaterials. Proceedings of the National Academy of Sciences 115 (23), 5944–5949.

Dahoumane, S.A., Mechouet, M., Wijesekera, K., Filipe, C.D., Sicard, C., Bazylinski, D.A., et al., 2017. Algae-mediated biosynthesis of inorganic nanomaterials as a promising route in nanobiotechnology—a review. Green Chemistry 19 (3), 552–587.

Danwittayakul, S., Jaisai, M., Dutta, J., 2015. Efficient solar photocatalytic degradation of textile wastewater using ZnO/ZTO composites. Applied Catalysis B: Environmental 163, 1–8.

Das, V.L., Thomas, R., Varghese, R.T., Soniya, E.V., Mathew, J., Radhakrishnan, E.K., 2014. Extracellular synthesis of silver nanoparticles by the *Bacillus* strain CS 11 isolated from industrialized area. 3 Biotech 4 (2), 121–126.

Dhanker, R., Hussain, T., Tyagi, P., Singh, K.J., Kamble, S.S., 2021. The emerging trend of bio-engineering approaches for microbial nanomaterial synthesis and its applications. Frontiers in Microbiology 12, 638003.

Dhas, T.S., Kumar, V.G., Karthick, V.A.K.J., Angel, K.J., Govindaraju, K., 2014. Facile synthesis of silver chloride nanoparticles using marine alga and its antibacterial efficacy. Spectrochimica Acta Part A: Molecular and Biomolecular Spectroscopy 120, 416–420.

Dhillon, G.S., Brar, S.K., Kaur, S., Verma, M., 2012. Green approach for nanoparticle biosynthesis by fungi: current trends and applications. Critical Reviews in Biotechnology 32 (1), 49–73.

Di Gregorio, S., Lampis, S., Vallini, G., 2005. Selenite precipitation by a rhizospheric strain of *Stenotrophomonas* sp. isolated from the root system of *Astragalus bisulcatus*: a biotechnological perspective. Environment International 31 (2), 233–241.

Dondi, R., Su, W., Griffith, G.A., Clark, G., Burley, G.A., 2012. Highly size-and shape-controlled synthesis of silver nanoparticles via a templated Tollens reaction. Small (Weinheim an der Bergstrasse, Germany) 8 (5), 770–776.

Du, L., Xu, Q., Huang, M., Xian, L., Feng, J.X., 2015. Synthesis of small silver nanoparticles under light radiation by fungus *Penicillium oxalicum* and its application for the catalytic reduction of methylene blue. Materials Chemistry and Physics 160, 40–47.

Ebadi, M., Zolfaghari, M.R., Aghaei, S.S., Zargar, M., Noghabi, K.A., 2022. *Desertifilum* sp. EAZ03 cell extract as a novel natural source for the biosynthesis of zinc oxide nanoparticles and antibacterial, anticancer and antibiofilm characteristics of synthesized zinc oxide nanoparticles. Journal of Applied Microbiology 132 (1), 221–236.

El-Moslamy, S., Kabeil, S., Hafez, E.J.O., 2016. Bioprocess development for *Chlorella vulgaris* cultivation and biosynthesis of anti-phytopathogens silver nanoparticles. Journal of Nanomaterials & Molecular Nanotechnology 9, 2.

El-Shanshoury, A.E.R.R., ElSilk, S.E., Ebeid, M.E., 2011. Extracellular biosynthesis of silver nanoparticles using *Escherichia coli* ATCC 8739, *Bacillus subtilis* ATCC 6633, and *Streptococcus thermophilus* ESh1 and their antimicrobial activities. International Scholarly Research Notices 2011.

Erasmus, M., Cason, E.D., van Marwijk, J., Botes, E., Gericke, M., van Heerden, E., 2014. Gold nanoparticle synthesis using the thermophilic bacterium *Thermus scotoductus* SA-01 and the purification and characterization of its unusual gold reducing protein. Gold Bulletin 47 (4), 245–253.

Francavilla, M., Pineda, A., Romero, A.A., Colmenares, J.C., Vargas, C., Monteleone, M., et al., 2014. Efficient and simple reactive milling preparation of photocatalytically active porous ZnO nanostructures using biomass derived polysaccharides. Green Chemistry 16 (5), 2876–2885.

Gahlawat, G., Shikha, S., Chaddha, B.S., Chaudhuri, S.R., Mayilraj, S., Choudhury, A.R., 2016. Microbial glycolipoprotein-capped silver nanoparticles as emerging antibacterial agents against cholera. Microbial Cell Factories 15 (1), 1–14.

Ghorbani, H.R., Rashidi, R., 2013. Biosynthesis of gold nanoparticles by *Escherichia coli*. Minerva Biotecnologica 25, 161–164.

Gole, A., Dash, C., Soman, C., Sainkar, S.R., Rao, M., Sastry, M., 2001. On the preparation, characterization, and enzymatic activity of fungal protease – gold colloid bioconjugates. Bioconjugate Chemistry 12 (5), 684–690.

Golmohammadi, H., Morales-Narváez, E., Naghdi, T., Merkoçi, A., 2017. Nanocellulose in sensing and biosensing. Chemistry of Materials 29 (13), 5426–5446.

Grasso, G., Zane, D., Dragone, R., 2019. Microbial nanotechnology: challenges and prospects for green biocatalytic synthesis of nanoscale materials for sensoristic and biomedical applications. Nanomaterials 10 (1), 11.

Guan, Z., Ying, S., Ofoegbu, P.C., Clubb, P., Rico, C., He, F., et al., 2022. Green synthesis of nanoparticles: current developments and limitations. Environmental Technology & Innovation 102336.

Guo, H., White, J.C., Wang, Z., Xing, B., 2018. Nano-enabled fertilizers to control the release and use efficiency of nutrients. Current Opinion in Environmental Science & Health 6, 77–83.

Gwala, M., Dutta, S., Ghosh Chaudhuri, R., 2021. Microalgae mediated nanomaterials synthesis. In *algae*. Springer, Singapore, pp. 295–324.

Han, X., Xu, K., Taratula, O., Farsad, K., 2019. Applications of nanoparticles in biomedical imaging. Nanoscale 11 (3), 799–819.

Handford, C.E., Dean, M., Henchion, M., Spence, M., Elliott, C.T., Campbell, K., 2014. Implications of nanotechnology for the agri-food industry: opportunities, benefits and risks. Trends in Food Science & Technology 40 (2), 226–241.

Hasan, S., 2015. A review on nanoparticles: their synthesis and types. Research Journal of Recent Sciences 2277, 2502.

He, X., Hwang, H.M., 2016. Nanotechnology in food science: functionality, applicability, and safety assessment. Journal of Food and Drug Analysis 24 (4), 671–681.

He, S., Guo, Z., Zhang, Y., Zhang, S., Wang, J., Gu, N., 2007. Biosynthesis of gold nanoparticles using the bacteria *Rhodopseudomonas capsulata*. Materials Letters 61 (18), 3984–3987.

Hergt, R., Dutz, S., 2007. Magnetic particle hyperthermia-biophysical limitations of a visionary tumour therapy. Journal of Magnetism and Magnetic Materials 311 (1), 187–192.

Hergt, R., Hiergeist, R., Zeisberger, M., Schüler, D., Heyen, U., Hilger, I., et al., 2005. Magnetic properties of bacterial magnetosomes as potential diagnostic and therapeutic tools. Journal of Magnetism and Magnetic Materials 293 (1), 80–86.

Hu, Y., Wang, Y., Jiang, J., Han, B., Zhang, S., Li, K., et al., 2016. Preparation and characterization of novel perfluorooctyl bromide nanoparticle as ultrasound contrast agent via layer-by-layer self-assembly for folate-receptor-mediated tumor imaging. BioMed Research International 2016.

Husain, S., Verma, S.K., Azam, M., Sardar, M., Haq, Q.M.R., Fatma, T., 2021. Antibacterial efficacy of facile cyanobacterial silver nanoparticles inferred by antioxidant mechanism. Materials Science and Engineering: C 122, 111888.

Husen, A., 2017. Gold nanoparticles from plant system: synthesis, characterization and their application. Nanoscience and Plant–soil Systems. Springer, Cham, pp. 455–479.

Ibrahim, E., Zhang, M., Zhang, Y., Hossain, A., Qiu, W., Chen, Y., et al., 2020. Green-synthesization of silver nanoparticles using endophytic bacteria isolated from garlic and its antifungal activity against wheat Fusarium head blight pathogen *Fusarium graminearum*. Nanomaterials 10 (2), 219.

Iravani, S., 2014. Bacteria in nanoparticle synthesis: current status and future prospects. International Scholarly Research Notices 2014.

Iravani, S., Varma, R.S., 2019. Biofactories: engineered nanoparticles via genetically engineered organisms. Green Chemistry 21 (17), 4583–4603.

Jain, N., Bhargava, A., Majumdar, S., Tarafdar, J.C., Panwar, J., 2011. Extracellular biosynthesis and characterization of silver nanoparticles using *Aspergillus flavus* NJP08: a mechanism perspective. Nanoscale 3 (2), 635–641.

Jain, N., Bhargava, A., Tarafdar, J.C., Singh, S.K., Panwar, J., 2013. A biomimetic approach towards synthesis of zinc oxide nanoparticles. Applied Microbiology and Biotechnology 97 (2), 859–869.

Jayaseelan, C., Rahuman, A.A., Kirthi, A.V., Marimuthu, S., Santhoshkumar, T., Bagavan, A., et al., 2012. Novel microbial route to synthesize ZnO nanoparticles using *Aeromonas hydrophila* and their activity against pathogenic bacteria and fungi. Spectrochimica Acta Part A: Molecular and Biomolecular Spectroscopy 90, 78–84.

Jena, J., Pradhan, N., Nayak, R.R., Dash, B.P., Sukla, L.B., Panda, P.K., et al., 2014. Microalga *Scenedesmus* sp.: a potential low-cost green machine for silver nanoparticle synthesis. Journal of Microbiology and Biotechnology 24 (4), 522–533.

Jose, A., Radhakrishnan, E.K., 2018. Applications of nanomaterials in agriculture and food industry. Green and Sustainable Advanced Materials: Applications 2, 343–375.

Kanchi, S., Kumar, G., Lo, A.Y., Tseng, C.M., Chen, S.K., Lin, C.Y., et al., 2018. Exploitation of de-oiled jatropha waste for gold nanoparticles synthesis: a green approach. Arabian Journal of Chemistry 11 (2), 247–255.

Kanhed, P., Birla, S., Gaikwad, S., Gade, A., Seabra, A.B., Rubilar, O., et al., 2014. In vitro antifungal efficacy of copper nanoparticles against selected crop pathogenic fungi. Materials Letters 115, 13–17.

Kaur, P., Thakur, R., Duhan, J.S., Chaudhury, A., 2018. Management of wilt disease of chickpea in vivo by silver nanoparticles biosynthesized by rhizospheric microflora of chickpea (*Cicer arietinum*). Journal of Chemical Technology & Biotechnology 93 (11), 3233–3243.

Keat, C.L., Aziz, A., Eid, A.M., Elmarzugi, N.A., 2015. Biosynthesis of nanoparticles and silver nanoparticles. Bioresources and Bioprocessing 2 (1), 1–11.

Khan, S.A., Ahmad, A., 2013. Fungus mediated synthesis of biomedically important cerium oxide nanoparticles. Materials Research Bulletin 48 (10), 4134–4138.

Kharissova, O.V., Dias, H.R., Kharisov, B.I., Pérez, B.O., Pérez, V.M.J., 2013. The greener synthesis of nanoparticles. Trends in Biotechnology 31 (4), 240–248.

Kolahalam, L.A., Viswanath, I.K., Diwakar, B.S., Govindh, B., Reddy, V., Murthy, Y.L.N., 2019. Review on nanomaterials: synthesis and applications. Materials Today: Proceedings 18, 2182–2190.

Kominkova, M., Milosavljevic, V., Vitek, P., Polanska, H., Cihalova, K., Dostalova, S., et al., 2017. Comparative study on toxicity of extracellularly biosynthesized and laboratory synthesized CdTe quantum dots. Journal of Biotechnology 241, 193–200.

Koul, B., Poonia, A.K., Yadav, D., Jin, J.O., 2021. Microbe-mediated biosynthesis of nanoparticles: applications and future prospects. Biomolecules 11 (6), 886.

Kumar, R., Lal, S., 2014. Synthesis of organic nanoparticles and their applications in drug delivery and food nanotechnology: a review. Journal of Nanomaterials & Molecular Nanotechnology 3: 4. of 11, 2.

Kumar, V., Guleria, P., Mehta, S.K., 2017. Nanosensors for food quality and safety assessment. Environmental Chemistry Letters 15 (2), 165–177.

Li, S., Shen, Y., Xie, A., Yu, X., Qiu, L., Zhang, L., et al., 2007. Green synthesis of silver nanoparticles using *Capsicum annuum L.* extract. Green Chemistry 9 (8), 852–858.

Li, W.L., Li, X.H., Zhang, P.P., Xing, Y.G., 2011a. Development of nano-ZnO coated food packaging film and its inhibitory effect on *Escherichia coli* in vitro and in actual tests. In *Advanced Materials Research*, Vol. 152. Trans Tech Publications Ltd, pp. 489–492.

Li, X., Xu, H., Chen, Z.S., Chen, G., 2011b. Biosynthesis of nanoparticles by microorganisms and their applications. Journal of Nanomaterials 2011.

Liang, X., Perez, M.A.M.J., Nwoko, K.C., Egbers, P., Feldmann, J., Csetenyi, L., et al., 2019. Fungal formation of selenium and tellurium nanoparticles. Applied Microbiology and Biotechnology 103 (17), 7241–7259.

Liu, L., Choi, S., 2021. Enhanced biophotoelectricity generation in cyanobacterial biophotovoltaics with intracellularly biosynthesized gold nanoparticles. Journal of Power Sources 506, 230251.

Liu, J., Zhang, X., Yu, M., Li, S., Zhang, J., 2012. Photoinduced silver nanoparticles/nanorings on plasmid DNA scaffolds. Small (Weinheim an der Bergstrasse, Germany) 8 (2), 310–316.

Liu, G., Fu, M., Li, F., Fu, W., Zhao, Z., Xia, H., et al., 2020. Tissue-engineered PLLA/gelatine nanofibrous scaffold promoting the phenotypic expression of epithelial and smooth muscle cells for urethral reconstruction. Materials Science and Engineering: C 111, 110810.

Lukman, A.I., Gong, B., Marjo, C.E., Roessner, U., Harris, A.T., 2011. Facile synthesis, stabilization, and anti-bacterial performance of discrete Ag nanoparticles using *Medicago sativa* seed exudates. Journal of Colloid and Interface Science 353 (2), 433–444.

Lv, Q., Zhang, B., Xing, X., Zhao, Y., Cai, R., Wang, W., et al., 2018. Biosynthesis of copper nanoparticles using *Shewanella loihica* PV-4 with anti-bacterial activity: novel approach and mechanisms investigation. Journal of Hazardous Materials 347, 141–149.

Mahdavi, M., Namvar, F., Ahmad, M.B., Mohamad, R., 2013. Green biosynthesis and characterization of magnetic iron oxide (Fe_3O_4) nanoparticles using seaweed (*Sargassum muticum*) aqueous extract. Molecules (Basel, Switzerland) 18 (5), 5954–5964.

Malandrakis, A.A., Kavroulakis, N., Chrysikopoulos, C.V., 2019. Use of copper, silver and zinc nanoparticles against foliar and soil-borne plant pathogens. Science of the Total Environment 670, 292–299.

Manti, A., Boi, P., Falcioni, T., Canonico, B., Ventura, A., Sisti, D., et al., 2008. Bacterial cell monitoring in wastewater treatment plants by flow cytometry. Water Environment Research 80 (4), 346–354.

Mejias, J.H., Salazar, F., Pérez Amaro, L., Hube, S., Rodriguez, M., Alfaro, M., 2021. Nanofertilizers: a cutting-edge approach to increase nitrogen use efficiency in grasslands. Frontiers in Environmental Science 52.

Mendrek, B., Chojniak, J., Libera, M., Trzebicka, B., Bernat, P., Paraszkiewicz, K., et al., 2017. Silver nanoparticles formed in bio-and chemical syntheses with biosurfactant as the stabilizing agent. Journal of Dispersion Science and Technology 38 (11), 1647–1655.

Mochochoko, T., Oluwafemi, O.S., Jumbam, D.N., Songca, S.P., 2013. Green synthesis of silver nanoparticles using cellulose extracted from an aquatic weed; water hyacinth. Carbohydrate Polymers 98 (1), 290–294.

Momeni, S., Nabipour, I., 2015. A simple green synthesis of palladium nanoparticles with *Sargassum* alga and their electrocatalytic activities towards hydrogen peroxide. Applied Biochemistry and Biotechnology 176 (7), 1937–1949.

Naeem, G.A., Jaloot, A.S., Owaid, M.N., Muslim, R.F., 2021. Green synthesis of gold nanoparticles from *Coprinus comatus*, agaricaceae, and the effect of ultraviolet irradiation on their characteristics. Walailak Journal of Science and Technology (WJST) 18 (8), 9396. 12.

Nahvi, I., Belkahla, S., Asiri, S.M., Rehman, S., 2021. . Overview and prospectus of algal biogenesis of nanoparticles. In *Microbial Nanotechnology: Green Synthesis and Applications*. Springer, Singapore, pp. 121–134.

Naikoo, G.A., Mustaqeem, M., Hassan, I.U., Awan, T., Arshad, F., Salim, H., et al., 2021. Bioinspired and green synthesis of nanoparticles from plant extracts with antiviral and antimicrobial properties: a critical review. Journal of Saudi Chemical Society 25 (9), 101304.

Nance, E., 2019. Careers in nanomedicine and drug delivery. Advanced Drug Delivery Reviews 144, 180–189.

Narayanan, K.B., Sakthivel, N., 2010. Biological synthesis of metal nanoparticles by microbes. Advances in Colloid and Interface Science 156 (1–2), 1–13.

Nnaji, C.O., Jeevanandam, J., Chan, Y.S., Danquah, M.K., Pan, S., Barhoum, A., 2018. Engineered nanomaterials for wastewater treatment: current and future trends. Fundamentals of Nanoparticles 129–168.

Nogueira, V., Lopes, I., Rocha-Santos, T., Santos, A.L., Rasteiro, G.M., Antunes, F., et al., 2012. Impact of organic and inorganic nanomaterials in the soil microbial community structure. Science of the Total Environment 424, 344–350.

Ovais, M., Khalil, A.T., Ayaz, M., Ahmad, I., Nethi, S.K., Mukherjee, S., 2018. Biosynthesis of metal nanoparticles via microbial enzymes: a mechanistic approach. International Journal of Molecular Sciences 19 (12), 4100.

Partlow, K.C., Chen, J., Brant, J.A., Neubauer, A.M., Meyerrose, T.E., Creer, M.H., et al., 2007. 19F magnetic resonance imaging for stem/progenitor cell tracking with multiple unique perfluorocarbon nanobeacons. The FASEB Journal 21 (8), 1647–1654.

Parveen, K., Banse, V., Ledwani, L., 2016. April. Green synthesis of nanoparticles: their advantages and disadvantages, In *AIP Conference Proceedings*, Vol. 1724. AIP Publishing LLC, p. 020048, No. 1.

Peiris, M.M.K., Gunasekara, T.D.C.P., Jayaweera, P.M., Fernando, S.S.N., 2018. TiO2 nanoparticles from Baker's yeast: a potent antimicrobial. Journal of Microbiology and Biotechnology 28 (10).

Polyak, A., Ross, T.L., 2018. Nanoparticles for SPECT and PET imaging: towards personalized medicine and theranostics. Current Medicinal Chemistry 25 (34), 4328–4353.

Ponmurugan, P., Manjukarunambika, K., Elango, V., Gnanamangai, B.M., 2016. Antifungal activity of biosynthesised copper nanoparticles evaluated against red root-rot disease in tea plants. Journal of Experimental Nanoscience 11 (13), 1019–1031.

Prasad, R., Bhattacharyya, A., Nguyen, Q.D., 2017. Nanotechnology in sustainable agriculture: recent developments, challenges, and perspectives. Frontiers in Microbiology 8, 1014.

Qi, P., Zhang, D., Zeng, Y., Wan, Y., 2016. Biosynthesis of CdS nanoparticles: a fluorescent sensor for sulfate-reducing bacteria detection. Talanta 147, 142–146.

Raj, R., Dalei, K., Chakraborty, J., Das, S., 2016. Extracellular polymeric substances of a marine bacterium mediated synthesis of CdS nanoparticles for removal of cadmium from aqueous solution. Journal of Colloid and Interface Science 462, 166–175.

Raliya, R., Tarafdar, J.C., 2014. Biosynthesis and characterization of zinc, magnesium and titanium nanoparticles: an eco-friendly approach. International Nano Letters 4 (1), 1–10.

Raliya, R., Biswas, P., Tarafdar, J.C., 2015. TiO_2 nanoparticle biosynthesis and its physiological effect on mung bean (Vigna radiata L.). Biotechnology Reports 5, 22–26.

Raliya, R., Saharan, V., Dimkpa, C., Biswas, P., 2017. Nanofertilizer for precision and sustainable agriculture: current state and future perspectives. Journal of Agricultural and Food Chemistry 66 (26), 6487–6503.

Roopan, S.M., Bharathi, A., Kumar, R., Khanna, V.G., Prabhakarn, A., 2012. Acaricidal, insecticidal, and larvicidal efficacy of aqueous extract of *Annona squamosa L* peel as biomaterial for the reduction of palladium salts into nanoparticles. Colloids and Surfaces B: Biointerfaces 92, 209–212.

Roy, N., Gaur, A., Jain, A., Bhattacharya, S., Rani, V., 2013. Green synthesis of silver nanoparticles: an approach to overcome toxicity. Environmental Toxicology and Pharmacology 36 (3), 807–812.

Sahoo, P.C., Kausar, F., Lee, J.H., Han, J.I., 2014. Facile fabrication of silver nanoparticle embedded $CaCO_3$ microspheres via microalgae-templated CO_2 biomineralization: application in antimicrobial paint development. RSC Advances 4 (61), 32562–32569.

Sahoo, C.R., Maharana, S., Mandhata, C.P., Bishoyi, A.K., Paidesetty, S.K., Padhy, R.N., 2020. Biogenic silver nanoparticle synthesis with cyanobacterium *Chroococcus minutus* isolated from Baliharachandi sea-mouth, Odisha, and in vitro antibacterial activity. Saudi Journal of Biological Sciences 27 (6), 1580–1586.

Salem, D.M., Ismail, M.M., Tadros, H.R., 2020. Evaluation of the antibiofilm activity of three seaweed species and their biosynthesized iron oxide nanoparticles (Fe3O4-NPs). *The Egyptian*. Journal of Aquatic Research 46 (4), 333–339.

Salunke, B.K., Sawant, S.S., Lee, S.I., Kim, B.S., 2016. Microorganisms as efficient biosystem for the synthesis of metal nanoparticles: current scenario and future possibilities. World Journal of Microbiology and Biotechnology 32 (5), 1–16.

Salvadori, M.R., Ando, R.A., Oller Nascimento, C.A., Correa, B., 2015. Extra and intracellular synthesis of nickel oxide nanoparticles mediated by dead fungal biomass. PLoS One 10 (6), e0129799.

San Chan, Y., Don, M.M., 2013. Biosynthesis and structural characterization of Ag nanoparticles from white rot fungi. Materials Science and Engineering: C 33 (1), 282–288.

Sanaeimehr, Z., Javadi, I., Namvar, F., 2018. Antiangiogenic and antiapoptotic effects of green-synthesized zinc oxide nanoparticles using *Sargassum muticum* algae extraction. Cancer Nanotechnology 9 (1), 1–16.

Santhoshkumar, T., Rahuman, A.A., Jayaseelan, C., Rajakumar, G., Marimuthu, S., Kirthi, A.V., et al., 2014. Green synthesis of titanium dioxide nanoparticles using *Psidium guajava* extract and its antibacterial and antioxidant properties. Asian Pacific Journal of Tropical Medicine 7 (12), 968–976.

Saxena, J., Sharma, P.K., Sharma, M.M., Singh, A., 2016. Process optimization for green synthesis of silver nanoparticles by *Sclerotinia sclerotiorum* MTCC 8785 and evaluation of its antibacterial properties. SpringerPlus 5 (1), 1–10.

Sharma, H., Mondal, S., 2020. Functionalized graphene oxide for chemotherapeutic drug delivery and cancer treatment: a promising material in nanomedicine. International Journal of Molecular Sciences 21 (17), 6280.

Sharma, N., Pinnaka, A.K., Raje, M., Fnu, A., Bhattacharyya, M.S., Choudhury, A.R., 2012. Exploitation of marine bacteria for production of gold nanoparticles. Microbial Cell Factories 11 (1), 1–6.

Sharma, A., Sharma, S., Sharma, K., Chetri, S.P., Vashishtha, A., Singh, P., et al., 2016. Algae as crucial organisms in advancing nanotechnology: a systematic review. Journal of Applied Phycology 28 (3), 1759–1774.

Sharma, D., Kanchi, S., Bisetty, K., 2019. Biogenic synthesis of nanoparticles: a review. Arabian Journal of Chemistry 12 (8), 3576–3600.

Shedbalkar, U., Singh, R., Wadhwani, S., Gaidhani, S., Chopade, B.A., 2014. Microbial synthesis of gold nanoparticles: current status and future prospects. Advances in Colloid and Interface Science 209, 40–48.

Shivaji, S., Madhu, S., Singh, S., 2011. Extracellular synthesis of antibacterial silver nanoparticles using psychrophilic bacteria. Process Biochemistry 46 (9), 1800–1807.

Sicard, C., Brayner, R., Margueritat, J., Hémadi, M., Couté, A., Yéprémian, C., et al., 2010. Nano-gold biosynthesis by silica-encapsulated microalgae: a "living" bio-hybrid material. Journal of Materials Chemistry 20 (42), 9342–9347.

Silva, L.P., Bonatto, C.C., Polez, V.L.P., 2016. Green synthesis of metal nanoparticles by fungi: current trends and challenges. Advances and Applications Through Fungal Nanobiotechnology 71–89.

Singh, P., Kim, Y.J., Zhang, D., Yang, D.C., 2016. Biological synthesis of nanoparticles from plants and microorganisms. Trends in Biotechnology 34 (7), 588–599.

Singh, T., Shukla, S., Kumar, P., Wahla, V., Bajpai, V.K., Rather, I.A., 2017. Application of nanotechnology in food science: perception and overview. Frontiers in Microbiology 8, 1501.

Sonbol, H., AlYahya, S., Ameen, F., Alsamhary, K., Alwakeel, S., Al-Otaibi, S., et al., 2021. Bioinspired synthesize of CuO nanoparticles using *Cylindrospermum stagnale* for antibacterial, anticancer and larvicidal applications. Applied Nanoscience 1–11.

Soni, N., Prakash, S., 2015. Antimicrobial and mosquitocidal activity of microbial synthesized silver nanoparticles. Parasitology Research 114 (3), 1023–1030.

Srivastava, N., Mukhopadhyay, M., 2014. Biosynthesis of SnO_2 nanoparticles using bacterium *Erwinia herbicola* and their photocatalytic activity for degradation of dyes. Industrial & Engineering Chemistry Research 53 (36), 13971–13979.

Srivastava, S.K., Constanti, M., 2012. Room temperature biogenic synthesis of multiple nanoparticles (Ag, Pd, Fe, Rh, Ni, Ru, Pt, Co, and Li) by *Pseudomonas aeruginosa*. Journal of Nanoparticle Research 14 (4), 1–SM10.

Stankic, S., Suman, S., Haque, F., Vidic, J., 2016. Pure and multi metal oxide nanoparticles: synthesis, antibacterial and cytotoxic properties. Journal of Nanobiotechnology 14 (1), 1–20.

Sudha, S.S., Rajamanickam, K.., Rengaramanujam, J., 2013. Microalgae mediated synthesis of silver nanoparticles and their antibacterial activity against pathogenic bacteria. Indian Journal of Experimental Biology 51 (5).

Suganya, K.U., Govindaraju, K., Kumar, V.G., Dhas, T.S., Karthick, V., Singaravelu, G., et al., 2015. Blue green alga mediated synthesis of gold nanoparticles and its antibacterial efficacy against Gram positive organisms. Materials Science and Engineering: C 47, 351–356.

Sunkar, S., Nachiyar, C.V., 2012. Biogenesis of antibacterial silver nanoparticles using the endophytic bacterium *Bacillus cereus* isolated from *Garcinia xanthochymus*. Asian Pacific Journal of Tropical Biomedicine 2 (12), 953–959.

Tania, I.S., Ali, M., 2021. Coating of ZnO nanoparticle on cotton fabric to create a functional textile with enhanced mechanical properties. Polymers 13 (16), 2701.

Thomas, R., Mathew, S., Nayana, A.R., Mathews, J., Radhakrishnan, E.K., 2017. Microbially and phytofabricated AgNPs with different mode of bactericidal action were identified to have comparable potential for surface fabrication of central venous catheters to combat *Staphylococcus aureus* biofilm. Journal of Photochemistry and Photobiology B: Biology 171, 96–103.

Ulbrich, K., Hola, K., Subr, V., Bakandritsos, A., Tucek, J., Zboril, R., 2016. Targeted drug delivery with polymers and magnetic nanoparticles: covalent and noncovalent approaches, release control, and clinical studies. Chemical Reviews 116 (9), 5338–5431.

Velhal, S.G., Kulkarni, S.D., Latpate, R.V., 2016. Fungal mediated silver nanoparticle synthesis using robust experimental design and its application in cotton fabric. International Nano Letters 6 (4), 257–264.

Vigneshwaran, N., Ashtaputre, N.M., Varadarajan, P.V., Nachane, R.P., Paralikar, K.M., Balasubramanya, R.H., 2007. Biological synthesis of silver nanoparticles using the fungus *Aspergillus flavus*. Materials Letters 61 (6), 1413–1418.

Vijayaraghavan, K., Nalini, S.K., 2010. Biotemplates in the green synthesis of silver nanoparticles. Biotechnology Journal 5 (10), 1098–1110.

Wang, Y., Weinstock, I., 2013. Polyoxometalate-protected metal nanoparticles: synthesis, structure and catalysis. *Polyoxometalate*. Chemistry: Some Recent Trends 8, 1.

Weissig, V., Pettinger, T.K., Murdock, N., 2014. Nanopharmaceuticals (part 1): products on the market. International Journal of Nanomedicine 9, 4357.

Xiang, L., Bin, W., Huali, J., Wei, J., Jiesheng, T., Feng, G., et al., 2007. Bacterial magnetic particles (BMPs)-PEI as a novel and efficient non-viral gene delivery system. Journal of Gene Medicine 9 (8), 679–690.

Xie, J., Lee, J.Y., Wang, D.I., Ting, Y.P., 2007. Identification of active biomolecules in the high-yield synthesis of single-crystalline gold nanoplates in algal solutions. Small (Weinheim an der Bergstrasse, Germany) 3 (4), 672–682.

Xue, B., He, D., Gao, S., Wang, D., Yokoyama, K., Wang, L., 2016. Biosynthesis of silver nanoparticles by the fungus *Arthroderma fulvum* and its antifungal activity against genera of *Candida, Aspergillus* and *Fusarium*. International Journal of Nanomedicine 11, 1899.

Yang, H., Zhu, S., Pan, N., 2004. Studying the mechanisms of titanium dioxide as ultraviolet-blocking additive for films and fabrics by an improved scheme. Journal of Applied Polymer Science 92 (5), 3201–3210.

Zahra, Z., Arshad, M., Rafique, R., Mahmood, A., Habib, A., Qazi, I.A., et al., 2015. Metallic nanoparticle (TiO_2 and Fe_3O_4) application modifies rhizosphere phosphorus availability and uptake by *Lactuca sativa*. Journal of Agricultural and Food Chemistry 63 (31), 6876–6882.

Zonooz, N.F., Salouti, M., 2011. Extracellular biosynthesis of silver nanoparticles using cell filtrate of *Streptomyces sp.* ERI-3. Scientia Iranica 18 (6), 1631–1635.

Chapter 7

Sustainable valorization of food waste for the biogeneration of nanomaterials

Uroosa Noor[1], Shashi Soni[1], Shalini Purwar[2] and Ena Gupta[1]
[1]*Department of Family and Community Sciences, University of Allahabad, Prayagraj, Uttar Pradesh, India,* [2]*Banda University of Agriculture and Technology, Banda, Uttar Pradesh, India*

7.1 Introduction to food waste and nanomaterials

In the current scenario, nanotechnology with increased investment in it has come into existence due to its positive effects on wide-ranging research areas with multiple applications such as in electronics, medicinal, environmental, communication, energy, material science, and many more. Numerous consumer products containing nanomaterials are also available in the market including cosmetics, antimicrobial and stain-resistant fabrics, electronic elements, cigarette filters, cleaning products, sunscreen, etc. By considering this beneficial and emerging field of interest, it was suggested by researchers to use some waste but worthy by-products of food in the formulation of nanomaterials.

The discarded, defective, or spoiled part of any material that has no further use or is worthless is categorized as waste. It could be of numerous types, but here the discussion is only restricted to food waste (Mahanthesh et al., 2022). According to United Nations Environment Programme (UNEP), "food waste" can be elaborated as food and its inedible parts (related with particular food such as rind, crown, core, bones, pits/stones, etc.), which are intentionally removed from the human food supply chain at various food sectors, namely household, retail, and food service area and finally reached to its end destination such as sewer, litter, or discarded or refused; controlled combustion; landfills; anaerobic or aerobic digestion; or composted. Food waste index report 2021 given by UNFP estimated that in 2019, around 931 million tonnes of food waste was generated globally, which accounts for around 17% of total global food production. This huge quantity of food waste could be perfectly used in the generation of a number of nanomaterials (Zhongming et al., 2021). Food wastes are the potential sources of carbohydrates with other organic compounds and multifunctional groups such as polymeric proteins, which makes it suitable for the biogeneration of environment-friendly nanomaterials (Mahanthesh et al., 2022). Classification of food waste is illustrated in Fig. 7.1.

The research and development of nanoscale materials, measuring ≤ 100 nm, is termed as nanotechnology that includes characterization, manipulation, and fabrication of nanoparticles. Nanoparticles exhibit at least one dimension of 5100 nm. These are formed in the environment through various sources such as natural (volcanic eruptions, forest fire); stationary (incinerators, combustion systems, industrial processes); medical (drug delivery); and mobile (transportation vehicles). These emitted nanoparticles get deposited in the environment such as in air, water and soil, so that their physical and chemical characteristics together with health and environmental impacts are distorted. Varied applications of nanotechnology also include detection and monitoring of environmental pollutants for improving the quality of water, air, and soil (Chen et al., 2011).

These nanomaterials possess exceptional functional properties, which are capable of being used as an active ingredient for industrial purposes including food and agricultural sectors (Bandyopadhyay et al., 2013). The major objective of this article is to elaborate the findings of research done in the field of nanomaterials and utilization of food waste for the formulation of various nanoparticles. This article starts with the overview of nanotechnology and types of food waste and then synthesis of nanomaterials through various techniques defined in the subsections, which is followed by structural characterization and types of nanoparticles synthesized from food waste. Finally, a comprehensive note is given for the application and future perspective of nanoparticles at different sectors.

Green and Sustainable Approaches Using Wastes for the Production of Multifunctional Nanomaterials. DOI: https://doi.org/10.1016/B978-0-443-19183-1.00022-2

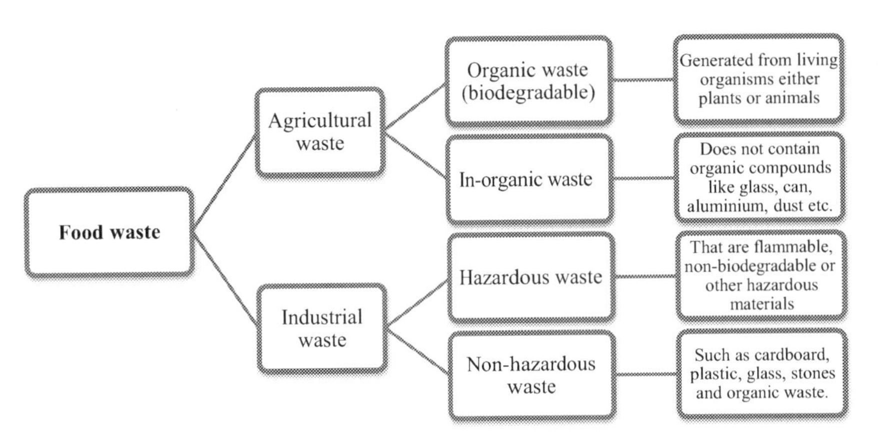

FIGURE 7.1 Schematic diagram of classification of food waste (Abdelbary and Abdelfattah, 2020).

7.2 Elucidation of food waste

Unwanted material discarded after the utilization of valuable parts is known as waste. Food waste is associated with the unwanted use of land, water, and energy and also responsible for the unnecessary generation of greenhouse gases through the production and delivery operation. On the contrary, worldwide nearly one-third of the food production is lost or wasted from farm to fork (Göbel et al., 2015; Raak et al., 2017). In the developed countries, food waste is mainly related to the consumers, due to the awareness of product quality, food wastes also created through farmers, processors, and the retailers and the items that are assumed to be undesired are sorted. Next to developing countries, food waste is related to poor technology, infrastructure and rapid spoilage due to the climatic conditions (Raak et al., 2017). Agricultural, industrial, and urban activities are the major factors for the pollution such as pesticide residue; heavy toxic metals and quality deterioration during storage are the main reasons for the generation of agricultural waste across the globe (Romanovski, 2021). In addition to these, there are many other reasons such as: postproduction, manufacturing, handling, storage, spoilage, and pest infections are also responsible for the production of agricultural waste after harvesting. Then, the processing industries produce food loss by peeling, slicing, and washing, which creates by-products such as peels, pomace, and spent grains. (Ravindran and Jaiswal, 2016).

Agricultural solid waste is mainly generated by activities such as farming, which comprises the generation of agricultural waste till land clearing to harvesting, due to poor network of road for transporting the harvested product resulting in large quantities of waste generation, due to lack of electricity, which produce wastes because of the lack of cold storage, improper techniques of drying or moisture monitoring resulting in the formation of aflatoxin in grains before storage, causes food spoilage, and leads to annual crop losses worldwide, and some of the solid waste is also generated in family level through kitchen garden after the consumption of agricultural products such as fruits and vegetable peels, which leads to microbial spoilage (Adejumo and Adebiyi, 2020). The function of agricultural-based industries is to increase the importance of raw agricultural products through the downstream processing to produce marketable, consumable products that provide profit to producers and engender their income. The upstream and downstream processing generates waste from agricultural industry that adversely affects the air, soil, and water quality. Besides this, some industrial wastes are nontoxic and nonhazardous such as organic waste, for example, sugarcane bagasse, sugar palm waste, wheat straw fiber, soy hull fiber, banana fiber, sugar beet fiber, and corncob are nonhazardous and can be used for other purposes (Ilyas et al., 2019). Agricultural food wastes are high in carbohydrate content in addition to other multifunctional groups and bioactive compounds, which makes agricultural waste an interesting area for research to overcome the environmental hazards through the production of high-value products such as nanomaterials (Mahanthesh et al., 2022). Recent research studies focused on to obtain nontoxic adsorption materials such as nonocelluleose and microcellulose from agricultural waste are deliberated as the major method for the removal of contamination caused by agricultural products without causing any harmful effects on human health and environment (Romanovski, 2021).

7.3 Synthesis of nanomaterials through different techniques

Nanomaterials can be synthesized by different synthesis methods, which are as follows:

1. Biological methods
2. Chemical methods
3. Physical methods

7.3.1 Biological methods

This is the simplest and easy method to generate nanomaterials by natural sources such as plant products, algae, fungi, bacteria, yeast, and viruses, which could be involved in the production of metal nanomaterials. Cellular, biochemical, and molecular mechanisms are the most common methods that facilitate the rate of synthesis and improve the properties of nanoparticles. Biological synthesis of nanoparticles is rapidly growing because it is an eco-friendly and reliable method with the controlled toxicity and with better understanding of biochemical and molecular mechanism of nanoparticle synthesis (Bhardwaj and Naraian, 2021; Chen et al., 2018; Kolahalam et al., 2019). Different biological methods for the synthesis of nanomaterials are as follows:

7.3.1.1 Microbial route for the synthesis of nanomaterials

Microorganisms are the essential microfactories that contain enormous potential such as it is eco-friendly, cost-efficient, evading toxic, hash chemicals, and the high energy requirement for the synthesis of nanomaterials. Due to the presence of some reductases enzyme, microorganisms have the capacity to accumulate and detoxify heavy metals, which in turn reduce the metal salts to metal nanoparticles through a narrow size distribution, consequently less polydispersity. Synthesis of nanomaterial through microorganisms is discussed below:

There are two mechanisms for the synthesis of nanomaterials through microorganisms: one is extracellular and another is intracellular. In extracellular synthesis, microorganisms are culturing for 1–2 days in a rotating shaker under favorable conditions (pH, medium components, temperature, etc.) and after the culturing, centrifuge it to remove the biomass, collected supernatant is used for the synthesis of nanomaterials with the addition of filtered and sterilized metal salts solutions and then incubate. Monitor the synthesis of nanomaterials by observing the change in color in the culture medium, for example, color change to deep brown indicates silver nanoparticle, and color change to ruby red to deep purple indicates gold nanoparticle. After incubation, nanomaterials again centrifuge at various speeds to remove the large particles or medium components and finally centrifuge with a density gradient or at high speed, washed vigorously with water or solvents (ethanol, methanol, etc.), and collect the nanomaterial in the form of bottom pallet. In the intracellular synthesis, microorganisms are cultured for a specific optimal period, centrifuge it to collect the biomass, and wash rigorously through sterile water, then mix it with sterilized-filtered solution of metal salts in sterile water. Monitoring of reaction mixture is similar as in the extracellular synthesis through the inspection of change in color. After the period of incubation, repeated cycle of ultrasonication, washing, and centrifugation is done to remove the biomass; these steps facilitate the breakdown of cell wall and thus allow to release the nanoparticles (Das et al., 2017; Marooufpour et al., 2019; Singh et al., 2016).

7.3.1.1.1 By the use of bacteria

Bacteria are the best aspirant for the production of nanoparticles because it holds the astonishing capacity to reduce the heavy and toxic metal ions into nanoparticles. Interactions of metal–microbes provide an important role along with enchanting applications such as biomineralization, bioremediation, bioleaching, and microbial corrosion. Previous research reported that *Delftia acidovorans* bacterium produced gold nanoparticle through the formation of small nonribosomal peptide, *dilfibactin*. *Rhodopseudomonas capsulate* bacterium also produces gold nanoparticles; besides this, *Morganella morganii* synthesizes the copper nanoparticles, and it was also documented that zero-valent palladium nanoparticles were produced by the bacteria present in Alpine sites (Anwar, 2018).

7.3.1.1.2 By the use of fungi

Nanoparticles are synthesized by fungi by a process known as mycosynthesis, and it comes under myconanotechnology. It was reported that using oyster mushroom fungi is more beneficial as compared with another biological agents such as bacteria because they are rich in tolerance capacity against heavy metals, and fungi are also easy to culture in bunches, which reduces the cost of downstreaming (Bhardwaj et al., 2018). Production of nanomaterial through fungi is one of the popular methods because its biomass is easy to handle and economically affordable. Along with this, it is effective for the production of enzymes in large scale through extracellular synthesis. These nanoparticles are stable for a very long time due to the activity of the enzymes, mainly NADH-reductase. It is observed that protein secretion is higher in fungal cells as compared with bacterial cells. Conversely, it might create genetic manipulation of organisms and lead to contamination (Devatha and Thalla, 2018; Kolahalam et al., 2019).

7.3.1.1.3 By the use of algae

Nanoparticles are produced through algae by the help of multiple applications, which have different characteristics, and these productions are growing rapidly. Some research revealed that *Sargassum wightii* algae are used to produce gold nanoparticles through extracellular synthesis, and 95% production was achieved within 12 hours. Limited research is explored regarding the synthesis of nanoparticles by the use of algae (Devatha and Thalla, 2018; Kolahalam et al., 2019). Two types of nanoparticles can be synthesized by algae: one is organic and another is inorganic. Organic nanopartiles comprise poly-ε-lysine (ε-PL), chitosan (CS), cationic quaternary polyelectrolytes, and quaternary ammonium compounds. These nanoparticles are not stable at high temperature as compared with inorganic ones; these properties of inorganic nanoparticle support it as antimicrobial polymer (Uzair et al., 2020).

7.3.1.2 Plant route for the synthesis of nanomaterial

Undeniably, a number of bacteria, yeast, and fungi have been recognized for the synthesis of nanomaterials, but still it is not feasible because they require extremely expensive medium and maintenance of high aseptic conditions. Therefore, plants and their extracts are investigated as potential biofactories, which are gaining attention for the production of nanoparticles without using any toxic chemicals and are eco-friendly, stable, simple, rapid, and low-cost method because it reduces the metal salt through phytochemicals such as flavones, organic acids, and quinines are present in it, which acts as a stabilizing agent in the formulation of nanoparticles (Kolahalam et al., 2019; Singh et al., 2016). Previous research revealed that many plant varieties such as *Sesbania grandiflora, Tribulus terrestris*, and *Rosmarinus officinalis* were used in the synthesis of gold and silver nanoparticles (Mustapha et al., 2022). Along with these, extracts of green tea leaf, *Terminalia chebula* fruit, Oolong and backtea leaf, banana peel, *Colocasiae sculenta* leaves, sorghum bran, eucalyptus leaf, *Azadirachta indica*, and *Tridax procumbens* had also been used for the preparation of nanoparticles (Devatha and Thalla, 2018).

7.3.2 Physical route for the synthesis of nanomaterials

Physical methods are categorized in two approaches: "top-down" and "bottom-up." In top-down approach, mechanical milling techniques are used by which the large materials are pulverized into small particles. Major disadvantage of this approach is the hardness of finding the desired particle shape and size due to the milling process. In the bottom-up approach, liquid and gaseous phases are used to form larger material by the chemical combination of smaller ions. Nanoparticles synthesized by this method are condensed (Kolahalam et al., 2019). Some of the physical methods are described below:

7.3.2.1 Laser evaporation method

This method is considered in the bottom-up approach used in the formation of magnetic nanopowders. In this method, laser beam is used to evaporate the raw metal oxides from solid or sometimes in liquid surface by irradiation, therefore the formation of nanoparticles transpires outside the evaporation zone through rapid condensation and nucleation as a result of steep temperature gradient. The particle size and magnetic phase can be changed via adjustment in the laser power and composition of the atmosphere in the evaporation chamber (Kolahalam et al., 2019; Rane et al., 2018).

7.3.2.2 RF plasma method

This method requires extremely high temperatures due to high-voltage RF coils, shielded around the evacuated system that heats the metals beyond its evaporation point. In this system, helium gas is passed, resulting in the generation of high temperature in the coils region, which causes the nucleation of metal vapors. After that by the help of diffusion process, it is passed to the colder collector rod and formed the nanoparticles (Károly et al., 2011; Kolahalam et al., 2019).

7.3.3 Chemical route of synthesis of nanomaterials

In the chemical method, various bottom-up approaches are used for the synthesis of nanoparticles. This method is mostly applicable for liquid and gas phases and produces pure and controlled particle-sized nanomaterials. There are so many techniques involved in chemical method such as sol–gel method, coprecipitation, hydrothermal technique, solvothermal, sonochemical, pyrolysis, vapor deposition, microemulsion, microwave-assisted, intercalation, ion-exchange, and reflux, some of which are discussed below:

7.3.3.1 Coprecipitation method

Preparations of nanoparticles through this method are the simplest, common, and widely used method that requires aqueous medium and can produce uniform nanoparticles. In coprecipitation method, a mix of two or more water-soluble salts generally of divalent and trivalent metal ions is used, in which mainly trivalent ions contain the soluble salts. Water-soluble salts are going through the reaction and reduced to form at least one water-insoluble salt for precipitation. This reaction and reducing agent may or may not follow the heat condition but require continuous stirring. The nature of particles prepared by this method is less crystalline, heat energy conceivably used to increase the crystallinity of the particle. The entire process is done under alkaline medium through the addition of common alkaline solutions such as ammonia, sodium hydroxide to maintain the essential pH. The size of nanoparticles revolves around some aspects such as the pH of the solution, ratio of the salt selected, type of base used, and the maintenance of the temperature for reaction medium. At the end by the help of filtration and centrifugation, solvent is extracted out, and the supernatants are further purified and dried and then after doping, diverse unusual earth metals in the ferrites are produced (Kolahalam et al., 2019).

7.3.3.2 Sol–gel method

This method involves low temperature and was initially developed for the production of glass and ceramic materials in which small molecules are used to produce solid materials. This chemical method involves the gradual development of a gel-like diphasic system having both liquid and solid phases and their morphology ranging from the detached particles to constant polymer networks. In the sol–gel method, by the presence of acid or base afterwardpolycondensation, the metal alkoxide solution is hydrolyzed into water and alcohol, after that polycondensation helps to change the liquid phase into gel phase through the elimination of water molecules in the solution and increment in the viscosity of the solution, when all the liquid phase or water molecules are condensed; then the gel phase changes into the powder phase, and further heat is required to produce fine crystalline powder (Kolahalam et al., 2019; Rane et al., 2018).

7.3.3.3 Hydrothermal method

In this method, solutions are exposed to the high temperature and pressure in a sealed chamber where chemical reaction takes place, which leads to the formation of good-quality crystals with controlled composition. Some important solids such as superionic conductors, electronically conducting solids, magnetic materials, luminescence phosphors, chemical sensing oxides, microporous crystals, complex oxide ceramic, and fluorides are prepared with this technique. In the hydrothermal synthesis, homogeneous mixture is prepared by mixing divalent and trivalent transition metal salts in 1:2 ratio and stirring constantly. After that the prepared solution is placed in the sealed vessel commonly known as autoclave, by heating, the pressure of the autoclave increases, which direly raises the solvent temperature above its boiling points; to form different types of nanoparticles, adjustment of time and temperature depends on the type of nanoparticles synthesized (Rane et al., 2018; Kolahalam et al., 2019).

7.3.3.4 Microwave-assisted synthesis

In this method, the materials are directly heated by microwave radiation with portable electric charges using electromagnetic resonance (EMR), where the EMR energy is converted into thermal energy, by frequencies ranging between 1 and 2.5 GHz, which develops the temperature between 100°C and 200°C. By this method, narrow-size small nanoparticles can be produced. Some of the common solids such as ferrites, oxides, selenides, and colloidal metals are synthesized by the microwave synthesis (Kolahalam et al., 2019)

7.4 Types of nanomaterials synthesized from food waste

7.4.1 Metal oxide nanoparticles

Silver nanoparticles (AgNPs) exhibit a great potential to inhibit the growth of wide-ranging bacteria of spoilage and pathogenic origin, thus it is frequently used in the food industries. By the use of agri-food waste, environment-friendly, green synthesis of AgNPs can be done. This method is more sustainable because it reduces environmental pollution, while AgNPs are produced. The flowchart for the synthesis of metal oxide nanoparticles is shown in Fig. 7.1. Uniform and spherical AgNPs of 8.67–4.7 nm diameter were synthesized with good antibacterial activity at the lowest concentration of 0.9 μg/mL against both types of bacteria, which is suitable for using in food and medicine industry. In a related study (El-Desouky et al., 2022), AgNPs were fabricated through biogenic processes with the utilization of coffee waste extract (CWE). AgNPs were stabilized in the CWE solution because it contains many bioactive compounds, which reduces cap. Monodispersed small size around 20 nm as-synthesized AgNPs were produced, which exhibited in-plane dipole plasmon resonance of hexagonal nanoplates. Physical and spectroscopic examinations of nanoparticles confirmed the hexagonal shape and nanoscale characteristics of AgNPs. These are also of superior quality in the means of antibacterial potential against *B. subtilis, Pseudomonas aeruginosa, S. aureus*, and *E. coli*. In the presence of citric acid with a removal efficiency of $\sim$94.6%, as-prepared AgNPs (12 mg) enabled the photodegradation of phenol compounds (20 mL) in less time. Thus, it was concluded that CWE is an environment-friendly extract for the manufacturing of low-cost AgNPs, useful in many areas such as phenol compounds removal, preparing functional nanodevices, and photocatalytic activities. Ahmad and Sharma (2012) also synthesized AgNPs with the use of *Ananas comosus* aqueous extract. Various physical parameters confirmed its crystalline nature along with spherical shape of 12 nm diameter. The different antioxidants present in the *A. comosus* juice were responsible for the simple reaction process of quite stable (even upto 4 months of incubation) AgNPs formation. Basavegowda and Lee (2013) also used aqueous extract of Satsuma mandarin peel for the preparation of AgNPs. When the peel was exposed to the silver ion, reduction reaction was started, which resulted into the formation of silver nanoparticles at 5–20 nm range of size. This study revealed the viability of fruit peel waste for the synthesis of AgNPs. Peel extract of *Nephelium lappaceum* L. (Rambutan) was used to obtain AgNPs. At 370 and 495 nm, two peaks were noticed on UV-Vis spectrophotometer and its triangular, hexagonal, and truncated triangular shapes of 132.6 ± 42 nm diameter were confirmed through TEM (transmission electron microscopy) technology. As well as AgNPs face-centered cubic symmetry along with crystalline nature was accomplished through X-ray diffraction (XRD) and selected area electron diffraction (SAED) analysis (Kumar et al., 2015). Under the optimum conditions such as 45°C temperature, pH 9, 1 mm $AgNO_3$ concentration, and 40 g/L (4% w/v) of extract concentration, AgNPs were prepared with a biosynthetic approach through oak fruit hull extract. AgNPs of an average of 40 nm size and spherical shape were prepared. Cytotoxicity potential of AgNPs was checked against MCF-7 (human breast cancer cell line) cell line with a mild positive result, and also long-term stability was shown through zeta potential analysis (Heydari and Rashidipour, 2015). In the meantime, Velmurugan et al. (2015) also developed cost-effective and eco-friendly AgNPs of 10–50 nm size from the extract of defatted cashew nut shell (CNS). Starch of this CNS was helpful in reducing silver ion particles. On comparing with commercially available AgNPs, it was shown that both are identical. In another study, Velu et al. (2015) prepared AgNPs of 20–50 nm diameter by reacting $AgNO_3$ solution with the aqueous extract of groundnut (*Arachis hypogaea)* for 1 hour at 28°C. The responsible elements for this reaction were carboxyl group, amino acids, and polyphenolic residues, and the new silver nanoparticles were exhibiting face-centered cubic structure with a crystalline nature. They also showed larvicidal activity against larvae of malaria (*Anopheles stephensi*) and dengue (*Aedes aegypti*) vectors. In a related study, seed shell of *Cola Nitida* was applied to prepare silver nanoparticles of 5–40 nm size with a spherical shape. The responsible mechanism of action behind this reaction was catalytic activity of alkaloids and proteins in the aqueous extract. Its antimicrobial activity was also checked against three strains *E. coli, Klebsiella granulomatis*, and *P. aeruginosa* with an MIC value of 50 μg/mL. Cocoa pod husk was used for the green synthesis of spherical AgNPs of 4–32 nm. Proteins and phenolics were envisaged as responsible elements for the development of nanoparticles. Good antibacterial activity (42.9%–100%) was observed when used synergistically altogether with cefuroxime and ampicillin against *E. coli* and *K. pneumonia*. When this nanoparticle was used as additive in emulsion paint, it totally inhibited the growth of *E. coli, Streptococcus pyogenes, S. aureus, A. flavus, A. niger, A. fumigates, P. aeruginosa*, and *K. pneumonia* (Lateef et al., 2016; Patil et al., 2022).

Dubey et al. (2010) investigated the green synthesis of gold nanoparticles (AuNPs) and AgNPs through the reduction process of chloroauric acid and silver nitrate solutions isolated from *Tanacetum vulgare* plant. Formation of these two nanoparticles was confirmed through the surface plasmon spectra, and peaks of absorbance were noticed at 546 and 452 nm. In this queue under optimum conditions, starch-supported green silver nanoparticles were prepared with the use of aqueous extract of orange peel waste generated from the food processing industry. Its reductive potential was predicted as useful due to the presence of flavonoids, carotenoids, ascorbic acid, pectins, and other flavones. On surface plasmon resonance at 404 nm. a peak was noticed within a narrow range spectrum of 3–12 nm. Furthermore, antilipid peroxidation, free-radical scavenging potential, antimicrobial, and cytocompatibility tests were done with positive results on goat liver homogenate, DPPH scavenging test, synergistic effect with rifampicin against *Bacillus subtilis* MTCC 736 and THP-1 cell-line (human leukemic monocytic cell line), respectively.

A huge quantity of diverse food source wastes are generated from agriculture and horticulture processes. The renewable and waste utilization approach has led to the opportunity to produce eco-friendly gold nanoparticles from this food waste. Chums-ard et al. (2019) biologically synthesized gold nanoparticles (AuNPs) from watermelon (*Citrullis lanatus var*) waste, both green and red part. With the help of numerous advanced characterization techniques, AuNPs were evaluated, and further antibacterial activity of nanoparticles was checked through the Kirby–Bauer sensitivity method. Results of the study showed that 100 nm–2.5 μm size of AuNPs with different shapes as green watermelon extract exhibited triangular-shaped nanoparticle, and on the other hand, red watermelon extract exhibited hexagonal and spherical shape particles with similar antibacterial properties against *Staphylococcus epidermidis* and *E. coli*. Peel of mango was utilized as raw material for the formulation of AuNPs of 6.03 ± 2.77 to 18.01 ± 3.67 nm size that was biocompatible at a higher concentration of 160 μg/mL with wI-38 (normal human fetal lung fibroblast cells) and CV-1 (African green monkey kidney normal cells) (Yang et al., 2014). Several bioactive compounds such as anthocyanidin, epicatechin, catechin, proanthocyanidins, and condensed tannins present in the grape processing wastes are responsible for the successful production of single-step AuNPs of 20–50 nm size at room temperature within 5 minutes. Maximal absorbance of stable AuNPs was recorded at 536–538 nm (Krishnaswamy et al., 2014).

In the group of metal oxide nanoparticles, zinc oxide nanoparticles are also synthesized through various biogenic methods by the use of food wastes generated from processing industries. In this manner, Yuvakkumar et al. (2014) synthesized biogenic ZnO NPs from *Nephelium lappaceum* L. (rambutan) peels. The action of the mechanism was reacting $Zn(NO_3)_2.6H_2O$ with peel extract of rambutan at 80°C for 2h for the formation of zinc-ellagate complex. This complex was then dried in oven at 40°C for 8h, then calcified in muffle furnace at 450°C to obtain the nanocrystals of zinc oxide due to the presence of polyphenols, which was characterized by several physical techniques. Remarkable antibacterial activity was observed against *S. aureus* (23 mm) and *E. coli* (18.5 mm) in cotton fabric coated with ZnONPs. Karnan and Selvakumar (2016) also developed ZnONPs from the rambutan peels extract as a natural ligation agent through the entire process. Moreover, tea solid waste was mixed with zinc acetate at 80°C to formulate ZnONPs. This mixture was then left overnight for drying at 70°C. Then obtained green ZnONPs was confirmed through several advanced parameters. Results showed an average 19.5 nm of particles with good stability (−41.3) mV through zeta potential analysis (Hassan et al., 2019).

As the zinc oxide nanoparticles were obtained in the same manner, NiNPs are also developed through the nickel ellagate complex formation through the use of rambutan peels extract. Polyphenols present in the extract were responsible for the generation of nickel oxide nanoparticles through $Ni(NO_3)_2.6H_2O$. Strong antibacterial activity was observed against *S. aureus* (35 mm) and *E. coli* (25 mm) in the cotton fabric coated with NiO nanocrystals (Yuvakkumar et al., 2014).

At 60°C, 80 mL of 1 mm palladium acetate Pd $(OAc)_2$ was reacted with 10 mL of aqueous extract of *Annona squamosa* for 4h for the synthesis of palladium nanoparticles (PdNPs). Physical characterization of these NPs revealed the spherical shape of nanoparticles with an average size of 100 nm. The functional group OH present in the extract showed that it was the reason for capping of synthesized nanoparticles (Roopan et al., 2012). In a similar study by Lakshmipathy et al. (2015), peels of watermelon were used as reducing and capping agent in the synthesized palladium nanoparticles with an average size of 96.4 nm. Another study done by Bankar et al. (2010) synthesized PdNPs by using banana peel extract to reduce Pd chloride. Average sizes of nanoparticles were noted as 50 nm with a crucial role of amine, hydroxyl, and carboxyl groups during the whole synthesis process. In addition, Coccia et al. (2012) and Nadagouda and Varma (2008) also synthesized spherical PdNPs of 5–100 nm size using lignin and commercially available tea and coffee waste extracts, respectively.

By the use of *Punica granatum* peel extract, spherical crystalline platinum nanoparticles (PtNPs) with an average size of 16–23 nm diameter were developed by Dauthal and Mukhopadhyay (2015). High stability of the nanoparticles was indicated by negative zeta potential. Results of the Fourier transform infrared spectroscopy (FTIR) analysis showed the major role of hydroxyl and carbonyl groups in the form of polyphenolic compounds present in the peel for the capping and stabilization of PtNPs.

Rice straw, an agricultural food waste, has been used as a lignocellulosic solid waste to prepare titanium dioxide nanoparticles TiO_2 NPs through sol–gel method (Ramimoghadam et al., 2014). In a related study, *Annona squmosa* L. peel was used as an agent for the formulation of spherical shape polydispersed TiO_2 NPs with an average size of 23 ± 2 nm diameter (Roopan et al., 2012).

7.4.2 Other nanoparticles

Cellulose is an extensively available biomass present in nature, which can be isolated from regular filament, due to their structural and dynamic connections, they allow to form several types of nanostructures cellulose, which is termed as nanocrystalline cellulose (NCC). From the literature survey it is detected that cellulose nanoparticles are separated and used for many purposes in different fields; besides this, nanocrystalline cellulose can be used as a filling agent to remove the barrier and improve the mechanical properties of biocomposites. NCCs comprise an important role in various fields such as paper products, fibers, energy crops, biofuel, consumables/tablets, building materials for sprays and insulations, thin transparent films for movies, cellophane, dry and wet compound as supportive polymer, computer parts, loudspeaker films, tobacco filter additive, antimicrobial films, coating, packaging, as a flavor preservative agent, as stabilizer, and also used in medical, pharmaceutical, and cosmetic fields (Ukkund et al., 2018). Ilyas et al. (2019) stated that NCC contains numerous prominent chemical, electrical, and ocular properties because of its needle-like shape; high surface area, high aspect ratio (length/diameter), high crystallinity, high strength and stiffness, nanoscale size along with this low density and highly negative charge that leads it to a distinctive behavior in solution. Due to their high chemical reactivity of surface making it personalized for innumerable applications, it also allows high-temperature application due to its heat stability. Additionally, they also contains high surface OH group that gives the active site to hydrogen bonds by interlocking with nonpolar matrices. Isolation of NCC widely needs hydrolysis chemical synthesis, but according to the type of raw material and degree of processing, physical, chemical, enzymatic, and ionic pretreatments are also required before the synthesis of nanocrystalline cellulose. A research conducted by Johar et al. (2012) isolated nanocrystallized cellulose and nanocellulose fiber from rice husk through the alkali (NaOH) and bleaching treatment ($NaCl_2O$) followed by acid hydrolysis (H_2SO_4) respectively, which is shown in Fig. 7.2. Morphological characterization was done by using SEM, TEM, FTIR, and XRD, resulting in the successful treatment, which would significantly increase the crystallinity of NCC. Arnata et al. (2020) conducted a study in which nanocrystalline cellulose was isolated from sago fronds, which are considered as waste material by using acid hydrolysis (sulfuric acid) with some variation in hydrolysis time and cationic modification in the surface of NNC. The samples underwent through Fourier transform infrared spectroscopy (FTIR), scanning electron microscopy (SEM), and XRD and showed that lignin content, hemicellulose, and fiber dimensions were decreased, followed by increase in cellulose content and in degree of crystallinity during the delignification, bleaching, and hydrolysis process. Result indicated that the longer the hydrolysis time, thermal stability, dimension, and yield of nanocrystalline cellulose were decreased; additionally, the degree of crystallinity and surface charge were increased, which gives it moderate stability, and were potential as wider-scaled material processing application (Fig. 7.3).

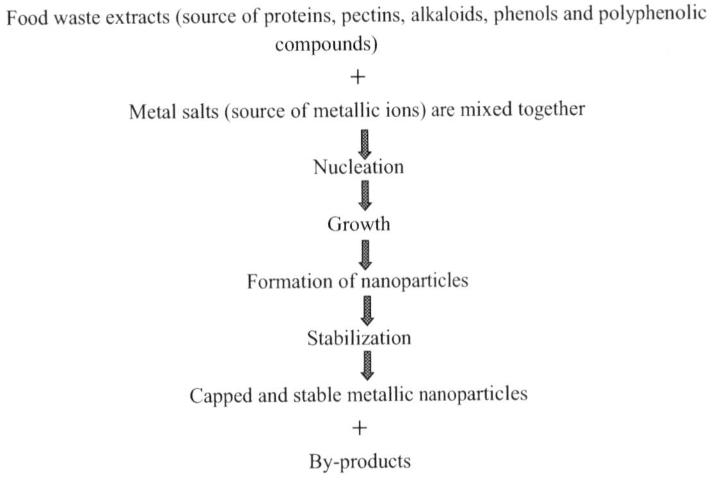

Food waste extracts (source of proteins, pectins, alkaloids, phenols and polyphenolic compounds)
+
Metal salts (source of metallic ions) are mixed together
↓
Nucleation
↓
Growth
↓
Formation of nanoparticles
↓
Stabilization
↓
Capped and stable metallic nanoparticles
+
By-products

FIGURE 7.2 Flowchart for the mechanism of action for the synthesis of metal oxide nanoparticles. *Reproduced from Ghosh, P.R., Fawcett, D., Sharma, S. B., & Poinern, G. E. 2017. Production of high-value nanoparticles via biogenic processes using aquacultural and horticultural food waste. Materials, 10 (8), 852.*

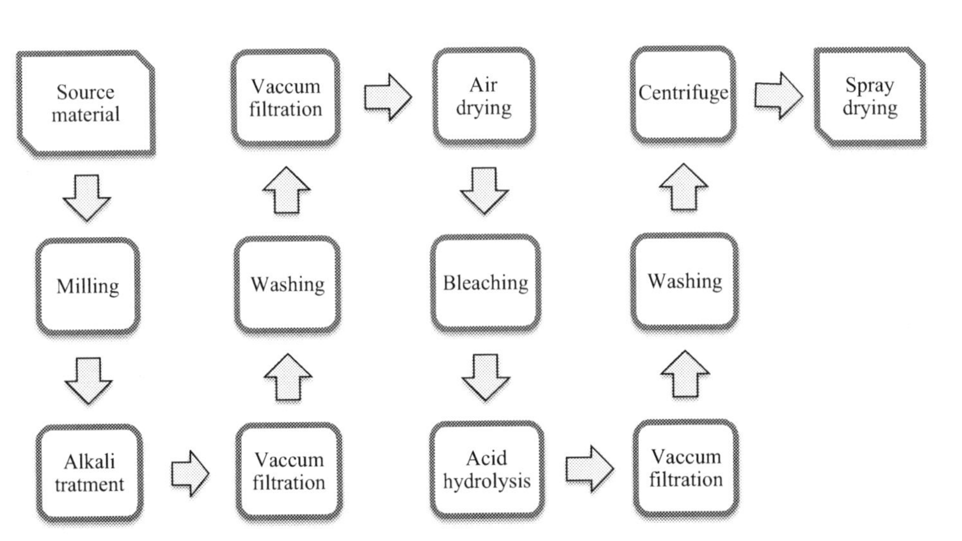

FIGURE 7.3 Flowchart for the extraction of nanocrystalline cellulose.

7.5 Applications and future perspective of nanomaterials in various areas

Nanoparticles are synthesized through various physical, chemical, and conventional methods; these are the novel particles to be used in various areas at commercial or industrial levels. Food waste generated from the food processing industries or household wastes are the major source of environmental pollution. Thus, the approach to convert this waste into value-added products has led to the formulation of nanoparticles. This has been reported as a very important antimicrobial agent, photocatalysts, pharmaceuticals, food packaging applications, and biosensors and as cancer therapy agent (Ghosh et al., 2017).

Through the green-chemistry approach, biogenic synthesized nanoparticles achieved great interest in the conversion of food waste into novel nanoparticles. Also its crucial role could be seen in the biomedicinal field due to less utilization of chemicals and ultimately minimal hazards (Elsaesser and Howard, 2012). In the area of cosmetics, pharmaceuticals, and biomedical products, metal oxide nanoparticles such as AgNPs, AuNPs, PdNPs, and PtNPs play a crucial role. Majorly AuNPs are applied in the biomedical field, diagnosis of the disease, and pharmaceutical sciences. In contrast, AgNPs are commonly used as antibacterial and antiinflammatory agents in the field of biomedicine for improving wound healing capacity. Other than these two, various other metal oxides are also used for wastewater treatment and remediation with the mechanism of action including adsorption and removal of heavy metals and other pollutants such as pathogens. In future, many more research studies could be done in the synthesis and applications of nanoparticles with easier techniques along with cost-effective and eco-friendly approaches (Ghosh et al., 2017).

7.6 Conclusion

Generation of food waste is inevitable during agricultural practices and food processing, which may lead to the generation of greenhouse gases and water contamination through food waste decomposition. These food wastes are the promising source for the synthesis of nanomaterials through various techniques, which are used to reduce these wastes; consequently, it may cause many hazardous effects when it is left untreated. This chapter emphasizes the utilization of waste generated through the agricultural and industrial food processing techniques. Several methods such as biological, physical, and chemical are used to synthesize value-added nanomaterials such as metal oxide nanomaterials (Ag, Au, Pd, Pt, ZnO, NiO_2, TiO_2, etc.) and nanocrystalline cellulose. Each of these techniques used for the preparation of nanomaterials has its own advantage and has many applications in the pharmaceutical, biomedicinal, cosmetics industries, and for waste water treatment. Many further studies may be accomplished in this field.

Acknowledgment

None declared.

Disclosure statement

The authors have declared no conflict of interest.

References

Abdelbary, S., Abdelfattah, H., 2020. Modern trends in uses of different wastes to produce nanoparticles and their environmental applications. Nanotechnology and the Environment . Available from: https://doi.org/10.5772/intechopen.93315.

Adejumo, I.O., Adebiyi, O.A., 2020. Agricultural solid wastes: Causes, effects, and effective management. Strategies of Sustainable Solid Waste Management 8.

Ahmad, N., Sharma, S., 2012. Green synthesis of silver nanoparticles using extracts of *Ananas comosus*. Chem 2, 141–147.

Anwar, S.H., 2018. A brief review on nanoparticles: Types of platforms, biological synthesis and applications. Research & Reviews: Journal of Material Sciences 6, 109–116.

Arnata, I.W., Suprihatin, S., Fahma, F., Richana, N., Sunarti, T.C., 2020. Cationic modification of nanocrystalline cellulose from sago fronds. Cellulose 27 (6), 3121–3141.

Bandyopadhyay, S., Peralta-Videa, J.R., Gardea-Torresdey, J.L., 2013. Advanced analytical techniques for the measurement of nanomaterials in food and agricultural samples: A review. Environmental Engineering Science 30 (3), 118–125.

Bankar, A., Joshi, B., Kumar, A.R., Zinjarde, S., 2010. Banana peel extract mediated novel route for the synthesis of palladium nanoparticles. Materials Letters 64 (18), 1951–1953.

Basavegowda, N., Lee, Y.R., 2013. Synthesis of silver nanoparticles using *Satsuma mandarin* (Citrus unshiu) peel extract: A novel approach towards waste utilization. Materials Letters 109, 31–33.

Bhardwaj, A.K., Naraian, R., 2021. Cyanobacteria as biochemical energy source for the synthesis of inorganic nanoparticles, mechanism and potential applications: A review. 3 Biotech 11 (10), 445.

Bhardwaj, A.K., Shukla, A., Maurya, S., Singh, S.C., Uttam, K.N., Sundaram, S., et al., 2018. Direct sunlight enabled photo-biochemical synthesis of silver nanoparticles and their bactericidal efficacy: Photon energy as key for size and distribution control. Journal of Photochemistry and Photobiology B: Biology 188, 42–49.

Chen, Z., Yadghar, A.M., Zhao, L., Mi, Z., 2011. A review of environmental effects and management of nanomaterials. Toxicological & Environmental Chemistry 93 (6), 1227–1250.

Chen, Y., Fan, Z., Zhang, Z., Niu, W., Li, C., Yang, N., et al., 2018. Two-dimensional metal nanomaterials: Synthesis, properties, and applications. Chemical Reviews 118 (13), 6409–6455.

Chums-ard, W., Fawcett, D., Fung, C.C., Poinern, G.E.J., 2019. Biogenic synthesis of gold nanoparticles from waste watermelon and their antibacterial activity against *Escherichia coli* and *Staphylococcus epidermidis*. International Journal of Research in Medical Sciences 7 (7), 2499–2505.

Coccia, F., Tonucci, L., Bosco, D., Bressan, M., d'Alessandro, N., 2012. One-pot synthesis of lignin-stabilised platinum and palladium nanoparticles and their catalytic behaviour in oxidation and reduction reactions. Green Chemistry 14 (4), 1073–1078.

Das, R.K., Pachapur, V.L., Lonappan, L., Naghdi, M., Pulicharla, R., Maiti, S., et al., 2017. Biological synthesis of metallic nanoparticles: Plants, animals and microbial aspects. Nanotechnology for Environmental Engineering 2 (1), 1–21.

Dauthal, P., Mukhopadhyay, M., 2015. Biofabrication, characterization, and possible bio-reduction mechanism of platinum nanoparticles mediated by agro-industrial waste and their catalytic activity. Journal of Industrial and Engineering Chemistry 22, 185–191.

Devatha, C.P., Thalla, A.K., 2018. Green synthesis of nanomaterials. Synthesis of inorganic nanomaterials. Woodhead Publishing, pp. 169–184.

Dubey, S.P., Lahtinen, M., Sillanpää, M., 2010. Tansy fruit mediated greener synthesis of silver and gold nanoparticles. Process Biochemistry 45 (7), 1065–1071.

El-Desouky, N., Shoueir, K., El-Mehasseb, I., El-Kemary, M., 2022. Synthesis of silver nanoparticles using bio valorization coffee waste extract: Photocatalytic flow-rate performance, antibacterial activity, and electrochemical investigation. Biomass Conversion and Biorefinery 1–15.

Elsaesser, A., Howard, C.V., 2012. Toxicology of nanoparticles. Advanced Drug Delivery Reviews 64 (2), 129–137.

Ghosh, P.R., Fawcett, D., Sharma, S.B., Poinern, G.E., 2017. Production of high-value nanoparticles via biogenic processes using aquacultural and horticultural food waste. Materials 10 (8), 852.

Göbel, C., Langen, N., Blumenthal, A., Teitscheid, P., Ritter, G., 2015. Cutting food waste through cooperation along the food supply chain. Sustainability 7 (2), 1429–1445.

Hassan, S.S., Abdel-Shafy, H.I., Mansour, M.S., 2019. Removal of pharmaceutical compounds from urine via chemical coagulation by green synthesized ZnO-nanoparticles followed by microfiltration for safe reuse. Arabian Journal of Chemistry 12 (8), 4074–4083.

Heydari, R., Rashidipour, M., 2015. Green synthesis of silver nanoparticles using extract of oak fruit hull (Jaft): Synthesis and in vitro cytotoxic effect on MCF-7 cells. International Journal of Breast Cancer 2015, 1–7.

Ilyas, R.A., Sapuan, S.M., Ibrahim, R., Atikah, M.S.N., Atiqah, A., Ansari, M.N.M., et al., 2019. Production, processes and modification of nanocrystalline cellulose from agro-waste: A review. Nanocrystalline Materials 3–32.

Johar, N., Ahmad, I., Dufresne, A., 2012. Extraction, preparation and characterization of cellulose fibres and nanocrystals from rice husk. Industrial Crops and Products 37 (1), 93–99.

Karnan, T., Selvakumar, S.A.S., 2016. Biosynthesis of ZnO nanoparticles using rambutan (*Nephelium lappaceum* L.) peel extract and their photocatalytic activity on methyl orange dye. Journal of Molecular Structure 1125, 358–365.

Károly, Z., Mohai, I., Klébert, S., Keszler, A., Sajó, I.E., Szépvölgyi, J., 2011. Synthesis of SiC powder by RF plasma technique. Powder Technology 214 (3), 300–305.

Kolahalam, L.A., Viswanath, I.K., Diwakar, B.S., Govindh, B., Reddy, V., Murthy, Y.L.N., 2019. Review on nanomaterials: Synthesis and applications. Materials Today: Proceedings 18, 2182–2190.

Krishnaswamy, K., Vali, H., Orsat, V., 2014. Value-adding to grape waste: Green synthesis of gold nanoparticles. Journal of Food Engineering 142, 210–220.

Kumar, B., Smita, K., Cumbal, L., Angulo, Y., 2015. Fabrication of silver nanoplates using *Nephelium lappaceum* (Rambutan) peel: A sustainable approach. Journal of Molecular Liquids 211, 476–480.

Lakshmipathy, R., Palakshi Reddy, B., Sarada, N.C., Chidambaram, K., Khadeer Pasha, S.K., 2015. Watermelon rind-mediated green synthesis of noble palladium nanoparticles: Catalytic application. Applied Nanoscience 5 (2), 223–228.

Lateef, A., Azeez, M.A., Asafa, T.B., Yekeen, T.A., Akinboro, A., Oladipo, I.C., et al., 2016b. Cocoa pod husk extract-mediated biosynthesis of silver nanoparticles: Its antimicrobial, antioxidant and larvicidal activities. Journal of Nanostructure in Chemistry 6 (2), 159–169.

Mahanthesh, A.B., Haldar, S., Banerjee, S., 2022. Biogeneration of valuable nanomaterials from food and other wastes. Biotechnology for Zero Waste: Emerging Waste Management Techniques 361–368.

Marooufpour, N., Alizadeh, M., Hatami, M., Asgari Lajayer, B., 2019. Biological synthesis of nanoparticles by different groups of bacteria. Microbial nanobionics. Springer, Cham, pp. 63–85.

Mustapha, T., Misni, N., Ithnin, N.R., Daskum, A.M., Unyah, N.Z., 2022. A review on plants and microorganisms mediated synthesis of silver nanoparticles, role of plants metabolites and applications. International Journal of Environmental Research and Public Health 19 (2), 674.

Nadagouda, M.N., Varma, R.S., 2008. Green synthesis of silver and palladium nanoparticles at room temperature using coffee and tea extract. Green Chemistry 10 (8), 859–862.

Patil, S.P., Chaudhari, R.Y., Nemade, M.S., 2022. *Azadirachta indica* leaves mediated green synthesis of metal oxide nanoparticles: A review. Talanta Open 100083.

Raak, N., Symmank, C., Zahn, S., Aschemann-Witzel, J., Rohm, H., 2017. Processing-and product-related causes for food waste and implications for the food supply chain. Waste Management 61, 461–472.

Ramimoghadam, D., Bagheri, S., Abd Hamid, S.B., 2014. Biotemplated synthesis of anatase titanium dioxide nanoparticles via lignocellulosic waste material. BioMed Research International 2014.

Rane, A.V., Kanny, K., Abitha, V.K., Thomas, S., 2018. Methods for synthesis of nanoparticles and fabrication of nanocomposites. Synthesis of inorganic nanomaterials. Woodhead publishing, pp. 121–139.

Ravindran, R., Jaiswal, A.K., 2016. Exploitation of food industry waste for high-value products. Trends in Biotechnology 34 (1), 58–69.

Romanovski, V., 2021. Agricultural waste based-nanomaterials: Green technology for water purification. Aquananotechnology. Elsevier, pp. 577–595.

Roopan, S.M., Bharathi, A., Kumar, R., Khanna, V.G., Prabhakarn, A., 2012. Acaricidal, insecticidal, and larvicidal efficacy of aqueous extract of *Annona squamosa* L peel as biomaterial for the reduction of palladium salts into nanoparticles. Colloids and Surfaces B: Biointerfaces 92, 209–212.

Singh, P., Kim, Y.J., Zhang, D., Yang, D.C., 2016. Biological synthesis of nanoparticles from plants and microorganisms. Trends in Biotechnology 34 (7), 588–599.

Ukkund, S.J., Taqui, S.N., Mayya, D.S., Prasad, P., 2018. Nanocrystalline cellulose from agricultural waste: An overview. International Journal of Nanoparticles 10 (4), 284–297.

Uzair, B., Liaqat, A., Iqbal, H., Menaa, B., Razzaq, A., Thiripuranathar, G., et al., 2020. Green and cost-effective synthesis of metallic nanoparticles by algae: Safe methods for translational medicine. Bioengineering 7 (4), 129.

Velmurugan, P., Park, J.H., Lee, S.M., Jang, J.S., Yi, Y.J., Han, S.S., et al., 2015. Reduction of silver (I) using defatted cashew nut shell starch and its structural comparison with commercial product. Carbohydrate Polymers 133, 39–45.

Velu, K., Elumalai, D., Hemalatha, P., Janaki, A., Babu, M., Hemavathi, M., et al., 2015. Evaluation of silver nanoparticles toxicity of *Arachis hypogaea* peel extracts and its larvicidal activity against malaria and dengue vectors. Environmental Science and Pollution Research 22 (22), 17769–17779.

Yang, N., WeiHong, L., Hao, L., 2014. Biosynthesis of Au nanoparticles using agricultural waste mango peel extract and its in vitro cytotoxic effect on two normal cells. Materials Letters 134, 67–70.

Yuvakkumar, R., Suresh, J., Nathanael, A.J., Sundrarajan, M., Hong, S.I., 2014. Rambutan (*Nephelium lappaceum* L.) peel extract assisted biomimetic synthesis of nickel oxide nanocrystals. Materials Letters 128, 170–174.

Zhongming, Z., Linong, L., Xiaona, Y., Wangqiang, Z., Wei, L., 2021. UNEP food waste index report 2021. UNEP: Nairobi, Kenya.

Chapter 8

Industrial wastes and their suitability for the synthesis of nanomaterials

Shikha Baghel Chauhan[1], Shikha Saxena[1] and Abhishek Kumar Bhardwaj[2]
[1]*Amity Institute of Pharmacy, Amity University, Noida, Uttar Pradesh, India,* [2]*Department of Environmental Science, Amity School of Life Sciences, Amity University, Gwalior, Madhya Pradesh, India*

8.1 Introduction

Numerous nanoparticles have been created over the past several decades using a variety of techniques, and they have been used to enhance technology for environmental applications such as water treatment, the detection of persistent contaminants, soil/water remediation, and many others. The study of alternate inputs for nanoparticle manufacturing as well as the use of green synthesis methodologies are driven by the area of materials science and engineering's growing interest in enhancing the efficiency of the processes involved in their manufacturing (Bhardwaj and Naraian 2020; Bhardwaj et al., 2018). Here, we begin by outlining the fundamental principles of creating nanoparticles from plastic, electrical, and industrial waste (UNEP, 2020).

Annual production of billion tons of hazardous waste is a result of industrialization and urbanization producing a severe environmental strain. Due to the earth's limited resources, the scarcity of land for waste disposal, and the negative effects on the ecosystem, it is imperative to create effective treatment technologies and conserve precious resources. There are significant amounts of secondary raw materials found in the trash produced by numerous industrial sectors that may be utilized in a variety of ways. Waste reduction, separation, and recycling procedures should be created in order to protect the environment. Industrial wastes may collect in people's bodies and lead to major illnesses and physical problems because they include heavy metals, toxic compounds, and dyes (Zhang et al., 2013). The most dangerous of them were thought to be heavy metals. For instance, short-term exposure to high doses of chromium can irritate the skin at the point of contact, harm the nasal mucosa, and result in skin ulcers. The mining, electroplating, leather tanning, metal processing, textile, dyeing, and steel fabrication sectors are some of the causes of chromium contamination. Other chromium-using industries include the paint pigment and paper industries. Steel waste, battery waste, and gasoline all contribute to lead contamination, which damages the kidneys, brain, muscles, and bones as well as causing hypertension and renal and brain damage. An efficient method of trash treatment and recycling that adheres to the ideas of minimal waste creation and riches from waste is nanomaterial synthesis from industrial waste (Koolivand et al., 2017).

The traditional techniques of treating electroplating waste currently include landfilling, manufacturing bricks, stabilizing with cement matrix, etc. Similar to this, fly ash is a complex and common solid waste that is created after the burning of coal in thermal power plants. Another example is red mud, which is made from small particles of iron, aluminum, silicon, titanium oxides, and hydroxides and is formed when alumina is made from bauxite ores. With Si, Al, Mn, and Cr only being utilized in very small quantities, calcium is the main ingredient used in the manufacturing of paper. If industrial waste is not properly processed, it might affect both the ecosystem and humans. Traditional techniques including landfilling, incineration, and stabilization are currently used to handle or dispose of the industrial waste. All of these technologies, meanwhile, have certain drawbacks. For instance, the leakage of organics and heavy metals from landfills has the potential to contaminate soil and groundwater; garbage incineration merely decreases the amount of waste that pollutes the air; and costly waste stabilization costs finding a way to utilize industrial waste is urgently needed in accordance with the "3R" criteria (Reduce, Reuse, and Recycle). In the long run, landfilling is not an environmentally friendly approach since it can result in significant secondary contamination from metal leaching and

Green and Sustainable Approaches Using Wastes for the Production of Multifunctional Nanomaterials. DOI: https://doi.org/10.1016/B978-0-443-19183-1.00016-7

resource loss. So, maximizing resource recovery and economically reusing resources is the promising alternative for industrial waste treatment. Numerous studies have been conducted to find ways to turn industrial waste materials into useful goods as ceramic, pigment, catalysts, etc. (Soliman and Moustafa 2020).

Due to its huge surface area and unique features, the creation of nanomaterials from trash and their use in the removal of pollutants have recently attracted attention. A review of the literature shows that more people are starting to take an interest in this topic (Bhardwaj et al., 2021). In addition to resolving issues with waste disposal, the use of waste-derived nanomaterials for industrial applications in pollution removal aids in regulating the exploitation of the natural resources. Additionally, replacing virgin resources with waste-derived materials improves the process economics, as valuable materials are used in the production process. As a result, there is growing interest in the use of waste-derived nanoparticles as an effective and affordable material for removing contaminants from the environment (Biswas et al., 2020).

This chapter details important facets of waste identification as a practical contribution to the treatment and recovery of metal- and carbon-based nanoparticles. In our subsequent discussion, we highlight potential conclusions drawn from the many control frameworks involved in the synthesis of nanoparticles. Then, we give some instances of waste-derived nanoparticles that were used in a proof-of-concept experiment to show how certain technologies may be used to improve the safety and quality of drinking water. We wrap off by going through current issues from the toxicological and existence perspectives that need to be taken into account before waste-derived nanoparticle production is scaled up and put into use.

8.1.1 Waste as starting materials for the production of nanoparticles

Depending on the source of emission, highly hazardous waste products that are generated in significant numbers can be separated into wastes produced by industry and wastes produced by consumers (Bhardwaj et al., 2022). Wastes generated by people and by businesses alike can be utilized as raw materials in processes that produce commodities with additional value. Significant sources of carbon, lead, and palladium in the manufacturing industries include large batteries, rubber tyres, wastewater, and biosolids. Massive amounts of plastic and electronic garbage are produced by consumer products that have reached the end of their useful life. E-waste contains a sizable amount of recoverable valuable and semiprecious metals (Soliman and Moustafa, 2020).

8.2 Industrial waste

Batteries, tires, wastewater, and sludge have all been investigated as possible low-cost and widely accessible introductory materials for the production of nanoparticles up to this point (Samaddar et al., 2018).

8.2.1 Types of industrial waste

8.2.1.1 Biodegradable waste

The term "biodegradable wastes" refers to industrial wastes that can be converted by the activity of certain microbes into nonpoisonous substances. Even household rubbish can be compared to them. These wastes are produced by the dairy, textile, and food processing sectors, among others. Wheat, leather, and paper are a few examples. They don't have a harmful nature and don't need any particular care either. Combustion and other methods are used in their treatment.

8.2.1.2 Nonbiodegradable waste

Nonbiodegradable waste cannot break down into nontoxic chemicals. Plastics and fly ash are a few examples. They are produced by factories that produce iron and steel, fertilizer, chemicals, pharmaceuticals, and dyes. According to the estimates, between 11% and 15% of all industrial wastes are nonbiodegradable and toxic, and the pace of growth in this waste category only continues to rise. These wastes can't be simply broken down and made safer.

For instance, slag and fly ash produced as waste by the steel industries are used in the cement sector. Other techniques used for hazardous trash include landfilling and incineration.

8.2.1.2.1 Lead batteries

China provides greater than 50 million units of used lead batteries per year; given the toxicological consequences of lead exposure and its ability for bioaccumulation, recycling these batteries is an ethical course of action that prevents

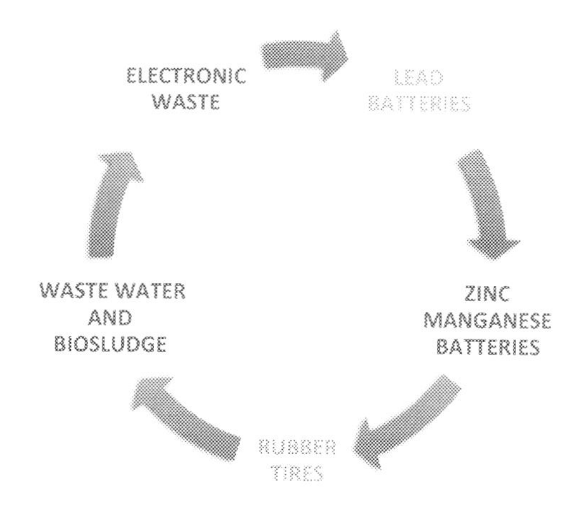

FIGURE 8.1 Types of industrial waste.

environmental and public health issues (Zhou et al., 2017). Metal oxide, dioxide, and sulfate forms, which are present in lead pastes, are insoluble and challenging to recover and repurpose. However, lead may be extracted from used lead-acid battery paste using pyrometallurgical techniques. For instance, Zhou et al. (2017) developed a synthesis method to create Pb nanopowder through the pyrolytic pathway from a model of wasted lead paste (Fig. 8.1). The thermal procedure used by the authors to create submicron lead oxide particles using povidone aids in particle size reduction (Zhou et al., 2017). The approach resolves the problem of the insolubility of lead paste in aqueous media, however pyrometallurgical operations may have adverse impacts due to their high energy requirements as well as their undesirable releases of air contaminants like SO_2 and lead dust.

8.2.1.2.2 Zinc manganese batteries

Similar to this, moveable batteries in industrialized countries are largely made of zinc-manganese (Zn-Mn) composites. China produces more than 10 billion units annually. In both people and other species, exposure to hazardous quantities of Zn can lead to decreased calcium absorption and metabolic. As a result, recovering Zn from recycled Zn-Mn batteries is an efficient way to allow the metal's usage in other technologies while limiting its uncontrolled environmental dispersion. Recycled Zn-Mn batteries have garnered a lot of attention recently as a starting material for the production of Zn nanomaterials such as NPs and flaky NPs.

Zinc separation from used batteries was made simple and effective by Xiang et al. (2015). Recycled zinc cathode samples were inserted in a corundum crucible and heated to an internal temp of 700°C, then cooled to an external temperature of 200°C (Fig. 8.1). Up to 99.68% of zinc may be separated using the straightforward method. Structured nanoparticles with various morphologies, such as hexagonal prisms, fibers, and sheets, may be produced by introducing a N_2 gas flow into the heat exchanger after the separation process.

8.2.1.2.3 Rubber tires

Each year, around one billion million trash tires are produced worldwide (Goldstein Research, 2020). A significant portion is burned outside or discarded, which causes air ecological damage harm (Moghaddasi et al., 2013). The majority of modern nations restrict the disposal of scrap rubber in dumps as a result. Finding environmentally friendly ways to handle rubber tire trash is therefore important. Samaddar et al. (2018) tires typically include 45%–47% natural rubber, 21%–22% carbon black, 12%–25% metals, 5%–10% textile, 6%–7% additives, 1%–2% zinc, and 1% sulfur. An appropriate option with the potential for value addition is the production of zinc NPs from old tyres. NPs from discarded tyres might be produced by ball milling for five hours nonstop. The process produced rubber ash particles that were smaller than 400 nm; with the help of silicon waste, the particles' size was further decreased to 50 nm. They were just as effective in supplying plants with zinc as commercial $ZnSO_4$ fertilizers. Since waste tyres are mostly made of carbon, they are found to be a potential source of carbon for a number of purposes.

In a nutshell, under optimal conditions, thermal transformation at high temperatures (1000°C) and self-pressurized conditions created carbon nanoparticles (22 nm) with a high yield (approximately 81%). The chemical examination of the chain-like agglomerated nanoparticles revealed that they were somewhat oxidized (C, 84.9%; O, 4.9%; S, 10.21%), and they showed excellent thermal conductivity and stability (Fig. 8.1). Numerous studies have argued in favor of using unorthodox hydrocarbon waste, including tyres and plastics, as low-cost raw materials for CNT manufacturing

(Sathiskumar and Karthikeyan 2019). In order to create carbon nanotubes on a quartz substrate, low boiling point hydrocarbon tire pyrolysis oil generated from used tire material was combined with ferrocene as the catalyst and circulated at a rate of 21 mL/min at 940°C.

8.2.1.2.4 Wastewater and bio sludge

As they were found to be rich sources of rare earth elements and metal oxides, industrial sludge and wastewater effluents were extensively investigated for the recovery of materials such as PbO, Ag, and SiO_2. While the quantity of trace elements in waste effluents varies greatly, there are correlations between the particular processes carried out in the emitting source and the variety of the mineral composition within a single effluent source. A novel method for recovering metal from aqueous waste, such as that produced during bioremediation or biomining operations, was demonstrated by Urbina et al. (2019). Functionalizing fungus mycelia is accomplished using metal-binding peptides. For predicting the binding affinity between metals and organic ligand-binding motifs, in silico models have been created (Fig. 8.1). These models are beneficial because they make it possible to estimate binding characteristics based on free-access protein databases to characterize the geometric properties of a cognate metal in a metalloprotein. The proof-of-concept results for high affinity and suitability for Cu show a strong agreement between the motifs found in nature and those created in silica (Elsayed et al., 2020).

A novel method for recovering metal from aqueous waste, such as that produced during bioremediation or biomining operations, was demonstrated by Urbina et al. (2019). Functionalizing fungus mycelia is accomplished using metal-binding peptides. In-silico models have been developed to predict the binding affinity between metals and organic ligand-binding motifs. These models can characterize the geometrical characteristics of a cognate metal in a metalloprotein using free-access protein sequences, which makes them helpful. The proof-of-concept results for Cu binding affinity and selectivity reveal a close correspondence between in-silico-created patterns and motifs observed in nature (Tran et al., 2017).

8.2.1.2.5 Electronic waste

As a consequence of the enormous amount of electronic items and equipment in the United States and Europe, the bulk of the e-waste receiving nations, which are in West Africa and Asia, are dealing with serious environmental and public health issues. With regard to the recovery of metallic components in particular, the United States and China have been in the forefront of the technological latest of effective e-waste recycling procedures over the last few decades. However, recycling e-waste is getting more and harder since electronic devices are made of complex mixtures of various metals, and even recently available nanomaterials. Additionally, traditional mechanical separation methods used in underdeveloped nations are having less success as a result of the continued downsizing of electronic components (Baldé et al., 2017). Despite the practical challenges, the significant economic worth of precious metals like gold drives e-waste mining. (Abdelbasir et al., 2018). The largest concentrations of precious metals and a sizeable fraction of base metals are found in printed circuit boards and mobile phones, as illustrated in Table 8.1.

The metal content in e-waste is frequently significantly higher than that of land-mined ores. For instance, virgin ore has a maximal copper content of 3%, but e-waste has a weight content of 10%–20% copper.

TABLE 8.1 Metal ratio in various e-waste scraps (Hsu et al., 2019).

Types of e-waste	Weight (%)				
	Cu	Fe	Al	Ni	Pb
Printed circuit board	21	5	3	1	2
Mobile phone	12	4	1	1	0.2
TV Board	11	27	8	0.03	1
Portable audio	22	22	1.0	0.3	0.15
DVD player	4	61	3	0.5	0.2
Calculator	3	4	5	0.5	0.1

Traditional methods for recovering metals from e-waste involve first converting the metallic components to oxide forms, then repeatedly elevating the element's pure form. E-waste processing methods that are effective, such pyro- and hydrometallurgy, have recently attracted a lot of interest. Crushed electronic debris is melted down in pyrometallurgy, where the refractories are expelled as slag while the plastic breaks down and releases energy. Following processing, basic or extraction of precious metal is done from this slag. As an alternative, hydrometallurgy processing is more dependable and manageable, but it is also quite expensive since it calls for the restoration and therapy of used liquids as well as the creation of new solutions (Shokri et al., 2017).

8.2.1.2.6 Plastic waste

Undoubtedly, the most important and troublesome class of waste materials in the world has resulted from the widespread use of nonbiodegradable plastics. Since nearly a century ago, plastic polymers have been extensively employed to create a vast array of items.

In reality, there has been a significant increase in the manufacture of synthetic polymers during the last few decades. Because they were made to be durable and withstand changes in external environmental conditions, traditional plastics continue to exist and accumulate in the environment as shown in Table 8.2. Currently, the bulk of solid waste generated worldwide is composed of plastics (Ritchie and Roser, 2020). There are alarmingly high levels of plastic waste and microplastics in the water. Through trophic chain pollution, these particles have a deleterious influence on people in addition to aquatic life and birds. PWs are dangerous because they include pigments with various very toxic trace elements that might migrate out of the polymeric matrix and into the environment, in addition to the issues with their slow degradation. The environmental harm caused by synthetic PWs is currently regarded as one of the most severe and probably permanent impacts of modern human activity (Zeng and Bai, 2016).

In conclusion, the methods shown have considerable potential for lowering the quantity of waste that is dumped in landfills and toxic substances that are released into the environment. It is important to keep in mind that most of the methods under consideration are still in the research and development phase. However, the suggested procedures act as a proof-of-concept illustration of how to depart from the conventional course of waste management. Perhaps the reduction of dangers to public health and the environment and the offsetting of processing, and disposal costs, will be enough incentives to encourage the use of these alternative methods.

8.3 Advantages of synthesis of nanomaterial from industrial waste

8.3.1 Nanomaterial recover/synthesis from industrial waste

The process of turning frequently produced garbage into valuable nanomaterial is typically referred to as development that is sustainable. Numerous industrial wastes have undergone thermochemical analysis and are processed to extract priceless nanomaterials. Various industrial processes have been used to create nanomaterials with different morphologies and sizes. For the manufacture of materials, traditional methods employ a top-down approach, where bulk using

TABLE 8.2 Properties and composition of polymers frequently included in plastic garbage (Banu et al., 2020).

Plastic	Molecular weight (g/mol)	Density (g/cm^{-3})	Crystallinity
Polystyrene (PS)	104.1	1.04–1.11	Highly amorphous (atactic), highly crystalline (syndiotactic)
Polyvinylchloride (PVC)	62.50	1.10–1.45	Highly amorphous
polyethylene terephthalate (PET)	192.17	1.38–1.40	Semicrystalline
Polypropylene (PP)	42.08	0.855, amorphous 0.946, crystalline	Highly crystalline and amorphous
(HDPE)	10^3–10^7	0.93–0.97	High crystalline, low amorphous
(LDPE)	8.8×10^4 to 4.5×10^5	0.910–0.940	Low crystalline, high amorphous

TABLE 8.3 Waste nanoparticles used in environmental applications.

Method	nanoparticles	Application
Wastewater treatment and water remediation	Iron oxide $CaCo_3$	Dye removal Lead adsorption
Monitoring of water pollution	Carbon	Detection of Hg
Air pollution capture	Nano cuprous	CO_2 capture

mechanical, chemical, or other types of energy and materials are broken down into tiny bits as shown in Table 8.3. Nevertheless, in the bottom-up strategy, the ions are chemically mixed to create the particles (Hou et al., 2021).

Coprecipitation, the hydrothermal method, microwave irradiation, and the ball milling techniques are examples of bottom-up methods for synthesizing industrial waste-derived nanomaterials. Cr, Cu, Ni, and nanomaterials have been formed from electroplating waste, steel industry waste, paper mill sludge, and phosphor gypsum waste, among other industrial wastes. MFe_2O_4 and zeolite are a few examples of recovered nanomaterials (Liao et al., 2017).

8.3.2 Prior treatment of industrial waste

Pretreating industrial waste is an essential step in the creation of nanomaterial from the trash. Physical and chemical methods are subcategories of preparation procedures based on the makeup of the waste sample. The physicochemical characteristics and surface functional group of synthesized nanomaterials are influenced by the chemical makeup of the waste and the treatment process. Temperature, the acid used for treatment, and the compounds utilized for activation are process factors that have an impact on the synthesis. Industrial waste was initially utilized as a catalyst immediately, without any processing of SO_4 and OH. Then, in order to improve the activity, the trash was pretreated using physical and chemical activation. One physical treatment approach that decreases particle size to produce a homogeneous mixture for additional use is ball milling industrial waste. In order to recover calcium from steelmaking slag, Lee et al. investigated the impact of ball milling on particle size. The reaction between the components is easier for finer particle size (Lee et al., 2017).

The ball milling impact boosted calcium extraction effectiveness (360 rpm for 72 hours). Therefore, the primary goal of chemical pretreatment is to eliminate any pollutants in the waste sample that may be made soluble via heating or chemical treatment. When soluble salts like chlorides, sulfates, and carbonates are present, they can dissociate and produce flaws in the products when they are fired at high temperatures. Due to the need for large amounts of water and a longer exposure period, Abreu et al. found that the waste's particle size distribution affected the extraction of the salt. Strong acids, such as HNO_3 and HCl, are concentrated and used in strong acid hydrolysis, a type of chemical pretreatment, to treat waste materials. For instance, to create Al nanopowder from recycled aluminum foils, waste materials were first processed in aqua regia, and then an ammonia solution was added to maintain the solution's pH in the alkaline range (Hadi et al., 2015).

8.3.3 Nanoparticle synthesis from recycled material

8.3.3.1 Metal and metal oxide nanoparticles

Following appropriate chemical and physical preparation, a variety of processes have been employed to create various types of nanomaterials from the trash. Grinding and milling are the most used physical preparation techniques. By heating or treating with chemicals like strong acids (such as H_2SO_4, and HCl), chemical pretreatment is used to remove any pollutants contained in the waste sample (Samaddar et al., 2018).

Electronic trash from computer circuit boards, cell phones, and supercapacitors can be recycled into metal nanoparticles and polymers, such as Pb, Hg, Pt, and Rh (Vermisoglou et al., 2016). The majority of efforts so far have been concentrated on recovering metal nanoparticles from discarded batteries. A number of valuable elements may also be found in other types of e-waste, though (Fig. 8.2) some nanoparticles are synthesized through recycled material. For instance, using microemulsion techniques, Cu nanoparticles were created from acidic $CuCl_2$ waste etchants produced during the manufacturing process for printed circuit boards.

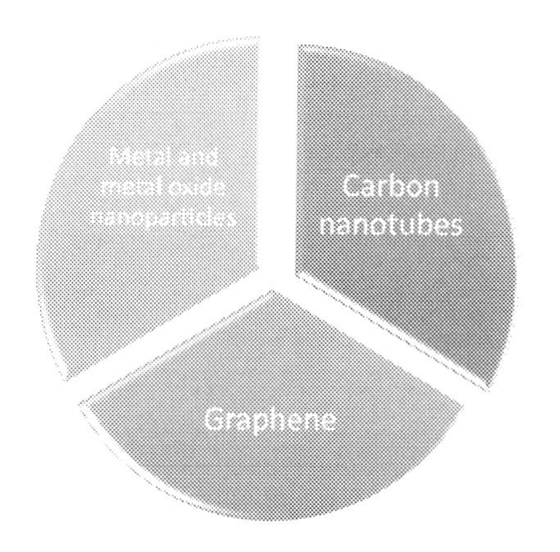

FIGURE 8.2 Nanoparticles synthesized through recycled material.

8.3.3.2 Carbon nanoparticles

Different technologies, including autoclaves and muffle furnaces, are used to create carbon nanotubes (CNTs) from waste plastic. Each technique uses pyrolysis to break down solid-state polymers into carbon-based building blocks. The bulk of plastic garbage is made of polyolefin, which has an 85 wt.% carbon content and is a polymer made from olefin monomer (CnH_2 n). CNTs can also be created using other polymers that have been merged with extra components, such as PVA and PET (Bazargan and McKay 2012).

A brand-new layer-by-layer assembly mechanism was created by Gong et al. This process involved catalyzing waste PP in a quartz tube reactor with activated carbon and Ni_2 O_3. PP was successfully broken down into light hydrocarbons by activated carbon. By promoting dehydrogenation and aromatization with the creation of various aromatic groups, activated carbon also offered very effective catalytic conversion with Ni. Based on benzene rings, CNT growth was made possible (Fig. 8.2). The recovery of Ni_2O_3 could be an alternative for more ecologically responsible scaleup production given the high energy need of this process (Bajad et al., 2015).

8.3.3.3 Graphene

There are several ways to produce graphene from discarded polymers. High yields of graphene flakes may be produced, according to Gong et al. Their process made use of waste polypropylene (PP) and montmorillonite that had undergone organic modification. To create the carbonized char, a homogeneous combination of modified montmorillonite, PP, and talcum was mixed in a crucible and heated to 700°C for 15 minutes. The carbonized char was submerged in HF and HNO_3 after cooling. The impurities were removed by HF, and the amorphous carbon was oxidized by HNO_3. Graphene flakes were produced after centrifuging and isolating from solution steps (Strudwick et al., 2015).

In the first furnace, a ceramic boat containing over 4 mg of plastic trash was heated to around 500°C. Secondly, Cu foil was used as the substrate which was maintained at a temperature of around 1020°C. The degraded carbonaceous chemicals were then put into the following furnace with a gas combination after pyrolysis. After that, the products reacted with a Cu substrate to promote the development of graphene (Fig. 8.2). The pace at which the disintegrated products were injected in this approach was essential for the production of graphene crystals. At a modest rate of pyrolysis and injection, large hexagonal single-layered graphene was created. On the other hand, bilayer graphene was created when the injection rate was high (Yang et al., 2016).

8.3.4 Nanomaterial synthesis from waste methods

Industrial waste-derived nanoparticles can be produced via a variety of thermochemical processes, including coprecipitation, hydrothermal, carbothermal, solvothermal, and microwave-assisted techniques. Due to their properties, the production of nanoparticles made from waste is a difficult procedure. Surface modification of nanoparticles is thought to be an important factor in terms of selectivity and water stability for applications involving pollution removal. Below is a discussion of the benefits and drawbacks of the most popular techniques for creating nanomaterials (Kefeni et al., 2017).

8.3.4.1 Coprecipitation method

One of the most popular wet chemical techniques for producing waste-derived nanoparticles is coprecipitation. Due to its straightforward operation and high production of nanomaterials, this approach is one of the most used. Acids were

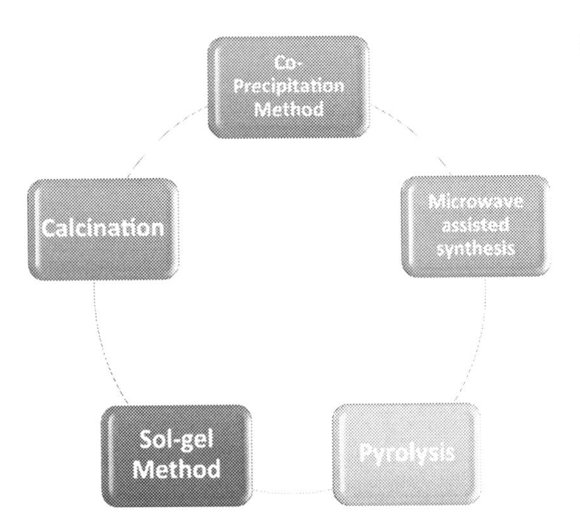

FIGURE 8.3 Nanomaterial synthesis from waste methods.

used to leach the metal from the industrial waste. The pH of the acid-extracted metal solutions is raised between 7 and 12 by adding alkali. The kind of extractant, the molar ratio of the metal, reaction temperatures, PH, and other reaction parameters, such as stirring rate and basic solution lowering speed, all affect the size and form of the nanoparticle (Fig. 8.3). By precipitating silica, Liang et al. created silica nanoparticles, which were then aged for 12 hours and dried at 100°C after alkali extraction and acid treatment to raise the pH to 7. 44.42% to 93.63% of the nanoparticles produced by precipitating silica from biomass fly ash were pure. The sole drawback to the precipitation approach is that the final goods appear to be less pure than those made using other techniques (Abdel-aal et al., 2016).

By heating at high temperatures (120°C–550°C) and pressures (20–150 bar) with oxygen or an inert gas like nitrogen, hydrothermal processing may transform waste into products with value added. The sole distinction between the hydro-thermal methods is that in the latter, other solvents are employed in place of water. Many oxygen-containing functional groups, such as C = O, C–O and O–H groups, are present in the materials produced using this approach. Here, C = O might produce oxygen species that, by resolving the environmental issue of carbon-rich waste, break down the organic pollution. The hydrothermal method was used to create ferrite catalyst from electroplating sludge, manganic biochar from ferric and biological sludge, sludge carbon/TiO_2 nanocomposite from anaerobic sludge, TiO_2-FA nanocomposites, and tungsten oxide FA nanocomposite. Because it offers more benefits than high-temperature pyrolysis, it is regarded as an effective way to manage wet solid waste that is high in carbon. Low energy requirements, moderate reaction conditions, and no requirements for raw material moisture content all contribute to low energy use (Liang et al., 2020).

8.3.4.2 Microwave-assisted synthesis

Various inorganic materials, including pure and metal oxides, amorphous and nonporous materials, core-shell particles, semiconductors, and metallic nanoparticles, have been synthesized using the microwave-assisted synthesis approach. Microwave radiation is applied to the material throughout the synthesis process through interactions between molecules and ions in the precursor solvents and reducing agents. The detoxification of sediment sludge and the immobilization of metal ions within the sediment have both been accomplished using microwave radiation (Mirzaei and Neri, 2016). This approach is one of the most successful methods for nanoparticle production because it has numerous benefits over other procedures, including as rapid heating, quick response, high yield, and great thermal stability (Fig. 8.3). For the purpose of removing Cr using a microwave chemical reactor, iron sludge and cotton stalks were used by Duan et al. to make magnetic biochar. Zhang et al. successfully produced hydroxyapatite nanoparticles from phosphor gypsum waste produced by phosphorus fertilizer companies in order to investigate its usage in the removal of fluoride from aqueous solutions. This approach is superior to previous synthesis techniques because it produces crystalline nanomaterials quickly and operates at low temperatures (Mondal et al., 2020).

8.3.4.3 Pyrolysis

One of the best processes for creating materials with a good surface area, high ion exchange capacity, and value-added surface functional groups like –OH, and C–O/C = O is pyrolysis. The temperature and length of the reaction during pyrolysis have a significant impact on how the material's surface chemistry changes (Chen Y di 2018). It alters the surface's propensity to adsorb water, which is a crucial element in the adsorption reaction. In a research, rise in temperature from 650°C to 950°C led to an increase in surface pH. As a result, altering the pyrolysis conditions can change the surface attributes including porosity, selectivity, or catalytic activity (Fig. 8.3). The novel surface chemistry created by solid-state reactions during pyrolysis directly influences the product's catalytic activity. Toxic gas emissions during the material process and significant energy usage are drawbacks of this approach (Bandosz and Block, 2006).

8.3.4.4 Sol–gel method

Since the early 1800s, the sol–gel process has been utilized to generate materials for catalysis, membranes, and fibers, as well as in the domains of science and engineering. In this method, the precursors are hydrolyzed, polycondensed, aged, dried, and thermally decomposed to create a colloidal suspension. Compared to materials made using the hydrothermal approach under identical reaction circumstances, the materials made using the sol–gel method were crystalline and had smaller crystallite sizes (Khan, 2020). The most often employed precursors in this process are metal-organic compounds or inorganic metal salts. Precursor concentration and temperature are variables that affect the gel formation's characteristics (Fig. 8.3). The main benefit of this technology is the formation of a nanomaterial at the conclusion of the process that has a large surface area and stable surfaces. In other words, utilizing the sol–gel method and chromium-containing tannery waste with an initial solution pH of 4.0 that was then calcined at 1100°C, single-phase chromium-doped alumina nanoparticles were produced (Chen Y di, 2018).

8.3.4.5 Calcination

Calcination is used to structurally convert waste, with an emphasis on the final oxide products (Fig. 8.3). The simple metal oxide was created when the calcination temperature was lower than 400°C. However, spinel oxide was produced at calcination temperatures of more than 600°C. Interesting functional materials might be produced by varying the calcination temperature. The precursor becomes layered double hydroxide when it is calcined at a lower temperature (Hou et al., 2021). The "memory effect" refers to the reconstruction that occurred after this product was submerged in water or exposed to wet gas. A list of industrial waste-derived nanomaterial synthesis is shown in Table 8.4.

TABLE 8.4 Industrial waste-derived nanomaterial synthesis method and properties (Boruah et al., 2020).

Waste-derived nanoparticle	Recyclable waste	Recycling method	Properties of nanoparticle
Iron oxide waste nanomaterial with 7.6 wt.% of Titanium dioxide	Iron oxide waste	Planetary wet ball milling technique	Particle size 10 nm
$MnFe_2O_4$ and $CuFe_2O_4$	Pickling waste liquor (PWL) + electroplating wastewater (EPW)	Coprecipitation	$MnFe_2O_4$ 30–50 nm
$CuFe_2O_4$	$CuFe_2O_4$-Printed circuit boards sludge	Acid leaching, chemical exchange, and ferrite process	Particle size 20 to 120 nm
Magnetic nickeliferous pyrrhotite	Nickel sulfide mine tailing	Crushed and pulverized	Specific surface area 1.00 m^2/g, average pore width 14.08 nm
Alum based adsorbent Aluminum oxide	Aluminum oxide Alum sludge	Calcination	BET surface area 70–181 m^2/g, Pore size 4.75–11.17 nm
Ceramic nanopigments $CrAl_2O_3$ nanoparticles	Tannery waste	Modified sol–gel method	Particle size 56 ± 4 nm
Magnetic bio char n ZVIWSBC	Persulfate-ZVI dewatered waste activated sludge	Pyrolysis	Particle size 20–50 nm saturation magnetization 54 emu/g
Zeolite	Fly Ash	Hydrothermal process	Pore diameter 14.16 nm, surface area 52.49 m^2/g

(Continued)

TABLE 8.4 (Continued)

Waste-derived nanoparticle	Recyclable waste	Recycling method	Properties of nanoparticle
Bio silica nanoparticles	Biomass power plant fly ash	Alkaline extraction method	Particle size 20 and 40 nm, surface area 115 m^2/g, pore diameter 3 nm, pore volume 0.13 cm^3/g
Zeolites fly ash bead/TiO_2 composite	Fly ash	Fly ash Alkali activation and mixing	Zeolite Fly ash bead particle size ~ 2 μm, TiO_2 particle size 20 nm, pore volume 5 ~ 50 nm
Activated carbon	Biological, chemical, and hybrid sludge	Acid-treated and pyrolyzed	Pore diameter 20–1000 Å BET surface area 89.5–1114.7 m^2/g
Tungsten oxide—fly ash composite	Fly Ash	Hydrothermal method	Surface area 8.606–31.189 m^2/g, Pore diameter 12.67–27.2 nm
Fe_3O_4	Spent pickling liquors	Ultrasonic-assisted chemical coprecipitation	Particle size 13–23 nm

8.4 Challenges

Researchers have been driven to focus on efficient trash usage in order to realize the notion of waste treatment and riches from waste due to the rising production of industrial garbage. Nanomaterial synthesis from waste is thought to be an efficient approach that offers important element recovery and financial advantages because nanotechnology is a rapidly expanding science with applications in many different fields. However, research must be conducted to enhance and strengthen the recovery process in order to address the issue brought on by growing waste creation and resource exploitation. Industrial waste has been used to synthesize waste-derived nanomaterials, and each technique of synthesis has its own benefits. Examples of industrial waste include waste from tanneries, electroplating industries, paper mills, steel industries, etc.

More precisely, by altering the surface chemistry, waste pretreatment has an impact on the characteristics of the nanomaterial. It is obvious that throughout the synthesis process, physical and chemical parameters have an impact on how nanomaterial is formed. Therefore, it is important to investigate whether synthesis processes are practical for producing waste-derived products on a big scale. More research must be done, namely, to regulate the hazardous gases generated during the thermal synthesis process. Environmental remediation uses the physicochemical-produced nanoparticles obtained from trash extensively. MFe_2O_4 (M-Mn, Cu, Zn, Ag), magnetic biochar, $MgCr_2O_4$, Cr_2O_3, Cu, copper ferrite, and others are among the nanomaterials created from industrial waste (Visa et al., 2012).

For larger-scale applications of waste-derived nanomaterials, however, LCA must be carried out with an emphasis on energy use, pollution reduction, and cost effectiveness. Sustainable zero waste may also be accomplished by boosting the nanoparticles' activity and stability.

8.5 Future opportunities

The most thoroughly investigated nanomaterials, including carbon nanotubes (CNTs), zero-valent metal nanoparticles (Ag, Fe, and Zn), metal oxide nanoparticles (TiO_2, ZnO, and iron oxides), and nanocomposites, were featured in this research. Additionally, their applications in the treatment of wastewater and water were thoroughly covered. Nanomaterials appear to be particularly promising for the treatment of water and wastewater given the present rate of research and application.

This review's discussion of research has highlighted and addressed several approaches to using waste materials as inputs for the production of NPs. Even while turning garbage into cutting-edge innovations for environmental applications can first seem like a good full-circle approach, there are still a lot of engineered nanomaterials gaps in knowledge that need to be filled before these concepts can be applied outside of the lab. Particularly addressing energy usage, the creation of secondary wastes, destiny, exposure routes in varied habitats, and toxicity levels, there are information gaps in these areas. When practicable methods get past the proof-of-concept stage, comprehensive cost-benefit analyses must be carried out.

However, just because NPs may be created using environment-friendly methods, it doesn't mean they are inherently safe or that their usage or release into the environment won't have a negative impact. Studies conducted in the past ten years have revealed that engineered nanoparticles (engineered NPs) used in a variety of settings, such as food production and packaging, have the potential to significantly disrupt epigenetic mechanisms and raise several concerns about their potential to cause disease. One of the more well-known classes of engineered NPs may be fullerenes. Numerous in vitro and in vivo studies have examined the effects and toxicity of carbon nanostructures with various biological systems. The same properties that make CNTs so desirable for application areas are also connected to inflammatory reactions and fibrogenic processes that reduce lung function in mice and increase their susceptibility to respiratory infections. In terms of how they impact the body after being breathed, CNTs have been compared to asbestos fibers, which are considered to be cytotoxic and cancer-causing particles.

In conclusion, a complex web of influencing elements must be recognized in order to reap the potential advantages of the application of waste-derived NPs, and judgments should lean on the side of prudence and ethics. Regional definitions of the scale and breadth of these NP applications, as well as plans for closed-loop management of the engineered nanoparticles, should be made, and management techniques should be based on health and environmental science.

8.6 Conclusion

Researchers have been driven to focus on efficient trash usage in order to realize the notion of waste treatment and riches from waste due to the rising production of industrial garbage. Nanomaterial synthesis from waste is thought to be an efficient approach that offers important element recovery and financial advantages because nanotechnology is a rapidly expanding science with applications in many different fields. However, research must be conducted to enhance and strengthen the recovery process in order to address the issue brought on by growing waste creation and resource exploitation. Industrial waste, including fly ash, silicon waste, paper mill waste, electroplating industry waste, and waste from steel mills, among others, has been used to create waste-derived nanomaterials.

More precisely, by altering the surface chemistry, waste pretreatment has an impact on the characteristics of the nanomaterial. It is obvious that throughout the synthesis process, physical and chemical parameters have an impact on how nanomaterial is formed. Therefore, it is important to investigate whether synthesis processes are practical for producing waste-derived products on a big scale. More research must be done, namely, to regulate the hazardous gases generated during the thermal synthesis process. Environmental remediation uses the physiochemically produced nanoparticles obtained from trash extensively.

Despite its usage in the remediation of contaminants, byproducts from the removal or leaching back of the original pollutants should be studied for potential hazards. However, LCA must be carried out with an emphasis on energy consumption, pollution reduction, and cost-effectiveness before further widespread deployment of waste-derived nanomaterials. Furthermore, sustained zero waste may be accomplished by enhancing the stability and activity of the nanomaterials.

References

Abdel-aal, E.A., Farghaly, F.E., Tahawy, R., El-shahat, M.F., 2016. A comparative study on recovery of chromium from tannery wastewater as nano magnesium chromite. Physicochemical Problems of Mineral Processing 52, 821–834. Available from: https://doi.org/10.5277/ppmp160224.

Abdelbasir, S.M., Hassan, S.S.M., Kamel, A.H., El-Nasr, R.S., 2018. Status of electronic waste recycling techniques: a review. Environmental Science and Pollution Research 25, 16533–16547.

Bajad, G.S., Tiwari, S.K., Vijayakumar, R.P., 2015. Synthesis and characterization of CNTs using polypropylene waste as precursor. Materials Science & Engineering B: Solid-State Materials for Advanced Technology 194, 68–77. Available from: https://doi.org/10.1016/j.mseb.2015.01.004.

Baldé, C.P., Forti, V., Gray, V., Kuehr, R., Stegmann, P., 2017. Quantities, flows, and resources the global E-waste. Available from: http://www.unu.edu (accessed 10.05.20).

Bandosz, T.J., Block, K., 2006. Effect of pyrolysis temperature and time on catalytic performance of sewage sludge/industrial sludge-based composite adsorbents. Applied Catalysis B: Environmental 67, 77–85.

Banu, J.R., Sharmila, V.G., Ushani, U., Amudha, V., Kumar, G., 2020. Impervious and influence in the liquid fuel production from municipal plastic waste through thermo-chemical biomass conversion technologies - a review. The Science of the Total Environment 718, 137287. Available from: https://doi.org/10.1016/j.scitotenv.2020.137287.

Bazargan, A., McKay, G., 2012. A review - synthesis of carbon nanotubes from plastic wastes. Chemical Engineering Journal 195–196, 377–391. Available from: https://doi.org/10.1016/j.cej.2012.03.077.

Bhardwaj, A.K., Naraian, R., 2020. Green synthesis and characterization of silver NPs using oyster mushroom extract for antibacterial efficacy. Journal of Chemistry, Environmental Sciences and its Applications 7 (1), 13–18.
Bhardwaj, A.K., Naraian, R., 2021. Cyanobacteria as biochemical energy source for the synthesis of inorganic nanoparticles, mechanism and potential applications: a review3 Biotech 11 (10), 445.
Bhardwaj, A.K., Naraian, R., Sundaram, S., Kaur, R., 2022. Biogenic and non-biogenic waste for the synthesis of nanoparticles and their applications. Bioremediation. CRC Press, pp. 207–218.
Bhardwaj, A.K., Shukla, A., Maurya, S., Singh, S.C., Uttam, K.N., Sundaram, S., et al., 2018. Direct sunlight enabled photo-biochemical synthesis of silver nanoparticles and their bactericidal efficacy: Photon energy as key for size and distribution control. Journal of Photochemistry and Photobiology B: Biology 188, 42–49.
Biswas, A., Patra, A.K., Sarkar, S., Das, D., Chattopadhyay, D., De, S., 2020. Synthesis of highly magnetic iron oxide nanomaterials from waste iron by one-step approach. Colloids and Surfaces A: Physicochemical and Engineering Aspects 589, 124420.
Boruah, P.K., Yadav, A., Das, M.R., 2020. Magnetic mixed metal oxide nanomaterials derived from industrial waste and its photocatalytic applications in environmental remediation. Journal of Environmental Chemical Engineering 8, 104297.
Chen, Y.di, Ho, S.H., Wang, D., Wei, Z.Su, Chang, J.S., Ren, N.Qi, 2018. Lead removal by a magnetic biochar derived from persulfate-ZVI treated sludge together with one-pot pyrolysis. Bioresource Technology 247, 463–470.
Elsayed, D.M., Abdelbasir, S.M., Abdel-Ghafar, H.M., Salah, B.A., Sayed, S.A., 2020. Silverandcoppernanostructuredparticlesrecoveredfrommetalized plastic waste for antibacterial applications. Journal of Environmental Chemical Engineering 8, 103826. Available from: https://doi.org/10.1016/j.jece.2020.103826.
GoldsteinResearch, 2020. Global tire recycling market report - edition 2020. Available from: https://www.goldsteinresearch.com/report/globaltire-recycling-industry-market-trends-analysis (accessed 10.05.20).
Hadi, P., Xu, M., Ning, C., Sze Ki Lin, C., McKay, G., 2015. A critical review on preparation, characterization and utilization of sludge-derived activated carbons for wastewater treatment. Chemical Engineering Journal 260, 895–906.
Hou, H., Liu, Z., Zhang, J., Zhou, J., Qian, G., 2021. A review on fabricating functional materials by heavy metal–containing sludges. Environmental Science and Pollution Research 28, 133–155.
Hsu, E., Barmak, K., West, A.C., Park, A.H.A., 2019. Advancements in the treatment and processing of electronic waste with sustainability: a review of metal extraction and recovery technologies. Green Chemistry: An International Journal and Green Chemistry Resource: GC 21, 919–936. Available from: https://doi.org/10.1039/C8GC03688H.
Kefeni, K.K., Mamba, B.B., Msagati, T.A.M., 2017. Application of spinel ferrite nanoparticles in water and wastewater treatment: A review. Separation and Purification Technology 188, 399–422.
Khan, F.A., 2020. Applications of Nanomaterials in Human Health, first ed. Springer, pp. 15–21.
Koolivand, A., Mazandaranizadeh, H., Binavapoor, M., Mohammadtaheri, A., Saeedi, R., 2017. Hazardous and industrial waste composition and associated management activities in Caspian industrial park. Iran. Environmental Nanotechnology, Monitoring & Management 7, 9–14.
Lee, S.H.M., Lee, S.H.M., Jeong, S.K., et al., 2017. Calcium extraction from steelmaking slag and production of precipitated calcium carbonate from calcium oxide for carbon dioxide fixation. Journal of Industrial and Engineering Chemistry 53, 233–240.
Liang, G., Li, Y., Yang, C., et al., 2020. Production of biosilica nanoparticles from biomass power plant fly ash. Waste Management (New York, N. Y.) 105, 8–17.
Liao, C.Z., Tang, Y., Lee, P.H., Liu, C., Shih, K., Li, F., 2017. Detoxification and immobilization of chromite ore processing residue in spinel-based glass-ceramic. Journal of Hazardous Materials 321, 449–455.
Mirzaei, A., Neri, G., 2016. Microwave-assisted synthesis of metal oxide nanostructures for gas sensing application: a review. Sensors and Actuators. B, Chemical 237, 749–775.
Moghaddasi, S., Khoshgoftarmanesh, A.H., Karimzadeh, F., Chaney, R.L., 2013. Preparation of nano-particles from waste tire rubber and evaluation of their effectiveness as zinc source for cucumber in nutrient solution culture. Scientia Horticulturae 160, 398–403. Available from: https://doi.org/10.1016/j.scienta.2013.06.028.
Mondal, P., Anweshan, A., Purkait, M.K., 2020. Green synthesis and environmental application of iron-based nanomaterials and nanocomposite: a review. Chemosphere 259, 127509.
Ritchie, H., Roser, M., 2020. Plastic pollution. Available from: https://ourworldindata.org/plastic-pollution.
Samaddar, P., Ok, Y.S., Kim, K.H., Kwon, E.E., Tsang, D.C.W., 2018. Synthesis of nanomaterials from various wastes and their new age applications. Journal of Cleaner Production 197, 1190–1209.
Sathiskumar, C., Karthikeyan, S., 2019. Recycling of waste tires and its energy storage application of by-products – a review. Sustainable Materials and Technologies 22.
Shokri, A., Pahlevani, F., Levick, K., Cole, I., Sahajwalla, V., 2017. Synthesis of copper-tin nanoparticles from old computer printed circuit boards. Journal of Cleaner Production 142, 2586–2592.
Soliman, N.K., Moustafa, A.F., 2020. Industrial solid waste for heavy metals adsorption features and challenges; a review. Journal of Materials Research and Technology 9, 10235–10253.
Strudwick, A.J., Weber, N.E., Schwab, M.G., Kettner, M., Weitz, R.T., Wünsch, J.R., et al., 2015. Chemical vapor deposition of high-quality graphene films from carbon dioxide atmospheres. ACS Nano 9, 31–42.
Tran, D.T., Joubert, A., Venditti, D., Durecu, S., Meunier, T., Le Bihan, O., et al., 2017. Characterization of polymer waste containing nano-fillers prior its end-of-life treatment. Waste Biomass Valorization 8, 2463–2471.

UNEP, 2020. Solid waste management | UNEP - UN environment programme. Available from: https://www.unenvironment.org/explore-topics/resourceefficiency/what-we-do/cities/solid-waste-management (accessed 10.05.20).

Urbina, J., Patil, A., Fujishima, K., Paulino-Lima, G., Saltikov, C., Rothschild, L.J., 2019. A new approach to biomining: bioengineering surfaces for metal recovery from aqueous solutions. Scientific Reports (Nature Publ. Group) 9, 16422.

Vermisoglou, E.C., Giannouri, M., Todorova, N., Giannakopoulou, T., Lekakou, C., Trapalis, C., 2016. Recycling of typical super capacitor materials. Waste Management & Research: The Journal of the International Solid Wastes and Public Cleansing Association, ISWA 34, 337–344.

Visa, M., Isac, L., Duta, A., 2012. Fly ash adsorbents for multi-cation wastewater treatment. Applied Surface Science 258, 6345–6352.

Xiang, X., Xia, F., Zhan, L., Xie, B., 2015. Preparation of zinc nano structured particles from spent zinc manganese batteries by vacuum separation and inert gas condensation. Separation and Purification Technology 142, 227–233. Available from: https://doi.org/10.1016/j.seppur.2015.01.014.

Yang, R.X., Chuang, K.H., Wey, M.Y., 2016. Carbon nanotube and hydrogen production from waste plastic gasification over Ni/AlSBA-15 catalysts: effect of aluminum content. RSC Advances 6, 40731–40740. Available from: https://doi.org/10.1039/C6RA04546D.

Zeng, W., Bai, H., 2016. High-performance CO_2 capture on amine-functionalized hierarchically porous silica nanoparticles prepared by a simple template-free method. Adsorption 22, 117–127. Available from: https://doi.org/10.1007/s10450-015-9698-0.

Zhang, T., Gao, P.P., Gao, P.P., Wei, J., Yu, Q., 2013. Effectiveness of novel and traditional methods to incorporate industrial wastes in cementitious materials - an overview. Resources, Conservation and Recycling 74, 134–143.

Zhou, H., Su, M., Lee, P.H., Shih, K., 2017. Synthesis of submicron lead oxide particles from the simulated spent lead paste for battery anodes. Journal of Alloys and Compounds 690, 101–107. Available from: https://doi.org/10.1016/j.jallcom.2016.08.094.

Chapter 9

Scope to improve the synthesis of nanomaterial's using industrial waste

Ajay Kumar Tiwari[1], Saket Jha[2], Mohee Shukla[3], Rohit Shukla[3], Ravikant Singh[4], Abhimanyu Kumar Singh[5], Ashok Kumar Pathak[6], Rudra Prakash Ojha[7] and Anupam Dikshit[3]

[1]*Department of Physics, Nehru Gram Bharati (Deemed to be University), Prayagraj, Uttar Pradesh, India,* [2]*Department of Surgery, College of Medicine, University of Illinois, Chicago, IL, United States,* [3]*Biological Product Laboratory, Department of Botany, University of Allahabad, Prayagraj, Uttar Pradesh, India,* [4]*Department of Biotechnology, Swami Vivekanand University, Sagar, Madhya Pradesh, India,* [5]*Department of Physics, Shyama Prasad Mukherjee Govt. Degree College, University of Allahabad, Prayagraj, Uttar Pradesh, India,* [6]*Department of Physics, Ewing Christian College, University of Allahabad, Prayagraj, Uttar Pradesh, India,* [7]*Department of Zoology, Nehru Gram Bharati (Deemed to be University), Prayagraj, Uttar Pradesh, India*

9.1 Introduction

Industrial waste management has now become a major concern, and several new approaches are being taken to reduce the waste. Recent advancements in dealing with this waste, such as the synthesis of nanomaterials from these industrial waste materials, are now attracting great interest due to their extensive use in the fields of medicine, electronics, space, and pharmaceutical industries. Several reports and ancient manuscripts suggest that these nanomaterials were synthesized in earlier times. However, advancement in the synthesis of nanocomposites has flourished in recent decades. These nanomaterials are formed into different shapes and structures because of peculiar properties as well as having more surface area to volume ratio (Teague, 2005). A variety of nanoparticles synthesis methods are known today, all of which are classified into three major categories: (1) physical, (2) chemical, and (3) biological. For the chemical and physical synthesis, several toxic compounds are used, which directly or indirectly cause harmful effects on human health as well as aquatic ecosystems. Thus the third method (biologically) is much eco-friendlier and less toxic to the environment (Salunke et al., 2016). Even for the minimum waste production and wealth from the industrial waste, biological synthesis via green root may approved as an effective tool for recycling (Biswas et al., 2020). Currently, preparation of nanomaterials from industrial waste is in starting phase still it has returned a good wealth and also moved toward the sustainable development (Dutta et al., 2018). Now a huge amount of production of nonbiodegradable waste materials from industries has become a challenge for the environment and human being. The conventional methods, which are utilized for the treatments of industrial waste, have some disadvantages. Synthesis of nanomaterials from industrial waste has become lately interested, and it can be utilized as a contaminant subtraction due to the unique properties of nanomaterials. Nowadays, the nanomaterials synthesized from the industrial waste are utilized not only as a pollution removal but also waste disposal problems can also be resolved. Not only this, economical growth can also be increased by improving the substituting materials from the industrials waste instead of virgin materials. Therefore researchers have developed a new method to synthesize the nanomaterials from industrial waste that are cost-effective and environmental-friendly. Still, there is a need for more simple, reliable, and eco-friendly nanocomposites synthesis, which has thus led to an interest in synthesizing these nanocomposites from industrial waste. This will not only reduce the earth's waste burden, but also make it more reusable, with more potential applications (Salunke et al., 2016).

9.2 Industrial wastes

The wastes produced by manufacturing or industrial processes, such as polymer batteries, wastewater, and tyres, are utilized to prepare the various types of nanomaterials. China alone produced the lead batteries approximately 50–60

Green and Sustainable Approaches Using Wastes for the Production of Multifunctional Nanomaterials. DOI: https://doi.org/10.1016/B978-0-443-19183-1.00011-8

million yearly (Zhou et al., 2019). The nano-size of lead oxide particles was produced using thermal degradation technique mixing with poly-N-vinyl-2-pyrrolidone (PVP). The addition of PVP increases the reduction in particle size, changing microns to nano-sized. These nano-synthesized lead oxides help reduce the toxic overload of these lead compounds, which were previously insoluble in water.

Exposure to high amounts of Zn can decrease the calcium intake in the body and lead to a low content of calcium in the human body. Recycling zinc–manganese (Zn–Mn) battery waste has been discovered to be a good source for producing various forms of nanomaterial, such as zinc nanoparticles and nanofibers. Zinc nanoparticles can be produced by zinc cathodes. (Jha et al., 2018; Abdelbasir et al., 2020). Water hydrolyzes lithium salts in general, but they are thermally unstable above 80°C. Therefore the evaporation temperature should not exceed above this (Wang et al., 2020a). But H_2SO_4 is not introduced for such applications. Several reports revealed the toxicity of leaching compounds and metal-associated particles during recycling of lithium-ion battery wastes. The excessive usage of chemical polymers gives rise to an abrupt increase in the waste material burden. This waste is not biologically degradable and can be further reused in making plastic bags, bottles, etc., but still remains as solid waste on earth. Recent advances provide a novel method for producing ultrafine nanotubes and carbon nanotubes from polyethylene terephthalate (PET) waste (Berkmans et al., 2014). This generates the various shapes and sizes of carbon nanotubes, which can be useful in various drug delivery applications and other means of protein and amino acid targeting pathways (Moghaddasi et al., 2013). Tire disposal to landfills is now banned in several European nations. Thus it required an innovative technique for reprocessing of rubber tire waste. In the presence of silicon waste, fabricated from rubber tire waste by the ball milling process was obtained in the range of 50–500 nm (Moghaddasi et al., 2013). Furthermore, cathode-ray tube funnel glass having a lead-rich amount (20%–30%) is also a harmful waste. The conventional leaching method is not sufficient for the extraction of all lead (Pb) from the funnel glass. Therefore a unified technique of vacuum carbon–thermal minimization and inert gas consolidation is used to obtain the Pb NPs in place of the conventional leaching method. Pb mostly exists in the form of lead oxides in funnel glasses and has much higher melting points as compared with metallic lead under ambient conditions (Yot and Méar, 2009). Lead oxide is obtained from metallic lead using an abbreviation process. The truncation and lead evaporation rates are incremented expeditiously with the enhancement of the temperature (Gaines, 2014). Some industrial waste types are described as below.

9.2.1 Biodegradable industrial waste

These can be easily degraded by microbial inoculants, which are nonpoisonous but accumulate on earth. They are even comparable with house wastes, which mainly include newsletters, wood, etc., mainly generated from beverage industries, food industries, slaughterhouses, milk factories, etc. As these are nontoxic to the environment and can be easily degradable as well as reusable, thus these do not need any specific treatment (Gaines, 2014).

9.2.2 Nonbiodegradable industrial waste

Industrial waste is mostly nonbiodegradable, which means it cannot be degraded or digested by microbial or biological components, and thus becomes toxic to the environment. The fertilizer, dye, chemical, and drug industries, as well as iron and steel plants, produce various types of nonbiodegradable waste. These industrial contaminants cannot be easily broken down into simpler components, which makes them more hazardous to living organisms and pollutes the ecosystem. These accumulated directly or indirectly in mammals, including humans, and cause several diseases. Several studies are still being conducted, with the goal of developing novel methods to remediate this waste rather than disposing of it in landfills (Gaines, 2014).

9.3 Effect of industrial waste material on the environment

Industrialization on a large scale has generated a billion metric tons of toxic waste, which has put a significant burden on the environment. Massive secondary raw materials exist in various types of industrial waste. These raw materials could be utilized in various applications (Schwarz et al., 2012). The minimization of waste and its recycling processes act as a safeguard for the environment (Koolivand et al., 2017). The heavy metals originating from industrial wastes, toxic chemicals, and dyes cause a variety of serious diseases and physical disorders. Lead pollution generated from industrial waste affects the human organs and bones (Soliman and Moustafa, 2020; Biswas et al., 2020).

9.4 Synthesis of nanomaterials from industrial waste

Synthesis of nanomaterials (NPs) from urban waste is found to be more effectual method to treat these contaminants and make them reusable materials (Biswas et al., 2020). Currently, the synthesizing of nanocomposites from industrial waste materials gained much interest because of the captivating the financial and sustainable approaches (Dutta et al., 2018, Figs. 9.1 and 9.2). Electroplating methods are widely considered and utilized for the fabrication of nanocomposite from industrial waste materials, as these are mostly metals, metal oxides, or metal salts. The nanocompounds derived from these wastes are found to be more durable and stable, with a high content of organic matter and several metals

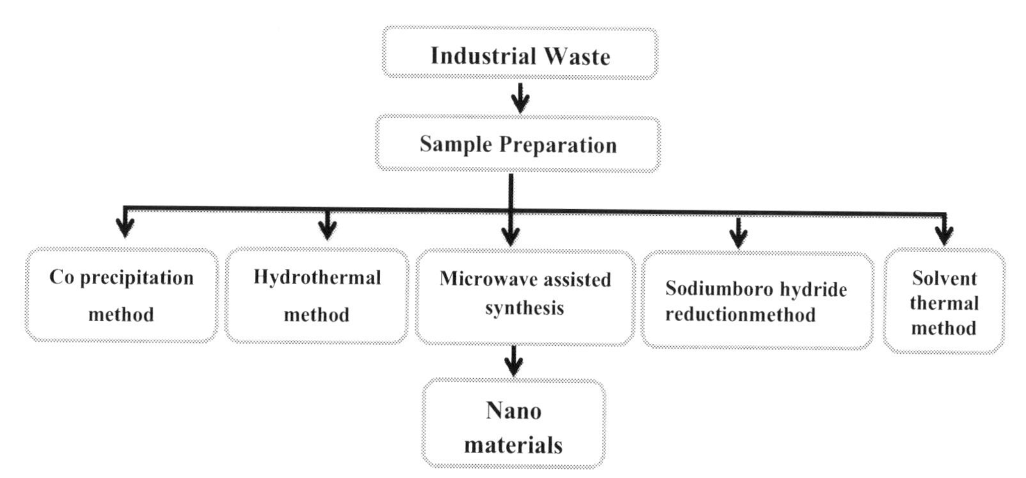

FIGURE 9.1 Graphical presentation of synthesis method of nanomaterials from industrial wastes.

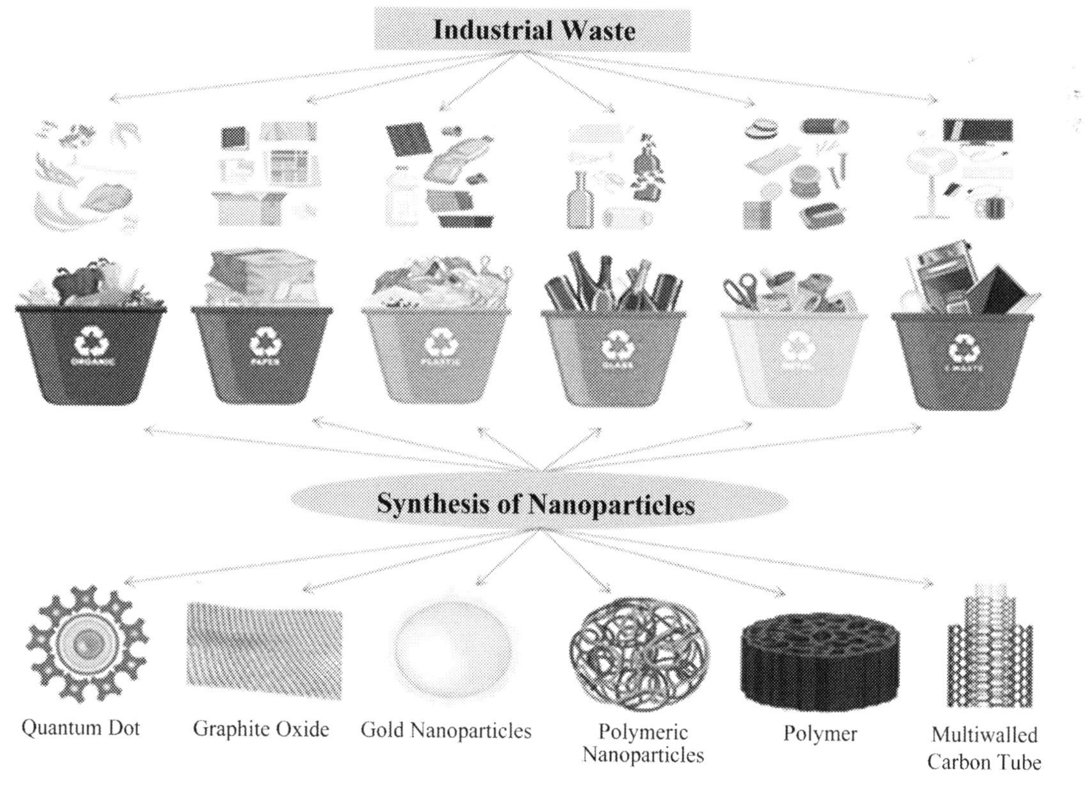

FIGURE 9.2 Glimpse of industrial wastes utilized for nanomaterial production.

and others, in several natures and crystal structures. Still, science faces a more astronomically massive challenge: developing nanoparticles with the desired properties to achieve the desired goal using congruous applications. These were dependent on the size, structure, and optical properties of the synthesized nanoparticles, which primarily depend on the synthesis process (Bhardwaj et al., 2022). This can be achieved by utilizing biological methods as an alternative to conventional approaches (Wang et al., 2020b).

The transformation of conventionally generated waste materials into nanomaterials resulted in long-term development (Saini et al., 2019). A wide range of industrial waste is being converted into usable nanomaterials using various methods (Table 9.1) to produce nanoparticles of various sizes and structures. Bulk materials or materials at their microscale can be transformed into their nanoscale by utilizing various forms of energy such as chemical and mechanical in the top-down method. Even in bottom-up, ions are fused together to compose the particles at their nanoscale (Kefeni

TABLE 9.1 List of industrial waste–derived nanomaterial with their particle size.

S. no.	Industrial waste used to prepare NPs	Synthesis technique	Prepared NPs and size	Application of the nanoparticles	References
1	Iron oxide waste	Planetary wet ball milling technique	Iron oxide (10 nm)	Drug delivery Biology and medicine	Boruah et al. (2020), Ali et al. (2016)
2	Waste sludge	Hydrothermal carbonization	Fe-doped biochar (10–20 nm)	To reduced trace element buildup in soil and plants	Zang et al. (2020), Naeem et al. (2022)
3	Marble waste	Nitric acid treatment	Nano-$CaCO_3$ (70 nm)	Antimicrobial agent, drug and gen delivery, agent in therapeutic purpose	Yang et al. (2020)
4	Waste iron	Hydrothermal method	Magnetic iron oxide Nanomaterial (60–100 nm)	MRI, hyperthermia, drug Delivery in vitro bio-separation	Biswas et al. (2020), Vangijzegem et al. (2019)
5	Biomass power plant fly ash	Alkaline extraction method	Biosilica nanoparticles (20–40 nm)	Advanced catalysis, drug delivery and applications in biomedical, environmental remediation, and wastewater treatment	Liang et al. (2020), Nayl et al. (2022)
6	Printed circuit board waste etchants	Microemulsion process	Copper nanoparticles (20–50 nm)	To increase "contact killing," as well as to develop antibacterial and antiviral combination effects	Mdlovu et al. (2018), Xu et al. (2022)
7	Ferric + biological sludge	Hydrothermal carbonization	Magnetic biochar (200 nm)	To remediate water bodies with pollutants	Zhang et al. (2017), Silva et al. (2020)
8	Waste ferrous sulfate and pyrite were	Solid-phase reduction reaction	Magnetite nanoparticle (25–50 nm)	Medicine, cancer theragnostic, biosensing, catalysis, agriculture, and the environment.	Ali et al. (2021), Ren et al. (2017)
9	Co-fired ceramic waste	Leaching and wet chemical reduction	Ag nanoparticles (100 nm)	Antimicrobial, anticancer larvicidal wound healing, medicinal textiles and devices	Swain et al. (2017), Khan et al. (2020)
10	Aluminum foil waste	Coprecipitation using NH_4OH as a precipitant	Nanoscale single crystalline γ- and α-Al powder (36–200 nm)	Various industrial application	El-Amir et al. (2016)

(Continued)

TABLE 9.1 (Continued)

S. no.	Industrial waste used to prepare NPs	Synthesis technique	Prepared NPs and size	Application of the nanoparticles	References
11	Phosphogypsum waste	Calcination	Ca hydroxyapatite nanomaterial (50–57 nm)	Bone tissue engineering and drug delivery	Mousa et al. (2016), Halim et al. (2021)
12	Fly Ash	Hydrothermal process	Zeolite 3 (14–16 nm)	Catalysis, gas separation and ion exchange.	Visa et al. (2016)
13	Tannery wastewater	Precipitation and calcination	Nanomagnesium chromite, chromium oxide (20–400 nm)	Automobile and aero plane	Ali et al. (2016), Duan and Liu (2019)
14	Synthetic wastewater representing	Coprecipitation	Copper oxide Nanoparticles (5–50 nm)	As an Antimicrobial agent	Heuss-Aßbichler et al. (2016), Verma and Kumar (2019)
15	Tire waste	Chemical activation	Nano activated carbon (2–12 nm)	Drug delivery, medical implants, tissue engineering, wound biosensing, bioimaging, vaccination, and photodynamic therapy	Al-Rahbiand Williams (2016), Gaur et al. (2021)
16	Steel industry waste	Acid leaching and precipitation	Zero valent iron (ZVI, 303 and 1660 nm)	As a reductant	Mukherjee et al. (2015), Plessl et al. (2022)
17	Iron ore tailings	Ball milling, acid leaching and urea hydrolysis	Ag-Fe_3O_4 (<100 nm)	As an antimicrobial	Kumar et al. (2015), Taufiq et al. (2020)
18	Zn–Mn battery waste	Vacuum evaporation	Zn nanoparticles (100–300 nm)	As an antimicrobial and antioxidant	Xiang et al. (2015), Tiwari et al. (2022a,b)
19	Cathode-ray tube funnel glass waste	Vacuum carbon–thermal reduction and inert gas consolidation procedures	Pb nanoparticles (4–34 nm)	Magnetic data storage, magnetic resonance imaging (MRI)	Xiang et al. (2015), Bratovcic (2020)
20	Anaerobic sludge	Hydrothermal deposition, chemical treatment	Sludge carbon/TiO_2 nanocomposites (2–22 nm)	As a catalytic oxidative	Athalathil et al. (2015), Ramankutty (2015)
21	Papermill sludge	Calcination	PMS Fe_3O_4 nanocomposite (18 nm)	Wastewater	Zhou et al. (2015), Ghanbari et al. (2019)
22	Aluminum and magnesium scrap from foundries	Coprecipitation	Magnesium aluminate spinel (<50 nm)	Optically transparent ceramic, Neutron radiation resistance, humidity sensor	Ewais et al. (2015), Shi et al. (2020)
23	Waste silicon sludge	Alkali dissolution and acid precipitation	Silica nanoparticles (20–45 nm)	Advanced catalysis, drug delivery and biomedical applications, environmental remediation applications, and wastewater treatment	Ding et al. (2015), Nayl et al. (2022)

(Continued)

TABLE 9.1 (Continued)

S. no.	Industrial waste used to prepare NPs	Synthesis technique	Prepared NPs and size	Application of the nanoparticles	References
24	Steel pickling waste liquor	Coprecipitation method	Fe_3O_4 (20–50 nm)	As biomedical applications, biosensing, environmental applications for the removal of heavy metals and organic pollutants, and applications in energy storage devices.	Huang et al. (2014), Nguyen et al. (2021)
25	Synthetic tin containing wastewater spherical shape	Hydrothermal method	Tin sulfide nanomaterial (<100 nm)	Photodetectors, solar cells, photochemical cells, battery anodes, and gas sensors	Bhardwaj et al. (2014), Koteeswara Reddy et al. (2015)
26	Printed circuit boards sludge	Acid leaching, chemical exchange, and ferrite process	$CuFe_2O_4$ (20–120 nm)	Magnetic resonance imaging, drug delivery, and cell labeling	Tu et al. (2013), López-Moreno et al. (2016)
27	Sulfate leach liquor of red mud	Chemical precipitation	TiO_2 nanoparticle (200 nm)	Toothpaste, pharmaceuticals, coatings, papers, inks, plastics, food products, cosmetics and textile	Tsakiridis et al. (2011), Waghmode et al. (2019)
28	Prepared Sulfate leach liquor of red mud	Chemical precipitation	TiO_2 nanopowders (200 nm)	Reducing toxicity of dyes and pharmaceutical drugs; wastewater treatment; reproduction of silkworm	Tsakiridis et al. (2011), Waghmode et al. (2019)
29	Steel pickling waste liquor	Sodium borohydride reduction	Nano-Ni/Fe particles (20–50 nm)	Nitrate removal from water as an alternative to biological processes	Fang et al. (2011), Valiyeva et al. (2019)
30	PCB sludge	Microwave hydrothermal	CuO NPs (13–30 nm)	Antimicrobial agents	Wu et al. (2009), Grigore et al. (2016)

et al., 2017). Coprecipitation (Fan et al., 2017; Zhang et al., 2017), the hydrothermal method (Zhang et al., 2017), microwave irradiation (Zhang et al., 2012), and the top-down approach, including the ball milling technique (Kumar et al., 2015; Lu et al., 2017), are examples of these methods (Mallampati et al., 2018).

9.5 Various types of nanomaterial synthesis

Various synthesis techniques, such as coprecipitation methods, hydrothermal methods, carbothermal, solvothermal, and microwave-assisted methods are utilized in the formation of the nanoparticles from environmental pollutant (Kefeni et al., 2017). The synthesized nanomaterials from industrial waste materials show more complex features (Tripathi et al., 2021; Shukla et al., 2022; Jha et al., 2022). Surface modifications of nanoparticles are becoming more important for industrial water waste remediation due to the critical role of having both selectivity and aqueous stability (Iravani and Varma, 2022). Detailed information about various methods used for the preparation of nanoparticles with respect to their pros and cons is discussed below.

9.5.1 Coprecipitation method

It is a type of wet chemical method that is mostly used in the synthesis of the nanoparticles from industrial waste. This technique is utilized frequently because of its modest procedure and great production of nanomaterials (Wang et al., 2020a).

9.5.2 Hydrothermal method

The hydrothermal method is an exquisite technique to convert waste into wealth. In this process, high temperatures between 120°C and 550°C and pressures 20–150 bar in the existence of oxygen or nitrogen like inert gases are used to make a useful product from waste (Munir et al., 2018). The hydrothermal technique of processing is analogous to the solvothermal technique, but in this technique, water is used in place of various solvents (Khan, 2020). The hydrothermal technique is very significant for degrading carbon-rich waste, which is a major cause of the environmental problem. The product developed by this technique contains a C = O like functional group that has the potential to produce nascent oxygen species that are able to reduce organic pollutants (Zang et al., 2020).

The hydrothermal technique is considered more effective than high-temperature pyrolysis for the treatment of carbon-rich wet solid waste because it is a mild reaction and doesn't require drying the waste raw substances, along with low energy consumption (Zang et al., 2020). One of the merits of this technique is that size of materials, morphology of materials, crystalline phase, and surface chemistry can be controlled by directing temperature, pressure, and solvent properties of the reaction; however, high temperature and pressure for the reaction are a demerit of this technique as well.

9.5.3 Microwave-supported synthesis

In this method of synthesis, microwave energy is utilized, which strikes the material through the molecular and ionic interactions of the precursor solvents and reducing agents (Mondal et al., 2020). The treatment of microwave radiation can also be used to immobilize metal ions within the sediment as well as detoxify sludge (Wu et al., 2009). As it is a thermally more stable and high-yielding technique, it is therefore considered an efficient method for the synthesis of nanomaterials (Zhang et al., 2012). Magnetic biochar was prepared with iron sludge and cotton stalk in another study of Duan et al. (2017). This is used for the removal of Cr with the help of microwave chemical reactor (Zhang et al., 2012) and successfully prepared hydroxyapatite nanoparticles from the waste of phosphor gypsum. This method has gained a lot of attention as it is a quick method for the manufacture of crystalline nanoparticles (Khan, 2020).

9.5.4 Pyrolysis

One of the most effective methods for better surface area management with stable configurations and value-added superficial functional groups such as hydroxyl and ketone is pyrolysis (Ho et al., 2018). A change in the surface chemistry of desired materials can be done easily due to the temperature and reaction time used in pyrolysis. The material gains affinity for adsorbing water due to a change in its surface. The discharge of highly toxic gases and extraordinary energy consumption are some of the drawbacks of this method (Wang et al., 2020b).

9.5.5 The sol–gel method

The sol–gel method is another well-known technique for the synthesis of NPs with a smaller crystallite size. Precursor molecules are mostly inorganic or organic compounds. Some important factors that readily impact the properties of gel are the solvent type, acid or base content, temperature, etc. Final products in the form of NPs are highly useful as they have a high surface area and a stable surface (Yilmaz and Soylak, 2020). Cr-doped alumina nanocompounds were formed by using sol–gel integration from tannery waste (da Costa Cunha et al., 2016).

9.5.6 Calcination

Calcinations can be used to establish the transformation of waste into oxides and mixed oxides. Hou et al. (2021) found that for the simple metal oxides were the result when the process of calcination, temperature is below 400°C, whereas for spinel oxide nanocomposites formation, temperature much higher to 600°C. As a result of controlling the calcination temperature, the desired NPs were successfully synthesized.

9.6 Mechanism of nanomaterial synthesis from industrial waste

Numerous types of NPs can be synthesized by a number of procedures after apt pretreatment. Grinding and milling are two of the main common physical pretreatment methodologies (Samaddar et al., 2018). A sodium borohydride reduction method was utilized for the preparation of metallic NPs. For the production of NPs, freshly prepared sodium borohydride is quickly added in the solution of waste material. These are then splashed cyclically with milli Q water and ethanol for the elimination of surplus sodium borohydride (Fang et al., 2011). Solvent thermal methods are another method in which chemical reactions occur in a closed autoclave reactor containing solvent in which chemical reactions occur on heating at a critical temperature (Dubin et al., 2010). In the solvent thermal method, hematite NPs (Fe_2O_3) were synthesized with the help of ethanol, and the solution was heated with the material combination ($FeCl_3$ and NaOH) at a temperature of 150°C for 2 hours to obtain nanoparticles (Tang et al., 2011). Electrode position is another procedure for the synthesis of NPs that is used to endorse chemical reactions in aqueous solutions. With the help of this method, nanoporous materials, nanowires, and nanocylinders are prepared. Apart from these, green synthesis is a method for the synthesis of NPs that is eco-friendly and biocompatible. In green synthesis, nanoparticles are synthesized in various ways, such as with the help of plant extracts, fungi, bacteria, etc. From electronic waste such as printed circuit board (PCB), cell phones, automobiles, and laptops, metal nanoparticles of Pb, Au, Fe, Cu, and Pt, etc., can be synthesized (Singh and Lee, 2016; Vermisoglou et al., 2016). Other types of e-waste materials are also being utilized for valuable materials.

The acidic cupric chloride waste materials were used to prepared the Cu nanoparticles that are utilized in the formation of printed circuit boards by the process of microemulsion (Mdlovu et al., 2018). Such as aqueous solutions of copper sulfate from the boards of printed circuit that were chemically reduced to synthesized organically stabilized nanoparticles of Cu. Wastes of printed circuit boards (PCB) are also reprocessed to found metals such as Au Fe, Pb, Cu, and Hg (Cui and Anderson, 2016). The metals obtained from printed circuit boards can be utilized for the synthesis of nanoparticles. Cu obtained from WPCBs acts like initial material for the preparation of Cu nanoparticles (El-Nasr et al., 2020). Waste of printed circuit boards can also be directly used to synthesize nanoparticles. Copper oxide photocatalysts can be recuperated using electrokinetic mechanism. Cooper oxide nanocompounds of varying sizes can be prepared by supercritical water oxidation methods and electrokinetic mechanism (Xiu and Zhang, 2012).

The fabrication of copper–tin (Cu–Sn) NPs from the WPCBs of spent computers was confirmed with the help of selective thermal transformation and separation of toxic lead (Pb) and antimony (Shokri et al., 2017). Lead nanoparticles were also prepared from WPCB solders under vacuum conditions with forced flow inert gas condensation. Some other types of e-waste can also be synthesized into nanoparticles such as Cu_2O nanoparticles, purified carbon nanotubes, and Cu_2O catalysts (Søndergaard et al., 2016). A leaching solution can be used to create bimetallic nanoparticles such as Ag/Cu, Ag, and Cu nanoparticles (Elsayed et al., 2020).

This method was also used for reusing nanoparticles, such as functionalized nanoparticles that can be successfully recycled for capturing toxin materials from spiked blood plasma samples with the help of glycine buffer for freeing up after using the NPs (Hassanain et al., 2017). And the obtained ZnO NPs are spherical in shape and approximately 50 nm (Farzana et al., 2018). According to Dutta et al. (2018), the recycling of waste electrical devices into synthesized nanocompounds is a new period in the field of nanotechnology and environmental research.

9.6.1 Carbon nanotubes

In diverse systems such as quartz tubes, crucibles, and autoclaves, the synthesis of carbon nanotubes (CNTs) is carried out (Bazargan and McKay, 2012). The aim of each methodology is to break down solid-state polymers in the form of their carbon precursors through the method of pyrolysis (Zhuo and Levendis, 2014). Two waste polyolefin materials, namely polypropylene (PP) and polyethylene (PE), can be utilized for the synthesis of carbon materials. Some other waste plastic materials, integrated with some added materials such as polyvinyl alcohol and PET, can also play very vital role for the preparation of CNTs (Deng et al., 2016). Gong et al. established a fresh mechanism for formation of nanotube, which is a layer-by-layer gathering mechanism, in which leftover polypropylene was catalyzed in a quartz tube reactor with the aid of activated carbon and Ni Polypropylene and was effortlessly broken down into its light hydrocarbon constituents using activated carbon. Activated carbon also plays an efficient role in catalytic conversion with Ni via increasing dehydrogenation and aromatization process through the synthesis of various aromatic groups (Gong et al., 2012).

9.6.2 Graphene

The atoms of graphene are arranged into a honeycomb-like lattice structure. Novoselov et al. (2012) reported that the high electrical conductivity and good chemical constancy are characteristics of graphene. Due to optical and electromechanical characteristics of graphenes (Marinho et al., 2012), it plays a significant role in all major sectors such as health, energy, and the environment (Surwade et al., 2015). Graphite and organic molecules are two of the essential raw materials from which graphene is manufactured. In the bottom-up approach, chemical vapor deposition is deposited on metallic films, while in the top-down approach, graphite crystal liquid exfoliation takes place (Coleman, 2013). HF and HNO_3 were immersed after cooling. HF has the ability to dissolve the impurities, and HNO_3 is capable of oxidizing the amorphous carbon (Gong et al., 2014).

9.7 Summary and future prospects

As discussed in this chapter, the production of industrial waste materials is indispensable, and if not controlled or try to reduce, the environment and human health are widely affected by it. Regenerative, recyclable, and reused industrial waste materials have made them an alternative source of economy. Various types of synthesis methods are discussed. Among them, hydrothermal synthesis via green root may be approved as a tool kit for the preparation of nanomaterials by using industrial waste. The chemically synthesized nanomaterials have antibacterial, antimicrobial, and antioxidant properties, but still, they have a toxic nature due to chemical reactions. Therefore biological or green synthesis can play an important role to synthesize nanomaterials using industrial waste due to their toxic-free nature and also the least energy demands. Two-dimensional nanomaterials such as graphene and corban nanotube were also synthesized by using industrial's waste materials. The study needs more improvement for the fabrication/synthesis of various types of nanomaterials with their specific shape and size without expending a huge amount of energy. Further study is needed to synthesize nanomaterials having negotiable toxic properties. There is one drawback in the synthesis of nanomaterials the chemical composition of various types of industrial waste, collected from different places around the globe will vary as a result the synthesis process at a large scale is not uniform. Further studies are also needed for the degradation time of various types of industrial waste materials so that they can collect timely. However, the earlier incredible effort has been kept for the synthesis of nanoparticles from industrial waste materials. Industrial waste–derived nanomaterials have the potential to be utilized broadly in the medical field, for therapeutic drugs, antimicrobial activity, antioxidant activity, water treatments, and food packaging. The syntheses of nanoparticles from industrial waste materials have lined a pathway to develop an eco-friendly and cost-effective process that confines the usage of hazardous chemical substances. With the extensive availability of industrial waste materials and biologically active biomolecules, the synthesis of nanoparticles from industrial waste may act as an alternative source.

9.8 Conclusion

These waste materials are not easily degradable and thus lead to all types of pollution as well as several diseases. This chapter highlights the negative aspects of industrial waste and problems regarding the remediation of these waste materials. The chapter also discusses the use of nanotechnology in the degradation of these waste materials, with a particular emphasis on developing novel nanomaterials from this industrial waste. Different types of waste materials can be converted into nanomaterials by different methods. These methods of synthesizing nanoparticles from waste can be more favorable as this waste can be reusable in several applications. These novel synthesized nanomaterials are less toxic and eco-friendly and are useful in electronics, space technology, packaging, etc. The use of biological approaches in production of nanoparticles provides a broad range of applications such as drug loading and medicine.

Acknowledgments

The authors are thankful to Head, Department of Botany, University of Allahabad, Prayagraj.

References

Abdelbasir, S.M., McCourt, K.M., Lee, C.M., Vanegas, D.C., 2020. Waste-derived nanoparticles: synthesis approaches, environmental applications, and sustainability considerations. Frontiers in Chemistry 8, 782.

Ali, A., Shah, T., Ullah, R., Zhou, P., Guo, M., Ovais, M., et al., 2021. Review on recent progress in magnetic nanoparticles: synthesis, characterization, and diverse applications. Frontiers in Chemistry 9, 629054.

Ali, A., Zafar, H., Zia, M., Ul Haq, I., Phull, A.R., Ali, J.S., et al., 2016. Synthesis, characterization, applications, and challenges of iron oxide nanoparticles. Nanotechnology, Science and Applications 9, 49.

Al-Rahbi, A.S., Williams, P.T., 2016. Production of activated carbons from waste tyres for low temperature NOx control. Waste Management 49, 188–195.

Athalathil, S., Font, J., Fortuny, A., Stüber, F., Bengoa, C., Fabregat, A., 2015. New sludge-based carbonaceous materials impregnated with different metals for anaerobic azo-dye reduction. Journal of Environmental Chemical Engineering 3 (1), 104–112. Available from: https://doi.org/10.1016/j.jece.2014.07.002.

Bazargan, A., McKay, G., 2012. A review–synthesis of carbon nanotubes from plastic wastes. Chemical Engineering Journal 195, 377–391.

Berkmans, A.J., Jagannatham, M., Priyanka, S., Haridoss, P., 2014. Synthesis of branched, nano channeled, ultrafine and nano carbon tubes from PET wastes using the arc discharge method. Waste Management 34 (11), 2139–2145.

Bhardwaj, R.K., Bharti, V., Sharma, A., Mohanty, D., Agrawal, V., Vats, N., et al., 2014. Green approach for in-situ growth of CdS nanorods in low band gap polymer network for hybrid solar cell applications. Advances in Nanoparticles 2014.

Bhardwaj, A.K., Naraian, R., Sundaram, S., Kaur, R., 2022. Biogenic and non-biogenic waste for the synthesis of nanoparticles and their applications. Bioremediation. CRC Press, pp. 207–218.

Biswas, A., Patra, A.K., Sarkar, S., Das, D., Chattopadhyay, D., De, S., 2020. Synthesis of highly magnetic iron oxide nanomaterials from waste iron by one-step approach. Colloids and Surfaces A: Physicochemical and Engineering Aspects 589, 124420.

Boruah, P.K., Yadav, A., Das, M.R., 2020. Magnetic mixed metal oxide nanomaterials derived from industrial waste and its photocatalytic applications in environmental remediation. Journal of Environmental Chemical Engineering 8 (5), 104297.

Bratovcic, A., 2020. Synthesis, characterization, applications, and toxicity of lead oxide nanoparticles. Lead Chem 6, 66.

Coleman, J.N., 2013. Liquid exfoliation of defect-free graphene. Accounts of Chemical Research 46 (1), 14–22.

Cui, H., Anderson, C.G., 2016. Literature review of hydrometallurgical recycling of printed circuit boards (PCBs). Journal of Advanced Chemical Engineering 6 (1), 142–153.

da Costa Cunha, G., Peixoto, J.A., de Souza, D.R., Romão, L.P.C., Macedo, Z.S., 2016. Recycling of chromium wastes from the tanning industry to produce ceramic nano-pigments. Green Chemistry 18 (19), 5342–5356.

Deng, J., You, Y., Sahajwalla, V., Joshi, R.K., 2016. Transforming waste into carbon-based nanomaterials. Carbon 96, 105–115.

Ding, H., Li, J., Gao, Y., Zhao, D., Shi, D., Mao, G., et al., 2015. Preparation of silica nanoparticles from waste silicon sludge. Powder Technology 284, 231–236.

Duan, S., Ma, W., Pan, Y., Meng, F., Yu, S., Wu, L., 2017. Synthesis of magnetic biochar from iron sludge for the enhancement of Cr (VI) removal from solution. Journal of the Taiwan Institute of Chemical Engineers 80, 835–841.

Duan, X., Liu, N., 2019. Magnesium for dynamic nano-plasmonics. Accounts of Chemical Research 52 (7), 1979–1989.

Dubin, S., Gilje, S., Wang, K., Tung, V.C., Cha, K., Hall, A.S., et al., 2010. A one-step, solvothermal reduction method for producing reduced graphene oxide dispersions in organic solvents. ACS Nano 4 (7), 3845–3852.

Dutta, T., Kim, K.H., Deep, A., Szulejko, J.E., Vellingiri, K., Kumar, S., et al., 2018. Recovery of nanomaterials from battery and electronic wastes: a new paradigm of environmental waste management. Renewable and Sustainable Energy Reviews 82, 3694–3704.

El-Amir, A.A., Ewais, E.M., Abdel-Aziem, A.R., Ahmed, A., El-Anadouli, B.E., 2016. Nano-alumina powders/ceramics derived from aluminum foil waste at low temperature for various industrial applications. Journal of Environmental Management 183, 121–125.

El-Nasr, R.S., Abdelbasir, S.M., Kamel, A.H., Hassan, S.S., 2020. Environmentally friendly synthesis of copper nanoparticles from waste printed circuit boards. Separation and Purification Technology 230, 115860.

Elsayed, D.M., Abdelbasir, S.M., Abdel-Ghafar, H.M., Salah, B.A., Sayed, S.A., 2020. Silver and copper nanostructured particles recovered from metalized plastic waste for antibacterial applications. Journal of Environmental Chemical Engineering 8 (4), 103826.

Ewais, E.M., Besisa, D.H., El-Amir, A.A., El-Sheikh, S.M., Rayan, D.E., 2015. Optical properties of nanocrystalline magnesium aluminate spinel synthesized from industrial wastes. Journal of Alloys and Compounds 649, 159–166.

Fan, S., Wang, Y., Li, Y., Tang, J., Wang, Z., Tang, J., et al., 2017. Facile synthesis of tea waste/Fe_3O_4 nanoparticle composite for hexavalent chromium removal from aqueous solution. RSC Advances 7 (13), 7576–7590.

Fang, Z., Qiu, X., Chen, J., Qiu, X., 2011. Degradation of the polybrominated diphenyl ethers by nanoscale zero-valent metallic particles prepared from steel pickling waste liquor. Desalination 267 (1), 34–41.

Farzana, R., Rajarao, R., Behera, P.R., Hassan, K., Sahajwalla, V., 2018. Zinc oxide nanoparticles from waste Zn-C battery via thermal route: characterization and properties. Nanomaterials 8 (9), 717.

Gaines, L., 2014. The future of automotive lithium-ion battery recycling: charting a sustainable course. Sustainable Materials and Technologies 1, 2–7.

Gaur, M., Misra, C., Yadav, A.B., Swaroop, S., Maolmhuaidh, F.Ó., Bechelany, M., et al., 2021. Biomedical applications of carbon nanomaterials: fullerenes, quantum dots, nanotubes, nanofibers, and graphene. Materials 14 (20), 5978.

Ghanbari, F., Ahmadi, M., Gohari, F., 2019. Heterogeneous activation of peroxymonosulfate via nanocomposite CeO_2-Fe_3O_4 for organic pollutants removal: the effect of UV and US irradiation and application for real wastewater. Separation and Purification Technology 228, 115732.

Gong, J., Liu, J., Wan, D., Chen, X., Wen, X., Mijowska, E., et al., 2012. Catalytic carbonization of polypropylene by the combined catalysis of activated carbon with Ni_2O_3 into carbon nanotubes and its mechanism. Applied Catalysis A: General 449, 112–120.

Gong, J., Liu, J., Wen, X., Jiang, Z., Chen, X., Mijowska, E., et al., 2014. Upcycling waste polypropylene into graphene flakes on organically modified montmorillonite. Industrial & Engineering Chemistry Research 53 (11), 4173–4181.

Grigore, M.E., Biscu, E.R., Holban, A.M., Gestal, M.C., Grumezescu, A.M., 2016. Methods of synthesis, properties and biomedical applications of CuO nanoparticles. Pharmaceuticals 9 (4), 75.

Halim, N.A.A., Hussein, M.Z., Kandar, M.K., 2021. Nanomaterials-upconverted hydroxyapatite for bone tissue engineering and a platform for drug delivery. International Journal of Nanomedicine 16, 6477.

Hassanain, W.A., Izake, E.L., Schmidt, M.S., Ayoko, G.A., 2017. Gold nanomaterials for the selective capturing and SERS diagnosis of toxins in aqueous and biological fluids. Biosensors and Bioelectronics 91, 664–672.

Heuss-Aßbichler, S., John, M., Klapper, D., Bläß, U.W., Kochetov, G., 2016. Recovery of copper as zero-valent phase and/or copper oxide nanoparticles from wastewater by ferritization. Journal of Environmental Management 181, 1–7.

Ho, S.H., Wang, D., Wei, Z.S., Chang, J.S., Ren, N.Q., 2018. Lead removal by a magnetic biochar derived from persulfate-ZVI treated sludge together with one-pot pyrolysis. Bioresource Technology 247, 463–470.

Hou, H., Liu, Z., Zhang, J., Zhou, J., Qian, G., 2021. A review on fabricating functional materials by heavy metal–containing sludges. Environmental Science and Pollution Research 28 (1), 133–155.

Huang, R., Fang, Z., Fang, X., Tsang, E.P., 2014. Ultrasonic Fenton-like catalytic degradation of bisphenol A by ferroferric oxide (Fe_3O_4) nanoparticles prepared from steel pickling waste liquor. Journal of Colloid and Interface Science 436, 258–266.

Iravani, S., Varma, R.S., 2022. Genetically engineered organisms: possibilities and challenges of heavy metal removal and nanoparticle synthesis. Clean Technologies 4 (2), 502–511.

Jha, S., Singh, R., Pandey, A., Bhardwaj, M., Tripathi, S.K., Mishra, R.K., et al., 2018. Bacterial toxicological assay of calcium oxide nanoparticles against some plant growth-promoting rhizobacteria. International Journal for Research in Applied Science and Engineering Technology 6 (11), 460–466.

Jha, S., Singh, R., Jha, G., Singh, P., Dikshit, A., 2022. Contamination and impacts of metals and metalloids on agro-environment. Metals Metalloids Soil Plant Water Systems. Academic Press, pp. 111–130.

Kefeni, K.K., Mamba, B.B., Msagati, T.A., 2017. Application of spinel ferrite nanoparticles in water and wastewater treatment: a review. Separation and Purification Technology 188, 399–422.

Khan, F.A., 2020. Synthesis of nanomaterials: methods & technology. Applications of Nanomaterials in Human Health. Springer, Singapore, pp. 15–21.

Khan, M., Khan, A.U., Alam, M.J., Park, S., Alam, M., 2020. Biosynthesis of silver nanoparticles and its application against phytopathogenic bacterium and fungus. International Journal of Environmental Analytical Chemistry 100 (12), 1390–1401.

Koolivand, A., Mazandaranizadeh, H., Binavapoor, M., Mohammadtaheri, A., Saeedi, R., 2017. Hazardous and industrial waste composition and associated management activities in Caspian industrial park, Iran. Environmental Nanotechnology, Monitoring & Management 7, 9–14.

Koteeswara Reddy, N., Devika, M., Gopal, E.S.R., 2015. Review on tin (II) sulfide (SnS) material: synthesis, properties, and applications. Critical Reviews in Solid State and Materials Sciences 40 (6), 359–398.

Kumar, R., Sakthivel, R., Behura, R., Mishra, B.K., Das, D., 2015. Synthesis of magnetite nanoparticles from mineral waste. Journal of Alloys and Compounds 645, 398–404.

Liang, G., Li, Y., Yang, C., Zi, C., Zhang, Y., Hu, X., et al., 2020. Production of biosilica nanoparticles from biomass power plant fly ash. Waste Management 105, 8–17.

López-Moreno, M.L., Avilés, L.L., Pérez, N.G., Irizarry, B.Á., Perales, O., Cedeno-Mattei, Y., et al., 2016. Effect of cobalt ferrite ($CoFe_2O_4$) nanoparticles on the growth and development of *Lycopersicon lycopersicum* (tomato plants). Science of the Total Environment 550, 45–52.

Lu, X., Ning, X.A., Lee, P.H., Shih, K., Wang, F., Zeng, E.Y., 2017. Transformation of hazardous lead into lead ferrite ceramics: crystal structures and their role in lead leaching. Journal of Hazardous Materials 336, 139–145.

Mallampati, S.R., Lee, B.H., Mitoma, Y., Simion, C., 2018. Sustainable recovery of precious metals from end-of-life vehicles shredder residue by a novel hybrid ball-milling and nanoparticles enabled froth flotation process. Journal of Cleaner Production 171, 66–75.

Marinho, B., Ghislandi, M., Tkalya, E., Koning, C.E., de With, G., 2012. Electrical conductivity of compacts of graphene, multi-wall carbon nanotubes, carbon black, and graphite powder. Powder Technology 221, 351–358.

Mdlovu, N.V., Chiang, C.L., Lin, K.S., Jeng, R.C., 2018. Recycling copper nanoparticles from printed circuit board waste etchants via a microemulsion process. Journal of Cleaner Production 185, 781–796.

Moghaddasi, S., Khoshgoftarmanesh, A.H., Karimzadeh, F., Chaney, R.L., 2013. Preparation of nano-particles from waste tire rubber and evaluation of their effectiveness as zinc source for cucumber in nutrient solution culture. ScientiaHorticulturae 160, 398–403.

Mondal, P., Anweshan, A., Purkait, M.K., 2020. Green synthesis and environmental application of iron-based nanomaterials and nanocomposite: a review. Chemosphere 259, 127509.

Mousa, S.M., Ammar, N.S., Ibrahim, H.A., 2016. Removal of lead ions using hydroxyapatite nano-material prepared from phosphogypsum waste. Journal of Saudi Chemical Society 20 (3), 357–365.

Mukherjee, R., Sinha, A., Lama, Y., Kumar, V., 2015. Utilization of zerovalent iron (ZVI) particles produced from steel industry waste for in-situ remediation of ground water contaminated with organo-chlorine pesticide heptachlor. International Journal of Environmental Research 9 (1), 19–26.

Munir, M.T., Mansouri, S.S., Udugama, I.A., Baroutian, S., Gernaey, K.V., Young, B.R., 2018. Resource recovery from organic solid waste using hydrothermal processing: opportunities and challenges. Renewable and Sustainable Energy Reviews 96, 64–75.

Naeem, M.A., Abdullah, M., Imran, M., Shahid, M., Abbas, G., Amjad, M., et al., 2022. Iron oxide nanoparticles doped biochar ameliorates trace elements induced phytotoxicity in tomato by modulation of physiological and biochemical responses: implications for human health risk. Chemosphere 289, 133203.

Nayl, A.A., Abd-Elhamid, A.I., Aly, A.A., Bräse, S., 2022. Recent progress in the applications of silica-based nanoparticles. RSC Advances 12 (22), 13706–13726.

Nguyen, M.D., Tran, H.V., Xu, S., Lee, T.R., 2021. Fe_3O_4 nanoparticles: structures, synthesis, magnetic properties, surface functionalization, and emerging applications. Applied Sciences 11 (23), 11301.

Novoselov, K.S., Colombo, L., Gellert, P.R., Schwab, M.G., Kim, K., 2012. A roadmap for graphene. Nature 490 (7419), 192–200.

Plessl, K., Russ, A., Vollprecht, D., 2022. Application and development of zero-valent iron (ZVI) for groundwater and wastewater treatment. International Journal of Environmental Science and Technology 1–16.

Ramankutty, S.A., 2015. TiO_2-Sludge carbon enhanced catalytic oxidative reaction in environmental wastewaters applications. Conversion of Sludge B into Catalysts: Environ Treatment Appl 147.

Ren, G., Yang, L., Zhang, Z., Zhong, B., Yang, X., Wang, X., 2017. A new green synthesis of porous magnetite nanoparticles from waste ferrous sulfate by solid-phase reduction reaction. Journal of Alloys and Compounds 710, 875–879. Available from: https://doi.org/10.1016/j.jallcom.2017.03.337.

Saini, D., Aggarwal, R., Anand, S.R., Sonkar, S.K., 2019. Sunlight induced photodegradation of toxic azo dye by self-doped iron oxide nano-carbon from waste printer ink. Solar Energy 193, 65–73.

Salunke, B.K., Sawant, S.S., Lee, S.I., Kim, B.S., 2016. Microorganisms as efficient biosystem for the synthesis of metal nanoparticles: current scenario and future possibilities. World Journal of Microbiology and Biotechnology 32 (5), 1–16.

Samaddar, P., Ok, Y.S., Kim, K.H., Kwon, E.E., Tsang, D.C., 2018. Synthesis of nanomaterials from various wastes and their new age applications. Journal of Cleaner Production 197, 1190–1209.

Schwarz, M., Veverka, M., Michalková, E., Lalík, V., Veverková, D., 2012. Utilisation of industrial waste for ferrite pigments production. Chemical Papers 66 (4), 248–258.

Shi, Z., Zhao, Q., Guo, B., Ji, T., Wang, H., 2020. A review on processing polycrystalline magnesium aluminate spinel ($MgAl_2O_4$): sintering techniques, material properties and machinability. Materials & Design 193, 108858.

Shokri, A., Pahlevani, F., Levick, K., Cole, I., Sahajwalla, V., 2017. Synthesis of copper-tin nanoparticles from old computer printed circuit boards. Journal of Cleaner Production 142, 2586–2592.

Shukla, M., Shukla, R., Jha, S., Singh, R., Dikshit, A., 2022. Myco-remediation: a sustainable biodegradation of environmental pollutants. Sustainable Management of Environmental Contaminants: Eco-friendly Remediation Approaches. Springer International Publishing, Cham, pp. 425–449.

Silva, T.C.F., VergÜtz, L., Pacheco, A.A., Melo, L.F., Renato, N.S., Melo, L.C., 2020. Characterization and application of magnetic biochar for the removal of phosphorus from water. Anais da Academia Brasileira de Ciências 92.

Singh, J., Lee, B.K., 2016. Recovery of precious metals from low-grade automobile shredder residue: a novel approach for the recovery of nanozerovalent copper particles. Waste management 48, 353–365.

Soliman, N.K., Moustafa, A.F., 2020. Industrial solid waste for heavy metals adsorption features and challenges; a review. Journal of Materials Research and Technology 9 (5), 10235–10253.

Søndergaard, R.R., Zimmermann, Y.S., Espinosa, N., Lenz, M., Krebs, F., 2016. Incineration of organic solar cells: efficient end of life management by quantitative silver recovery. Energy & Environmental Science 9 (3), 857–861.

Surwade, S.P., Smirnov, S.N., Vlassiouk, I.V., Unocic, R.R., Veith, G.M., Dai, S., et al., 2015. Water desalination using nanoporous single-layer graphene. Nature Nanotechnology 10 (5), 459–464.

Swain, B., Shin, D., Joo, S.Y., Ahn, N.K., Lee, C.G., Yoon, J.H., 2017. Selective recovery of silver from waste low-temperature co-fired ceramic and valorization through silver nanoparticle synthesis. Waste Management 69, 79–87.

Tang, W., Li, Q., Gao, S., Shang, J.K., 2011. Arsenic (III, V) removal from aqueous solution by ultrafine α-Fe_2O_3 nanoparticles synthesized from solvent thermal method. Journal of Hazardous Materials 192 (1), 131–138.

Taufiq, A., Saputro, R.E., Susanto, H., Hidayat, N., Sunaryono, S., Amrillah, T., et al., 2020. Synthesis of Fe_3O_4/Ag nanohybrid ferrofluids and their applications as antimicrobial and antifibrotic agents. Heliyon 6 (12), e05813.

Teague, C., 2005. United States national nanotechnology initiative. Epidemiology (Cambridge, Mass.) 16 (5), S153.

Tiwari, A.K., Jha, S., Agrawal, R., Mishra, S.K., Pathak, A.K., Singh, A.K., et al., 2022a. Anti-bacterial efficacy of phyto-synthesized zinc oxide nanoparticles using *Murraya paniculata* l. leaf extract. Sciences 4 (7).

Tiwari, A.K., Jha, S., Singh, A.K., Mishra, S.K., Pathak, A.K., Ojha, R.P., et al., 2022b. Innovative investigation of zinc oxide nanoparticles used in dentistry. Crystals 12 (8), 1063.

Tripathi, A., Singh, R., Jha, S., Pandey, A., Dikshit, A., 2021. Nano-biorem: a new concept toward remedial study. Heavy Metal Toxicity in Plants. CRC Press, pp. 181–192.

Tsakiridis, P.E., Oustadakis, P., Katsiapi, A., Perraki, M., Agatzini-Leonardou, S., 2011. Synthesis of TiO_2 nano-powders prepared from purified sulphate leach liquor of red mud. Journal of Hazardous Materials 194, 42–47.

Tu, Y.J., You, C.F., Chang, C.K., Wang, S.L., Chan, T.S., 2013. Adsorption behavior of As (III) onto a copper ferrite generated from printed circuit board industry. Chemical Engineering Journal 225, 433–439.

Valiyeva, G.G., Bavasso, I., Di Palma, L., Hajiyeva, S.R., Ramazanov, M.A., Hajiyeva, F.V., 2019. Synthesis of Fe/Ni bimetallic nanoparticles and application to the catalytic removal of nitrates from water. Nanomaterials 9 (8), 1130.

Vangijzegem, T., Stanicki, D., Laurent, S., 2019. Magnetic iron oxide nanoparticles for drug delivery: applications and characteristics. Expert Opinion on Drug Delivery 16 (1), 69–78.

Verma, N., Kumar, N., 2019. Synthesis and biomedical applications of copper oxide nanoparticles: an expanding horizon. ACS Biomaterials Science & Engineering 5 (3), 1170–1188.

Vermisoglou, E.C., Giannouri, M., Todorova, N., Giannakopoulou, T., Lekakou, C., Trapalis, C., 2016. Recycling of typical supercapacitor materials. Waste Management & Research 34 (4), 337–344.

Visa, M., Andronic, L., Enesca, A., 2016. Behavior of the new composites obtained from fly ash and titanium dioxide in removing of the pollutants from wastewater. Applied Surface Science 388, 359–369.

Waghmode, M.S., Gunjal, A.B., Mulla, J.A., Patil, N.N., Nawani, N.N., 2019. Studies on the titanium dioxide nanoparticles: biosynthesis, applications and remediation. SN Applied Sciences 1 (4), 1–9.

Wang, M., Tian, Y., Liu, W., Zhang, R., Chen, L., Luo, Y., et al., 2020a. A moving urban mine: the spent batteries of electric passenger vehicles. Journal of Cleaner Production 265, 121769.

Wang, S., Yan, W., Zhao, F., 2020b. Recovery of solid waste as functional heterogeneous catalysts for organic pollutant removal and biodiesel production. Chemical Engineering Journal 401, 126104.

Wu, C., Kuo, C., Lo, S., 2009. From waste to resource: a case study of heavy-metal sludge by microwave treatment. Journal of Environmental Engineering Management 19, 119–125.

Xiang, X., Xia, F., Zhan, L., Xie, B., 2015. Preparation of zinc nano structured particles from spent zinc manganese batteries by vacuum separation and inert gas condensation. Separation and Purification Technology 142, 227–233.

Xiu, F.R., Zhang, F.S., 2012. Size-controlled preparation of Cu_2O nanoparticles from waste printed circuit boards by supercritical water combined with electrokinetic process. Journal of Hazardous Materials 233, 200–206.

Xu, V.W., Nizami, M.Z.I., Yin, I.X., Yu, O.Y., Lung, C.Y.K., Chu, C.H., 2022. Application of copper nanoparticles in dentistry. Nanomaterials 12 (5), 805.

Yang, H., Yan, Y., Hu, Z., 2020. The preparation of nano calcium carbonate and calcium silicate hardening accelerator from marble waste by nitric acid treatment and study of early strength effect of calcium silicate on C30 concrete. Journal of Building Engineering 32, 101507. Available from: https://doi.org/10.1016/j.jobe.2020.101507.

Yilmaz, E., Soylak, M., 2020. Functionalized nanomaterials for sample preparation methods. Handbook of Nanomaterials in Analytical Chemistry. Elsevier, pp. 375–413. Available from: https://doi.org/10.1016/B978-0-12-816699-4.00015-3.

Yot, P.G., Méar, F.O., 2009. Influence of AlN, TiN and SiC reduction on the structural environment of lead in waste cathode-ray tubes glass: an x-ray absorption spectroscopy study. Journal of Physics: Condensed Matter 21 (28), 285104. Available from: https://doi.org/10.1088/0953-8984/21/28/285104.

Zang, T., Wang, H., Liu, Y., Dai, L., Zhou, S., Ai, S., 2020. Fe-doped biochar derived from waste sludge for degradation of rhodamine B via enhancing activation of peroxymonosulfate. Chemosphere 261, 127616.

Zhang, D., Luo, H., Zheng, L., Wang, K., Li, H., Wang, Y., et al., 2012. Utilization of waste phosphogypsum to prepare hydroxyapatite nanoparticles and its application towards removal of fluoride from aqueous solution. Journal of Hazardous Materials 241, 418–426.

Zhang, H., Liu, J., Ou, C., Shen, J., Yu, H., Jiao, Z., et al., 2017. Reuse of Fenton sludge as an iron source for $NiFe_2O_4$ synthesis and its application in the Fenton-based process. Journal of Environmental Sciences 53, 1–8.

Zhou, G., Fang, F., Chen, Z., He, Y., Sun, H., Shi, H., 2015. Facile synthesis of paper mill sludge-derived heterogeneous catalyst for the Fenton-like degradation of methylene blue. Catalysis Communications 62, 71–74.

Zhou, N., Price, L., Yande, D., Creyts, J., Khanna, N., Fridley, D., et al., 2019. A roadmap for China to peak carbon dioxide emissions and achieve a 20% share of non-fossil fuels in primary energy by 2030. Applied Energy 239, 793–819.

Zhuo, C., Levendis, Y.A., 2014. Upcycling waste plastics into carbon nanomaterials: a review. Journal of Applied Polymer Science 131 (4).

Chapter 10

Application and characterization of nonbiogenic synthesized nanomaterials

Devi Selvaraj[1] and Tharmaraj Vairaperumal[2,3]

[1]PG Department of Chemistry, Cauvery College for Women (Autonomous), Tiruchirappalli, Tamil Nadu, India, [2]Institute of Clinical Medicine, College of Medicine, National Cheng Kung University, Tainan, Taiwan ROC, [3]Environmental Science and Technology Laboratory, Department of Chemical Engineering, SRM Institute of Science and Technology, Kattankulathur, Tamil Nadu, India

10.1 Introduction

Nonbiogenic waste belongs to any waste material that does not contain organic matter or other than biologically derived substances such as plastic, glass, metals, and electronics (Gabryszewska and Gworek, 2020). Nonbiogenic waste causes a significant environmental issue, more years to take to decompose leads to long-lasting pollution and environmental damage (Manfredi et al., 2010). Improper disposal of nonbiogenic wastes also has serious issues for human health (Rada et al., 2018). Therefore, proper nonbiogenic waste management including proper disposal, recycling, and reducing waste is significantly protecting the environment and human health. Various technologies have been developed to reuse, reduce, and recycle nonbiogenic waste materials by converting them into new materials without producing hazardous materials (Alfarisa et al., 2015).

Recently, researchers are exploiting the promising approach of nanotechnology to provide a sustainable solution in the environment (Ibrahim et al., 2016). Nanomaterials have unique optical, chemical, and physical properties that enhanced their reactivity compared to bulk materials due to the high surface-to-volume ratio (Patil and Burungale, 2020). Nanomaterials can be classified into different types based on their size, morphology, and the properties such as metal nanoparticles (Rao et al., 2000), metal oxide nanoparticles (Chavali and Nikolova, 2019), carbon nanomaterials (Choudhary et al.,2014), semiconductor nanomaterials (Kumar and Nann, 2006), and organic nanoparticles (Horn and Rieger, 2001). The most commonly used nanomaterials' synthesis approaches are biogenic (biological) and nonbiogenic (physical and chemical) methods (Huston et al., 2021.; Hachem et al., 2022.; Arole and Munde, 2014). The biological method has been used for the synthesis of nanomaterials using various microorganisms such as bacteria, fungi, algae, and plant extracts (Jadoun et al., 2021, 2022). The physical approach to the synthesis of nanomaterials can be done by various methods such as laser ablation, physical vapor deposition (PVD), mechanical milling, thermal evaporation, sonochemical, etc. (Kulkarni, 2015; Krishnia et al., 2022). The chemical methods for the synthesis of nanomaterials have utilized varieties of techniques including chemical reduction (Tharmaraj and Pitchumani, 2011, 2013; Tharmaraj and Yang, 2014), sol–gel, solvothermal, hydrothermal, and reverse micelle (Reverberi et al.,2016).

Nanomaterials' characterization has been employed in a variety of techniques which are based on their physical, chemical, and optical properties including size, shapes, functional group, and elemental composition (Mourdikoudis et al., 2018). The most common characterization techniques are UV-Visible spectroscopy (Quevedo et al., 2021), Fourier transforms infrared spectroscopy (Eid, 2022), powder X-ray diffraction (XRD) (Cervellino et al., 2016), scanning electron microscopy (SEM) (Vladár and Hodoroaba, 2020), and transmission electron microscopy (TEM) (Kumar, 2013) with energy-dispersive X-ray spectroscopy (EDAX) (Iqbal et al., 2019).

Nonbiogenic waste materials such as e-waste, glass waste, industrial waste, and mining waste can be used as source materials for the synthesis of various types of nanomaterials such as metal, metal oxide, and composite materials (Bhattacharya et al., 2021; Samaddar et al., 2018; Hou et al.,2020). For instance, one-step synthesis of magnetic Fe_2O_3 nanomaterials from iron waste (Biswas et al., 2020), fibrous carbon nanomaterial from waste polyethylene terephthalate (Zhou et al., 2020), TiO_2 nanoparticles from industrial wastewater (Chowdhury et al., 2016), thermal method synthesis

Green and Sustainable Approaches Using Wastes for the Production of Multifunctional Nanomaterials. DOI: https://doi.org/10.1016/B978-0-443-19183-1.00024-6

of ZnO_2 nanoparticles from waste Zn-C battery (Farzana et al., 2018), carbon nanomaterial from plastic wastes (Zhuo and Levendis, 2014), graphene nanoplatelets from waste palm oil (Azam et al.,2018), solar grade SiO_2 nanoparticles from agro-wastes (Adebisi et al., 2017), Ag NPs from agro-industrial waste (Tade et al., 2020), and mesoporous silica nanoparticles synthesized from agricultural and industrial waste (Razak et al., 2022) have been achieved. These waste-derived nanomaterials have been utilized for several applications such as environmental pollutant remediation (Vinitha Judith and Vasudevan, 2021), heterogeneous photocatalysis (Oviedo et al., 2022), antibacterial activity (Tade et al., 2020), water purification (Romanovski, 2021), dye degradation (Singh et al., 2014), biodiesel production (Zhong et al., 2020), energy storage, catalysis, and sensing (Bhattacharya et al., 2021).

In this chapter, we focused on the nonbiogenic synthesis approaches of nanomaterials from various nonbiogenic waste products and their potential applications as seen in Fig. 10.1. The detailed discussion has been made about the synthesis of nanomaterials using nonbiogenic methods of chemical reduction, sol–gel, hydrothermal, solvothermal, pulsed laser ablation, electrochemical, microwave irradiation, UV irradiation, photochemical, sonochemical, and mechanical milling. Then, the characterization techniques of UV-Vis spectroscopy, Fourier transforms infrared spectroscopy (FT-IR), powder XRD, SEM, and TEM were discussed. In addition, recent advances in nonbiogenic synthesized nanomaterials are utilized for various potential applications such as water remediation, sensor, electrocatalyst, supercapacitor, and antifungal activity. Finally, we summarized and concluded the research challenges and future directions of nonbiogenic synthesized nanomaterials.

10.2 Impacts of nonbiogenic wastes

Globally, a huge amount of industrial waste, approximately 7.6 billion tons, is generated every year. These industrial wastes such as plastics, tires, and e-waste have negative impacts on the environment such as being highly reactive, radioactive, flammable, toxic, and corrosive. It is necessary to issue a solution for the recycling or conversion of waste into wealth products for the sustainable development of the environment (EPA's Guide, 2016). Plastic is one of the nonbiogenic wastes which is composed of many toxic chemicals that leach into the environment and pollute water, air, and land. The burning of plastics causes respiratory problems in humans and animals (Veksha et al., 2018; Verma et al., 2016; Thompson et al.,2009). The recycling of electronic waste such as various parts of mobile, laptops, etc., and exposure to electroplating, leather tanning, textile, steel fabrication, paint, paper industries, and batteries have released toxic chemicals such as dioxins, polyaromatic hydrocarbons, chromium, metals like zinc, manganese, and waste alkali, which damage muscles, kidneys, brain, and affect human health in various aspects (Percival et al., 2021; Robinson, 2009).

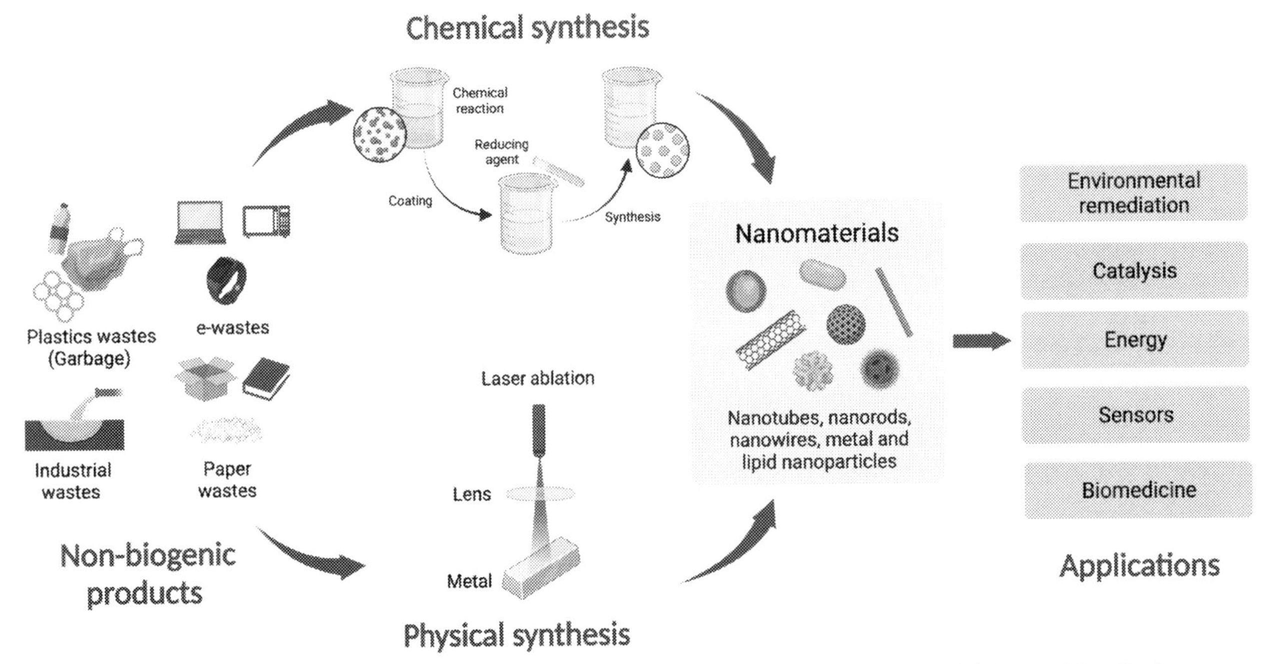

FIGURE 10.1 Nonbiogenic synthesis approaches of nanomaterials from various nonbiogenic waste products and their potential applications.

10.3 Nonbiogenic methods for the synthesis of nanomaterials

The synthesis of nanomaterials from nonbiogenic waste can be achieved by using physical and chemical methods that depend on the type of waste and the desired properties of the nanomaterials (Samaddar et al., 2018). Mainly three steps are involved in synthesizing nanomaterials from nonbiogenic waste including (1) collection of waste materials from nonbiogenic waste, (2) pretreatment of collected nonbiogenic waste, involving cleaning *via* chemical or physical methods such as leaching, acid digestion, or roasting and then grinding the waste material to a fine powder (Singh et al., 2021), and (3) finally, pretreated nonbiogenic waste can be used for the synthesis of nanomaterials by physical and chemical methods such as sol–gel, chemical vapor deposition, electrodeposition, ball milling, PVD, and spray pyrolysis. The whole process of nanomaterials synthesis from nonbiogenic waste was shown in Fig. 10.2.

10.3.1 Physical methods

The pulsed laser ablation is one of the simplest physical methods based on the PVD process in which a high power short plus laser radiation is targeted as solid materials (Ashfold et al., 2004). The pulsed laser ablation technique has been utilized for the synthesis of nanomaterials including thin films, metal, oxide, semiconductors, and nitrides (Yogesh et al., 2021). Ultraviolet irradiation (UV) is one of the important techniques in the synthesis of nanomaterials because this technique was used without affecting any characteristics or properties of nanomaterials (Abbas Syed et al., 2021). The photochemical method has obtained significant attention in the synthesis of nanomaterials which avoid the use of toxic chemical compounds and it can be carried out at room temperature (Jara et al., 2021). The microwave irradiation technique is considered a greener technique for the synthesis of nanomaterials because it produced clean products without any toxic residuals (Tiwari and Talreja, 2022). The electrochemical method allows a synthesis of different types of nanomaterials for electrocatalysts and sensors applications (Rocco et al., 2023 and Stulov et al., 2023). Ultrasonic energy was utilized in sonochemical approach for the synthesis of nanomaterials with any agglomeration. This method generates hydrogen-free radicals during the sonication process from water that acts as a reducing agent for the formation of nanoparticles (Kumar et al., 2023; Ghotekar et al., 2023). The mechanical milling method also produced well-dispersed nanoparticles by grinding the precursor-activated solid-state chemical reaction that leads to the formation of nanomaterials (Smaoui et al., 2023 and Dubadi et al., 2023). The thermal decomposition method allowed the synthesis

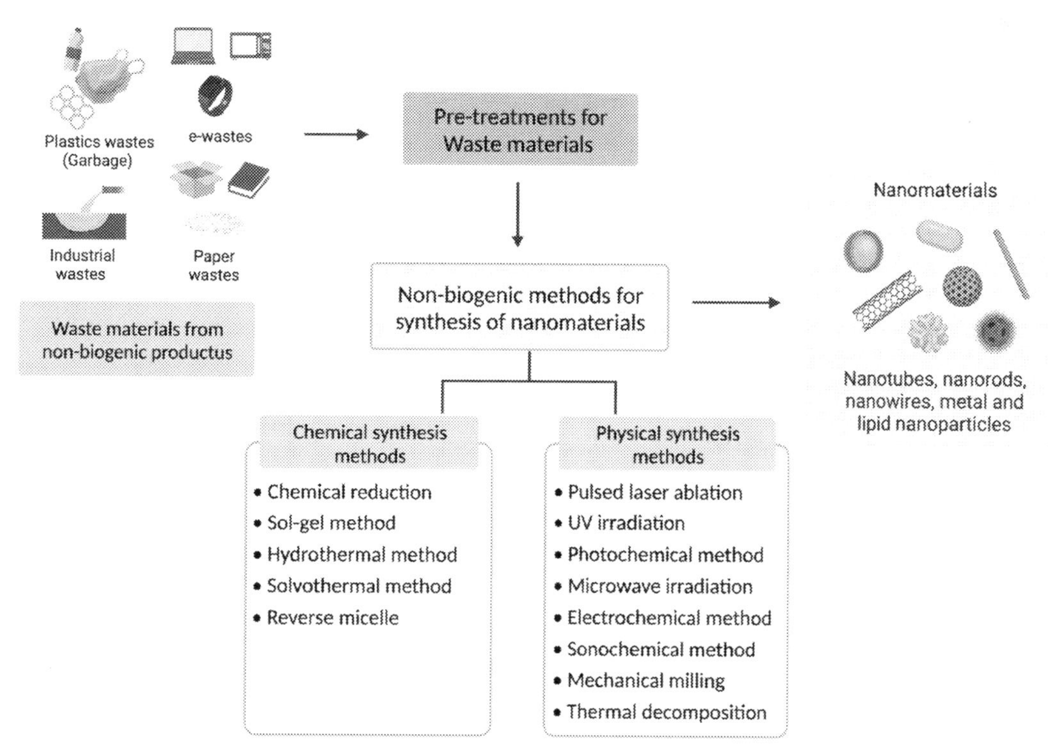

FIGURE 10.2 Schematic representation for the synthesis of nanomaterials by nonbiogenic methods using nonbiogenic waste products.

of metal oxide nanomaterials using corresponding metal complexes at high temperatures (Dogan et al., 2023 and Singh et al., 2023). Table 10.1 shows various types of nanomaterials synthesized using different types of physical synthesis approaches.

10.3.2 Chemical methods

The chemical reduction method employed for the synthesis of nanoparticles using reducing agents, such as borohydrides, citric and oxalic acids, polyols, hydrogen peroxide, and sulfites, etc., can provide an electron to reduce the metal ions into metal nanoparticle (e.g., Au^{+} to Au^{0}) (De Souza et al., 2019). Sometimes stabilizing agent also acts as a reducing agent (Kimling et al., 2006). Sol–gel route is one of the efficient methods for the synthesis of metal oxide nanoparticles because it produced nanoparticles with high homogeneity at lower temperature and low cost (Parashar et al., 2020). A hydrothermal method is an interesting chemical method used to directly prepare various ceramic nanomaterials with suitable sizes and shapes that have been achieved by controlling reaction time, optimum temperature, pH, and solute concentration (Deus et al., 2013). Generally, solvothermal processes have been carried out using autoclave systems with different metal salts (eg., Ag^{+}, Au^{+}, and Zn^{2+}, etc.,) precursors with water or organic solvents at a temperature between 100°C and 350°C to obtain nanomaterials (Navas et al., 2020). Variety of nanomaterials have been synthesized using reversed micelles process, a self-assembly surfactant-induced aggregation which was known as a reverse micelles (Eastoe et al., 2006). Table 10.2 shows various types of nanomaterials synthesized using different types of chemical synthesis approaches.

10.4 Synthesis of nanomaterials from nonbiogenic wastes

10.4.1 Electronic waste (e-waste)-derived nanomaterials

E-waste or waste electrical and electronic equipment are one of the fastest rising nonbiogenic waste streams, causing environmental problems globally. E-waste is categorized into three types: (1) information technology and telecom

TABLE 10.1 Various type of nanomaterials synthesized by using physical methods.

S. No	Nanomaterial	Physical approach	Applications	References
1	SnO_2	Pulsed laser ablation	Photocatalytic	Al Baroot et al. (2023)
2	Ag/Ta_2O_5 nanocomposite	Pulsed laser ablation	Antibacterial activity	Alheshibri et al. (2023)
3	AgNPs	UV irradiation	Antibacterial activity	Rheima et al. (2019)
4	Cr_2O_3 nanoparticles	UV irradiation	Removal of zinc ions	Mohammed et al. (2020)
5	Ag and Au NPs	Photochemical	Catalytic	Saha et al. (2010)
6	Pt NPs	Microwave irradiation	Fuel cell	Chen et al. (2002)
7	Se NPs	Microwave irradiation	Solar cell	Panahi-Kalamuei et al. (2014)
8	Au NPs	Electrochemical	Sensors	Lin et al. (2016)
9	$NiFe_2O_4$	Electrochemical	Catalytic	Galindo et al. (2012)
10	$CdWO_4$	Sonochemical	Photocatalytic	Hosseinpour-Mashkani and Sobhani-Nasab (2016)
11	Eu-doped ZnO NPs	Sonochemical	Catalytic	Khataee et al. (2020)
12	TiO_2	Thermal decomposition	Photocatalytic	Chin et al. (2010)

TABLE 10.2 Various type of nanomaterials synthesized by using chemical methods.

S. No	Nanomaterial	Chemical approach	Applications	References
1	Ag NPs	Chemical reduction	Biomedical	Zahra et al. (2016)
2	Ag NPs	Chemical reduction	Antibacterial activity	Guzmán et al. (2009)
3	SiO_2 NPs	Sol–gel	Industrial	Dubey et al. (2015)
4	$ZnTiO_3$ NPs	Sol–gel	Photocatalytic	Salavati-Niasari et al. (2016)
5	Ag NPs	Hydrothermal	Sensors	Lu et al. (2011)
6	Fe_3O_4 NPs	Hydrothermal	Lithium-ion battery	Ni et al. (2009)
7	ZnO NPs	Solvothermal	Antiinfection	Bai et al. (2015)
8	Fe_3O_4	Solvothermal	Magnetic hyperthermia	Fotukian et al. (2020)
9	ZnO NPs	Reversed micelles	Antibacterial	Çakır et al. (2012)
10	Fe-doped TiO_2 NPs	Reversed micelles	Photocatalytic	Mancuso et al. (2023)

equipment such as personal computers, laptops, and monitors; (2) large household appliances like washing machines and refrigerators; (3) consumer equipment such as televisions, mobile phones, DVDs, and MP3 players. Even though the recycling of e-wastes provided a way to reduce the generation of e-waste, due to the high cost rendered to the use of the recycling process in developing countries. There is an urgent need to develop a technology to convert waste materials into valuable products. The synthesis of nanomaterials from e-waste provides a pathway for reducing the dumping of e-waste in the environment.

Photocopier selenium drum is an electronic waste used for the synthesis of selenium nanoparticles (SeNPs). Bacillus thuringiensis was used during the synthesis of SeNPs (Eswarapriya and Jegatheesan, 2015). The surface sterilized layer was scrapped from the e-waste and made into fine powder through the grounding of the scrapped layer. An enrichment medium consists of approximately 80 mg of lyophilized e-waste and 100 μL of inoculated Bacillus thuringiensis cell suspension incubated at 37°C for 24 hours to 48 hours. The change in color from light yellow to red indicated the formation of SeNPs. Dried SeNPs were obtained through the centrifugation of solution followed by washing with deionized distilled water and dried on Lyo Guard Chamber.

Manganese oxide nanomaterials (Mn_3O_4) are synthesized from waste batteries such as spent Zn-C batteries through thermal methods (Farzana et al., 2019). The powdered materials obtained after the removal of the zinc casing, carbon rod, sealing, and outer metal shell were used as raw materials for synthesizing Mn_3O_4 nanoparticles. Manganese dioxide was used as cathode materials in Zn-C batteries. The moisture content in the battery was removed by drying powder in an oven at 90°C for two hours. The agglomeration of nanoparticles can be reduced by pulverizing dried powder in a ball-milling machine. The experiment equipment consists of a graphite rod to allow the sample into a furnace, a quartz tube containing a horizontal tube, and a gas flow system. The furnace temperature was maintained at 900°C. The dried battery powder was taken in a ceramic crucible and placed on a graphite rod followed by pushing it into the furnace. The powder was kept in the furnace for 1 hour under an argon atmosphere. The formation of the greenish-colored powder indicated the formation of Mn_3O_4 nanoparticles.

Similarly, nano-zinc oxide (ZnO) has been synthesized from waste zinc-manganese batteries through high thermal evaporation - separation and oxygen control oxidation. Therein, air acts as both an oxidizing agent and carrier gas for the formation of ZnO nanomaterials (Zhan et al., 2018a,b). The vacuum metallurgy method was mostly adopted for the synthesis of nano-ZnO due to the low boiling points of zinc under vacuum. The flowing of oxygen was controlled by using both physical and chemical methods separately for the formation of nano ZnO. The copper cap, carbon stick, manganese dioxide, electrolyte, and zinc hull were removed from waste Zn-Mn batteries by hand picking. The zinc hull was taken in a corundum crucible and heated by placing it in the first heating zone of the furnace. The second and third zones of the furnace behave as condensation chambers consisting of silicon steel sheets fixed through a wire. The prepared zinc oxide nanomaterials were collected in the condensation area after cooling down the furnace.

Silicon nitride nanowire (Si_3N_4) was prepared from waste obsolete computers (Maroufi et al., 2018). Si_3N_4 has unique physicochemical properties such as corrosion resistance, chemical inertness, and high thermal conductivity. It is used as ceramic components for structural applications such as automobiles, chemical reactors, truck engines, and gas turbine components. It is also used in catalysis, light emission devices, nanoelectronics, and laser fields. The waste computer was dismantled manually to separate the plastic shells of computers as carbon sources and the glass fraction of monitors as silica sources. These carbon and silica source materials were used for the synthesis of Si_3N_4 nanowires through the carbothermal nitridation technique. The crushed glass fraction and plastic shells were mixed in the 3:2 ratio with a total mass of one gram. The thermal reaction was taken between the mixture at 1550°C in a hot tubular furnace under atmospheric pressure and the purging of nitrogen. The reaction was run up to 130 minutes. Then the sample was cooled down to get a powdered nanowire. The nanowire was also prepared by the addition of waste toner powder as an iron catalyst to study the effect of the catalyst on the morphology of the nanowire.

Fluorescent carbon quantum dots (CQDs) are synthesized from electronic waste like batteries (Devi et al., 2020). CQDs was synthesized from Duracell AA batteries by collecting carbon black followed by heating at high temperatures. These CQDs are mixed with the solution of 1 mM auric chloride under stirring to prepare gold-modified CQDs. The formation of gold-modified CQDs can be identified by the changing of color from pale yellow to pinkish red.

10.4.2 Industrial waste-derived nanomaterials

Nowadays, industrialization has generated a huge number of wastes such as organic matter and metals from electroplating industries, dyes from textile industries, fly ash from thermal power plants, and metals particularly calcium from paper industries. The synthesis of nanomaterials from the wastes gained great attention due to its unique physicochemical properties and large surface-to-volume ratio. Bottom-up approaches such as microwave irradiation, coprecipitation method, and hydrothermal synthesis, and top-down approaches such as ball milling have been used for the synthesis of nanomaterials from industrial wastes (Zhang et al., 2012, 2017; Chen et al., 2016; Visa et al., 2015; Kumar et al., 2015).

The refining of crude oil generates petrochemicals along with carbon black waste as a byproduct. The generated carbon black waste contains vanadium metal that leaches from crude oil. Carbon black waste is very harmful to human health because vanadium damages the antioxidant enzymatic actions of human cell lines. Converting carbon black waste into valuable vanadate nanomaterials reduces its hazardous effects. Vanadate nanomaterials have plentiful applications in the field of batteries, chemical sensors, biological imaging, and therapy. At first, the retrieval of vanadium was achieved by the complete dissolution of metals from carbon black waste through leaching processes. Three types of alkaline earth vanadate nanomaterials namely barium vanadate polyhedral nanoparticles, strontium vanadate nanorods, and calcium vanadate nanorods were synthesized from the leaching solutions of carbon black wastes through hydrothermal synthesis (Zhan et al., 2018a,b). Vanadate materials have good magnetic and electrical properties due to the presence of magnetic V^{4+} ions. The first step of the leaching process is the drying of the wet carbon black waste sample, carried out in an oven at 150°C for 12 hours. The dried samples were dispersed for 1 hour in 50 mL of sodium hydroxide or nitric acid using a sonicator. The leaching solution was separated and converted into an alkaline leaching solution by adjusting the pH of the solution. The solution was mixed with alkaline earth metal salts by stirring followed by hydrothermal treatment at 200°C for 12 hours. Vanadium nanomaterials were collected as a white solid material.

10.4.3 Plastic waste-derived nanomaterials

Carbon nanotube (CNT) and graphene were synthesized from plastic wastes (Bazargan and McKay, 2012). Plastic waste-derived CNTs were prepared using various reactor systems such as a muffle furnace, autoclave, quartz tube, and crucible. The objective of these methods was the breaking of large solid polymers into small carbon precursors through the thermal method (Zhuo and Levendis, 2014). Waste plastics mainly consisted of polyolefins such as polyethylene (PE) and polypropylene (PP) (Gong et al., 2012). CNT has been synthesized in quartz tubes through chemical vapor deposition techniques (Mishra et al., 2012) PP and Ni as catalysts were placed in two different quartz boats. PP decomposed into carbon atoms during the pyrolysis of plastic waste at 400°C followed by dissolving into a hot area of nickel. Carbon atom gets precipitated in a cold area of nickel, which served as nucleating sites for the growth of CNT. Solid waste plastics were used as a source of carbon for the synthesis of single-crystal graphene on polycrystalline copper foil (Sharma et al., 2014). Similarly, CNT was also prepared from polymer waste (Bajad et al., 2015). Synthesized carbon nanoparticles were also produced from plastic bag waste through thermooxidative degradation, polymerization, carbonization, and passivation methods (Hu et al., 2014).

10.4.4 Tires waste-derived nanomaterials

The manufacture of tires for automobiles has been increasing every year. The tires are made up of natural and synthetic rubber along with carbon black. The tire waste is generated due to their irreparable damage. The recycling of tires is very difficult due to the thermosetting properties of cross-linked molecular structures. Recently, research is going on for the synthesis of carbon nanomaterials by recovering carbon components from tires and destroying hazardous components present within the tires. Tire waste is used as a starting precursor of carbon source for the preparation of carbon nanomaterials due to their low cost, high carbon content, and high abundance (Zhang et al., 2016; Zhang and Williams, 2016). Carbon nanostructures are synthesized by mixing residual solid waste of waste tires with ferrocene (Mis-Fernández et al., 2012). The mixture under evacuation and heating followed by cooling produced carbon nanomaterials. Carbon nanoparticles are synthesized from waste tires through the heating of ground waste tires in a high-temperature furnace for 4 hours (Gómez-Hernández et al., 2019). They recovered 81% of carbon nanoparticles from tire waste with an average size of 22 nm. Carbon nanosheets doping with heteroatoms as electrode materials were synthesized from the pyrolysis of waste tires through chemical vapor deposition (Veksha et al., 2018). Nickel oxide-loaded calcium carbonate was used as a catalyst for the preparation of carbon nanosheets. They have studied the influences of ammonia and water on the formation of carbon nanosheets from waste tires.

Core-shell carbon black nanoparticles use the reactive extrusion process. The extrusion process takes place at two different temperatures: 300°C and 140°C (Song et al., 2018). At low temperatures, the degradation of ground tire rubber was done by using a green plasticizer and soybean oil as a swelling agent. The chemical vapor deposition method was used for the preparation of multiwall carbon nanotubes from waste tires (Mwafy, 2020). This reaction was carried out in a quartz tube reactor. Graphene has also been prepared (Wang et al., 2019) from waste tires through a pyrolysis process in an alkaline medium. Simple one-step processes, no requirement for complex equipment, and expensive chemicals are the advantages of the pyrolysis process. The arrangement of carbon atoms into a graphene-like structure was achieved by the induction of potassium metal vapor. Waste tires into carbon nanospheres use ferrocene as a catalyst through a chemical vapor deposition (CVD) reaction (Heidari and Younesi, 2020). There are two steps involved in the synthesis of carbon nanospheres. The first step involved the carbonization of feedstock and the second step was the mixing of carbonized feedstock with a ferrocene catalyst. Similarly, waste tires were converted into carbon nanotubes (Yang et al., 2012) using cobalt as a catalyst through CVD. The researchers have also synthesized CNT from scrap tires over various catalysts such as nickel, cobalt, iron, and copper with aluminum oxide. The waste tire was converted into carbon nanoparticles (Maroufi et al., 2017).

10.4.5 Paper waste-derived nanomaterials

Paper waste was used for the synthesis of two-dimensional graphene oxide quantum dots (GOQD). GOQD is a graphene oxide sheet with a size of less than 100 nm. Recently, carbon nanostructures could be synthesized from waste paper through microwave-aided hydrothermal degradation processes (Adolfsson et al., 2015). The synthesized carbon nanostructures are further converted into two-dimensional GOQD. This aided the conversion of waste resources into value-added products. The degradation of paper was done in sulfuric acid solution in Teflon vessels at a temperature of 160°C to prepare carbon nanostructures. The carbon nanostructures were further reacted with nitric acid by heating to give an orange/red solid of GOQD.

10.4.6 Mining waste-derived nanomaterials

The disposal of mining waste is a challenging task in the environment. Two types of mining wastes have been generated during the extraction of ore. (1) Mining tailings such as chemical residues, mud, and sandy materials produced during ore concentration. (2) The waste earth materials such as soil, rocks, and sediment are separated through the excavation process. Among these, the mining tailings have entered into waterways thereby causing pollution. It is an urgent need to reconvert mining tailings into high value-added materials. For example, synthesized a nanoadsorbent from bauxite tailings for the removal of water pollutants (Nascimento et al., 2022). A carbon nanotube can be obtained from active metals present in mining waste. Iron ore is one of the raw mineral materials obtained in the largest quantity from the earth's crust. It plays a vital role in many industries for making steel. However, a huge amount of hazardous wastes containing sulfide minerals and toxic metals was produced during iron mining activities. Magnetic nanomaterials have been synthesized through the recycling of iron mining wastes (Cruz et al., 2021).

10.5 Characterization techniques of nonbiogenic synthesized nanomaterials

There are several spectroscopy and microscopy techniques used to characterize the nanomaterial properties including size, shape, crystal structure, and composition (Shaheen et al., 2023). Herein, we have discussed the optical, structural, and morphological characterization of nanomaterials such as UV-visible spectroscopy, FTIR, XRD, SEM, and TEM with EDAX.

10.5.1 UV-Vis spectroscopy

UV-Vis. spectroscopy is used to analyze the optical properties of surface plasmon resonance (SPR) of nanomaterials quantitatively at the range of near ultraviolet 180–390 nm and visible 390–780 nm radiation. These particular radiation ranges cause electronic transitions of the molecules so that the amount of absorbed radiation is directly proportional to the analyte concentration in the solution (Beer-lambert law) (Bagoria and Kumar, 2023). The inorganic transition metal ions and their aqueous complexes are generally colored and absorbed in the near-ultraviolet region because of $\sigma \rightarrow \sigma^*$, $n \rightarrow \sigma$, $n \rightarrow \pi^*$, and $\pi \rightarrow \pi^*$ transitions, leading to the formation of sharp absorption peaks (Worsfold et al., 2019). Inorganic transition metal nanoparticles such as Ag and Au NPs absorbed light radiation at around 420 and 520 nm, respectively (Abdulhameed et al., 2014).

10.5.2 Fourier transform infrared spectroscopy

FTIR spectroscopy has been used to identify the functional groups present in the nanomaterials' surfaces (Zhang and Yan, 2010). For instance, FTIR was used to identify the conformation of nanomaterial conjugates with proteins (Jiang et al., 2005), and the covalent attachment of functional molecule with carbon nanotubes (Baudot et al., 2010).

10.5.3 X-ray diffraction

XRD technique has been extensively used to identify the crystalline structure of the nanomaterial, the nature of the nanoparticle phase, nanoparticle surface lattice parameters, and crystalline particle grain size (Kumar et al., 2022). XRD analysis of silver nanoparticles showed 2θ positions of hkl values corresponding to FCC silver and the estimated particle size (Bykkam et al., 2015).

10.5.4 Scanning electron microscopy

SEM analysis provided the particle size and morphology of nanomaterials (Buhr et al., 2009). Nonconductive nanomaterials' samples during the drying process may cause alteration in the morphology of nanomaterials because of shrinking (Bootz et al., 2004). So, ultrathin layer coating is required for nonconductive nanomaterials during the SEM sample preparation (Suzuki, 2002).

10.5.5 Transmission electron microscopy

TEM coupled with EDX is a widely used technique for the direct imaging analysis of nanomaterials which provides size, morphology, and chemical information of nanomaterials with high resolution up to 1 nm (Wang, 1999). EDX elemental analysis can provide the quantitative chemical composition of the nanomaterials (Hodoroaba, 2020). Here, the characterizations of some nonbiogenic synthesized nanomaterials were discussed in detail.

The field emission scanning electron microscopy (FE-SEM) and TEM studies of selenium nanoparticles synthesized from e-waste have shown spherical shape with size in the range from 50 to 500 nm. EDAX analysis showed the peak of selenium, confirming the formation of SeNPs (Eswarapriya et al.,2015). Mn_3O_4 was synthesized from e-waste and characterized by using microscopic and spectroscopic methods. The residue of manganese oxide was obtained from waste Zn-C batteries through evaporation and condensation methods. The residue of manganese dioxide was converted into Mn_3O_4 by oxidation at 800°C under an air atmosphere (Farzana et al., 2019).

FE-SEM and TEM images of Mn_3O_4 have shown the morphology of nanoparticles as spherical and cubic shapes. Mn_3O_4 NPs have shown lattice fringes with d-spacing of 0.31 and 0.21 nm, corresponding to (112) and (211) planes of crystalline nanoparticles, respectively. The polycrystalline nature of Mn_3O_4 NPs was confirmed by the selected area electron diffraction (SAED) pattern. The diameter of nanoparticles was calculated as around 15 nm from XRD data using the Scherrer formula. The XRD pattern of Mn_3O_4 NPs corresponded to the Hausmannite phase of tetragonal

Mn_3O_4 and space group of 141/AMD. The FT-IR spectra of nanoparticles showed the characteristic peaks of metal-oxygen bonding and stretching vibration of spinal metal oxide. The absorption peaks at 607 and 527 cm^{-1} corresponded to Mn^{2+} ions occupied in the tetrahedral sites and Mn^{3+} occupied in the octahedral sites of Mn_3O_4 NPs. The tetragonal Hausmannite with spinal structure was further confirmed from the characteristic bands obtained from Raman spectra. An energy dispersive spectroscopy (EDS) mapping study has shown the uniform distribution of Mn and O in nanoparticles. The oxidation state and composition of synthesized manganese oxide nanoparticles were confirmed from X-ray photoelectron spectroscopy (XPS) analysis. The thermal gravimetric analysis (TGA) of the synthesized nanoparticles has shown the good thermal stability of manganese oxide nanoparticles. The powder XRD, Raman spectroscopy, FTIR, and SEM with EDS of manganese oxide nanomaterials are shown in Fig. 10.3.

Nano ZnO as the white powder was obtained from waste zinc manganese batteries (Zhan et al., 2018a,b). The morphology of nano-ZnO was analyzed at various operating parameters. The XRD pattern of nano-ZnO indicated the complete crystallization of nanomaterials with no impurities. SEM images showed the tetrapod-shaped spatial structure of nano-ZnO as seen in Fig. 10.4.

Si_3N_4 nanowire was prepared from waste computers through the carbothermal nitridation technique (Maroufi et al., 2018). The morphology of nanowires was studied to study the effect of iron catalysts on nanowire formation. The sharp diffraction peaks in the XRD pattern of nonbiogenic waste-derived Si_3N_4 nanowire showed α and β crystalline phases of Si_3N_4. There were no diffraction peaks observed for unreduced SiO_2 and unreacted carbon, indicating the complete

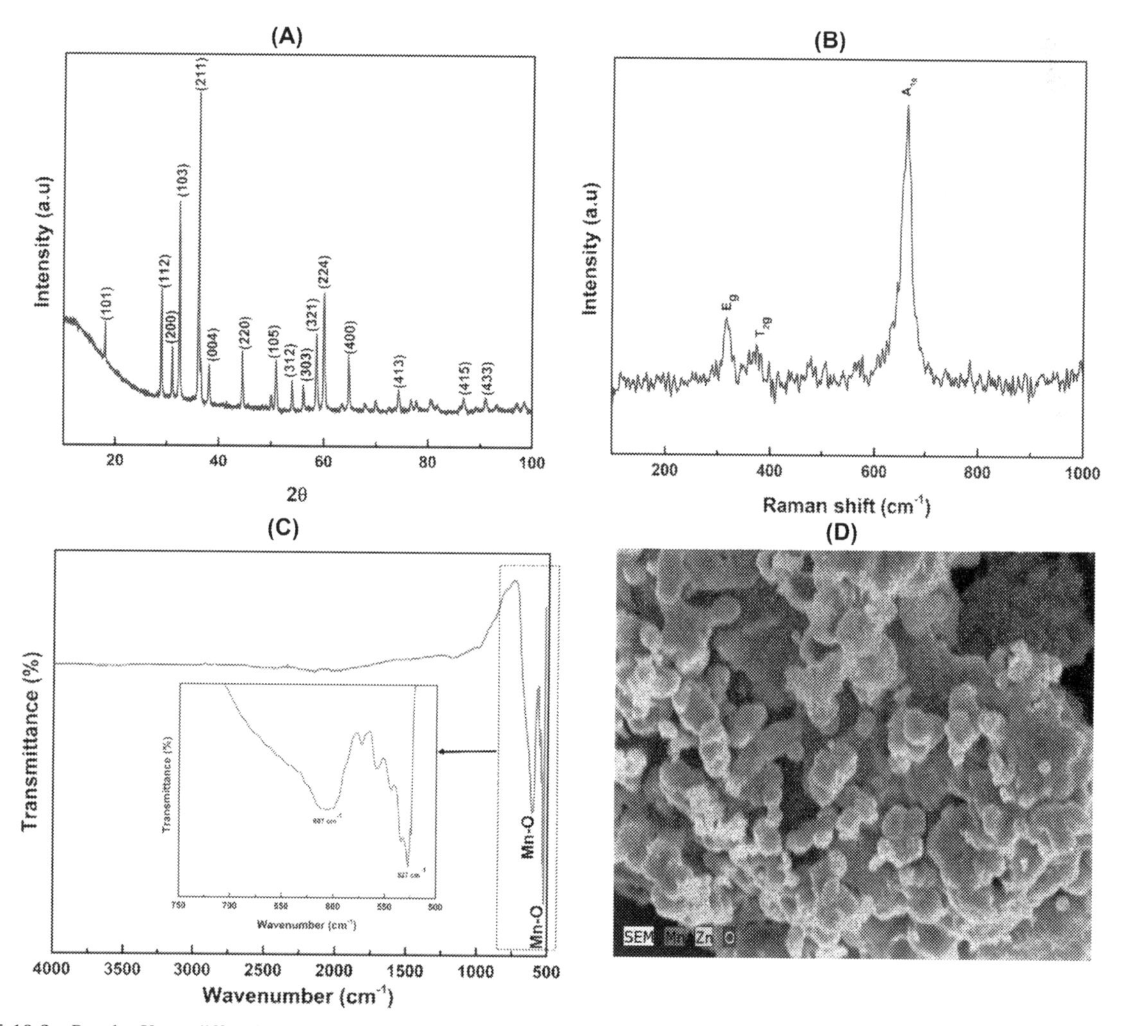

FIGURE 10.3 Powder X-ray diffraction (A), Raman (B), Fourier transform infrared spectroscopy spectra (C) (inset: the region of 500–750 cm^{-1}), and scanning electron microscopy with energy dispersive spectroscopy mapping (D) for the synthesized Mn_3O_4 nanoparticles from spent Zn-C battery. *This figure is reproduced from Farzana, R., Hassan, K. and Sahajwalla, V., 2019. Manganese oxide synthesized from spent Zn-C battery for supercapacitor electrode application. Scientific Reports, 9 (1), 1–12 with permission of Springer Nature.*

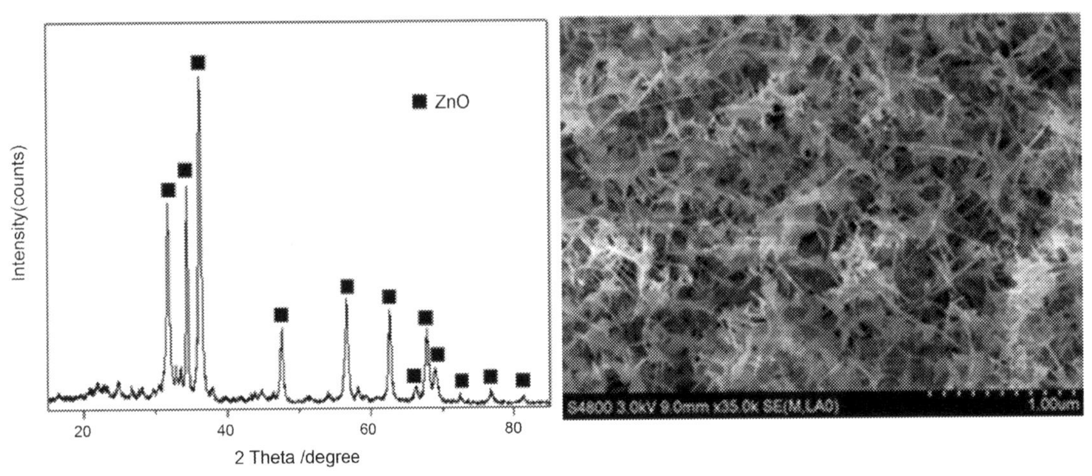

FIGURE 10.4 Powder X-ray diffraction pattern and scanning electron microscopy image of nano-zinc oxide synthesis from waste zinc-manganese batteries. *This figure is reproduced from Zhan, L., Li, O., Wang, Z., Xie, B., 2018a. Recycling zinc and preparing high-value-added nanozinc oxide from waste zinc-manganese batteries by high-temperature evaporation-separation and oxygen control oxidation. ACS Sustainable Chemistry & Engineering, 6 (9), 12104–12109.*

reduction of silicon dioxide. The IR spectrum of the nanowire showed the absorption peaks corresponding to Si-Si stretching mode and Si-N stretching vibration mode. SEM images of the nanowire showed T-shaped and Y-shaped morphologies with twisted surfaces. The optical properties of e-waste-derived CQDs and gold-functionalized CQDs have been investigated by using absorption and emission spectroscopy. The increased absorption and emission intensity indicated the increased degree of carbonization. The absorption peak of CQDs at 350 nm was due to the molecular transition of C = C from n to π^* and π to π^*. The peak at 520 nm indicated the SPR band of gold functionalized with CQDs. The existence of uniform sp^2 carbonaceous structure of CQDs could be confirmed by their photoluminescence spectrum. The D and G bands of the Raman spectrum confirmed the uniform sp^2 carbonaceous structure consisting of sp^3 defects. The HRTEM analyses of CQDs and Au-functionalized CQDs showed the formation of spherical particles with uniform particle distribution and size of 5 and 20 nm, respectively. The crystalline nature of both compounds was confirmed through XRD studies (Devi et al., 2020).

Three different vanadium nanomaterials were prepared from carbon waste generated by petrochemical industries. XPS studies revealed the oxidation state of vanadium as V, which is the fully oxidized state of vanadium. NMR analysis showed the presence of vanadium as VO_4^{3-} in an alkaline leaching solution. The acid solution was added to alkaline to get an acid-leaching solution that showed an absorption band at ~405 nm in the UV-Visible spectrum. The color of the leaching solution was changed from white to yellow while adding acid (blue color) to the alkaline (colorless) leaching solution. The blue color of the acid-leaching solution indicated the presence of Ni^{2+} and Fe^{2+} ions, showing intensive absorption of visible light. SEM analysis of calcium vanadate, strontium vanadate, and barium vanadate showed the morphology of vanadate as nanorods with a bush-like assembly, ellipsoids-like structure, and polyhedral nanoparticles. The XRD pattern of all three vanadates showed the hexagonal phases of vanadate nanomaterials (Zhan et al., 2018a,b). The formation of graphene from solid waste plastics was characterized by spectroscopic and chromatographic techniques. Chromatographic analysis showed the decomposition of polymers into graphene at different temperatures. Optical microscopic images showed the formation of a round shaped and large hexagonal single-crystal graphene through the pyrolysis of polymer with a controlled rate.

The TEM images of carbon nanosheets synthesized from waste tires have shown the predominant carbon nanosheet structure at all temperatures. However, some carbon black nanoparticle impurities were observed along with carbon nanosheets when the temperatures were in the range of 700°C–800°C. This was attributed to the carbon structure encapsulated with a catalyst. The morphology of carbon nanomaterials depends on the type of starting precursors and the noncondensable pyrolysis gases whether it contains sulfur or nitrogen. For example, plastic pyrolysis gave multiwall carbon nanotubes whereas noncondensable gas pyrolysis of tires produced carbon nanosheets using the same catalyst and temperatures. The carbon growth increased with increasing temperatures up to 750°C in catalytic chemical vapor deposition and then decreased. TEM images also showed the addition of nickel oxide to calcium carbonate only increasing the yield without altering the morphology of the carbon nanosheet. The ICP-OES measurements showed that there was no calcium and nickel present on the surface of the carbon nanosheet. XPS measurements confirmed the presence of heteroatoms such as nitrogen and sulfur in carbon nanosheets. The intensive D-band was observed in XPS, indicating the presence of a defect in carbon nanosheets (Veksha et al., 2020).

The powder-XRD pattern of tire waste-derived multiwall carbon nanotube (MWCNT) has shown the diffraction peaks corresponding to the nature of the graphitic structure. The morphology analysis confirmed the formation of MWCNT with an opening at the end of the tube with a diameter of 40 nm. The SEM images of pristine CNT have shown the distribution of iron catalysts as tiny particles through the CNT network. EDAX analysis of pristine and purified CNT confirmed the presence and absence of iron, respectively. The percentage composition of carbon and oxygen was 95.54% and 4.06%, respectively, indicating no impurities on MWCNT. The appearance of G and D bands in Raman spectroscopy indicated the presence of SP^3 hybridized carbon atom on the CNT skeleton and SP^2 hybridized carbon atom as defects on CNT, respectively. The absence of a band at nearly 200 cm^{-1} indicated no formation of a single-walled carbon nanotube. TGA analysis showed the thermal decomposition of pure MWCNT started at 500°C. These results indicated the successful formation of MWCNT from tire waste (Yang et al., 2012).

The morphology of tire waste-derived graphene was similar to that of nanoflake structure in which the vertical carbon nanosheets were interconnected with cross-linking among them. HRTEM images of 3D graphene have shown a sharp texture on a monolithic plate, suggesting high crystal quality of graphene structure with a thickness of less than 10 nm. TEM images at different temperatures and concentrations of KOH indicated the various stages for the conversion of zero-dimensional carbon particles into three-dimensional graphene. The image of graphitic structure undergoes some changes with pyrolysis time. The hexagonal diffraction pattern on SAED further confirmed the high pure crystallinity of the graphene structure. Atomic force microscopy analysis indicated the size of graphene as several micrometers with a thickness of 4 nm. The peaks with high intensity in the XRD pattern of KOH-activated samples were due to the existence of more micropore structures. The peaks obtained at high temperatures corresponded to both amorphous and graphitic structures, indicating the destruction of graphitic configuration by potassium hydroxide. Elemental analysis showed that the oxygen content decreased with increasing pyrolytic temperature and time. The presence of micropores was further confirmed by the existence of a typical IV isotherm with H1 type hysteresis loop in nitrogen adsorption/desorption isotherms. This method indicated the conversion of tire waste into worthwhile products (Veksha et al., 2020).

The TEM image of core-shell carbon nanoparticles formed in a two-extrusion process was compared with commercial carbon black by Song et al. (2018). The waste tires are made up of commercial carbon black. TEM images of core-shell carbon nanoparticles formed at high-temperature extrusion have shown that the largest aggregation of core-shell carbon nanoparticles was due to the presence of rubber gel on its top, making the recombination of individual core-shell carbon nanoparticles. The core-shell carbon nanoparticles obtained at warm degradation have similar morphology as carbon black except for the roughness of the surface. The surface roughness of core-shell materials was due to the adsorption of rubber on the surface of the core-shell materials. SEM and TEM images of carbon nanospheres (CNS) synthesized from tire waste have shown that CNS have smooth surfaces with no porous and spherical shape with narrow size distribution. The average diameter of spherical-shaped CNS was 38–40 nm. XRD pattern indicated the presence of a highly ordered crystalline nature of graphitic carbon. The graphitic character was confirmed by the interlayer d-spacing value of CNS. The G- and D-bands observed in the Raman spectrum were attributed to the sp^2 hybridized carbon of graphite and carbon atoms, respectively. The degree of graphitization can be determined from the ratio of intensities of D- and G-bands. The functional group present on the surface of the CNS was investigated through FT-IR analysis. BET analysis indicated the isotherm of CNS follows type III, indicating the nonporous nature of CNS. These analyses indicated the successful formation of CNS from tire waste (Song et al., 2018).

SEM images of carbon nanostructures synthesized from waste paper showed the change in morphology of untreated paper to spherical-shaped particles, indicating the formation of carbon nanospheres. BET analysis indicated the presence of low porosity in carbon structures. The graphitic flakes and carbon nanosheets appeared on carbon nanostructures. The formation of GOQD from carbon nanostructures was monitored by the dynamic light scattering technique. This showed that the particle size was decreased as compared to that of carbon nanostructures, indicating the formation of GOQD. Graphene oxide was formed with a desirable size by controlling the reaction conditions. FT-IR studies showed that more defects were observed in the carbon nanostructure with increasing reaction time. XPS analysis indicated the increase in oxygen content in GOQDs compared to that of carbon nanostructures. The peaks observed in UV-Visible and emission spectroscopies further confirmed the formation of GOQD. The synthesis of GOQD from waste paper provided a novel methodology for the conversion of waste paper into wealth products (Adolfsson et al., 2015).

10.6 Application of nonbiogenic waste-derived nanomaterials

10.6.1 Antifungal activity

The disk diffusion method was used for analyzing the antifungal activity of SeNPs synthesized from e-waste (Eswarapriya and Jegatheesan, 2015). Two fungi such as *Candida albicans* and *Aspergillus niger* were taken for testing the antifungal activity of the SeNPs. The antifungal activity studies showed that SeNPs exhibited good antifungal activity against *C. albicans* as compared with *A. niger*. SeNPs have better antifungal activity and low toxicity than other inorganic and organic forms of selenium because of their high surface-to-volume ratios.

10.6.2 Supercapacitors

Recently, supercapacitors have attracted much attention in energy-related applications such as electronics and automobiles due to their high specific power, fast charge-discharge rate, and long-life cycle. An electrode material made up of nanostructured metal oxide has played a vital role in various applications such as high-density magnetic storage media and catalysis because of its properties like high specific capacitance, flexibility, and low resistance. Manganese oxide nanomaterial was used as supercapacitor electrode materials due to its low cost, environmental compatibility, and good electrochemical performance. The waste batteries were transformed again for energy applications instead of burying the waste in a landfill for avoiding environmental and health threats. The synthesized nanomaterials have shown the highest capacitance value of 125 F/g at a scan rate of 5 mV/s. The electrodes were very stable after 2100 cycles, confirming the good cycling stability, good reversibility, and high efficiency of waste-derived nanoparticles as supercapacitors. The whole experimental setup and different steps involved in the synthesis of Mn_3O_4 with the steps of electrode fabrication are shown in Fig. 10.5. Vanadate nanomaterials synthesized from leaching solutions of carbon black waste were employed as potential semiconductors with wide band gaps. These materials were used in applications such as electronic devices and semiconducting glasses (Zhan et al., 2018a,b). Tire waste-derived graphene material was used as a supercapacitor electrode due to its high excellent capacitive, rate performance, and high conductivity.

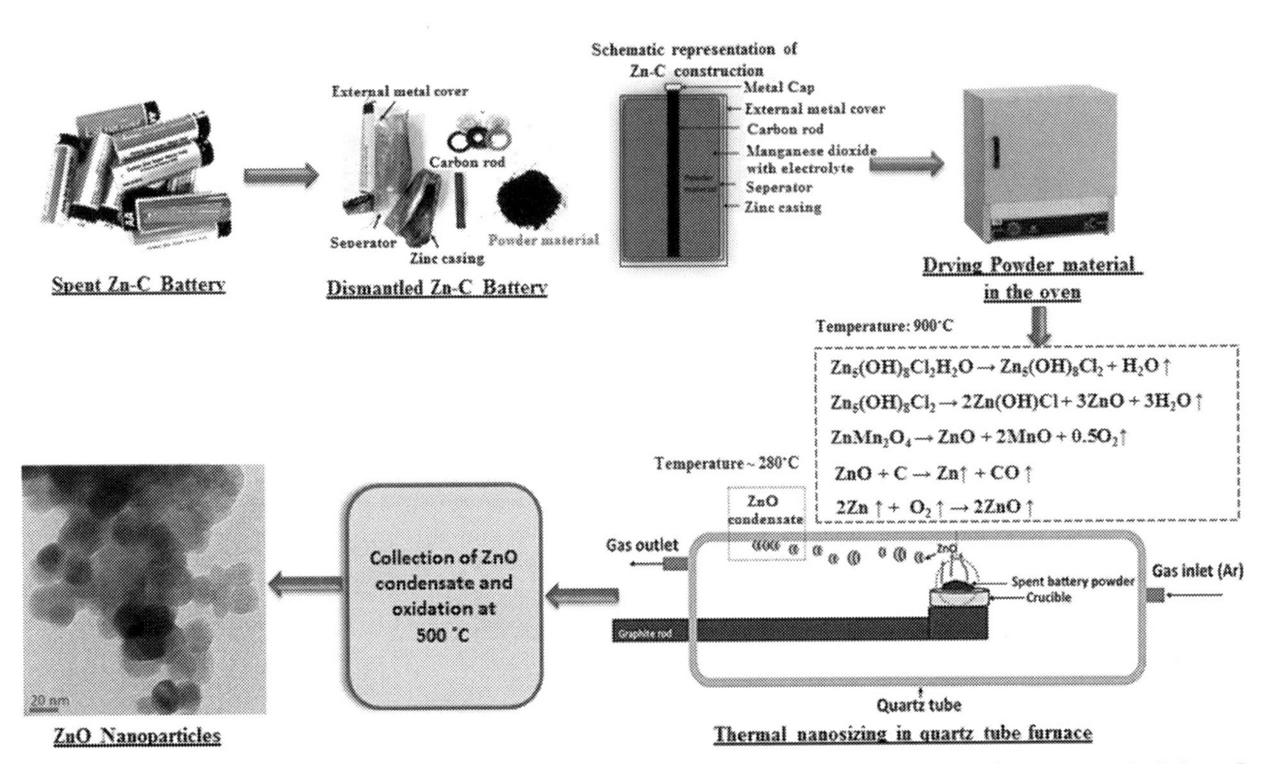

FIGURE 10.5 Schematic illustration of specific capacitor process of device fabrication. *This figure is reproduced from (Farzana, R., Rajarao, R., Behera, P.R., Hassan, K., Sahajwalla, V., 2018. Zinc oxide nanoparticles from waste Zn-C battery via thermal route: characterization and properties. Nanomaterials, 8 (9), 717) with permission of nature.*

10.6.3 Electrocatalyst

Carbon nanomaterials such as graphene and carbon nanosheets as electrocatalysts have attracted great attention in oxygen reduction reactions. This vital reaction has been used in rechargeable metal air batteries and low-temperature fuel cells. Carbon nanomaterials have certain advantages such as inexpensive, readily available sources, large-scale synthesis, and high durability over other metal atoms. The electrocatalytic property was increased by the doping of carbon nanomaterials with heteroatoms such as nitrogen, sulfur, and oxygen. Carbon nanosheets have been prepared from waste tires as electrocatalysts for the oxygen reduction reaction (Yang et al., 2012). The electrocatalytic oxidation reaction was carried out in a three-electrode electrochemical cell using platinum, glassy carbon, and silver/silver chloride electrodes as counter, working, and reference electrodes. The linear sweep voltammetry experiment was done in both air-saturated and argon-saturated solutions. The peaks obtained in linear sweep voltammetry correspond to the oxygen reduction reaction of the air-saturated solution. There were no peaks observed in argon-saturated solutions. The increased catalytic activity of carbon nanosheet-modified glassy carbon electrodes compared with bare glassy carbon electrodes was due to the increased content of sulfur on the surface of carbon nanosheets. The confinement effect of sulfur-doped carbon nanosheets lowered the adsorption and dissociation energies of oxygen, thereby increasing the catalytic activity. The addition of nickel oxide over calcium carbonate catalyst on the carbon surface only increased the carbon content in nanomaterials. It did not influence the electrocatalytic activity of carbon nanosheets due to the shielding effect of nickel particles by graphene layers. The high stability and electrocatalytic activity of carbon nanosheets during oxygen reduction reaction imply their use as a potential electrocatalyst compared with other metals such as platinum, ruthenium, and iridium.

10.6.4 Sensor

CQDs have been used in various applications such as drug delivery, energy harvesting, and chemical and biosensing due to their unique physiochemical properties such as low photobleaching, chemical internet, low cytotoxicity, ease of synthesis, and biocompatibility. CQD functionalized with gold was used as a surface-enhanced Raman scattering (SERS)-based sensor for the detection of organic dyes such as methylene blue. This substance was coated on a screen-printed electrode through the drop-casting method. A Raman peak of methylene blue at $1620\ cm^{-1}$ was chosen to study the SERS. The intensity of the peak was increased for gold-functionalized CQDs, indicating that the incorporation of gold into CQDs increased the roughness of molecules with an increase in the number of SERS spots. The construction of paper-based SERS substrates using gold-functionalized CQDs showed high sensitivity toward target molecules (Devi et al., 2020).

10.6.5 Water remediation

Magnetic nanomaterials, cobalt ferric oxide nanoparticles, were synthesized from mining waste using natural organic matter as a solvent. The nanomaterials were used as both catalysts and adsorbents for the removal of nitrophenol and polycyclic aromatic hydrocarbons from wastewater, respectively, thereby minimizing environmental pollution (Cruz et al., 2021). The application of various nanomaterials derived from nonbiogenic waste is tabulated in Table 10.3.

10.7 Summary and conclusion

In this chapter, we have attempted to provide an overview of the recent development of nonbiogenic synthesized nanomaterials from various nonbiogenic waste products and their potential useful applications in the environment and human health. In recent years, nanomaterials derived from nonbiogenic waste materials such as e-waste, industrial waste, plastic waste, tire waste, paper waste, mining waste, and glass waste by nonbiogenic synthesis methods such as physical and chemical approaches have been discussed. The size, shapes, functionality, and chemical compositions of nonbiogenic synthesized nanomaterials characterized by using spectroscopic and microscopic techniques including UV-Vis., FTIR, XRD, SEM, and TEM with EDX have also been summarized. Finally, a detailed discussion has been made on their potential applications including antifungal activity, supercapacitors applications, electrocatalysts, sensors, and water remediation. E-waste-derived SeNPs showed better antifungal activity and low toxicity compared to other forms of selenium (Eswarapriya and Jegatheesan, 2015). Tire waste-derived graphene nanomaterials have been used as supercapacitor electrodes because they provided high conductivity and large capacitive with excellent performance (Zhan et al., 2018a,b). Also, carbon nanosheets were prepared from waste tires and showed an excellent electrocatalytic

TABLE 10.3 Nanomaterials derived from nonbiogenic waste and their applications.

S. No.	Nonbiogenic waste	Nanomaterials	Applications	References
1	Iron ore tailing	Iron oxide nanoparticles (Fe_3O_4 NPs)	Adsorption of dye molecules	Giri et al. (2011)
2	Polyethylene terephthalate	Graphene	Dye removal	El Essawy et al. (2017)
3	Printed circuit boards	Nanocomposites of zinc oxide and copper oxide	Photocatalyst	Nayak et al. (2019)
4	Electrical waste	Nano cuprous oxide	Electrochemical sensing of dopamine and mercury	Abdelbasir et al. (2018)
5	Pickling lime of a steel plant	Zero-valent iron nanoparticles	Removal of nitrobenzene	Lee et al. (2015)
6	Fabric filter dust	Metal-doped zinc oxide nanoparticles	Photodegradation of phenol	Wu et al. (2014)
7	Aluminum foil	Nano alumina powder	Abrasives	El-Amir et al. (2016)
8	Stainless steel slags	Platinum nanoparticles/Silica oxide	The catalyst for carbon monoxide oxidation	Domínguez et al. (2008)
9	Plastic bag wastes	Carbon nanoparticles	Cellular imaging and iron sensing	Hu et al. (2014)
10	Mining waste	Magnetic nanomaterials	Water remediation	Cruz et al. (2021)
11	Camphor oil	CNT	Energy storage material	TermehYousefi et al. (2014)
12	E-waste	SeNPs	Antifungal activity	Eswarapriya and Jegatheesan (2015)
13	Paper cups	Graphene sheets	Fuel cell	Zhao and Zhao (2013)
14	Glass fiber waste	Zeolite like nanomaterials	Dye adsorption	Tsai and Horng (2021)

performance of oxygen reduction reactions because of the presence of various heteroatoms like nitrogen, sulfur, and oxygen for doping with various functional materials (Yang et al., 2012). Waste paper-derived CQDs were used for bio/chemical sensor applications and drug delivery because of their biocompatibility, less toxicity, and low photobleaching properties. Magnetic $CoFe_2O_4$ NPs synthesis is done from mining waste and is used for the removal of nitrophenol and polycyclic aromatic hydrocarbons from wastewater (Cruz et al., 2021). Nonbiogenic waste-derived nanomaterial's approach is one of the key factors to minimize the generation of hazardous materials in the environment. In the future, research will involve the fabrication of low-cost sensing devices using waste-derived nanomaterials for the removal of various organic pollutions from the environment.

References

Abbas Syed, Q., Hassan, A., Sharif, S., Ishaq, A., Saeed, F., Afzaal, M., et al., 2021. Structural and functional properties of milk proteins as affected by heating, high pressure, Gamma and ultraviolet irradiation: a review. International Journal of Food Properties. 24 (1), 871–884.

Abdelbasir, S.M., El-Sheikh, S.M., Morgan, V.L., Schmidt, H., Casso-Hartmann, L.M., Vanegas, D.C., et al., 2018. Graphene-anchored cuprous oxide nanoparticles from waste electric cables for electrochemical sensing. ACS Sustainable Chemistry & Engineering 6 (9), 12176–12186.

Abdulhameed, N.R., Salih, H.A., Hassoon, K.I., Ali, A.K., 2014. Plasmonic absorption of gold and silver nanoparticles in water. Journal of Engineering Technology 32, 6 Part (B) Scientific).

Adebisi, J.A., Agunsoye, J.O., Bello, S.A., Ahmed, I.I., Ojo, O.A., Hassan, S.B., 2017. Potential of producing solar grade silicon nanoparticles from selected agro-wastes: a review. Solar Energy 142, 68–86.

Adolfsson, K.H., Hassanzadeh, S., Hakkarainen, M., 2015. Valorization of cellulose and waste paper to graphene oxide quantum dots. RSC Advances 5 (34), 26550–26558.

Al Baroot, A., Elsayed, K.A., Haladu, S.A., Magami, S.M., Alheshibri, M., Ercan, F., et al., 2023. One-pot synthesis of SnO_2 nanoparticles decorated multi-walled carbon nanotubes using pulsed laser ablation for photocatalytic applications. Optics & Laser Technology 157, 108734.

Alfarisa, S.U.H.U.F.A., Abu Bakar, S., Mohamed, A.Z.M.I., Hashim, N.O.R.H.A.Y.A.T.I., Kamari, A.Z.L.A.N., Md Isa, I., et al., 2015. Carbon nanostructures production from waste materials: a review. Advanced Materials Research 1109, 50–54.

Alheshibri, M., Kotb, E., Haladu, S.A., Al Baroot, A., Drmosh, Q.A., Ercan, F., et al., 2023. Synthesis of highly stable Ag/Ta_2O_5 nanocomposite by pulsed laser ablation as an effectual antibacterial agent. Optics & Laser Technology 162, 109295.

Arole, V.M., Munde, S.V., 2014. Fabrication of nanomaterials by top-down and bottom-up approaches-an overview. Journal of Materials Science 1, 89–93.

Ashfold, M.N., Claeyssens, F., Fuge, G.M., Henley, S.J., 2004. Pulsed laser ablation and deposition of thin films. Chemical Society Reviews 33 (1), 23–31.

Azam, M.A., Abd Mudtalib, N.E.S.A., Seman, R.N.A.R., 2018. Synthesis of graphene nanoplatelets from palm-based waste chicken frying oil carbon feedstock by using catalytic chemical vapour deposition. Materials Today Communications 15, 81–87.

Bagoria, R., Kumar, M., 2023. Nanomaterial characterization techniques. Nanomaterials in Manufacturing Processes. CRC Press, pp. 219–234.

Bai, X., Li, L., Liu, H., Tan, L., Liu, T., Meng, X., 2015. Solvothermal synthesis of ZnO nanoparticles and anti-infection application in vivo. ACS Applied Materials & Interfaces 7 (2), 1308–1317.

Bajad, G.S., Tiwari, S.K., Vijayakumar, R.P., 2015. Synthesis and characterization of CNTs using polypropylene waste as precursor. Materials Science and Engineering: B 194, 68–77.

Baudot, C., Tan, C.M., Kong, J.C., 2010. FTIR spectroscopy as a tool for nano-material characterization. Infrared Physics and Technology 53 (6), 434–438.

Bazargan, A., McKay, G., 2012. A review-synthesis of carbon nanotubes from plastic wastes. Chemical Engineering Journal 195, 377–391.

Bhattacharya, G., Fishlock, S.J., McLaughlin, J.A., Roy, S.S., 2021. Metal-oxide nanomaterials recycled from E-waste and metal industries: a concise review of applications in energy storage, catalysis, and sensing. International Journal of Energy Research 45 (5), 8091–8102.

Biswas, A., Patra, A.K., Sarkar, S., Das, D., Chattopadhyay, D., De, S., 2020. Synthesis of highly magnetic iron oxide nanomaterials from waste iron by one-step approach. Colloids and Surfaces. A, Physicochemical and Engineering Aspects 589, 124420.

Bootz, A., Vogel, V., Schubert, D., Kreuter, J., 2004. Comparison of scanning electron microscopy, dynamic light scattering and analytical ultracentrifugation for the sizing of poly (butyl cyanoacrylate) nanoparticles. European Journal of Pharmaceutics and Biopharmaceutics: Official Journal of Arbeitsgemeinschaft fur Pharmazeutische Verfahrenstechnik e.V 57 (2), 369–375.

Buhr, E., Senftleben, N., Klein, T., Bergmann, D., Gnieser, D., Frase, C.G., et al., 2009. Characterization of nanoparticles by scanning electron microscopy in transmission mode. Measurement Science & Technology 20 (8), 084025.

Bykkam, S., Ahmadipour, M., Narisngam, S., Kalagadda, V.R., Chidurala, S.C., 2015. Extensive studies on X-ray diffraction of green synthesized silver nanoparticles. Advances Nanoparticles 4 (1), 1–10.

Çakır, B.A., Budama, L., Topel, Ö., Hoda, N., 2012. Synthesis of ZnO nanoparticles using PS-b-PAA reverse micelle cores for UV protective, self-cleaning and antibacterial textile applications. Colloids and Surfaces. A, Physicochemical and Engineering Aspects 414, 132–139.

Cervellino, A., Frison, R., Masciocchi, N., Guagliardi, A., 2016. X-ray powder diffraction characterization of nanomaterials. X-ray and neutron techniques for nanomaterials characterization, pp. 545–608.

Chavali, M.S., Nikolova, M.P., 2019. Metal oxide nanoparticles and their applications in nanotechnology. SN Applied Sciences 1 (6), 607.

Chen, W.X., Lee, J.Y., Liu, Z., 2002. Microwave-assisted synthesis of carbon supported Pt nanoparticles for fuel cell applications. Chemical Communications 21, 2588–2589.

Chen, D., Li, Q., Shao, L., Zhang, F., Qian, G., 2016. Recovery and application of heavy metals from pickling waste liquor (PWL) and electroplating wastewater (EPW) by the combination process of ferrite nanoparticles. Desalination and Water Treatment 57 (60), 29264–29273.

Chin, S., Park, E., Kim, M., Jurng, J., 2010. Photocatalytic degradation of methylene blue with TiO_2 nanoparticles prepared by a thermal decomposition process. Powder Technology 201 (2), 171–176.

Choudhary, N., Hwang, S., Choi, W., 2014. Carbon nanomaterials: a review. Handbook of Nanomaterials Properties 709–769.

Chowdhury, S.U.J.A.N., Mandal, P.C., Zulfiqar, M.U.H.A.M.M.A.D., Subbarao, D.U.V.V.U.R.I., 2016. Development of ionothermal synthesis of titania nanomaterial for waste-water treatment, Advanced Materials Research, Vol. 1133. Trans Tech Publications Ltd., pp. 537–541.

Cruz, D.R., Silva, I.A., Oliveira, R.V., Buzinaro, M.A., Costa, B.F., Cunha, G.C., et al., 2021. Recycling of mining waste in the synthesis of magnetic nanomaterials for removal of nitrophenol and polycyclic aromatic hydrocarbons. Chemical Physics Letters 771, 138482.

De Souza, C.D., Nogueira, B.R., Rostelato, M.E.C., 2019. Review of the methodologies used in the synthesis gold nanoparticles by chemical reduction. Journal of Alloys and Compounds 798, 714–740.

Deus, R.C., Cilense, M., Foschini, C.R., Ramirez, M.A., Longo, E., Simões, A.Z., 2013. Influence of mineralizer agents on the growth of crystalline CeO_2 nanospheres by the microwave-hydrothermal method. Journal of Alloys and Compounds 550, 245–251.

Devi, P., Hipp, K.N., Thakur, A., Lai, R.Y., 2020. Waste to wealth translation of e-waste to plasmonic nanostructures for surface-enhanced Raman scattering. Applied Nanoscience 10 (5), 1615–1623.

Dogan, N., Ozel, F., Koten, H., 2023. Structural, morphological, and magnetic characterization of iron oxide nanoparticles synthesized at different reaction times via thermal decomposition method. Current Nanoscience 19 (1), 33–38.

Domínguez, M.I., Barrio, I., Sánchez, M., Centeno, M.Á., Montes, M., Odriozola, J.A., 2008. CO and VOCs oxidation over Pt/SiO_2 catalysts prepared using silicas obtained from stainless steel slags. Catalysis Today 133, 467–474.

Dubadi, R., Huang, S.D., Jaroniec, M., 2023. Mechanochemical synthesis of nanoparticles for potential antimicrobial applications. Materials 16 (4), 1460.

Dubey, R.S., Rajesh, Y.B.R.D., More, M.A., 2015. Synthesis and characterization of SiO2 nanoparticles via sol-gel method for industrial applications. Materials Today: Proceedings 2 (4–5), 3575–3579.

Eastoe, J., Hollamby, M.J., Hudson, L., 2006. Recent advances in nanoparticle synthesis with reversed micelles. Advances in Colloid and Interface Science 128, 5–15.

Eid, M.M., 2022. Characterization of nanoparticles by FTIR and FTIR-microscopy. Handbook of Consumer Nanoproducts. Springer, Singapore, pp. 1–30, Singapore.

El Essawy, N.A., Ali, S.M., Farag, H.A., Konsowa, A.H., Elnouby, M., Hamad, H.A., 2017. Green synthesis of graphene from recycled PET bottle wastes for use in the adsorption of dyes in aqueous solution. Ecotoxicology and Environmental Safety 145, 57–68.

El-Amir, A.A., Ewais, E.M., Abdel-Aziem, A.R., Ahmed, A., El-Anadouli, B.E., 2016. Nano-alumina powders/ceramics derived from aluminum foil waste at low temperature for various industrial applications. Journal of Environmental Management 183, 121–125.

Eswarapriya, B., Jegatheesan, K.S., 2015. Antifungal activity of biogenic selenium nanoparticles synthesized from electronic waste. International Journal of PharmTech Research 8 (3), 383–386.

Farzana, R., Rajarao, R., Behera, P.R., Hassan, K., Sahajwalla, V., 2018. Zinc oxide nanoparticles from waste Zn-C battery via thermal route: characterization and properties. Nanomaterials 8 (9), 717.

Farzana, R., Hassan, K., Sahajwalla, V., 2019. Manganese oxide synthesized from spent Zn-C battery for supercapacitor electrode application. Scientific Reports 9 (1), 1–12.

Fotukian, S.M., Barati, A., Soleymani, M., Alizadeh, A.M., 2020. Solvothermal synthesis of $CuFe_2O_4$ and Fe_3O_4 nanoparticles with high heating efficiency for magnetic hyperthermia application. Journal of Alloys and Compounds 816, 152548.

Gabryszewska, M., Gworek, B., 2020. Impact of municipal and industrial waste incinerators on PCBs content in the environment. PLoS One 15 (11), e0242698.

Galindo, R., Mazario, E., Gutiérrez, S., Morales, M.P., Herrasti, P., 2012. Electrochemical synthesis of $NiFe_2O_4$ nanoparticles: characterization and their catalytic applications. Journal of Alloys and Compounds 536, S241–S244.

Ghotekar, S., Pansambal, S., Lin, K.Y.A., Pore, D., Oza, R., 2023. Recent advances in synthesis of $CeVO_4$ nanoparticles and their potential scaffold for photocatalytic applications. Top Catalysis 66 (1–4), 89–103.

Giri, S.K., Das, N.N., Pradhan, G.C., 2011. Synthesis and characterization of magnetite nanoparticles using waste iron ore tailings for adsorptive removal of dyes from aqueous solution. Colloids and Surfaces. A, Physicochemical and Engineering Aspects 389 (1–3), 43–49.

Gómez-Hernández, R., Panecatl-Bernal, Y., Méndez-Rojas, M.Á., 2019. High yield and simple one-step production of carbon black nanoparticles from waste tires. Heliyon 5 (7), e02139.

Gong, J., Liu, J., Wan, D., Chen, X., Wen, X., Mijowska, E., et al., 2012. Catalytic carbonization of polypropylene by the combined catalysis of activated carbon with Ni_2O_3 into carbon nanotubes and its mechanism. Applied Catalysis 449, 112–120.

Guzmán, M.G., Dille, J., Godet, S., 2009. Synthesis of silver nanoparticles by chemical reduction method and their antibacterial activity. International Journal of Chemical and Molecular Engineering 2 (3), 104–111.

Hachem, K., Ansari, M.J., Saleh, R.O., Kzar, H.H., Al-Gazally, M.E., Altimari, U.S., et al., 2022. Methods of chemical synthesis in the synthesis of nanomaterial and nanoparticles by the chemical deposition method: a review. BioNanoSci 12 (3), 1032–1057.

Heidari, A., Younesi, H., 2020. Synthesis, characterization and life cycle assessment of carbon nanospheres from waste tires pyrolysis over ferrocene catalyst. Journal of Environmental Chemical Engineering 8 (2), 103669.

Hodoroaba, V.D., 2020. Energy-dispersive X-ray spectroscopy (EDS). Characterization of Nanoparticles. Elsevier, pp. 397–417.

Horn, D., Rieger, J., 2001. Organic nanoparticles in the aqueous phase—theory, experiment, and use. Angewandte Chemie (International Ed. in English) 40 (23), 4330–4361.

Hosseinpour-Mashkani, S.M., Sobhani-Nasab, A., 2016. A simple sonochemical synthesis and characterization of $CdWO_4$ nanoparticles and its photocatalytic application. Journal of Materials Science Materials 27, 3240–3244.

Hou, S., Duan, Z., Ma, Z., Singh, A., 2020. Improvement on the properties of waste glass mortar with nanomaterials. Construction and Building Materials 254, 118973.

Hu, Y., Yang, J., Tian, J., Jia, L., Yu, J.S., 2014. Green and size-controllable synthesis of photoluminescent carbon nanoparticles from waste plastic bags. RSC Advances 4 (88), 47169–47176.

Huston, M., DeBella, M., DiBella, M., Gupta, A., 2021. Green synthesis of nanomaterials. Nanomaterials 11 (8), 2130.

Ibrahim, R.K., Hayyan, M., AlSaadi, M.A., Hayyan, A., Ibrahim, S., 2016. Environmental application of nanotechnology: air, soil, and water. Environmental Science and Pollution Research 23, 13754–13788.

Iqbal, T., Ejaz, A., Abrar, M., Afsheen, S., Batool, S.S., Fahad, M., et al., 2019. Qualitative and quantitative analysis of nanoparticles using laser-induced breakdown spectroscopy (LIBS) and energy dispersive x-ray spectroscopy (EDS). Laser Physics 29 (11), 116001.

Jadoun, S., Arif, R., Jangid, N.K., Meena, R.K., 2021. Green synthesis of nanoparticles using plant extracts: a review. Environmental Chemistry Letters 19, 355–374.

Jadoun, S., Chauhan, N.P.S., Zarrintaj, P., Barani, M., Varma, R.S., Chinnam, S., et al., 2022. Synthesis of nanoparticles using microorganisms and their applications: a review. Environmental Chemistry Letters 20 (5), 3153–3197.

Jara, N., Milán, N.S., Rahman, A., Mouheb, L., Boffito, D.C., Jeffryes, C., et al., 2021. Photochemical synthesis of gold and silver nanoparticles-A review. Molecules (Basel, Switzerland) 26 (15), 4585.

Jiang, X., Jiang, J., Jin, Y., Wang, E., Dong, S., 2005. Effect of colloidal gold size on the conformational changes of adsorbed cytochrome c: probing by circular dichroism, UV − Visible, and Infrared Spectroscopy. Biomacromol 6 (1), 46–53.

Khataee, A., Karimi, A., Zarei, M., Joo, S.W., 2020. Eu-doped ZnO nanoparticles: sonochemical synthesis, characterization, and sonocatalytic application. Ultrasonics Sonochemistry 67, 102822.

Kimling, J., Maier, M., Okenve, B., Kotaidis, V., Ballot, H., Plech, A., 2006. Turkevich method for gold nanoparticle synthesis revisited. The Journal of Physical Chemistry. B 110 (32), 15700–15707.

Krishnia, L., Thakur, P., Thakur, A., 2022. Synthesis of nanoparticles by physical route. In: Thakur, A., Thakur, P., Khurana, S.P. (Eds.), Synthesis and Applications of Nanoparticles. Springer, Singapore. Available from: https://doi.org/10.1007/978-981-16-6819-7_3.

Kulkarni, S.K., 2015. Synthesis of nanomaterials—I (physical methods). Nanotechnology: Principles and Practices. Springer, Cham. Available from: https://doi.org/10.1007/978-3-319-09171-6_3.

Kumar, C.S. (Ed.), 2013. Transmission Electron Microscopy Characterization of Nanomaterials. Springer Science & Business Media. Available from: https://doi.org/10.1007/978-3-642-38934-4.

Kumar, S., Nann, T., 2006. Shape control of II–VI semiconductor nanomaterials. Small (Weinheim an der Bergstrasse, Germany) 2 (3), 316–329.

Kumar, R., Sakthivel, R., Behura, R., Mishra, B.K., Das, D., 2015. Synthesis of magnetite nanoparticles from mineral waste. Journal of Alloys and Compounds 645, 398–404.

Kumar, M., Ranjan, R., Dandapat, S., Srivastava, R., Sinha, M.P., 2022. XRD analysis for characterization of green nanoparticles: a mini review. Brazilian Journal of Pharmaceutical Sciences 58, e19173.

Kumar, V.B., Gedanken, A., Ze'ev, I.P., 2023. Sonochemistry of molten metals. Nanoscale.

Lee, H., Kim, B.H., Park, Y.K., Kim, S.J., Jung, S.C., 2015. Application of recycled zero-valent iron nanoparticle to the treatment of wastewater containing nitrobenzene. Nanomaterials 2015.

Lin, P., Chai, F., Zhang, R., Xu, G., Fan, X., Luo, X., 2016. Electrochemical synthesis of poly (3, 4-ethylenedioxythiophene) doped with gold nanoparticles, and its application to nitrite sensing. Microchimica Acta 183, 1235–1241.

Lu, W., Liao, F., Luo, Y., Chang, G., Sun, X., 2011. Hydrothermal synthesis of well-stable silver nanoparticles and their application for enzymeless hydrogen peroxide detection. Electrochimica Acta 56 (5), 2295–2298.

Mancuso, A., Blangetti, N., Sacco, O., Freyria, F.S., Bonelli, B., Esposito, S., et al., 2023. Photocatalytic degradation of crystal violet dye under visible light by Fe-doped TiO_2 prepared by reverse-micelle sol–gel method. Nanomaterials 13 (2), 270.

Manfredi, S., Tonini, D., Christensen, T.H., 2010. Contribution of individual waste fractions to the environmental impacts from landfilling of municipal solid waste. Waste Management 30 (3), 433–440.

Maroufi, S., Mayyas, M., Sahajwalla, V., 2017. Nano-carbons from waste tyre rubber: an insight into structure and morphology. Waste Management 69, 110–116.

Maroufi, S., Mayyas, M., Nekouei, R.K., Assefi, M., Sahajwalla, V., 2018. Thermal nanowiring of e-waste: a sustainable route for synthesizing green Si_3N_4 nanowires. ACS Sustainable Chemistry & Engineering 6 (3), 3765–3772.

Mis-Fernández, R., Rios-Soberanis, C.R., Arenas-Alatorre, J., Azamar-Barrios, J.A., 2012. Synthesis of carbon nanostructures from residual solids waste tires. Journal of Applied Polymer Science 123 (4), 1960–1967.

Mishra, N., Das, G., Ansaldo, A., Genovese, A., Malerba, M., Povia, M., et al., 2012. Pyrolysis of waste polypropylene for the synthesis of carbon nanotubes. Journal of Analytical and Applied Pyrolysis 94, 91–98.

Mohammed, M.A., Rheima, A.M., Jaber, S.H., Hameed, S.A., 2020. The removal of zinc ions from their aqueous solutions by Cr_2O_3 nanoparticles synthesized via the UV-irradiation method. Egyptian Journal of Chemistry 63 (2), 425–431.

Mourdikoudis, S., Pallares, R.M., Thanh, N.T., 2018. Characterization techniques for nanoparticles: comparison and complementarity upon studying nanoparticle properties. Nanoscale 10 (27), 12871–12934.

Mwafy, E.A., 2020. Eco-friendly approach for the synthesis of MWCNTs from waste tires via chemical vapor deposition. Environmental Nanotechnology, Monitoring & Management 14, 100342.

Nascimento, R.S., Correa, J.A.M., Figueira, B.A.M., Pinheiro, P.A., Silva, J.H., Freire, P.T.C., et al., 2022. From mining waste to environmental remediation: a nanoadsorbent from amazon bauxite tailings for the removal of erythrosine B dye. Applied Clay Science 222, 106482.

Navas, D., Ibañez, A., González, I., Palma, J.L., Dreyse, P., 2020. Controlled dispersion of ZnO nanoparticles produced by basic precipitation in solvothermal processes. Heliyon 6 (12), e05821.

Nayak, P., Kumar, S., Sinha, I., Singh, K.K., 2019. ZnO/CuO nanocomposites from recycled printed circuit board: preparation and photocatalytic properties. Environmental Science and Pollution Research 26 (16), 16279–16288.

Ni, S., Wang, X., Zhou, G., Yang, F., Wang, J., Wang, Q., et al., 2009. Hydrothermal synthesis of Fe_3O_4 nanoparticles and its application in lithium ion battery. Materials Letters 63 (30), 2701–2703.

Oviedo, L.R., Muraro, P.C.L., Pavoski, G., Espinosa, D.C.R., Ruiz, Y.P.M., Galembeck, A., et al., 2022. Synthesis and characterization of nanozeolite from (agro) industrial waste for application in heterogeneous photocatalysis. Environmental Science and Pollution Research 29, 3794–3807.

Panahi-Kalamuei, M., Salavati-Niasari, M., Hosseinpour-Mashkani, S.M., 2014. Facile microwave synthesis, characterization, and solar cell application of selenium nanoparticles. Journal of Alloys and Compounds 617, 627–632.

Parashar, M., Shukla, V.K., Singh, R., 2020. Metal oxides nanoparticles via sol–gel method: a review on synthesis, characterization and applications. Journal of Materials Science Materials 31, 3729–3749.

Patil, S.P., Burungale, V.V., 2020. Physical and chemical properties of nanomaterials. In Nanomedicines for Breast Cancer Theranostics. Elsevier, pp. 17–31.

Percival, R.V., Schroeder, C.H., Miller, A.S., Leape, J.P., 2021. Environmental Regulation: Law, Science, and Policy [Connected EBook with Study Center]. Wolters Kluwer Law & Business.

Quevedo, A.C., Guggenheim, E., Briffa, S.M., Adams, J., Lofts, S., Kwak, M., et al., 2021. UV-Vis spectroscopic characterization of nanomaterials in aqueous media. Journal of Visualized Experiments 176, e61764.

Rada, E.C., Ionescu, G., Conti, F., Cioca, U.I., Torretta, V., 2018. Energy from municipal solid waste: some considerations on emissions and health impact. Calitatea 19 (167), 118–122.

Rao, C.R., Kulkarni, G.U., Thomas, P.J., Edwards, P.P., 2000. Metal nanoparticles and their assemblies. Chemical Society Reviews 29 (1), 27–35.

Razak, N.A.A., Othman, N.H., Shayuti, M.S.M., Jumahat, A., Sapiai, N., Lau, W.J., 2022. Agricultural and industrial waste-derived mesoporous silica nanoparticles: a review on chemical synthesis route. Journal of Environmental Chemical Engineering 107322.

Reverberi, A., Kuznetsov, N., Meshalkin, V., Salerno, M., Fabiano, B., 2016. Systematical analysis of chemical methods in metal nanoparticles synthesis. Theoretical Foundations of Chemical Engineering 50 (1).

Rheima, A.M., Mohammed, M.A., Jaber, S.H., Hameed, S.A., 2019. Synthesis of silver nanoparticles using the UV-irradiation technique in an antibacterial application. Journal of Southwest Jiaotong University 54 (5).

Robinson, B.H., 2009. E-waste: an assessment of global production and environmental impacts. The Science of the Total Environment 408 (2), 183–191.

Rocco, D., Moldoveanu, V.G., Feroci, M., Bortolami, M., Vetica, F., 2023. Electrochemical synthesis of carbon quantum dots. ChemElectroChem e202201104.

Romanovski, V., 2021. Agricultural waste based-nanomaterials: green technology for water purification. In Aquananotechnology. Elsevier, pp. 577–595.

Saha, S., Pal, A., Kundu, S., Basu, S., Pal, T., 2010. Photochemical green synthesis of calcium-alginate-stabilized Ag and Au nanoparticles and their catalytic application to 4-nitrophenol reduction. Langmuir: The ACS Journal of Surfaces and Colloids 26 (4), 2885–2893.

Salavati-Niasari, M., Soofivand, F., Sobhani-Nasab, A., Shakouri-Arani, M., Faal, A.Y., Bagheri, S., 2016. Synthesis, characterization, and morphological control of $ZnTiO_3$ nanoparticles through sol-gel processes and its photocatalyst application. Advances Powder Technology 27 (5), 2066–2075.

Samaddar, P., Ok, Y.S., Kim, K.H., Kwon, E.E., Tsang, D.C., 2018. Synthesis of nanomaterials from various wastes and their new age applications. Journal of Cleaner Production 197, 1190–1209.

Shaheen, I., Khalil, A., Shaheen, R., Tahir, M.B., 2023. A review on nanomaterials: types, synthesis, characterization techniques, properties and applications. International Journal of Innovative Science and Research Technology 2 (1), 56–62.

Sharma, S., Kalita, G., Hirano, R., Shinde, S.M., Papon, R., Ohtani, H., et al., 2014. Synthesis of graphene crystals from solid waste plastic by chemical vapor deposition. Carbon 72, 66–73.

Singh, S., Srivastava, V.C., Mandal, T.K., Mall, I.D., 2014. Synthesis of different crystallographic Al_2O_3 nanomaterials from solid waste for application in dye degradation. RSC Advances 4 (92), 50801–50810.

Singh, M.P., Bhardwaj, A.K., Bharati, K., Singh, R.P., Chaurasia, S.K., Kumar, S., et al., 2021. Biogenic and non-biogenic waste utilization in the synthesis of 2D materials (graphene, h-BN, g-C2N) and their applications. Frontiers in Nanotechnology 3, 685427.

Singh, A., Gogoi, H.P., Barman, P., 2023. Synthesis of metal oxide nanoparticles by facile thermal decomposition of new Co (II), Ni (II), and Zn (II) Schiff base complexes-optical properties and photocatalytic degradation of methylene blue dye. Inorganica Chimica Acta 546, 121292.

Smaoui, S., Chérif, I., Hlima, H.B., Khan, M.U., Rebezov, M., Thiruvengadam, M., et al., 2023. Zinc oxide nanoparticles in meat packaging: a systematic review of recent literature. Food Packaging and Shelf Life 36, 101045.

Song, P., Zhao, X., Cheng, X., Li, S., Wang, S., 2018. Recycling the nanostructured carbon from waste tires. Composites Communications 7, 12–15.

Stulov, Y., Dolmatov, V., Dubrovskiy, A., Kuznetsov, S., 2023. Electrochemical synthesis of functional coatings and nanomaterials in molten salts and their application. Coatings 13 (2), 352.

Suzuki, E., 2002. High-resolution scanning electron microscopy of immunogold-labelled cells by the use of thin plasma coating of osmium. Journal of Microscopy 208 (3), 153–157.

Tade, R.S., Nangare, S.N., Patil, P.O., 2020. Agro-industrial waste-mediated green synthesis of silver nanoparticles and evaluation of its antibacterial activity. Nano Biomedicine and Engineering 12 (1), 57–66.

TermehYousefi, A., Bagheri, S., Shinji, K., Rouhi, J., Rusop Mahmood, M., Ikeda, S., 2014. Fast synthesis of multilayer carbon nanotubes from camphor oil as an energy storage material. BioMed Research International 2014.

Tharmaraj, V., Pitchumani, K., 2011. Alginate stabilized silver nanocube-Rh6G composite as a highly selective mercury sensor in aqueous solution. Nanoscale 3 (3), 1166–1170. Available from: https://doi.org/10.1039/c0nr00749h.

Tharmaraj, V., Pitchumani, K., 2013. A highly selective ratiometric fluorescent chemosensor for Cu(ii) based on dansyl-functionalized thiol stabilized silver nanoparticles. Journal of Materials Chemistry B 1 (14), 1962–1967. Available from: https://doi.org/10.1039/c3tb00534h.

Tharmaraj, V., Yang, J., 2014. Sensitive and selective colorimetric detection of Cu(2 +) in aqueous medium via aggregation of thiomalic acid functionalized Ag nanoparticles. Analyst 139 (23), 6304–6309. Available from: https://doi.org/10.1039/c4an01449a.

Thompson, R.C., Moore, C.J., Vom Saal, F.S., Swan, S.H., 2009. Plastics, the environment and human health: current consensus and future trends. Philosophical Transactions of the Royal Society 364 (1526), 2153–2166.

Tiwari, S., Talreja, S., 2022. Green chemistry and microwave irradiation technique: a review. Journal of Pharmaceutical Research International 34 (39A), 74–79.

Tsai, C.K., Horng, J.J., 2021. Transformation of glass fiber waste into mesoporous zeolite-like nanomaterials with efficient adsorption of methylene blue. Sustainability 13 (11), 6207.

Veksha, A., Giannis, A., Oh, W.D., Chang, V.W.C., Lisak, G., 2018. Upgrading of non-condensable pyrolysis gas from mixed plastics through catalytic decomposition and dechlorination. Fuel Processing Technology 170, 13–20.

Veksha, A., Latiff, N.M., Chen, W., Ng, J.E., Lisak, G., 2020. Heteroatom doped carbon nanosheets from waste tires as electrode materials for electrocatalytic oxygen reduction reaction: effect of synthesis techniques on properties and activity. Carbon 167, 104–113.

Verma, R., Vinoda, K.S., Papireddy, M., Gowda, A.N.S., 2016. Toxic pollutants from plastic waste-a review. Procedia Environmental Sciences 35, 701–708.

Vinitha Judith, J., Vasudevan, N., 2021. Synthesis of nanomaterial from industrial waste and its application in environmental pollutant remediation. Environmental Engineering Research .

Visa, M., Isac, L., Duta, A., 2015. New fly ash TiO_2 composite for the sustainable treatment of wastewater with complex pollutants load. Applied Surface Science 339, 62–68.

Vladár, A.E., Hodoroaba, V.D., 2020. Characterization of nanoparticles by scanning electron microscopy. In Characterization of Nanoparticles. Elsevier, pp. 7–27.

Wang, Z.L., 1999. Transmission electron microscopy and spectroscopy of nanoparticles. Characterization of Nanophase Materials 37–80.

Wang, C., Li, D., Zhai, T., Wang, H., Sun, Q., Li, H., 2019. Direct conversion of waste tires into three-dimensional graphene. Energy Storage Materials 23, 499–507.

Worsfold, P., Townshend, A., Poole, C.F., Miró, M., 2019. Encyclopedia of Analytical Science. Elsevier.

Wu, Z.J., Huang, W., Cui, K.K., Gao, Z.F., Wang, P., 2014. Sustainable synthesis of metals-doped ZnO nanoparticles from zinc-bearing dust for photodegradation of phenol. Journal of Hazardous Materials 278, 91–99.

Yang, W., Sun, W.J., Chu, W., Jiang, C.F., Wen, J., 2012. Synthesis of carbon nanotubes using scrap tyre rubber as carbon source. Chinese Chemical Letters 23 (3), 363–366.

Yogesh, G.K., Shukla, S., Sastikumar, D., Koinkar, P., 2021. Progress in pulsed laser ablation in liquid (PLAL) technique for the synthesis of carbon nanomaterials: a review. Applied Physics A 127, 1–40.

Zahra, Q., Fraz, A., Anwar, A., Awais, M., Abbas, M., 2016. A mini review on the synthesis of Ag-nanoparticles by chemical reduction method and their biomedical applications. NUST Journal of Natural Sciences 9 (1), 1–7.

Zhan, L., Li, O., Wang, Z., Xie, B., 2018a. Recycling zinc and preparing high-value-added nanozinc oxide from waste zinc-manganese batteries by high-temperature evaporation-separation and oxygen control oxidation. ACS Sustainable Chemistry & Engineering 6 (9), 12104–12109.

Zhan, G., Ng, W.C., Koh, S.N., Wang, C.H., 2018b. Template-free synthesis of alkaline earth vanadate nanomaterials from leaching solutions of oil refinery waste. ACS Sustainable Chemistry & Engineering 6 (2), 2292–2301.

Zhang, B., Yan, B., 2010. Analytical strategies for characterizing the surface chemistry of nanoparticles. Analytical and Bioanalytical Chemistry 396, 973–982.

Zhang, D., Luo, H., Zheng, L., Wang, K., Li, H., Wang, Y., et al., 2012. Utilization of waste phosphogypsum to prepare hydroxyapatite nanoparticles and its application towards removal of fluoride from aqueous solution. Journal of Hazardous Materials 241, 418–426.

Zhang, Y., Williams, P.T., 2016. Carbon nanotubes and hydrogen production from the pyrolysis catalysis or catalytic-steam reforming of waste tyres. Journal of Analytical and Applied Pyrolysis 122, 490–501.

Zhang, Y., Wu, C., Nahil, M.A., Williams, P., 2016, August. High-value resource recovery products from waste tyres. In: Proceedings of the Institution of Civil Engineers-Waste and Resource Management (Vol. 169, No. 3, pp. 137–145). Thomas Telford Ltd.

Zhang, H., Liu, J., Ou, C., Shen, J., Yu, H., Jiao, Z., et al., 2017. Reuse of Fenton sludge as an iron source for $NiFe_2O_4$ synthesis and its application in the Fenton-based process. Journal of Environmental Sciences 53, 1–8.

Zhao, H., Zhao, T.S., 2013. Graphene sheets fabricated from disposable paper cups as a catalyst support material for fuel cells. Journal of Materials Chemistry 1 (2), 183–187.

Zhong, L., Feng, Y., Wang, G., Wang, Z., Bilal, M., Lv, H., et al., 2020. Production and use of immobilized lipases in/on nanomaterials: a review from the waste to biodiesel production. International Journal of Biological Macromolecules 152, 207–222.

Zhou, X., Deng, J., Yang, R., Zhou, D., Fang, C., He, X., et al., 2020. Facile preparation and characterization of fibrous carbon nanomaterial from waste polyethylene terephthalate. Journal of Waste Management 107, 172–181.

Zhuo, C., Levendis, Y.A., 2014. Upcycling waste plastics into carbon nanomaterials: a review. Journal of Applied Polymer Science 131 (4).

Chapter 11

Nanomaterial synthesis using tire and plastic

Sarfaraz Ahmed Mahesar[1], Muhammad Saqaf Jagirani[1,2,3], Aamna Balouch[1], Aftab Hussain Khuhawar[4], Abdul Hameed Kori[1] and Syed Tufail Hussain Sherazi[1]

[1]*National Centre of Excellence in Analytical Chemistry, University of Sindh, Jamshoro, Pakistan,* [2]*Institute of Green Chemistry and Chemical Technology, School of Chemistry & Chemical Engineering, Jiangsu University, Zhenjiang, P.R. China,* [3]*School of Materials Science & Engineering, Jiangsu University, Zhenjiang, P.R. China,* [4]*Chemical Engineering and Technology, Ocean University of China, Qingdao, P.R. China*

11.1 Introduction

Nanotechnology (NT) is the science of manipulating materials into nano size and control of matter at the nanoscale (between 1 and 100 nm); due to unique phenomena, NT can be used in new applications. When the matter is controlled at the nanoscale, the processes make more efficient and get to places that are hard to reach. NT has a broad spectrum of applications, and it is emerging science. Due to the high surface-to-mass ratio, nanoscale materials are highly reactive. At a smaller scale, the materials need less energy to proceed with the reaction, and the nanomaterials are more likely to be sustainable (Recordati et al., 2015). The properties of bulk and nanoscale materials differ, and the properties of nanoscale materials themselves can also vary. For example, it was found that nanoparticles (NPs) made of the same material but different sizes had different absorption wavelengths (Chen et al., 2019a). The size and shape of the NPs also found to be a factor in how well they are absorbed and reacted (Ahmed and Ali, 2020). This difference in properties based on the shape and size of the NPs could be used as a regulating factor for new applications. Nanomaterials have a wide range of magnetic, electrical, mechanical, and catalytic properties. In recent years, several engineered nanomaterials have demonstrated impressive potential for use in high-tech fields such as energy conversion and storage, pollution inspecting, and smart packing sustainability approaches, frameworks, and metrics are increasing (Yang et al., 2019; Falinski et al., 2018). This growing trend will likely continue over the next few years in line with current worldwide patterns regarding sustainable development. In this chapter, we discuss the history, problems, and potential solutions for recycling industrial waste (rubber tires, large batteries, and wastewater), plastics such as polyvinyl chloride, polyethylene (PE), polypropylene (PP), and polystyrene and polyvinyl alcohol (PVA) waste (printed circuit boards, copper cables, and electronic equipment), and as suitable inputs for making NPs that have added value such as monitoring, treatment, and cleanup of water pollution. We end with some comments about sustainability that highlight the most important things that need to be done to make and use nanomaterials made from waste on a larger scale. The NT plays an essential role in many fields, including energy, transportation, agriculture, healthcare, communication, and information, to achieve new development and sustainability. For example, nanocapsules can be used to deliver drugs and make medicines, while quantum dots are used in biosensors to take pictures of cells. NT could also help make agriculture and food systems more sustainable and help make waste into valuable products, as plastic waste has been used to prepare different types of nanomaterials, such as carbon-based and plastic-based metal nanomaterials. This could be done in several ways, including using NT to make fertilizers and treat water for agricultural fields (Das et al., 2019). Carbon nanoparticles (CNPs) can easily pass through cell membranes because their sizes are in the nanoscale range. This is true even if they bind to drug molecules, proteins, or genes. This benefit and biocompatible CNPs make them a good choice as carriers. Also, the fact that CNPs are used in photo- and electrocatalysis is not a big surprise (Kim et al., 2016). Because CNPs create an adjustable band gap, we can use solar energy in the visible light and near-infrared ranges to speed up the catalysis process. Some methods that have been tried to make CNPs are microwave irradiation, pulsed laser irradiation, hydrothermal and ultrasonic treatment, etc. (Bhardwaj et al., 2022).

Green and Sustainable Approaches Using Wastes for the Production of Multifunctional Nanomaterials. DOI: https://doi.org/10.1016/B978-0-443-19183-1.00023-4

11.2 Tire and plastic-based preparation of nanomaterials

Several methods have been reported for the preparation of tire and plastic waste–based nanomaterials such as quartz tubes, autoclaves, muffle furnaces, and crucibles (Bazargan, McKay, 2012). Fig. 11.1 denotes the general schematic view of the fabrication of nanomaterials (NMs).

11.3 Quartz tube

Russian researchers reported a Quartz tube–based process for synthesizing (polyethylene) plastic waste–based NPs such as CNTs. In these experiments, crooked CNTs with sizes ranging from 25 to 65 nm were produced at a temperature of 800 °C (Chernozatonskii et al., 1998). Granular PE was pyrolyzed in a chamber. The resulting products were sent through a quartz tube reactor at a pressure of four atmospheres of hydrocarbons and helium (He) gas while passing over a nickel plate. The temperature at which it was conducted is the primary factor that impacts the process. Jalili et al. reported the crooked method-based preparation of CNTs at 420 °C–450 °C temperature range, which led to the synthesis of crooked CNTs with diameters ranging from 10 to 40 nm. The reaction lasted under an hour, producing approximately 103 g/cm^2 per hour of CNTs. The time required for the reaction was about 59 minutes. After an investigation into the product's resistance to heat, it was discovered that the filamentous material begins to oxidize at approximately 420 °C (Jalili et al., 2008).

11.4 Autoclave

Kong and Zhang (2007) reported a new autoclave-based method for preparing plastic waste–based CNTs. During the preparation of CNTs using an autoclave under high temperatures of up to 700 °C, the proposed method obtained high yields of CNTs with 20–60 nm. They also successfully produced helical CNTs at a rate of 5% throughout the process. CNTs can be found in various shapes in addition to the straight form, including spiral, curved, and planar-spiral configurations. In the nanoelectromechanical domains, helical CNTs are of particular interest because it has been demonstrated that they possess unique electrical, magnetic, and mechanical properties (Volodin et al., 2004).

CNTs yields, on the other hand, increased as temperatures rose to 700 °C up to 80%. It was also discovered that in the absence of maleated polypropylene (MA-PP), the iron would congregate at the base of the autoclave, and the number of CNTs produced was very low. MA-PP performed the function of a compatibilizer to improve the dispersion of iron in the substrate. Zhang et al. (2008) published yet another study in which PP, rather than PE, was used as the reagent in this study, and Ni, rather than ferrocene, was utilized. The yield remained at a high level of up to 80%

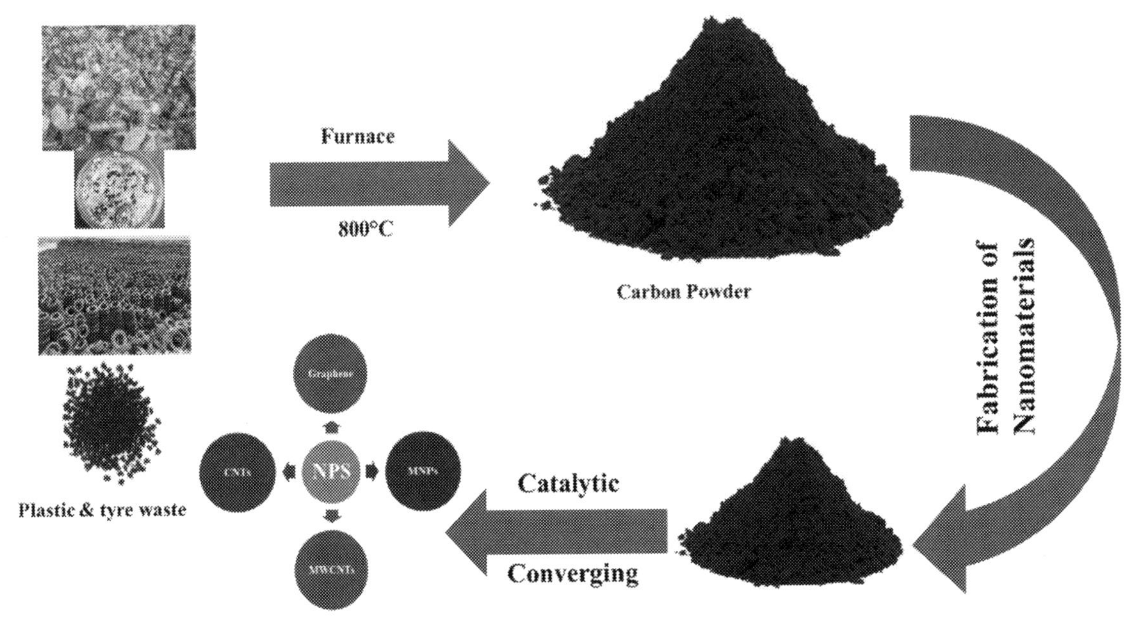

FIGURE 11.1 General schematic view of fabrication of NMs.

(at 800 °C optimal conditions). Pyrolysis of PP and ferrocene [$Fe(C_5H_5)_2$] in an autoclave was used by Zhang et al. (2010) to produce clusters of CNTs in the form of microspheres with sizes ranging from 5.5 to 7.5 μm. CNTs synthesized at 700 °C exhibited better crystallinity and fewer defects than those produced at lower temperatures. At 600 °C and 700 °C temperatures, the BET surface areas of the material were respectively 140.6 and 74.5 m^2/g. Compared with other processes that produce spheres, this method has several advantages, and one is that it does not require the premixing of ingredients. Pol and Thackeray (2011) produced high-purity nanomaterials using their autogenic method. These nanomaterials were made from a variety of PE products. The electrodes made from the synthesized materials performed exceptionally well when used in lithium half cells (Masarapu et al., 2009).

11.5 Crucible

Song et al. (2009) used two types of zeolites (HZMS-5 or H-Beta) as synergistic additives in a simple reactor crucible to help PP burn in the presence of Ni_2O_3 catalyst and make multiwalled carbon nanotubes (MWCNTs). Then, SEM and TEM were used to look at the shapes and structures, and it was found that the products made from H-Beta zeolite (PP/H-Beta/Ni_2O_3: 90/5/5) were better and showed hair-like fibers. The better performance of H-Beta can be explained by its large-pored, strong-acid molecular sieve that breaks apart at high temperatures, exposing more acid sites that help MWCNTs form. The products' thermal gravimetric (TG) analysis showed that the breakdown process started after 500 °C and peaked at 590 °C, which means they were stable. Tang et al. (2005) prepared MWCNTs by Brabender mixer in the presence of PP using modified clay, a Ni catalyst, and maleated PP at 100 rpm and 190 °C for 10 minutes. The mixture was then heated to 600 °C in a crucible. After completing the reaction, the resultant char was cooled and cleaned. When the shapes of the cleaning products were looked at, many hollow tube-like structures with diameters of 20–40 nm were found. It has been thought that having clay in the burning process would help MWCNTs formation.

11.6 Muffle furnace

Song and Ji (2011) proposed a catalytic-based method that employs molybdenum (Mo) and magnesium (Mg) as the catalyst, significantly improving upon their previous use of Ni for producing CNTs from PP. The catalyst was then mixed with the substrate in a Brabender mixer to make a composite, placed in a crucible, and heated for 10 minutes in a muffle furnace at 850 °C. After the black powder had cooled to room temperature, it could be seen without further treatment or cleaning. After figuring out how many tubes were made, the catalyst combination that made the most tubes was chosen for the rest of the work. Conventional CNTs comprised 95% of the product, and their diameters were less than 30 nm. About 5% of the CNTs had structures made up of two helices with diameters of 60 nm. It is important to note that the quality of the CNTs was very good, and TEM results indicated the formation of amorphous carbon. TGA analysis also showed that the degradation range of the synthesized products was 600 °C–800 °C, with a peak at 730 °C. These values were similar to CNTs made in the old way, using the arc process. Also, when the temperature exceeded 800 °C, only 0.3% of the material was left, which shows purity. Other studies have found that using Mo catalysts results in the formation of helical products. Therefore, the researchers speculated that Mo was responsible for the observed spiral patterns (Somanathan and Pandurangan, 2010). On the other hand, if too much Mo were used in a catalyst that contained Co and Mg to make CNTs from methane, the graphene walls would be less graphitized, and the amount of CNTs made would be less (Yeoh et al., 2010).

11.7 Plastic and tire waste–based nanomaterials

CNTs can be made from WP in different ways, including autoclaves, quartz tubes, muffle furnaces, and crucibles. Polyolefin is a polymer made from an olefin monomer (CnH_2n), has a carbon content of about 85.7% by weight and is found in most plastic waste (Gong et al., 2012). PP and PE are two types of waste polyolefins used to make carbon materials. Other plastics such as PVA and PET that have been added can also be used to make CNTs (Deng et al., 2016). Gong and his colleagues developed a novel method of layer-by-layer fabrication. In a quartz tube reactor, activated carbon and Ni_2O_3 were used to catalyze used PP. The PP could be broken down into light hydrocarbons with the help of activated carbon. The activated carbon promoted dehydrogenation and aromatization, forming different aromatic groups and a very efficient catalytic conversion when Ni was used. The growth of CNTs was made possible by benzene rings. Synergistic catalysis was also made possible by the carboxylic parts of activated carbon and Ni_2O_3. At 820 °C and a ratio of PP:10Ni_2O_3:8AC (wt.%) for the raw materials, the highest yield (50 wt.%) of carbon was reached (Gong et al., 2012). As this method uses a lot of energy, recovering Ni_2O_3 could be a better way to manufacture it on a larger scale with less energy. Zhou et al. developed a new way of breaking down PE by combustion using a stainless steel wire mesh as a catalyst and a base. The resultant CNTs weighed more than 10% (Zhuo et al., 2010). For another method, Zhang et al. utilized an autoclave to combine PP (2 g), maleated PP (0.5 g), and Ni catalysis powders (0.5 g). The mixture was then heated on an electric stove for 12 hours until it reached 700 °C. Then it was left to cool down to the temperature of the room. Ni plays a key role during the preparation of CNTs from the PP. MA-PP helped CNTs grow in two ways: first, it made it easier for Ni to spread out in PP, and then it made it easier for carbon atoms and Ni catalysts to work together as one system. Ni particles were separated to make them surrounded by carbon, and a high surface packing density of Ni particles made it possible for the nanotubes to grow in a straight line and 80% product of CNT was obtained (Zhang et al., 2008). Bajad et al. made MWNTs by burning PP waste in the presence of Ni/Mo/MgO catalyst, yielding about 45.8%. The size and number of nanotubes depended on the ratio of Ni to Mo. The high-resolution transmission electron microscopy (HRTEM) images showed that the ratio of Ni to Mo controlled the CNTs' size and shape. The CNTs grew in size as the Mo content increased, while the yield increased when the Mo content decreased and the CNTs shrank. The authors used response surface methodology (RSM) to find the best Ni/Mo mole ratio. RSM suggested that a Ni/Mo mole ratio of 22.04 would produce 394% carbon. While 514% of CNTs would be produced over a $Ni_4Mo.2MgO$ catalyst at 800 °C, 5 g of polymer, 150 mg of catalyst, and 10 minutes of burning (Bajad et al., 2015). CNTs have a unique structure that makes them very stable and good at conducting electricity. Because of this, they can be used to reinforce structures for other electrochemical purposes, such as making supercapacitors (Borsodi et al., 2016). Veksha et al. (2018) tried some electrocatalytic applications, and they found that CNT electrocatalysts made directly from plastic are not as good as commercial Pt/C catalysts. Carbon Black (CB) is an alternative nanomaterial often used to strengthen the structure of rubber products such as car tires. It has also been used to support catalysts and make conductive inks (Noked et al., 2011). Catalysis-pyrolysis of old plastics has been used to make carbon nanomaterials such as CNTs, Zhang and Williams (2016). In this process, the extraordinary thing was that instead of unwanted coke, CNTs were made, which could seriously damage the catalyst. Also, low-cost CNTs made from used plastics have been used to reinforce materials with good results in tensile and flexural strength, showing that they have a lot of potential for use in the industry (Borsodi et al., 2016). It is known that methane and ethylene, which come from the oil industry, are usually used as the carbon precursors for making CNTs with the chemical vapor deposition (CVD) method, which has been the most common way to produce large quantities of CNTs (Shah and Tali, 2016). As the small gases needed to make CNTs can also be made from pyrolyzing WP, this is an attractive way to make CNTs from WP. Zhang et al. (2015) found that Ni/Al_2O_3 was more active at making MWCNTs and gave off more H_2 than Co/Al_2O_3 and Cu/Al_2O_3 when reforming old tires catalytically. Yang et al. (2015) made CNTs with a diameter of 20–30 using an HNi/Al_2O_3 catalyst in a pilot-scale system. This showed that Ni-based catalysts could continuously treat plastics to make high-value CNTs. From the catalytic decomposition of methane with a Ni/La_2O_3 catalyst. Pudukudy et al. (2016) discovered bulk carbon deposition of rather homogeneous CNTs and reported a yield of 55% hydrogen. Fe-based catalysts are also a good choice for making CNTs because they are cheap and good for the environment. Shen et al. (2008) used a two-stage fixed-bed reactor to study the catalytic pyrolysis of WP in the presence of a bimetallic Ni-Fe catalyst to make H_2 and CNTs (Mishra et al., 2012). WP is made of organic polymers with chains of hydrocarbons, which were thought to be the source of carbon atoms for the growth of carbon-based materials. Quan et al. (2010) also reported that printed circuit board (PCB) waste could be used to make CNTs and porous carbon. Electrical and electronic waste makes it easy to collect PCB waste. It was first pyrolyzed to produce oil, then polymerized at 95 °C for 4 hours with a formaldehyde solution (6.5 g, 37%) and an ammonium hydroxide catalyst. The resin was heated for 2 hours at 60 °C and 12 hours at 120 °C. This process made PCB waste pyrolysis oil-based resin, which was used as a precursor. Using a long pyrolysis process at 900 °C on oil-based resin and ferrocene catalyst, 56.82% of CNTs with an outer diameter of 338 nm, and a wall thickness of 86 nm were made. The porous carbon was made by heating resin treated with KOH to 700 °C.

11.8 Graphene-based nanomaterials

Graphene is a single atom with a honeycomb-like structure with sp^2-hybridized carbon atoms. This material has important qualities such as a large surface area, good chemical stability, high electrical conductivity, and high mechanical strength. Health, energy, and the environment have all been changed by graphene (Yang et al., 2016). Graphite and organic molecules are the only things used to make graphene. Chemical reduction of graphene oxide is all common way to make graphene (Abdolhosseinzadeh et al., 2018). CVD is used on metallic films in the bottom-up method, while the top-down method uses the liquid exfoliation of graphite crystals. Graphene can be made from different types of WP in several ways. Gong and his colleagues made many graphene flakes using waste PP and montmorillonite that had been changed organically. Graphene flakes were made by repeatedly spinning down a solution and separating it from the rest (Gong et al., 2014b). Manukyan et al. found a way to burn waste silicon carbide (SiC) and polytetrafluoroethylene using less energy to make graphene sheets. The process was similar to epitaxial growth on SiC, where C_2F_4 took out Si in an exothermic reaction (Manukyan et al., 2013). Graphene has high electric conductivity and a large surface area. This makes it a good candidate for electrochemical energy storage. As an anode material in lithium-ion batteries (LIBs), it has shown great electrochemical performance (Min et al., 2019). But graphene rearranges strongly due to the interactions between carbon layers (Li et al., 2015).

11.9 Metal and metal oxide nanoparticles

Several procedures have been used to synthesize various types of NPs from wastes after appropriate pretreatment, either physically or chemically, or a combination of both. Grinding and milling are two of the most common physical pretreatment methods. Chemical pretreatment is used to eliminate any contaminants in a waste sample. This is done by heating or treating the sample with strong acids such as H_2SO_4, HNO_3, and HCl (Samaddar et al., 2018). The production of ZnNPs from used tires is a good alternative that can add value. Moghaddasi et al. (2013) ball-milled old tires for 5 hours to make NPs. The process resulted in rubber ash particles less than 500 nm in size. Si ash further reduced the particle size to below 50 nm. As a source of Zn for plants, rubber ash and ground tire rubber particles were just as good as a commercial $ZnSO_4$ fertilizer. Maroufi et al. (2017a) used high temperatures to make high-value CNPs from waste tire rubber. Gómez-Hernández et al. (2019) described another easy and inexpensive way to make CBNPs from old tires. In short, thermal transformation at 1000 °C and self-pressured conditions produced carbon NPs (22 nm) with a high yield (about 81%) when the conditions were optimized. Gómez-Hernández et al. (2019) did a chemical analysis and found that the chain-like agglomerated NPs were partly oxidized ($C = 84.9\%$, $O = 4.9\%$, and $S = 10.21\%$). Many studies have reported that tires and plastics are good low-cost starting materials for preparing CNTs because they are made of unconventional hydrocarbon waste.

11.10 Applications of plastic and tire waste–based nanomaterials

Pollutants in industrial wastewater are specific to the industries that generate them. The textile industry, for instance, generates dye-tainted waste streams. Dye-tainted water must be purified for environmental and human health (Arslan et al., 2016). Several research groups are trying to develop new processes to treat industrial wastewater with NT and recycled NPs. This helps meet goals for circularity and sustainability. Fig. 11.2 represents the general applications of plastic and tire waste–derived nanomaterials in different fields.

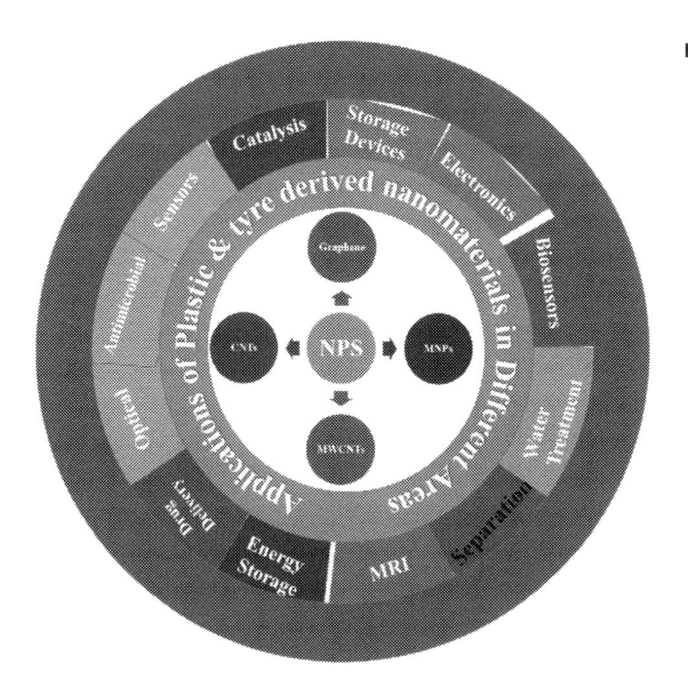

FIGURE 11.2 Applications of plastic and tire waste–derived NMs.

Abdul Rahim Arifin et al. (2017) used modified Fe NPs that could absorb more than 99% of the dye wastewater, with the 53.76 nm particles absorbing the most. El Essawy et al. looked into another way to get dye out of water. They used recycled PET as a starting material for making NP and turning PET into graphene. The ability to stick with methylene blue and acid blue 25 dyes was checked. This graphene did an excellent job absorbing methylene blue at a pH of 12, and within 30 minutes, the dye had reached a stable state. Acidic solutions were the best places for the acid blue 25 dye to stick, and after about 50 minutes, the attached dye reached a state of equilibrium. PET-based graphene could remove both dyes from solutions (El Essawy et al., 2017).

Thai et al. (2019) looked at how highly porous aerogels can be made from recycled materials such as plastic bottles. These aerogels made from waste are especially useful for cleaning up oil spills in water bodies because they are very light and can soak up a lot of liquid. For example, cotton aerogels can absorb more than 100 g of motor oil per g of aerogel, much more than most commercial sorbents can do. Zhang et al. tested another way to clean up dyewater waste using Fe to make saccharin. The difficult-to-treat saccharin wastewater was first used to get the $NiFe_2O_4$NPs utilized as a catalyst. Then, a hydrothermal method was used to make a $NiFe_2O_4$/ZnCuCr-LDH compost at a mesoporous magnet treating dye water (Zhang et al., 2020).

Rovani et al. (2018) devised ways to make dye-absorbing Si NPs from agricultural waste. In this study, authors used the ash from burned sugar cane to make high-purity SiO_2NPs. The NPs were tested to see how well they could soak acid orange 8 (AO8) dye. The Si NPs could absorb 230 mg/g and be used repeatedly up to five times. Nayak et al. employed electronic waste, specifically PCBs, to generate a nanocomposite of Zn and Cu oxide. With visible light and hydrogen peroxide, the NPs worked well as a photocatalyst to break down the methyl orange dye (Nayak et al., 2019). Wu et al. looked into another way to eliminate contaminants in wastewater using photocatalysis. Wu and his colleagues used fabric filter dust to make a metal-doped ZnO (M-ZnO) nanomaterial. Fe, Mg, Ca, and Al were used as dopants, and Ca, Al, and Zn were used as elemental sources. All of these were found in fabric filter dust. The NPs of doped M-ZnO were made by combining solvolysis and coprecipitation processes. Under visible light, the NPs work as a photocatalyst to break down organic substances, particularly phenol (Wu et al., 2014).

Rubber is a versatile material utilized in producing a wide variety of goods, including but not limited to tires, rubber–shell products, hoses, sealing strips, and so on (Acevedo and Barriocanal, 2015). The most recent report from the International Rubber Research Group (IRSG) predicted that the total amount of rubber consumed worldwide would rise by 2.5% to 30.03 million tonnes (Chen et al., 2019b). The ever-increasing demand for rubber will inevitably bring about an acceleration in the production of rubber waste. One of the most common types of rubber waste is scrap tire, which is resistant to the biological breakdown that occurs during the natural elimination process and can take decades to complete (Angin, 2014). Many recycling strategies have been proposed to reutilize it in various ways, such as derived fuel, road paving, and floor mats. It is essential to broaden the application of this technology, although a significant quantity of used tires cannot be processed appropriately. CB accounts for 22% of a tire's rubber, while raw rubber makes up 45%–48% (Alexandre-Franco et al., 2008) because it contains a high percentage of carbon—somewhere between 68 and 75 wt.%, it can be used as a raw material in producing carbon materials as absorbents for the heavy metal ion in an aqueous solution.

Quek and Balasubramanian used a mixture of nitrogen and oxygen to activate the carbonaceous char. The adsorbent reached its maximum adsorption capacity of 20 mg/g for lead and 15 mg/g for Cu ions. Following the pyrolysis of the tire rubber at 900 °C, the carbonaceous char was treated with hydrogen peroxide (10%) and nitric acid (Quek and Balasubramanian, 2009). Gupta et al. reported carbon-based nanomaterials used for the extraction of Ni^{2+}. The proposed method obtained an adsorption capacity of approximately 25.45 mg/g. Based on the findings presented above, it was clear that the pyrolysis conditions significantly impacted the adsorption capacity of the carbonaceous char for various heavy metals (Gupta et al., 2014). In addition, it was hypothesized that the low density of the carbonaceous absorbents would cause them to float to the surface of the wastewater. This would make it difficult to disperse the absorbents and recycle them. The magnetic NPs are designed to be incorporated into the carbonaceous absorbents to compensate for the earlier shortage. In recent years, biochar, a type of adsorbent derived from a wide range of biomass, including algae, manure, crops, and forest residues (Wan et al., 2020), has been the subject of significant research. In addition, magnetic particles (MPs) have been the subject of extensive research for adsorbing heavy metal ions from an aqueous solution (Mashile et al., 2020). Table. 11.1 represents the preparation method and applications of plastic and tire-derived nanomaterials.

Fe_3O_4 was purchased from the market to adsorb heavy metal ions from an aqueous solution without further modification. About 15 mg/g was the maximum amount of Pb^{2+} that the Fe_3O_4 could adsorb before it became saturated (Nassar, 2010). Wang et al. modified the Fe_3O_4 by coating it with a layer of SiO_2 and adding –SH to the mixture. The maximum adsorption capacity for Hg^{2+} demonstrated by the $Fe_3O_4@SiO_2$–SH was approximately 90 mg/g (Wang et al., 2016).

TABLE 11.1 Preparation method and applications of plastic and tire-derived nanomaterials.

Nanomaterial	Sources	Methods	Applications	References
Heteroatom-doped carbon nanosheets	Waste tires	Pyrolysis	Electrocatalytic oxygen reduction	Veksha et al. (2020)
Carbon black nanoparticles		Lab-made cylindrical stainless steel reactor	Flexible electronic Devices	Gómez-Hernández et al. (2019)
Mesoporous silicon carbide (SiC) nanofiber/particle composite		Carbothermal reduction		Maroufi et al. (2017b)
Core–shell structured carbon nanoparticles		Light Pyrolysis		Li et al. (2016)
Magnetic adsorbents		One-pot pyrolysis	Removal of Pb (II)	Ji et al. (2020)
MWCNTs	Waste plastics	Quartz tube		Mishra et al. (2012)
MWCNTs		Quartz tube		Gong et al. (2012).
CNTs		Quartz tube		Gong et al. (2014a)
CNTs		Quartz tube		Zhuo et al. (2010)
CNTs		Quartz tube		Zhang et al. (2008)
CNTs		Muffle furnace		Bajad et al. (2015)
CNTs		Crucible		Gong et al. (2014b)
Graphene		Crucible		Ruan et al. (2011)
Graphene		Quartz tube		Sharma et al. (2014)
Graphene		Stainless steel reactor		Manukyan et al. (2013)
Graphene				Takami et al. (2014)
Graphene				Sun et al. (2013)
Fullerenes			Dermatology and Cosmetics	Mousavi et al. (2017)
Nano-SiO_2			Pest control, Medicine, Concrete mixtures	Kerboua et al. (2010)
Nanoporous Carbon			Supercapacitance	Cheng et al. (2015)
Graphene flakes			Transistors, Conductive Inks	Liu et al. (2007)
Carbon nanotubes		Pyrolysis-catalysis process		Yao et al. (2021)
Magnetic nanocomposites		Catalytic conversion	Chrome (VI) detoxification	Xu et al. (2021)
CNTs		Catalytic pyrolysis		Yao et al. (2017)

(Continued)

TABLE 11.1 (Continued)

Nanomaterial	Sources	Methods	Applications	References
Carbon nanosheets	Mixed plastic		Degradation of dyes	Gong et al. (2015)
Nano-clay/Nanocomposites			Improved mechanical properties of the polymer after recycling	Wen et al. (2019)
Carbon nanosheets			Supercapacitance	Xu et al. (2015)
Carbon nanospheres			Energy storage	Gong et al. (2014b)
Nanofoams	Plastic foam		Degradation of dyes	De Assis et al. (2018)

In conclusion, coated and functionally modified MPs exhibited high selectivity for target metal ions and were easily recoverable from aqueous solutions; however, preparing adsorbents required excessive care and effort. This combination was carried out to create novel magnetic carbonaceous adsorbents (MCAs). This technique lessened the negative effects of waste rubber on the surrounding ecosystem and improved the properties of the rubber's use as an adsorbent for heavy metals. It was determined by studying a variety of preparation conditions which one was the most effective. In addition to this, the prepared MCA's structural and adsorption properties on heavy metals were thoroughly researched. In the end, it investigated the desorption of heavy metals and the recycling of the MCA (Ji et al., 2020) as shown in Table 11.1.

11.11 Conclusion

Plastics have been used as carbonaceous feeds in several techniques for manufacturing nanomaterials. Creating these high-value items from the ostensibly useless waste stream has become even more appealing due to the growing issue of waste streams and the possible advantages and profits associated with nanomaterials. The literature results indicate that such a project is feasible using various reactor types and process setups. This chapter is the first review to compile all relevant literature. A wide range of topics has not yet been explored due to the shortage of studies in this area. However, the most encouraging results are highlighted, and the outcomes of several processes are contrasted. Additionally, topics that seem to need more research are touched upon. Future study in this area is still very much possible to obtain high-value nanomaterial from less expensive sources and use easier processes. The use of waste materials serves as a substitute for using waste materials, the majority of which are hazardous to the environment, as well as a way to discover a plentiful and affordable source for creating carbon materials.

References

Abdolhosseinzadeh, S., Sadighikia, S., Alkan Gürsel, S., 2018. Scalable synthesis of sub-nanosized platinum-reduced graphene oxide composite by an ultraprecise photocatalytic method, ACS Sustainable Chemistry & Engineering, 6. pp. 3773–3782.

Abdul Rahim Arifin, A.S., Ismail, I., Abdullah, A.H., Shafiee, F.N., Nazlan, R., Ibrahim, I.R., 2017. Iron Oxide Nanoparticles Derived From Mill Scale Waste as Potential Scavenging Agent in Dye Wastewater Treatment For Batik Industry. Solid State Phenomena. Trans Tech Publications Ltd., pp. 393–398.

Acevedo, B., Barriocanal, C., 2015. Texture and surface chemistry of activated carbons obtained from tyre wastes. Fuel Processing Technology 134, 275–283.

Ahmed, S., Ali, W., 2020. s. Green Nanomaterials: Processing, Properties, and Application. Springer Nature.

Alexandre-Franco, M., Fernández-González, C., Macías-García, A., Gómez-Serrano, V., 2008. Uptake of lead by carbonaceous adsorbents developed from tire rubber. Adsorption 14, 591–600.

Angin, D., 2014. Utilization of activated carbon produced from fruit juice industry solid waste for the adsorption of Yellow 18 from aqueous solutions. Bioresource Technology 168, 259–266.

Arslan, S., Eyvaz, M., Gürbulak, E., Yüksel, E., 2016. A review of state-of-the-art technologies in dye-containing wastewater treatment–the textile industry case. Textile Wastewater Treatment 1–29.

Bajad, G.S., Tiwari, S.K., Vijayakumar, R., 2015. Synthesis and characterization of CNTs using polypropylene waste as precursor. Materials Science and Engineering: B 194, 68–77.

Bazargan, A., McKay, G., 2012. A review—synthesis of carbon nanotubes from plastic wastes. Chemical Engineering Journal 195, 377–391.

Bhardwaj, A.K., Naraian, R., Sundaram, S., Kaur, R., 2022. Biogenic and non-biogenic waste for the synthesis of nanoparticles and their applications. In Bioremediation. CRC Press, pp. 207–218.

Borsodi, N., Szentes, A., Miskolczi, N., Wu, C., Liu, X., 2016. Carbon nanotubes synthesized from gaseous products of waste polymer pyrolysis and their application. Journal of Analytical and Applied Pyrolysis 120, 304–313.

Chen, M., He, Y., Ye, Q., Wang, X., Hu, Y., 2019a. Shape-dependent solar thermal conversion properties of plasmonic Au nanoparticles under different light filter conditions. Solar Energy 182, 340–347.

Chen, R., Li, Q., Zhang, Y., Xu, X., Zhang, D., 2019b. Pyrolysis kinetics and mechanism of typical industrial non-tyre rubber wastes by peak-differentiating analysis and multi kinetics methods. Fuel 235, 1224–1237.

Cheng, L.X., Zhang, L., Chen, X.Y., Zhang, Z.J., 2015. Efficient conversion of waste polyvinyl chloride into nanoporous carbon incorporated with MnOx exhibiting superior electrochemical performance for supercapacitor application. Electrochimica Acta 176, 197–206.

Chernozatonskii, L., Kukovitskii, E., Musatov, A., Ormont, A., Izraeliants, K., L'vov, S., 1998. Carbon crooked nanotube layers of polyethylene: synthesis, structure and electron emission. Carbon 36, 713–715.

Das, G., Patra, J.K., Paramithiotis, S, Shin, H.-S., 2019. The sustainability challenge of food and environmental nanotechnology: Current status and imminent perceptions. International Journal of Environmental Research and Public Health 16, 4848.

De Assis, G.C., Skovroinski, E.B., Leite, V.D., Rodrigues, M.O., Galembeck, A., Alves, M.C., et al., 2018. Conversion of "Waste Plastic" into photocatalytic nanofoams for environmental remediation. ACS Applied Materials & Interfaces 10, 8077–8085.

Deng, J., You, Y., Sahajwalla, V., Joshi, R.K., 2016. Transforming waste into carbon-based nanomaterials. Carbon 96, 105–115.

El Essawy, N.A., Ali, S.M., Farag, H.A., Konsowa, A.H., Elnouby, M., Hamad, H.A., 2017. Green synthesis of graphene from recycled PET bottle wastes for use in the adsorption of dyes in aqueous solution. Ecotoxicology and Environmental Safety 145, 57–68.

Falinski, M.M., Plata, D.L., Chopra, S.S., Theis, T.L., Gilbertson, L.M., Zimmerman, J.B., 2018. A framework for sustainable nanomaterial selection and design based on performance, hazard, and economic considerations. Nature Nanotechnology 13, 708–714.

Gómez-Hernández, R., Panecatl-Bernal, Y., Méndez-Rojas, M.Á., 2019. High yield and simple one-step production of carbon black nanoparticles from waste tires. Heliyon 5, e02139.

Gong, J., Liu, J., Wan, D., Chen, X., Wen, X., Mijowska, E., et al., 2012. Catalytic carbonization of polypropylene by the combined catalysis of activated carbon with Ni_2O_3 into carbon nanotubes and its mechanism. Applied Catalysis A: General 449, 112–120.

Gong, J., Liu, J., Jiang, Z., Feng, J., Chen, X., Wang, L., et al., 2014a. Striking influence of chain structure of polyethylene on the formation of cup-stacked carbon nanotubes/carbon nanofibers under the combined catalysis of CuBr and NiO. Applied Catalysis B: Environmental 147, 592–601.

Gong, J., Liu, J., Wen, X., Jiang, Z., Chen, X., Mijowska, E., et al., 2014b. Upcycling waste polypropylene into graphene flakes on organically modified montmorillonite. Industrial & Engineering Chemistry Research 53, 4173–4181.

Gong, J., Liu, J., Chen, X., Jiang, Z., Wen, X., Mijowska, E., et al., 2015. Converting real-world mixed waste plastics into porous carbon nanosheets with excellent performance in the adsorption of an organic dye from wastewater. Journal of Materials Chemistry A 3, 341–351.

Gupta, V.K., Nayak, A., Agarwal, S., Chaudhary, M., Tyagi, I., 2014. Removal of Ni (II) ions from water using scrap tire. Journal of Molecular Liquids 190, 215–222.

Jalili, S., Jafari, M., Habibian, J., 2008. Effect of impurity on electronic properties of carbon nanotubes. Journal of the Iranian Chemical Society 5, 641–645.

Ji, J., Chen, G., Zhao, J., Wei, Y., 2020. Efficient removal of Pb (II) by inexpensive magnetic adsorbents prepared from one-pot pyrolysis of waste tyres involved magnetic nanoparticles. Fuel 282, 118715.

Kerboua, N., Cinausero, N., Sadoun, T., Lopez-Cuesta, J., 2010. Effect of organoclay in an immiscible poly (ethylene terephtalate) waste/poly (methyl methacrylate) blend. Journal of Applied Polymer Science 117, 129–137.

Kim, Y.K., Sharker, S.M., In, I., Park, S.Y., 2016. Surface coated fluorescent carbon nanoparticles/TiO2 as visible-light sensitive photocatalytic complexes for antifouling activity. Carbon 103, 412–420.

Kong, Q., Zhang, J., 2007. Synthesis of straight and helical carbon nanotubes from catalytic pyrolysis of polyethylene. Polymer Degradation and Stability 92, 2005–2010.

Li, Q., Yao, K., Zhang, G., Gong, J., Mijowska, E., Kierzek, K., et al., 2015. Controllable synthesis of 3D hollow-carbon-spheres/graphene-flake hybrid nanostructures from polymer nanocomposite by self-assembly and feasibility for lithium-ion batteries. Particle & Particle Systems Characterization 32, 874–879.

Li, S., Wan, C., Wu, X., Wang, S., 2016. Core-shell structured carbon nanoparticles derived from light pyrolysis of waste tires. Polymer Degradation and Stability 129, 192–198.

Liu, P.S., Li, L., Zhou, N.L., Zhang, J., Wei, S.H., Shen, J., 2007. Waste polystyrene foam-graft-acrylic acid/montmorillonite superabsorbent nanocomposite. Journal of Applied Polymer Science 104, 2341–2349.

Manukyan, K.V., Rouvimov, S., Wolf, E.E., Mukasyan, A.S., 2013. Combustion synthesis of graphene materials. Carbon 62, 302–311.

Maroufi, S., Mayyas, M., Sahajwalla, V., 2017a. Nano-carbons from waste tyre rubber: an insight into structure and morphology. Waste Management 69, 110–116.

Maroufi, S., Mayyas, M., Sahajwalla, V., 2017b. Waste materials conversion into mesoporous silicon carbide nanocermics: nanofibre/particle mixture. Journal of Cleaner Production 157, 213–221.

Masarapu, C., Subramanian, V., Zhu, H., Wei, B., 2009. Long-cycle electrochemical behavior of multiwall carbon nanotubes synthesized on stainless steel in Li Ion batteries. Advanced Functional Materials 19, 1008–1014.

Mashile, G.P., Mpupa, A., Nqombolo, A., Dimpe, K.M., Nomngongo, P.N., 2020. Recyclable magnetic waste tyre activated carbon-chitosan composite as an effective adsorbent rapid and simultaneous removal of methylparaben and propylparaben from aqueous solution and wastewater. Journal of Water Process Engineering 33, 101011.

Min, J., Zhang, S., Li, J., Klingeler, R., Wen, X., Chen, X., et al., 2019. From polystyrene waste to porous carbon flake and potential application in supercapacitor. Waste Management 85, 333–340.

Mishra, N., Das, G., Ansaldo, A., Genovese, A., Malerba, M., Povia, M., et al., 2012. Pyrolysis of waste polypropylene for the synthesis of carbon nanotubes. Journal of Analytical and Applied Pyrolysis 94, 91–98.

Moghaddasi, S., Khoshgoftarmanesh, A.H., Karimzadeh, F., Chaney, R.L., 2013. Preparation of nanoparticles from waste tire rubber and evaluation of their effectiveness as zinc source for cucumber in nutrient solution culture. Scientia Horticulturae 160, 398–403.

Mousavi, S.Z., Nafisi, S., Maibach, H.I., 2017. Fullerene nanoparticle in dermatological and cosmetic applications. Nanomedicine: Nanotechnology, Biology and Medicine 13, 1071–1087.

Nassar, N.N., 2010. Rapid removal and recovery of Pb (II) from wastewater by magnetic nanoadsorbents. Journal of Hazardous Materials 184, 538–546.

Nayak, P., Kumar, S., Sinha, I., Singh, K.K., 2019. ZnO/CuO nanocomposites from recycled printed circuit board: preparation and photocatalytic properties. Environmental Science and Pollution Research 26, 16279–16288.

Noked, M., Soffer, A., Aurbach, D., 2011. The electrochemistry of activated carbonaceous materials: past, present, and future. Journal of Solid State Electrochemistry 15, 1563–1578.

Pol, V.G., Thackeray, M.M., 2011. Spherical carbon particles and carbon nanotubes prepared by autogenic reactions: Evaluation as anodes in lithium electrochemical cells. Energy & Environmental Science 4, 1904–1912.

Pudukudy, M., Yaakob, Z., Takriff, M.S., 2016. Methane decomposition into COx free hydrogen and multi-walled carbon nanotubes over ceria, zirconia and lanthana supported nickel catalysts prepared via a facile solid state citrate fusion method. Energy Conversion and Management 126, 302–315.

Quan, C., Li, A., Gao, N., 2010. Synthesis of carbon nanotubes and porous carbons from printed circuit board waste pyrolysis oil. Journal of Hazardous Materials 179, 911–917.

Quek, A., Balasubramanian, R., 2009. Low-energy and chemical-free activation of pyrolytic tire char and its adsorption characteristics. Journal of the Air & Waste Management Association 59, 747–756.

Recordati, C., De Maglie, M., Bianchessi, S., Argentiere, S., Cella, C., Mattiello, S., et al., 2015. Tissue distribution and acute toxicity of silver after single intravenous administration in mice: nano-specific and size-dependent effects. Particle and Fibre Toxicology 13, 1–17.

Rovani, S., Santos, J.J., Corio, P., Denise, A.F., 2018. Highly pure silica nanoparticles with high adsorption capacity obtained from sugarcane waste ash. ACS Omega 3, 2618–2627.

Ruan, G., Sun, Z., Peng, Z., Tour, J.M., 2011. Growth of graphene from food, insects, and waste. ACS Nano 5, 7601–7607.

Samaddar, P., Ok, Y.S., Kim, K.-H., Kwon, E.E., Tsang, D.C., 2018. Synthesis of nanomaterials from various wastes and their new age applications. Journal of Cleaner Production 197, 1190–1209.

Shah, K.A., Tali, B.A., 2016. Synthesis of carbon nanotubes by catalytic chemical vapour deposition: a review on carbon sources, catalysts and substrates. Materials Science in Semiconductor Processing 41, 67–82.

Sharma, S., Kalita, G., Hirano, R., Shinde, S.M., Papon, R., Ohtani, H., et al., 2014. Synthesis of graphene crystals from solid waste plastic by chemical vapor deposition. Carbon 72, 66–73.

Shen, W., Huggins, F.E., Shah, N., Jacobs, G., Wang, Y., Shi, X., et al., 2008. Novel Fe–Ni nanoparticle catalyst for the production of CO-and CO_2-free H_2 and carbon nanotubes by dehydrogenation of methane. Applied Catalysis A: General 351, 102–110.

Somanathan, T., Pandurangan, A., 2010. Helical multi-walled carbon nanotubes synthesized by catalytic chemical vapor deposition. Carbon 13, 3974.

Song, R., Ji, Q., 2011. Synthesis of carbon nanotubes from polypropylene in the presence of Ni/Mo/MgO catalysts via combustion. Chemistry Letters 40, 1110–1112.

Song, R., Li, B., Zhao, S., Li, L., 2009. Transferring polypropylene into carbon nanotubes via combustion of PP/zeolites (H-ZSM-5 or H-beta)/Ni_2O_3. Journal of Applied Polymer Science 112, 3423–3428.

Sun, L., Tian, C., Li, M., Meng, X., Wang, L., Wang, R., et al., 2013. From coconut shell to porous graphene-like nanosheets for high-power supercapacitors. Journal of Materials Chemistry A 1, 6462–6470.

Takami, T., Seino, R., Yamazaki, K., Ogino, T., 2014. Graphene film formation on insulating substrates using polymer films as carbon source. Journal of Physics D: Applied Physics 47, 094015.

Tang, T., Chen, X., Meng, X., Chen, H., Ding, Y., 2005. Synthesis of multi-walled carbon nanotubes by catalytic combustion of polypropylene. Angewandte Chemie International Edition 44, 1517–1520.

Thai, Q.B., Le, D.K., Luu, T.P., Hoang, N., Nguyen, D., Duong, H., 2019. Aerogels from wastes and their applications. JOJ Materials Science 5, 10.19080.

Veksha, A., Giannis, A., Oh, W.-D., Chang, V.W.-C., Lisak, G., 2018. Upgrading of non-condensable pyrolysis gas from mixed plastics through catalytic decomposition and dechlorination. Fuel Processing Technology 170, 13–20.

Veksha, A., Latiff, N.M., Chen, W., Ng, J.E., Lisak, G., 2020. Heteroatom doped carbon nanosheets from waste tires as electrode materials for electrocatalytic oxygen reduction reaction: effect of synthesis techniques on properties and activity. Carbon 167, 104–113.

Volodin, A., Buntinx, D., Ahlskog, M., Fonseca, A., Nagy, J.B., Van Haesendonck, C., 2004. Coiled carbon nanotubes as self-sensing mechanical resonators. Nano Letters 4, 1775–1779.

Wan, J., Liu, L., Ayub, K.S., Zhang, W., Shen, G., Hu, S., et al., 2020. Characterization and adsorption performance of biochars derived from three key biomass constituents. Fuel 269, 117142.

Wang, Z., Xu, J., Hu, Y., Zhao, H., Zhou, J., Liu, Y., et al., 2016. Functional nanomaterials: study on aqueous Hg (II) adsorption by magnetic Fe_3O_4@ SiO_2-SH nanoparticles. Journal of the Taiwan Institute of Chemical Engineers 60, 394–402.

Wen, Y., Kierzek, K., Chen, X., Gong, J., Liu, J., Niu, R., et al., 2019. Mass production of hierarchically porous carbon nanosheets by carbonizing "real-world" mixed waste plastics toward excellent-performance supercapacitors. Waste Management 87, 691–700.

Wu, Z.-J., Huang, W., Cui, K.-K., Gao, Z.-F., Wang, P., 2014. Sustainable synthesis of metals-doped ZnO nanoparticles from zinc-bearing dust for photodegradation of phenol. Journal of Hazardous Materials 278, 91–99.

Xu, F., Tang, Z., Huang, S., Chen, L., Liang, Y., Mai, W., et al., 2015. Facile synthesis of ultrahigh-surface-area hollow carbon nanospheres for enhanced adsorption and energy storage. Nature Communications 6, 1–12.

Xu, D., Yang, S., Su, Y., Xiong, Y., Zhang, S., 2021. Catalytic conversion of plastic wastes using cost-effective bauxite residue as catalyst into H_2-rich syngas and magnetic nanocomposites for chrome (VI) detoxification. Journal of Hazardous Materials 413, 125289.

Yang, R.-X., Chuang, K.-H., Wey, M.-Y., 2015. Effects of nickel species on Ni/Al_2O_3 catalysts in carbon nanotube and hydrogen production by waste plastic gasification: bench-and pilot-scale tests. Energy & Fuels 29, 8178–8187.

Yang, K., Feng, L., Liu, Z., 2016. Stimuli responsive drug delivery systems based on nano-graphene for cancer therapy. Advanced Drug Delivery Reviews 105, 228–241.

Yang, Z., Tian, J., Yin, Z., Cui, C., Qian, W., Wei, F., 2019. Carbon nanotube-and graphene-based nanomaterials and applications in high-voltage supercapacitor: a review. Carbon 141, 467–480.

Yao, D., Wu, C., Yang, H., Zhang, Y., Nahil, M.A., Chen, Y., et al., 2017. Co-production of hydrogen and carbon nanotubes from catalytic pyrolysis of waste plastics on Ni-Fe bimetallic catalyst. Energy Conversion and Management 148, 692–700.

Yao, D., Yang, H., Hu, Q., Chen, Y., Chen, H., Williams, P.T., 2021. Carbon nanotubes from post-consumer waste plastics: investigations into catalyst metal and support material characteristics. Applied Catalysis B: Environmental 280, 119413.

Yeoh, W.-M., Lee, K.-Y., Chai, S.-P., Lee, K.-T., Mohamed, A.R., 2010. The role of molybdenum in Co-Mo/MgO for large-scale production of high quality carbon nanotubes. Journal of Alloys and Compounds 493, 539–543.

Zhang, J., Li, J., Cao, J., Qian, Y., 2008. Synthesis and characterization of larger diameter carbon nanotubes from catalytic pyrolysis of polypropylene. Materials Letters 62, 1839–1842.

Zhang, J., Du, J., Qian, Y., Xiong, S., 2010. Synthesis, characterization and properties of carbon nanotubes microspheres from pyrolysis of polypropylene and maleated polypropylene. Materials Research Bulletin 45, 15–20.

Zhang, Y., Wu, C., Nahil, M.A., Williams, P., 2015. Pyrolysis–catalytic reforming/gasification of waste tires for production of carbon nanotubes and hydrogen. Energy & Fuels 29, 3328–3334.

Zhang, Y., Williams, P.T., 2016. Carbon nanotubes and hydrogen production from the pyrolysis catalysis or catalytic-steam reforming of waste tyres. Journal of Analytical and Applied Pyrolysis 122, 490–501.

Zhang, H., Xia, B., Wang, P., Wang, Y., Li, Z., Wang, Y., et al., 2020. From waste to waste treatment: mesoporous magnetic $NiFe_2O_4$/ZnCuCr-layered double hydroxide composite for wastewater treatment. Journal of Alloys and Compounds 819, 153053.

Zhuo, C., Hall, B., Richter, H., Levendis, Y., 2010. Synthesis of carbon nanotubes by sequential pyrolysis and combustion of polyethylene. Carbon 48, 4024–4034.

Chapter 12

Emerging biowaste-derived surfaces to support redox-sensitive nanoparticles: applications in removal of synthetic dyes

Nitin Khandelwal and Gopala Krishna Darbha
Environmental Nanoscience Laboratory, Department of Earth Sciences & Centre for Climate and Environmental Studies, Indian Institute of Science Education and Research Kolkata, Mohanpur, West Bengal, India

12.1 Introduction

Every day of our life starts with a cup of tea or coffee and proceeds with many fruits, vegetables, and other food items. Before consumption, all these products travel through a lot of processing resulting in several byproducts. Think about it, after eating an orange, what do we generally do with its peel? Similarly, for shell and coir after drinking coconut water or eating coconut, shell, and skins after eating dry-fruits, etc. Similarly, we are served with the final food products, but the kitchen holds many processed materials in the bins. We presume them as waste, and in this form, being unattended, they result in a substantial ecological burden.

Let's try to change the approach and consider the unique qualities of biowastes. In general, agro/biowaste is very light in weight due to its loose porous structure. It contains lignin and cellulose and is a rich source of porous carbon (Soffian et al., 2022). In addition, several functional groups on the waste biomass surface can bind the toxicants. Researchers have tried to enhance the capacity of this waste to capture pollutants by pyrolyzing it at high temperatures of 350°C–550°C (Yaashikaa et al., 2020). Pyrolysis results in the loss of volatile organic compounds with the retention of a porous carbon matrix known as biochar, having intact functional groups. Biochar has applications in water retention, soil fertilization, mineralization, etc. (Ippolito et al., 2020).

In a similar line, nanomaterials are being continuously engineered and explored for their applications in various sectors, including environmental sustainability (Khan et al., 2019). Eco-friendly nanomaterials satisfy SE3 criteria of being scalable, efficient, economical, and environmentally friendly, and are finding their applications in removal of various contaminants from soils and water bodies (Khandelwal and Darbha, 2022).

While biowaste-derived carbon can be obtained in nano-size range and serve as an eco-friendly nanomaterial, various forms of biowastes also help in synthesizing and preserving nanomaterials and composites (Ram and Abhishek, 2020, Singh et al., 2021). This in turn enhances their efficiency and reactivity, and decreases their cost to make the process and material sustainable for environmental applications (Samaddar et al., 2018). In summary, this zero-cost agro/biowaste can be a feedstock for innovative nano-adsorbents, nano-catalysts, nano-disinfectants, etc. (Omran and Baek, 2022).

Industrial expansion to fulfill the demands of the exponentially rising population around the world has led to the release of various contaminants in soils and water bodies (Liyanage and Yamada, 2017). Dye molecules are synthetic aromatic compounds consisting of several organic functional groups, and some of them may contain toxic metals in their basic structure (Sarkar et al., 2017). Apart from textile industries holding the largest market for synthetic dyes utilization, pulp mills, paper printing, dyeing, painting, plastics manufacturing, and leather tanneries, etc. contribute significantly in dyes' utilization and release (Hunger, 2007). Based on chemical properties and applications, dyes can be classified in different groups holding considerable chemical diversity, as shown in Fig. 12.1 (Benkhaya et al., 2020).

During the dyeing process, not all the dye molecules bind entirely to the target fabric, leading to their effluent loss (Lellis et al., 2019). This loss or release of dyes as a contaminant depends on the chemical characteristics of dyes and

Green and Sustainable Approaches Using Wastes for the Production of Multifunctional Nanomaterials. DOI: https://doi.org/10.1016/B978-0-443-19183-1.00014-3

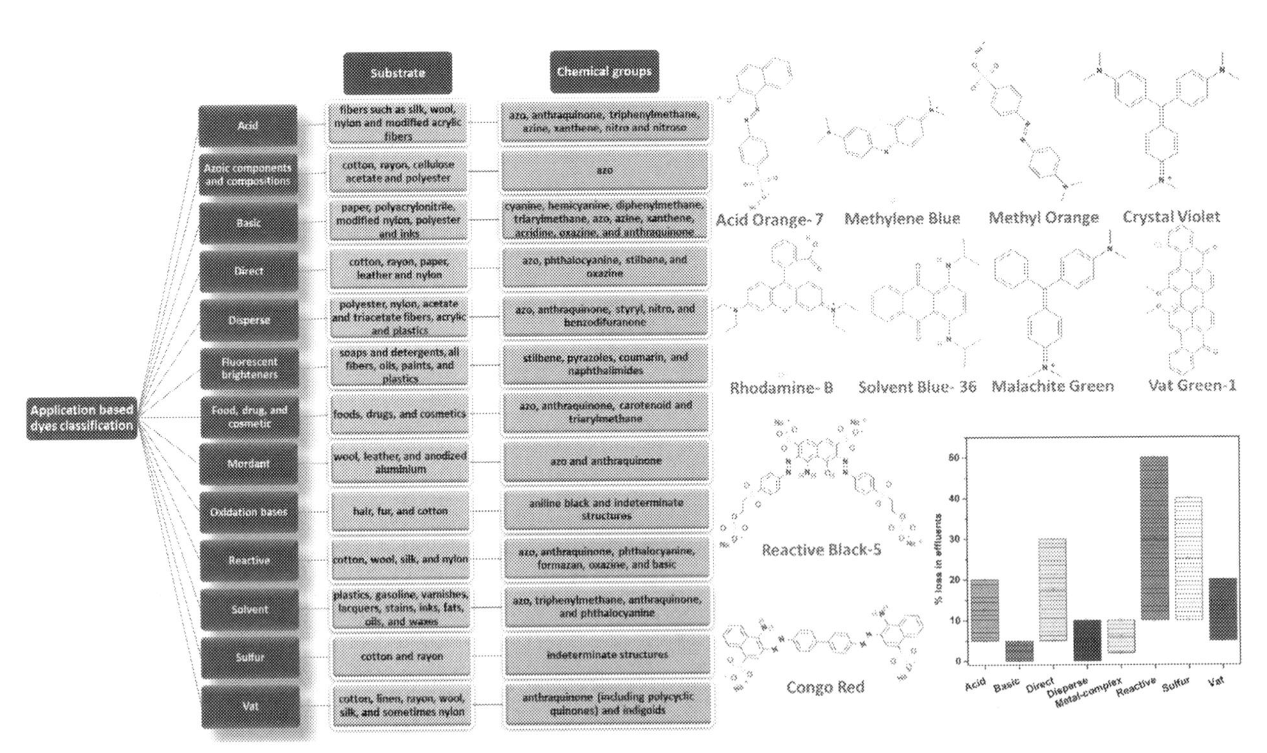

FIGURE 12.1 Dyes' classification and chemical structures of commonly explored dyes for dyes removal experiments and range of % loss of various dyes during dying due to varying binding affinity for fabrics (right).

can range as high as >50% for reactive dyes (Mattioli et al., 2005). The range of loss in the effluent for different dyes has been provided in Fig. 12.1 (right). A dye processing unit generally consumes about 80 L of water/1 kg of product. Data reveals that nearly 10^4 tonnes/year of dyes are discharged into wastewater streams by the textile industries (Chequer et al., 2013). Other than textile industries, industrial effluents from newspaper processing, pulp mills, painting and dying, and leather tanning industries contain dyes in significant amounts (Ejder Korucu et al., 2015). A small concentration of these dyes, even <1 ppm results in colored water, which reflects or adsorbs the sunlight entering the water, causes an imbalance in the aquatic ecosystem by hindering photosynthesis (Samchetshabam et al., 2017). Dyes also increase the biological oxygen demand which reduces reoxygenation in the aqueous reservoir and therefore affects the growth of photoautotrophic organisms (Ghoreishi and Haghighi, 2003). Based on the concentration and time of exposure, dyes can have both acute and chronic effects on different organisms (Verma, 2008).

Azo dyes contributing to more than 50% of total production have aromatic amines in them, which increases the risk of bladder cancer (Chung, 2016). Other health impacts include acute tubular necrosis supervene, blindness, chemosis, contact dermatitis, exophthalmos, hypertension, lacrimation, skin irritation, etc. (Chung, 2016). Toxic heavy metals present in the framework of some dyes can result in chronic diseases or may cause kidney failure (Jaishankar et al., 2014). Such potential health threats clearly show that the release of dyes contaminated water, which poses an alarming threat to the aquatic ecosystem and environment, requiring the scientific community's immediate attention for innovative and eco-friendly solutions. In this chapter, we have integrated the existing knowledge on the development of biowaste-derived and other redox-active iron nanocomposites and their applications in removal of synthetic dyes. The chapter focuses in detail on the removal mechanisms, existing challenges, and future perspectives.

12.2 Available dyes removal techniques

Plenty of methods have been explored to remediate dyes contaminated water and can be classified into three major groups- chemical, biological, and physical treatment methods (Sharma and Bhattacharya, 2017). Wet oxidation, ozonation, and coagulation-flocculation are chemical methods, while adsorption and membrane filtration belong to physical treatment processes (Yaseen and Scholz, 2019). Whereas microbial degradation, aerobic, and anaerobic processes are biological means of dyes degradation and removal (Khan et al., 2013). Setup and operational costs, energy consumption, environmental impacts, maintenance, secondary contamination, removal efficiency, kinetics, selectivity, behavior in complex matrices, and scalability are some of the factors to be reviewed to comment on the applicability and suitability of any technique. For example, coagulation-flocculation generally forms colloidal aggregates through electrostatic destabilization and allows the contaminants to settle with the coagulants (Tetteh and Rathilal, 2019). Generally used coagulants include iron salts, alum, lime, magnesium salts, etc. (Gregory and Duan, 2001). Higher chemical cost, pH adjustment, high sludge generation, and limited floc settling are some of this technique's major drawbacks (Irfan et al., 2017). Other chemical oxidation techniques face high operational costs; secondary toxic aromatic amines generation and scalability are some of the major disadvantages (Katheresan et al., 2018). Ozonation shows promising results for the removal of several dyes; and owing to its application in the gaseous stage, it has no volume increase, but a shorter half-life of ozone requires continuous ozonation in the system, making the process cost-intensive (Yerramilli et al., 2005). Various biological treatments on the other hand are widely applied due to their cost-effectiveness, scalability, and high dyes degradation efficiency but slower removal kinetics, toxicity to microorganisms, and limited removal of nonbiodegradable contaminants limits its applicability. Considering both advantages and disadvantages, utilizing eco-friendly adsorbents can be a promising solution due to their low power consumption, environmentally friendly nature, minimal setup and operational costs, tunable nature for removal efficiency and selectivity, etc. (Bolisetty et al., 2019).

Most of the conventional natural adsorbents are very economical to use due to their abundance like clays, zeolite, etc. but have shown limited removal efficiency due to their low surface area and reactivity. In the series of advancing the adsorbents, researchers have been exploring the nano-based materials for their potential as adsorbents for various contaminants (Sadegh et al., 2017, Anjum et al., 2019). Pertaining to their high specific surface area, tuneable functionality, and reactivity, nano-adsorbents have shown their candidature as adsorbents for various contaminants (El-Sayed, 2020), ranging from toxic metals, metalloids and metal oxides, pharmaceuticals, and personal care products, to micro- and nano-sized plastics (Tiwari et al., 2020), etc. Further scientific advancement resulted in the development of redox-sensitive nanoparticles (RSNP), for example, reduced metal nanoparticles of iron, aluminum, selenium, and copper, etc., bimetallic core-shell nanoparticles, for example, Cu/Ni, Ni/Fe nanoparticles, etc. (Sharma et al., 2019), superoxidized ferrate and reduced metal oxides like magnetite, etc. (Xu and Fu, 2020). Apart from having physical adsorption of contaminants on the surface, such nanoparticles have the electron transferring ability on the interface contributing to interfacial chemical transformation or degradation of the contaminants (Khan et al., 2019). But stabilization or controlling their nano-size range along with the preservation of their redox-sensitive nature in the natural environment is quite critical and requires further modifications (Camargo et al., 2009).

12.3 Redox-sensitive iron nanoparticles

Out of several explored metals and metal oxide nanoparticles, iron-based materials, mainly nanoscale zerovalent iron (nZVI) particles, are the most eco-friendly due to easily scalable synthesis and formation of insoluble iron-oxyhydroxides after interaction with contaminants resulting in the minimal secondary release of contaminants (Kharisov et al., 2012). nZVI or Fe^0 nanoparticles can release electrons via oxidizing themselves, which can simultaneously lead to either reduction and coprecipitation of dyes or to the formation of free radicals causing degradation of dyes in the presence of dissolved oxygen. Example of the reactions 12.1–12.7 are provided below (Guo et al., 2016b).

$$Fe^0 + O_2 + 2H^+ \rightarrow Fe^{2+} + H_2O_2 \quad (12.1)$$

$$Fe^{2+} + H_2O_2 \rightarrow Fe^{3+} + _OH + OH^- \quad (12.2)$$

$$2Fe^{3+} + Fe^0 \rightarrow 3Fe^{2+} \quad (12.3)$$

$$Fe^0 + 2H^+ \rightarrow Fe^{2+} + H_2 \quad (12.4)$$

$$Fe^{2+} + O_2 \rightarrow Fe^{3+} + O_2^- \quad (12.5)$$

$$O_2^- + H^+ \leftrightarrow HO_2 \quad (12.6)$$

$$HO_2 + HO_2 \rightarrow H_2O_2 + O_2 \quad (12.7)$$

Similar to other RSNPs, nZVI generally oxidizes in the natural environment and forms aggregates. The thick oxide layer formed on the outer shell of the nZVI particles inhibits electron transfer and limits their applicability. Further, the generation of soluble Fe^{2+} as an intermediate product during the iron oxidation process can cause the desorption of adsorbates, which may lead to secondary contamination (Lefevre et al., 2016).

12.3.1 Synthesis, characterization, and modifications of nZVI

Both top-down and bottom-up approaches of particle synthesis have been explored to prepare nZVI particles and composites (Mukherjee et al., 2015). Top-down processes include standard grinding, precision milling, and lithography, which are relatively inexpensive methods but have shown limited control over particle size distribution and morphology. Bottom-up approaches include the most famous and commonly used wet chemical reduction using $NaBH_4$, carbothermal reduction, ultrasound-assisted chemical reduction, electrochemical, and some of the explored green synthesis protocols using plant extracts, etc. (Fang et al., 2018a). The wet chemical reduction method shows reasonable control over the particles' size (1–100 nm) and has also been well explored for the development of modified nanocomposites. This method can be combined with the assistance of ultrasound to generate particles of even smaller size ($\approx$ 10 nm) (Lu et al., 2016). Green synthesis involving plant extracts is an eco-friendly and inexpensive process but generates irregularly shaped particles (Guan et al., 2015). Whereas the electrochemical method generally results in the formation of aggregated clusters of nZVI particles. Apart from the utilization of very high temperatures, the carbothermal method doesn't involve the use of any toxic chemical reductant and generates spherical nanoparticles (20–150 nm) in the presence of gaseous reducing agents like H_2, CO_2, etc. (Mackenzie and Georgi, 2019).

To enhance the efficiency of nZVI and overcome earlier-stated demerits, pretreatment with acid (Lai and Lo, 2008) or ultrasound (Dutta et al., 2016a), chemical treatment with ethylene diamine tetraacetic acid (EDTA), rhamnolipids, carboxymethyl cellulose (CMC), etc. (Xue et al., 2018, Adusei-Gyamfi and Acha, 2016), supporting with other sorbents like clays, zeolite, graphene, activated carbon, etc. (Lv et al., 2017), are some of the countermeasures experimented so far. Comparative details of various pretreatment methods have been provided in Table 12.1. Considering the involvement of comparatively no significant secondary cost for most of the supporting surfaces like natural clays or waste to resource conversion (through agricultural waste generated biochar-based surfaces) along with their efficiency in preventing nZVI aggregation and preserving iron oxidation (Fig. 12.2A) and added minimal secondary release of contaminants due to their adsorbing nature make surface-supported redox-sensitive iron nanocomposites as promising candidates for aqueous decontamination.

TABLE 12.1 Commonly performed modifications for nZVI stabilization (Guan et al., 2015, Lu et al., 2016).

Explored pretreatments	Preparation/Major role	Associated challenges
Bimetallic nZVI formation		
	Particles of nZVI that have been coated with a catalyst, such as platinum, gold, nickel, copper, palladium, etc., to enhance the reduction reaction	Short lifetime due to generation of oxide shell, loss of the loosely bound second metal particles, frequency of replacement, and the high cost of noble metals. Possible eco-toxicity
Surface coated		
	Particles that have been coated with surface stabilizers (e.g., polyelectrotyte, surfactant, and biopolymer) in order to increase the suspension stability and therefore particle mobility.	Ability of ligands to mobilize toxic metals, potential ecotoxicity (e.g., EDTA), use of toxic chemicals, tedious synthesis procedure
Emulsified		
	Emulsified nZVI particles are nZVI particles that have been coated with a membrane made from biodegradable oil and water to facilitate the treatment of chlorinated hydrocarbons by making the particles more hydrophobic	Not generally cost-effective for dispersed plumes, viscosity issue
Surface supported		
	In situ growth of nZVI particles on different surfaces like clays, biochar, carbon-based nanoadsorbents, zeolite, etc. to prevent aggregation and oxidation along with minimal secondary release of contaminants	May involve more than one synthesis step

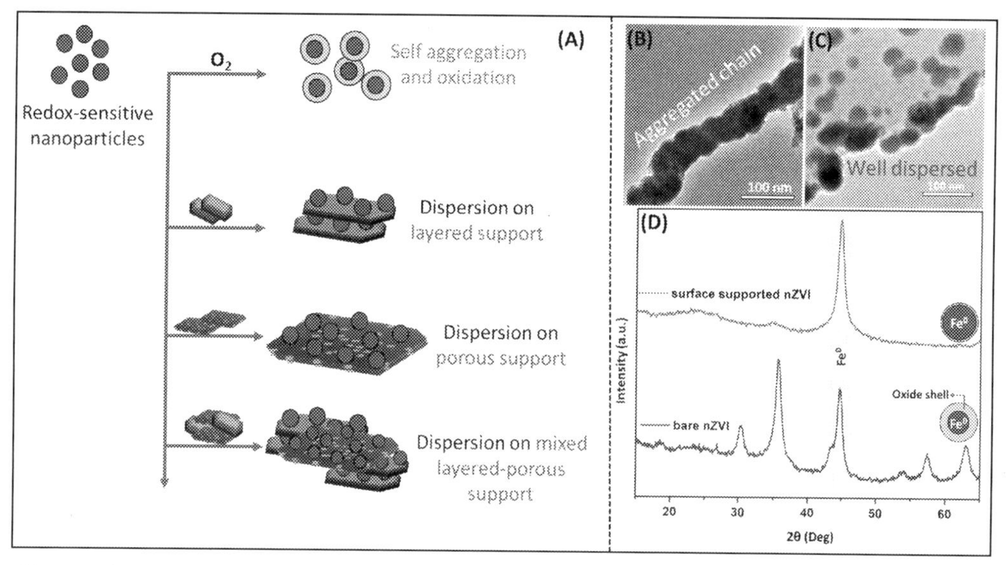

FIGURE 12.2 (A) supporting surfaces for redox-sensitive nanoparticles, scanning electron microscope (SEM) images of (B) bare nZVI, (C) surface supported nZVI, and (D) respective pXRD spectra of bare and supported nZVI composites.

12.3.2 Surface-supported redox-sensitive iron nanocomposites

As described earlier, bare nZVI particles are not stable in the environment and tend to self-aggregate and form chains (Fig. 12.2B) like morphology (Phenrat et al., 2007). Whereas, supporting surfaces (Fig. 12.2C) clearly show well-dispersed spherical nZVI particles of size <100 nm. Characteristic nZVI peak corresponding to α-Fe° at $2\theta = 44.8$ degrees was observed in the pXRD spectra of both bare and supported nZVI particles (Fig. 12.2D), but other peaks around 30 degrees and 35 degrees that belong to various iron oxides were only visible in bare nZVI, suggesting the formation of very thick iron oxide layer on the surface of nZVI (Khandelwal et al., 2020).

So far, researchers have used either natural or synthetically developed surfaces to preserve the redox-sensitive nature of nZVI and prevent their aggregation. On the basis of morphological characteristics, these surfaces can be divided into three major domains- (1) layered surfaces, (2) porous surfaces, and (3) polymer-based entrapment.

Natural surfaces include various clays (layered), zeolite and starch (porous), etc. whereas synthetically developed surfaces hold vastness, that is, layered double hydroxides, graphene, boron nitride, Mxene, etc. as layered surfaces and activated charcoal, agricultural waste-derived biochar, metal and covalent organic frameworks (MOFs & COFs), and mesoporous silica as porous supports (Fang et al., 2018a). Supporting nZVI on such adsorbent surfaces not only shows enhanced reactivity due to their excellent dispersion and electron transferring ability but also results in the minimal secondary release of adsorbed contaminants or soluble iron (Fe^{2+}) along with ease of complete separation. Interfacial wet chemical reduction with borohydride in the presence of supporting materials is the easiest and most common synthesis protocol for supported nZVI nanocomposites preparation (Yan et al., 2013). These surfaces also help control the smaller size of nanoparticles via fast nucleation and limited growth in porous materials like charcoal. Researchers have also explored possibilities of utilizing surface-mediated in situ thermal reduction of iron in the presence of gaseous-reducing agents like generated CO_2 for synthesizing-supported iron nanocomposites (Stefaniuk et al., 2016). Green synthesis protocols involving leaf extracts as reducing agents in the presence of supporting surfaces and precursor Fe^{2+}/Fe^{3+} ions have also been investigated for preparing such nanocomposites (Ahamed et al., 2016). Recently, novel surfaces have also been developed by combining layered and porous materials to support nZVI for enhanced removal and degradation of contaminants (Khandelwal et al., 2020). These surface-supported nZVI nanocomposites have been explored as either adsorbents for removal of toxic metals, radionuclides, organic contaminants, and other ions or as catalysts for the degradation mainly of dyes and other organic contaminants (Fang et al., 2018b).

12.4 Use of biowastes in designing and preserving reactivity of redox-sensitive nanoparticles

Biowaste constitutes of biodegradable materials that can no longer be suitable for direct utilization and therefore may pose as an ecological burden (Maraveas, 2020). It can range from typical household and kitchen waste to agricultural residues, sewage sludge, processed wood, natural textiles, and papers (Lohri et al., 2017). In the field of redox-sensitive nanomaterials, such wastes have found multiple applications, that is (1) green synthesis, (2) functionalization and stabilization, and (3) supporting surface for preserved reactivity. All these applications have been explained in detail in the following section.

12.4.1 Use in green synthesis of redox-sensitive nanoparticles

As described earlier, RSNPs synthesis involves either the use of toxic chemical reducing agents such as borohydride or may involve high temperatures, making the process cost-intensive and less environmentally friendly (Sravanthi et al., 2018). Researchers have shown that the extracts from natural products, such as plants leaf/bark, tea, coffee, fruits, etc., are enriched in polyphenols and flavonoids that can serve as reducing and capping agents for nanoparticles (Kozma et al., 2016).

Such synthesis can be performed at room temperature and generally doesn't require use of anaerobic/inert environment. For example, epicatechin polyphenols from tea that has higher reduction potential were also explored for iron reduction and formation of RSNPs as shown in Fig. 12.3A (Wang et al., 2017a). However, solo use of green reducing agents leads to limited or incomplete reduction of iron ions and therefore can have limited degradation of organic compounds including dyes (Kozma et al., 2016). Combined use of these green RSNPs with other catalysts has shown promising organic degradation applications (Bao et al., 2019).

12.4.2 Use in functionalization and stabilization of redox-sensitive nanoparticles

Stabilization of RSNPs can prevent their agglomeration and may also preserve their reactivity. Researchers have explored various bio-derived polymers and surfactants for the stabilization of RSNPs (Singh and Misra, 2016). These biosurfactants and biopolymers also help in surface functionalization of RSNPs by capping them (Moura et al., 2022).

Carboxymethylcellulose, starch, gelatin, alginate, guar gum, chitosan, rhamnolipids, etc. are some of the biopolymers and biosurfactants that have been used for RSNPs' stabilization. For example, carboxylate functional groups of CMC allow complex formation with iron ions which leads to the coverage of ions with CMC network (He et al., 2007). Later, addition of reducing agent results in rapid nucleation and nanoparticles' growth. CMC also helps in controlling the size of nanoparticles due to charge-induced steric hindrance (Fatisson et al., 2010). Therefore, these bio-derived stabilizers can be helpful in designing reactive RSNPs for efficient environmental applications.

12.4.3 Use as supporting surfaces

Recently, scientific community has started looking at conversion of such biowastes to resources that can have multiple environmental benefits. Loose porous structure and versatile functional groups on surface make them a perfect material

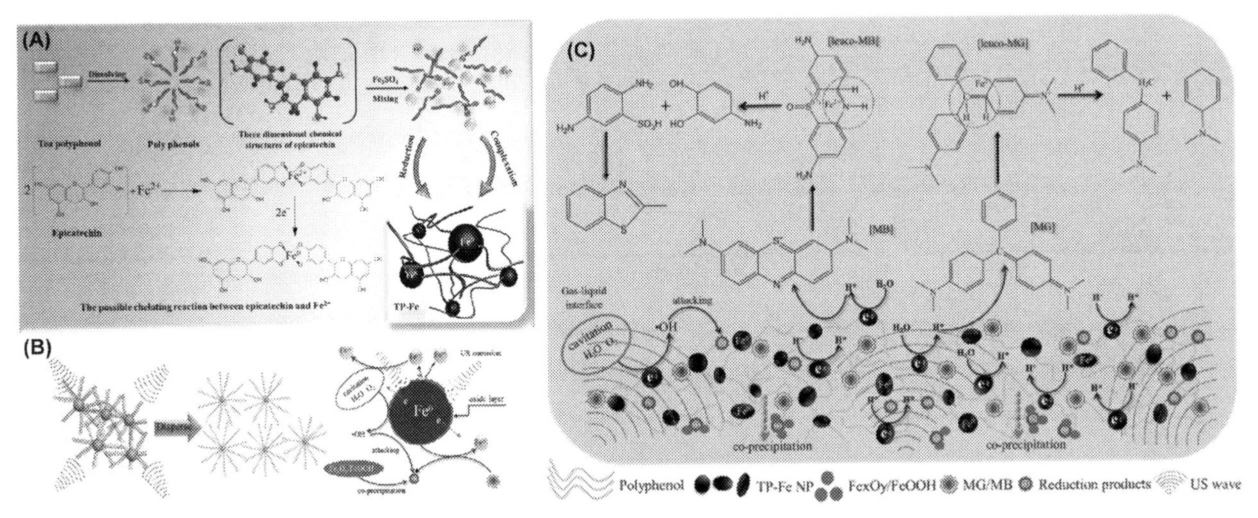

FIGURE 12.3 (A) Schematic representation of tea-polyphenols extraction and nZVI composite synthesis, (B, C) dyes degradation mechanisms. *Modified and republished with permission (Wang, X. Y., Wang, A. Q., Ma, J., Fu, M. L., 2017a. Facile green synthesis of functional nanoscale zero-valent iron and studies of its activity toward ultrasound-enhanced decolorization of cationic dyes. Chemosphere, 166, 80–88).*

to grow RSNPs. Low-temperature pyrolysis (350°C−550°C) of such wastes results in biochar development with very high surface area and porosity along with preserved multifunctionality. At the same time, high-temperature pyrolysis leads to the development of charcoal, a more carbon-enriched material with higher porosity but limited functional groups on the surface. Multifunctional surface with high surface area allows initial interaction and sorption of constituent metal ions on the surface followed by rapid nucleation and growth of nanoparticles in the pores, which results in the prevented agglomeration and preserved reactivity of RSNPs. Such biowaste-derived nanocomposites have also been designed and explored for the removal of synthetic dyes.

Similarly, textile natural fibers, papers, and process wood are among rich sources of cellulose that can be polymerized and entrap nZVI particles helping in their well dispersion and preserved reactivity on the surface. Researchers have also explored the applicability of such cellulose-modified RSNPs in the removal of dyes (Wang et al., 2015).

Direct growth of RSNPs on biomass surface such as seaweeds/microalgae (*Sargassum swartzii*) through liquid phase reduction method was also explored (Jerold et al., 2017).

Therefore, used biowastes can range from pomegranate peel, oak leaves, tree seed pod fibers, to biochar generated from various feedstocks and fossilized algae rocks such as diatomite, to grow RSNPs. Using biowaste-derived coating agents and porous carbon surfaces to preserve RSNPs through either a two-step wet-chemical approach or one-step copyrolysis technique has shown promising environmental applications of derived redox-sensitive nanocomposites (Khandelwal and Darbha, 2021).

In summary, various biowastes, derived surfaces, and waste-extracted polymers/surfactants can be used to support or modify nZVI for their enhanced reactivity for removal of various pollutants.

12.5 Application of surface-supported redox-sensitive iron nanocomposites in dyes removal and prevailing mechanisms

The presence of supporting surfaces generally leads to synergistic removal of various synthetic dyes due to their overall porosity, surface functionality, good particle dispersion, and redox-nature preservation of nZVI particles on the surface. These supported nanocomposites can remove dyes from aqueous solutions through various mechanisms including (1) enhanced complexation and sorption, (2) reductive degradation and coprecipitation, (3) oxidative degradation in Fenton system, (4) advanced oxidation through peroxymonosulfate (PMS) activation, (5) organic degradation through radical generation in the presence of dissolved oxygen (no involvement of compounds like H_2O_2), and (6) coagulation/sedimentation with generated iron-oxy-hydroxides. Therefore, if we exclude the use of other external oxidative agents, supported nZVI can remove dyes through four different mechanisms, as illustrated in Fig. 12.4, and can also handle secondary pollutants via sorption on the adsorbing surfaces (Ezzatahmadi et al., 2017). Chemical structures of commonly investigated dyes for removal experiments have been shown in Fig. 12.1. Adsorbing surfaces cause a faster transfer of dye molecules on the surface, which can be degraded further via interfacial electron transfer, causing cleavage of bonds and resulting in the deactivation of available chromophores which ultimately leads to the decolorization of dyes (Dutta et al., 2016b).

Decolorization can make the water look transparent but can't assure complete decontamination of dyes, which requires higher removal of total organic carbon removal in the system (Zhang et al., 2012). Free radical generation, either in the presence of oxidative agents like H_2O_2 or via iron oxidation and simultaneous in situ generation of H_2O_2

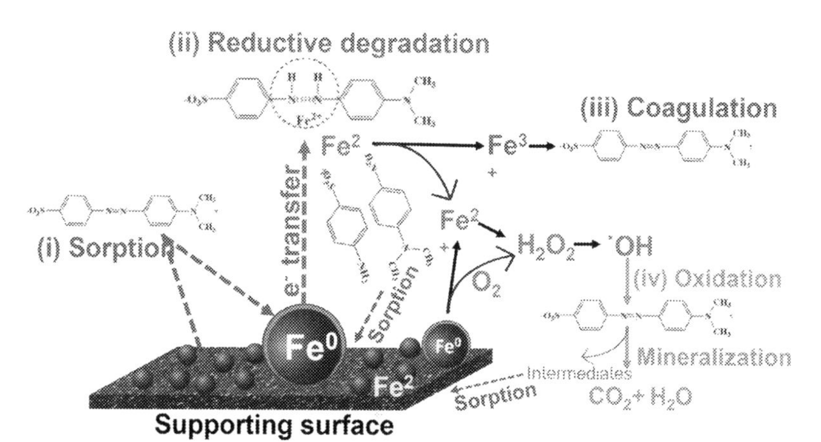

FIGURE 12.4 Various dye removal mechanisms for supported nZVI nanocomposites.

in the presence of dissolved oxygen (reaction 1–7), can further degrade dyes and may ultimately result in complete mineralization (Tokumura et al., 2011). In the past decade, researchers have explored several surfaces that are dominantly layered or porous in nature to support nZVI particles and investigated their application as both catalyst and adsorbent to remove dyes (Sun et al., 2015). For example: in 2008, Zhao et al. supported nZVI particles on cation exchange resin and observed efficient removal of various water-soluble azo dyes (Zhao et al., 2008).

Removal of dyes was limited (70%–85%) at pH > 6.2, whereas it reached as high as 100% in acidic conditions; results show nZVI-assisted electron transfer and simultaneous reduction degradation of dye molecules via cleavage of azo (–N = N–) bond. Apart from supporting nZVI on the resin (Shu et al., 2010), researchers have explored various clay-supported nZVI particles for reductive degradation of dyes such as: bentonite-nZVI composite for methyl orange (Chen et al., 2011), acid violet red-B (Lin et al., 2014), kaolinite supported nZVI for crystal violet (Chen et al., 2012), direct black-G (Liu et al., 2014), etc., and montmorillonite-nZVI for rhodamine 6G removal (Rao et al., 2018). Major removal mechanisms can be summarized as: oxidation of iron, adsorption of dyes onto clay-nZVI composites, forming Fe(II)–dye complex, and reduction of the –N = N– bond at the iron surface (Li et al., 2017). Carbon-based adsorbents have also been utilized as a support for nZVI particles via different synthesis routes. For instance: the liquid phase reduction method was used to synthesize biochar-nZVI nanocomposite, which showed 98.3% removal of acid orange-7 but at a very acidic pH of 2 (Quan et al., 2014). Chitosan-modified biochar-nZVI was utilized for the removal of methylene blue, which showed 68% removal of methylene blue (MB) via reduction to leuco-MB and its precipitation (Zhou et al., 2014). Using biochar as a supporting surface can be a promising option for dyes removal due to its generation through pyrolysis of agricultural biomass and its extensive functionality. Synthesis of activated carbon, fly-ash, and carbon nanofiber-supported nZVI nanocomposites was achieved through thermal reduction via in situ generation of CO_2 as a reducing agent (Wang et al., 2017b, Du et al., 2020).

Microalgae supported nZVI showed 142.8 mg/g removal capacity for malachite green dye at a pH of 10. Solutions containing 100 mg/L of dyes and 0.1 g of composite showed an increase in the malachite green removal capacity from 11.3 to 89.6 mg/g due to increased deprotonation and enhanced electrostatic attraction (Jerold et al., 2017). Apart from seaweeds or microalgae, researchers have also used yeast to modify nZVI particles (Guler and Kundakci, 2018). Green-synthesized nZVI composites using tea polyphenols also showed efficient dye removal (> 90%) of methylene blue and malachite green within 30 minutes of interaction. Electron transfer from composite and generation of active radicals helped in degradation of dyes. Cleavage of –C = N– bond in MB was major dye discoloration mechanism as summarized in Fig. 12.3B and C (Wang et al., 2017a).

Carbonized pomegranate peel-supported nZVI showed 99.7% removal of malachite green dye at an adsorbent dose of 0.15 g within 30 minutes of interaction with a maximum sorption capacity of 32.5 mg/g (Gunduz and Bayrak, 2018). Kecić et al. showed 92% removal of water-based magenta flexographic dye at a concentration of 180 mg/L, adsorbent dose of 60 mg/L and 11 mM H_2O_2 at pH = 2. Structural transformation and destruction of azo bond were major dye discoloration mechanisms (Kecic et al., 2018). nZVI loaded on anaerobic sludge has showed 99% removal of methyl orange with a capacity of 168 mg/g, whereas sludge-derived biochar combined with bacteria has showed efficient continuous removal (up to 99.6%) of methylene blue dye with a sorption capacity of 77.5 mg/g in batch mode (Ahmad et al., 2021). nZVI immobilized on sycamore (*Platanus occidentalis*) tree seed pod fibers showed efficient removal of cationic dyes including malachite green oxalate (92.6 mg/g), methyl violet (92.6 mg/g), and methylene blue (140.8 mg/g) dyes (Parlayici and Pehlivan, 2019).

Apart from reductive degradation of dyes, redox-sensitive iron nanocomposites have also been explored as a catalyst for the activation of H_2O_2, PMS and PDS, etc. which further leads to enhanced organic degradation and mineralization of dyes like direct black G(597 mg/g), etc. (Mossmann et al., 2019, Lin et al., 2017). Granular red mud-supported nZVI (Fe@GRM) was synthesized through thermal reduction and has shown efficient degradation (97.8%) of acid orange-7 through generated sulfate radicals via PS activation. Such methods, where the further use of other chemical reagents is involved, may have a higher cost of operation and can also cause secondary contamination. To enhance the dye degradation and removal efficiency even more, surface supported bimetallic nanoparticles have been prepared. For example: clay-supported bimetallic nZVI/Pd nanoparticles have shown ultraefficient removal of methyl orange (985.56 mg/g) due to enhanced reduction and sorption along with >90% removal within 10 minutes of interaction and nearly 94% removal of MO in wastewater solutions (Wang et al., 2013). Oxidation of nZVI can result in the formation of Fe^{2+} in the system as a corrosion product, which, in acidic conditions, can lead to the formation of H_2O_2 in the presence of dissolved oxygen in the water, causing in situ generation of reactive hydroxyl radicals in the system (reaction 1–7). For example: polyaniline-supported nZVI was used for the removal of rhodamine B (RB). Results showed 83.31% removal of RB (0.02 mM, pH = 6.5) compared to 11.1% with bare nZVI within 2 hours of interaction (Guo et al., 2016a). Results also showed efficient RB removal and degradation in acidic conditions via radicals generated through activation of molecular oxygen by PANI/nZVI in a heterogeneous Fenton-like system. Recently, a combined layered-porous surface was prepared by wet chemical mixing of layered bentonite and porous charcoal and utilized for supporting nZVI particles. Such clay-charcoal-nZVI system was used for the removal of cationic methylene blue and zwitterionic rhodamine B from complex aqueous matrices like wastewater. Spectroscopic observations, in combination with dyes' recovery in an organic solvent, suggested a combination of sorption and organic degradation of both the dyes in the absence of additional chemical reagents. Results also showed efficient removal of dyes, that is, 112.3 mg/g (MB) and 58.7 mg/g (RB) and significant removal in wastewater. It was also observed that nZVI-supported pm surfaces enriched with porous charcoal were able to organically degrade the dyes. In contrast, clay-enriched layered surfaces have shown dominant sorption and minimal degradation. Organic degradation at circum-neutral pH without utilizing any other chemical reagent shows the promising candidature of the charcoal-enriched clay-charcoal-nZVI composites to remove organic dyes from complex aqueous matrices (Khandelwal et al., 2020). Interestingly, to limit the use of toxic chemicals in the system, nanoscale zerovalent iron particles supported on multiwalled carbon nanotubes/attapulgite were synthesized via reduction using *Ruellia tuberosa* leaf extract as a reducing agent, and the composite had shown good removal efficiencies for both methylene blue (149.9 mg/g) and malachite green (177.9 mg/g) due to sorption-enhanced reductive degradation (Zhang et al., 2020).

Apart from living and fresh biomass, fossilized algae rocks such as diatomite were also used to support nZVI particles which showed up to 98% removal of acid blue dye which was higher than raw diatomite (37%) and nZVI (40%) components (Flores-Rojas et al., 2021). Use on newer feedstocks for biochar generation and utilization of greener reducing agents such as tea-leaf extracts or extracted polyphenols are some of the frontiers in the field of waste-derived redox-active nanocomposites for environmental applications. Recently, El-Monaem et al. utilized lemon-derived biochar to support nZVI particles which resulted in 1959.9 mg/g methylene blue sorption capacity within 5 minutes of interaction (Abd El-Monaem et al., 2022). Researchers are further moving toward more complex systems including the discoloration of real wastewater solutions and removal of mixture of dyes from aqueous solutions. For example, recently Eddy et al. showed the efficiency of tea-leaf extract-generated nZVI in degrading mixed rhodamine B and methyl orange dyes through Fenton process. At a ratio of 1:1, 100% discoloration was observed with 100% and 66.4% degradation of rhodamine B and methyl orange, respectively, whereas COD values were decreased by 92.1% (Eddy et al., 2022).

12.6 Conclusion: current challenges and future perspectives

Industrial wastewater containing a considerable amount of colored water possesses a serious threat to the environment, and conventional methods, photocatalytic degradation, or advanced processes face several limitations due to either cost-intensive nature or effectiveness. Surface supported RSNPs, and specifically nZVI nanocomposites, have gathered a lot of attention due to their eco-friendly nature, interfacial electron transferring ability and high reactivity. Researches have shown that both synthetic and natural adsorbing surfaces or both layered and porous surfaces are efficient in preserving nZVI particles from self-aggregation and oxidation. From the reviewed literature, various dyes removal mechanisms can be summarized as follows- (1) surface-enhanced complexation, sorption, and secondary contamination control, (2) reductive degradation and coprecipitation, (3) oxidative degradation in Fenton-like systems via activation of oxidative chemical reagents like H_2O_2, (4) advanced oxidation through sulfate radicals generated via PMS activation, and most promising (5) organic degradation through radical generation in the presence of dissolved oxygen (no involvement of compounds like H_2O_2). In situ generation of reactive hydroxyl radicals in supported nZVI system was found to be surface-mediated; therefore, further research is critically required to optimize the surface characteristics in order to achieve efficient dyes degradation. In this regard, biowaste-derived surfaces and other materials including biopolymers and biosurfactants can be a useful entity due to zero cost and waste-to-resource conversion. In addition, green extracts such as polyphenols from tea leaves can be an alternate eco-friendly option to prepare RSNPs compared to conventional liquid phase reduction using borohydride.

Furthermore, in situ and on-site generation of supported nZVI composites, continuous regeneration of adsorbent, and condition optimization are also critical challenges to be explored. Most of the dye removal studies reviewed here have been performed in the laboratory. Only some of them investigated behavior in wastewater like matrices, which makes it challenging to predict the behavior of the composite in the natural environment and highly complex wastewater matrices. It is also essential to understand and explore the toxicity of the final dyes containing composites, economical ways of recovering adsorbed dye fractions, and simultaneous composite regeneration. Overall, it can be concluded that surface-supported redox-sensitive iron-based nanocomposites can be optimized for their promising candidature as adsorbent and catalyst for the degradation of most of the synthetic dyes in complex aqueous matrices.

References

Abd El-Monaem, E.M., Omer, A.M., El-Subruiti, G.M., Mohy-Eldin, M.S. & Eltaweil, A.S., 2022. Zero-valent iron supported-lemon derived biochar for ultra-fast adsorption of methylene blue. *Biomass Conversion and Biorefinery*.

Adusei-Gyamfi, J., Acha, V., 2016. Carriers for nano zerovalent iron (nZVI): synthesis, application and efficiency. RSC Advances 6, 91025–91044.

Ahamed, N., Anbu, S., Vikraman, G., Nasreen, S., Muthukumari, M., Kumar, M., 2016. Green synthesis of nano zerovalent iron particles (Nzvi) for environmental remediation. Life Science Archives *2454-1354*.

Ahmad, A., Singh, A.P., Khan, N., Chowdhary, P., Giri, B.S., Varjani, S., et al., 2021. Bio-composite of Fe-sludge biochar immobilized with Bacillus Sp. in packed column for bio-adsorption of Methylene blue in a hybrid treatment system: isotherm and kinetic evaluation. Environmental Technology & Innovation 23.

Anjum, M., Miandad, R., Waqas, M., Gehany, F., Barakat, M.A., 2019. Remediation of wastewater using various nano-materials. Arabian Journal of Chemistry 12, 4897–4919.

Bao, T., Jin, J., Damtie, M.M., Wu, K., Yu, Z.M., Wang, L., et al., 2019. Green synthesis and application of nanoscale zero-valent iron/rectorite composite material for P-chlorophenol degradation via heterogeneous Fenton reaction. Journal of Saudi Chemical Society 23, 864–878.

Benkhaya, S., M'rabet, S., El Harfi, A., 2020. A review on classifications, recent synthesis and applications of textile dyes. Inorganic Chemistry Communications 115, 107891.

Bolisetty, S., Peydayesh, M., Mezzenga, R., 2019. Sustainable technologies for water purification from heavy metals: review and analysis. Chemical Society Reviews 48, 463–487.

Camargo, P.H.C., Satyanarayana, K.G., Wypych, F., 2009. Nanocomposites: synthesis, structure, properties and new application opportunities. Materials Research 12, 1–39.

Chen, Z.X., Jin, X.Y., Chen, Z.L., Megharaj, M., Naidu, R., 2011. Removal of methyl orange from aqueous solution using bentonite-supported nanoscale zero-valent iron. Journal of Colloid and Interface Science 363, 601–607.

Chen, Z.X., Cheng, Y., Chen, Z.L., Megharaj, M., Naidu, R., 2012. Kaolin-supported nanoscale zero-valent iron for removing cationic dye-crystal violet in aqueous solution. Journal of Nanoparticle Research 14.

Chequer, F.D., De Oliveira, G.A.R., Ferraz, E.A., Cardoso, J.C., Zanoni, M.B., De Oliveira, D.P., 2013. Textile dyes: dyeing process and environmental impact. Eco-friendly Textile Dyeing and Finishing 6, 151–176.

Chung, K.-T., 2016. Azo dyes and human health: a review. Journal of Environmental Science and Health. Part C, Environmental Carcinogenesis & Ecotoxicology Reviews 34.

Du, Y.F., Dai, M., Cao, J.F., Peng, C.S., Ali, I., Naz, I., et al., 2020. Efficient removal of acid orange 7 using a porous adsorbent-supported zero-valent iron as a synergistic catalyst in advanced oxidation process. Chemosphere 244.

Dutta, S., Ghosh, A., Satpathi, S., Saha, R., 2016a. Modified synthesis of nanoscale zero-valent iron and its ultrasound-assisted reactivity study on a reactive dye and textile industry effluents. Desalination and Water Treatment 57, 19321–19332.

Dutta, S., Saha, R., Kalita, H., Bezbaruah, A.N., 2016b. Rapid reductive degradation of azo and anthraquinone dyes by nanoscale zero-valent iron. Environmental Technology & Innovation 5, 176–187.

Eddy, D.R., Nursyamsiah, D., Permana, M.D., Solihudin, S., Noviyanti, A.R., Rahayu, I., 2022. Green production of zero-valent iron (ZVI) using tea-leaf extracts for fenton degradation of mixed rhodamine B and methyl orange dyes. Materials 15.

Ejder Korucu, M., Gurses, A., Doğar, Ç., Sharma, S. & Açikyildiz, M. 2015. Removal of organic dyes from industrial effluents: an overview of physical and biotechnological applications.

El-Sayed, M.E.A., 2020. Nanoadsorbents for water and wastewater remediation. Science of The Total Environment 739, 139903.

Ezzatahmadi, N., Ayoko, G.A., Millar, G.J., Speight, R., Yan, C., Li, J.H., et al., 2017. Clay-supported nanoscale zero-valent iron composite materials for the remediation of contaminated aqueous solutions: a review. Chemical Engineering Journal 312, 336–350.

Fang, Y., Wen, J., Zeng, G., Shen, M., Cao, W., Gong, J., et al., 2018a. From nZVI to SNCs: development of a better material for pollutant removal in water. Environmental Science and Pollution Research 25, 6175–6195.

Fang, Y., Wen, J., Zeng, G.M., Shen, M.C., Cao, W.C., Gong, J.L., et al., 2018b. From nZVI to SNCs: development of a better material for pollutant removal in water. Environmental Science and Pollution Research 25, 6175–6195.

Fatisson, J., Ghoshal, S., Tufenkji, N., 2010. Deposition of carboxymethylcellulose-coated zero-valent iron nanoparticles onto silica: roles of solution chemistry and organic molecules. Langmuir: The ACS Journal of Surfaces and Colloids 26, 12832–12840.

Flores-Rojas, E., Schnabel, D., Justo-Cabrera, E., Solorza-Feria, O., Poggi-Varaldo, H.M., Breton-Deval, L., 2021. Using nano zero-valent iron supported on diatomite to remove acid blue dye: synthesis, characterization, and toxicology test. Sustainability 13.

Ghoreishi, S.M., Haghighi, R., 2003. Chemical catalytic reaction and biological oxidation for treatment of non-biodegradable textile effluent. Chemical Engineering Journal 95, 163–169.

Gregory, J., Duan, J., 2001. Hydrolyzing metal salts as coagulants. Pure and Applied Chemistry 73, 2017–2026.

Guan, X., Sun, Y., Qin, H., Li, J., Lo, I.M., He, D., et al., 2015. The limitations of applying zero-valent iron technology in contaminants sequestration and the corresponding countermeasures: the development in zero-valent iron technology in the last two decades (1994–2014). Water Research 75, 224–248.

Guler, U.A., Kundakci, O., 2018. Batch and fixed-bed column adsorption of crystal violet and safranin by zero-valent iron (nZVI) and nZVI-modified S. cerevisiae. Desalination and Water Treatment 106, 292–304.

Gunduz, F., Bayrak, B., 2018. Synthesis and performance of pomegranate peel-supported zero-valent iron nanoparticles for adsorption of malachite green. Desalination and Water Treatment 110, 180–192.

Guo, W., Hao, F., Yue, X., Liu, Z., Zhang, Q., Li, X., et al., 2016a. Rhodamine B removal using polyaniline-supported zero-valent iron powder in the presence of dissolved oxygen. Environmental Progress & Sustainable Energy 35, 48–55.

Guo, W.L., Hao, F.F., Yue, X.X., Liu, Z.H., Zhang, Q.Y., Li, X.H., et al., 2016b. Rhodamine B removal using polyaniline-supported zero-valent iron powder in the presence of dissolved oxygen. Environmental Progress & Sustainable Energy 35, 48–55.

He, F., Zhao, D., Liu, J., Roberts, C.B., 2007. Stabilization of Fe – Pd nanoparticles with sodium carboxymethyl cellulose for enhanced transport and dechlorination of trichloroethylene in soil and groundwater. Industrial & Engineering Chemistry Research 46, 29–34.

Hunger, K., 2007. Industrial Dyes: Chemistry, Properties, Applications. John Wiley & Sons.

Ippolito, J.A., Cui, L., Kammann, C., Wrage-Mönnig, N., Estavillo, J.M., Fuertes-Mendizabal, T., et al., 2020. Feedstock choice, pyrolysis temperature and type influence biochar characteristics: a comprehensive meta-data analysis review. Biochar 2, 421–438.

Irfan, M., Butt, T., Imtiaz, N., Abbas, N., Khan, R.A., Shafique, A., 2017. The removal of COD, TSS and colour of black liquor by coagulation–flocculation process at optimized pH, settling and dosing rate. Arabian Journal of Chemistry 10, S2307–S2318.

Jaishankar, M., Tseten, T., Anbalagan, N., Mathew, B.B., Beeregowda, K.N., 2014. Toxicity, mechanism and health effects of some heavy metals. Interdisciplinary Toxicology 7, 60–72.

Jerold, M., Vidya, E.V., Sankar, R., Arun, N., Sivasubramanian, V., 2017. Nanoscale zerovalent iron-Sargassum swartzii biocomposites for the removal of malachite green from an aqueous solution. Separation Science and Technology 52, 965–974.

Katheresan, V., Kansedo, J., Sie Yon, J.L., 2018. Efficiency of various recent wastewater dye removal methods: a review. Journal of Environmental Chemical Engineering 6.

Kecic, V., Kerkez, D., Prica, M., Luzanin, O., Becelic-Tomin, M., Pilipovic, D.T., et al., 2018. Optimization of azo printing dye removal with oak leaves-nZVI/H_2O_2 system using statistically designed experiment. Journal of Cleaner Production 202, 65–80.

Khan, R., Bhawana, P., Fulekar, M.H., 2013. Microbial decolorization and degradation of synthetic dyes: a review. Reviews in Environmental Science and Bio/Technology 12, 75–97.

Khan, I., Saeed, K., Khan, I., 2019. Nanoparticles: properties, applications and toxicities. Arabian Journal of Chemistry 12, 908–931.

Khandelwal, N., Darbha, G.K., 2021. Combined antioxidant capped and surface supported redox-sensitive nanoparticles for continuous elimination of multi-metallic species. Chemical Communications 57, 7280–7283.

Khandelwal, N., Darbha, G.K., 2022. Sorption and continuous filtration of heavy metals and radionuclides using novel nano-farringtonite: mechanisms delineation using EXAFS. Chemosphere 308, 136376.

Khandelwal, N., Tiwari, E., Singh, N., Marsac, R., Schafer, T., Monikh, F.A., et al., 2020. Impact of long-term storage of various redox-sensitive supported nanocomposites on their application in removal of dyes from wastewater: mechanisms delineation through spectroscopic investigations. Journal of Hazardous Materials 401, 123375.

Kharisov, B.I., Rasika Dias, H.V., Kharissova, O.V., Manuel Jiménez-Pérez, V., Olvera Pérez, B., Muñoz Flores, B., 2012. Iron-containing nanomaterials: synthesis, properties, and environmental applications. RSC Advances 2, 9325–9358.

Kozma, G., Rónavári, A., Kónya, Z., Kukovecz, Á., 2016. Environmentally benign synthesis methods of zero-valent iron nanoparticles. ACS Sustainable Chemistry & Engineering 4, 291–297.

Lai, K.C.K., Lo, I.M.C., 2008. Removal of chromium (VI) by acid-washed zero-valent iron under various groundwater geochemistry conditions. Environmental Science & Technology 42, 1238–1244.

Lefevre, E., Bossa, N., Wiesner, M.R., Gunsch, C.K., 2016. A review of the environmental implications of in situ remediation by nanoscale zero valent iron (nZVI): behavior, transport and impacts on microbial communities. The Science of the Total Environment 565, 889–901.

Lellis, B., Fávaro-Polonio, C.Z., Pamphile, J.A., Polonio, J.C., 2019. Effects of textile dyes on health and the environment and bioremediation potential of living organisms. Biotechnology Research and Innovation 3, 275–290.

Li, X.G., Zhao, Y., Xi, B.D., Meng, X.G., Gong, B., Li, R., et al., 2017. Decolorization of Methyl Orange by a new clay-supported nanoscale zero-valent iron: synergetic effect, efficiency optimization and mechanism. Journal of Environmental Sciences 52, 8–17.

Lin, Y.M., Chen, Z.X., Chen, Z.L., Megharaj, M., Naidu, R., 2014. Decoloration of acid violet red B by bentonite-supported nanoscale zero-valent iron: reactivity, characterization, kinetics and reaction pathway. Applied Clay Science 93–94, 56–61.

Lin, J.J., Sun, M.Q., Liu, X.W., Chen, Z.L., 2017. Functional kaolin supported nanoscale zero-valent iron as a Fenton-like catalyst for the degradation of Direct Black G. Chemosphere 184, 664–672.

Liu, X.W., Wang, F.F., Chen, Z.L., Megharaj, M., Naidu, R., 2014. Heterogeneous Fenton oxidation of Direct Black G in dye effluent using functional kaolin-supported nanoscale zero iron. Environmental Science and Pollution Research 21, 1936–1943.

Liyanage, C.P., Yamada, K., 2017. Impact of population growth on the water quality of natural water bodies. Sustainability 9, [Online].

Lohri, C.R., Diener, S., Zabaleta, I., Mertenat, A., Zurbrügg, C., 2017. Treatment technologies for urban solid biowaste to create value products: a review with focus on low- and middle-income settings. Reviews in Environmental Science and Bio/Technology 16, 81–130.

Lu, H.-J., Wang, J.-K., Ferguson, S., Wang, T., Bao, Y., Hao, H.-X., 2016. Mechanism, synthesis and modification of nano zerovalent iron in water treatment. Nanoscale 8, 9962–9975.

Lv, X.S., Zhang, Y.L., Fu, W.Y., Cao, J.Z., Zhang, J., Ma, H.B., et al., 2017. Zero-valent iron nanoparticles embedded into reduced graphene oxide-alginate beads for efficient chromium (VI) removal. Journal of Colloid and Interface Science 506, 633–643.

Mackenzie, K., Georgi, A., 2019. NZVI synthesis and characterization. In: Phenrat, T., Lowry, G.V. (Eds.), Nanoscale Zerovalent Iron Particles for Environmental Restoration: From Fundamental Science to Field Scale Engineering Applications. Springer International Publishing, Cham.

Maraveas, C., 2020. Production of sustainable and biodegradable polymers from agricultural waste. Polymers (Basel) 12.

Mattioli, D., De Florio, L., Giordano, A., Tarantini, M., Scalbi, S., Aguado, M., et al., 2005. Efficient use of water in the textile finishing industry.

Mossmann, A., Dotto, G.L., Hotza, D., Jahn, S.L., Foletto, E.L., 2019. Preparation of polyethylene-supported zero-valent iron buoyant catalyst and its performance for Ponceau 4R decolorization by photo-Fenton process. Journal of Environmental Chemical Engineering 7.

Moura, C.C., Salazar-Bryam, A.M., Piazza, R.D., Carvalho Dos Santos, C., Jafelicci Jr, M., Marques, et al., 2022. Rhamnolipids as green stabilizers of nZVI and application in the removal of nitrate from simulated groundwater. Front Bioeng Biotechnol 10, 794460.

Mukherjee, R., Kumar, R., Sinha, A., Lama, Y., Saha, A., 2015. A review on synthesis, characterization and applications of nano-zero valent iron (nZVI) for environmental remediation. Critical Reviews in Environmental Science and Technology 46, 00.

Omran, B.A., Baek, K.-H., 2022. Valorization of agro-industrial biowaste to green nanomaterials for wastewater treatment: approaching green chemistry and circular economy principles. Journal of Environmental Management 311, 114806.

Parlayici, S., Pehlivan, E., 2019. Fast decolorization of cationic dyes by nano-scale zero valent iron immobilized in sycamore tree seed pod fibers: kinetics and modelling study. International Journal of Phytoremediation 21, 1130–1144.

Phenrat, T., Saleh, N., Sirk, K., Tilton, R.D., Lowry, G.V., 2007. Aggregation and sedimentation of aqueous nanoscale zerovalent iron dispersions. Environmental Science & Technology 41, 284–290.

Quan, G.X., Sun, W.J., Yan, J.L., Lan, Y.Q., 2014. Nanoscale zero-valent iron supported on biochar: characterization and reactivity for degradation of acid orange 7 from aqueous solution. Water, Air, and Soil Pollution 225.

Ram, N., Abhishek, A.K.B., 2020. Green synthesis and characterization of silver NPs using oyster mushroom extract for antibacterial efficacy. *Journal of Chemistry*. Environmental Sciences and its Applications 7, 13–18.

Rao, W.X., Liu, H., Lv, G.C., Wang, D.Y., Liao, L.B., 2018. Effective degradation of Rh 6G using montmorillonite-supported nano zero-valent iron under microwave treatment. Materials 11.

Sadegh, H., Ali, G.A.M., Gupta, V.K., Makhlouf, A.S.H., Shahryari-Ghoshekandi, R., Nadagouda, M.N., et al., 2017. The role of nanomaterials as effective adsorbents and their applications in wastewater treatment. Journal of Nanostructure in Chemistry 7, 1–14.

Samaddar, P., Ok, Y.S., Kim, K.-H., Kwon, E.E., Tsang, D.C.W., 2018. Synthesis of nanomaterials from various wastes and their new age applications. Journal of Cleaner Production 197, 1190–1209.

Samchetshabam, G., Ajmal, H., Choudhury, T.G., 2017. Impact of textile dyes waste on aquatic environments and its treatment. Environment and Ecology 35, 2349–2353.

Sarkar, S., Banerjee, A., Halder, U., Biswas, R., Bandopadhyay, R., 2017. Degradation of synthetic azo dyes of textile industry: a sustainable approach using microbial enzymes. Water Conservation Science and Engineering 2, 121–131.

Sharma, S., Bhattacharya, A., 2017. Drinking water contamination and treatment techniques. Applied Water Science 7, 1043–1067.

Sharma, G., Kumar, A., Sharma, S., Naushad, M., Prakash Dwivedi, R., Alothman, Z.A., et al., 2019. Novel development of nanoparticles to bimetallic nanoparticles and their composites: a review. Journal of King Saud University - Science 31, 257–269.

Shu, H.Y., Chang, M.C., Chen, C.C., Chen, P.E., 2010. Using resin supported nano zero-valent iron particles for decoloration of Acid Blue 113 azo dye solution. Journal of Hazardous Materials 184, 499–505.

Singh, R., Misra, V., 2016. Stabilization of zero-valent iron nanoparticles: role of polymers and surfactants. In: Aliofkhazraei, M. (Ed.), Handbook of Nanoparticles. Springer International Publishing, Cham.

Singh, M.P., Bhardwaj, A.K., Bharati, K., Singh, R.P., Chaurasia, S.K., Kumar, S., et al., 2021. Biogenic and non-biogenic waste utilization in the synthesis of 2D materials (graphene, h-BN, g-C2N) and their applications. 3.

Soffian, M.S., Abdul Halim, F.Z., Aziz, F., Rahman, M.A., Mohamed Amin, M.A., Awang Chee, D.N., 2022. Carbon-based material derived from biomass waste for wastewater treatment. Environmental Advances 9, 100259.

Sravanthi, K., Ayodhya, D., Yadgiri Swamy, P., 2018. Green synthesis, characterization of biomaterial-supported zero-valent iron nanoparticles for contaminated water treatment. Journal of Analytical Science and Technology 9, 3.

Stefaniuk, M., Oleszczuk, P., Ok, Y.S., 2016. Review on nano zerovalent iron (nZVI): from synthesis to environmental applications. Chemical Engineering Journal 287, 618–632.

Sun, X., Kurokawa, T., Suzuki, M., Takagi, M., Kawase, Y., 2015. Removal of cationic dye methylene blue by zero-valent iron: effects of pH and dissolved oxygen on removal mechanisms. Journal of Environmental Science and Health. Part A, Toxic/Hazardous Substances & Environmental Engineering 50, 1057–1071.

Tetteh, E.K., Rathilal, S., 2019. Application of Organic Coagulants in Water and Wastewater Treatment. IntechOpen.

Tiwari, E., Singh, N., Khandelwal, N., Monikh, F.A., Darbha, G.K., 2020. Application of Zn/Al layered double hydroxides for the removal of nano-scale plastic debris from aqueous systems. Journal of Hazardous Materials 397, 122769.

Tokumura, M., Morito, R., Hatayama, R., Kawase, Y., 2011. Iron redox cycling in hydroxyl radical generation during the photo-Fenton oxidative degradation: dynamic change of hydroxyl radical concentration. Applied Catalysis B: Environmental 106, 565–576.

Verma, Y., 2008. Acute toxicity assessment of textile dyes and textile and dye industrial effluents using Daphnia magna bioassay. Toxicology and Industrial Health 24, 491–500.

Wang, T., Su, J., Jin, X.Y., Chen, Z.L., Megharaj, M., Naidu, R., 2013. Functional clay supported bimetallic nZVI/Pd nanoparticles used for removal of methyl orange from aqueous solution. Journal of Hazardous Materials 262, 819–825.

Wang, X.Y., Wang, P., Ma, J., Liu, H.L., Ning, P., 2015. Synthesis, characterization, and reactivity of cellulose modified nano zero-valent iron for dye discoloration. Applied Surface Science 345, 57–66.

Wang, X.Y., Wang, A.Q., Ma, J., Fu, M.L., 2017a. Facile green synthesis of functional nanoscale zero-valent iron and studies of its activity toward ultrasound-enhanced decolorization of cationic dyes. Chemosphere 166, 80–88.

Wang, Y.M., Lopez-Valdivieso, A., Zhang, T., Mwamulima, T., Zhang, X.L., Song, S.X., et al., 2017b. Synthesis of fly ash and bentonite-supported zero-valent iron and its application for removal of toxic cationic dyes from aqueous solutions. Environmental Engineering Science 34, 740–751.

Xu, L., Fu, F.L., 2020. Se(IV) oxidation by ferrate(VI) and subsequent in-situ removal of selenium species with the reduction products of ferrate(VI): performance and mechanism. Journal of Environmental Science and Health Part a-Toxic/Hazardous Substances & Environmental Engineering .

Xue, W.J., Huang, D.L., Zeng, G.M., Wan, J., Zhang, C., Xu, R., et al., 2018. Nanoscale zero-valent iron coated with rhamnolipid as an effective stabilizer for immobilization of Cd and Pb in river sediments. Journal of Hazardous Materials 341, 381–389.

Yaashikaa, P.R., Kumar, P.S., Varjani, S., Saravanan, A., 2020. A critical review on the biochar production techniques, characterization, stability and applications for circular bioeconomy. Biotechnology Reports 28, e00570.

Yan, W., Lien, H.-L., Koel, B.E., Zhang, W.-X., 2013. Iron nanoparticles for environmental clean-up: recent developments and future outlook. Environmental Science: Processes & Impacts 15, 63–77.

Yaseen, D.A., Scholz, M., 2019. Textile dye wastewater characteristics and constituents of synthetic effluents: a critical review. International Journal of Environmental Science and Technology 16, 1193–1226.

Yerramilli, A., Chary, N., Dasary, S., 2005. Decolourization of industrial effluents – available methods and emerging technologies – a review. Reviews in Environmental Science and Bio/Technology 4, 245–273.

Zhang, Q., Li, C., Li, T., 2012. Rapid photocatalytic degradation of methylene blue under high photon flux UV irradiation: characteristics and comparison with routine low photon flux. International Journal of Photoenergy 2012, 398787.

Zhang, Y.J., Xu, H., Tang, J., Tian, L., Chen, M.Z., Tian, W.G., et al., 2020. Simultaneous removal of binary cationic dyes from wastewater by nanoscale zero-valent iron particles supported on multi-walled carbon nanotubes/attapulgite. Desalination and Water Treatment 187, 410–426.

Zhao, Z.S., Liu, J.F., Tai, C., Zhou, Q.F., Hu, J.T., Jiang, G.B., 2008. Rapid decolorization of water soluble azo-dyes by nanosized zero-valent iron immobilized on the exchange resin. Science in China Series B-Chemistry 51, 186–192.

Zhou, Y.M., Gao, B., Zimmerman, A.R., Chen, H., Zhang, M., Cao, X.D., 2014. Biochar-supported zerovalent iron for removal of various contaminants from aqueous solutions. Bioresource Technology 152, 538–542.

Chapter 13

Nanomaterials synthesis from the industrial solid wastes

Leow Hui Ting Lyly[1], Zhen Hong Chang[2] and Yeit Haan Teow[1,3]
[1]*Research Centre for Sustainable Process Technology (CESPRO), Faculty of Engineering and Built Environment, Universiti Kebangsaan Malaysia, Bangi, Selangor, Malaysia,* [2]*Department of Chemical and Petroleum Engineering, Faculty of Engineering, Technology and Built Environment, UCSI University, Kuala Lumpur, Malaysia,* [3]*Department of Chemical and Process Engineering, Faculty of Engineering and Built Environment, Universiti Kebangsaan Malaysia, Bangi, Selangor, Malaysia*

13.1 Introduction

Industrial waste is defined as waste produced during product manufacturing, extraction of natural resources, and agricultural production (Osorio et al., 2021). Industrial waste is divided into two types, which are solid waste and liquid waste because both need to be handled by different approaches and both are generated at a different scale. Most of the industrial waste released has shown harmful properties that could increase the environmental burden due to containing heavy metals such as arsenic, cadmium, chromium, copper, lead, mercury, nickel, and zinc (Singh et al., 2021; Obasi and Akudinobi, 2020). Some examples of industries that consist of heavy metal waste include battery manufacturing, chemical manufacturing, construction, electroplating, metal processing, mining, paper and pulp, paint manufacturing, petroleum refining, textiles, and tanneries industry (Ishchenko, 2018; Yue et al., 2019). Since the industrial revolution, industrial waste has been an issue because most industrial wastes are corrosive, ignitable reactive, or toxic (Godswill et al., 2020). If these released wastes are not properly managed with extreme care, the released waste could bring an impact on the environment and human health. To reduce the waste output, strategies such as maximizing waste recovery, reuse of resources economically, and recycling waste are promising options for industrial waste treatment.

According to life cycle assessments studied by Hertwich et al. (2015), the production of nanomaterials needs more natural resources and energy when they are generated from an eco-friendly process. Therefore, the employment of waste as raw materials for synthesizing nanomaterial is attractive due to low-priced, readily available, and its pathway approaching sustainable development. Waste-derived nanomaterials have been actively studied by researchers in recent years as an integrated method for recycling and waste treatment (Abdelbasir et al., 2020). Additionally, strict environmental regulation has resulted in the rising demand for the recycling of waste (Tam et al., 2018). However, there are still challenges encountered in the aspect of economic, environmental, social, and technical on the large-scale production of nanomaterials from waste. For instance, during the metal and plastic recycling process, harmful substances were utilized for extraction and purification, which could increase the health risk for both the environment and humans (Kral et al., 2019). In addition, some chemical-based path has been reported in the literature to dispose of or recycle solid waste, but it is often time-consuming and harmful to the environment (Abdel-Shafy and Mansour, 2018). Thus, developing a simple and green route for the preparation of nanomaterials with desired properties from solid waste remains a challenge to scientists and environmentalists.

A variety of nanomaterials have been produced through various techniques and demonstrated potential for advanced applications, which include water/soil reclamation (Linley and Thomson, 2021), water/wastewater treatment (Mohammed et al., 2018), membrane technology (Lyly et al., 2021), bone and tissue engineering (Eivazzadeh-Keihan et al., 2019), energy transformation and storage (Yang et al., 2019), drug delivery (Singh et al., 2019), intelligent packaging (Bumbudsanpharoke and Ko, 2019), pollution monitoring (Xue et al., 2017), etc. The physicochemical properties of nanomaterials such as reactivity and high specific area have acquired great attention owing to their wide-ranging application in different fields (Xu et al., 2012). Hence, the preparation of nanomaterials with unique functional

Green and Sustainable Approaches Using Wastes for the Production of Multifunctional Nanomaterials. DOI: https://doi.org/10.1016/B978-0-443-19183-1.00021-0

properties is being pursued by researchers. Fig. 13.1 presents the statistic of publications related to green synthesis of nanomaterials over the past 10 years (2013–23), retrieved from SCOPUS database on February 18, 2023. The number of publications shows an increasing trend over the years with a total number of publications of 12,024, including articles, reviews, books, book chapters, conference papers, and conference reviews.

Different and various techniques have been employed to synthesize nanomaterials with different properties. There are two approaches to the synthesis of nanomaterials from industrial solids, which are the top-down approach and the bottom-up approach (Abid et al., 2021). The top-down approach is the broking down of bulk materials into smaller or fine pieces through chemical, mechanical, and other types of energy (Low et al., 2017). Techniques involved in the top-down approach are grinding and milling. The bottom-up approach involves the combination of ions chemically to form particles (Ahmadi Tehrani et al., 2019). Some of the techniques in the bottom-up approach are calcination, coprecipitation, hydrothermal, pyrolysis, sol–gel, and chemical vapor deposition. Herein, this chapter provides an overview of the generation of nanomaterials from solid industrial waste. The chapter starts with the source of industrial solid waste as an input for the synthesis of nanomaterials as added-value products and their potential application. This is followed by an outline of the methods to derive nanomaterials from industrial. Lastly, the sustainability consideration of the utilization of industrial waste and its future perspective are discussed.

13.2 Industrial solid waste for the synthesis of nanomaterials

Due to the rapid growth in industrialization and urbanization, the amount of waste produced increased sharply all over the world. This phenomenon becomes deteriorated as the waste keeps accumulating without proper management, resulting in severe impacts on the environment and humans. Solid waste is either managed by the third party or disposed of directly. Hence, the mismanaged solid waste will not only cause pollution to the soil and water bodies but also contribute to a large fraction's loss of valuable resources. This section highlights the various sources of industrial waste, which could be a source for the production of value-added nanomaterials.

13.2.1 Generation of waste for production of nanomaterials and their potential application

Depending on the types of industrial sectors, the waste contains both organic and inorganic compounds in different compositions. By having proper solid waste management, these compounds could be recovered and acted as the feedstock to produce valuable nanomaterials. With the rising concerns about sustainability in terms of resource recovery, researchers have studied ways to retrieve useful materials from industrial solid waste and transform them into nanomaterials that can be employed in a wide range of applications (Bhardwaj and Naraian, 2021). Table 13.1 presents a summary of various nanomaterials that are synthesized from industrial wastes together with their synthesis methods, sizes, and applications.

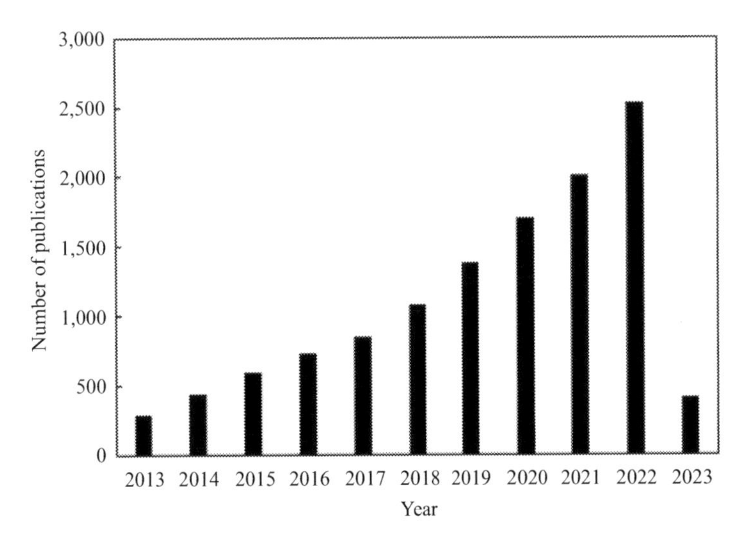

FIGURE 13.1 Number of publications related to green synthesis of nanomaterials over the past 10 years as retrieved from SCOPUS database.

TABLE 13.1 Summary of nanomaterials retrieved from industrial wastes and their potential application.

Industry	Waste materials	Nanomaterials	Synthesis methods	Size (nm)	Applications	References
Metallurgical	High-magnesium nickel waste	$Mg(OH)_2$ nanosheet	Hydrothermal	~13	Catalyst	Qian et al. (2022)
	Nickel-laden electroplating sludge	Nickel ferrite nanoparticles	Regulator-assisted hydrothermal	50–80	Anode in Lithium-ion battery	Weng et al. (2020)
Coal mining	Iron mile tailing sludge (waste iron salts)	Cobalt ferrite nanomaterials	Mixing with Cobalt(II) chloride in natural organic matter-rich water	N/A	Catalyst for reducing nitrophenol and adsorbents for polycyclic aromatic hydrocarbons	Cruz et al. (2021)
	Maghara coal	Nano-activated carbon	Carbonization	~38	Adsorbent for dyes	Shokry et al. (2019)
Power plant	Brown coal fly ash	$Mg(OH)_2$ nanosheet	Hydrothermal	~18	Catalyst for water-gas shift	Qian et al. (2021)
		Al^{3+}-defected iron oxide nanoflakes	Hydrothermal	~50	Adsorbent for heavy metal and dyes	(Qian et al., 2020)
Pulp and paper	Eucalyptus scraps	Nanofibrillar-biochar	Carbonization	<100	Conductive hosting network for electrochemical and biosensoristic field	Fiori et al. (2023)
	Waste lime sludge	Nano-sized waste lime sludge particles	Drying and milling	~39[a]	Flocculent for microalgae harvesting	Kumar et al. (2022)
Food and beverage	Sugar beet processing residuals	CaO nanoparticles	Drying and milling	~15	Adsorbent for heavy metals	Lashen et al. (2022)
		CaO nanorod	Drying and milling	~10	Adsorbent for heavy metals	Lashen et al. (2022)
Construction	Glass wastes	Glass nanoparticles	Milling	80	Cement replacement material in self-compacted concrete	Hussien et al. (2022)
	Ceramic wastes	Ceramic nanoparticles	Milling	60	Cement replacement material in self-compacted concrete	Hussien et al. (2022)
	Brick factory residuals	SiO_2 nanoparticles	Drying and milling	~18	Adsorbent for heavy metals	Lashen et al. (2022)
Textile	Bleached cotton and dyed denim	Nanocellulose-reinforced polypropylene composites	Integrated nanofibrillation and compounding	100–500	Interior components of automobiles and aircrafts	Liang et al. (2023)
	Chrome-tanned buffing dust	Nanofibrous carbon	Pulse pyrolysis	~40	Bitumen modifier	Murugan et al. (2020)

N/A, Not available.
[a]*Crystallite size.*

13.2.1.1 Metallurgical industry

Slag is a solid waste that is mainly generated during the distilling and refining of metals and nonmetals in the metallurgical industry (Habib et al., 2020). For that reason, it contains several important metals and is therefore regarded as a secondary source of metals. Nickel slag is produced during the air-cooling and water-quenching of a nickel melt (Qian et al., 2022). When high magnesium garnierite is used as a feedstock for nickel production, it results in the generation of high-magnesium nickel slag. According to Qian et al. (2022), the elemental composition of MgO in high-magnesium nickel slag accounts for the second largest proportion (28.61 wt.%). They then separated the Mg^{2+} from the slag by leaching and precipitation. Then, hydrothermal treatment was employed to produce the $Mg(OH)_2$ nanosheets. The produced nickel could be used for catalyst application.

Electroplating sludge is the by-product from the electroplating process in metallurgical industry, containing multiple metallic components (i.e., Ni, Cu, Zn, Cr, etc.) (Xia et al., 2020). The complex composition of electroplating sludge makes the selective recovery of metal resources become harder. For that reason, a regulator was introduced during the hydrothermal process of electroplating sludge as reported by Weng et al. (Weng et al., 2020). In their study, they extracted Ni from nickel-laden electroplating sludge and converted it into nickel ferrite nanomaterials (50–80 nm) using a regulator-assisted hydrothermal acid-washing technique. The $NiFe_2O_4$ nanoparticles obtained demonstrated a long cycle life (> 100 cycles) with insignificant capacity decay, suggesting that it is a potential material used for electrochemical Li-storage.

13.2.1.2 Mining and power plant industry

Mining industry involves the extraction and processing of minerals from the earth surface, such as iron ore and coal. The iron mining activity produces high contents of sulfide minerals and toxic metals, which are hazardous to the environment. Proper waste management system is required not only to prevent the leaching of hazardous waste to the environment but also to develop strategy to utilize this waste and transform it into valuable products. Cruz et al. (2021) extracted iron precursors from iron mine tailings and used them to synthesize magnetic hybrid cobalt ferrite nanomaterial. This nanomaterial was proved to have excellent catalytic activity (99% conversion) in reducing 4-nitrophenol and have high adsorption ability (~ 80%) toward polycyclic aromatic hydrocarbons.

Apart from iron ore, coal is another commonly used raw material in the metals industry. It can be also used as a fuel for power generation in most countries, depending on its ranking. Generally, coal can be ranked based on the amount of carbon present in the coal (Cheng et al., 2017). For that reason, Shokry et al. (2019) utilized the Maghara coal obtained from coal mining as the feedstock to synthesize nano-activated carbon (38 nm) by carbonization and NaOH activation. The obtained nano-activated carbon proved to achieve high methylene blue removal.

Coal fly ash, which is the by-product generated during the burning of coal, is produced in large quantities in most countries. Similar to the production of $Mg(OH)_2$ from nickel slag where Qian et al. (2021) conducted a series of leaching-precipitation-hydrothermal processes using brown coal fly ash to produce $Mg(OH)_2$ nanosheet. In addition, they further converted the $Mg(OH)_2$ to MgO and impregnated Fe^{3+} on it, forming a Fe-MgO catalyst nanoparticle. They then proved the Fe-MgO for its excellent catalysis for high-temperature water–gas shift reactions. Apart from Mg $(OH)_2$ nanosheet, coal fly ash is also a feedstock for the synthesis of iron oxide (α-Fe_2O_3) nanoparticles. Qian et al. (2021) recovered the α-Fe_2O_3 from the brown coal fly ash and further doped it with Al^{3+}, producing in-situ Al^{3+}-defected iron oxide nanoflakes. The nanoflakes were proved to exhibit high adsorption efficiency.

13.2.1.3 Pulp and paper industry

Lime sludge is the common residual produced from the pulp and paper industry, with calcium carbonate accounting for its largest proportion followed by silica and alumina (Singh et al., 2020). Previous study had proven the efficiency of calcium carbonate as a flocculant in water treatment (Andreoli and Sabogal-Paz, 2019). Thereupon, Amit et al. (Kumar et al., 2022) utilized the waste lime sludge by turning it into nano-sized flocculent through drying and milling processes. The results also revealed that the nanoparticles could achieve up to 86% of biomass harvesting under optimal conditions without affecting the biodiesel profile of the microalgae *Tetraselmis indica*.

13.2.1.4 Food and beverage industry

Food and beverage waste, on the other hand, is the residual generated as the result of coordination impairment between different activities and high consumer expectations for quality (Kwan et al., 2018; Fao, 2017). Despite the high sugar content in the food and beverage waste, it also contains unprocessed raw materials, which are carbonaceous

(Kwan et al., 2018). For instance, Lashen et al. (2022) used the sugar beet processing residuals, which were the by-products of the purification of beetroot juice, to synthesize nano-sized adsorbent. The waste was first air-dried, followed by mechanically grounded into nanoparticles. They found that the nano-adsorbent derived from the sugar beet residual could be removed up to 91% of Cd and 99% of Cu in water in both acidic and neutral conditions (pH 3–7). This is due to the presence of high composition of CaO (36.04%) in the nanomaterials, making it a potent metal sorbent.

13.2.1.5 Construction industry

Glass and ceramic are both common materials that are produced in a huge quantity to meet the needs of the domestic market for construction and decoration. However, both glass and ceramic will not decompose over time, resulting in the production of waste. Hussien et al. (2022) came out with the ideas of replacing the small proportion of cement used in the self-compacted concrete with the nano-sized glass and ceramic particles. The glass and ceramic wastes were mechanically grounded into nano-sized particles (80 and 60 nm, respectively). They found that the mechanical properties of self-compacted concrete improved by replacing 3% of cement with glass or ceramic nanoparticles.

Fired clay is produced as the residuals during the manufacturing of brick used for construction purposes. Generally, it contains a high amount of clay, silt, Si, and heavy metals (i.e., Al, Fe and Mn), which can be recovered and formed into nanomaterials (Shaheen et al., 2017; Lashen et al., 2022). For example, Lashen et al. (2022) mechanically grounded the brick factory residuals into nano-sized particles (18 nm) and utilized them as metal adsorbents for Cu and Cd.

13.2.1.6 Textile industry

Cotton waste is one of the major solid wastes produced in textile industry, from process of apparel cuttings, rejection for faulty products, cuttings of excess fabric (Kar et al., 2022). However, cotton is a good source of cellulose (made up of 95% of cellulose). For that reason, Liang et al. (2023) proposed to extract the cellulose from the cotton and incorporated into polypropylene to improve the strength of the polymer. The waste cotton was first pretreated mechanically to reduce the fiber size followed by nanofibrillation and compounding. The resulted cellulose-reinforced polypropylene nanocomposites showed improved properties, in terms of tensile strength, flexure strength, and impact resistance.

Chromium tanning is the common method used during the processing of leather, which resulted in the formation of chrome-tanned buffing dust (Prochon et al., 2020). Apart from chrome content, the buffing dust contains plenty of organics, proteins, and nutrients, which could be an alternative source for the production of carbon-based nanomaterials (Hittini et al., 2019; Swarnalatha et al., 2008). Murugan et al. (2020) synthesized nanofibrous carbon from chrome-tanned buffing dust by a pulse pyrolysis process. The pyrolysis was conducted in three stages with controlled temperature, heating rate, and oxygen purging to ensure stable combustion of the organic material in the solid waste without causing the conversion of Cr(III) to Cr(IV). They then incorporate the nanofibrous carbon into bitumen through the microwave heat mixing method and found that the stability and physical properties of the bitumen increased as well.

13.3 Synthesis of nanomaterials from solid waste

Various nanomaterials could be synthesized from industrial waste residues such as carbon or metallic nanoparticles, carbon nanosheets or nanotubes, nano-sized activated carbon, graphene, and nanofibers.

13.3.1 Pretreatment of industrial waste

Pretreatment is the first step before the synthesis of nanomaterials from waste. Due to different types of industrial waste being used for recycling, thus, different pretreatment methods have been introduced. Besides the chemical composition of the waste itself, the pretreatment method could affect the surface functional group and also the physicochemical properties of the final nanomaterial product. The pretreatment method can be categorized into chemical and physical methods.

13.3.1.1 Chemical method for pretreatment of waste

Chemical pretreatment is defined as the separation of compounds by heating or using strong acid, alkali, or oxidant. The purpose of using chemical pretreatment is to eliminate the presence of contaminants in the existing waste sample. For instance, some soluble salts that are present in the waste could interfere with or cause defects in final products due to their dissociation at high temperatures. Strong acid hydrolysis where concentrated acids such as hydrochloride acid (HCl), sulfuric acid (H_2SO_4), and nitric acid (HNO_3) were applied to treat solid waste materials. For instance, waste aluminum foil waste was first dissolved into aqua regia (a mixture of HNO_3 and HCl in a 1:3 molar ratio) to maintain its solution pH prior to the further process to produce nano-α-alumina powder (El-Amir et al., 2016). However, unless it is necessary, it is not recommended to use strong acid due to its corrosive nature, which needs extra care during handling, and it is necessary to neutralize and recycle the acids. Some pretreatment does not require the use of chemicals but only water. For example, ferrous sulfate waste produced by the titanium oxide (TiO_2) industry was pretreated by dissolving the waste in distilled water before it was further processed to α-Fe_2O_3 nanoparticles (Li et al., 2016). The addition of solvent into the waste materials is also one of the pretreatment methods. In some of the industrial waste such as polymer waste and plastics waste, they possessed high viscosity, which then hampered the mass and heat transfer during the process. The addition of solvent into the waste material could reduce the viscosity.

13.3.1.2 Physical method for pretreatment of waste

Physical pretreatment is defined as the breakdown of substances or particles to a smaller size. The two most easy and common pretreatment methods are grinding and milling. The grinding process involves simple tools such as a mortar and pestle. On the other hand, the milling process involves a ball mill. Ball milling process has been utilized for the production of nanomaterials from industrial waste. It could reduce the particle size to achieve a homogeneous mixture size for further processes or other applications owing to the particle size could affect the reaction between components. In the work of Moghaddasi et al. (2013), waste tires were utilized from the rubber industry for the synthesis of zinc nanoparticles. Before the preparation of zinc nanoparticles, a few times of pretreatment milling process successfully produced the rubber ash with a size less than 50 nm. In another work of synthesis of lead particles from glass waste by Xing and Zhang (2011), the glass waste was first pretreated by grounding them into small pieces, which have a size of ≤ 10 mm. The glass particles were undergoing a ball milling and sieving process to obtain the size ≤ 74 μm after wet scrubbing and ultrasonic cleaning.

13.3.2 Methods to synthesis nanomaterials from waste

Various techniques are available for the generation of nanomaterials from industrial waste. Methods that are discussed in this section are calcination, coprecipitation, hydrothermal, leaching, pyrolysis, sol–gel, catalytic carbonization, chemical vapor deposition, and exfoliation. The general steps for each method are briefly introduced followed by the pro and cons of each method and related examples that have been studied by other researchers.

13.3.2.1 Calcination method

Calcination focuses on the final oxide product, which is a process to change its structure. Different structures could form with different calcination temperatures. Fig. 13.2A shows the calcination method conducted in a lab furnace by introducing heat to remove pollutants and contaminants, to oxidize organic fractions, to eliminate moisture content, and to decalcify at desired high temperature. A simple metal oxide was formed when the calcination temperature is less than 400°C, whereas spinel oxide was formed when the calcination temperature is more than 600°C. Therefore, different structures of nanomaterials could be formed by manipulating the calcination temperature. At lower calcined temperatures, materials tend to convert into layered double hydroxide. However, when the material is soaked in water, the material will be reconstructed. On the other hand, at higher calcined temperatures, the formation of spinel is irrecoverable. For example, the preparation of aluminum oxide (Al_2O_3) from sludge (Singh et al., 2014) and the recovery of chromium (Cr) from electroplating sludge (Wang et al., 2022). Fig. 13.2B shows the pilot-scale installation for flash calcination method and the enlargement on calciner unit. In this technique, the materials are exposed rapidly to air under extreme high temperatures (600°C–1200°C) in the calciner unit. The flash process is instantaneous; therefore it could cause the appearance of glassy phase, which is at amorphous state.

13.3.2.2 Coprecipitation method

The coprecipitation method is commonly used for nanomaterial synthesis due to its high yield and its simplicity during operation. However, the purity of nanomaterial produced using coprecipitation is less compared with using other methods. Briefly, industrial waste that contains metal was first extracted using acid to obtain the acid-extracted metal solution. It was treated with an alkali to increase the pH to 7–12. The extracted nanoparticles' morphology and size are affected by a few factors such as the molar ratio of metal, type of extractant used, stirring speed, falling speed of alkali solution, pH, and reaction temperature. Silica nanoparticles were synthesized from biomass fly ash via a coprecipitation process. The waste was undergoing alkali extraction followed by acid treatment to neutralize the alkali-extracted silica

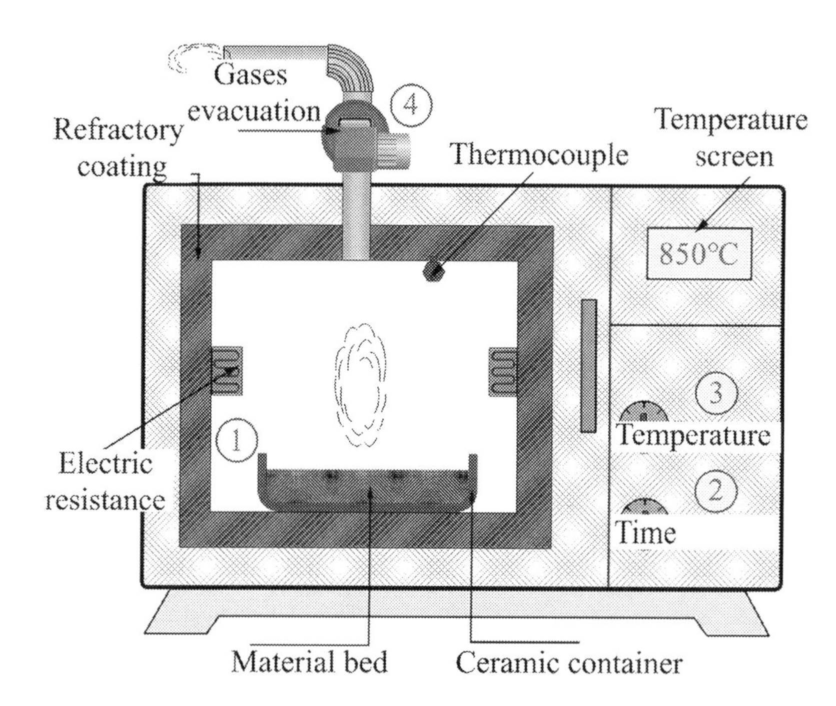

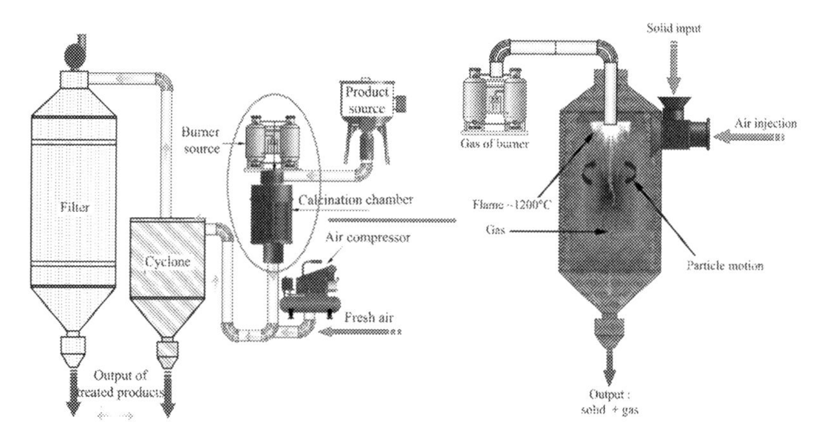

FIGURE 13.2 Schematic diagram of (A) lab furnace for calcination and (B) pilot-scale installation for flash calcination and zoomed in for calciner unit (Amar et al., 2021). *Adapted with permission from Amar, M., Benzerzour, M., Kleib, J., Abriak, N.-E., 2021. From dredged sediment to supplementary cementitious material: characterization, treatment, and reuse. International Journal of Sediment Research 36, 92–109. Copyright 2021 Elsevier.*

solution. The extracted silica solution that has a pH value of 7 was left aging for 12 hours and dried at a temperature of 100°C. The purity of the final silica nanoparticles obtained was in the range of 44%–94% (Razak et al., 2022). Besides, El-Amir et al. (2016) have successfully developed a green routes synthesis of nano-α-alumina powder using aluminum foil waste as raw material via the coprecipitation method.

13.3.2.3 Hydrothermal method

The hydrothermal method involves heating industrial waste at an elevated temperature at a range of 120°C–550°C and applying high pressure at a range of 20–150 bar with inert gas or oxygen. For the hydrothermal method, water was used for the process; however, when a different solvent was used instead of water, it is called the solvent thermal method where a chemical reaction is involved in the solvent-contained reactor. The hydrothermal method is efficient toward the solid waste that is plenteous in carbon. The nanomaterials synthesized by this method from waste often consist of oxygen functional groups such as the carbonyl group and hydroxyl group. The hydrothermal process has a mild reaction condition, which has no specification on the moisture content of waste material. Hence, drying samples prior to hydrothermal treatment is not necessary. Besides, the hydrothermal method involves low energy consumption. The properties of the synthesized nanomaterials method such as morphology, size, surface chemistry, and crystalline phase could be manipulated by controlling the solvent properties, reaction temperature, and pressure during hydrothermal treatment. Some examples of the hydrothermal process of converting waste to nanomaterials are the preparation of carbon-TiO_2 nanocomposite from anaerobic sludge waste (Vinitha Judith and Vasudevan, 2021), WO_3-FA nanocomposite from fly ash and semiconductors (Maria et al., 2016), and magnetic biochar from ferric and biological sludge waste (Zhang et al., 2018).

13.3.2.4 Sol–gel method

The sol–gel method is employed to produce solid material from small molecules. It was commonly applied to the synthesis of metal oxide nanomaterial from waste especially silicon and titanium oxide. In a sol–gel method, colloidal solutions (sol) are formed and act as a precursor for network polymer (gel). The typical precursors used in this method are metal–organic compounds and an inorganic metal salt. A few factors should be considered in the sol–gel method such as type of solvent, acid/alkali content, water content, temperature, and concentration of precursor as it could affect the gel properties. Sol–gel method is favorable in the synthesis of nanomaterial because the end product has a stable surface with a high surface area. Under a similar reaction condition, the nanomaterial synthesis via the sol–gel method has a smaller crystallite size, and it is highly crystalline compared with nanomaterial produced by the hydrothermal method. Some examples of the production of nanomaterials from industrial waste by the sol–gel method are the preparation of nano-calcium oxide (CaO) from eggshell waste (Habte et al., 2019), extraction of silica (Si) nanoparticles from palm kernel shell ash (Imoisili et al., 2020), and synthesis of MgO-fly ash composite nanomaterial from fly ash (Kumar et al., 2018). In addition, the synthesis of silica particles from several industrial solid waste with different approach such as hydrothermal, coprecipitation, and sol–gel methods has been illustrated in Fig. 13.3.

13.3.2.5 Pyrolysis method

Synthesis of nanomaterial from waste via pyrolysis route could render nanomaterial with stable structure, good surface area, excellent ion exchange capacity, and surface functional groups such as hydroxyl, carbonyl, and carboxyl groups. Changing the pyrolysis condition could change the surface properties of nanomaterials such as catalytic activity, porosity, and selectivity where the catalytic activity of nanomaterials is highly dependent on the new surface properties formed during pyrolysis treatment. For instance, the increase in temperature from 450°C to 950°C could affect the surface pH of nanomaterials (He et al., 2021). Therefore controlling the temperature and reaction time during the pyrolysis method is a key to alternating the surface chemistry of nanomaterial produced where the surface chemistry could give an impact on the water adsorption reaction. Nonetheless, during the synthesis of nanomaterial by the pyrolysis process,

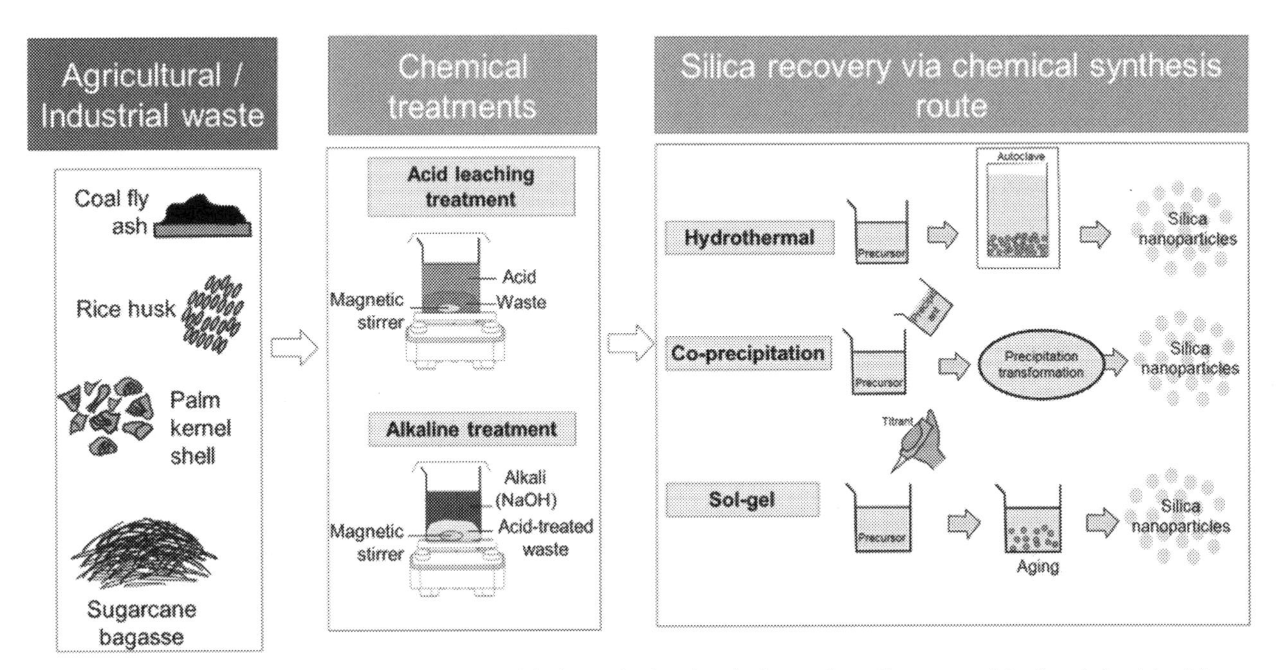

FIGURE 13.3 Typical procedure for hydrothermal, coprecipitation, and sol–gel method to produce silica nanoparticles from industrial solid waste. *Adapted with permission from Razak, N.A.A., Othman, N.H., Shayuti, M.S.M., Jumahat, A., Sapiai, N., Lau, W. J., 2022. Agricultural and industrial waste-derived mesoporous silica nanoparticles: a review on chemical synthesis route. Journal of Environmental Chemical Engineering 10, 107322. Copyright 2022 Elsevier.*

undesirable toxic gases such as lead dust and sulfur dioxide are emitted into the air, which could give an impact on the environment. The pyrolysis process system is shown in Fig. 13.4. Additionally, due to the high reaction temperature (above 600°C), the energy consumption is high for pyrolysis treatment compared with hydrothermal treatment. An example of nanomaterial synthesis using pyrolysis method is the production of zinc oxide (ZnO) from spent battery at 900°C under an argon atmosphere. The produced ZnO has a size of 50 nm with a spherical shape (Farzana et al., 2018).

13.3.2.6 Leaching

Leaching is the process of extracting materials from a solid by dissolving them in liquid. The synthesis method is popular for the synthesis of metallic nanoparticles. In the study by Fang et al. (2011), the freshly prepared sodium borohydride solutions were quickly added to the steel pickling waste materials for reaction. Then, the produced nanoparticles are washed a few times with water and alcohol to remove the excess sodium borohydride and finally iron (Fe) nanoparticles were obtained as the end product.

13.3.2.7 Catalytic carbonization

This novel method was proposed by Gong et al. (2014) to obtain a high yield of graphene by employing waste polypropylene (PP) as a carbon source as shown in Fig. 13.5. In this method, PP waste is first catalyzed with organic modified montmorillonite (OMMT), then, the mixture was heated in a furnace at 700°C to obtain carbonized char. The char is then immersed in acid to dissolve impurities and oxidized carbon. A high yield of graphene flakes was obtained after repeated centrifugation and isolation process.

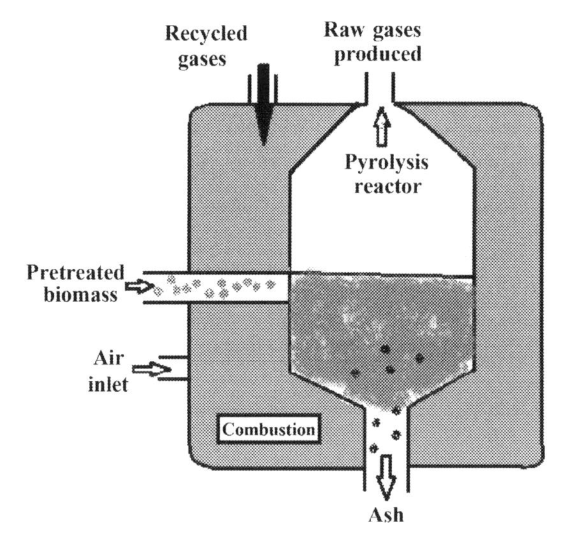

FIGURE 13.4 Schematic diagram of pyrolysis reactor for pyrolysis process (Yaashikaa et al., 2020). *Reprinted with permission from Yaashikaa, P.R., Kumar, P.S., Varjani, S., Saravanan, A., 2020. A critical review on the biochar production techniques, characterization, stability and applications for circular bioeconomy. Biotechnology Reports 28, e00570. Copyright 2020 Elsevier.*

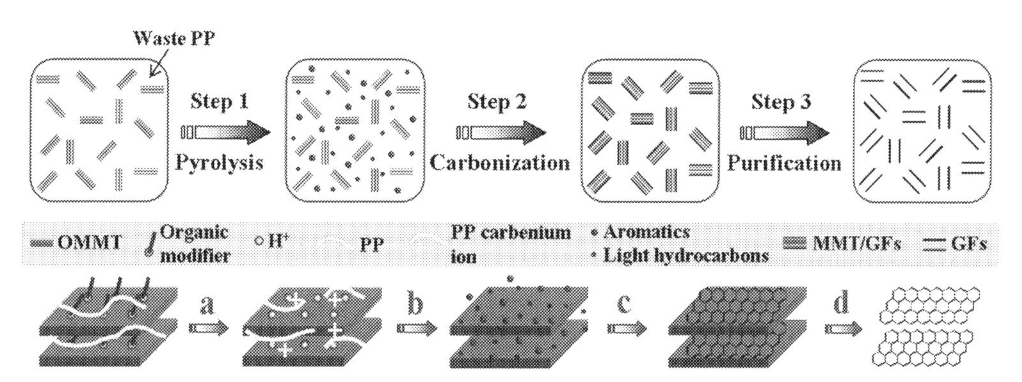

FIGURE 13.5 Schematic diagram illustrates the steps of synthesizing graphene nanoparticles via catalytic carbonization. *Reprinted with permission from Gong, J., Liu, J., Wen, X., Jiang, Z., Chen, X., Mijowska, E., et al., 2014. Upcycling waste polypropylene into graphene flakes on organically modified montmorillonite. Industrial & Engineering Chemistry Research 53, 4173–4181. Copyright 2014 American Chemical Society.*

13.3.2.8 Chemical vapor deposition method

Graphene nanomaterial can be prepared via a bottom-up approach that used the chemical vapor deposition CVD method or a top-down approach that uses liquid exfoliation. Plastic waste, cellulose waste, and agriculture biomass waste are commonly utilized as feedstock to synthesize graphene nanomaterials. The CVD method was used to produce good-quality solid material. It is widely used to deposit nanomaterials in numerous forms such as amorphous, epitaxial, monocrystalline, and polycrystalline. In a typical CVD process, the substrate is exposed to volatile precursors where a reaction takes place on the substrate surface to obtain the desired deposit. Consistently, the volatile by-product is produced during the process and the flowing gas in the reaction chamber could remove it. High purity of crystalline nanomaterial is obtained via the CVD method. The process of CVD is shown in Fig. 13.6. The unique properties of graphene such as strong mechanical strength, high electrical conductivity, high surface area, and good chemical stability (Tiwari and Syväjärvi, 2015) have made graphene suitable for any application. However, the production of graphene involves toxic reaction conditions, high cost, and nanosheet agglomeration. Hence, the usage of waste to synthesize graphene is encouraged. For instance, in the synthesis of graphene using waste plastic via chemical vapor deposition, high purity of graphene crystal was obtained. The injection rate of the polymer component of plastic waste into the furnace affects the growth of crystals in graphene. Low injection rate results in large crystals while high injection rate results in few-layer of graphene crystals (Sharma et al., 2014).

13.3.2.9 Exfoliation

Exfoliation is a process where materials enlarge by factors up to the hundreds with a unique axis, resulting in larger materials with high-temperature resistance and low density. Electrochemical exfoliation is a promising method to produce graphene from graphite. In brief, voltage is applied to drive ionic species to intercalate into graphite. Graphite will then expand and exfoliate graphene sheets. Graphene was produced using mosquito-repellent refill as a precursor via electrochemical exfoliation by extracting the anode and cathode of the graphite rod in an aqueous electrolyte (Udhaya Sankar et al., 2018). On the other hand, graphene oxide was produced using pencil cores as a precursor via electrochemical exfoliation in a sulfuric acid electrolyte solution (Liu et al., 2013). Liquid-phase exfoliation is a process to produce a single layer of graphene from graphite by overcoming the van der Waals force. For instance, graphene sheets were produced via liquid-phase exfoliation utilizing polyethylene terephthalate waste (Ko et al., 2020). Besides, liquid-phase exfoliation could produce nitrogen-doped carbon sheets by using sugarcane waste (Babu and Ramesha, 2019). To produce reduced graphene oxide, a chemical exfoliation method was used using a waste dry battery (Roy et al., 2016).

13.4 Sustainability consideration and future outlook

In the previous section, the source of industrial solid waste generation that has the potential for nanomaterials synthesis and application is reviewed. This is followed by a discussion of methods used in the preparation of nanomaterials from industrial waste. The concept of utilizing industrial waste for nanotechnology applications in various fields looks like an interesting approach, yet, fundamental concepts and knowledge gaps about waste-derived nanomaterials are significant before turning the ideas and research results from lab to industrial scale. Some researchers have proposed novel methods that are practical for the synthesis of nanomaterials from waste. Thus consideration should be taken for large-scale production. Besides, it is important to highlight the study of life cycle assessment for the development of new processes and technology that focus on aging transformation, pollution, energy consumption, and cost efficiency in the large application of waste-derived nanomaterials. By conducting the life cycle assessment, unintended consequences in the future could be minimized. Experimental data are required to perform the study of this life cycle assessment; however, most data available in the literature are results from simulation and modeling. Although utilization of waste for the production of waste is a sustainable approach, however, concerns regarding energy consumption and secondary waste

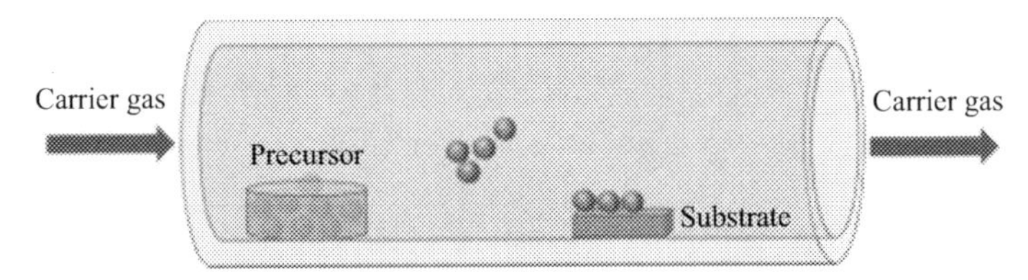

FIGURE 13.6 Schematic diagram of chemical vapor deposition method (Mittal et al., 2021). *Reprinted with permission from Mittal, M., Sardar, S., Jana, A., 2021. Chapter 7 – Nanofabrication techniques for semiconductor chemical sensors. In: Hussain, C.M., Kailasa, S.K. (Eds.), Handbook of Nanomaterials for Sensing Applications. Elsevier. Copyright 2021 Elsevier.*

generation need to be considered to avoid causing harm to the environment and human health. For example, by employing the thermal synthesis method, toxic gasses are released. It is crucial to further study the hazard and the management of the toxic gases released into the environment. Besides, the quality of final waste-derived nanoparticles such as purity and properties are often varied from one batch to another due to the different compositions and conditions of the waste material collected. This could affect the application of nanomaterials in terms of efficiency and effectiveness. For commercialization purposes, high-quality nanomaterials at low cost and reproducible are needed. Therefore, the inconsistent quality of waste-derived nanomaterials is hard to commercialize. To minimize this situation, pretreatment of the waste material is crucial to ensure the composition and condition of the waste before the process of synthesis of nanomaterials from waste. In short, researchers need to understand the complex web of influencing factors, management of the waste-derived nanomaterials, and the scope and scale of application before the implementation of waste-derived nanomaterials for benefit.

13.5 Conclusion

In this age of rapid technological advance, the development of industrialization and urbanization has made human life better than before. However, large-scale production from industries is always associated with an increased amount of waste generated. On the other hand, the nanotechnology field is growing rapidly due to its application in various fields, hence, the utilization of waste as feedstock to synthesize nanomaterials is an effective way to recover valuable compounds from waste to achieve the concept of wealth from waste. Throughout the chapter, various industrial solid wastes as starting materials for nanomaterials production have been discussed. Industrial solid wastes from different industries such as electroplating waste, steel waste, fly ash, paper mill waste, and tannery waste contain a greater amount of metals, which could be utilized and transformed into waste-derived nano-sized material. Pretreatment of waste is necessary either chemically, physically, or the combination of both prior synthesis of nanomaterial because it could affect the physicochemical properties of the nanomaterials. Different approaches to the synthesis of nanomaterials from industrial waste are also discussed in this chapter. Lastly, most of the waste-derived nanomaterials were reported along with their application, including catalysts for various chemical reactions, adsorbents for dye and heavy metal removal, antibacterial agent as well as electrode.

Acknowledgement

The authors are thankful to the financial support provided by Geran Translasional UKM (TR-UKM), UKM-TR-009.

References

Abdelbasir, S.M., Mccourt, K.M., Lee, C.M., Vanegas, D.C., 2020. Waste-derived nanoparticles: synthesis approaches, environmental applications, and sustainability considerations. Frontiers in Chemistry 8, 782.

Abdel-Shafy, H.I., Mansour, M.S., 2018. Solid waste issue: sources, composition, disposal, recycling, and valorization. Egyptian Journal of Petroleum 27, 1275–1290.

Abid, N., Khan, A.M., Shujait, S., Chaudhary, K., Ikram, M., Imran, M., et al., 2021. Synthesis of nanomaterials using various top-down and bottom-up approaches, influencing factors, advantages, and disadvantages: a review. Advances in Colloid and Interface Science 102597.

Ahmadi Tehrani, A., Omranpoor, M.M., Vatanara, A., Seyedabadi, M., Ramezani, V., 2019. Formation of nanosuspensions in bottom-up approach: theories and optimization. DARU Journal of Pharmaceutical Sciences 27, 451–473.

Amar, M., Benzerzour, M., Kleib, J., Abriak, N.-E., 2021. From dredged sediment to supplementary cementitious material: characterization, treatment, and reuse. International Journal of Sediment Research 36, 92–109.

Andreoli, F.C., Sabogal-Paz, L.P., 2019. Coagulation, flocculation, dissolved air flotation and filtration in the removal of Giardia spp. and Cryptosporidium spp. from water supply. Environmental Technology 40, 654–663.

Babu, D.B., Ramesha, K., 2019. Melamine assisted liquid exfoliation approach for the synthesis of nitrogen doped graphene-like carbon nano sheets from bio-waste bagasse material and its application towards high areal density Li-S batteries. Carbon 144, 582–590.

Bhardwaj, A.K., Naraian, R., 2021. Cyanobacteria as biochemical energy source for the synthesis of inorganic nanoparticles, mechanism and potential applications: a review. 3 Biotech 11, 445.

Bumbudsanpharoke, N., Ko, S., 2019. Nanomaterial-based optical indicators: promise, opportunities, and challenges in the development of colorimetric systems for intelligent packaging. Nano Research 12, 489–500.

Cheng, Y., Jiang, H., Zhang, X., Cui, J., Song, C., Li, X., 2017. Effects of coal rank on physicochemical properties of coal and on methane adsorption. International Journal of Coal Science & Technology 4, 129–146.

Cruz, D.R.S., Silva, I.a A., Oliveira, R.V.M., Buzinaro, M.a P., Costa, B.F.O., Cunha, G.C., et al., 2021. Recycling of mining waste in the synthesis of magnetic nanomaterials for removal of nitrophenol and polycyclic aromatic hydrocarbons. Chemical Physics Letters 771.

Eivazzadeh-Keihan, R., Maleki, A., De La Guardia, M., Bani, M.S., Chenab, K.K., Pashazadeh-Panahi, P., et al., 2019. Carbon based nanomaterials for tissue engineering of bone: building new bone on small black scaffolds: a review. Journal of advanced research 18, 185–201.

El-Amir, A.a M., Ewais, E.M.M., Abdel-Aziem, A.R., Ahmed, A., El-Anadouli, B.E.H., 2016. Nano-alumina powders/ceramics derived from aluminum foil waste at low temperature for various industrial applications. Journal of Environmental Management 183, 121–125.

Fang, Z., Qiu, X., Chen, J., Qiu, X., 2011. Degradation of the polybrominated diphenyl ethers by nanoscale zero-valent metallic particles prepared from steel pickling waste liquor. Desalination 267, 34–41.

Fao, F., 2017. The future of food and agriculture–Trends and challenges. Annual Report 296, 1–180.

Farzana, R., Rajarao, R., Behera, P.R., Hassan, K., Sahajwalla, V., 2018. Zinc oxide nanoparticles from waste Zn-C battery via thermal route: characterization and properties. Nanomaterials 8, 717.

Fiori, S., Della Pelle, F., Silveri, F., Scroccarello, A., Cozzoni, E., Del Carlo, M., et al., 2023. Nanofibrillar biochar from industrial waste as hosting network for transition metal dichalcogenides. Novel sustainable 1D/2D nanocomposites for electrochemical sensing. Chemosphere 317, 137884.

Godswill, A.C., Gospel, A.C., Otuosorochi, A.I., Somtochukwu, I.V., 2020. Industrial and community waste management: global perspective. American Journal of Physical Sciences 1, 1–16.

Gong, J., Liu, J., Wen, X., Jiang, Z., Chen, X., Mijowska, E., et al., 2014. Upcycling waste polypropylene into graphene flakes on organically modified montmorillonite. Industrial & Engineering Chemistry Research 53, 4173–4181.

Habib, A., Bhatti, H.N., Iqbal, M., 2020. Metallurgical processing strategies for metals recovery from industrial slags. Zeitschrift für Physikalische Chemie 234, 201–231.

Habte, L., Shiferaw, N., Mulatu, D., Thenepalli, T., Chilakala, R., Ahn, J.W., 2019. Synthesis of nano-calcium oxide from waste eggshell by sol-gel method. Sustainability 11, 3196.

He, M., Xu, Z., Sun, Y., Chan, P.S., Lui, I., Tsang, D.C.W., 2021. Critical impacts of pyrolysis conditions and activation methods on application-oriented production of wood waste-derived biochar. Bioresource Technology 341, 125811.

Hertwich, E.G., Gibon, T., Bouman, E.A., Arvesen, A., Suh, S., Heath, G.A., et al., 2015. Integrated life-cycle assessment of electricity-supply scenarios confirms global environmental benefit of low-carbon technologies. Proceedings of the National Academy of Sciences 112, 6277–6282.

Hittini, W., Mourad, A.-H.I., Abu-Jdayil, B., 2019. Cleaner production of thermal insulation boards utilizing buffing dust waste. Journal of Cleaner Production 236, 117603.

Hussien, R.M., Abd El-Hafez, L.M., Mohamed, R.A.S., Faried, A.S., Fahmy, N.G., 2022. Influence of nano waste materials on the mechanical properties, microstructure, and corrosion resistance of self-compacted concrete. Case Studies in Construction Materials 16.

Imoisili, P.E., Ukoba, K.O., Jen, T.-C., 2020. Green technology extraction and characterisation of silica nanoparticles from palm kernel shell ash via sol–gel. Journal of Materials Research and Technology 9, 307–313.

Ishchenko, V., 2018. Environment contamination with heavy metals contained in waste. Environmental Problems 3 (1), 21–24.

Kar, S., Santra, B., Kumar, S., Ghosh, S., Majumdar, S., 2022. Sustainable conversion of textile industry cotton waste into P-dopped biochar for removal of dyes from textile effluent and valorisation of spent biochar into soil conditioner towards circular economy. Environmental Pollution (Barking, Essex: 1987) 312, 120056.

Ko, S., Kwon, Y.J., Lee, J.U., Jeon, Y.-P., 2020. Preparation of synthetic graphite from waste PET plastic. Journal of Industrial and Engineering Chemistry 83, 449–458.

Kral, U., Morf, L.S., Vyzinkarova, D., Brunner, P.H., 2019. Cycles and sinks: two key elements of a circular economy. Journal of Material Cycles and Waste Management 21, 1–9.

Kumar, T.V., Sivasankar, V., Fayoud, N., Abou Oualid, H., Sundramoorthy, A.K., 2018. Synthesis and characterization of coral-like hierarchical MgO incorporated fly ash composite for the effective adsorption of azo dye from aqueous solution. Applied Surface Science 449, 719–728.

Kumar, N., Verma, S., Park, J., Jaiswal, A.K., Ghosh, U.K., Gautam, R., 2022. Utilization of nano-sized waste lime sludge particles in harvesting marine microalgae for biodiesel feedstock production. Nanotechnology for Environmental Engineering 7, 99–107.

Kwan, T.H., Ong, K.L., Haque, M.A., Tang, W., Kulkarni, S., Lin, C.S.K., 2018. High fructose syrup production from mixed food and beverage waste hydrolysate at laboratory and pilot scales. Food and Bioproducts Processing 111, 141–152.

Lashen, Z.M., Shams, M.S., El-Sheshtawy, H.S., Slaný, M., Antoniadis, V., Yang, X., et al., 2022. Remediation of Cd and Cu contaminated water and soil using novel nanomaterials derived from sugar beet processing-and clay brick factory-solid wastes. Journal of Hazardous Materials 128205.

Li, D., Jia, J., Zhang, Y., Wang, N., Guo, X., Yu, X., 2016. Preparation and characterization of Nano-graphite/TiO_2 composite photoelectrode for photoelectrocatalytic degradation of hazardous pollutant. Journal of Hazardous Materials 315, 1–10.

Liang, D., Liu, W., Zhong, T., Liu, H., Dhandapani, R., Li, H., et al., 2023. Nanocellulose reinforced lightweight composites produced from cotton waste via integrated nanofibrillation and compounding. Sci Rep 13, 2144.

Linley, S., Thomson, N.R., 2021. Environmental applications of nanotechnology: nano-enabled remediation processes in water, soil and air treatment. Water, Air, & Soil Pollution 232, 1–50.

Liu, J., Yang, H., Zhen, S.G., Poh, C.K., Chaurasia, A., Luo, J., et al., 2013. A green approach to the synthesis of high-quality graphene oxide flakes via electrochemical exfoliation of pencil core. Rsc Advances 3, 11745–11750.

Low, Z.H., Chen, S.K., Ismail, I., Tan, K.S., Liew, J., 2017. Structural transformations of mechanically induced top-down approach $BaFe_{12}O_{19}$ nanoparticles synthesized from high crystallinity bulk materials. Journal of Magnetism and Magnetic Materials 429, 192–202.

Lyly, L.H.T., Chang, Y.S., Ng, W.M., Lim, J.K., Derek, C.J.C., Ooi, B.S., 2021. Development of membrane distillation by dosing SiO_2-PNIPAM with thermal cleaning properties via surface energy actuation. Journal of Membrane Science 636, 119193.

Maria, V., Andreea, C., Anca, D., 2016. Nanocomposite WO_3-TiO_2/fly ash with dual functionality in simultaneous removal of pollutants from wastewater. In: 12th International Conference on Colloid and Surface Chemistry. Lasi Romania.

Mittal, M., Sardar, S., Jana, A., 2021. Chapter 7 – Nanofabrication techniques for semiconductor chemical sensors. In: Hussain, C.M., Kailasa, S.K. (Eds.), Handbook of Nanomaterials for Sensing Applications. Elsevier.

Moghaddasi, S., Khoshgoftarmanesh, A.H., Karimzadeh, F., Chaney, R.L., 2013. Preparation of nano-particles from waste tire rubber and evaluation of their effectiveness as zinc source for cucumber in nutrient solution culture. Scientia Horticulturae 160, 398–403.

Mohammed, N., Grishkewich, N., Tam, K.C., 2018. Cellulose nanomaterials: promising sustainable nanomaterials for application in water/wastewater treatment processes. Environmental Science: Nano 5, 623–658.

Murugan, K.P., Balaji, M., Kar, S.S., Swarnalatha, S., Sekaran, G., 2020. Nano fibrous carbon produced from chromium bearing tannery solid waste as the bitumen modifier. Journal of Environmental Management 270, 110882.

Obasi, P.N., Akudinobi, B.B., 2020. Potential health risk and levels of heavy metals in water resources of lead–zinc mining communities of Abakaliki, southeast Nigeria. Applied Water Science 10, 1–23.

Osorio, L.L.D.R., Flórez-López, E., Grande-Tovar, C.D., 2021. The potential of selected agri-food loss and waste to contribute to a circular economy: applications in the food, cosmetic and pharmaceutical industries. Molecules (Basel, Switzerland) 26, 515.

Prochon, M., Marzec, A., Dzeikala, O., 2020. Hazardous waste management of buffing dust collagen. Materials 13, 1498.

Qian, B., Liu, C., Lu, J., Jian, M., Hu, X., Zhou, S., et al., 2020. Synthesis of in-situ Al3 + -defected iron oxide nanoflakes from coal ash: a detailed study on the structure, evolution mechanism and application to water remediation. Journal of Hazardous Materials 395, 122696.

Qian, B., Zhang, J., Zhou, S., Lu, J., Liu, Y., Dai, B., et al., 2021. Synthesis of (111) facet-engineered MgO nanosheet from coal fly ash and its superior catalytic performance for high-temperature water gas shift reaction. Applied Catalysis A: General 618, 118132.

Qian, B., Liu, H., Ma, B., Wang, Q., Lu, J., Hu, Y., et al., 2022. Bulk trash to nano treasure: synthesis of two-dimensional brucite nanosheet from high-magnesium nickel slag. Journal of Cleaner Production 333, 130196.

Razak, N.A.A., Othman, N.H., Shayuti, M.S.M., Jumahat, A., Sapiai, N., Lau, W.J., 2022. Agricultural and industrial waste-derived mesoporous silica nanoparticles: a review on chemical synthesis route. Journal of Environmental Chemical Engineering 10, 107322.

Roy, I., Sarkar, G., Mondal, S., Rana, D., Bhattacharyya, A., Saha, N.R., et al., 2016. Synthesis and characterization of graphene from waste dry cell battery for electronic applications. RSC Advances 6, 10557–10564.

Shaheen, S.M., Shams, M.S., Khalifa, M.R., Mohamed, A., Rinklebe, J., 2017. Various soil amendments and environmental wastes affect the (im) mobilization and phytoavailability of potentially toxic elements in a sewage effluent irrigated sandy soil. Ecotoxicology and Environmental Safety 142, 375–387.

Sharma, S., Kalita, G., Hirano, R., Shinde, S.M., Papon, R., Ohtani, H., et al., 2014. Synthesis of graphene crystals from solid waste plastic by chemical vapor deposition. Carbon 72, 66–73.

Shokry, H., Elkady, M., Hamad, H., 2019. Nano activated carbon from industrial mine coal as adsorbents for removal of dye from simulated textile wastewater: operational parameters and mechanism study. Journal of Materials Research and Technology 8, 4477–4488.

Singh, S., Srivastava, V.C., Mandal, T.K., Mall, I.D., 2014. Synthesis of different crystallographic Al_2O_3 nanomaterials from solid waste for application in dye degradation. RSC Advances 4, 50801–50810.

Singh, A.P., Biswas, A., Shukla, A., Maiti, P., 2019. Targeted therapy in chronic diseases using nanomaterial-based drug delivery vehicles. Signal Transduction and Targeted Therapy 4, 1–21.

Singh, S., Singh, A., Singh, B., Vashistha, P., 2020. Application of thermo-chemically activated lime sludge in production of sustainable low clinker cementitious binders. Journal of Cleaner Production 264, 121570.

Singh, M.P., Bhardwaj, A.K., Bharati, K., Singh, R.P., Chaurasia, S.K., Kumar, S., et al., 2021. Biogenic and non-biogenic waste utilization in the synthesis of 2D materials (graphene, h-BN, g-C2N) and their applications. Frontiers in Nanotechnology 3.

Swarnalatha, S., Srinivasulu, T., Srimurali, M., Sekaran, G., 2008. Safe disposal of toxic chrome buffing dust generated from leather industries. Journal of Hazardous Materials 150, 290–299.

Tam, V.W., Soomro, M., Evangelista, A.C.J., 2018. A review of recycled aggregate in concrete applications (2000–2017). Construction and Building materials 172, 272–292.

Tiwari, A., Syväjärvi, M., 2015. Graphene Materials: Fundamentals and Emerging Applications, pp. 1-23.

Udhaya Sankar, G., Ganesa Moorthy, C., Rajkumar, G., 2018. Synthesizing graphene from waste mosquito repellent graphite rod by using electrochemical exfoliation for battery/supercapacitor applications. Energy Sources, Part A: Recovery, Utilization, and Environmental Effects 40, 1209–1214.

Vinitha Judith, J., Vasudevan, N., 2021. Synthesis of nanomaterial from industrial waste and its application in environmental pollutant remediation. Environmental Engineering Research .

Wang, Y., Xu, Y., Lin, S., Liu, W., Wang, Y., 2022. Recycling of chromium electroplating sludge using combined calcination-hydrothermal treatment: a risk-reducing strategy for separation of Cr(III) from solid waste. Surfaces and Interfaces 29, 101750.

Weng, C., Sun, X., Han, B., Ye, X., Zhong, Z., Li, W., et al., 2020. Targeted conversion of Ni in electroplating sludge to nickel ferrite nanomaterial with stable lithium storage performance. Journal of Hazardous Materials 393, 122296.

Xia, M., Muhammad, F., Li, S., Lin, H., Huang, X., Jiao, B., et al., 2020. Solidification of electroplating sludge with alkali-activated fly ash to prepare a non-burnt brick and its risk assessment. RSC Advances 10, 4640–4649.

Xing, M., Zhang, F.-S., 2011. Nano-lead particle synthesis from waste cathode ray-tube funnel glass. Journal of Hazardous Materials 194, 407–413.

Xu, P., Zeng, G.M., Huang, D.L., Feng, C.L., Hu, S., Zhao, M.H., et al., 2012. Use of iron oxide nanomaterials in wastewater treatment: a review. Science of the Total Environment 424, 1–10.

Xue, X.-Y., Cheng, R., Shi, L., Ma, Z., Zheng, X., 2017. Nanomaterials for water pollution monitoring and remediation. Environmental Chemistry Letters 15, 23–27.

Yaashikaa, P.R., Kumar, P.S., Varjani, S., Saravanan, A., 2020. A critical review on the biochar production techniques, characterization, stability and applications for circular bioeconomy. Biotechnology Reports 28, e00570.

Yang, Z., Tian, J., Yin, Z., Cui, C., Qian, W., Wei, F., 2019. Carbon nanotube-and graphene-based nanomaterials and applications in high-voltage supercapacitor: a review. Carbon 141, 467–480.

Yue, Y., Zhang, J., Sun, F., Wu, S., Pan, Y., Zhou, J., et al., 2019. Heavy metal leaching and distribution in glass products from the co-melting treatment of electroplating sludge and MSWI fly ash. Journal of Environmental Management 232, 226–235.

Zhang, H., Xue, G., Chen, H., Li, X., 2018. Magnetic biochar catalyst derived from biological sludge and ferric sludge using hydrothermal carbonization: preparation, characterization and its circulation in Fenton process for dyeing wastewater treatment. Chemosphere 191, 64–71.

Chapter 14

Nanomaterials' synthesis from the industrial solid wastes

Brahim Achiou, Doha El Machtani Idrissi, Ahlam Essate, Zakariya Chafiq Elidrissi, Youness Kouzi, Majda Breida, Mohamed Ouammou and Saad Alami Younssi

Laboratory of Membranes, Materials and Environment, Faculty of Sciences and Technologies of Mohammedia, Hassan II University of Casablanca, Mohammedia, Morocco

14.1 Introduction

Nanomaterials (NMs) are an increasingly important product of nanotechnology and they are used for a range of practical applications (Fig. 14.1). The terms nanometer and nanotechnology were firstly used in 1914 and 1959 respectively by Richard Adolf Zsigmondy and Richard Feynman (Baig et al., 2021). The British Standards Institution defined nanotechnology as the manipulation and control of matter on the nanoscale dimension. While NMs are defined as materials with any internal or external structures in the nanoscale range from 1 to 100 nm (Samaddar et al., 2018). The small size gives NMs more significant surface area and high reactivity, while it simultaneously enhances their physical, chemical, and biological properties. These special properties make them suitable for several fields such as energy, electronics, sensors, biotechnology, as well as cleanup and recycling technologies (Baig et al., 2021). NMs can have multifunctional properties depending on their composition and preparation approaches.

The synthesis of NMs includes different strategies involving bottom-up and top-down approaches (Fig. 14.2). In the top-down approach, the bulk materials are divided by means of reduction. This approach includes mechanical milling, electrospinning, lithography, sputtering, arc discharge, and laser ablation methods. The bottom-up approach involves manipulating individual atoms and molecules using several techniques, including chemical vapor deposition (CVD) as well as solvothermal, hydrothermal, sol–gel, and reverse micelle methods (Baig et al., 2021). Regardless of the selected approach, most of these synthesis methods are complicated, and use toxic precursors making the process costly, inefficient, and harmful to the environment.

The recovery and the synthesis of NMs from wastes offer benefits such as high availability, reduced cost, promotion of circular economy concept, and environmental remediation. Vast energy savings and economic benefits have been estimated from the recuperation of metals from wastes compared to conventional ores extraction. Specific examples of comparative cost-benefits are 95% for aluminum (Al), 85% for copper (Cu), 75% for zinc (Zn), and 90% for nickel (Ni) (Elsayed et al., 2020).

Waste materials applied to NMs can be classified into two major categories notably organic sources and inorganic sources. The residues have been employed for the synthesis of metallic/carbon nanoparticles (NPs), carbon nanotubes (CNTs)/nanosheets (NSs), nanoactivated carbon, nanofibers (NFs), nanocrystals (NCs), and nanowires (NWs) (Brar et al., 2022; Samaddar et al., 2018).

The chapter focuses on the methods for synthesizing NMs from industrial waste generated from mining, thermal power, battery, electronic, glass, plastic, and food industries. The findings can drive future innovations for optimizing the recent technologies to produce NMs at a marginal cost with potential properties in green and sustainable applications. Finally, potentialities and challenges of NMs' synthesis from industrial wastes were highlighted to pave way for future research.

Green and Sustainable Approaches Using Wastes for the Production of Multifunctional Nanomaterials. DOI: https://doi.org/10.1016/B978-0-443-19183-1.00003-9

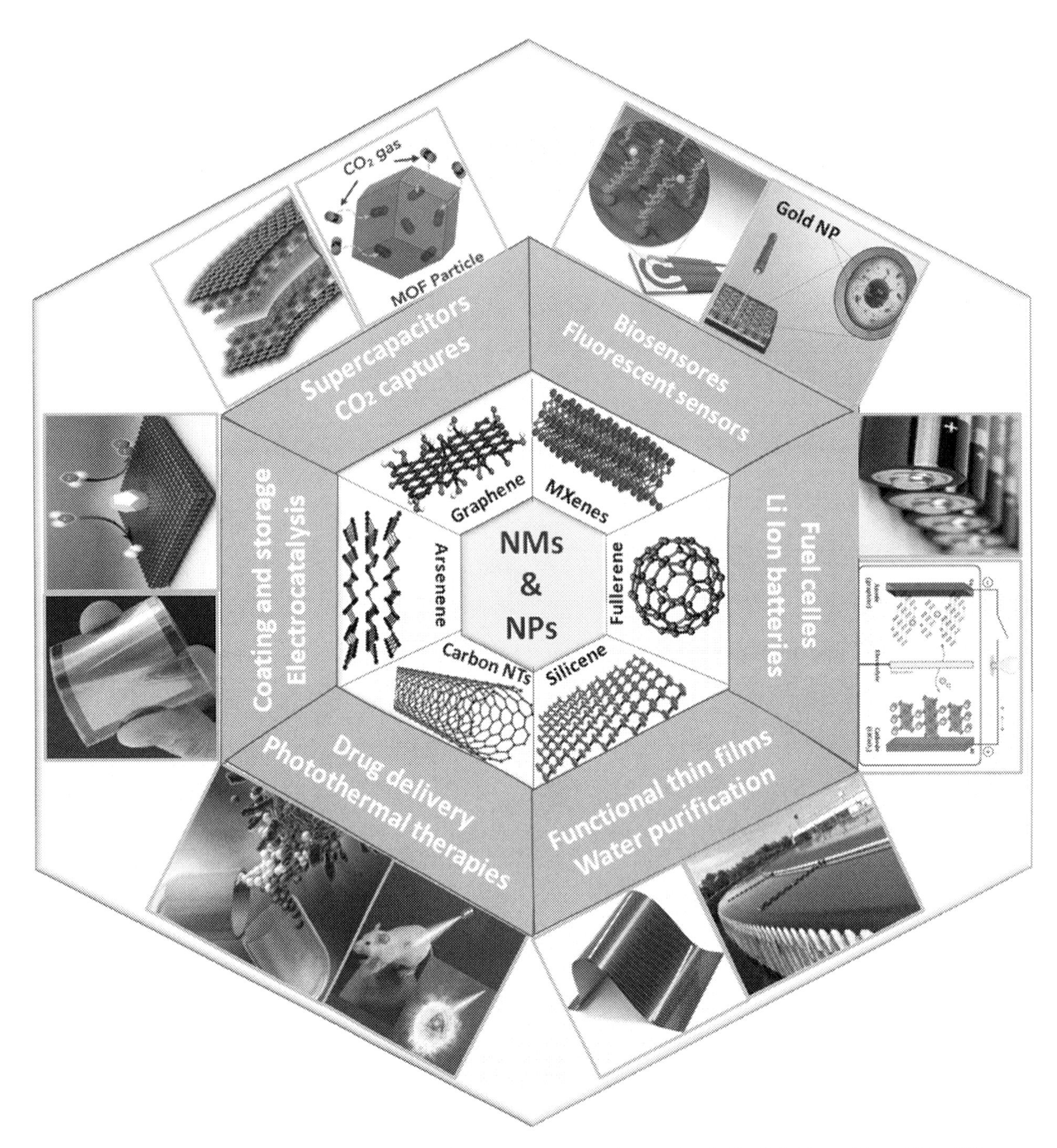

FIGURE 14.1 Nanoparticles and their applications. *Adapted from Baig, N., Kammakakam, I., Falath, W., 2021. Mater. Adv. 2, 1821–1871.*

14.2 Synthesis processes

14.2.1 Physical processes

Physical techniques are used for NMs' recovery from wastes due to their noncomplication, absence of chemical compounds, and relative selectivity to produce pure materials.

In NMs' synthesis, milling or grinding techniques are frequently employed as a physical pretreatment step and as a method in top-down approach. The method consists in the physical breakdown of big particles to directly form NMs with no further steps. This approach employs mechanical energy generated by the impact of two balls that crush the

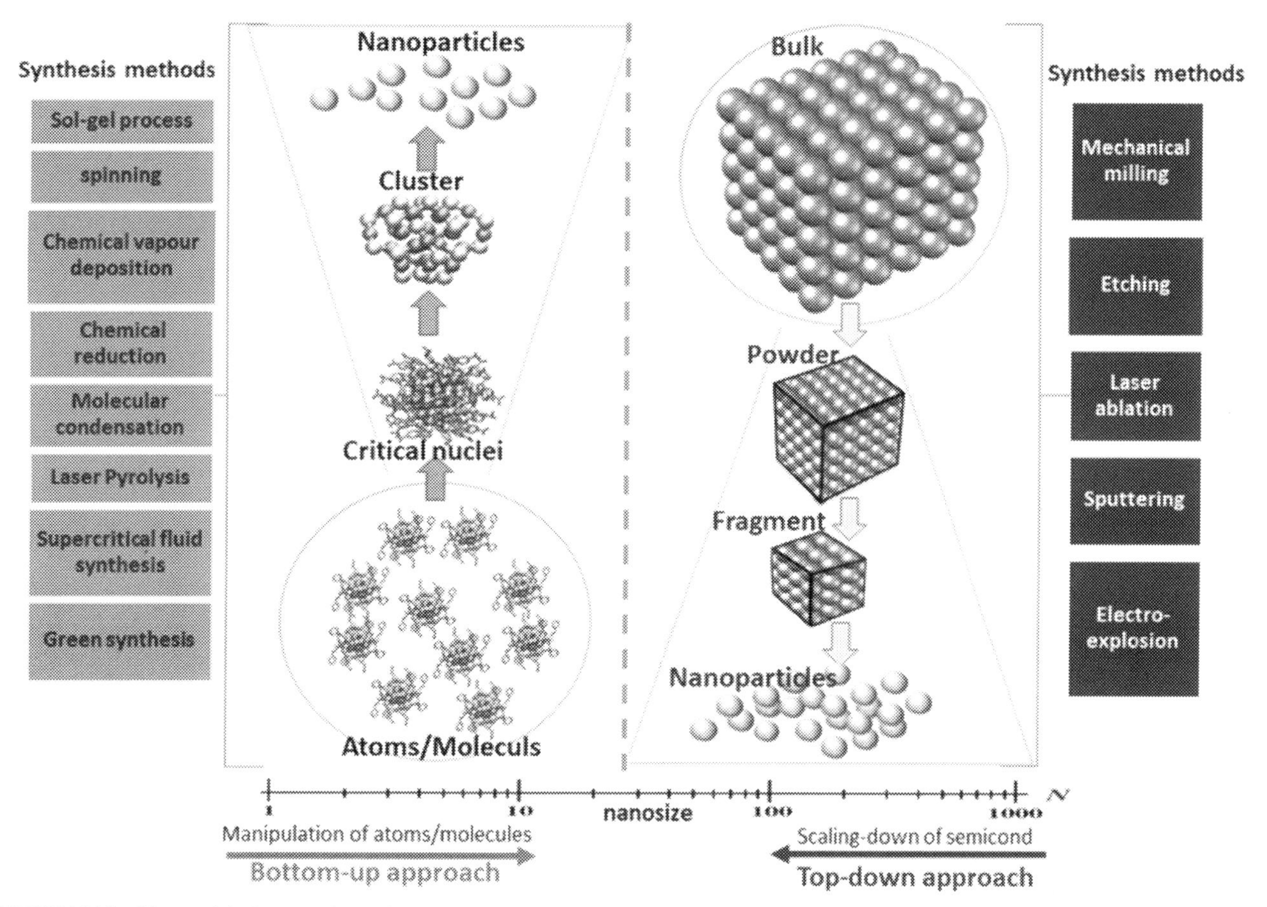

FIGURE 14.2 Nanoparticles' approaches and synthesis methods.

particles upon impact, resulting in size reduction. Generally, it is followed by a separation technique such as magnetic separation to increase purity. This method is inexpensive and simple to implement and was used for preparing nanostructured duplex and ferritic stainless steel powders with crystallite size below 10 nm via ball milling of an elemental mixture of iron (Fe), chrome (Cr), and Ni (Shashanka and Chaira, 2015).

The electric arc discharge is a long-lasting electrical discharge caused by the electrical breakdown of inert gas. The plasma is generated when a relatively strong current at a relatively low voltage runs through a nonconductive substance, such as air. This process is commonly used to generate carbon-based NMs such as carbon nanodots, CNTs, and graphene (Gr), with little mention of metal oxides. The process requires evaporating the material with an electrical arc created between graphite electrodes in a nitrogen environment, followed by vapor cooling over a substrate. The voltage could be adjusted to achieve the greatest results by altering the distance between the electrodes. This process was used to create homogeneous pure copper oxide (CuO) NPs with a diameter of 67 nm from Cu scrap by evacuating the chamber to a base pressure of 2×10^{-2} mbar and then backfilled with ambient air to 1 atm (Tharchanaa et al., 2021).

Sonication, often known as ultrasonic irradiation, is another low-cost technology for degrading particles in suspension employing high-frequency irradiation to break weak connections between the atoms. The waves that traverse through the liquid in suspension create microbubbles with local temperatures as high as 4800°C and pressures up to 500 atm. The bubbles develop and collapse, releasing smaller particles in their surroundings. This technology has been extensively utilized to extract NMs from wastes such as spherical cellulose with a diameter under 100 nm produced from cotton waste (Pandi et al., 2021).

Furthermore, inert gas condensation is a process for producing high-purity NMs, that involves evaporation and condensation phases. It generally consists in heating the metal at higher temperatures until evaporation, then transferring the vapor using inert gas to the condensation chamber where the metal particles are condensed by cold inert gas to produce pure NCs. The heating and condensing temperatures, distance between condensation and evaporation source, inert gas pressure, and condensation substrate are the primary parameters that impact the final NMs' characteristics. This technique was applied to produce Ge-monopnictides (GeAs) NMs with 55.15 nm from coal fly ash (CFA) (Zhang and Xu, 2017).

14.2.2 Chemical processes

The chemical processes are considered very efficient for the synthesis of NMs with different sizes and shapes that have resulted in a wide range of synthesis methods. For large-scale NMs' production, chemical processes produce materials with high purity and consume less energy compared to physical processes, however they require the use of hazardous chemicals (Brar et al., 2022).

CVD is considered as one of the most used methods in which vapor-phase substances condense and form a solid-phase material. This method can be split into two steps: (1) thin-film growth that involves a catalyst dispersed on support and (2) floating catalyst method in which volatile compounds are used as catalysts. For example, Sivamaran et al. synthesized successfully CNTs by CVD process using a combination of catalyst materials ($NiO/CuO/Al_2O_3$). The average particle size was varying from 50 to 80 nm for CNTs diameter (Sivamaran et al., 2022).

On the other hand, the hydrothermal method permits the preparation of materials and crystals with high performances in numerous technology fields (Feng and Li, 2017). This method involves chemical reactions between the reactant ions and/or molecules in a solution where the mixture is placed in autoclave under controlled conditions of temperature and pressure. For instance, Bensalah et al. adopted a surfactant-assisted hydrothermal reaction to synthesize nanohydroxyapatite (HAp) with a diameter of 20 nm from phosphogypsum (PG) (Bensalah et al., 2018).

Nowadays, the sol–gel method has been frequently used for NPs' synthesis due to its low cost, easy to use, and ability to produce particles with high quality (Fig. 14.3A). In sol–gel, the colloid is fabricated by precursors containing metal elements encircled by ligands.

Depending on the hydrolysis rate, the sol–gel processes are classified into two categories, (1) the polymeric and (2) the colloidal routes. The polymeric route is the most used method in which the sol is formed in alcoholic media by successive reactions of hydrolysis and polycondensation reactions. The transition of the sol–gel leads to the gel state which undertakes a drying step followed by a thermal treatment to obtain a sintered material. In contrast, hydrolysis occurs rapidly when the amount of water is large and causes the formation of a large precipitate that breaks down into small units. This technique represents the colloidal route. As an example, Sapawe et al. prepared a green silica (SiO_2) from agricultural waste via sol–gel method (Sapawe et al., 2018).

The sonochemical synthesis is one of the first approaches to easily prepare NMs with green properties (Kamali et al., 2021). The method uses a hot-spot mechanism caused by ultrasonic irradiation of a specific liquid. The mechanism involved in primary sonication is the sonochemical destruction of weak bonds within the bubbles. While the secondary sonication englobes the reactions between chemically activated species when diffusing in the precursor medium. The process was used by Sharififard et al. to prepare chitosan NMs from shrimp shells which were then used to synthesize chitosan/activated carbon/iron bionanocomposites (Sharififard et al., 2018).

The microemulsions (MEs) are thermodynamic stable systems in which nano-sized water droplets are dispersed in a continuous oil medium stabilized by surfactant and cosurfactant molecules. Further, MEs have shown interesting applications because of their low interfacial tensions, large interfacial areas, high thermodynamic stability, and ability to solubilize immiscible liquids. Mdlovu et al. recovered Cu NPs from printed circuit board wastes by the reduction of copper chloride ($CuCl_2$) in nonionic water/oil MEs with sodium borohydride ($NaBH_4$) under different temperature and pressure conditions. They obtained Cu NPs with spherical shape and diameters of 20–50 nm (Mdlovu et al., 2018).

Finally, the coprecipitation technique is widely used for the preparation of NMs. This technique is very simple, uses less energy, and has good control over the size and the homogeneity of the NMs (Abdul Razak et al., 2022). The reactions of the coprecipitation perform when nucleation, growth, coarsening, and/or agglomeration processes occur at the same time.

14.2.3 Biological processes

The biological processes are highly recommended compared to physical and chemical processes. The synthesis of NMs via biosynthesis requires less energy consumption but they are relatively slower. Waste materials containing different functional groups such as alcohol, aldehyde, and ketone are largely employed in the green fabrication of many types of metal oxide NMs (García-Quintero and Palencia, 2021). Mechanisms like reduction, enzymatic oxidation, nucleation, and peptides/polysaccharides chelation are processes adapted for the green fabrication of NMs.

The synthesis of the NMs by microorganisms (MOs) (e.g., microbes, viruses, and bacteria) is one of the most used biosynthesis processes because of the significant role of MOs in the reduction of metal ions (into different sizes) intracellularly or extracellularly (Yang et al., 2022). Different MOs are implicated in NPs biosynthesis, for instance, yeast, fungi, and algae (Brar et al., 2022). The process starts with capturing the metal ions within microbial cells or on the surface. The metals are

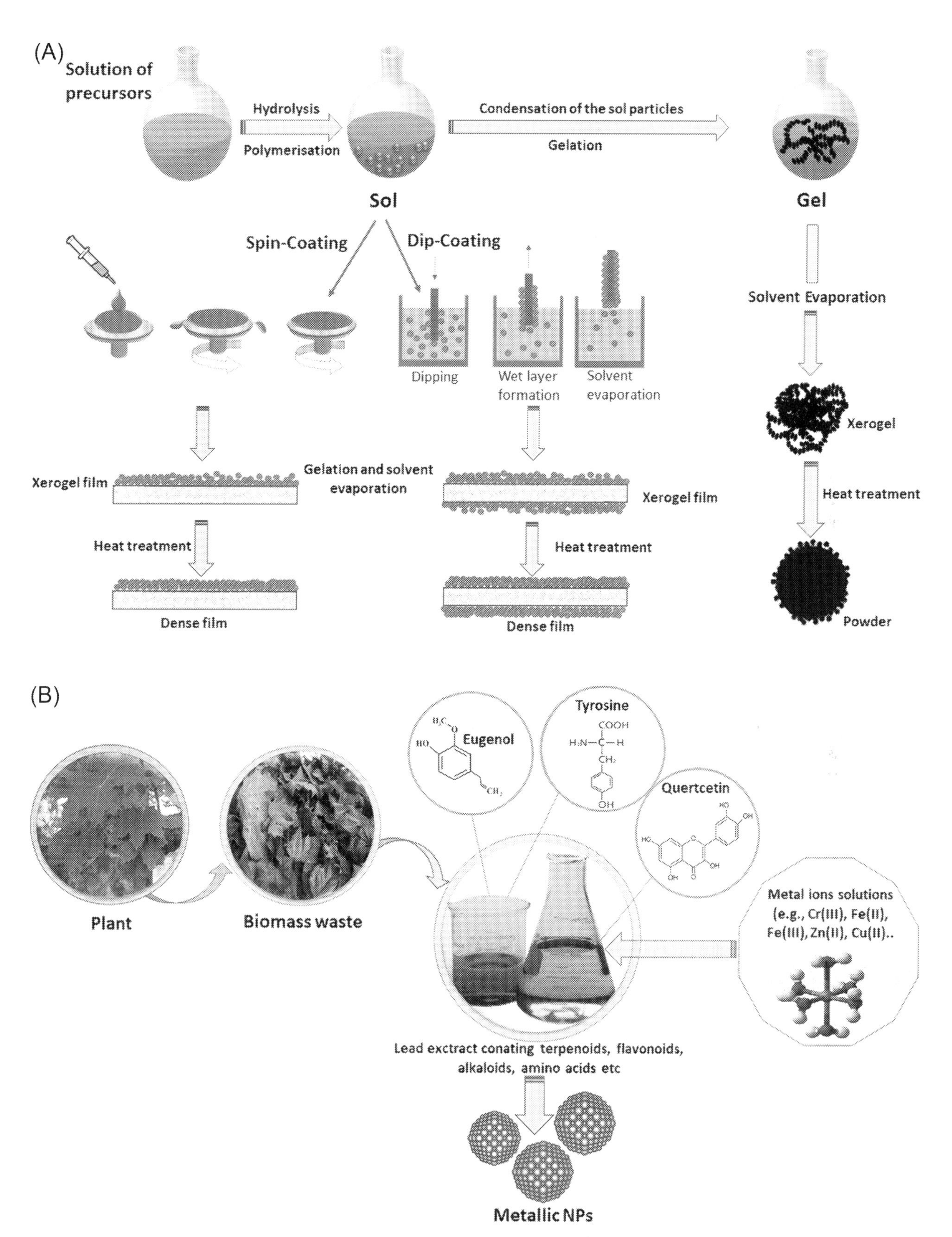

FIGURE 14.3 An overview showing (A) two sol–gel methods, (B) biological waste synthesis examples (Pandi et al, 2021). *Adapted from Baig, N., Kammakakam, I., Falath, W., 2021. Mater. Adv. 2, 1821–1871, Pandi, N., Sonawane, S.H., Anand Kishore, K., 2021. Ultrasonics Sonochemistry 70, 105353.*

then reduced/oxidized, followed by shaping, sizing, and capping of the obtained NPs with agents. These agents prevent NPs from aggregation and are bound to peptides. The final step of NPs' synthesis is the precipitation (e.g., in form of sulfides, carbonates, and phosphates) followed by volatilization through methylation, ethylation or both (Brar et al., 2022). NPs' biosynthesis using yeast can cause metals' precipitation and transformation mainly in elemental form or metal sulfides. While the majority of the fungal strains generate extracellular NPs' synthesis which makes the recovery practice eco-friendly. It is worth mentioning that the size and shape of the synthesized NPs using fungi are regulated by the variation of pH, temperature, and the presence of ammonium. The latter limits the synthesis of the NPs by preventing the release of nitrate reductase. On the other hand, low pH induces the production of Ag NPs in small size while high pH produces larger size (Brar et al., 2022).

Plant extracts present another source for NPs' synthesis with their ability for metal reduction and chelation (Fig. 14.3B). The natural-rich biowastes that consist mainly of food, agricultural byproducts, and households (leftover) are considered the most used plants (Ashrafi et al., 2022). Families of plants including terpenoids and flavonoids ensure the synthesis of NPs by the release of reactive hydrogen or by the dissociation of protons that are capable of the generation of resonance structures to assure further oxidation (García-Quintero and Palencia, 2021). Extraction of polyphenols from the peels of lemon, citrus paradise, and pomegranate fruit was applied as reducing, stabilizing, and capping agents to reduce the metal ions into metal oxide NMs. Further, agricultural wastes such as crop residues namely rice husks, corncob, and sugarcane bagasse were used as an oxide precursor giving the high content of SiO_2 (Sarkar et al., 2021). The use of plant-based synthesis is more efficient and simpler in terms of setup conditions compared to MOs. These experiments are carried out under ambient temperature and atmospheric pressure without complicated steps such as the determination of optimal nutrition and cell culture growth conditions.

In turn, due to the wide chemical compounds' variety of plants, it is challenging to determine the nature of the used biomolecules which limits the reproducibility of the process (García-Quintero and Palencia, 2021). Additionally, animal wastes are also categorized as one of the principal sources of NMs' synthesis. Eggshells and seashells are the most employed wastes in terms of agglomeration reduction and NMs' stabilization giving their high composition of calcium carbonate ($CaCO_3$) and organic matter as well as their porous structure.

14.2.4 Hybrid processes

Hybrid processes mainly involve the adoption of two or more processes in order to obtain efficient results. The synthesis of NMs with high purity and specific particle properties (e.g., morphology, size, porosity, crystallinity, colloidal stability, and surface chemistry) requires the employment of more than one process. For example, Kumar et al. performed both physical and chemical processes to synthesize magnetite NPs, using mechanical grinding followed by acid digestion. The combination of the chemical route with the mechanical milling helped to inhibit the side reaction of SiO_2 produced during the grinding. The prepared NPs have a size in the range of 4–5 nm with high purity (Kumar et al., 2015). In addition, Kumari et al. combined the carbonization and chemical activation processes for the synthesis of carbon fibers, the employment of both routes led to producing NMs with high adsorption capacity along with long thermic stability (Kumari et al., 2022). While Zhan et al. integrated the evaporation-separation and oxygen oxidation processes to recycle Zn, both chemical and physical methods were explored in this work for the purpose of achieving satisfying oxidation rate of Zn (Zhan et al., 2018).

In order to develop an environmentally clean production of NMs, the NPs synthesized by the biosynthesis route and sol–gel method are gaining wide attention. The use of bioreducing agents extracted from plants instead of chemical-reducing agents (e.g., acids and alcohol) in the sol–gel chemical process is environmentally friendly and cost-effective. For instance, Rambabu et al. succeeded to synthesize zinc oxide (ZnO) NPs using biosynthesis sol–gel method while performing ultrasonication to achieve the desired chelation temperature of the Zn ions and phytochemicals of the plants (Rambabu et al., 2021).

14.3 Inorganic waste-based nanomaterials

14.3.1 Fly ash

Fly ash (FA) is a byproduct of coal-burning in thermal power stations. The world produces about 780 million tons of FA each year (such as 39% in the United States, 47% in Europe, and 43% in India) (Gollakota et al., 2019). Worldwide, only about 10%–30% of FA is mainly used as an additive for cement concrete, filling materials, etc. The rest has to be stored in ponds or deposited in old mines and it remains a source of air, water and soil pollution. Major constituents of FA can be presented as silica (SiO_2), alumina (Al_2O_3), ferric oxide (Fe_2O_3), calcium oxide (CaO) and other impurities (Gollakota et al., 2019). Therefore, FA is recommended in a large spectrum of applications (e.g., cement concrete).

In this context, several innovative efforts were taken in the recent past for the synthesis of FA-based NMs. For instance, Yang and coworkers prepared NPs of Al_2O_3/SiO_2 from thermal power plant FA and scrap Al foil by using alkali extraction, followed by precipitation and calcination. FA was washed and digested with NaOH (2N) for 24 hours. Afterward, the solid residues were separated and registered to maximize Al recovery and silicon compounds. The filtrate was then treated with sulfuric acid (H_2SO_4) (1N) to form gelatinous precipitate of Al and silicon hydroxide. The slurry was kept for 24 hours for complete settling, then washed with water. Finally, the white gels were calcined at 973K for 4 hours to obtain NPs in the range of 6.16 to 61.14 nm (Chatterjee et al., 2019). Nanocrystalline zeolites were also prepared from FA. Murukutti et al. synthesized pure nanocrystalline zeolite type A and zeolite type X using CFA obtained from Indian thermal power plants by employing alkali fusion method followed by hydrothermal technique. The preparation of nanocrystalline zeolite type A and zeolite type X was started by adding FA to Na_2CO_3 and fused at 800°C for 2 hours. The fused sample was homogenized with 3 M NaOH, then the mixture was transferred to an autoclave for hydrothermal reaction at 100°C for 8 hours to synthesize zeolite X and at 100°C for 6 hours to obtain zeolite A. It is worthy of notice that all the zeolites present a particle size varying from 42 to 82 nm (Murukutti and Jena, 2022).

Moreover, Visa et al. fabricated FA-TiO_2 nanocomposite using the hydrothermal method (Visa et al., 2015). To remove soluble compounds, FA was washed under mechanical stirring (48 hours), filtered, and dried at 115°C. The dry FA was then mixed with nano-sized powder of TiO_2 and hexadecyltrimethyl-ammonium bromide (HTAB) as a surface-controlling agent. The resulting NMs have a size less than 74.65 nm.

Additionally, in a recent study, FA-based CNT was successfully prepared following alkali activation and film casting method by Singh et al. The starting FA was milled for 10 hours utilizing ball milling. The raw material was treated with 2 M NaOH under stirring at 80°C for 24 hours. The gained material was then grounded into fine powder and used as precursors for the fabrication of biocomposite films of polyvinyl alcohol. The obtained biocomposite was pyrolyzed at 500°C for 20 minutes. After the pyrolysis, the material was converted to a graphitized structure (CNT) with particle size varying from 20 to 30 nm (Singh et al., 2020).

14.3.2 Phosphogypsum

Recently, the quantity of PG waste increased around the world and created serious problems, especially environmental ones. Extensive research on the potential of producing NPs using PG has been conducted (Bensalah et al., 2018).

Lu and coworkers prepared nano-$CaCO_3$ using PG waste as a source of calcium by the dissolution of $CaSO_4$ (Lu et al., 2016). The preparation was done by the reactive crystallization process in which three phases are mixed—solid (PG), liquid ($NH_3{\bullet}H_2O$), and gas (CO_2)—resulting in $CaCO_3$ crystallization. PG and deionized water were mixed according to a specific weight ratio and turned into a slurry. The latter was then blended with ammonium hydroxide and a given flow rate of CO_2 was piped under setting temperature and continuous stirring. Once the pH was decreased to 8, $CaCO_3$ was collected, filtered, and dried. Moreover, the obtained $CaCO_3$ particles were amorphous with an aggregate structure in the range of 86−104 nm. In another research work, nano-$CaSO_4$ was prepared from PG through a hydrothermal process. A pretreatment consisting of washing (removal of floating organic matter), drying, and grounding of PG was done. A mixture of glycerol, ethanol, and CTAB was prepared and stirred under specific conditions. On the other hand, a mixture containing PG, nitric acid (HNO_3), and H_2SO_4 was prepared. The first mixture was slowly added to the PG mixture and stirred. Afterward, the mixture was put in a reaction kettle under a 140°C air-blast drying box for 24 hours. The effect of guiding agent on the morphology of the $CaSO_4$ crystals was investigated and the optimal conditions were found at temperature of 140°C, volume ratio of glycerol to ethanol of 2:1, and solid-liquid ratio of PG: CTAB: HNO_3:H_2SO_4 of 4:1:4:4.

The obtained NCs have a good cubic structure with sizes below 100 nm. Further, the application of the nano $CaSO_4$ in the reinforcement of linear high-density polyethylene (HDPE) indicates that the addition of 3% nano-$CaSO_4$ has a good effect (2 times higher) on the melt strength of the substrate of HDPE. Moreover, the elongation and tensile strength were increased by 31.43 and 9.16%, respectively (Gong et al., 2020).

Lately, it has been demonstrated that HAp is considered as the principal constituent of human hard tissues. The HAp was investigated in many works as constituent of bone scaffolds and teeth implants (You et al., 2022). In this way, Mousa et al. prepared successfully a nanocalcium HAp from PG waste using wet method (Mousa et al., 2016).

This is a very simple approach in which an amount of PG reacts with H_2PO_4 in an alkaline medium at room temperature and the expected reactions occurr according to Eqs. (14.1)–(14.2).

$$5CaSO_4 \cdot 2H_2O + 3H_3PO_4 \rightarrow Ca_5(PO_4)_3OH + 5H_2SO_4 + 9H_2O \quad (14.1)$$

$$10NH_4OH + 5H_2SO_4 \rightarrow 5(NH_4)_2SO_4 + 10H_2O \quad (14.2)$$

The prepared HAp showed desultory structure and agglomerated particles of rod-like morphology. The size of these particles was found in the range of 50–57 and 5–7 nm in width.

Another work has been carried out to prepare the nanocrystalline HAp by a low-cost surfactant-assisted hydrothermal method (Bensalah et al., 2018). The preparation of HAp was studied under the variation of multiple parameters such as pH (5–11), temperature (100°C–200°C), time (1–15 hours), and surfactant concentrations (0.003–0.01 mol). In this method, a defined quantity of PG was added to potassium dihydrogen phosphate (KH_2PO_4) under stirring. Afterward, nonionic Brij-93 ($C_{18}H_{35}(OCH_2CH_2)_2OH$) was added (as surfactant) to achieve the growth of crystals in one direction and to prevent the agglomeration of nanorods. Finally, the obtained mixture was put into Teflon-lined steel autoclave and hydrothermally treated under different temperatures and times. The HAp was obtained according to Eq. (14.3).

$$10(CaSO_4, \tfrac{1}{2}H_2O) + 6KH_2PO_4 \rightarrow Ca_{10}(PO_4)_6(OH)_2 + 3K_2SO_4 + 7H_2SO_4 + 3H_2O \quad (14.3)$$

The prepared HAp consists of uniform rod-like NPs with a diameter and length of 18 and 63 nm, respectively. In addition, HAp crystals have multiadsorbing sites, a large surface area, and high biocompatibility.

An attempt has been made by Ennaciri et al. in order to produce nanocalcium fluoride (CaF_2) from PG waste using recycled hydrofluoric acid (HF) and ammonia (NH_3) (Ennaciri et al., 2020). The authors found that the total conversion of PG is achieved using stoichiometric proportions of PG, HF, and sodium fluoride (NaF) during a time reaction of 1.5 hours at room temperature. The CaF_2 was obtained following Eq. (14.4).

$$CaSO_4 \cdot 2H_2O + NH_3 + 2HF \rightarrow CaF_2 + (NH_4)HSO_4 + 2H_2O \quad (14.4)$$

The characterization of the CaF_2 nanocrystalline powder confirms that the particles size is greater than 20 nm.

14.3.3 Glass wastes

In the past decade, the waste generated from glass occupies large areas and volumes and it is accounted to represent 5% of total global waste (Dadsetan et al., 2022). Glass waste has been widely exploited because of its rich SiO_2 and CaO composition and recently many researchers have been largely interested in the utilization of this silica-based waste in the synthesis of NPs, especially nanosilica synthesis. Asadi and Norouzbeigi applied the glass waste powder as an inexpensive and available SiO_2 source for the synthesis of colloidal nanosilica (Asadi and Norouzbeigi, 2018). The process of the nanosilica preparation consists mainly of reactions in the acidic and basic media, and gel preparation.

The glass powder is mixed with acid solution in a reflux system (at 80°C for 5 hours). Afterward, the settled mixture is separated through filtration and dried at 110°C for 12 hours. These residues reacted with alkaline solution of NaOH to remove impurities and separate sodium silicate (Na_2SiO_3) through a second filtration. Finally, the addition of H_2SO_4 (2.5 M) to Na_2SiO_3 solution allows the formation of SiO_2 gel. The gel was washed and distilled respectively with H_2SO_4 and water to remove sodium sulfate. The three main chemical reactions taking place are described by Eqs. (14.5)–(14.7).

$$\text{Glass powder} + HCl \rightarrow M(Cl)n + SiO_2 \quad (14.5)$$

$$SiO_2 + 2NaOH \rightarrow Na_2SiO_3 + H_2O \quad (14.6)$$

$$Na_2SiO_3 + H_2SO_4 \rightarrow SiO_2 + Na_2SO_4 + H_2O \quad (14.7)$$

The production of the optimized silica gel was possible at 4 M acid concentration, 2.5 M alkaline concentration, 2 hours stabilization duration and 90°C. The silica gel had 98.50% content of SiO_2 with an average particle size of 21.9 nm.

Additionally, a study by Owoeye et al. reutilized the waste container glasses as a precursor in the preparation of nanosized silica xerogel (Owoeye et al., 2020). It should be noted that the nano-SiO_2 synthesis in this work was realized by applying two distinct processes. The synthesis of Na_2SiO_3 was done through hydrothermal process in which a mixture of recycled waste glasses and NaOH solution (3 M) was heated up to 200°C for 3 hours under stirring. While the synthesis of silica gel was done via the sol–gel method by adding HCl (3 M) into the prepared Na_2SiO_3 solution under stirring. Particles in the range between 31 and 51 nm were successfully obtained.

On the other hand, different types of waste glasses, like conventional bottles, have been widely employed to produce NMs in order to enhance the mechanical permanence of traditional cement in building industries and thus reducing CO_2 emissions. Hamzah et al. succeeded to achieve a high porosity and durability of the mortar after replacing ground furnace blast slag with physically treated glass waste nanopowders (80 nm). The optimum nanoscale distribution of obtained glass waste particles was achieved by grinding for 7 hours using ball mill machine (Hamzah et al., 2021).

Display glass wastes like cathode ray-tube funnel glass were reutilized as a precursor in the synthesis process of NMs. Xing and Zhang proposed a novel approach that consists of combining vacuum carbon-thermal reduction with inert gas consolidation for the synthesis of nanolead particles from cathode ray-tube CRT funnel glass waste (Xing and Zhang, 2011). In the proposed process, the lead oxide (PbO) in the funnel glass was initially reduced to elemental Pb, then evaporated in circumstance and quenched to form NPs with a size ranging between 4 and 34 nm.

14.3.4 Mining wastes

In the recent years, great efforts have been made to use mining waste as a potential source for NMs' synthesis. For instance, Shirmehenji et al. proposed an eco-friendly and nontoxic process to produce selenium (Se) NPs using anode slimes collected from a local Cu-electrorefining plant. The anode slimes contain significant amounts of Se (6.11%) along with impurities [e.g., Ag, Pb, and barium sulfate ($BaSO_4$)] (Shirmehenji et al., 2021). Se synthesis process comprises consecutive steps starting with a gravity separation to remove barium sulfate, leaching of the concentrate by HNO_3 (3 M), precipitation of Ag and Pb as insoluble chloride salts, and finally purification of Se. The purified solution of Se was then used as a precursor for the green preparation of Se NPs using different fruit extracts (e.g plum, grape, orange, pomegranate, tomatoes, and lime) as reducing agents as well as stabilizers. This approach exhibits a high Se yield of 68% and a particle size varying from 70 to 80 nm.

Ewais et al. succeeded in synthesizing magnesium aluminate ($MgAl_2O_4$) spinel NPs, with high purity, through coprecipitation using industrial wastes of Al and Mg scraps (Ewais et al., 2015). The synthesis process is based on the preparation of magnesium nitrate ($Mg(NO_3)_2$) and aluminum nitrate ($Al_2(NO_3)_3$) by dissolving Mg and Al scarp respectively in HNO_3 and acidic solution containing 20% HCl and 20% HNO_3. The aqueous suspensions were formed by dropwise addition of NH_4OH for 15 minutes under stirring while pH was maintained at 10 during the precipitation. The obtained product was filtered, washed with water, and dried in an oven overnight. The dry precursor was calcined at 650°C in order to obtain ultrafine spinel powder (<50 nm).

Iron ore tailings (IOTs) are a form of solid waste produced by the mining industry and have been effectively used for the synthesis of magnetite NPs. Kumar and coworkers employed mechanical milling and chemical routes to convert IOTs into magnetite NPs' particles with size less than 100 nm (Kumar et al., 2015). Firstly, an amount of iron powder was added to IOTs (the molar concentration ratio was 1:4) in order to convert the hematite portion of the tailings into magnetite. Afterward, the mixed sample was milled for 2 hours to uniformly disperse the iron in the tailings. Finally, the acid digestion of the mixture sample was realized at 95°C.

In addition, Boruah et al. used a simple and low-cost planetary wet ball milling technique to synthesize iron oxide NPs containing 7.6 wt.% titanium dioxide (IOW) from raw waste iron oxide (IOW-R) (Boruah et al., 2020). Initially, IOW-R powder was intensively washed with deionized water to remove the impurities and inorganic ions, then dried at 120°C. To synthesize IOW NPs, the washed IOW-R powder was placed in a steel vessel with ultrapure water and an equal amount of the sample mixture was placed in another steel vessel. The two sample mixtures were then milled in a planetary ball mill for 9 hours. IOW NPs with an average size distribution of 10 nm were obtained without using any chemical treatment.

14.3.5 Batteries and electronic wastes

Electronic waste (e-waste) and end-of-life batteries have been studied as alternative low-cost and widespread source materials for NPs synthesis. The value of metal in e-waste was estimated at 52 billion U.S. dollars in 2014. Further, nearly 95 wt.% of used Pb-acid batteries are recovered for recycling Pb for new laboratory production in China (Liu et al., 2018).

Liu et al. used vacuum chlorinating and hydrothermal synthesis to produce nano-Pb sulfide dendrites (PbS) from the Pb paste in spent Pb-acid batteries (Liu et al., 2018). The waste paste contains PbO, lead dioxide (PbO_2), and lead sulfate ($PbSO_4$). The Pb paste was firstly crushed and sorted. Then, it was mixed with $CaCl_2$ and transferred to a vacuum tube furnace at a temperature ranging from 400°C to 800°C to produce lead chloride ($PbCl_2$). The author suggested the following reactions for $PbCl_2$ recovery [Eqs. (14.8)–(14.10)]. The prepared PbCl was mixed with thiourea (CH_4N_2S) as sulfide agent in water and placed in an autoclave for solvothermal synthesis at 120°C for 24 hours. The PbS recovery of 94.9% with size of 10 nm was obtained with the advantage of zero waste as all the byproducts were recycled and reused in the process again.

$$Pb + PbO_2 \rightarrow 2PbO \tag{14.8}$$

$$PbO + CaCl_2 \rightarrow PbCl_2 + CaO \tag{14.9}$$

$$PbSO_4 + CaCl_2 \rightarrow PbCl_2 + CaSO_4 \tag{14.10}$$

Zhan et al. synthesized zinc oxide (ZnO) NFs from zinc-manganese (Zn-Mg) waste batteries using high-temperature evaporation and oxygen oxidation (Zhan et al., 2018). The Zn hull part of the battery was used as Zn source and placed into a three-chamber furnace. In the first chamber, the waste was heated at 850°C to produce Zn vapor. Then, the evaporated Zn particles were transferred to the other chambers while being oxidized by the airflow. Finally, the condensation occurred in the last furnace chamber producing ZnO. The authors reported that pure ZnO NFs were prepared with less than 80 nm in diameter.

Farzana et al. produced nanospherical and cubic manganese oxide (Mn_3O_4) from zinc-carbon (Zn-C) waste batteries using heat pulverization (Farzana et al., 2019). The powder was recovered from the Zn-C waste, dried at 90°C, and pulverized in a ball milling to minimize agglomeration. The powder was then moved to a furnace under an argon atmosphere at 900°C to remove Zn impurities according to Eq. (14.11). Thereafter, the material was then cooled for 1 hour before returning to the furnace at 800°C under an air environment for oxidation [Eq. (14.12)]. The characterization confirmed the synthesis of Mn_3O_4 with a mean particle diameter of 15 nm.

$$ZnMn_2O_4 \rightarrow ZnO + 2MnO + 0.5O_2 \tag{14.11}$$

$$3MnO + 0.5O_2 \rightarrow 2Mn_3O_4 \tag{14.12}$$

Liu et al. employed multilayer ceramic capacitors (MLCCs) to electrochemically deposit nanopalladium (Pd) (Liu et al., 2020). Initially, the waste was crushed and enriched in Cu, C, SiO_2, and Al_2O_3 producing Cu-Pd-Ag-Sn allow. This latter was dissolved in HNO_3, then the insoluble tin oxide (SnO_2) was removed by filtration and Ag was transformed to AgCl by adding NaCl. Palladium nitrate ($Pd(NO_3)_2$) was obtained and used for Pd electrochemical deposition. Titanium was employed as the cathode. The deposition was done at E^0 of -0.25 V [Eq. (14.13)]. The characterization showed uniformly Pd NPs with size around 20 nm, purity higher than 99%, and recovery rate of 99.02%.

$$2Pd^{2+} + 2H_2O \rightarrow 2Pd + O_2 + 4H^+ \tag{14.13}$$

Many authors have employed hydrometallurgy for its high selectivity and excellent purity. He et al. used this technique to synthesize gold nanoparticles (AuNPs) from computer central processing units (CPUs) (He et al., 2022). To eliminate Ni and Cu, CPUs were treated with HNO_3 in an oil bath at 90°C. The solution was then filtered and the residue containing Au was dissolved in aqua regia solution. After 20 hours, the pH was adjusted to 3 and amidoximated fabrics were used to extract AuNPs from the leachate solution. The recovery was 84.32% with AuNPs' diameter of 4.41 nm.

14.4 Organic waste-based nanomaterials

14.4.1 Carbonaceous industrial waste

Carbonaceous wastes are well known for their high carbon content, making them an excellent source of carbon-derived NMs. Saikia et al. produced carbon quantum dots (CQDs) and Gr from different carbonaceous wastes (Saikia et al., 2020). Before the synthesis, waste samples were pretreated by drying, grounding, and sieving to get particles with sizes below 0.211 mm. Then, the hydrothermal technique was used for CQDs and Gr preparation. Firstly, an amount of crushed and sieved powder waste was dispersed in ultrapure water and transferred to an autoclave, and kept at 200°C for 2 hours. Afterward, the mixture was filtered by an UF membrane followed by a dialysis process to ensure the elimination of impurities. Finally, rotary evaporation was used to further concentrate the carbon NMs. The products obtained from all wastes were analyzed and showed that the CQDs produced are homogenous with globular shapes and have particle size between 3.2 and 6.5 nm and graphitization degree of around 89%. The author used fluorescence quantum yield to determine the optical characteristics of the products, the value reached 21.95% with photocatalytic degradation of 2-nitrophenols higher than 80.79%.

In a related study, Kumar et al. extracted nanodiamonds (NDs) from smoke deposits of Hawan Kund (Kumar et al., 2016). The waste was first crushed into a powder and an acid treatment using a mixture of sulfuric and fuming nitric acids (at 250°C in air for 3 hours) was done to eliminate metallic impurities. This technique is considered a risk-free economical procedure. The obtained powder was further treated by fuming HNO_3 and hydrogen peroxide mixture and kept at 150°C for 3 hours to remove carbonated compounds and isolate NDs. The characterization proved the synthesis of NDs with particles size between 4 and 5 nm. To demonstrate the environmental benefits, the authors performed a toxicity test on blood cells, and the findings demonstrated that NDs had low cytotoxicity.

Polymer wastes are a great source of carbonaceous. Mir et al. used propylene waste to produce nitrogen incorporated carbon-coated molybdenum (Mo2C@C/N) by hydrothermal route (Mir et al., 2020). The propylene sample was washed and dried before it was placed in the autoclave along with ammonium heptamolybdate tetrahydrate as a source for molybdenum/nitrogen (Mo/N) and magnesium powder used as a reducing agent. The authors found that the increase in temperature leads to a better carbonization at 800°C, resulting in more uniform and nonagglomerated particles with size more than 44 nm. The materials were used for electrochemical hydrogen production (hydrogen evolution reaction).

Furthermore, Zhang et al. yielded multiwallet carbon nanotubes (MWCNTs) while producing H_2 from truck tires waste by pyrolysis and steam reforming method (Zhang et al., 2017). Firstly, metals were removed, and then tire rubbers were shredded and transferred to a reactor containing two heating chambers. The first one is for pyrolysis where an amount of the waste is evaporated by heating the sample at 600°C for 20 minutes. The vapor was then transferred to the second chamber for steam reforming. Al_2O_3-SiO_2 prepared via sol–gel was used as a catalyst with different ratios and the reforming occurred at 800°C where H_2 was produced and then the MWCNTs were condensed with water injection. The diameter of the produced MWCNTs ranged from 11 to 36 nm with higher crystallinity using 3/5 ratio.

14.4.2 Food wastes

In the past decade, several processes including physical, chemical, biological, and hybrid were widely applied for the NMs' synthesis from food wastes. Among these processes, acid pretreatment followed by precipitation is largely reported by researchers, especially for the synthesis of silica and magnetite NPs. Recently, Rangaraj and Venkatachalam produced silica NPs from food and agroindustrial waste such as bamboo leaves characterized by high compositions of silica 50.2 wt.% (Rangaraj and Venkatachalam, 2017). A pretreatment with HCl to eliminate the impurities was realized at first, followed by alkali extraction using NaOH to obtain the SiO_2 precursor (Na_2SiO_3). After preparing the precursor of the reaction, concentrated H_2SO_4 was added to produce the SiO_2 NPs with high purity (99%) and an average particle size of 25 nm. Porrang et al. also synthesized the mesoporous nanosilica from rice and wheat husk following the same approach (Porrang et al., 2021). After performing an acid pretreatment on the selected wastes, alkali treatment was followed in order to obtain a Na_2SiO_3 solution. Finally, CTAB was added as a template to create a mesoporous structure. After aging the mixture for 24 hours and adjusting the pH by adding the NaOH, the obtained precipitation exhibited particles with an average particle size of 41.8 nm. For the synthesis of magnetite NPs from food, Akpomie and Conradie applied the thermo-chemical precipitation technique that consists of adding the washed peels of Musa acuminata to a solution of dissolved iron chloride ($FeCl_3$) and iron sulfate ($FeSO_4$) (Akpomie and Conradie, 2020). The reaction mixture was realized under constant stirring and followed by the addition of NaOH in order to obtain a nanosized particles' magnetite (25.9 nm).

Besides precipitation method, silica NPs could be synthesized via cocondensation that includes hydrolysis and condensation in one step. Rovani et al. produced the SiO_2 NPs from sugarcane waste ash based on the production of silanol groups (hydrolysis) and formation of siloxane (condensation) using H_2SO_4 in the presence of micelle maker CTAB (Rovani et al., 2018). The production of silanol groups was achieved by adding NaOH at temperature of 400 °C. This method generates silica NPs with high purity and particle size less than 20 nm.

The citrate combustion process was explored as well. Zhang et al. fabricated copper ferrite ($CuFe_2O_4$) nanocomposite from the eggshell by dispersing the prepared eggshell powder in a mixture of $Cu(NO_3)_2$, ferric nitrate along with citric acid. The obtained gel was then dried in order to obtain powder and calcinated at 500°C (Zhang et al., 2021). The obtained nanocomposite exhibits a multichannel architecture morphology and pore with size of 20 nm.

14.4.3 Plastic waste

The production of plastic waste is estimated to be more than 150 million metric tons per year and microplastics have been found in high concentration in the ocean (Abdelbasir et al., 2020). Current approaches target the insertion of plastic waste into construction materials, fiber, and paper-like composites fabrics as well as the production of new polymers.

Kumari et al. proposed an efficient methodology for the preparation of three types of low-cost activated carbon fibers (ACs) from plastic waste through carbonization and chemical activation processes (Kumari et al., 2022). Plastic wastes (in the form of plastic cups, bottles, and polybags) were cleaned with water, placed in a SiO_2 crucible, then heated in a muffle furnace at 400°C for 3 hours. The chemical activation of plastic was done by soaking an amount of waste powder in basic solution (8 M KOH) for 24 hours at ambient temperature. The obtained polybags-ACs, cups-ACs, and bottles-ACs are characterized by crystalline sizes of 13.14, 10.9, and 18.3 nm, respectively. In addition, ACs were found to possess multiple functionalities like −OH and −COOH over the adsorbent surface. The formed ACs were found to be water stable at different pH values in the range of 2−12 nm.

In another work, Veksha et al. converted flexible plastic packaging into MWCNT and pyrolysis oil (Veksha et al., 2020). Plastics pyrolysis, catalytic cracking of volatile hydrocarbons, and catalytic chemical vapor deposition (CCVD) of noncondensable pyrolysis gas represent the three successive steps in the NPs synthesis. The plastic waste along with Fe-ZSM and Ni-Ca catalysts were put into 3 reactors (a stainless-steel pyrolysis reactor, a catalytic cracking reactor, and a quartz CCVD reactor). The reactors were then purged with nitrogen (for 30 minutes) to remove air, in parallel both the catalytic cracking and CCVD reactors were respectively heated to 400°C and 700°C. To prevent condensation of pyrolysis products between the reactors, the setup connections and pipe were preheated, and the pyrolysis reactor was then heated to 550°C. The pyrolysis vapor that has gone through the catalytic cracking reactor is cooled to 0−5°C allowing the collection of oil. On the other hand, the noncondensable pyrolysis gas was used for MWCNT's synthesis in CCVD reactor. The NMs resulting have a size between 5 and 30 nm.

Catalytic pyrolysis of low-density polyethylene (LDPE) waste was used for MWCNT's synthesis (Aboul-Enein et al., 2021). The LDPE pyrolysis and CNTs growth were performed within two separated reactors in different positions (vertical and horizontal). A defined amount of LDPE and zeolite was placed in the vertical reactor while the monometallic Co/MgO catalyst was placed in the horizontal reactor. Initially, Co/MgO was reduced with H_2 gas for 1 hour then N_2 gas was used as a carrier gas. The production of noncondensable and condensable vapors inside the vertical reactor was performed at different temperatures (350°C and 600°C). Afterward, the noncondensable gases were decomposed to CNT and H_2 gas in the horizontal reactor over Co/MgO catalyst at 700°C. Finally, the carbon product was cooled to room temperature under N_2 gas flow and then purified from both catalyst and any impurities with concentrated HCl. Moreover, the size of produced CNTs is in the range of 7−40 nm.

A convenient method was adopted by Dai et al. to synthesize molybdenum carbide (Mo_2C) NPs using hydrothermal method (Dai et al., 2019). The NMs preparation involves the thermal treatment of a mixture of polyvinyl chloride (PVC) (as a carbon source), molybdenum sulfide, and metallic Na in a stainless-steel autoclave. The autoclave was heated for 10 hours under 600°C. After cooling at room temperature, the product was collected and treated respectively with dilute acid (HCl at 1 mol/L), distilled water, and alcohol to eliminate impurities. Finally, the obtained product was dried under vacuum at 60°C for 5 hours. The NMs have a particle size in the range of 50 nm and the formation of MO_2C depends on the temperature.

Another sustainable method was performed to synthesize nanostructured particles (Ag and Cu) from metalized acrylonitrile butadiene styrene plastic wastes (Elsayed et al., 2020). For the Ag NPs' synthesis, a mixture containing acidic leachate and chitosan solution was heated at 60°C. Afterward, L-ascorbic acid solution was added and stirred at 60°C. The prepared NPs were centrifuged and intensively washed with pure water and ethyl alcohol. On the other hand, bimetallic "Ag-Cu" NPs were synthesized from the acidic leachate after adjusting the pH solution to 1 using NaOH. To this solution, a mixture of L-ascorbic acid and chitosan solution was added. Ag−Cu NPs were obtained after centrifugation and water/ethanol treatment. The prepared particles have a particle size around 50−80 nm. Moreover, the synthesized NPs exhibited good antibacterial activities for both gram-positive and gram-negative bacteria.

14.5 Challenges and recommendations

A well-thought-out roadmap for the development of nanotechnology is being established by researchers and industrial partners. Future technologies depend upon how effectively materials can be manipulated on the nanoscale for various applications. Consequently, the development and effective utilization of NMs involves many challenges (Fig. 14.4).

The potential threat of NMs to both human health and environment—caused by their long-term persistence and toxicity—must be evaluated versus the beneficial opportunities of nanotechnology products. The release of NMs into the environment is more likely to happen as more products are manufactured. One of the greatest dangers associated to NMs is their accumulation and persistence in different media. However, the key parameters affecting toxicity and its behavior in environmental system (the extent to which NMs may bioaccumulate in plants or animals and thus food chain) are still uncertain.

FIGURE 14.4 Challenges in nanomaterials' synthesis.

The agglomeration of NPs is an inherent issue that causes damage of performances. This agglomeration occurs through physical entanglement, electrostatic interaction, or high surface energy. For instance, CNTs are difficult to properly align or disperse in polymer matrices as they undergo van der Waals interactions. Additionally, CNTs are associated with inflammatory and fibrogenic responses.

Similarly, Gr features and high surface areas are compromised due to severe agglomeration. The preparation of Gr hybrid materials faces challenges which lie in achieving a large contact area and short distance between materials enhancing carrier transfer. The efficiency of NMs could be enhanced via the development of 2D ultrathin materials and 3D architecture. These architectures have been applied to improve materials' inherent features, such as Gr.

Another challenge involves the production of NMs on a large scale. The presence of defects can affect the performance and the integral characteristics of NMs. In effect, NMs with high purity and free defects are produced through sophisticated devices and under harsh conditions at high prices. For this reason, efforts are required to develop new, safe, and eco-friendly synthesis techniques that overcome the challenges of conventional methods.

Many plastics and metals used for NMs' synthesis originate from waste recycling process which involves the use of hazardous substances for extraction and purification, thereby resulting in new health risks. Further, the recycling of plastic waste is challenging giving the difficulties caused by the chemical heterogeneity of packaging waste with the presence of polymer and Al together.

One of the challenges during the production of NFs using plastic wastes is impurities and contamination present in the recycled materials which induce major variation in the fiber properties. The use of polymer-based plastic wastes for carbon NPs' synthesis is challenging due to toxicity concern especially in occupational settings where the NMs are manufactured and handled in large quantities. Other wastes such as LDPE and HDPE were chemically treated and recycled to produce new polymers. However, it was found that the mechanical properties of the new polymer do not meet the conventional requirements, in addition the purification of monomers is expensive.

Printed circuit boards (PCBs) and mobile phones contain a high quantity of precious metals and a significant number of base metals. The PCBs' composition is estimated to be 40% metal, 30% organic, and 30% persistent noxious materials and they represent 3% of the total mass of electric and e-waste. However, e-waste recycling is challenging giving the complex mixture of materials (such as metals, plastic, and glass) as well as small size of components which diminishes the efficiency of conventional mechanical separation. Additionally, the high toxicity of materials (e.g., Pb and brominated flame retardants) present in PCBs is a major concern in the process of characterization and waste treatment.

The development of technology is proportionally linked with the improvement of nanotechnology. This has been proven possible by the different approaches applied for NMs' synthesis based on waste valorization. These materials are characterized by a wide range of size, properties, and demonstrated promising results in various applications.

14.6 Conclusion

By 2050, the World Bank predicts a growth of 70% of the global annual waste due to population growth, urbanization, and industrialization affecting thereby human health and environment. As a result, several studies have been directed toward the valorization of waste into NMs' synthesis via multiple processes. The selection of the synthesis process for NMs is critical and offers a great potential for diverse fields of applications. Throughout this chapter, the preparation of NMs from wastes generated from industrial activities such as thermal power, mining, plastics, fertilizers, glass, food, battery, and electronic has been exemplified and discussed. Further, various procedures including physical, chemical, and biological approaches were studied. Based on the type of waste as well as the NM's targeted parameters (e.g., size and purity), the approaches can be divided into multiple methods according to their selectivity, cost-effectiveness, and ease of use.

The chapter also aims to point out the challenges in the recovery and synthesis of highly pure NMs from industrial wastes such as the use of expensive/complicated devices, toxic chemicals, and generation of byproducts especially at high-scale production. In this context, it is expected that new synthesis processes (e.g., green approaches) will be introduced allowing the use of less energy, less toxic materials, and cost effectiveness.

References

Abdelbasir, S.M., McCourt, K.M., Lee, C.M., Vanegas, D.C., 2020. Frontiers in Chemistry 8, 782.

Abdul Razak, N.A., Othman, N.H., Mat Shayuti, M.S., Jumahat, A., Sapiai, N., Lau, W.J., 2022. Journal of Environmental Chemical Engineering 10, 107322.

Aboul-Enein, A.A., Arafa, E.I., Abdel-Azim, S.M., Awadallah, A.E., 2021. Fullerenes, Nanotubes and Carbon Nanostructures. 29, 46–57.

Akpomie, K.G., Conradie, J., 2020. The Arabian Journal of Chemistry 13, 7115–7131.

Asadi, Z., Norouzbeigi, R., 2018. Ceramics International 44, 22692–22697.

Ashrafi, G., Nasrollahzadeh, M., Jaleh, B., Sajjadi, M., Ghafuri, H., 2022. Advances in Colloid and Interface Science 301, 102599.

Baig, N., Kammakakam, I., Falath, W., 2021. Advanced Materials 2, 1821–1871.

Bensalah, H., Bekheet, M.F., Alami Younssi, S., Ouammou, M., Gurlo, A., 2018. Journal of Environmental Chemical Engineering 6, 1347–1352.

Boruah, P.K., Yadav, A., Das, M.R., 2020. Journal of Environmental Chemical Engineering. 8, 104297.

Brar, K.K., Magdouli, S., Othmani, A., Ghanei, J., Narisetty, V., Sindhu, R., et al., 2022. Environmental Research 207, 112202.

Chatterjee, A., Basu, J.K., Jana, A.K., 2019. Powder Technology 354, 792–803.

Dadsetan, S., Siad, H., Lachemi, M., Mahmoodi, O., Sahmaran, M., 2022. Journal of Cleaner Production. 342, 130931.

Dai, W., Lu, L., Han, Y., Wang, L., Wang, J., Hu, J., et al., 2019. ACS Omega 4, 4896–4900.

Elsayed, D.M., Abdelbasir, S.M., Abdel-Ghafar, H.M., Salah, B.A., Sayed, S.A., 2020. Journal of Environmental Chemical Engineering. 8, 103826.

Ennaciri, Y., Bettach, M., El Alaoui-Belghiti, H., 2020. Journal of Material Cycles and Waste Management. 22, 2039–2047.

Ewais, E.M.M., Besisa, D.H.A., El-Amir, A.A.M., El-Sheikh, S.M., Rayan, D.E., 2015. Journal of Alloys and Compounds 649, 159–166.

Farzana, R., Hassan, K., Sahajwalla, V., 2019. Scientific Reports 9, 8982.

Feng, S.-H., Li, G.-H., 2017. Hydrothermal and solvothermal syntheses. Modern Inorganic Synthetic Chemistry. Elsevier, pp. 73–104.

García-Quintero, A., Palencia, M., 2021. Science of the Total Environment. 793, 148524.

Gollakota, A.R.K., Volli, V., Shu, C.-M., 2019. Science of the Total Environment 672, 951–989.

Gong, S., Li, X., Song, F., Lu, D., Chen, Q., 2020. ACS Sustainable Chemistry & Engineering. 8, 4511–4520.

Hamzah, H.K., Huseien, G.F., Asaad, M.A., Georgescu, D.P., Ghoshal, S.K., Alrshoudi, F., 2021. Case Studies in Construction Materials. 15, e00775.

He, Y., Wang, M., Mao, X., Zhang, M., Feng, X., Ji, Z., et al., 2022. Radiation Physics and Chemistry 194, 110006.

Kamali, M., Dewil, R., Appels, L., Aminabhavi, T.M., 2021. Chemosphere 276, 130146.

Kumar, R., Sakthivel, R., Behura, R., Mishra, B.K., Das, D., 2015. Journal of Alloys and Compounds 645, 398–404.

Kumar, V., Srivastava, A.K., Toyoda, S., Kaur, I., 2016. Fullerenes, Nanotubes and Carbon Nanostructures 24, 190–194.

Kumari, M., Chaudhary, G.R., Chaudhary, S., Umar, A., 2022. Chemosphere 294, 133692.

Liu, K., Liang, S., Wang, J., Hou, H., Yang, J., Hu, J., 2018. ACS Sustainable Chemistry & Engineering 6, 17333–17339.

Liu, Y., Zhang, L., Song, Q., Xu, Z., 2020. Journal of Cleaner Production 257, 120370.

Lu, S.Q., Lan, P.Q., Wu, S.F., 2016. Industrial & Engineering Chemistry Research 55, 10172–10177.

Mdlovu, N.V., Chiang, C.-L., Lin, K.-S., Jeng, R.-C., 2018. Journal of Cleaner Production. 185, 781–796.

Mir, R.A., Kaur, G., Pandey, O.P., 2020. International Journal of Hydrogen Energy 45, 23908–23919.

Mousa, S.M., Ammar, N.S., Ibrahim, H.A., 2016. Journal of Saudi Chemical Society 20, 357–365.

Murukutti, M.K., Jena, H., 2022. Journal of Hazardous Materials. 423, 127085.

Owoeye, S.S., Jegede, F.I., Borisade, S.G., 2020. Materials Chemistry and Physics. 248, 122915.

Pandi, N., Sonawane, S.H., Anand Kishore, K., 2021. Ultrasonics Sonochemistry 70, 105353.

Porrang, S., Rahemi, N., Davaran, S., Mahdavi, M., Hassanzadeh, B., 2021. The European Journal of Pharmaceutical Sciences 163, 105866.

Rambabu, K., Bharath, G., Banat, F., Show, P.L., 2021. Journal of Hazardous Materials. 402, 123560.

Rangaraj, S., Venkatachalam, R., 2017. Applied Nanoscience 7, 145–153.

Rovani, S., Santos, J.J., Corio, P., Fungaro, D.A., 2018. ACS Omega 3, 2618–2627.

Saikia, M., Das, T., Dihingia, N., Fan, X., Silva, L.F.O., Saikia, B.K., 2020. Diamond and Related Materials 106, 107813.

Samaddar, P., Ok, Y.S., Kim, K.-H., Kwon, E.E., Tsang, D.C.W., 2018. Journal of Cleaner Production 197, 1190–1209.

Sapawe, N., Surayah Osman, N., Zulkhairi Zakaria, M., Amirul Shahab Syed Mohamad Fikry, S., Amir Mat Aris, M., 2018. Materials Today: Proceedings 5, 21861–21866.

Sarkar, Jit, Mridha, D., Sarkar, Joy, Orasugh, J.T., Gangopadhyay, B., Chattopadhyay, D., et al., 2021. Biocatalysis and Agricultural Biotechnology. 37, 102175.

Sharififard, H., Shahraki, Z.H., Rezvanpanah, E., Rad, S.H., 2018. Bioresource Technology 270, 562–569.

Shashanka, R., Chaira, D., 2015. Powder Technology 278, 35–45.

Shirmehenji, R., Javanshir, S., Honarmand, M., 2021. Journal of Cluster Science 32, 1311–1323.

Singh, R., Volli, V., Lohani, L., Purkait, M.K., 2020. Waste and Biomass Valorization. 11, 4957–4966.

Sivamaran, V., Balasubramanian, V., Gopalakrishnan, M., Viswabaskaran, V., Rao, A.G., 2022. Chemical Physics Impact 4, 100072.

Tharchanaa, S.B., Priyanka, K., Preethi, K., Shanmugavelayutham, G., 2021. Material Technology. 36, 97–104.

Veksha, A., Yin, K., Moo, J.G.S., Oh, W.-D., Ahamed, A., Chen, W.Q., et al., 2020. Journal of Hazardous Mater. 387, 121256.

Visa, M., Andronic, L., Duta, A., 2015. Journal of Environmental Management 150, 336–343.

Xing, M., Zhang, F.-S., 2011. Journal of Hazardous Materials. 194, 407–413.

Yang, Y., Waterhouse, G.I.N., Chen, Y., Sun-Waterhouse, D., Li, D., 2022. Biotechnology Advances 55, 107914.

You, B.C., Meng, C.E., Mohd Nasir, N.F., Mohd Tarmizi, E.Z., Fhan, K.S., Kheng, E.S., et al., 2022. Journal of Materials Research and Technology. 18, 3215–3226.

Zhan, L., Li, O., Wang, Z., Xie, B., 2018. ACS Sustainable Chemistry & Engineering. 6, 12104–12109.

Zhang, Y., Tao, Y., Huang, J., Williams, P., 2017. Waste Management & Research: The Journal for a Sustainable Circular Economy. 35, 1045–1054.

Zhang, L., Xu, Z., 2017. Scientific Reports 7, 3641.

Zhang, Y., Chen, Y., Kang, Z.-W., Gao, X., Zeng, X., Liu, M., et al., 2021. Colloids and Surfaces A: Physicochemical and Engineering Aspects 612, 125874.

Chapter 15

Green synthesis of nanomaterials from plant resources: its properties and applications

Rajashree Bhuyan[1], Palakshi Bordoloi[2], Jitendra Singh Verma[2], Kulbhushan Samal[3] and Sachin Rameshrao Geed[2]

[1]*Kalinga Institute of Industrial Technology, Bhubaneswar, Odisha, India,* [2]*CSIR-North East Institute of Science and Technology, Jorhat, Assam, India,* [3]*CSIR- Central Mechanical Engineering Research Institute, Durgapur, West Bengal, India*

15.1 Introduction

Nanomaterials play a crucial role in the sustainable development of technologies for socioeconomic benefit. The definition of nanomaterials is the material whose average size is smaller than 100 nm, and at the atomic/molecular level, at least one dimension can be manipulated or synthesized. Nanomaterials with specific proposed properties are synthesized using physical and chemical processes (Korde et al., 2020; Singh et al., 2021). Nanotechnology is referred to as the application of science at the molecular level. In the past few decades, a tremendous expansion of the nanotechnology field has been seen, particularly in the synthesis and applications of nanomaterials in various sectors. Nanoparticles have a large specific surface area and are smaller in size, showing physiochemical properties such as structural, thermal, mechanical, electrical, biological, catalytic activity, melting point, and optical properties that differ from the bulk of the same material. The development and production of defined nanoscale structures of nanomaterials play a critical role in many developing technologies related to chemical industries, biomedical industries, pharmaceutical industries, optical devices, aerospace industries, electronics, energy, and the environment (Devatha and Thalla, 2018; Singh et al., 2021).

Initially, nanomaterials were synthesized using a mechanical method (grinding) that started with solid mass, made it smaller, and stabilized into the required size. In this method, it was not easy to achieve the predictable narrow size of nanomaterials. Chemical methods such as the sol−gel, hydrothermal, gas phase, and hydrolysis are used to synthesize the nanomaterials starting with the atomic size and fabricating into the nanoscale (Korde et al., 2020). Though, aerosol technologies, laser ablation, UV irradiation, and photochemical processes synthesize nanomaterials. These methods are costly and release harmful toxic byproducts. In addition, these methods make it difficult to control the nanomaterial's surface chemistry in terms of size, shape, and structure (Kharissova et al., 2013).

Nanomaterial biosynthesis is an approach that is ecological, biocompatible, commercial, and safe. A plant extract−mediated nanoparticle biosynthetic approach primarily reduces hazardous by-products and helps tune the size of nanomaterials. These plants' phytochemicals take part in a crucial job in nanomaterial synthesis (Devatha and Thalla, 2018). Flavonoids are the most frequent photochemical found in all plants, and they are thought to be responsible for the nanoparticles to the huge number of phytochemicals found in plant extracts. Many researchers believe those flavonoid bioactive molecules may help form metal oxide nanoparticles because nanomaterials produced using a biomaterials approach have better catalytic activity and less costly and toxic chemicals (Kharissova et al., 2013; Saravanakumar et al., 2018).

15.2 Plant-extracted bioactive molecules involved in the synthesis of nanomaterials

In general, the bioactive molecules are obtained from the plant, and plant extract involves the extraction in the water. The plant parts are first dried, ground, and cut into pieces, then deep into hot water, filtered, and stored for some time.

Green and Sustainable Approaches Using Wastes for the Production of Multifunctional Nanomaterials. DOI: https://doi.org/10.1016/B978-0-443-19183-1.00017-9

This filter has many different biological molecules depending on plant sources. Usually, phenols, terpenoids, and flavonoids are majorly present in plant extracts (Saravanakumar et al., 2018; Singh et al., 2021). The synthesis of nanoparticles from plant extracts has a large number of bioactive molecules, which indicate the reducing properties. The potential reducing agents present in plant extract are alkaloids, alcoholic compounds, amino acids, chlorophyll, enzymes, polysaccharides, proteins, and quinol (Bhardwaj and Naraian, 2021a,b). By contrast, glucosides, proteins, and polysaccharides also play an essential function in the biosynthesis of nanomaterials (Fardsadegh and Jafarizadeh-Malmiri, 2019; Yap et al., 2020). Active components in these bioactive molecules behave as regulators to diminish and stabilize nanoparticle precursors. Much literature is available on the synthesis of Ag, Cu, Au, Se, etc., nanomaterials from various plant parts among stabilizing, capping, and reducing agents found in the extract. Bioactive molecules are present in vegetable oil, tea polyphenols, Cannabis sativa and black currant, and Carpesium cernuum, among many others. Metallic nanomaterials are synthesized using free radicals generally present in vegetable oils, including cashew oil (Kumar et al., 2008). These bioactive radicals exchange through oxidative drying of oils to reduce $C_7H_5AgO_2$ and then convert to Ag nanoparticles. Ag-based nanoparticles are synthesized using polyester resin as a protection as well as fatty acids and aldehydes are stabilizing agents (Kumar et al., 2008). Table 15.1 shows the bioactive molecules present in plant leaves of various species. The fruit of the Myristica fragrans tree comes from an evergreen tree species. Metallic nanoparticles confirm that quercetin, flavonoids, and phenols, nanomaterials when the non-seed fruit part, or pericarp, was crushed, dried, water boiled with CuO or $AgNO_3$ (Bhardwaj et al., 2017; Sasidharan et al., 2020).

15.3 Plant resource for nanomaterials synthesis

In the last 30 years, the continuous development of nanoparticles from metal salt using plant extract has gained significant attention due to the simplicity of the method. The whole plant extract, extract of a different part of the plants, plant tissue, microorganisms, and marine algae have been employed in nano-based material synthesis (Makarov et al., 2014). Due to the simplicity of plant extract–mediated nanomaterial synthesis, it has gained more attention for producing nanoparticles economically than a relatively expensive whole plant extract and microorganism-based processes. The significant advantage of plant extract–mediated nanoparticle synthesis is the reduction of metal ions and the stabilization of nanoparticles using the same plant extract (Makarov et al., 2014). The plant extract source influences the characteristics of the nanoparticles due to the difference in reducing agent concentration in the extract. The plant extract–based synthesis involves mixing the plant extract with aqueous metal salt at room temperature.

The different available plants are used to synthesize nanomaterials, as given in Fig. 15.1 and Table 15.2. *Ixora brachypoda* DC leaf extract is used for Ag nanomaterial production (Bhat et al., 2021). These Ag-based nanomaterials were produced through green synthesis, 18–50 nm in size, and mostly spherical in shape. Das et al. (2019) have used the *Morus alba* L. (Mulberry) leaf extract for green Ag-based nanomaterials production. These nanoparticles can be

TABLE 15.1 Bioactive molecules are extracted from plant and plant extracts used for the nanomaterial synthesis.

Bioactive molecule	Nanoparticles	Plant	Size (nm)	Reference
Flavonoids phenols quercetin	Ag, Cu	*Myristica fragrans*	10–50	Sasidharan et al. (2020)
Flavonoids glucosides quercetin	Ag	*Allium cepa L.*	12.5	Yap et al. (2020)
Terpenoids carboxylic acid flavonoids	Ag	*Berberis vulgaris*	30–70	Behravan et al. (2019)
Phenols flavonoids proteins	Ag	*Cannabis sativa*	26.52	Chouhan and Guleria (2020)
Phenols proteins terpenoids	Ag	*Mentha pulegium*	5–50	Kelkawi Abd et al. (2017)
Phenols flavonoids terpenoids	Au	*Ribes nigrum*	6–44	Stozhko et al. (2019)
Phenols, flavonoids, terpenoids	Au	*Ribes uvacrispa*	8–47	Stozhko et al. (2019)
Polysaccharides phenols flavonoids proteins	Se	*Aloe Vera*	50	Fardsadegh and Jafarizadeh-Malmiri (2019)

FIGURE 15.1 Plants used for the synthesis of nanomaterials.

varied from 12 to 39 nm in size and spherical. Ag, Cu, and Pd metallic nanoparticles can be synthesized from *M. alba* L. fruit extract. The mulberry leaf extract was also used to synthesize Ag nanoparticles of 15–53 nm and nearly spherical particles (Adavallan and Krishnakumar, 2014). According to Keshari et al. (2021), the size of green-synthesized Ag nanoparticles was 5–40 nm when the leaf extract of the *Catharanthus roseus* (Periwinkle) plant (Fig. 15.1) was used. The Ag nanomaterial was spherical. They used leaf extract for Ag nanoparticle synthesis. Umadevi et al. (2013) used the *Lycopersicon esculentum* (Tomato) fruit extract to synthesize metallic nanoparticles. The synthesis of ZnO nanoparticles using extract from tomatoes has also been reported by Sutradhar and Saha (2016), who used the microwave-assisted process for the skin removal of red tomatoes. Then the whole mass was ground to make juice. The resulting ZnO nanomaterials were spherical in shape and 40–100 nm in size (Sutradhar and Saha, 2016). Another report by Muhsen Mohamed et al. (2020) shows the average size of synthesized Ag nanomaterials from tomato extract after applying green chemistry was found to be 9.58–72.69 nm, and they were irregular in shape (Table 15.2). Santiago et al. (2019) have reported Ag nanoparticle synthesis of <87 nm-sized from tomato leaf extract. Pods of *the Papaver somniferum* plant were powdered and soaked in methanol for 2 days. From analytical-grade gold chloride ($HAuCl_4$), gold nanoparticles were reported to form in a greener way. The morphology was spherical, and the average size of Au nanoparticles was 77 nm (Muhammad et al., 2019).

Moodley et al.'s (2018) studies show that using *Moringa oleifera* L. plant leaf extract for Ag-based nanomaterial synthesis by sunlight as the main energy source resulted in the formation of 9 and 11 nm-sized Ag nanoparticles, respectively, from fresh and freeze-dried samples. Here, the reaction mixture of leaf extract and $AgNO_3$ solution was exposed to direct sunlight for Ag nanoparticle formation. Cu nanoparticles with amorphous nature using a hydroalcoholic extract of leaves of this plant are also reported by Das et al. (2020). Marfu'ah et al. (2020) studied the ZnO-based nanoparticles with an average size of 8–10 nm when they used banana extract peel as a reducing agent to produce nanomaterial based on ZnO. An easy method was reported by Ibrahim (2015) for synthesizing green nanoparticles from banana peel extract. They could produce 23.7 nm (average) sized Ag nanomaterials from boiled bananas. *Azadirachta indica* (Neem) leaf extract reduces $AgNO_3$ to nanosilver particles. Neem trees have been used in the past for various therapeutic uses.

Ag nanoparticles can be synthesized in a very simple way from the leaf powder of the neem tree. The Ag nanoparticles are formed in the mixture of *neem* leaf powder and $AgNO_3$ solution when heated up to 90°C–100°C (Ahmed et al., 2016). Crude extracts of the *Aegle marmelos* plant can be a source of important biomolecules as the methanolic

TABLE 15.2 Plant sources for the green synthesis of nanomaterials.

Plant	Plant part used	Type of extract	Average size of nanomaterial (nm)	Related metal	Morphology	Reference
Ixora brachypoda	Leaf extract	Aqueous	18–50	Ag	Spherical	Bhat et al. (2021)
Morus alba L.	Leaf extract	Aqueous	12–39	Ag	Spherical	Das et al. (2019)
M. alba L.	Fruit extract	Aqueous	80–150 AgNP 50–200 CuNP 50–100 PdNP	Ag, Cu, and Pd	Spherical, Non regular	Razavi et al. (2020)
M. alba L.	Leaf extract	Aqueous	15–53	Au	Nearly spherical	Adavallan and Krishnakumar, (2014)
Catharanthus roseus	Leaf extract	Aqueous	5–40	Ag	Spherical	Keshari et al. (2021)
Catharantus roseus	Leaf extract	Aqueous	49	Ag	Crystalline	Al-Shmgani et al. (2017)
Lycopersicon esculentum Mill.	Fruit extract	Aqueous	10	Ag	Spherical	Umadevi et al. (2013)
L. esculentum Mill.	Fruit extract	Aqueous	40–100	ZnO	Almost spherical	Sutradhar and Saha (2016)
L. esculentum Mill.	Fruit extract	Aqueous	9.58–72.69	Ag	Irregular	Mohamed et al. (2020)
L. esculentum Mill.	Leaf extract	Aqueous	< 87	Ag	Spherical, Oval	Santiago et al. (2019)
Moringa olifera	Leaf extract	Hydroalcoholic	35.8–49.2	Cu	Amorphous	Das et al. (2020)
M. olifera	Leaf extract	Methanolic leaf extract	15–20	Au	Crystalline	Boruah et al. (2021)
M. olifera	Leaf extract	Aqueous	9–11	Ag	Crystalline	Moodley et al. (2018)
M. olifera	Leaf extract	Aqueous	70	Ag	Spherical, cuboidal, triangular	Medda et al. (2015)
M. olifera	Leaf extract	Aqueous	20–60	Au	Spherical	Chakraborty et al. (2013)
Aloe barbadensis Miller	Leaf extract	Aqueous	30–80	Ag	Spherical	Arshad et al. (2022)
A. barbadensis Miller	Leaf extract	Aqueous	20–24	Ag	Spherical	Begum et al. (2020)
A. barbadensis Miller	Leaf extract	Aqueous	70	Ag	Spherical, cuiboidal, triangular	Medda et al. (2015)
A. barbadensis Miller	Leaf extract	Hot water extraction	70.70 ± 22–192.02 ± 53	Ag	Crystalline	Tippayawat et al. (2016)

(Continued)

TABLE 15.2 (Continued)

Plant	Plant part used	Type of extract	Average size of nanomaterial (nm)	Related metal	Morphology	Reference
A. barbadensis Miller	Leaf gel	Aqueous	Ag = 12 to 40 Au = < 15	Au, Ag	Crystalline	Kamala Nalini and Vijayaraghavan (2020)
Banana	Peel extract	Aqueous	23.7	Ag	Spherical	Ibrahim (2015)
Banana	Peel extract	Aqueous	8–10	ZnO		Marfu'ah et al. (2020)
Cinnamomum camphora	Leaf extract	Aqueous	55–80 Ag 80 Au Triangular 23.4 Au sperical	Au, Ag	Spherical, triangular	Huang et al. (2007)
Emblica officinalis	Leaf extract	Aqueous	16	ZnO	Nano flakes	Gnanasangeetha and Thambavani (2014)
Azadirachta indica	Leaf extract	Aqueous	51	ZnO	Nano flowers	Gnanasangeetha and Thambavani (2014)
A. indica	Leaf extract	-	5–50	Pt	Spherical	Thirumurugan et al. (2016)
A. indica	Leaf extract	Aqueous	20–50	Ag	Spherical, triangular, cuboidal	Asimuddin et al. (2020)
Aegle marmelos	Leaf extract	Aqueous	8–10	NiO	Crystalline	Ezhilarasi et al. (2018)
A. marmelos	Fruit extract	Methanolic fruit extract	159–181	Ag	Spherical	Devi et al. (2020)
Ocimum sanctum L.	Leaf extract	Aqueous	15.0 ± 12.34	Ag	Nearly spherical	Malapermal et al. (2017)
O. sanctum L.	Leaf extract	Aqueous	23	Pt	Irregular	Soundarrajan et al. (2012)
O. sanctum L.	Leaf extract	Aqueous	2	Pt	Irregular	Prabhu and Gajendran (2017)
O. sanctum L.	Leaf extract	Aqueous	12–36	Ni	Nearly spherical	Pandian et al. (2015)
Papaver somniferum L.	Pod extract	Aqueous	48	ZnO	Crystalline	Muhammad et al. (2019)

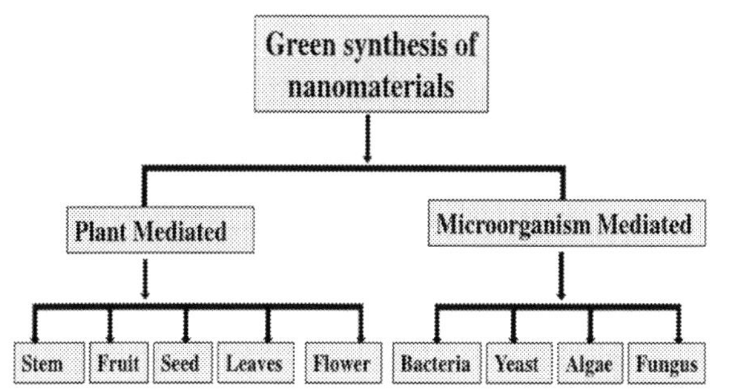

FIGURE 15.2 Green synthesis of nanomaterials from plant and microorganism mediated.

extract of fruits possessing silver nanoparticles could prevent the growth of some strong pathogenic bacteria (Devi et al., 2020). The *Ocimum sanctum* leaf extract can bind and stabilize the silver nanoparticles by amino acid residues or proteins present in the plant. Due to magnetic effects, differently shaped nanoparticles may be formed. Ni-based nanomaterials show another important use by adsorbing various dyes such as crystal violet and methyl orange and ions such as nitrate and sulfate from water (Malapermal et al., 2017). Ag nanoparticle production using *O. sanctum* aqueous leaf extract was also reported by Malapermal et al. (2017).

The group reported a stable, spherical shape of Ag nanoparticles with a size of 15.0 ± 12.34 nm; Pt-based nanoparticle synthesis using chloroplatinic acid as substrate with the *O. sanctum* plant extract was reported by Soundarrajan et al. (2012).

15.4 Green synthesis methods for nanomaterials

A green synthesis that has been common recently can be used to produce metallic nanoparticles of various dimensions, compositions, and physicochemical features. Yeasts, bacteria, molds, actinobacteria, plants, and algae, as well as their products, can all be employed to synthesize in one step. Proteins, phenolic chemicals, enzymes, alkaloids, amines, and pigments found in plants and microorganisms are reduced to form nanoparticles (Castro et al., 2014). Fig. 15.2 shows the green synthesis of nanomaterials (Peralta-Videa et al., 2016).

15.4.1 Plant-mediated synthesis

Plant extracts are employed to synthesize nanomaterials because they are easy to control and include a variety of metabolites required for reduction. By evaluating whole-plant extracts and plant tissues for manufacturing nanoparticles, plant extracts are found to be simpler, more scalable, and potentially less expensive (Castillo-Henríquez et al., 2020). Plants are known to have a variety of medicinal chemicals that have been utilized in traditional medicine since ancient times. Plant extracts have recently been mentioned in several reports for the production of nanoparticles because they are safe, easily obtainable, and encompass a wide range of biomolecules that can facilitate the synthesis of nanoparticles (such as alkaloids, quinines, phenols, flavonoids, terpenoids, tannins, and others). Secondary metabolites of plants have proven to be a valuable source of novel medicinal compounds. The basic types of plant-derived anticancer compounds include Vinca alkaloids, Camptothecin quinoline alkaloids, Taxane diterpenoids, and Epipodophyllotoxin lignans derivatives (Kuppusamy et al., 2016).

Parts of plants such as leaves, seeds, fruits, roots, and stems have also been used to synthesize nanoparticles because their extract contains phytochemicals that function as both a reducing and a stabilizing agent. After 24, 48, 72, 96, and 120 hours of nanoparticle production, UV-Vis spectrophotometer peaks of ZnO nanoparticles synthesized from Trifolium pratense flower extract were comparable, indicating that the nanomaterials were stable (Akhtar et al., 2013; Kuppusamy et al., 2016).

Plant biodiversity offers fascinating biochemical and yield-specific nanoparticle synthesis opportunities. Plants are essential sources because they include biomolecules including coenzyme and vitamin-based intermediates that can convert metal ions to nanoparticles. Like other biosynthesis approaches, plant extract–mediated synthesis was first investigated for Au and Ag nanoparticles (Parmar and Sanyal, 2022). *Acalypha indica*, Datura metel, *Jatropha curcas*, *Cassia auriculata*, etc., have all been used to make silver nanoparticles. *Aloe barbadensis* Miller, *Diopyros kaki, Medicago sativa, Pelargonium graveolens* leaves, and *Cinnamomum camphora* have also been used to make gold nanoparticles. Zn, Ni, Cu, and Co nanoparticles were produced in *M. sativa*, *Brassica juncea*, and *Helianthus annus* plants. *C. camphora, A. indica*, and *J. curcas* have also biologically generated Ag/Au bimetallic, Pd, and Pb nanoparticles (Parmar and Sanyal, 2022).

In addition to bacteria and fungi, plant extracts are great sources of metal oxide nanoparticle production. For example, plant-derived iron oxide nanoparticles have been widely manufactured. With *Camellia sinensis* (green tea) extracts, round and uneven cluster-structured Fe-based nanoparticles were synthesized (Ishak et al., 2019). Fe-based nanoparticles (unevenly shaped) of size 20–40 nm were obtained by employing extracts from three distinct species. Agricultural waste–based extracts can also be used to make iron oxide. Iron oxide nanoparticles were synthesized using *Sorghum* sp., grape seed proanthocyanidin, plantain peel extract, and banana peel ash extract (Parmar and Sanyal, 2022). Furthermore, leaf extracts of *Eucalyptus globulus, Dodonaea viscosa, pomegranate, vine leaf*, and *Tridax procumbens*, extract of pericarp from dry fruits of *Terminalia chebula*, and leaves of *Rosmarinus officinalis*, *Melaleuca nesophila*, and *Eucalyptus tereticornis* have all been employed in the production of iron oxide nanoparticles. ZnO nanoparticles have been made using extracts from Coriandrum sativum leaves, *A. indica, Calotropis gigantea, Hibiscus rosa-sinensis*, *aloe* leaf broth extract, *C. sinensis*, and *Calotropis procera* milky latex. *Neem* leaf extract, *J. curcas* latex, and *Eclipta prostrate* have also been used to make TiO_2 nanoparticles. Leaf extracts of *A. barbadensis Miller* and *Malva sylvestris* yielded CuO nanoparticles (Ishak et al., 2019; Parmar and Sanyal, 2022).

All studies involving plant leaf extracts were linked to the availability of various bioactive compounds, including alkaloids, alcoholic compounds, amino acids, and many other chelating proteins, all of which were thought to be accountable for the reduction of metal ions into nanoparticles. Extracts from leaves are a suitable source for nanoparticle manufacturing since they are high in these bioactive chemicals (Jeevanandam et al., 2016).

15.4.2 Plant parts sources for synthesis of metallic nanomaterials

15.4.2.1 Stem

Shameli et al. (2012) employed methanolic extract from the stem of *Callicarpa maingayito* to make silver nanoparticles. The aldehyde group plays an important role in the conversion of silver ions into metallic Ag-based nanomaterials. The molecular investigations on the synthesis of Ag crystal are difficult. Earlier research has provided model processes for nanoparticle interactions with pathogenic microorganisms. The Ag nanoparticles bind to the proteins present in the outer cell wall of the fungus, viral entities, and bacteria, which breaks the cell wall lipoproteins, thus halting cell division and causing the death of the cell. Vanaja et al. (2013) reported on Ag-based nanomaterials' photosynthesis using *Cissus quadrangularis* extracts at room temperature.

15.4.2.2 Fruits

Gopinath et al. (2012) employed fruit extracts of Tribulus terrestris with silver nitrate solution of varied molar concentrations to produce Ag nanomaterials. The extract's active phytochemical components are responsible for a one-step reduction reaction. T. terrestris extract produced spherical silver nanoparticles that demonstrated excellent antibacterial efficacy against multidrug-resistant human infections (Kuppusamy et al., 2016). A similar study used grape polyphenol to produce palladium nanoparticles that are beneficial to cure diseases caused by bacteria. Extract from *Rumex hymenosepalus* also serves as a reducing and stabilizing agent in producing Ag nanoparticles. In pharmaceuticals, to cure different endemic diseases, optimal physicochemical conditions to produce nanomaterials are highly successful (Kuppusamy et al., 2016).

15.4.2.3 Seeds

Flavonoids, as well as other natural bioactive compounds such as vitamins, lignin, and saponins, are abundant in fenugreek seed extract. Potent-reducing agents such as fenugreek seed extract reduce chloroauric acid function as a superior surfactant. The seed extract contains the CN, COO, and CC functional groups. The functional groups present in metabolites function as a surfactant for Au nanomaterials, and flavonoids help Au nanomaterials maintain their electrostatic stability (Vijayakumar et al., 2013; Bhardwaj and Naraian, 2021a,b).

15.4.2.4 Leaves

It was reported that extract from leaves was used as a mediator in the creation of nanoparticles. *Murraya koenigii, Centella asiatica, Alternanthera sessilis*, and a variety of other plant leaves have been researched. *P. nigrum leaves* have recently been discovered to have an important bioactive component that is used in the environmentally friendly manufacturing of nanoparticles. Ag nanoparticles are used in cancer medicine to treat a variety of cancers and other adverse disorders. Longumine and piper longminine, found in *P. nigrum* extracts, operate as a capping agent to create Ag nanoparticles and boost the tumor cell's cytotoxic effects. Vijayakumar et al. (2013) have studied the synthesis of Ag nanoparticles utilizing *Artemisia nilagirica* leaves. It demonstrates that antimicrobial drugs can be used effectively now and in the future. Similarly, Ag nanoparticles derived from plants may be used to treat several pathogenic conditions in humans.

15.4.2.5 Flowers

Noruzi et al. (2011) investigated an environmentally benign gold nanoparticle production method utilizing rose petals. Sugars and proteins abound in the extract media. The principal sources of tetrachloroaurate salt reduction into bulk Au nanoparticles are these functional compounds. Similarly, *C. roseus* and *Clitoria ternatea*, two different flower species, are employed to make metallic nanoparticles of various sizes and forms. Plant-derived nanoparticles have been shown to successfully inhibit hazardous pathogenic bacteria. Medicinally useful *Nyctanthes arbortristis* flowers of Au nanoparticles extract have also been created using a green chemistry method. The flowers extract of *Mirabilis jalapa* acts as a reducing agent, resulting in environmentally benign Au nanoparticles (Vijayakumar et al., 2013; Bhardwaj and Naraian, 2021a,b).

15.5 Advantages of green synthesis methods over chemical synthesis for nanomaterials

The advantage of green synthesis over physicochemical methods is that it is less expensive, more environmentally friendly, and easier to scale up for large-scale synthesis because it does not require high temperature, pressure, harmful chemicals, and energy. Because toxic chemicals are not used to produce nanoparticles, using environmentally benign components such as bacteria, fungi, plant extracts, and enzymes has various advantages as eco-friendly and compatible with pharmaceutical applications. These drawbacks necessitated the employment of unique and well-refined technologies, which opened the door to investigating safe and environmentally friendly nanoparticle syntheses (Ahmed et al., 2022). Biocompatibility, such as reduced metal cytotoxicity, is necessary for nanoparticles with biological uses. When compared with physicochemically produced nanoparticles, biogenic nanoparticles are devoid of hazardous contamination from by-products that attach to nanoparticles during physiochemical production, limiting the biomedical applications of the resulting nanoparticles. The advantages of biological nanoparticle synthesis include quick and environmentally friendly production methods (El-Gendy et al., 2019; Ahmed et al., 2022). Furthermore, no additional stabilizing agents are required because plant and microbe ingredients act as capping and stabilizing agents. Furthermore, when biological nanoparticles come into touch with complicated biological fluids, their surfaces gradually and selectively absorb biomolecules due to the attachment between biomolecule and nanoparticle surface. There are abundant metabolites with pharmacological action in medicinal plants that are thought to adhere to the produced nanoparticles, offering extra benefit by boosting the nanoparticles' efficacies. The biological synthesis of nanoparticles also reduces the number of steps necessary, such as attaching certain functional groups to the nanoparticle surface to make them physiologically active (Ahmed et al., 2017).

Nanoparticles have two primary basic features that make them advantageous for medication delivery. First, nanoparticles, due to their small size, may pass through capillaries and be taken up by cells, allowing for effective drug collection in target areas. Another one is that the use of biodegradable resources for nanomaterials manufacturing allows for long-term drug release to the target region. Nanoparticles are significant for a variety of reasons, not just pharmaceuticals (De Jong and Borm, 2008; Ahmed et al., 2022).

15.6 Properties of nanomaterials synthesized from plant resources

Microorganisms, plants, and other biological structures are involved in eliminating harmful and waste metals from the environment, which is accomplished by the reduction, oxidation, and catalytic action of nanoparticles (Nadaroglu et al., 2017). Nanomaterials having unique properties as optical, insulator, antimicrobial, stability, antioxidant, manipulability, biocompatibility, and biosynthesized nanomaterials are used in the biomedical field. Metallic nanoparticles with catalytic activity, which can be exploited in the industrial field, are essential nowadays (Nadaroglu et al., 2017).

15.6.1 Surface area

Nanomaterials' surface areas are often much larger than those of their bulk counterparts, and all nanomaterials share this trait.

15.6.2 Quantum effects

At the nanoscale, quantum effects are more evident. The size at which these effects arise is significantly influenced by the type of the semiconductor material.

15.6.3 Great catalyst support

2D sheets made of different nanomaterials allowed for effective dispersion of active catalyst nanoparticles, significantly improving catalytic performance. To improve performance, catalysts have recently been atomically disseminated on 2D sheets of nanomaterials.

15.6.4 Antimicrobial activity

Some nanomaterials contain antiviral, antibacterial, and antifungal characteristics, making them ideal for treating pathogen-related illnesses.

15.6.5 Electrical and optical properties

Nanoparticles' electrical and optical properties are significantly more interwoven. For instance, noble metallic nanoparticles possess optical properties based on size and a prominent UV–visible extinction band, which is lacking from the spectra of bulk metals. The LSPR (localized surface plasmon) range shows the highest wavelength depending on the dimensions and interparticle spacing of the nanomaterials, their dielectric characteristics, and those of their surroundings, which include the substrate adsorbate and solvents. Au and Ag-based nanoparticles show electrical and optical properties (Link and El-Sayed, 2003).

15.6.6 Mechanical properties

Scientists can use the unique mechanical properties of nanoparticles to use them in fields such as surface engineering, nonmanufacturing, tribology, and nanofabrication. Many mechanical metrics, such as adhesion, hardness, elastic modulus, friction, stress, strain, coagulation, surface coating, and lubrication, can be studied to find the real mechanical characteristics of nanoparticles in order to optimize their mechanical qualities. In the presence of elevated contact pressure, the rigidity difference between the nanoparticles and the outer surface in contact dictates if the nanoparticle in a lubricated touch is depressed into the plan surface, which could reveal nanoparticle behavior in a contact environment (Khan et al., 2019; Naraian and Abhishek, 2020).

15.6.7 Magnetic properties

Magnetic nanoparticles have attracted the attention of scientists working in diverse fields such as catalysis, magnetic fluids, MRI (magnetic resonance imaging), data storage, biomedicine, and remediation such as water purification. As per the research, nanoparticles operate best when their size is less than the critical range, that is, between 10 and 20 nm. The magnetic characteristics of nanomaterials are efficiently dominated at such low scales, rendering these particles worthy and applicable in various ways. The magnetism of nanoparticles is because of their unequal electrical distribution. These features are also influenced by artificial methods and procedures such as coprecipitation, solvothermal, microemulsion, flame spray, and thermal breakdown synthesis (Banejee et al., 2010; Bhardwaj and Naraian, 2021a,b).

15.6.8 Thermal properties

Metal nanoparticles possess superior thermal conductivities than solid-state fluids. At room temperature, Cu possesses thermal conductivity, 700 times that of water. Oxides such as Al_2O_3 have a higher thermal conductivity than water. Hence, the thermal conductivity of liquids with dispersed solid materials is much more excellent than those of typical heat transfer liquids. Dispersion of solid particles in a liquid such as water, oils, or ethylene glycol on the nanometric scale leads to the formation of nanofluids. Concerning features, nanofluids are expected to outperform standard heat transfer liquids and liquids having tiny particles. Since heat flow occurs at the material surface, materials with a wide total surface area are preferred, improving the suspension's stability. Nanofluids containing Al_2O_3 nanoparticles in water or ethylene have recently been shown to have superior thermal conductivity (Fig. 15.3) (Bhardwaj and Naraian, 2021a,b).

15.7 Applications of green synthesis nanomaterials

15.7.1 Pharmaceutical applications

15.7.1.1 Antibactericidal activities

In pathogenic species, Ag nanoparticles efficiently disrupted the cell membrane. Ag nanoparticles with higher concentrations exhibit higher membrane permeability than those with lesser concentrations, causing bacteria to breach their cell walls. When compared with AgNO3 treated cells, Ag nanoparticles reduced by *Rhizophora apiculata* showed less growth of bacteria in the plate, which could be owing to the particles' smaller size and more significant surface area,

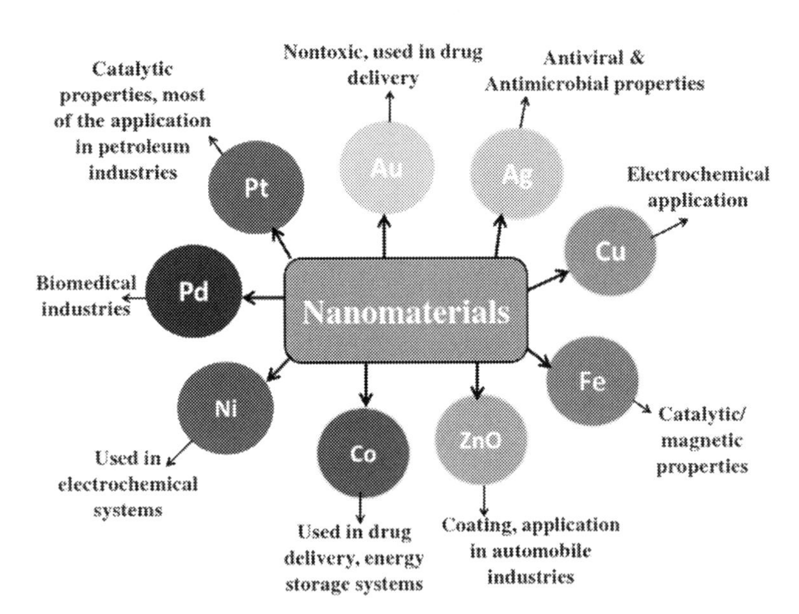

FIGURE 15.3 Properties and application of nanomaterials.

leading to increased membrane permeability and cell death. Bacterial interactions with metallic Ag and Au nanoparticles bind to the active site of the cell membrane, inhibiting cell cycle processes. In a one-step technique, extract from *Citrus sinensis* peel was applied as a reducing and capping agent to produce biosynthesized Ag nanoparticles. Ag nanoparticles were effectively decreased by *C. sinensis* peel extract, and action against *Pseudomonas aeruginosa*, *E. coli*, and *Staphylococcus aureus* was demonstrated (Paredes et al., 2014; Naraian and Abhishek, 2020).

15.7.1.2 Antifungicidal activity

Biosynthesized nanoparticle shows the greater fungicidal potential than commercially available antibiotics. Plant-extracted Ag nanomaterial damages the cell membrane of *Candida* sp. as well as intercellular mechanism, and ultimately lost cell activity. Most commercial antifungal medicines have limited clinical application, with more unpleasant effects and a slower recovery from disease. Nanoparticles were produced to provide new and effective antimicrobial medications. The fungal cell wall is made up of the fatty acid polymer and protein. The multifunctional Ag nanoparticles have potential antifungal action and effectively kill spore-producing fungus. By treating the fungal cell membrane with metallic nanoparticles, considerable alterations were detected (Khandel and Shahi, 2018; Naraian and Abhishek, 2020).

15.7.1.3 Anticancer activity

Ag-based nanomaterials produced from plants influence the cell cycle and enzymes in circulation. Furthermore, the plant-made nanomaterials limit the generation of free radicals in the cell. Free radicals are known to cause cell propagation and harm cell functions. A moderate Au concentration of nanomaterials promotes the apoptotic pathway in malignant cells. Similarly, MCF-7 is a human breast cancer cell treated with Ag nanoparticles that maintain biomolecule concentrations in the cells, resulting in cell metabolic regulation. Metallic nanomaterial has demonstrated unique medical applications in diagnosing and treating cancer illnesses. Bio-based nanomaterials have novel and revolutionary treatments for malignant deposits that do not interfere with normal cells. Compared with existing chemical-based synthetic pharmaceuticals, Suman et al. (2013) found that green manufacture of Ag nanomaterials had shown considerable cytotoxic activity in HeLa cell lines.

15.7.1.4 Antiviral activity

The medical application of Ag nanomaterials for treating and regulating viral pathogen development is as plant-mediated nanoparticles. Viruses enter a host with great care, and they translate speedily and grow their colonies. Ag-based nanomaterials can be biosynthesized to operate as effective broad-spectrum antiviral agents, limiting virus-cell activities. Suriyakalaa et al. (2013) investigated bio-based Ag nanomaterials with potent anti-HIV activity early in the reverse transcription pathway. Metallic nanomaterials are powerful antiviral agents that prevent viruses from entering the human system. Sun et al. (2005) found that biosynthesized-based nanoparticles show great potential against virucidal viruses. Furthermore, the Ag and Au nanoparticles continually block the HIV-1 life cycle's postentry stages. As a result, metallic nanoparticles will be an effective antiviral agent against retroviruses.

15.7.1.5 Antioxidant activity

Free radical generation is regulated by antioxidant agents, including both enzymatic and nonenzymatic compounds. Cellular damage caused by free radicals includes brain damage, atherosclerosis, and cancer. Reactive oxygen species produce free radicals that are regulated by bio-molecules. Ag-based nanomaterials have a stronger antioxidant impact than other synthetic compounds, such as ascorbic acid. Reichelt et al. (2012) found that nanoparticles have stronger antioxidant activity than tea leaf extract, with higher phenolic and flavonoid content.

15.7.2 Commercial applications

15.7.2.1 Elimination of organic and inorganic pollutants

Eutrophication in natural water resources produced by ammonia and phosphate can degrade pristine water habitats significantly; as a result, innovative and cost-effective restoration solutions are needed immediately. Fe-based nanoparticles synthesized by the green approach diffused onto zeolite by leaf extracts of eucalyptus were used to simultaneously remove phosphate and ammonia from aqueous solutions. Xu et al. (2020) reported the produced substance eliminated 43.3% and 99.8% of NH_4^+ and PO_4^{3-} respectively at initial concentrations of 10 mg/L for two coexisting ions. The maximal adsorption capacity of the generated material for NH_4^+ was 3.47 mg/g, and for PO_4^{3-} was 43 38.91 mg/g postoptimization. *Tamarindus indica* leaves are also used to produce water-soluble carbon quantum dots (QDs) with a wavelength of ~260–400 nm. The generated QDs can be utilized as a sensing probe for Hg^{2+} in a range of 0 to 0.1 μM, with a 6 nM of detection line; the device's feasibility was monitored by employing pond water samples for Hg^{2+} detection it could be customized for further study (Bano et al., 2018). Nanomaterials are also used for the treatment of industrial wastewater/effluents (Naraian and Abhishek, 2020; Bhardwaj and Naraian, 2021a,b).

Extracts from *Citrus limon*, *Cucumis sativus*, and *Vitis vinifera* were combined to create a new Fe_3O_4 nonabsorbent. The Fe-based nanosorbents were used to remove sulfamethoxazole, piperacillin, ampicillin, trimethoprim, tetracycline, erythromycin, and tazobactam from water systems. The optimum conditions for antibiotic elimination were evaluated using the Box–Behnken design concept. The Freundlich, Temkin, and Langmuir adsorption isotherm models were shown to be perfectly suited for the adsorption of chosen antibiotics, with an astounding 90% clearance rate for most of them.

15.7.2.2 Industrial application

Ag-based nanomaterials are employed in a variety of mechanical devices because Ag metal is a highly heat-conductive substance. It's mostly utilized in heat-sensitive instruments such as PCR lids and UV spectrophotometers. Ag nanomaterials are utilized as a coating material for sections of the apparatus. It is extremely stable at high temperatures and has no effect on the samples, because of multiple open-scale activities in the food industries, such as manufacturing, processing, and transportation of raw materials.

As indicated by their mechanical properties, nanomaterials can be employed in a number of mechanical applications, including lubricants, coatings, and adhesives. This property can also be exploited to make nanodevices that are mechanically stronger for a range of applications. Tribological qualities can be tweaked at the nanoscale by inserting nanomaterials in metal and polymer matrixes to increase their mechanical strength. The rolling motion of nanomaterials in the lubricated contact area allowed for extremely low friction and wear.

Furthermore, nanomaterials have strong sliding and delamination capabilities, which could reduce friction and wear while enhancing lubrication. Hardness and wear resistance are increased by coating, which can result in a range of mechanical properties (Bano et al., 2018).

As nanomaterials are easily manipulable and reversible, they can be integrated into electric, electronic, or optical devices using the "bottom-up" or "self-assembly" method, which is the gold standard in nanotechnology (Fig. 15.3).

15.8 Conclusion

Green synthesis technologies are thus being researched to close the gap and address the problems. Biological synthesis techniques are reasonably simple, nontoxic, and inexpensive. In the process of "green synthesis," the biological components are derived from plant extracts or tree parts. The biomolecules that come from plants are suitable for synthesis of metallic nanoparticles. Our focus is on the exploration of plant diversity, applications of green synthesis nanomaterials, characteristics of synthetic nanomaterials derived from plant sources are encountered for the green synthesis of nanomaterials from plant resources.

References

Adavallan, K., Krishnakumar, N., 2014. Mulberry leaf extract mediated synthesis of gold nanoparticles and its anti-bacterial activity against human pathogens. Advances in Natural Sciences: Nanoscience and Nanotechnology 5 (2), 025018.

Ahmed, S., Chaudhry, S.A., Ikram, S., 2017. A review on biogenic synthesis of ZnO nanoparticles using plant extracts and microbes: a prospect towards green chemistry. Journal of Photochemistry and Photobiology B: Biology 166, 272–284.

Ahmed, S., Saifullah, Ahmad, M., Swami, B.L., Ikram, S., 2016. Green synthesis of silver nanoparticles using Azadirachta indica aqueous leaf extract. Journal of Radiation Research and Applied Sciences 9 (1), 1–7.

Ahmed, S.F., Mofijur, M., Rafa, N., Chowdhury, A.T., Chowdhury, S., Nahrin, M., et al., 2022. Green approaches in synthesising nanomaterials for environmental nanobioremediation: Technological advancements, applications, benefits and challenges. Environmental Research 204, 111967.

Akhtar, M.S., Panwar, J., Yun, Y.S., 2013. Biogenic synthesis of metallic nanoparticles by plant extracts. ACS Sustainable Chemistry & Engineering 1 (6), 591–602.

Al-Shmgani, H.S., Mohammed, W.H., Sulaiman, G.M., Saadoon, A.H., 2017. Biosynthesis of silver nanoparticles from Catharanthus roseus leaf extract and assessing their antioxidant, antimicrobial, and wound-healing activities. Artificial cells, Nanomedicine, and Biotechnology 45 (6), 1234–1240.

Arshad, H., Saleem, M., Pasha, U., Sadaf, S., 2022. Synthesis of Aloe vera-conjugated silver nanoparticles for use against multidrug-resistant microorganisms. Electronic Journal of Biotechnology 55, 55–64.

Asimuddin, M., Shaik, M.R., Adil, S.F., Siddiqui, M.R.H., Alwarthan, A., Jamil, K., et al., 2020. Azadirachta indica based biosynthesis of silver nanoparticles and evaluation of their antibacterial and cytotoxic effects. Journal of King Saud University-Science 32 (1), 648–656.

Banejee, R., Katsenovich, Y., Lagos, L., McIintosh, M., Zhang, X., Li, C.Z., 2010. Nanomedicine: magnetic nanoparticles and their biomedical applications. Current Medicinal Chemistry 17 (27), 3120–3141.

Bano, D., Kumar, V., Singh, V.K., Hasan, S.H., 2018. Green synthesis of fluorescent carbon quantum dots for the detection of mercury (II) and glutathione. New Journal of Chemistry 42 (8), 5814–5821.

Begum, Q., Kalam, M., Kamal, M., Mahboob, T., 2020. Biosynthesis, characterization, and antibacterial activity of silver nanoparticles derived from aloe barbadensis miller leaf extract. Iranian Journal of Biotechnology 18 (2), e2383.

Behravan, M., Panahi, A.H., Naghizadeh, A., Ziaee, M., Mahdavi, R., Mirzapour, A., 2019. Facile green synthesis of silver nanoparticles using Berberis vulgaris leaf and root aqueous extract and its antibacterial activity. International Journal of Biological Macromolecules 124, 148–154.

Bhardwaj, A.K., Naraian, R., 2021a. Cyanobacteria as biochemical energy source for the synthesis of inorganic nanoparticles, mechanism and potential applications: a review. 3 Biotech 11 (10), 1–16.

Bhardwaj, A.K., Naraian, R., 2021b. Cyanobacteria as biochemical energy source for the synthesis of inorganic nanoparticles, mechanism and potential applications: a review. 3 Biotech 11 (10), 445.

Bhardwaj, A.K., Shukla, A., Mishra, R.K., Singh, S.C., Mishra, V., Uttam, K.N., et al., 2017. Power and time dependent microwave assisted fabrication of silver nanoparticles decorated cotton (SNDC) fibers for bacterial decontamination. Frontiers in Microbiology 8, 330.

Bhat, M., Chakraborty, B., Kumar, R.S., Almansour, A.I., Arumugam, N., Kotresha, D., et al., 2021. Biogenic synthesis, characterization and antimicrobial activity of Ixora brachypoda (DC) leaf extract mediated silver nanoparticles. Journal of King Saud University-Science 33 (2), 101296.

Boruah, J.S., Devi, C., Hazarika, U., Reddy, P.V.B., Chowdhury, D., Barthakur, M., et al., 2021. Green synthesis of gold nanoparticles using an antiepileptic plant extract: in vitro biological and photo-catalytic activities. RSC Advances 11 (45), 28029–28041.

Castillo-Henríquez, L., Alfaro-Aguilar, K., Ugalde-Álvarez, J., Vega-Fernández, L., Montes de Oca-Vásquez, G., Vega-Baudrit, J.R., 2020. Green synthesis of gold and silver nanoparticles from plant extracts and their possible applications as antimicrobial agents in the agricultural area. Nanomaterials 10 (9), 1763.

Castro, L., Blázquez, M.L., González, F.G., Ballester, A., 2014. Mechanism and applications of metal nanoparticles prepared by bio-mediated process. Reviews in Advanced Sciences and Engineering 3 (3), 199–216.

Chakraborty, A., Das, D.K., Sinha, M., Dey, S., Bhattacharjee, S., 2013. Moringa oleifera leaf extract mediated green synthesis of stabilized gold nanoparticles. Journal of Bionanoscience 7 (4), 415–419.

Chouhan, S., Guleria, S., 2020. Green synthesis of AgNPs using Cannabis sativa leaf extract: characterization, antibacterial, anti-yeast and α-amylase inhibitory activity. Materials Science for Energy Technologies 3, 536–544.

Das, D., Ghosh, R., Mandal, P., 2019. Biogenic synthesis of silver nanoparticles using S1 genotype of Morus alba leaf extract: characterization, antimicrobial and antioxidant potential assessment. SN Applied Sciences 1 (5), 1–16.

Das, P.E., Abu-Yousef, I.A., Majdalawieh, A.F., Narasimhan, S., Poltronieri, P., 2020. Green synthesis of encapsulated copper nanoparticles using a hydroalcoholic extract of Moringa oleifera leaves and assessment of their antioxidant and antimicrobial activities. Molecules (Basel, Switzerland) 25 (3), 555.

De Jong, W.H., Borm, P.J., 2008. Drug delivery and nanoparticles: applications and hazards. International Journal of Nanomedicine 3 (2), 133.

Devatha, C.P., Thalla, A.K., 2018. Green synthesis of nanomaterials. In Synthesis of Inorganic Nanomaterials. Woodhead Publishing, pp. 169–184.

Devi, M., Devi, S., Sharma, V., Rana, N., Bhatia, R.K., Bhatt, A.K., 2020. Green synthesis of silver nanoparticles using methanolic fruit extract of Aegle marmelos and their antimicrobial potential against human bacterial pathogens. Journal of Traditional and Complementary Medicine 10 (2), 158–165.

El-Gendy, N.S., El-Gendy, N.S., Omran, B.A., 2019. Green synthesis of nanoparticles for water treatment. Nano and Bio-Based Technologies for Wastewater Treatment: Prediction and Control Tools for the Dispersion of Pollutants in the Environment, pp. 205–263.

Ezhilarasi, A.A., Vijaya, J.J., Kaviyarasu, K., Kennedy, L.J., Ramalingam, R.J., Al-Lohedan, H.A., 2018. Green synthesis of NiO nanoparticles using Aegle marmelos leaf extract for the evaluation of in-vitro cytotoxicity, antibacterial and photocatalytic properties. Journal of Photochemistry and Photobiology B: Biology 180, 39–50.

Fardsadegh, B., Jafarizadeh-Malmiri, H., 2019. Aloe vera leaf extract mediated green synthesis of selenium nanoparticles and assessment of their in vitro antimicrobial activity against spoilage fungi and pathogenic bacteria strains. Green Processing and Synthesis 8 (1), 399–407.

Gnanasangeetha, D., Thambavani, S.D., 2014. Facile and eco-friendly method for the synthesis of zinc oxide nanoparticles using Azadirachta and Emblica. International Journal of Pharmaceutical Sciences and Research 5 (7), 2866.

Gopinath, V., MubarakAli, D., Priyadarshini, S., Priyadharsshini, N.M., Thajuddin, N., Velusamy, P., 2012. Biosynthesis of silver nanoparticles from Tribulus terrestris and its antimicrobial activity: a novel biological approach. Colloids and surfaces B: biointerfaces 96, 69–74.

Huang, J., Li, Q., Sun, D., Lu, Y., Su, Y., Yang, X., et al., 2007. Biosynthesis of silver and gold nanoparticles by novel sundried Cinnamomum camphora leaf. Nanotechnology 18 (10), 105104.

Ibrahim, H.M., 2015. Green synthesis and characterization of silver nanoparticles using banana peel extract and their antimicrobial activity against representative microorganisms. Journal of Radiation Research and Applied Sciences 8 (3), 265–275.

Ishak, N.M., Kamarudin, S.K., Timmiati, S.N., 2019. Green synthesis of metal and metal oxide nanoparticles via plant extracts: an overview. Materials Research Express 6 (11), 112004.

Jeevanandam, J., Chan, Y.S., Danquah, M.K., 2016. Biosynthesis of metal and metal oxide nanoparticles. ChemBioEng Reviews 3 (2), 55–67.

Kamala Nalini, S.P., Vijayaraghavan, K., 2020. Green synthesis of silver and gold nanoparticles using aloe vera gel and determining its antimicrobial properties on nanoparticle impregnated cotton fabric. Journal of Nanotechnology Research 2 (3), 42–50.

Kelkawi Abd, A.H., Kajani, A.A., Bordbar, A.K., 2017. Green synthesis of silver nanoparticles using Mentha pulegium and investigation of their antibacterial, antifungal and anticancer activity. IET Nanobiotechnology 11 (4), 370–376.

Keshari, A.K., Srivastava, A., Chowdhury, S., Srivastava, R., 2021. Antioxidant and antibacterial property of biosynthesised silver nanoparticles.

Khan, I., Saeed, K., Khan, I., 2019. Review nanoparticles: properties, applications and toxicities. Arabian Journal of Chemistry 12 (2), 908–931.

Khandel, P., Shahi, S.K., 2018. Mycogenic nanoparticles and their bio-prospective applications: current status and future challenges. Journal of Nanostructure in Chemistry 8 (4), 369–391.

Kharissova, O.V., Dias, H.R., Kharisov, B.I., 2013. BO P erez and VMJ P erez. Trends in Biotechnology 31, 240.

Korde, P., Ghotekar, S., Pagar, T., Pansambal, S., Oza, R., Mane, D., 2020. Plant extract assisted eco-benevolent synthesis of selenium nanoparticles-a review on plant parts involved, characterization and their recent applications. Journal of Chemical Reviews 2 (3), 157–168.

Kumar, A., Vemula, P.K., Ajayan, P.M., John, G., 2008. Nature 7, 236.

Kuppusamy, P., Yusoff, M.M., Maniam, G.P., Govindan, N., 2016. Biosynthesis of metallic nanoparticles using plant derivatives and their new avenues in pharmacological applications–An updated report. Saudi Pharmaceutical Journal 24 (4), 473–484.

Link, S., El-Sayed, M.A., 2003. Optical properties and ultrafast dynamics of metallic nanocrystals. Annual Review of Physical Chemistry 54 (1), 331–366.

Makarov, V.V., Love, A.J., Sinitsyna, O.V., Makarova, S.S., Yaminsky, I.V., Taliansky, M.E., et al., 2014. "Green" nanotechnologies: synthesis of metal nanoparticles using plants. Acta Naturae (англоязычная ФерФия) 6 (1), 35–44. 20.

Malapermal, V., Botha, I., Krishna, S.B.N., Mbatha, J.N., 2017. Enhancing antidiabetic and antimicrobial performance of Ocimum basilicum, and Ocimum sanctum (L.) using silver nanoparticles. Saudi Journal of Biological Sciences 24 (6), 1294–1305.

Marfu'ah, S., Rohma, S.M., Fanani, F., Hidayati, E.N., Nitasari, D.W., Primadi, T.R., et al., 2020, May. Green synthesis of ZnO nanoparticles by using banana peel extract as capping agent and its bacterial activity. In: IOP Conference Series: Materials Science and Engineering (Vol. 833, No. 1, p. 012076). IOP Publishing.

Medda, S., Hajra, A., Dey, U., Bose, P., Mondal, N.K., 2015. Biosynthesis of silver nanoparticles from Aloe vera leaf extract and antifungal activity against Rhizopus sp. and Aspergillus sp. Applied Nanoscience 5 (7), 875–880.

Mohamed, M., Faraj, K.M., Al-Jobori, K., 2020. Green synthesis of silver nanoparticles using tomato (Lycopersicon esculentum) extract and evaluation of their antifungal activitiesUniversity of Baghdad Plant Archives 20 (2), 5773–5786.

Moodley, J.S., Krishna, S.B.N., Pillay, K., Sershen, Govender, P., 2018. Green synthesis of silver nanoparticles from Moringa oleifera leaf extracts and its antimicrobial potential. Advances in Natural Sciences: Nanoscience and Nanotechnology 9 (1), 015011.

Muhammad, W., Ullah, N., Haroon, M., Abbasi, B.H., 2019. Optical, morphological and biological analysis of zinc oxide nanoparticles (ZnO NPs) using Papaver somniferum L. RSC Advances 9 (51), 29541–29548.

Nadaroglu, H., Güngör, A.A., Selvi, İ.N.C.E., 2017. Synthesis of nanoparticles by green synthesis method. International Journal of Innovative Research and Reviews 1 (1), 6–9.

Naraian, R., Abhishek, A.K.B., 2020. Green synthesis and characterization of silver NPs using oyster mushroom extract for antibacterial efficacy. Journal of Chemistry. Environmental Sciences and its Applications 7 (1), 13–18.

Noruzi, M., Zare, D., Khoshnevisan, K., Davoodi, D., 2011. Rapid green synthesis of gold nanoparticles using Rosa hybrida petal extract at room temperature. Spectrochimica Acta Part A: Molecular and Biomolecular Spectroscopy 79 (5), 1461–1465.

Pandian, C.J., Palanivel, R., Dhananasekaran, S., 2015. Green synthesis of nickel nanoparticles using Ocimum sanctum and their application in dye and pollutant adsorption. Chinese journal of Chemical Engineering 23 (8), 1307–1315.

Paredes, D., Ortiz, C., Torres, R., 2014. Synthesis, characterization, and evaluation of antibacterial effect of Ag nanoparticles against Escherichia coli O157: H7 and methicillin-resistant Staphylococcus aureus (MRSA). International Journal of Nanomedicine 9, 1717.

Parmar, M., Sanyal, M., 2022. Extensive study on plant mediated green synthesis of metal nanoparticles and their application for degradation of cationic and anionic dyes. Environmental Nanotechnology, Monitoring & Management 17, 100624.

Peralta-Videa, J.R., Huang, Y., Parsons, J.G., Zhao, L., Lopez-Moreno, L., Hernandez-Viezcas, J.A., et al., 2016. Plant-based green synthesis of metallic nanoparticles: scientific curiosity or a realistic alternative to chemical synthesis? Nanotechnology for Environmental Engineering 1 (1), 1–29.

Prabhu, N., Gajendran, T., 2017. Green synthesis of noble metal of platinum nanoparticles from Ocimum sanctum (Tulsi) Plant-extracts. IOSR Journal of Biochemistry and Biotechnology 3, 107–112.

Razavi, R., Molaei, R., Moradi, M., Tajik, H., Ezati, P., Shafipour Yordshahi, A., 2020. Biosynthesis of metallic nanoparticles using mulberry fruit (Morus alba L.) extract for the preparation of antimicrobial nanocellulose film. Applied Nanoscience 10 (2), 465–476.

Reichelt, K.V., Hoffmann-Luecke, P., Hartmann, B., Weber, B., Ley, J.P., Krammer, G.E., et al., 2012. Phytochemical characterization of South African bush tea (Athrixia phylicoides DC.). South African Journal of Botany 83, 1–8.

Santiago, T.R., Bonatto, C.C., Rossato, M., Lopes, C.A., Lopes, C.A., Mizubuti, E.S.G., et al., 2019. Green synthesis of silver nanoparticles using tomato leaf extract and their entrapment in chitosan nanoparticles to control bacterial wilt. Journal of the Science of Food and Agriculture 99 (9), 4248–4259.

Saravanakumar, K., Chelliah, R., Shanmugam, S., Varukattu, N.B., Oh, D.H., Kathiresan, K., et al., 2018. Green synthesis and characterization of biologically active nanosilver from seed extract of Gardenia jasminoides Ellis. Journal of Photochemistry and Photobiology B: Biology 185, 126–135.

Sasidharan, D., Namitha, T.R., Johnson, S.P., Jose, V., Mathew, P., 2020. Synthesis of silver and copper oxide nanoparticles using Myristica fragrans fruit extract: Antimicrobial and catalytic applications. Sustainable Chemistry and Pharmacy 16, 100255.

Shameli, K., Bin Ahmad, M., Jaffar Al-Mulla, E.A., Ibrahim, N.A., Shabanzadeh, P., Rustaiyan, A., et al., 2012. Green biosynthesis of silver nanoparticles using Callicarpa maingayi stem bark extraction. Molecules (Basel, Switzerland) 17 (7), 8506–8517.

Singh, M.P., Bhardwaj, A.K., Bharati, K., Singh, R.P., Chaurasia, S.K., Kumar, S., et al., 2021. Biogenic and non-biogenic waste utilization in the synthesis of 2D materials (graphene, h-BN, g-C2N) and their applications. Frontiers in Nanotechnology 53.

Soundarrajan, C., Sankari, A., Dhandapani, P., Maruthamuthu, S., Ravichandran, S., Sozhan, G., et al., 2012. Rapid biological synthesis of platinum nanoparticles using Ocimum sanctum for water electrolysis applications. Bioprocess and Biosystems Engineering 35 (5), 827–833.

Stozhko, N.Y., Bukharinova, M.A., Khamzina, E.I., Tarasov, A.V., Vidrevich, M.B., Brainina, K.Z., 2019. The effect of the antioxidant activity of plant extracts on the properties of gold nanoparticles. Nanomaterials 9 (12), 1655.

Suman, T.Y., Rajasree, S.R., Kanchana, A., Elizabeth, S.B., 2013. Biosynthesis, characterization and cytotoxic effect of plant mediated silver nanoparticles using Morinda citrifolia root extract. Colloids and surfaces B: Biointerfaces 106, 74–78.

Sun, R.W., Rong, C., Chung, N.P.Y., Ho, C.M., Lin, C.L.S., Che, C.M., 2005. Silver nanoparticles fabricated in Hepes buffer exhibit cytoprotective activities toward HIV-1 infected cells. Chemical Communications 28, 5059–5061.

Suriyakalaa, U., Antony, J.J., Suganya, S., Siva, D., Sukirtha, R., Kamalakkannan, S., et al., 2013. Hepatocurative activity of biosynthesized silver nanoparticles fabricated using Andrographis paniculata. Colloids and Surfaces B: Biointerfaces 102, 189–194.

Sutradhar, P., Saha, M., 2016. Green synthesis of zinc oxide nanoparticles using tomato (Lycopersicon esculentum) extract and its photovoltaic application. Journal of Experimental Nanoscience 11 (5), 314–327.

Thirumurugan, A., Aswitha, P., Kiruthika, C., Nagarajan, S., Christy, A.N., 2016. Green synthesis of platinum nanoparticles using Azadirachta indica—An eco-friendly approach. Materials Letters 170, 175–178.

Tippayawat, P., Phromviyo, N., Boueroy, P., Chompoosor, A., 2016. Green synthesis of silver nanoparticles in aloe vera plant extract prepared by a hydrothermal method and their synergistic antibacterial activity. PeerJ 4, e2589.

Umadevi, M., Bindhu, M.R., Sathe, V., 2013. A novel synthesis of malic acid capped silver nanoparticles using Solanum lycopersicums fruit extract. Journal of Materials Science & Technology 29 (4), 317–322.

Vanaja, M., Rajeshkumar, S., Paulkumar, K., Gnanajobitha, G., Malarkodi, C., Annadurai, G., 2013. Phytosynthesis and characterization of silver nanoparticles using stem extract of Coleus aromaticus. International Journal of Materials and Biomaterials Applications 3 (1), 1–4.

Vijayakumar, M., Priya, K., Nancy, F.T., Noorlidah, A., Ahmed, A.B.A., 2013. Biosynthesis, characterisation and anti-bacterial effect of plant-mediated silver nanoparticles using Artemisia nilagirica. Industrial Crops and Products 41, 235–240.

Xu, Q., Li, W., Ma, L., Cao, D., Owens, G., Chen, Z., 2020. Simultaneous removal of ammonia and phosphate using green synthesized iron oxide nanoparticles dispersed onto zeolite. Science of the Total Environment 703, 135002.

Yap, Y.H., Azmi, A.A., Mohd, N.K., Yong, F.S.J., Kan, S.Y., Thirmizir, M.Z.A., et al., 2020. Green synthesis of silver nanoparticle using water extract of onion peel and application in the acetylation reaction. Arabian Journal for Science and Engineering 45 (6), 4797–4807.

Chapter 16

Nanotechnology for sustainable development and future: a review

Rajat Tokas[1], Laxmi Kant Bhardwaj[1], Naresh Kumar[2] and Tanu Jindal[1]

[1]*Amity Institute of Environmental Toxicology, Safety and Management (AIETSM), Amity University, Noida, Uttar Pradesh, India,* [2]*Amity Institute of Environmental Sciences (AIES), Amity University, Noida, Uttar Pradesh, India*

16.1 Introduction

To live a healthy and a sustainable life, every individual needs food, house, water, energy, medical facilities, job, clothes, etc. (Diallo et al., 2013). The biggest downfall for humans is that we have failed to provide a good life with all basic necessities for better living. According to statistics, the world population will reach 8–10 billion by 2050. We need to lower the impact on the environment and change in climate because of human activities and race to development. Even though we are not able to meet the increasing demand for food, water, shelter, and electricity along with failure to lower the emission of gases like carbon dioxide (CO_2), methane (CH_4), nitrous oxide (N_2O), ozone (O_3), and carbon footprints on the environment (Godfray et al., 2010; Diallo and Brinker, 2011).

In the continuous progress of nanotechnology (NT), which refers to the construction and manipulation of matter at the nanoscale (1–100 nm) (Tahan, 2007), the researchers are focusing on two major issues related to sustainable development which may occur over a period of time:

1. Will NT be able to solve the issues related to sustainability in food, water, logistics, shelter, medical facilities, energy, and employment?
2. Will NT be able to create sustainable behavior amongst people for social benefit and less effect on climate change and global warming due to human activities?

16.1.1 Need toward sustainability

The whole world is suffering from the loss of ecosystem, cutting down trees, improper disposal of waste, air, water, and plastic pollution, inadequate drinking water availability, toxic harmful pollutants, climate change, and global warming. Less ease of availability is there for medical services, education, training for skill employment, etc. Humans are moving toward a dead-end for themselves, flora, and fauna. United Nations Sustainable Development Goals (UNSDGs) have been developed by the United Nations to address the important issues of the environment, economic, and political aspects (UNSDG's, 2021). These goals are shown in Fig. 16.1.

The goals have been designed in such a manner that they interconnect with each other, and one winning situation will surely be a win situation for all. Researchers have been continuously trying to advance in the path of drinking water, health, energy and environment, innovation, infrastructure, production, and consumption. These goals should help in reducing poverty and hunger, improvement in education, and equality along with economic progress. And this will turn into a peaceful life, justice for all, and powerful organization to help everyone.

16.1.2 How can nanotechnology contribute?

In lieu of step-by-step advancements to existing technologies, NT extends uncontrollable, innovative improvements that answer immediately with the best solutions for the benefit of mankind, the ecosystem, and our planet (Mercier-Laurent, 2015). The domain in which NT has shown improvement is in the sector of energy, protection of the environment,

Green and Sustainable Approaches Using Wastes for the Production of Multifunctional Nanomaterials. DOI: https://doi.org/10.1016/B978-0-443-19183-1.00012-X

FIGURE 16.1 Different types of sustainable development goals.

management of resources, and medical facility. NT has also developed skills related to scientific communication, medical engineering, etc. (Pokrajac et al., 2021). This undoubtedly makes NT an important enabler of sustainability in materials and resources, while enhancing the quality of life for a growing world population with increasing demand for energy and materials (Hyeon et al., 2015).

16.1.3 Teamwork, collaborative approach significant for consequential for Society 5.0

The investment by the government has put a lot of impact and may keep doing it in the future. A very impressive and effective example is the way mRNA vaccines for COVID-19 are processed with fatty nanoparticles (NPs), which has acted as a revolution which is on a very deep research done over the past many years in nanoscience and NT (Chauhan et al., 2020).

16.1.3.1 Society 5.0

The concept of Society 5.0 was introduced by the Japanese government from 2016 to 2020, which is vital for science & technology and the novelty of fundamental planning (Narvaez Rojas et al., 2021; Althabhawi et al., 2022) (Fig. 16.2). It is a central concept in the midterm science and focuses on mankind, which will help to achieve social concerns along with economic growth with a primitive combination of cyber as well as the substantial planet and in addition, it will contribute to achieving UNSDGs. Humans are now in the race to achieve and win in the concept of Society 5.0 after a long history of hunter-gatherer society and agricultural society, with the help of industrial society and informative society.

In Society 5.0, humans and materials are interconnected to the Internet of Things (IoT), where all data and information can be shared with Artificial Intelligence (AI) whenever required, resulting in a change of life (Rahmawati et al., 2021). A healthy life is expected from precautionary medical equipment and robots which can operate as doctors and take care of patients as a nurse at a low cost.

16.1.4 Nanotechnology for sustainable development

Nanotechnology involves in the research and design of materials or devices at the atomic and molecular levels (Navalakhe and Nandedkar, 2007). Nanometers, which are one billionth of a meter, contain about 10 atoms. However, developing an accurate definition of NT is a daunting task. Even scholars in this field say, "It depends on who you

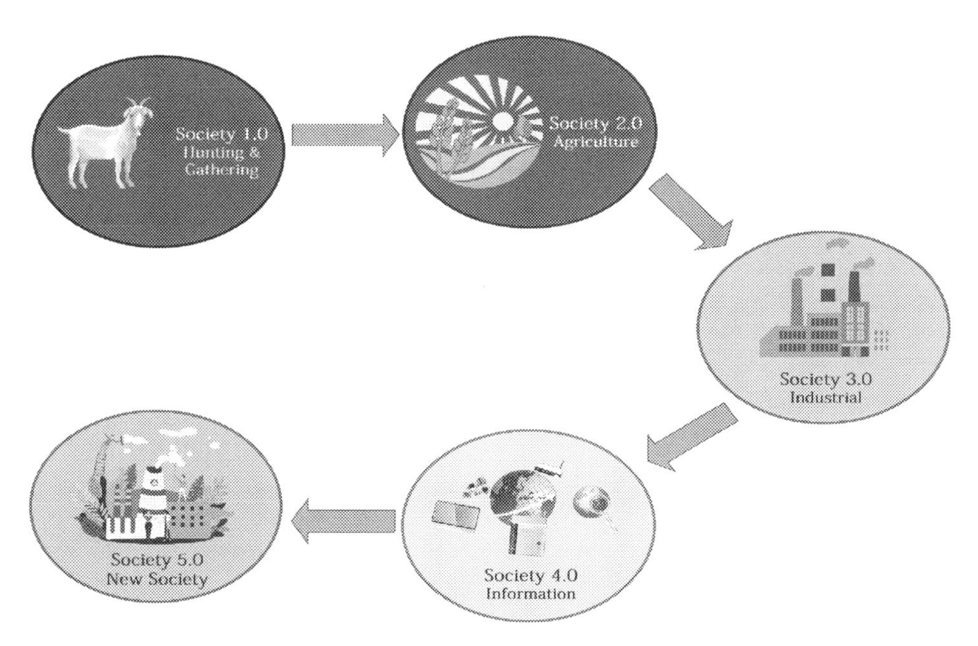

FIGURE 16.2 Diagrammatic representation of the Society 1.0 to Society 5.0.

ask." For example, some researchers use this term to describe almost all studies with a critical size of <1 micron (1000 nm). However, other researchers reserve this term for studies involving sizes from 1 to 100 nm (Bayisa et al., 2015).

There is also debate about whether naturally occurring NPs such as carbon black fall under the heading of NT (Uddin et al., 2017; Bhardwaj and Naraian, 2021; Bhardwaj et al., 2022). Finally, some reserve the term "NT" solely for manufacturing with atomic precision and use this term to describe the use of nanomaterials (NMs) to build materials, devices, and systems. Meanwhile, the National Science Foundation has described NT as "research and technological development at the atomic, molecular, or macromolecular level on a length scale of about 1–100 nm" (Kim et al., 2009). It is used to create and use structures, devices, and systems with new characteristics and capabilities due to the medium size of the NMs.

Nanomaterials have important physicochemical properties that make them particularly attractive as functional materials for sustainable technologies (Goesmann and Feldmann, 2010; Sun and Liao, 2020). On a bulk basis, they are much larger and have a more active surface area than bulk materials. They can be functioned with a variety of chemical groups to increase their affinity for specific compounds such as solutes and gases (Khin et al., 2012; Lu, 2014). They can also be functioned with key biochemical components and chemical groups that selectively target the metabolic/signaling networks of bacteria and viruses in water.

They also offer an unprecedented opportunity to develop functional materials with excellent electronic, optical, catalytic, and magnetic properties. These new functional materials can be processed into a variety of form factors such as water-soluble supramolecular hosts, particles, fibers, and membranes. Recent advances in the use of NT to address global sustainability challenges include:

- Water purification
- Clean energy technology
- Nanotechnology agricultural applications
- Greenhouse gas management
- Material supply and use
- Green manufacturing and chemistry
- Nanotechnology-based fiber applications

16.1.4.1 Green technologies

One of the burning issues and challenges the whole world is facing is climate change (Kumari et al., 2020). Over the past 20 years, the emission of greenhouse gases has increased the usage of fossil fuels such as coal and petroleum, which have come up as common factors to increase the emission (Yoro and Daramola, 2020). To meet the increasing consumption of energy along with a reduction in CO_2 emission, it needs the classification of measures to be taken for clean and renewable energy devices.

Nanotechnology opens vast opportunities to progress and innovate green technologies (Wiek et al., 2016; Di Sia, 2017; Palit and Hussain, 2018). Solar photovoltaic has come up as one of the popular sources of renewable electrical energy because it is available in plenty, is resourceful, and easy to implement with very less impact on the environment with respect to the usage of water and land (Jones and Olsson, 2017; Gulluce, 2021). Choi et al. (2012) have discussed the production of hydrogen by solar water breaking with the help of organic toxicity in wastewater as atoning electron donors. This study is very much important as solar radiation is isolated and the execution of solar power on a major scale needs an efficient device which converts solar energy into high-density chemical fuels.

16.1.4.2 *Purification of water*

In the 21st century, the whole world is facing the issue of clean drinking water, which is nowadays a big scarcity and a difficult situation to deal with. Most countries are now facing the issue of a sustainable supply of water in the fields of agriculture, processing food, producing energy, extraction of minerals, processing chemicals, and manufacturing of drinking water (Diallo and Brinker, 2011). As per the report of Bates et al. (2008), global warming and climate change will affect global clean water resources in the following ways:

- Raising of floods and droughts.
- Scarcity of water stored in glaciers and snowpacks.
- Lowering down of water quality because of the increase in salt, sediments, and pollutants in water resources all over the world.

Researchers have studied the quality of PET bottled drinking water (Bhardwaj and Sharma, 2021; Bhardwaj, 2022a) and stated that it is safe for human health. They also studied microplastics in drinking water in another study (Bhardwaj, 2022b) and stated that these tiny plastic particles came from human activities.

16.1.4.3 *A "formula" for implementation of sustainable nanotechnology*

How can we use this formula for data and information? A circular economy means unique supervision of the flow of energy and production by manufacturing and economic system with the channel to alleviate harm to the ecosystem and society which is known to support sustainability. We disagree that the circular formation in the flow of knowledge is dangerous to making vital advances in way of sustainability through NT. The old concept of translation of information and data commencing science to society is now out-of-date and is destructive to society as well as a scientific project. We would propose reinventing and reimagining of data and information, translation as a circular exchange between researchers and practitioners, and the flow of knowledge should not be one way.

Progress for sustainability will move fast in an efficient and meaningful manner with ongoing talks between researchers and practitioners, organizations, society, and groups of people who have knowledge of the ecosystem, also in schools and universities to motivate and to outline the research for a large impact. If the governance of data and information understand the needs of the market, industries, and partner's points of view for risk to health and the ecosystem, society only accepts innovative technology that outlines the design of research.

16.1.4.4 *Potential of sustainable nanotechnology in India*

In 2007, the Government of India launched a 5-year program called "Nano Mission" which allocated Rs 1000 crores budget (Deshpande Sarma and Anand, 2012; Purushotham, 2012; Ghosal and Chakraborty, 2021). The mission is to do basic research on NT, develop the infrastructure, and use skilled human resources for the identification of potential and challenges, taking global collaboration into account so that the best of minds can work for making sustainable earth to live in.

Based on the previous data, we have done some analysis on NT and would like to share these points for future research. These points are as follows:

- Innovation of sustainable NT will lead to a decrease in poverty as it will help in real-time solving issues related to basic necessities by identifying the exact need of the citizen.
- Needs can be served by boosting agricultural production artificially with NT if the requirement of food can't be achieved naturally.

- Science and technology are moving ahead, which can also make artificial clouds for rain in the area which is under drought. NT can help in achieving the water requirement for drinking as well as for daily purposes.
- Desalination of seawater can also be done if more effective research can be done in the field of NT.
- Job opportunities can be created in the field of research on sustainable NT which will lead to more research and employment.
- Sustainable NT may help in the economic growth of the country by creating sustainable education and industries.
- Land degradation and environmental degradation can be reduced with the help of sustainable NT. It will create clean air, safe drinking water, and soil full of nutrition for the production of food grains and vegetables.
- Marine resources can be saved and preserved by using NT to reduce plastic pollution and industrial waste dumped into the seawater.

16.1.5 Product transmission

The companies should focus and work keeping in mind the development of products according to their life cycle. Life cycle assessment should be considered in respect of economic, environmental, and governance constituents. We may understand with an example that the best quality mines are very less available and it results that statutory bodies should start developing mines of lower quality with less quantity target product and separating essential efficient minerals from toxic materials. Companies should start considering recycling of products with the very first step of product development along with the waste disposal factors for less impact on air and water with saving the environment and cost. If the product is recycled, then it will help to extend the supply by canceling the point of product unavailability which will ultimately reduce the price of the product. Raising concern is less availability of lithium (Li), nickel (Ni), and carbon monoxide (CO) along with increasing their price due to maximum usage in electric vehicles for batteries (Lu et al., 2016). Therefore, recycling the product can help in saving the above precious materials.

In this chapter, we discuss in detail the approaches, properties, and applications of NT in several sectors like clinical and medical, agriculture, energy and environment, electronics, food, textile, wastewater treatment, cosmetics, sports, furniture, etc.

16.2 Applications/uses of nanotechnology and its equipment

Several researchers have studied the different applications/uses of this technology in various sectors (Misra et al., 2010; Singh et al., 2017; Singh et al., 2021a) which are shown in Fig. 16.3 and Table 16.1. The improved equipment of NT is going to have a very vital role in the society (Phuyal et al., 2020). This will take up the technological breakthrough along with an improved transformation by supplying the nanodevices supposed to be used in society such as IoT sensors, self-driving vehicles, innovative robots, etc. Nanodevices are expected to take an important part in the realization of a sustainable society by providing water purification, reduction in CO_2 emissions, and boosting device supply with a recyclable viewpoint (Nagar and Pradeep, 2020).

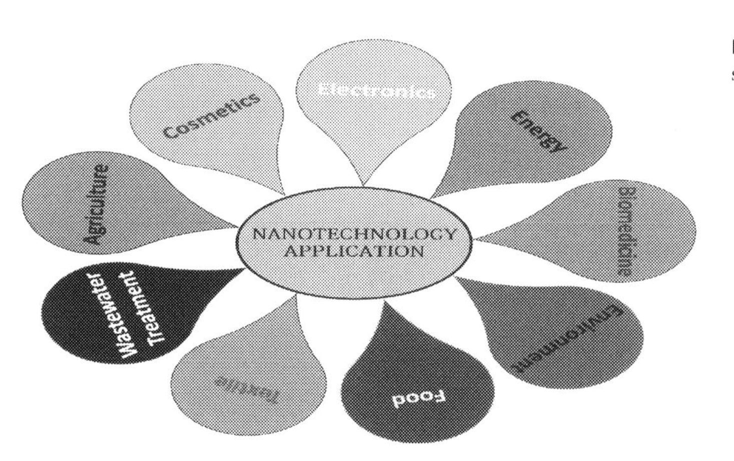

FIGURE 16.3 Applications/uses of nanotechnology in different sectors.

TABLE 16.1 Occurrence of different uses of nanotechnology/nanomaterials in different sectors.

S. no.	Sector	Nanomaterials	Uses of nanotechnology/nanomaterials
1	Medicine	Gold nanoparticles	Used in the detection of Alzheimer's disease
		Quantum dots	Used in the detection of cancer cells
		Nanofibers	Used in help in the artificially growing of heart muscles
		Nanocoatings	Used to stop viruses from binding to cells
		Nanomaterials	Used in eye drops to prevent blindness, used in a vaccine against hepatitis
2	Electronics	Nanorods	Used in computers for less consumption of electricity and less heat emission
		Carbon nanotubes	Used in the making of smaller, faster, and more efficient microchips and devices, used in making the flexible touchscreens
		Nanowires	Used in the fabrication of chips
		Nanoscale transistors	Used in the making of faster, more powerful and energy-efficient devices
		Nanostructured polymer films	Used in the formation of a brighter and wider image, lighter weight, better picture density, lower power consumption, and longer lifetime devices
		Flash memory chips	Used in iPod
		Antimicrobial/ Antibacterial coatings	Used in mouse, keywords, and cell phones
		Conductive inks	Used in printers and smart card
		Nanoscale fibers	Used in mobile to allow to bend, stretch, and fold
		Photovoltaic nanowire	Used in the charging of the device
3	Energy	Nanofabric materials	Used in the creation of nanoscale devices, capture, store, and transfer energy
		Nanostructured solar cells	Used in the conversion of solar energy to electric energy
4	Textiles	Carbon particles membrane	Used in the protection from electrostatic charges
		Smart fabrics	Used in the protection from stains and wrinkles, used to make stronger, lighter, and more durable materials
		Nano-sized whiskers	Used in making water and stain-repellent clothes, bullet-proof jackets, antimicrobial, and antibacterial clothes
		Nanotech sensor	Mostly used in the military for the detection of biological agents, protection from high temperatures and chemicals
5	Automobile	Nanopaint	Used in to make the water and dirt repellent, resistant to chipping and scratches
6	Environment	Nano ceramic particles	Used in the improvement of smoothness and heat resistance of common household equipment
		Nano paint	Used in buildings to reduce air pollution
		Nanobubbles or Nanofiltration systems	Used in water treatment plants to treat wastewater or removing of heavy metals
7	Food	Nanobiosensors	Used in the detection of the presence of pathogens, improves food production by increasing mechanical and thermal resistance, and decreasing oxygen transfer in packaged products

Many technical challenges are to be faced on the path of advancement of NT and devices. Below are the six major issues which can be faced:

1. IoT edge and AI chips and quantum devices for innovative computing.
2. Logistics with the best security and less impact on the environment.
3. Nanobiotechnology for medical facilities.
4. Robots to be used in the service sector and for helping human beings.
5. Smart devices for sustainable air, water, and other important necessities.
6. Energy equipment for renewable energy production and saving.

16.2.1 Clinical and medical

Remote clinical and medical services are open and warmly welcomed during the pandemic COVID-19 (Udugama et al., 2020; Bhalla et al., 2020). With the help of NT devices that is biosensors, the data was collected from patients by the doctors. Not only it was collected but also analyzed with the help of highly efficient computers equipped with NT. Nowadays, patients do not need to travel for diagnoses from one city to another as with the help of AI it can be done. Surgeries are going to be possible in the coming days with the help of nanotechnological devices such as robots, HD image sensors, 5G speed used for good communication, etc. (Mohan et al., 2021).

Nanotechnology may help in improving the health sector by providing biosensors to wear and biomaterials for reviving treatment. Wearable biosensors can record complex medical information about the patient in rural regions and can be sent to big hospitals directly. Doctors who are efficient and well-informed about this technology can do surgery in rural areas remotely with the help of robots (Pokrajac et al., 2021). However, various treatments and medicines for cancer and brain tumor may have side effects on the patients. Some NMs are used in the diagnosis and treatment of neurovegetative diseases or cancer (Valdiglesias et al., 2015). These particles attack only infected cells instead of the whole body. NMs are being used to enhance the efficiency of imaging devices (Wang et al., 2008). Nanotechnology is also used in gene therapy, wound treatment, etc. (Mozafari, 2018; Blanco-Fernandez et al., 2021). Nanotechnology can also help in the development of effective medical drugs for saving mankind and giving more quality of healthy life.

16.2.1.1 Against bacteria, viruses, and diseases

Nanobiotechnology is used in fighting against the upcoming and new evolving viruses like COVID-19 with the help of diagnoses through nanodevices, the discovery of new vaccinations, and treatment (Weiss et al., 2020; Singh et al., 2021b). Material and photonic technologies can also help effectively in the prevention of these kinds of viruses with the help of photocatalytic NPs, ultraviolet AlGaN light-emitting diodes, and air purification through virus capture with various membranes (Li et al., 2021; Minamikawa et al., 2021). It can also help to understand the interconnection of microorganisms that cause diseases with different types of materials (Konda et al., 2020). In the direction of progress, nanoscale simulations and data sciences can take a very important part (Mori et al., 2021). Silver nanoparticles, which were extracted from decorated cotton, were more effective against gram negative bacteria than gram positive bacteria (Bhardwaj et al., 2017a). Cuprous oxide (CuO-NPs) nanoparticles have the antibacterial properties and due to the lower cost, it may be the alternative to silver NPs (Bhardwaj et al., 2019).

16.2.2 Agriculture

Nanotechnology is used to increase food productivity and food quality. It has really attracted the agriculture industry due to its effectiveness and low-cost production with quality of food grains and vegetables. Nano-based agrochemicals can be used to provide the best nutrition to plants to grow. Farmers can also use nanobased fertilizers which help plants to grow faster without any harm to the soil and mankind. For example, nanocopper fertilizer gives plants nutrition and protects them from any type of disease (Elmer et al., 2018; Guha et al., 2020).

Gold-nanoparticles (Au-NPs)-based biosensors have been used for the detection of enzyme activity (Xiao-Ming et al., 2018). Carbon nanotubes provide strength to the crops and protect them from strong wind (Anand and Sinha, 2013; Yousefi et al., 2017). NPs are used in the agricultural field as an alternative to pesticides and other chemicals to resist several plant diseases (El-Moneim et al., 2021).

16.2.3 Energy and environment

The very important need of society for saving energy and the environment is to reduce CO_2 emissions, sustainable energy, and reduction of cost for production of energy (Razmjoo et al., 2021). Nanodevices can play a very important role in the sector of logistics and saving energy. These devices can also take part in the production of renewable energy with solar power, wind turbines, and fuel cells with green hydrogen. The progressive action of developing technologies is to produce hydrogen from solar energy and use it for carrying and storing of energy.

Nanotechnology has improved fuel production efficiency from raw petroleum materials by using better catalysis. Nanocatalysts are used to make chemical reactions more efficient. Carbon nanotubes are used in the production of windmill blades for the increasing the amount of electricity (Ma and Zhang, 2014; Elhenawy et al., 2021). Carbon nanotube scrubbers are being used to separate the power plant exhaust CO_2 (Irani et al., 2017).

16.2.3.1 Renewable energies

Nature and human, both act as an alarm when situation or needs are out of control. Renewable resources are somehow on the verge and will vanish in the coming years. Here comes the role of creating and developing sustainable NT to meet the requirement of mankind (Serrano et al., 2009). Not only it will fulfill the needs but also it will help in saving the ecosystem and will create a channel of using minimal energy in the most efficient manner (Badawy, 2015).

16.2.3.2 Solar energy

Solar energy can be conserved with the help of solar cells. But according to nanostructures, the efficiency of these cells depends on their design, shape, and type (Tsakalakos, 2012). Nanostructured cells are made with the help of nanowires, nanorods, and quantum dot structures enabling highly efficient and low-costing equipment. Solar panels convert sunlight into electricity (Xiang et al., 2019). There are two types of solar cells, which are very efficient.

16.2.3.2.1 Dye-sensitized solar cells

They are manufactured by placing dye-sensitized films between two transparent electrodes (Kyaw et al., 2011). Gratzel cell has transparent glass as an anode and platinum as a cathode with titanium dioxide film and also has an electrolyte between film and platinum.

16.2.3.2.2 Quantum dot solar cells

They are the nanosemiconductors which are clustered with high photoconductivity. They are competitive with nonrenewable sources like coal and petroleum because they are highly efficient and cost-effective.

16.2.3.3 Wind energy

Natural resources like wind energy are highly efficient and abundant in nature (Tejeda and Ferreira, 2014). But still, challenges like consideration of manufacturing cost, environmental issues, and construction of specific plants for conversion of wind into channelized energy are taken into account. The heavy turbines which are used to convert wind energy into conventional energy are also a challenge with transportation and installation. The cost of heavy turbines is high which leads to an increase in the overall production cost (Rathore et al., 2021). Sustainable NT needs to be identified to overcome these challenges (Li and Lu, 2014).

16.2.4 Electronics

Carbon nanotubes are used for making smaller, faster, and more efficient electric microchips and devices. Graphene is used in the development of flexible touchscreens (Vlasov et al., 2017; Bubnova, 2021).

16.2.5 Food

Nanobiosensors are used in the detection of the presence of pathogens in food (Vanegas et al., 2017; Bhardwaj et al., 2017b). Nanocomposites are used in the improvement of food production by increasing thermal and mechanical resistance.

16.2.6 Textile

Nanotechnology is used in the textile industry for the development of smart fabrics which are more durable (Shah et al., 2022). Nowadays, water and liquid replant clothes are in a fashion which are made of NPs like silica. Silica is sprayed on the fabric to make it waterproof (Mao et al., 2013).

16.2.7 Water treatment

Nanotechnology is used in the purification of water, and it is a highly effective and cheaper technique (Kumar et al., 2014). Unique NPs are used in filtration membranes to improve the quality of water by removing chemicals, heavy metals, and industrial waste. Photocatalytic degradation is the most popular technique in the treatment of wastewater (Melemeni et al., 2009). By the use of ions and nanobubbles, NT is used in the purification of wastewater (Atkinson et al., 2019).

16.2.8 Cosmetics

Titanium dioxide and zinc oxide are used in sunscreen for protection from UV rays (Morganti, 2010; Schneider and Lim, 2019). Liposome's NPs are used in antiaging skin creams (Thong et al., 2007). Nanoemulsions and liposomes are used in skin care products while liposomes and ethosomes are used in hair growth products. Iron oxide NPs are used as a pigment in some lipsticks (Wiechers and Musee, 2010; Ealia and Saravanakumar, 2017).

16.2.9 Sports

Nanotechnology is responsible for the bouncing nature of the tennis balls for a longer time (Hester and Harrison, 2016; Ćibo et al., 2019).

16.2.10 Furniture

To make nonflammable furniture, carbon nanofibers are coated over the furniture which reduces the flammability by up to 35% (Kim et al., 2011; Holder et al., 2016).

16.2.11 Adhesive

The adhesives or glues lose their stickiness due to high temperatures, but nanoglue helps to stay in high temperatures with stronger adhesiveness (Rai and Saraswat, 2022).

16.2.12 Safe and sustainable driving

Nanotechnology and its equipment can help to make a safe and environmentally friendly drive (Roco, 2005). AI chips, sensors, and other devices can really be useful and helpful for the safety of the driver as well as passengers.

16.3 Conclusions and recommendations

Nanotechnology has an important role to play in international efforts in sustainability. This emerging technology is used in several sectors. Several researchers are focusing on this technique and are making new NMs for keeping the product fresh with increased shelf life. At the beginning stage, there are numerous applications of NT and most of them require a high quality of research and development for sustainable development. The research regarding the application of NT is growing every day. An understanding of the safe application of NMs will help in sustainable development. The compulsory testing of nanomodified products should be performed before they are allowed to be introduced into the market. The toxicity of NPs is poorly understood due to the lack of validated test methods.

Based on the previous research, we would like to propose some recommendations in service to attain sustainable development.

- Further research for NT is required in-depth from a different perspective in life.
- Nanotechnology applications can be used in different sectors to make sustainable businesses along with protecting the environment.

- Society 5.0 concept can surely help to achieve social and economic goals with the help of nanodevices to reduce carbon footprint and focus on renewable energy with sustainability.
- Green technology can help in the production of sufficient energy and the reduction of gases which are the core reason for global warming and climate change.
- We need to work on the "formula" of a remodification of old methods to gain a circular economy and sustainable development with the help of NT.
- The policymakers and regulatory bodies should provide guidance documents for the validated protocols, safe uses, and the disposal of the NPs.
- The policymakers should make the procedures of filing patent easier so that it encourages the researcher/scientists/ product developer.
- Standard test procedures should be available to study the impact of NPs on humans.
- Government should promote and help in the research of more sustainable nanodevices by creating more opportunities in funding schemes.
- Programs should be launched as a mission to attain and sustain the process of sustainable development through NT by awarding the researchers through various schemes.
- Relaxation in taxes should be given by organizations for working and developing NT so that both society and organizations may be benefitted.
- More sessions should be organized by the institutes for society to create awareness on how we all can contribute by using nanodevices to achieve a sustainable future and development.
- More effective and efficient storage techniques should be developed through NT for saving and storing renewable energies.

Acknowledgments

The authors thank Amity University, Noida for providing the platform to do this study.

Conflict of interest

The authors declare that they have no competing financial interest or personal relationship that could have appeared to influence the work reported in this chapter.

References

Althabhawi, N.M., Zainol, Z.A., Bagherib, P., 2022. Society 5.0: a new challenge to legal norms. Sriwijaya Law Review 6 (1), 41–54.

Anand, A., Sinha, S.R.P., 2013. Performance evaluation of logic gates based on carbon nanotube field effect transistor. International Journal of Recent Technology and Engineering 2 (5).

Atkinson, A.J., Apul, O.G., Schneider, O., Garcia-Segura, S., Westerhoff, P., 2019. Nanobubble technologies offer opportunities to improve water treatment. Accounts of Chemical Research 52 (5), 1196–1205.

Badawy, W.A., 2015. A review on solar cells from Si-single crystals to porous materials and quantum dots. Journal of Advanced Research 6 (2), 123–132.

Bates, B., Kundzewicz, Z., Wu, S., 2008. Climate change and water. Intergovernmental Panel on Climate Change Secretariat.

Bayisa, T.K., Bule, M.H., Lenjisa, J.L., 2015. The potential of nano technology-based drugs in lung cancer management. The Pharma Innovation 4 (4, Part B), 77. Available from: https://www.thepharmajournal.com/vol4Issue4/Issue_jun_2015/4-4-16.1.pdf.

Bhalla, N., Pan, Y., Yang, Z., Payam, A.F., 2020. Opportunities and challenges for biosensors and nanoscale analytical tools for pandemics: COVID-19. ACS Nano 14 (7), 7783–7807.

Bhardwaj, L.K., 2022a. Evaluation of bis (2-ethylhexyl) phthalate (DEHP) in the PET bottled mineral water of different brands and impact of heat by GC–MS/MS. Chemistry Africa, pp. 1–14.

Bhardwaj, L.K., 2022b. Microplastics (MPs) in drinking water uses, sources and transport. Futuristic Trends in Agriculture Engineering & Food Science 2 (1), ISBN: 978-93-95632-65-2.

Bhardwaj, A.K., Naraian, R., 2021. Cyanobacteria as biochemical energy source for the synthesis of inorganic nanoparticles, mechanism and potential applications: a review. 3 Biotech 11 (10), 1–16.

Bhardwaj, L.K., Sharma, A., 2021. Estimation of physicochemical, trace metals, microbiological and phthalate in PET bottled water. Chemistry Africa 4 (4), 981–991.

Bhardwaj, A.K., Shukla, A., Mishra, R.K., Singh, S.C., Mishra, V., Uttam, K.N., et al., 2017a. Power and time dependent microwave assisted fabrication of silver nanoparticles decorated cotton (SNDC) fibers for bacterial decontamination. Frontiers in Microbiology 8, 330.

Bhardwaj, N., Bhardwaj, S.K., Nayak, M.K., Mehta, J., Kim, K.H., Deep, A., 2017b. Fluorescent nanobiosensors for the targeted detection of foodborne bacteria. TrAC Trends in Analytical Chemistry 97, 120–135.

Bhardwaj, A.K., Kumar, V., Pandey, V., Naraian, R., Gopal, R., 2019. Bacterial killing efficacy of synthesized rod shaped cuprous oxide nanoparticles using laser ablation technique. SN Applied Sciences 1 (11), 1–8.

Bhardwaj, L.K., Rath, P., Choudhury, M., 2022. A comprehensive review on the classification, uses, sources of nanoparticles (NPs) and their toxicity on health. Aerosol Science and Engineering 1–18.

Blanco-Fernandez, B., Castaño, O., Mateos-Timoneda, M.Á., Engel, E., Pérez-Amodio, S., 2021. Nanotechnology approaches in chronic wound healing. Advances in Wound Care 10 (5), 234–256.

Bubnova, O., 2021. A decade of R2R graphene manufacturing. Nature Nanotechnology 16 (10), 1050.

Chauhan, G., Madou, M.J., Kalra, S., Chopra, V., Ghosh, D., Martinez-Chapa, S.O., 2020. Nanotechnology for COVID-19: therapeutics and vaccine research. ACS Nano 14 (7), 7760–7782.

Choi, J., Qu, Y., Hoffmann, M.R., 2012. SnO_2, IrO_2, Ta_2O_5, Bi_2O_3, and TiO_2 nanoparticle anodes: electrochemical oxidation coupled with the cathodic reduction of water to yield molecular H2. In Nanotechnology for Sustainable Development. Springer, Cham, pp. 223–234.

Ćibo, M., Šator, A., Kazlagić, A., Omanović-Mikličanin, E., 2019. Application and impact of nanotechnology in sport. In Scientific-Experts Conference of Agriculture and Food Industry. Springer, Cham, pp. 349–362, September.

Deshpande Sarma, S., Anand, M., 2012. Status of nanoscience and technology in India. Proceedings of the National Academy of Sciences, India Section B: Biological Sciences 82 (1), 99–126. Available from: https://doi.org/10.1007/s40011-012-0077-2.

Di Sia, P., 2017. Nanotechnology among innovation, health and risks. Procedia-Social and Behavioral Sciences 237, 1076–1080.

Diallo, M., Brinker, C.J., 2011. Nanotechnology for sustainability: environment, water, food, minerals, and climate. In Nanotechnology Research Directions for Societal Needs in 2020. Springer, Dordrecht, pp. 221–259.

Diallo, M.S., Fromer, N.A., Jhon, M.S., 2013. Nanotechnology for sustainable development: retrospective and outlook. In Nanotechnology for Sustainable Development. Springer, Cham, pp. 1–16.

Ealia, S.A.M., Saravanakumar, M.P., 2017, November. A review on the classification, characterisation, synthesis of nanoparticles and their application. In: IOP Conference Series: Materials Science and Engineering (Vol. 263, No. 3, p. 032019). IOP Publishing.

Elhenawy, Y., Fouad, Y., Marouani, H., Bassyouni, M., 2021. Performance analysis of reinforced epoxy functionalized carbon nanotubes composites for vertical axis wind turbine blade. Polymers 13 (3), 422.

Elmer, W., Ma, C., White, J., 2018. Nanoparticles for plant disease management. Current Opinion in Environmental Science & Health 6, 66–70.

El-Moneim, D.A., Dawood, M.F., Moursi, Y.S., Farghaly, A.A., Afifi, M., Sallam, A., 2021. Positive and negative effects of nanoparticles on agricultural crops. Nanotechnology for Environmental Engineering 6 (2), 1–11.

Ghosal, M., Chakraborty, A., 2021, August. The growing use of nanotechnology in the built environment: a review. In: IOP Conference Series: Materials Science and Engineering (Vol. 1170, No. 1, p. 012007). IOP Publishing.

Godfray, H.C.J., Beddington, J.R., Crute, I.R., Haddad, L., Lawrence, D., Muir, J.F., et al., 2010. Food security: the challenge of feeding 9 billion people. Science (New York, N.Y.) 327 (5967), 812–818.

Goesmann, H., Feldmann, C., 2010. Nanoparticulate functional materials. Angewandte Chemie International Edition 49 (8), 1362–1395.

Guha, T., Gopal, G., Kundu, R., Mukherjee, A., 2020. Nanocomposites for delivering agrochemicals: a comprehensive review. Journal of Agricultural and Food Chemistry 68 (12), 3691–3702.

Gulluce, H., 2021. Using of nanotechnology for photovoltaic solar energy systems. NanoEra 1 (2), 39–44. Available from: https://dergipark.org.tr/en/pub/nanoera/issue/69475/1105623.

Hester, R.E., Harrison, R.M., 2016. Nanotechnology consequences for human health & the environment (issues in environmental science and technology).

Holder, K.M., Cain, A.A., Plummer, M.G., Stevens, B.E., Odenborg, P.K., Morgan, A.B., et al., 2016. Carbon nanotube multilayer nanocoatings prevent flame spread on flexible polyurethane foam. Macromolecular Materials and Engineering 301 (6), 665–673.

Hyeon, T., Manna, L., Wong, S.S., 2015. Sustainable nanotechnology. Chemical Society Reviews 44 (16), 5755–5757.

Irani, M., Jacobson, A.T., Gasem, K.A., Fan, M., 2017. Modified carbon nanotubes/tetraethylenepentamine for CO_2 capture. Fuel 206, 10–18.

Jones, L.E., Olsson, G., 2017. Solar photovoltaic and wind energy providing water. Global Challenges 1 (5), 1600022.

Khin, M.M., Nair, A.S., Babu, V.J., Murugan, R., Ramakrishna, S., 2012. A review on nanomaterials for environmental remediation. Energy & Environmental Science 5 (8), 8075–8109.

Kim, S., Kwon, K., Kwon, I.C., Park, K., 2009. Nanotechnology in drug delivery: past, present, and future. Nanotechnology in Drug Delivery. Springer, New York, NY, pp. 581–596. Available from: https://doi.org/10.1007/978-0-387-77668-2_19.

Kim, Y.S., Davis, R., Cain, A.A., Grunlan, J.C., 2011. Development of layer-by-layer assembled carbon nanofiber-filled coatings to reduce polyurethane foam flammability. Polymer 52 (13), 2847–2855.

Konda, A., Prakash, A., Moss, G.A., Schmoldt, M., Grant, G.D., Guha, S., 2020. Aerosol filtration efficiency of common fabrics used in respiratory cloth masks. ACS Nano 14 (5), 6339–6347.

Kumar, S., Ahlawat, W., Bhanjana, G., Heydarifard, S., Nazhad, M.M., Dilbaghi, N., 2014. Nanotechnology-based water treatment strategies. Journal of Nanoscience and Nanotechnology 14 (2), 1838–1858.

Kumari, B., Solanki, H., Kumar, A., 2020. Climate change: a burning issue for the world. Medicine 35, 501–507. Available from: http://www.capecomorinpublisher.com.

Kyaw, A.K.K., Tantang, H., Wu, T., Ke, L., Peh, C., Huang, Z.H., et al., 2011. Dye-sensitized solar cell with a titanium-oxide-modified carbon nanotube transparent electrode. Applied Physics Letters 99 (2), 021107. Available from: https://doi.org/10.1063/1.3610488.

Li, R., Cui, L., Chen, M., Huang, Y., 2021. Nanomaterials for airborne virus inactivation: a short review. Aerosol Science and Engineering 5 (1), 1–11. Available from: https://doi.org/10.1007/s41810-020-00080-4.

Li, Y., Lu, J., 2014. Lightweight structure design for wind energy by integrating nanostructured materials. Materials & Design 57, 689–696. Available from: https://doi.org/10.1016/j.matdes.2013.11.082.

Lu, B., 2014. Surface Reactivity of Hematite Nanoparticles. https://www.diva-portal.org/smash/get/diva2:728316/FULLTEXT01.pdf.

Lu, J., Chen, Z., Ma, Z., Pan, F., Curtiss, L.A., Amine, K., 2016. The role of nanotechnology in the development of battery materials for electric vehicles. Nature Nanotechnology 11 (12), 1031–1038. Available from: https://doi.org/10.1038/nnano.2016.207.

Ma, P.C., Zhang, Y., 2014. Perspectives of carbon nanotubes/polymer nanocomposites for wind blade materials. Renewable and Sustainable Energy Reviews 30, 651–660.

Mao, X., Chen, Y., Si, Y., Li, Y., Wan, H., Yu, J., et al., 2013. Novel fluorinated polyurethane decorated electrospun silica nanofibrous membranes exhibiting robust waterproof and breathable performances. RSC Advances 3 (20), 7562–7569.

Melemeni, M., Stamatakis, D., Xekoukoulotakis, N.P., Mantzavinos, D., Kalogerakis, N., 2009. Disinfection of municipal wastewater by TiO_2 phtocatalysis with UV-A, visible and solar irradiation and BDD electrolysis. Global Nest Journal 11 (3), 357–363.

Mercier-Laurent, E., 2015. The Innovation Biosphere: Planet and Brains in the Digital Era. John Wiley & Sons.

Minamikawa, T., Koma, T., Suzuki, A., Mizuno, T., Nagamatsu, K., Arimochi, H., et al., 2021. Quantitative evaluation of SARS-CoV-2 inactivation using a deep ultraviolet light-emitting diode. Scientific Reports 11 (1), 1–9.

Misra, R., Acharya, S., Sahoo, S.K., 2010. Cancer nanotechnology: application of nanotechnology in cancer therapy. Drug Discovery Today 15 (19–20), 842–850.

Mohan, A., Wara, U.U., Shaikh, M.T.A., Rahman, R.M., Zaidi, Z.A., 2021. Telesurgery and robotics: an improved and efficient era. Cureus 13 (3).

Morganti, P., 2010. Use and potential of nanotechnology in cosmetic dermatology. Clinical, Cosmetic and Investigational Dermatology: CCID 3, 5.

Mori, T., Jung, J., Kobayashi, C., Dokainish, H.M., Re, S., Sugita, Y., 2021. Elucidation of interactions regulating conformational stability and dynamics of SARS-CoV-2 S-protein. Biophysical Journal 120 (6), 1060–1071.

Mozafari, M., 2018. Nanotechnology in wound care: one step closer to the clinic. Molecular Therapy 26 (9), 2085–2086.

Nagar, A., Pradeep, T., 2020. Clean water through nanotechnology: needs, gaps, and fulfillment. ACS Nano 14 (6), 6420–6435.

Narvaez Rojas, C., Alomia Peñafiel, G.A., Loaiza Buitrago, D.F., Tavera Romero, C.A., 2021. Society 5.0: a Japanese concept for a superintelligent society. Sustainability 13 (12), 6567.

Navalakhe, R.M., Nandedkar, T.D., 2007. Application of nanotechnology in biomedicine. http://nopr.niscpr.res.in/handle/123456789/5224.

Palit, S., Hussain, C.M., 2018. Green sustainability, nanotechnology and advanced materials: a critical overview and a vision for the future. Green and Sustainable Advanced Materials 2, 1–18.

Phuyal, S., Bista, D., Bista, R., 2020. Challenges, opportunities and future directions of smart manufacturing: a state of art review. Sustainable Futures 2, 100023. Available from: https://doi.org/10.1016/j.sftr.2020.100023.

Pokrajac, L., Abbas, A., Chrzanowski, W., Dias, G.M., Eggleton, B.J., Maguire, S., et al., 2021. Nanotechnology for a sustainable future: addressing global challenges with the international network4sustainable nanotechnology. https://doi.org/10.1021/acsnano.1c10919.

Purushotham, H., 2012. Transfer of nanotechnologies from R&D institutions to SMEs in India. Carbon 16, 47.

Rahmawati, M., Ruslan, A., Bandarsyah, D., 2021. The era of Society 5.0 as the unification of humans and technology: a literature review on materialism and existentialism. Jurnal Sosiologi Dialektika 16 (2), 151–162.

Rai, J.P.N., Saraswat, S., 2022. Principles and potentials of nanobiotechnology. Nano-biotechnology for Waste Water Treatment. Springer, Cham, pp. 1–40.

Rathore, N., Yettou, F., Gama, A., 2021, October. Improvement in wind energy sector using nanotechnology. In: 2020 6th International Symposium on New and Renewable Energy (SIENR) (pp. 1–5). IEEE.

Razmjoo, A., Kaigutha, L.G., Rad, M.V., Marzband, M., Davarpanah, A., Denai, M., 2021. A Technical analysis investigating energy sustainability utilizing reliable renewable energy sources to reduce CO_2 emissions in a high potential area. Renewable Energy 164, 46–57.

Roco, M.C., 2005. Environmentally responsible development of nanotechnology. Environmental Science & Technology 39 (5), 106A–112A.

Schneider, S.L., Lim, H.W., 2019. A review of inorganic UV filters zinc oxide and titanium dioxide. Photodermatology, Photoimmunology & Photomedicine 35 (6), 442–446.

Serrano, E., Rus, G., Garcia-Martinez, J., 2009. Nanotechnology for sustainable energy. Renewable and Sustainable Energy Reviews 13 (9), 2373–2384.

Shah, M.A., Pirzada, B.M., Price, G., Shibiru, A.L., Qurashi, A., 2022. Applications of nanotechnology in smart textile industry: a critical review. Journal of Advanced Research . Available from: https://doi.org/10.1016/j.jare.2022.01.008.

Singh, T., Shukla, S., Kumar, P., Wahla, V., Bajpai, V.K., Rather, I.A., 2017. Application of nanotechnology in food science: perception and overview. Frontiers in Microbiology 8, 1501.

Singh, M.P., Bhardwaj, A.K., Bharati, K., Singh, R.P., Chaurasia, S.K., Kumar, S., et al., 2021a. Biogenic and non-biogenic waste utilization in the synthesis of 2D materials (graphene, h-BN, g-C2N) and their applications. Frontiers in Nanotechnology 53.

Singh, R., Behera, M., Kumari, N., Kumar, S., Rajput, V.D., Minkina, T.M., et al., 2021b. Nanotechnology-based strategies for the management of COVID-19: recent developments and challenges. Current Pharmaceutical Design 27 (41), 4197–4211.

Sun, Z., Liao, T. (Eds.), 2020. Responsive Nanomaterials for Sustainable Applications. Springer. Available from: https://doi.org/10.1007/978-3-030-39994-8.

Tahan, C., 2007. Identifying nanotechnology in society. Advances in Computers 71, 251–271.

Tejeda, J., Ferreira, S., 2014. Applying systems thinking to analyze wind energy sustainability. Procedia Computer Science 28, 213–220. Available from: https://doi.org/10.1016/j.procs.2014.03.027.

Thong, H.Y., Zhai, H., Maibach, H.I., 2007. Percutaneous penetration enhancers: an overview. Skin Pharmacology and Physiology 20 (6), 272–282. Available from: https://doi.org/10.1159/000107575.

Tsakalakos, L., 2012. Application of micro-and nanotechnology in photovoltaics. Comprehensive Renewable Energy 1.

Uddin, M., Rahman, M.M., Asmatulu, R., 2017. Recent progress on synthesis, characterization and applications of carbon black nanoparticles. Advances in Nanotechnology 39.

Udugama, B., Kadhiresan, P., Kozlowski, H.N., Malekjahani, A., Osborne, M., Li, V.Y., et al., 2020. Diagnosing COVID-19: the disease and tools for detection. ACS Nano 14 (4), 3822–3835. Available from: https://doi.org/10.1021/acsnano.0c02624.

United Nations Sustainable Development Goals (UNSDG's), 2021. Department of economic and social affairs, sustainable development. The 17 Goals. https://sdgs.un.org/goals (accessed 14.05.21).

Valdiglesias, V., Kiliç, G., Costa, C., Fernández-Bertólez, N., Pásaro, E., Teixeira, J.P., et al., 2015. Effects of iron oxide nanoparticles: cytotoxicity, genotoxicity, developmental toxicity, and neurotoxicity. Environmental and Molecular Mutagenesis 56 (2), 125–148.

Vanegas, D.C., Gomes, C.L., Cavallaro, N.D., Giraldo-Escobar, D., McLamore, E.S., 2017. Emerging biorecognition and transduction schemes for rapid detection of pathogenic bacteria in food. Comprehensive Reviews in Food Science and Food Safety 16 (6), 1188–1205.

Vlasov, A.I., Terent'ev, D.S., Shakhnov, V.A., 2017. Graphene flexible touchscreen with integrated analog-digital converter. Russian Microelectronics 46 (3), 192–199.

Wang, X., Yang, L., Chen, Z., Shin, D.M., 2008. Application of nanotechnology in cancer therapy and imaging. CA: A Cancer Journal for Clinicians 58 (2), 97–110.

Weiss, C., Carriere, M., Fusco, L., Capua, I., Regla-Nava, J.A., Pasquali, M., et al., 2020. Toward nanotechnology-enabled approaches against the COVID-19 pandemic. ACS Nano 14 (6), 6383–6406.

Wiechers, J.W., Musee, N., 2010. Engineered inorganic nanoparticles and cosmetics: facts, issues, knowledge gaps and challenges. Journal of Biomedical Nanotechnology 6 (5), 408–431.

Wiek, A., Foley, R.W., Guston, D.H., Bernstein, M.J., 2016. Broken promises and breaking ground for responsible innovation–intervention research to transform business-as-usual in nanotechnology innovation. Technology Analysis & Strategic Management 28 (6), 639–650. Available from: https://doi.org/10.1080/09537325.2015.1129399.

Xiang, C., Zhao, X., Tan, L., Ye, J., Wu, S., Zhang, S., et al., 2019. A solar tube: efficiently converting sunlight into electricity and heat. Nano Energy 55, 269–276. Available from: https://doi.org/10.1016/j.nanoen.2018.10.077.

Xiao-Ming, M.A., Mi, S., Yue, L., Yin-Jin, L., Fang, L.U.O., Long-Hua, G., et al., 2018. Progress of visual biosensor based on gold nanoparticles. Chinese Journal of Analytical Chemistry 46 (1), 1–10.

Yoro, K.O., Daramola, M.O., 2020. CO2 emission sources, greenhouse gases, and the global warming effect. Advances in Carbon Capture. Woodhead Publishing, pp. 3–28.

Yousefi, S., Kartoolinejad, D., Naghdi, R., 2017. Effects of priming with multi-walled carbon nanotubes on seed physiological characteristics of Hopbush (Dodonaeaviscosa L.) under drought stress. International Journal of Environmental Studies 74 (4), 528–539.

Chapter 17

Utilization of biogenic waste as a valuable calcium resource in the hydrothermal synthesis of calcium-orthophosphate nanomaterial

A.P. Bayuseno, R. Ismail, J. Jamari and S. Muryanto

Centre for Waste Management, Department of Mechanical Engineering, Diponegoro University, Semarang, Indonesia

17.1 Introduction

17.1.1 Recovery of biogenic waste calcium resources for hydroxyapatite production

Calcium orthophosphate (Ca-P)-based bioceramics, mainly hydroxyapatite (HA), can be derived from biogenic sources and become synthetic biomaterials for human cells because of their specific properties, such as bioactivity and biocompatibility with human bones. HA bioceramics are coating, artificial bone grafts, and porous bioceramic scaffolds relating to self-setting as a type of cement for filling out the damaged bone (Ribas et al., 2019; Turnbull et al., 2018). Here physiological environments are required to develop bioceramics based on Ca-P for bone healing (Baino et al., 2015; Oh et al., 2006; Tavoni et al., 2021). Bioceramics require bioactivity modification of the implanted bone surface, especially on surface-coating bone grafts (Kokubo et al., 2003). Furthermore, the implant materials would be subjected to loading under the influence of external stimulants for many years of use (Dorozhkin, 2013a,b). Ca-P material applications have now spread from orthopedics to dentistry (Abraham, 2014; Klein et al., 1985). Many Ca-P-based bioceramics are available for regenerative medicine with implant coatings and bone implants (Surmenev et al., 2014). The Ca-P material has shown clinical success as a biomedical material for bone regeneration and fracture curing (Havelin et al., 2000; Moroni et al., 2002; Larsson and Bauer, 2002). Ca-P bioceramics are widely used in medicine (Dorozhkin, 2013a) and may, for some reason, outperform autografts in bone regeneration (Russell and Leighton, 2008).

Since then, Ca-P materials research has expanded rapidly over the last 15 years. These materials are easily manufactured and clinically certified, despite their remarkable biological performance (Castro et al., 2022; Ferro and Guedes, 2019; Fihri et al., 2017; He et al., 2016; Pramanik et al., 2007; Yazdani et al., 2018; Zakaria et al., 2013). For example, Ca-P biomaterial scaffolds may promote gas, nutrient, and regulatory factor diffusion, including cell proliferation and differentiation (Heimann, 2002). In particular, porous bones are composed of 65%–70% minerals (hydroxyapatite and microminerals), 20%–25% organic compounds (primarily collagen), and 5.8%–10% H_2O (Jandt, 2007; Williams, 2009). Other elements found in HA that are necessary for bone growth include CO_3^{2-}, Zn^{2+}, Sr^{2+}, and F^- (Fihri et al., 2017; Gruselle, 2015; Hughes and Rakovan, 2015). As a result, inserting those ions into a synthetic HA crystal lattice may change its crystallinity and morphology, influencing its performance in human cells (Fihri et al., 2017). It is possible to synthesize human bone using a biocomposite containing hydroxyapatite and polymers, which provide mechanical strength, flexibility, and biodegradability, whereas HA provides bioactivity and osteoinductivity (Dorozhkin, 2013a).

However, porous polymer structures can induce bone tissue ingrowth and regeneration (Dorozhkin, 2019). Supposedly, chitosan biopolymers are composite scaffolds (CS) for bone tissue regeneration. Here, CS is naturally made of polysaccharides through the deacetylation of the biopolymer chitin, whereas marine crustacean shells have the potential contribution of nontoxic, biodegradable, and biocompatible chitin and chitosan biopolymers (Faridi and Arabhosseini, 2018; Krishna et al., 2007; Liu et al., 2003). Antibacterial and antimicrobial amino acids, in particular, do not cause

Green and Sustainable Approaches Using Wastes for the Production of Multifunctional Nanomaterials. DOI: https://doi.org/10.1016/B978-0-443-19183-1.00005-2

inflammatory reactions (Liu et al., 2003; Salma-Ancane et al., 2016). They have been widely used as scaffolding materials in medical technology (Krishna et al., 2007). Instead, nano-sized HA for biomedical engineering is advantageous for the surface and high catalytic activity and molecule absorbance. Synthetic HA stimulates antiinflammatory, immune, or irritating responses. Given an artificial cytoskeleton, HA can provide the necessary support while allowing cells to proliferate and differentiate functions.

Further, a green powder synthesis with the mechanochemical and hydrothermal techniques is of interest for creating nano-sized HA with diverse powder precursors containing calcium and phosphates. A freeze-drying technology with magnetic composite aerogels (SPIONS@nHA) assisted in nano-sized hydroxyapatite formed with a high-porous structure (Krishnan et al., 2021). As a result, the biosynthesis of nanoparticles (NPs) has become a concern for many researchers (Singh et al., 2021). Moreover, the biological approach has been adopted for synthesizing various nanomaterials (NMs) over the last few decades (Khan et al., 2019; Bhardwaj et al., 2021). Nanotechnology has emerged as a green synthesis route for producing products with excellent biomedical properties (Bhardwaj et al., 2018). Consequently, NPs innovation provided a range of products spreading from biosensing, medical, clinical, solar energy harnessing, and water treatment strategies for human benefit (Cohen-Karni et al., 2012).

Recently, recycling biogenic wastes into HA nanoparticles has gained popularity due to their remarkable biointeractive surfaces at the cell level with improved cell attachment and growth (Ghosh et al., 2011). By reducing and recycling biogenic waste, HA synthesis may promote an environmentally friendly approach (Hassan et al., 2016; Bhardwaj et al., 2021). Many biogenic wastes are high in calcium sources and chemical elements required for human bone (Castro et al., 2022; Ferro and Guedes, 2019). However, searching the terms "hydroxyapatite," "nanoparticles," and "green synthesis" in the ScienceDirect database yielded only publications of 94 documents from 2018 to 2023 (Fig. 17.1). A closer examination of the publication output revealed that publications mainly focused on practical insights into green synthesis for hydroxyapatite nanoparticles. There has been little research into recycling biogenic waste for the production of nanoparticle HA feedstock. The progress of powder green hydrothermal synthesis technology appears slower, particularly for HA materials for biomedical applications. Accordingly, this book chapter presents the progress of hydrothermal synthesis of calcium orthophosphate from biogenic sources, primarily nanocrystalline hydroxyapatite. This book chapter also covers hydroxyapatite chemistry, reaction thermodynamics, and kinetics.

17.2 Calcium orthophosphates phases

Ca- and P-ion molar ratios are set in practice to produce calcium-phosphate minerals (Angervall et al., 2009). Hydroxyapatite has calcium-to-phosphate molar ratios in the range from 1.5 to 1.9, whereas molar ratios between 1.5 and 1.67 may provide hydroxyapatite with a calcium deficit $[Ca_{10-X}(PO_4)_{6-X}(HPO_4)_X(OH)_{2-X}]$ (where x is a value between 0 and 1) (Ofudje et al., 2018). Fig. 17.2 shows the temperature-dependent (°K) calcium orthophosphate-based bioceramics of P_2O_5 and CaO compounds. Nonstoichiometric phases of hydroxyapatite may exist due to cation (Ca^{2+}) and anion (OH) release (Rey et al., 2011). As a result, blending calcium nitrate and di-ammonium phosphate at a Ca/P molar ratio of 1.6 may produce calcium-deficient phases (Raynaud et al., 2002) (Eq. 17.1).

$$9.6\,Ca(NO_3)_2 + 6(NH_4)_2HPO_4 \rightarrow Ca_{9.6}(PO_4)_{5.6}(HPO_4)_{0.4}(OH)_{1.6} \quad (17.1)$$

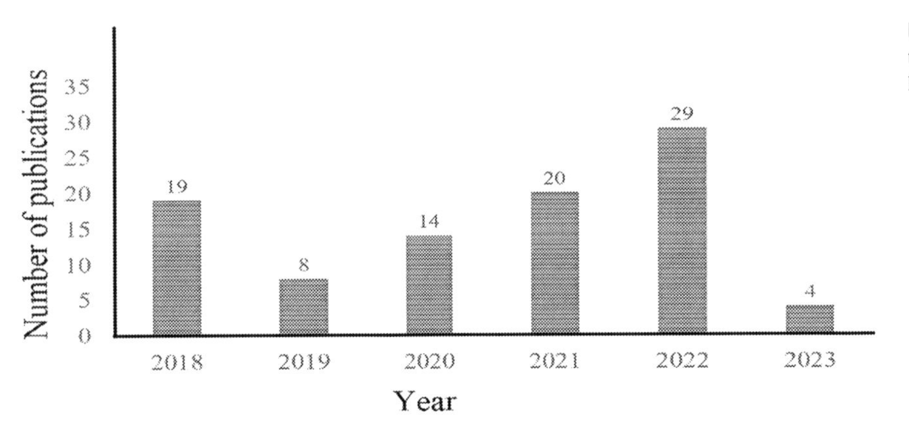

FIGURE 17.1 Trends in annual publication output on green synthesis of nanosized hydroxyapatite research from 2018 to 2023.

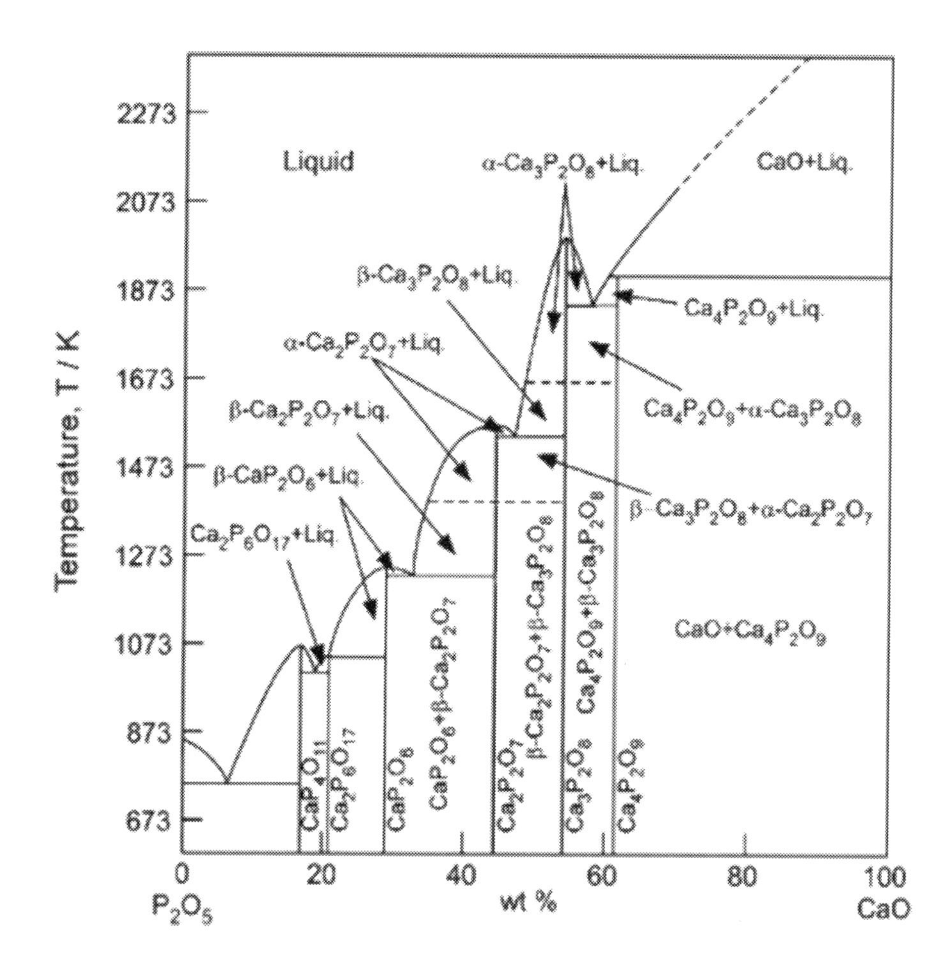

FIGURE 17.2 The phase diagram for the calcium phosphate system (Agbeboh et al., 2020). *Adapted with permission from Agbeboh, NI., Oladele, IO., Daramola, OO., Adediran, AA., Olasukanmi, OO., Tanimola, MO. (2020) Review article: environmentally sustainable processes for the synthesis of hydroxyapatite. Heliyon 6, e03765..*

Heating further causes tricalcium phosphate and hydroxyapatite to form calcium orthophosphate as presented in Eq. (17.2) (Valletregi, 1997).

$$Ca_{10-x}(PO_4)_{6-x}(HPO_4)x(OH)_{2-x} \rightarrow (1-x)Ca_{10}(PO_4)_6(OH)_2 + 3xCa_3(PO_4)_2 \tag{17.2}$$

Table 17.1 also shows the properties of calcium orthophosphates with varying Ca/P molar ratios in the range from 1.5 to 1.67 (Dorozhkin, 2013a). Calcium orthophosphates are practically insoluble in water. Cement can self-harden based on the solid-to-liquid (S/L) ratio, influencing bioresorbability and rheological properties. These common bioceramics can be used directly for bone regeneration, adapting to the shape of the porous bone. Furthermore, bioceramic cement pastes may harden at human body temperature. Correspondingly, a novel and innovative bone regeneration treatment may use various bioceramic cement phases. This condition, in turn, would promote improved osteoconductive properties of compressive strength and noncytotoxicity required for bone restoration. Finally, calcium orthophosphate cement can replace osteotransductively for new bone tissue after implantation.

17.3 The demand for nanoparticle hydroxyapatite biomaterial powder on a global scale

There has been an increase in demand for synthesizing HA powder using biogenic waste such as eggshells, green mussel shells, and fish bones (Dorozhkin and Dorozhkina, 2003). Cuttlefish (Geçer et al., 2008) and bovine bone are two additional biological sources of HA ceramics that provide low-cost raw materials for implant and coating materials (Silva et al., 2013; Wang et al., 2015; Zhang et al., 2010). Additionally, synthetic HA can be made from fish bones for biomedical material products (Park et al., 2017). Calcination, alkaline hydrolysis, and polymer-based methods are all used to extract HA from marine organisms (Dong et al., 2007; Geçer et al., 2008). Nanocrystalline HA has emerged as a promising biogenic waste-derived material, especially for self-setting cement (Simionescu et al., 2017), which provides a bioactive and biodegradable biomedical material (Liu et al., 2005). HA can be made by calcining Japanese sea

TABLE 17.1 Solubility property of specific Ca-P bioceramics (Dorozhkin, 2011).

Calcium/ phosphate molar ratio	Mineral phases	Mineral formula	Soluble at 25°C, g/L	Application
0.5	Monocalcium phosphate monohydrate	$Ca(H_2PO_4)_2 \cdot H_2O$	~18	Increase root fluoride absorption
0.5	Monocalcium phosphate (anhydrous)	$Ca(H_2PO_4)_2$	~17	A synthetic bone graft
1.5	α-tricalcium phosphate	α-$Ca_3(PO_4)_2$	~0.0025	bone repair biodegradable composite
1.5	β-tricalcium phosphate	β-$Ca_3(PO_4)_2$	~0.0025	Orthopedic procedures
1.2–2	Amorphous calcium orthophosphates	$Ca_X H_y(PO_4)_z \cdot n(H_2O)$; n: 3–4.5; 15%–20% H_2O	Not measured precisely	Dentistry
1.5–1.67	Calcium-deficient hydroxyapatite	$Ca_{10-X} (PO_4)_{6-X} (HPO_4)_X (OH)_{2-X}$; $0<x<1$	~0.0094	Bone transplantation
1.67	Hydroxyapatite	$Ca_{10}(PO_4)_2(OH)_2$	~0.0003	Bone regeneration and restoration
1.67	Fluorapatite (FA)	$Ca_{10}(PO_4)_2F$	~0.0002	Fluorine source in pharmaceutical products
1.67	Oxyapatite	$Ca_{10}(PO_4)_2O$	~0.087	Dentistry
1.67	Tricalcium phosphate	$Ca_4(PO_4)_2O$	~ 0.0007	Used in metallic implants as the type of cement and coatings.

bream, including the fish scales, at 1300°C (Fu et al., 2005). In addition, cuttlefish and coral have been investigated to determine the abundance of a calcium source for HA synthesis (Bisht et al., 2005; Kakizawa et al., 2004). HA products derived from this coral are now available with the brand names *ProOsten* and *Interpore200* for primarily biomedical purposes (Oyane et al., 2013). There is also biogenic opal (biosilica) as a mineral found in many marine organisms (Sporysh et al., 2010). Demosponges and hexactinellids are sponges that are high in silica. Calcarea also has a calcium carbonate skeleton, which the enzyme uses to control the formation of sponge biosilica. The axial filament is composed of silicate in its vicinity (Yin and Stott, 2005). Sponges are the only organisms capable of polymerizing silica enzymatically and producing siliceous structure components (Bigi et al., 2000). Also, an enzyme can dissolve nontoxic siliceous spicules (Astala et al., 2006; Boanini et al., 2006a,b; Ikawa et al., 2009; Norton et al., 2006). This biosilica sponge is valuable for bone healing (Astala et al., 2006; Fellah and Layrolle, 2009; Khalil et al., 2007). During bone healing, bioactive glasses (silica) can develop to strongly bind with new bone (Dorozhkin, 2015; Owens et al., 2019).

17.4 Synthesis methods of hydroxyapatite based on biogenic waste

Recycling waste materials into calcium orthophosphates is a more environmentally friendly manner. This synthetic approach's primary goals are to (1) reduce synthesis costs and (2) use fewer hazardous chemicals. Principally, biogenic wastes contain a significant calcium source as well as a variety of additional chemical elements required for HA synthesis (Eq. 17.3):

$$10\ Ca(OH)_2 + 6\ H_3PO_4 \rightarrow Ca_{10}(PO_4)_6(OH)_2 + 18\ H_2O \tag{17.3}$$

Furthermore, the synthesis of HA bioceramics may follow dry, wet, and high-temperature processing routes (Sadat-Shojai et al., 2013). Each of these methods may produce pure crystalline HA with varying crystallite sizes and morphologies influencing the bioactivity of human bone, according to Fig. 17.3 (Cox et al., 2014).

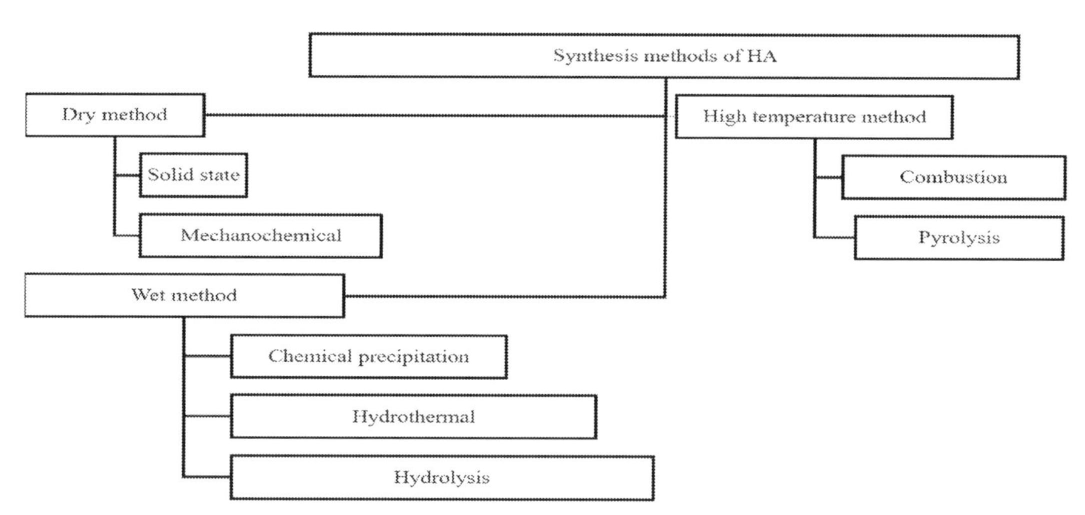

FIGURE 17.3 Synthesis method of hydroxyapatite bio ceramics.

17.5 Various syntheses for producing hydroxyapatite powder

Biocompatibility, physical and chemical properties of HA powder are necessary for implant coatings, dental cement, and dental toothpaste (Rey et al., 2007). The appropriate powder precursor and processing routes using different chemicals and reactants for Ca/P sources can contribute to these variables. Table 17.2 shows the various methods for synthesizing synthetic HA and the results obtained. Advanced techniques have been available to prepare nano-sized HA for human biomedical devices. A comparison of all synthesis methods revealed that changing the technology approach can result in different nanocrystalline HA and polymorphs and morphology of the crystal.

17.6 Hydrothermal synthesis of hydroxyapatite powders

Hydrothermal processing of HA powders uses reactants containing phosphate and calcium ions at specific pressures or the vapor pressure generated by itself at elevated temperatures of 275°C (Yoshimura and Suda, 1994). This method potentially produces nanocrystalline HA with a rod-like morphology and a hexagonal shape, allowing for effective stoichiometric HA control. In particular, the hydrothermal synthesis necessitates specialized autoclaves capable of heating aqueous solutions to temperatures as high as 200°C (Fig. 17.4) (Yoshimura and Suda, 1994). The HA particles of various shapes (sphere, cylinder) formed under these conditions by dissolving and recrystallizing a poorly soluble or insoluble substance. Fig. 17.5 depicts the general routes of hydrothermal processing.

17.7 Hydrothermal hydroxyapatite synthesis using biogenic waste shell sources

Starting materials for hydroxyapatite synthesis can be from biogenic wastes of eggshells, green mussel shells, snail shells, and seashells, which are low-cost, easily accessible natural calcium resources (Abdulrahman et al., 2014). This is an essential strategy for producing a valuable raw material in HA bioceramics (Akram et al., 2014; Oladele et al., 2018a,b). Powder processing routes, such as grinding, cleaning, and calcining, are followed for converting eggshells, seashells, and snail shells into Ca-sources for synthetic HA (Núnez et al., 2018). Cockle, clam, and eggshells, for example, are higher in calcium carbonate and converted to hydroxyapatite. Calcium sources could account for 11% of an eggshell's weight. Eggshells, in particular, are composed of 94% $CaCO_3$, 1% $Ca_3(PO_4)_2$, and 1% $MgCO_3$, with the remaining 4% made up of minute impurities (organic compounds and amino acids) (Li-Chan and Kim, 2008). The cuticle, lamellar, and spongeous layers are the three layers of the eggshell. The cuticle contains structural proteins and traces of needle-like hydroxyapatite. In contrast, calcium carbonate–reinforced fiber protein is found in the structure of the lamellar and spongeous layers (Jojor et al., 2015). Eggshells are oleophilic and necessary for bone tissue absorption. Eggshells and snail shells (Helix Aspersa) contain calcium carbonate skeletons (Jojor et al., 2015).

In particular, different powder processing approaches followed for shells with varying physical and mechanical strengths by typically ground and milled to the desired particle size. After being calcined at around 850°C, the powder is chemically treated with the appropriate chemicals to produce HA powder with specific chemical and biological properties. Among other shell sources, Mollusk shells (Pomacea lineata) are the strongest and contain approximately 97 wt.% of

TABLE 17.2 Synthesis of hydroxyapatite using various Ca/P powder precursor sources.

Synthesis technique	Calcium/phosphate ratio source	Result on mineralogy and morphology	Application	Reference
Solid state	γ-Fe_2O_3 and hydroxyapatite	γ-Fe_2O_3 and HA	Pharmaceutical applications	Azarifar et al. (2022)
	$CaCO_3$, $NH_4H_2PO_4$	91.23% Hydroxyapatite, 5.27% TCP, 3.50% CaO (1250°C for 2 h sintering)	Biomaterials	Punyanitya et al. (2019)
	$CaCO_3$, $CaHPO_4$	Hydroxyapatite (1300°C)	Biomedical; Pharmaceutical applications	Arkin et al. (2015)
	CaO, P_2O_5	Monetite and calcium-deficient hydroxyapatite	Biomedical; Pharmaceutical applications	Pramanik et al. (2007)
	β-TCP, $Ca(OH)_2$	Hydroxyapatite (Chemical ratio 3:2 and 3:3 at 1000°C)	Biomedical; Pharmaceutical applications	Rao et al. (1997)
Mechanochemical	$[Ca(NO_3)_2 \cdot 4H_2O]$, (Na_2HPO_4)	Fe_3O_4/HA/Au	Photocatalysis	Wu et al. (2023)
	Na_2HPO_4, $CaCl_2$ NaOH	HA- nanoparticles	Pharmaceutical applications; Drug-delivery	Ghate et al. (2022)
	Na_2HPO_4, $CaCl_2$ NaOH, $MgCl_2$	Mg/HA scaffold with Ag nanoparticles	Antibacterial properties	Calabrese et al. (2021)
	$FeCl_2 0.4H_2O$, $C_9H_{23}NO_3Si$, (NH4)2HPO4, methanol, tetraethyl orthosilicate (TEOS) and NaOH	Fe_3O_4/SiO_2-hydroxyapatite	Biomedical applications	Orooji et al. (2020)
	$AgNO_3$, $(Ca(NO_3)_2 \cdot 4H_2O)$, (P_2O_5), (C_2H_5OH) NH3 · H2O, NaCl	HA and HA/Ag nanocomposite	Antibacterial properties	Ni et al. (2018)
	$CaHPO_4 \cdot 2H_2O$, $CaCO_3$, CH_4N_2O	Carbonated HA (1 h) Size: 50–150 nm in length and 8 nm in width after milling for24 h	Biomedical applications	Shu et al. (2005)
	$Ca(OH)_2$, CaF_2, P_2O_5	Fluorapatite (6 h) with a size range from 35 to 65 nm	Material application for bioactivity	Fathi and Mohammadi Zahrani (2009)
	$Ca_2P_2O_7$, $CaCO_3$	HA	Biomedical applications	Rhee (2002)
	$CaHPO_4$, CaO	HA (>20–25 h) with a size of about 25 nm	Bioceramic material	Yeong et al. (2001)
Hydrothermal	$Ca(OH)_2$, HNO_3 $(NH4)_2HPO_4$: Temperature 50°C	Irregular agglomerate shape (5–40 mm) of carbonated HA -Ca/P ratio:1.67	Bioceramic material	Prihanto et al. (2023)
	NaH_2PO_4, $CaCl_2$	Hydroxyapatite nanowires	Organics absorbance	Li et al. (2022)
	$C_3H_6N_6$, EDTA, $C_{10}H_{16}N_2O_8$, Na_2HPO_4, NaOH deionized water	g-C_3N_4/HA nanocomposites	Photocatalysts	Mohammad et al. (2022)

(Continued)

TABLE 17.2 (Continued)

Synthesis technique	Calcium/phosphate ratio source	Result on mineralogy and morphology	Application	Reference
	$Ca(NO_3)_2$, H_3PO_4, NaOH and CH_3COONa, H_3PO_4	Template-free HA (TF-HA) and NaAc templated HA (HA-NCPs)	Bioceramic material	Ji et al. (2020)
	$Ca(NO_3)_2 \cdot 6H_2O$, $Tb(NO_3)_3$, Mg $(NO_3)_2$, $(NH_4)_2HPO_4$	HA/Tb/Mg	Bioceramic material	Chen et al. (2018)
	$Ca(OH)_2$, $(NH_4)_3PO_4$; Temperature 200°C	-Irregular agglomerate shape (5–40 mm) of carbonated HA -Ca/P ratio:1.86–2.08	Bioceramic material	Ortiz et al. (2017)
	$Ca(NO_3)_2 0.4H_2 0$, $(NH_4)_2HPO_4$ Temperatures of 25°C–180°C	HA (pH 10) with granule-like shape (30, 50, and 75 nm)	Bioceramic material	Nagata et al. (2013)
	$CaCl_2$, H_3PO_4; Temperature 100°C	HA with rod-like shape (80 nm length and 15 nm width)	Bioceramic material	Zhang et al. (2011)
	$CaCl_2$, K_2HPO_4 Temperature of 60°C–150°C	HA (120°C) with rod-like shape(15–20 nm diameter and 60–75 nm length)	Bioceramic material	Wang et al. (2006)
Hydrolysis	$Ni(NO_3)_2 0.6H_2O$, Ca $(NO_3)_2 0.4H_2O$, PEG, NH_3, and $(NH_4)_2HPO_4$.	HA (500°C, 2 h) nanoparticles	Antibacterial, antifungal, antioxidant, and anticancer	Sebastiammal et al. (2022)
	NaH_2PO_4, $CaCl_2$ NH_4OH	HA (750°C, 2 h) nanoparticles	Bioceramic material	Vaikundam et al. (2022)
	Ca(OH)2, H_3PO_4, NH_4OH	HA (800°C, 3 h) nanoparticles	Bioceramic material	Aguilar et al. (2021)
	$Ca(OH)_2$,$(NH_4)_2HPO_4$, ethanol, $HAuCl_4$	HA (700°C, 2 h) nanoparticles, Au/HA	Antibacterial and antioxidant nanocomposite	Fatimah et al. (2021)
	$Ca(NO_3)_2 \cdot 4H_2O$, $(NH_4)_2HPO_4$, $Ca(NO_3)_2 \cdot 4H_2O$, Zn $(CH_3COO)_2 \cdot 2H_2O$	HA (700°C, 2 h) nanoparticles HA/Zn particles	Bioceramic material	Beyene and Ghosh (2019)
	$CaCl_2 \cdot 2H_2O$ and (NH_4)2HPO_4,	PLA/nHAp/DCM composites	Biocomposite material	Shebi and Lisa (2018)
	DCPD NaOH + alcohol + (CTAB)	HA	Bioceramic material	Wang et al. (2015)
	$Ca(NO_3)_2$,$(NH_4)_2HPO_4$ Solvent: Propane diol (PD); Ethylene glycol (EG)	HA	Bioceramic material	Mechay et al. (2012)
	α-TCP, water	HA (48 h at 40°C) HA (3 h at 100°C)	Bioceramic material	Sinitsyna et al. (2005)
	α-TCP, and solvent: Water – Water + ethanol – Water + 1-butanol – Water + 1-hexanol – Water + 1-octanol	– HA (24 h) – HA + –α-TCP – HA (<72 h) – HA (24 h) – HA (<24 h)	Bioceramic material	Nakahira et al. (1999)
Combustion	NH_4-H_2PO_4, $Ca(NO_3)_2 \cdot 4H_2O$ Fuels: Lemon at Combustion temperature of 700°C	HA (major)	Anticancer activity against human cancer cell lines	Baladi et al. (2023)

(Continued)

TABLE 17.2 (Continued)

Synthesis technique	Calcium/phosphate ratio source	Result on mineralogy and morphology	Application	Reference
	$Ca(NO_3)_2 0.4H_2 0$, Fuels: Urea at Combustion temperature of 500°C	HA (major) and β-TCP (minor)	Bioceramic material	Canillas et al. (2017)
	$Ca(C_2H_3O_2)_2$, $(NH_4)_2HPO_4$ Fuels: Urea at a Combustion temperature of 500°C	HA (ignition temperature of 500°C at a pH of 7.4 with 30 min) Small nanorods with smooth edges of diameter 5 ± 2 nm and length – 17 nm	Bioceramic material	Kavitha et al. (2014)
	$Ca(NO_3)_2$, $NH_4H_2PO_4$, Fuels: Urea at Combustion temperature of 500°C; $Ca(OH)_2$	HA and CaO phases, Rectangular	Bioceramic material	Narayanan et al. (2009)
	$Ca(NO_3)_2$, $(NH_4)_2HPO_4$ Citric acid, succinic Acids, 185–425; Combustion temperature of 185–425	Ca/P ratio: 1.67 – TCP (major) in a mixture of fuel – Single fuel results in carbonated HA	Bioceramic material	Sasikumar, and Vijayaraghavan (2008)
Pyrolysis	H_3PO_4, HCL, NaOH, CaO	HA, Ca/P ratio: 1.67	Bioceramic material	Vinayagam et al. (2023)
	$Ca(C_2H_3 0_2)_2$, $(NH_4)_2HPO_4$	HA	Bioceramic material	Widiyastuti et al. (2014)
	$Ca_3(PO_4)_2$, $Ca(NO_3)_2 0.4H_2 0$, $(NH_4)_2HPO_4$	Hollow sphere with size 1–4 μm HA	Bioceramic material	Cho and Rhee (2012)
	$Ca(NO_3)_2$, $(NH_4)_2HPO_4$, HNO_3	HA, β-$Ca_2P_2O_7$	Bioceramic material	Nakazato et al. (2012)
	$Ca(C_2H_3O_2)_2$, $(CH_3)_3PO_4$	Irregular and flat shape smaller than 40 mm (1013 °K)	Bioceramic material	Nakazato et al. (2012)
		HA, β-$Ca_3(PO_4)_2$	Bioceramic material	Nakazato et al. (2012)
		HA, $CaCO_3$	Bioceramic material	Nakazato et al. (2012)
	$Ca(OH)_2$, H_3PO_4	Irregular and flat shape smaller than 10 mm (913 °K)	Bioceramic material	Kim and Saito (2001)

calcium carbonate (Parveen et al., 2020). Mollusk shells are boiled in water, thoroughly washed, and then pulverized in a ball mill before being calcined in a high-temperature furnace to yield calcium oxide. The calcined products follow the chemical processes routes for producing hydroxyapatite. FTIR and XRD analysis confirmed the formation of hydroxyapatite.

Instead, proteins and organic residues from eggshells were removed by heating at an increasing temperature of 5°C per min to 450°C and a holding time of 2 hours (Rivera et al., 1999). Hence, eggshells were calcined at 0.5°C/min and then kept at 900°C for a holding time of 2 hours, converting the eggshells to calcium oxide by exhausting CO_2, as needed for the subsequent hydrothermal process. In this case, the calcined shell powder dissolved in a phosphate solution and was then hydrothermally treated at an increasing temperature to 1050°C within completely sealed-close vessels. The precipitating solid was separated and collected before drying overnight at 100°C, and the dried powders evaluation uses analytical techniques (SEM, XRD, and FTIR). XRD analysis revealed that pure hydroxyapatite formed with minor impurities. Furthermore, FTIR spectra in the hydroxyapatite phase at 605 and 1050 cm^{-1} bands exhibit HA and TCP functional groups $(PO)_4$. As a result, hydroxyapatite is hydrothermally formed with spherically shaped morphology and has a size of 4.03–10.4 μm and a Ca/P ratio of less than 1.68.

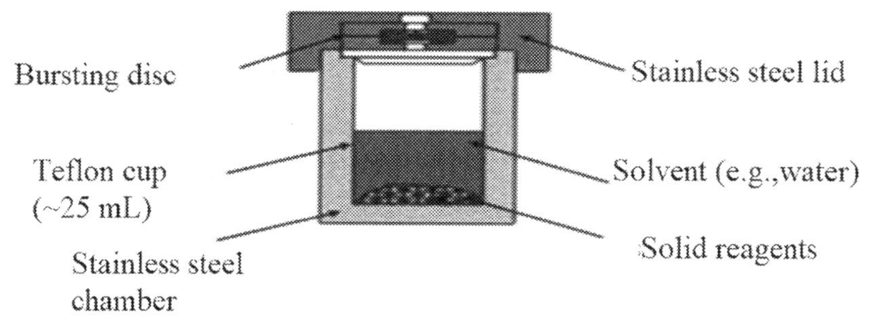

FIGURE 17.4 Autoclave hydrothermal reactor.

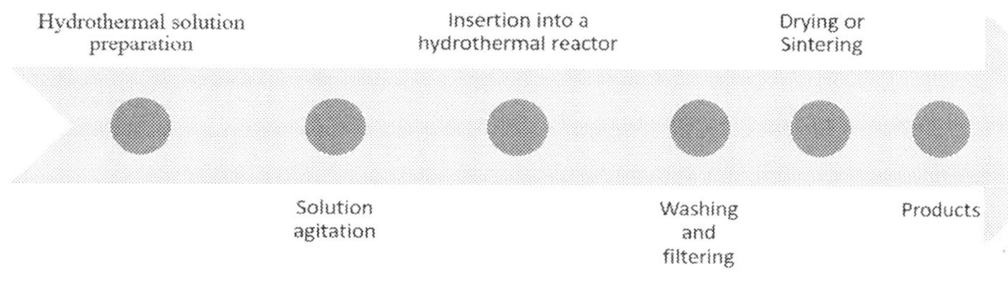

FIGURE 17.5 Hydrothermal procedures for powder synthesis.

Further, Otabil et al. (2018) previously reported work on hydrothermally hydroxyapatite synthesis using calcined eggshells and $(NH4)_2HPO_4$ at 100°C and 110°C for 2, 4, 6, and 8 hours, respectively. The hydrothermal hydroxyapatite product has a crystallite size of 0.702 nm, a fraction of crystal of 90.7%, and a HA phase content of 94.5%. This HA product is suitable for dental composite application and may render bacteria growth with an increased pH. Similarly, the calcined eggshells were obtained from the temperature of 200°C, 400°C, 600°C, 800°C and 900°C for a holding time of 1 hour and subsequently employed for hydrothermal synthesis in a solution containing $Ca_3(PO_4)$ (Hussain and Sabiruddin, 2021). The hydrothermal solution was thermally treated at 1050°C for 3 hours to give hydroxyapatite powder with traces of portlandite [$Ca(OH)_{2}$] and whitlockite [$Ca_9(PO_4)_6PO_3OH$].

On the other hand, the hydrothermal method of calcined mussel shell (*Perna viridis*) powder has been demonstrated for HA synthesis at varying temperatures and time durations (Ismail et al., 2021; Prihanto et al., 2023). After being calcined at 900°C for 5 hours, the powders were chemically treated with nitric acid [$Ca(NO_3)_2$] and by mineral carbonation with CO_2 gas. The precipitating calcium carbonate (PCC) from the pH 12 solution throughout this time was considered a viable process. Mixing PCC products with $(NH_4)_2HPO_4$ during the hydrothermal reaction resulted in high-purity HA. According to FTIR, SEM, TEM, and XRD analysis, this method produced all of the HA's characteristics. The low-temperature hydrothermal for HA synthesis using PCC powder feedstock-derived green mussel shells was demonstrated in our current study (Prihanto et al., 2023). Fig. 17.6 depicts the results of hydrothermal products heated at 110°C for a holding time of 24 hours in the pH solution of 10, as shown on an XRD Rietveld plot for pure hydroxyapatite (Rietveld, 1969). The XRD intensity plots, in this case, showed that the calculated and observed intensities were in agreement, supporting the conclusion that the synthesized powder product is pure hydroxyapatite.

Further, hydrothermal synthesis of HA bioceramics may yield a powder with varying point defects, which can be avoided by carefully controlling process parameters or using a solvent-thermal method. A hydrothermally induced

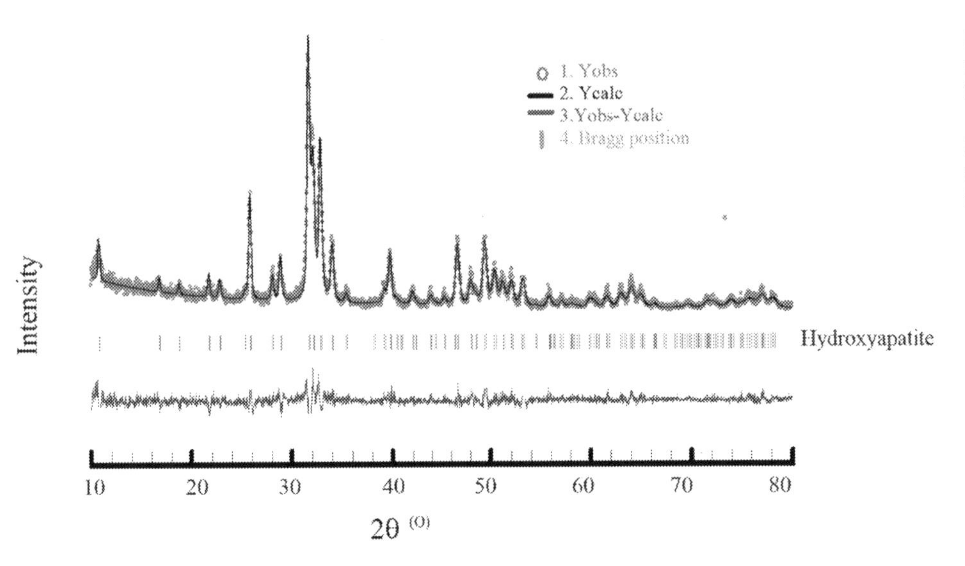

FIGURE 17.6 The XRD Rietveld refinement plot of hydrothermally hydroxyapatite powder formed at 110°C for a holding time of 24 hours in a hydrothermal pH solution of 10. Ycalc: calculated intensity; Yobs: observed intensity.

crystal growth can be observed in real time using various methods, allowing the hydrothermal synthesis to be better controlled. As a result, synthetic HA powder may have more bioactivity than biological apatite and conventional HA. The green mussel shell powder is the potential material to create a highly bioactive, nanosized, and thermo stable HA powder. Green mussel shell HA is a potential biomaterial for bone repair compared with other synthetic HA. The current hydroxyapatite synthesis research focused on waste materials (eggshells and green mussel shells) as feedstock powder for biomedical applications (Townsed et al., 2018). In particular, microwave, ultrasonic, mechanical, and electrochemical energies have improved the kinetics of hydrothermal reactions (Earl et al., 2006), making the experiment's length shorter and the technique more cost-effective for synthesizing HA powder because longer hydrothermal processing times may result in monetite (CaHPO4). Hydrothermal and nanotechnology are two complementary methods for producing nanosized HA particles. The crystallite sizes of HA products ranging from 10 to 60 nm are suitable for dental applications, including drug delivery systems, hyperthermia, neutron capture therapy, bio-imaging, and fluorescent labeling. The hydrothermal method can coat powdered metals, polymers, and ceramics with various compounds.

17.8 The current and future state of integrated calcium resource recovery for hydroxyapatite biomaterials

Emerging research topics in the hydrothermal synthesis of hydroxyapatite focused on developing a cost-effective method with nanoscale implications, accelerating synthesis kinetics, understanding material crystallization within hydrothermal solutions, and advancing in situ material characterization. The availability of calcium carbonate sources derived from biogenic wastes is currently a trend in synthetic HA for biomedical applications. Because of mineral source sustainability, obtaining calcium sources in HA synthesis from biogenic sources would provide the economic benefits of using a low-cost precursor. Biogenic wastes contain chemical elements similar to human cells (Townsed et al., 2018).

Further, calcium extraction from natural sources generally follows various stages of the powder processing method. Grinding, washing, and subsequent calcination at temperature and time may be followed to convert biogenic wastes into a powder precursor suitable for HA synthesis. Egg and mussel shell powder calcined products could mineralize CO_2 while producing valuable precipitated calcium carbonate (Prihanto et al., 2022). Calcium carbonates derived from biogenic materials, including corals, eggshells, and seashells, have been demonstrated to be used for hydrothermal HA synthesis. Moreover, the solid-to-liquid (S/L) ratio determines the quantity of HA powder with solution controlling the self-setting HA formulation.

Indeed, there are differences in microstructures of synthetic and nonsynthetic hydroxyapatites. For example, the HA structure derived from mammalian bones revealed a plate-like shaped morphology with B-type carbonated crystals with an average length of 50 nm, a thickness of 2–3 nm, and a width of 25 nm (Agbeboh et al., 2020). As a result, developing hydroxyapatite-like structures similar to those found in mammalian hard tissue is a research challenge. Also, the compatibility of HA must meet the requirements needed as a drug delivery system and cell carrier (Riman et al., 2002). An improvement in its osteoinductive properties through incorporating trace elements into the crystal structure of HA is a research challenge. The current use of hydroxyapatite in the development of biocomposites proposed for bone regeneration adds to its significance. Biocomposite materials to enhance their bioactivity blend with nanosized HA for bone replacement. As a result, much research has focused on using natural resources for hydrothermally nanosized HA synthesis.

17.9 Conclusion

This book chapter discusses some recent hydrothermal methods for the synthesis of HA using a powder precursor derived from waste biogenic shell sources. Some eggshells and green mussel shells contain a high concentration of $CaCO_3$, which is then processed further to yield a high purity of HA. Furthermore, extracting calcium from natural sources ensures sustainability allowing for waste material nutrient recovery to transform it into value-added material. This powder processing stage may include calcination, carbonation, and chemical precipitation leading to produce polymorphic calcium carbonate. The hydrothermal synthesis method yields a Ca/P ratio of hydroxyapatite as required for human bone regeneration. This hydrothermal technology was simple, effective, and inexpensive for converting calcium-derived biogenic wastes into hydroxyapatite. Increasing demand for homogeneous nanosized HA is essential for insight into reaction parameters such as temperature, time, and initial pH solution in hydrothermal synthesis. In addition, the most current study's HA product has limited osteoconductivity, relating to mineral components that match those found in mammalian hard tissue. Similarly, there is a need for future research conducted to improve the osteoinductive properties of HA by incorporating trace elements into the structure of the developed HA, as well as a thorough analytical evaluation of HA-derived properties and microstructure.

Acknowledgment

The authors would like to thank Diponegoro University Semarang in Indonesia for providing them with world-class university research support (WCUR). 118−32/UN7.6.1/PP/2022 Grant.

References

Abdulrahman, I., Tijani, H.I., Mohammed, B.A., Saidu, H., Yusuf, H., Jibrin, M.N., et al., 2014. From garbage to biomaterials: an overview on eggshell based hydroxyapatite. Journal of Materials 2014, 802467.

Abraham, C.M., 2014. A brief historical perspective on dental implants, their surface coatings, and treatments. Open Dentistry Journal 8 (Suppl 1-M2), 50−55.

Agbeboh, N.I., Oladele, I.O., Daramola, O.O., Adediran, A.A., Olasukanmi, O.O., Tanimola, M.O., 2020. Review article: environmentally sustainable processes for the synthesis of hydroxyapatite. Heliyon 6, e03765.

Aguilar, A.E.M., Fagundes, A.P., Macuvele, D.L.P., Cesca, K., Porto, L., Padoin, N., et al., 2021. Green synthesis of nano hydroxyapatite: morphology variation and its effect on cytotoxicity against fibroblast. Materials Letters 284 (2), 129013.

Akram, M., Ahmed, R., Shakir, I., Ibrahim, W.A.W., Hussain, R., 2014. Extracting hydroxyapatite and its precursors from natural resources. Journal of Materials Science 49 (4), 1461−1475.

Angervall, L., Berger, S., Rockert, H., 2009. A micro radiographic and X-ray crystallographic study of calcium in the pineal body and intracranial tumors. Acta Pathologica et Microbiologica Scandinavica 44 (2), 113−119.

Arkin, V.H., Lakhera, M., Manjubala, I., Narendra, U., Kumar, 2015. Solid state synthesis and characterization of calcium phosphate for biomedical application. International Journal of ChemTech Research 8, 264−267.

Astala, R., Calderin, L., Yin, X., Stott, M.J., 2006. Ab initio simulation of Si-doped hydroxyapatite. Chemistry of Materials: A Publication of the American Chemical Society 18, 413−422.

Azarifar, D., Ghaemi, M., Jaymand, M., Karamian, R., Asadbegy, M., Ghasemlou, F., 2022. Green synthesis and biological activities assessment of some new chromeno[2,3-b] pyridine derivatives. Molecular Diversity 26, 891−902.

Baino, F., Novajra, G., Vitale-Brovarone, C., 2015. Bioceramics and scaffolds: a winning combination for tissue engineering. Frontiers in Bioengineering and Biotechnology 3, 202.

Baladi, M., Amiri, M., Mohammadi, P., Mahdi, K.S., Golshani, Z., Razavi, R., et al., 2023. Green sol−gel synthesis of hydroxyapatite nanoparticles using lemon extract as capping agent and investigation of its anticancer activity against human cancer cell lines (T98, and SHSY5). Arabian Journal of Chemistry 16 (4), 104646.

Beyene, Z., Ghosh, R., 2019. Effect of zinc oxide addition on antimicrobial and antibiofilm activity of hydroxyapatite: a potential nanocomposite for biomedical applications. Materials Today Communications 21, 100612.

Bhardwaj, A.K., Shukla, A., Maurya, S., Singh, S.C., Uttam, K.N., Sundaram, S., et al., 2018. Direct sunlight enabled the photo-biochemical synthesis of silver nanoparticles and their bactericidal efficacy: photon energy as key for size and distribution control. Journal of Photochemistry and Photobiology B: Biology 188, 42−49.

Bhardwaj, A.K., Sundaram, S., Yadav, K.K., Srivastav, A.L., 2021. An overview of silver nano-particles as promising materials for water disinfection. Environmental Technology & Innovation 23, 101721.

Bigi, A., Boanini, E., Panzavolta, S., Roveri, N., 2000. Biomimetic growth of hydroxyapatite on gelatin films doped with sodium polyacrylate. Biomacromolecules 1, 752–756.

Bisht, S., Bhakta, G., Mitra, S., Maitra, A., 2005. DNA loaded calcium phosphate nanoparticles: highly efficient non-viral vector for gene delivery. International Journal of Pharmaceutics 288, 157–168.

Boanini, E., Fini, M., Gazzano, M., Bigi, A., 2006a. Hydroxyapatite nanocrystals modified with acidic amino acids. European Journal of Inorganic Chemistry 2006, 4821–4826.

Boanini, E., Torricelli, P., Gazzano, M., Giardino, R., Bigi, A., 2006b. Nanocomposites of hydroxyapatite with aspartic acid and glutamic acid and their interaction with osteoblast-like cells. Biomaterials 27, 4428–4433.

Calabrese, G., Petralia, S., Franco, D., Nocito, G., Claudia Fabbi, C., Forte, L., et al., 2021. A new Ag-nanostructured hydroxyapatite porous scaffold: antibacterial effect and cytotoxicity study. Materials Science and Engineering: C 118, 111394.

Canillas, M., Rivero, R., García-carrodeguas, R., Barba, F., Rodríguez, M.A., 2017. Processing of hydroxyapatite obtained by combustion synthesis. Boletín La Soc Española Cerámica y Vidr 56, 237.

Castro, M.A.M., Portelaa, T.O., Correaa, G.S., Oliveiraa, M.M., Rangel, J.H.G., Rodrigues, S.F., et al., 2022. Synthesis of hydroxyapatite by hydrothermal and microwave irradiation methods from biogenic calcium source varying pH and synthesis time. Boletín de la sociedad española de cerámica y vidrio 6 (1), 35–41.

Chen, L.-J., Chen, T., Cao, J., Liu, B.-L., Shao, C.-S., Zhou, K.-C., et al., 2018. Effect of Tb/Mg doping on composition and physical properties of hydroxyapatite nanoparticles for gene vector application. Transactions of Nonferrous Metals Society 28 (1), 125–136.

Cho, J.S., Rhee, S.-H., 2012. Preparation of submicron-sized hydroxyapatite powders by spray pyrolysis. Key Engineering Materials 493–494, 215–218.

Cohen-Karni, T., Langer, R., Kohane, D.S., 2012. The smartest materials: the future of nanoelectronics in medicine. ACS Nano 6, 6541–6545.

Cox, S.C., Mallick, K.K., Walton, R.I., 2014. Comparison of techniques for the synthesis of hydroxyapatite. Bioinspired, Biomimetic and Nanobiomaterials. 4, 37–47.

Dong, H., Ye, J.D., Wang, X.P., Yang, J.J., 2007. Preparation of calcium phosphate cement tissue engineering scaffold reinforced with chitin fiber. Journal of Inorganic Materials 22, 1007–1010.

Dorozhkin, S.V., 2011. Calcium orthophosphates: occurrence, properties, biomineralization, pathological calcification, and biomimetic applications. Biomatter 1, 121–164.

Dorozhkin, S.V., 2013a. A detailed history of calcium orthophosphates from the 1770s till 1950. Materials Science and Engineering: C 33 (6), 3085–3110.

Dorozhkin, S.V., 2013b. Calcium orthophosphate-based bioceramics. Materials 6 (9), 3840–3942.

Dorozhkin, S.V., 2015. Calcium orthophosphate-containing biocomposites and hybrid biomaterials for biomedical applications. Journal of Functional Biomaterials 6, 708–832.

Dorozhkin, S.V., 2019. Functionalized calcium orthophosphates ($CaPO_4$) and their biomedical applications. Journal of Materials Chemistry B 7, 7471.

Dorozhkin, S.V., Dorozhkina, E.I., 2003. The influence of bovine serum albumin on the crystallization of calcium phosphates from a revised simulated body fluid. Colloids Surface A 215, 191. 19.

Earl, J.S., Wood, D.J., Milne, S.J., 2006. Hydrothermal synthesis of hydroxyapatite. Journal of Physics: Conference Series 26, 268–271.

Faridi, H., Arabhosseini, A., 2018. Application of eggshell wastes as valuable and utilizable products: a review. Research in Agricultural Engineering 6, 4104–4114.

Fathi, M.H., Mohammadi Zahrani, E., 2009. Mechanical alloying synthesis and bioactivity evaluation of nanocrystalline fluoridated hydroxyapatite. Journal of Crystal Growth 311, 1392–1403.

Fatimah, I., Citradewi, P.W., Yahya, A., Nugroho, B.H., Hidayat, H., Purwiandono, G., et al., 2021. Biosynthesized gold nanoparticles-doped hydroxyapatite as antibacterial and antioxidant nanocomposite. Materials Research Express 8, 115003.

Fellah, B.H., Layrolle, P., 2009. Sol–gel synthesis and characterization of macroporous calcium phosphate bioceramics containing microporosity. Acta Biomaterialia 5, 735–742.

Ferro, A.C., Guedes, M., 2019. Mechanochemical synthesis of hydroxyapatite using cuttlefish bone and chicken eggshell as calcium precursors. Materials Science and Engineering: C 97, 124–140.

Fihri, A., Len, C., Varma, R.S., Solhy, A., 2017. Hydroxyapatite: a review of syntheses, structure, and applications in heterogeneous catalysis. Coordination Chemistry Reviews 347, 48–76.

Fu, H.H., Hu, Y.H., McNelis, T., Hollinger, J.O., 2005. A calcium phosphate-based gene delivery system. Journal of Biomedical Materials Research Part A 74A, 40–48.

Geçer, A., Yıldız, N., Erol, M., Çalimli, A., 2008. Synthesis of chitin calcium phosphate composite in different growth media. Polymer Composites 29, 84–91.

Ghate, P., Prabhu, D., Murugesan, G., Goveas, L.C., Varadavenkatesan, T., Vinayagam, R., et al., 2022. Synthesis of hydroxyapatite nanoparticles using Acacia falcata leaf extract and study of their anti-cancerous activity against cancerous mammalian cell lines. Environmental Research 214 (2), 113917.

Ghosh, S.K., Roy, S.K., Kundu, B., Datta, S., Basu, D., 2011. Synthesis of nano-sized hydroxyapatite powders through solution-combustion route under different reaction conditions. Materials Science and Engineering: B 176, 14–21.

Gruselle, M., 2015. Apatites: a new family of catalysts in organic synthesis. Journal of Organometallic Chemistry 793, 93–101.

Hassan, M.N., Mahmoud, M.M., El-Fattah, A.A., Kandil, S., 2016. Microwave-assisted preparation of nano-hydroxyapatite for bone substitutes. Ceramics International 42, 3725–3744.

Havelin, L.I., Engesaeter, L.B., Espehaug, B., Furnes, O., Lie, S.A., Vollset, S.E., 2000. The norwegian arthroplasty register: 11 years and 73,000 arthroplasties. Acta Orthopaedica Scandinavica 71, 337–353.

He, J., Zhang, K., Wu, S., Cai, X., Chen, K., Li, Y., et al., 2016. Performance of novel hydroxyapatite nanowires in treatment of fluoride contaminated water. Journal of Hazardous Materials 303, 119–130.

Heimann, R.B., 2002. Materials science of crystalline bioceramics: a review of basic properties and applications. CMU Journal 1, 23–46.

Hughes, J.M., Rakovan, J.F., 2015. Structurally robust, chemically diverse: apatite and apatite supergroup minerals. Elements 11 (3), 165–170.

Hussain, S., Sabiruddin, K., 2021. Synthesis of eggshell based hydroxyapatite using hydrothermal method. IOP Conference Series: Materials Science and Engineering 1189, 012024.

Ikawa, N., Kimura, T., Oumi, Y., Sano, T., 2009. Amino acid containing amorphous calcium phosphates and the rapid transformation into apatite. Journal of Materials Chemistry 19, 4906–4913.

Ismail, R., Laroybafih, M.B., Fitriyana, D.F., Nugroho, S., Santoso, Y.I., Hakim, A.J., et al., 2021.) The Effect of hydrothermal holding time on the characterization of hydroxyapatite synthesized from green mussel shells. Journal of Advanced Research in Fluid Mechanics and Thermal Sciences 80 (1), 84–93.

Jandt, K.D., 2007. Evolutions, revolutions and trends in biomaterials science —a perspective. Advanced Engineering Materials 9, 1035–1050.

Ji, X., Fang, Z., Fan, X., Deng, J., Shan, R., Li, Q., et al., 2020. Novel sodium acetate promoted, induced and stabilized mesoporous nanorods: hydroxyapatite nanocaterpillars. Materials Letters 271, 127857.

Jojor, L.M., Bambang, S., Decky, J.I., 2015. Characterization of hydroxyapatite derived from bovine. Asian Journal of Applied Sciences 3 (4), 758. ISSN: 2321 – 0893. Asian Online Journals. Available from: http://www.ajouronline.com.

Kakizawa, Y., Miyata, K., Furukawa, S., Kataoka, K., 2004. Size controlled formation of a calcium phosphate-based organic–inorganic hybrid vector for gene delivery using poly(ethylene glycol)-block-poly(aspartic acid). Advanced Materials 16, 699–702.

Kavitha, M., Subramanian, R., Narayanan, R., Udhayabanu, V., 2014. Solution combustion synthesis and characterization of strontium substituted hydroxyapatite nanocrystals. Powder Technology 253, 129–137.

Khalil, K.A., Kim, S.W., Dharmaraj, N., Kim, K.W., Kim, H.Y., 2007. Novel mechanism to improve the toughness of the hydroxyapatite bioceramics using high-frequency induction heat sintering. Journal of Materials Processing Technology 187–188, 417–420.

Khan, I., Saeed, K., Khan, I., 2019. Nanoparticles: properties, applications, and toxicities. Arabian Journal of Chemistry 12, 908–931.

Kim, W., Saito, F., 2001. Sonochemical synthesis of hydroxyapatite from H_3PO_4 solution with $Ca(OH)_2$. Ultrasonics Sonochemistry 8 (2), 85–88.

Klein, C.P.A.T.K., de Groot, A.A., Drissen, H.B.M.V.D.L., 1985. Interaction of biodegradable β-whitlockite ceramics with bone tissue: an in vivo study. Biomaterials 6 (3), 189–192.

Kokubo, T., Kim, H.M., Kawashita, M., 2003. Novel bioactive materials with different mechanical properties. Biomaterials 24 (13), 2161–2175.

Krishna, D.S.R., Siddharthan, A., Seshadri, S.K., Kumar, T.S.S., 2007. A novel route for the synthesis of nanocrystalline hydroxyapatite from eggshell waste. Journal of Materials Science: Materials in Medicine 18, 1735–1743.

Krishnan, G.K., Prabhakaran, K., George, B.K., 2021. Biogenic magnetic nano hydroxyapatite: sustainable adsorbent for the removal of perchlorate from water at near-neutral pH. Journal of Environmental Chemical Engineering 9, 106316.

Larsson, S., Bauer, T.W., 2002. Use of injectable calcium phosphate cement for fracture fixation: a review. Clinical Orthopaedics and Related Research 395, 23–32.

Li, B., Ren, J., Cheng, X., He, Y., Song, P., Wang, R., 2022. Hydroxyapatite nanowires-based Janus micro-rods for selective separation of organics. Colloids and Surfaces A: Physicochemical and Engineering 652, 129826.

Li-Chan, E.C., Kim, H.-O., 2008. Structure and chemical composition of eggs. Egg Bioscience and Biotechnology. John Wiley & Sons, Inc, pp. 1–95.

Liu, J., Ye, X., Wang, H., Zhu, M., Wang, B., Yan, H., 2003. The influence of pH and temperature on the morphology of hydroxyapatite synthesized by hydrothermal method. Ceramics International 29, 629–633.

Liu, T.Y., Chen, S.Y., Liu, D.M., Liou, S.C., 2005. On the study of BSA-loaded calcium-deficient hydroxyapatite nanocarriers for controlled drug delivery. Journal of Controlled Release 107, 112–121.

Mechay, A., Feki, H.E.L., Schoenstein, F., Jouini, N., 2012. Nanocrystalline hydroxyapatite ceramics prepared by hydrolysis in polyol medium. Chemical Physics Letters 541, 75–80.

Mohammad, I., Jeshurun, A., Ponnusamy, P., Reddy, B.M., 2022. Mesoporous graphitic carbon nitride/hydroxyapatite (g-C_3N_4/HAp) nanocomposites for highly efficient photocatalytic degradation of rhodamine B dye. Materials Today Communications 33, 104788.

Moroni, A., Faldini, C., Rocca, M., Stea, S., Giannini, S., 2002. Improvement of the bone–screw interface strength with hydroxyapatite-coated and titanium-coated AO/ASIF cortical screws. Journal of Orthopaedic Trauma 16 (4), 257–263.

Nagata, F., Yamauchi, Y., Tomita, M., Kato, K., 2013. Hydrothermal synthesis of hydroxyapatite nanoparticles and their protein adsorption behavior. Journal of the Ceramic Society of Japan 121, 797–801.

Nakahira, A., Sakamoto, K., Yamaguchi, S., Kaneno, M., Takeda, S., Okazaki, M., 1999. Novel synthesis method of hydroxyapatite whiskers by hydrolysis of alpha-tricalcium phosphate in mixtures of water and organic solvent. Journal of the American Ceramic Society 82, 2029–2032.

Nakazato, T., Tsukui, S., Nakagawa, N., Kai, T., 2012. Continuous production of hydroxyapatite powder by drip pyrolysis in a fluidized bed. Advanced Powder Technology 23, 632–639.

Narayanan, R., Singh, V., Kwon, T., Kim, K., 2009. Combustion synthesis of hydroxyapatite and hydroxyapatite (silver) powders. Key Engineering Materials 398, 411–419.

Ni, Z., Gu, X., He, Y., Wang, Z., Zou, X., Zhao, Y., et al., 2018. Synthesis of silver nanoparticle-decorated hydroxyapatite (HA@Ag) poriferous nanocomposites and the study of their antibacterial activities. RSC Advances 8 (73), 41722–41730.

Norton, J., Malik, K.R., Darr, J.A., Rehman, I., 2006. Recent developments in processing and surface modification of hydroxyapatite. Advances in Applied Ceramics 105, 113–139.

Núnez, D., Elgueta, E., Varaprasad, K., Oyarzún, P., 2018. Hydroxyapatite nanocrystals synthesized from calcium-rich bio-wastes. Materials Letters 230, 64–68.

Ofudje, E.A., Rajendran, A., Adeogun, A.I., Idowu, M.A., Kareem, S.O., Pattanayak, D.K., 2018. Synthesis of organic derived hydroxyapatite scaffold from pig bone waste for tissue engineering applications. Advanced Powder Technology 29 (1), 1–8.

Oh, S., Oh, N., Appleford, M., Ong, J.L., 2006. Bioceramics for tissue engineering applications – a review. American Journal of Biochemistry and Biotechnology 2 (2), 49–56.

Oladele, I.O., Akinola, O.S., Agbabiaka, O.G., Omotoyinbo, J.A., 2018a. A mathematical model for the prediction of impact energy of organic material-based hydroxyapatite (HAp) reinforced Epoxy composites. Fibers Polymers 19 (2), 452–459.

Oladele, I.O., Agbabiaka, O.G., Olasunkanmi, O.G., Balogun, A.O., Popoola, M.O., 2018b. Non-synthetic sources for the development of hydroxyapatite. Journal of Applied Biotechnology & Bioengineering 5 (2), 88–95.

Orooji, Y., Mortazavi-Derazkola, S., Ghoreishi, S.M., Amiri, M., Salavati-Niasari, M., 2020. Mesopourous Fe_3O_4@SiO_2-hydroxyapatite nanocomposite: green sonochemical synthesis using strawberry fruit extract as a capping agent, characterization and their application in sulfasalazine delivery and cytotoxicity. Journal of Hazardous Materials 400, 123140.

Ortiz, S.L., Avila, J.H., Gutierrez, M.P., Gomez-Pozos, H., Karthik, T.V.K., Lugo, V.R., 2017. Hydrothermal synthesis and characterization of hydroxyapatite microstructures. In: 14th International Conference on Electrical Engineering, Computing Science and Automatic Control, CCE, 0–3.

Otabil, A., Yeboah, F., Hou Gbologah, Y., 2018. Synthesis of hydroxyapatite from eggshells through hydrothermal process. EJUR 2 (1), 110. 101.

Owens, C.L., Nash, G.R., Hadler, K., Fitzpatrick, R.S., Anderson, C.G., Wall, F., 2019. Apatite enrichment by rare earth elements: a review of the effects of surface properties. Advances in Colloid and Interface Science 265, 14–28.

Oyane, H., Araki, H., Sogo, Y., Ito, A., Tsurushima, H., 2013. Coprecipitation of DNA and calcium phosphate using an infusion fluid mixture. Key Engineering Materials 529–530, 465–470.

Park, K.H., Kim, S.J., Hwang, M.J., Song, H.J., Park, Y.J., 2017. Biomimetic fabrication of calcium phosphate/chitosan nanohybrid composite in modified simulated body fluids. Express Polymer Letters 11, 14–20.

Parveen, S., Chakraborty, A., Chanda, D.K.R., Pramanik, S., Barik, A., Aditya, G., 2020. Microstructure analysis and chemical and mechanical characterization of the shells of three freshwater snails. ACS Omega 5 (40), 25757–25771.

Pramanik, S., Agarwal, A.K., Rai, K.N., Garg, A., 2007. Development of high strength hydroxyapatite by solid-state-sintering process. Ceramics International 33, 419–426.

Prihanto, A., Muryanto, S., Ismail, R., Jamari, J., Bayuseno, A.P., 2022. Practical insights into the recycling of green mussel shells (Perna-Viridis) for the production of precipitated calcium carbonate. Environmental Technology 25, 1–11.

Prihanto, A., Muryanto, S., Sancho Vaquer, A., Schmahl, W.W., Ismail, R., Jamari, J., et al., 2023. In-depth knowledge of the low-temperature hydrothermal synthesis of nanocrystalline hydroxyapatite from waste green mussel shell (*Perna viridis*). Environmental Technology 5, 1–13.

Punyanitya, S., Thiansem, S., Raksujarit, A., Chankachang, P., Sirisoam, T., Koonawoot, R., 2019. Fabrication and characterization of porous bioceramic made from bovine bone powder mixed calcium phosphate glass. Key Engineering Materials 803, 187–191.

Rao, R.R., Roopa, H.N., Kannan, T.S., 1997. Solid state synthesis and thermal stability of HAP and HAP - beta-TCP composite ceramic powders. Journal of Materials Science. Materials in Medicine 8, 511–518.

Raynaud, S., Champion, E., Bernache-Assollant, D., Thomas, P., 2002. Calcium phosphate apatites with variable Ca/P atomic ratio I. Synthesis, characterization, and thermal stability of powders. Biomaterials 23 (4), 1065–1072.

Rey, C., Combes, C., Drouet, C., Sfihi, H., Barroug, A., 2007. Physicochemical properties of nanocrystalline apatites: implications for biominerals and biomaterials. Materials Science and Engineering 27, 198–205.

Rey, C., Combes, C., Drouet, C., Grossin, D., 2011. Bioactive ceramics: physical chemistry. Ducheyne, Paul. Comprehensive Biomaterials. Elsevier, pp. 187–281, 1st.

Rhee, S.H., 2002. Synthesis of hydroxyapatite via mechanochemical treatment. Biomaterials 23, 1147–1152.

Ribas, R.G., Schatkoski, V.M., Montanheiro, T.L.D.-A., de Menezes, B.R.C., Stegemann, C., Leite, D.M.G., et al., 2019. Review article: current advances in bone tissue engineering concerning ceramic and bioglass scaffolds: a review. Ceramics International 45 (17), 21051–21061. Part A, 1.

Rietveld, H.M., 1969. A profile refinement method for nuclear and magnetic structures. Journal of Applied Crystallography 2, 65–71.

Riman, R.E., Suchanek, W.L., Byrappa, K., Chen, C.-W., Shuk, P., Oakes, C.S., 2002. Solution synthesis of hydroxyapatite designer particulates. Solid State Ion 151, 393–402.

Rivera, E.M., Araiza, M., Brostow, W., Castano, V.M., Dıaz-Estrada, J.R., Hernandez, R., et al., 1999. Synthesis of hydroxyapatite from eggshells. Materials Letters 41 (3), 128–134.

Russell, T.A., Leighton, R.K., 2008. Comparison of autogenous bone graft and endothermic calcium phosphate cement for defect augmentation in tibial plateau fractures. A multicenter, prospective, randomized study. Journal of Bone and Joint Surgery - Series A 90 (10), 2057–2061.

Sadat-Shojai, M., Khorasani, M.T., Dinpanah-Khoshdargi, E., Jamshidi, A., 2013. Synthesis methods for nanosized hydroxyapatite with diverse structures. Acta Biomaterialia 9, 7591–7621.

Salma-Ancane, K., Stipniece, L., Irbe, Z., 2016. Effect of biogenic and synthetic starting materials on the structure of hydroxyapatite bioceramics. Ceramics International 42, 9504–9510.

Sasikumar, S., Vijayaraghavan, R., 2008. Solution combustion synthesis of bioceramic calcium phosphates by single and mixed fuels- a comparative study. Ceramics International 34, 1373–1379.

Sebastiammal, S., Fathima, A.S.L., Henry, J., Wadaan, M.A., Mahboob, S., Wadaan, A.M., et al., 2022. Synthesis, characterization, antibacterial, antifungal, antioxidant, and anticancer activities of nickel-doped hydroxyapatite nanoparticles. Fermentation 8 (12), 677.

Shebi, A., Lisa, S., 2018. Pectin mediated synthesis of nano hydroxyapatite-decorated poly(lactic acid) honeycomb membranes for tissue engineering. Carbohydrate Polymers 201, 39–47.

Shu, C., Yanwei, W., Hong, L., Zhengzheng, P., Kangde, Y., 2005. Synthesis of carbonated hydroxyapatite nanofibers by mechanochemical methods. Ceramics International 31, 135–138.

Silva, S.S., Duarte, A.R.C., Oliveira, J.M., Mano, J.F., Reis, R.L., 2013. Alternative methodology for chitin–hydroxyapatite composites using ionic liquids and supercritical fluid technology. Journal of Bioactive and Compatible Polymers 28, 481–491.

Simionescu, B.C., Drobota, M., Timpu, D., Vasiliu, T., Constantinescu, C.A., Rebleanu, D., et al., 2017. Biopolymers/poly(e-caprolactone)/polyethylenimine functionalized nano-hydroxyapatite hybrid cryogel: synthesis, characterization and application in gene delivery. Materials Science and Engineering: C 81, 167–176.

Singh, M.P., Bhardwaj, A.K., Keval Bharati, K., Singh, R.P., Chaurasia, S.K., Kumar, S., et al., 2021. Biogenic and non-biogenic waste utilization in the synthesis of 2D materials (graphene, h-BN, g-C2N) and their applications. Frontiers in Nanoscience 3, 53.

Sinitsyna, O.V., Veresov, A.G., Kovaleva, E.S., Kolenko, Y.V., Putlyaev, V.I., Tretyakova, Y.D., 2005. Synthesis of hydroxyapatite by hydrolysis of α-$Ca_3(PO_4)_2$. Russian Chemical Bulletin International Edition 54, 79–86.

Sporysh, E., Shynkaruk, O., Lysko, O., Shynkaruk, A., Dubok, V., Buzaneva, E., et al., 2010. Biomimetic hydroxyapatite nanocrystals in composites with C60 and Au-DNA nanoparticles: IR-spectral study. Materials Science and Engineering: B 169, 128–133.

Surmenev, R.A., Surmeneva, M.A., Ivanova, A.A., 2014. Significance of calcium phosphate coatings for the enhancement of new bone osteogenesis - a review. Acta Biomaterialia 10 (2), 557–579.

Tavoni, M., Dapporto, M., Tampieri, A., Sprio, S., 2021. Bioactive calcium phosphate-based composites for bone regeneration. Journal of Composites Science 5 (9), 227.

Townsed, M., Davies, K., Hanley, N., Hewitt, J.E., Lundquist, C.J., Lohrer, A.M., 2018. The challenge of implementing the marine ecosystem service concept. Frontiers in Marine Science 5, 359.

Turnbull, G., Clarke, J., Picard, F., Riches, P., Jia, L., Han, F., et al., 2018. 3D bioactive composite scaffolds for bone tissue engineering. Bioactive Materials 3 (3), 278–314.

Vaikundam, M., Shanmugam, S., Aldawood, S., Arunkumar, P., Santhanam, A., 2022. Preparation of biopolymer pectin fascinate hydroxyapatite nanocomposite for biomedical applications. Applied Nanoscience 2022.

Valletregi, M., 1997. Synthesis and characterization of calcium-deficient apatite. Solid State Ion 101–103, 1279–1285.

Vinayagam, R., Sandhya Kandati, S., Murugesan, G., Goveas, L.C., Baliga, A., Shraddha Pai, A., et al., 2023. Bioinspiration synthesis of hydroxyapatite nanoparticles using eggshells as a calcium source: evaluation of Congo red dye adsorption potential. Journal of Materials Research and Technology 22, 169–180.

Wang, Y., Zhang, S., Wei, K., Zhao, N., Chen, J., Wang, X., 2006. Hydrothermal synthesis of hydroxyapatite nanopowders using cationic surfactant as a template. Materials Letters 60, 1484–1487.

Wang, M., Chen, H., Shih, W., Chang, H., Wang, M.-C., Chen, H.-T., et al., 2015. Crystalline size, microstructure, and biocompatibility of hydroxyapatite nanopowders by hydrolysis of calcium hydrogen phosphate dehydrate (DCPD). Ceramics International 41, 2999–3008.

Widiyastuti, W., Setiawan, A., Winardi, S., Nurtono, T., Setyawan, H., 2014. Particle formation of hydroxyapatite precursor containing two components in a spray pyrolysis process. Frontiers of Chemical Science and Engineering 8, 104–113.

Williams, D.F., 2009. On the nature of biomaterials. Biomaterials 30, 5897–5909.

Wu, R., Song, J., Lu, J., Ji, X., Tian, G., Zhang, F., 2023. Constructions of Fe_3O_4/HAp/Au nanohybrids with multifunctional structure for efficient photocatalysis and environmental remediation of organic dyes. Journal of Molecular Structure 1278, 134908.

Yazdani, J., Ahmadian, E., Sharifi, S., Shahi, S., Dizaj, S.M., 2018. A short view on nano-hydroxyapatite as coating of dental implants. Biomedicine & Pharmacotherapy = Biomedecine & Pharmacotherapie 105, 553–557.

Yeong, K.C.B., Wang, J., Ng, S.C., 2001. Mechanochemical synthesis of nanocrystalline hydroxyapatite from CaO and $CaHPO_4$. Biomaterials 22, 2705–2712.

Yin, X., Stott, M.J., 2005. Theoretical insights into bone grafting Si-stabilized β-tricalcium phosphate. The Journal of Chemical Physics 122, 024709.

Yoshimura, M., Suda, H., 1994. Hydrothermal processing of hydroxyapatite: past, present, and future. Hydroxyapatite and Related Compounds. CRC Press Inc., Boca Raton, FL, USA, pp. 45–72.

Zakaria, S.M., Zein, S.H.S., Othman, R., Yang, F., Jansen, J.A., 2013. Nanophase hydroxyapatite as a biomaterial in advanced hard tissue engineering: a review. Tissue Engineering Part B: Reviews 19, 431–441.

Zhang, Y., Reddy, V.J., Wong, S.Y., Li, X., Su, B., Ramakrishna, S., et al., 2010. Enhanced biomineralization in osteoblasts on a novel electrospun biocomposite nanofibrous substrate of hydroxyapatite/collagen/chitosan. Tissue Engineering. Part A 16, 1949–1960.

Zhang, G., Chen, J., Yang, S., Yu, Q., Wang, Z., Zhang, Q., 2011. Preparation of amino-acid-regulated hydroxyapatite particles by hydrothermal method. Materials Letters 65, 572–574.

Chapter 18

A review of plant-derived metallic nanoparticles synthesized by biosynthesis: synthesis, characterization, and applications

Banafsheh Haji Ali and Majid Baghdadi
School of Environment, College of Engineering, University of Tehran, Tehran, Iran

18.1 Introduction

As time progressed, nanoscale metals demonstrated greater performance than bulk metals in specific functions due to their unique physical and chemical properties (such as solubility, purity, chemical composition, and stability) (Adeyemi et al., 2022). Metallic nanoparticles (MNPs) are one of the most significant and practicable parts of nanomaterials, which means they are a major product of nanotechnology (Solgi and Taghizadeh, 2020). As a result of their unique characteristics, many metal-based nanomaterials (like Ag, Au, Cu, etc.) have received significant attention in a variety of fields, including medicine, biotechnology, catalysts, biosensors, cosmetics, electrochemistry, water treatment and purification, and agricultural and food industries. Therefore new MNPs have been rapidly developed and produced to meet the growing request for products and devices derived from nanotechnology (Ettadili et al., 2022; Yazdanian et al., 2022; Ying et al., 2022).

Metallic nanoparticles are a cluster of metal atoms ranging from 1 to 100 nm in size, making them highly effective in many profitable fields (Solgi and Taghizadeh, 2020; Jadoun et al., 2021; Adeyemi et al., 2022). MNPs are made with precursors of metals and also have a core of metals, metal oxides, metal sulfides, metal phosphates, etc.—which are usually covered with organic or inorganic materials (Ishtiaq et al., 2019). The most common methods for synthesizing MNPs are chemical, physical, and biological (Aboyewa, et al., 2021a,b). It should be mentioned that chemical and physical methods are not only costly and difficult but they also release toxic substances into the environment. Furthermore, high levels of radiation as well as high concentrations of reductants (such as sodium citrate, sodium borohydride ($NaBH_4$), and organic solvents) and stabilizers (such as polyethylene glycol) employed in these methods harm both the environment and the humans. Consequently, a new method has been developed for the synthesis of metallic nanoparticles called "green" synthesis, which is more environmentally friendly and green (Siakavella et al., 2020; Ettadili et al., 2022). Since green synthesis is based on clean-up, regulation/guideline, control, and remediation methods, it directly enhances eco-friendly materials and nanomaterials. In addition, green synthesis minimizes the amount of waste produced, uses renewable and environmentally friendly feedstocks and raw materials instead of harmful ones, and avoids the use of hazardous solvents, separation agents, and hazardous chemicals in the synthesis process—all of which reduce derivatives and contamination (Patel et al., 2020; Sidhu et al., 2022). In green synthesis, the utilization of safer solvents, such as water, acetone, ethanol, and methanol, is promoted (Nair et al., 2022). Moreover, green synthesis aims to generate nontoxic/less toxic chemicals and nanoparticles with the desired characteristics (Patel et al., 2020). MNPs can be synthesized at room temperature, natural pH neutrality, and at a low cost by using organic compounds, microbes, plants, and plant-derived materials (Ettadili et al., 2022; Nair et al., 2022).

In aqueous suspension, MNPs can be dispersed and transported based on their ability to form environmental stability. Metallic nanoparticles' stability can be estimated by assessing their tendency to aggregate or interact with the

Green and Sustainable Approaches Using Wastes for the Production of Multifunctional Nanomaterials. DOI: https://doi.org/10.1016/B978-0-443-19183-1.00001-5

environment. Generally, the collision rate of particles determines the rate of aggregation of particles over time, while the particle size and their interaction with the environment greatly determine the stability of suspensions (Singh et al., 2018). It has been shown that surface complexation can alter the intrinsic stability of the nanoparticles by controlling the colloidal stability. Colloidal stability or dissolution rate can be controlled according to the particle size and surface capping or functionalization (Pal et al., 2019). According to recent studies, MNPs synthesized by utilizing plant extracts and plant metabolites have higher stability and appropriate dimensions than those synthesized by other green synthetic methods (Ettadili et al., 2022; Nair et al., 2022). This chapter aims to provide a comprehensive overview of the most useful information regarding the biosynthesis of metallic nanoparticles derived from plant extracts and their application in biomedical, wastewater treatment, and agriculture.

18.1.1 The general procedure for preparing a plant extract

Simple MNP preparation from plants generally involves first collecting and washing desired plant parts using running tap water or distilled water. The plant part can be applied either fresh or dried. The fresh plant part is usually dried at room temperature and then crushed by applying a mortar and pestle. The prepared fresh or dried part is boiled for a short time, and then, using filter paper, the solution is pressed out to obtain a clear extract. In order to prepare nanoparticles, plant extracts can be stored at low temperatures or mixed directly with the corresponding metal solutions. Thermal degradation of phytochemicals may occur as a result of excessive heating. It is essential to adapt efficient procedures for the effective extraction of phytochemical compounds from plants. By immersing the crushed plant parts in the solvents for a relatively long period of time, the content of chemical compounds of plant extract may be increased. In addition, the extraction rate is influenced by the selection of the appropriate solvent. It was found that the phytochemical content was increased in the alcoholic and phenolic extracts. Furthermore, in this step, a number of optimization factors can be adjusted, including pH, temperature, and time. As a result of the incubation period, the mixture color changes, confirming that the MNPs have been synthesized (Nabi et al., 2018; Miu and Dinischiotu, 2022; Nair et al., 2022) (Fig. 18.1).

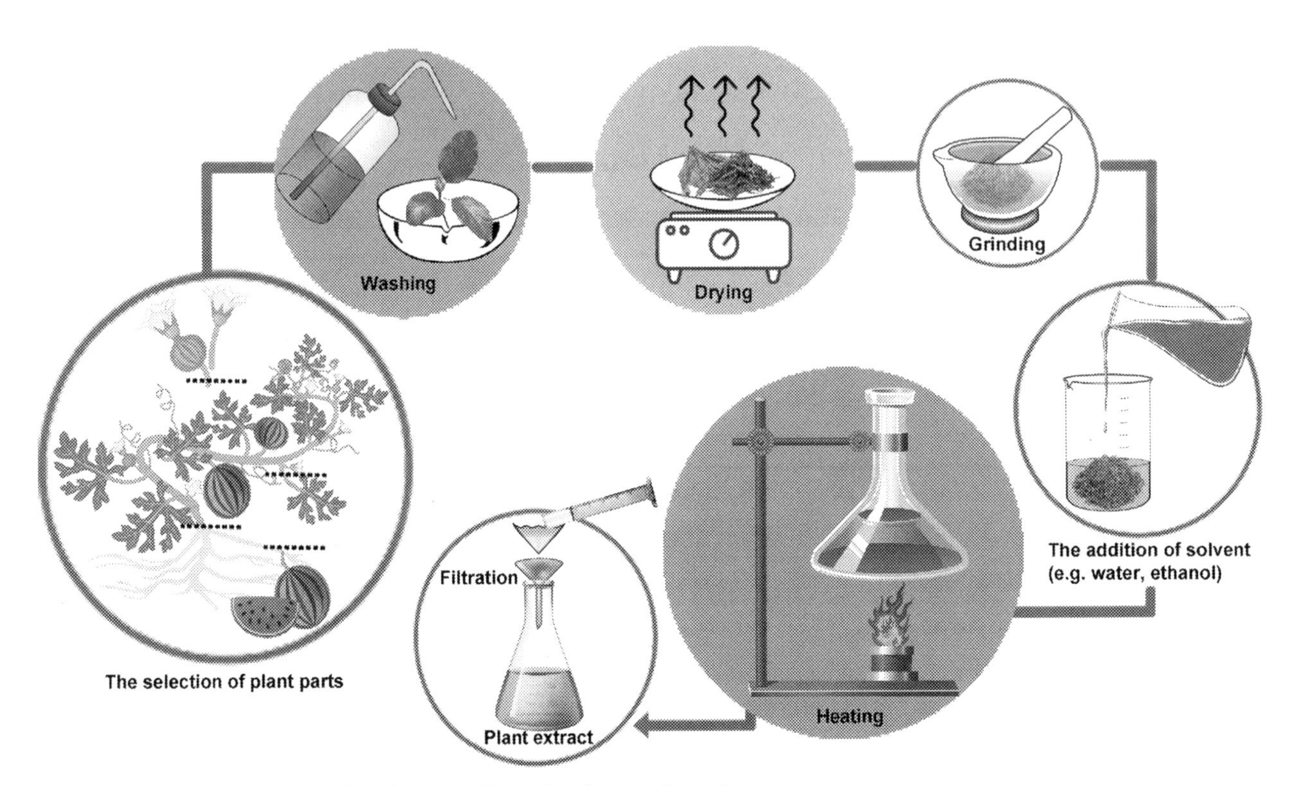

FIGURE 18.1 Schematic representation of the general procedure for preparing a plant extract.

18.2 Advantages of the plant extract-mediated synthesis of metallic nanoparticles

Plant extract-based biosynthesis methods owing to being easy, effective, economical, low cost, and practicable are ideal replacements for conventional techniques (Singh et al., 2018). Furthermore, the use of plant extract for "green" synthesis of MNPs is becoming increasingly important due to its ability to synthesize large amounts of nanoparticles, synthesize nanoparticles with different morphologies at a rapid rate, elimination of complicated cell culture maintenance, simplicity, nonpathogenicity, and cost-effectiveness compared to bacteria and/or fungi. In other words, the pathogenicity of microorganisms and the need for maintaining a large cell culture are significant disadvantages of using them to produce metallic nanoparticles (Aboyewa et al., 2021a,b; Kulkarni et al., 2021).

In the MNPs' synthesis, the biodiversity of plants has been taken into account owing to the presence of essential phytochemicals (including ketones, aldehydes, flavones, amides, terpenoids, carboxylic acids, phenols, and ascorbic acids) in extracts of different plant parts, particularly leaves (Alyamani et al., 2021; Vanitha, 2021). It should be mentioned that, due to the presence of biomolecules (such as carbohydrates, proteins, and coenzymes) with the exceptional ability for reducing metal salts into nanoparticles, plants are the best biological agents for preparing the metal and metal-oxide nanoparticles, and metal salts can be reduced to metal nanoparticles by such components (Singh et al., 2018). In addition, the possibility that plants can accumulate different amounts of heavy metals in different parts offers some interesting opportunities for further exploration of their applications in the preparation of metallic nanoparticles, which would allow for reduction and stabilization (Singh et al., 2018; Aboyewa et al., 2021a,b).

18.3 The role of plant extract in the synthesis of metallic nanoparticles

As noted above, plant extracts, due to their unique nontoxic phytochemical and bioactive components, are one of the most suitable candidates for metallic nanoparticle synthesis, since they have a great capacity for working as agents for capping/reducing/stabilizing in nontoxic conditions (Adeyemi et al., 2022; Ettadili et al., 2022). Regarding previous studies, many plants are really great sources of various bioactive compounds such as proteins, amino acids, polysaccharides, and phytochemicals (including flavonoids, polyphenols, polysaccharides, steroids, alkaloids, sapogenins, tannin, terpenoids, carbohydrates, proteins, amino acids, alcoholic compounds, organic acid, heterocyclic compounds, and vitamins). In general, based on the recent research, the existence of these derived substances in plant extracts has been proven much effective in converting metal ions to nanoparticles either by reducing or capping/stabilizing mechanisms (Jadoun et al., 2021; Ying et al., 2022).

As a first step in the synthesis of MNPs using plant extracts, phytochemical compounds have been found to be involved in the reduction of metal ions to metal/oxide metal NPs by phytochemical degradation. After that, metal-oxide ions are bound together by electrostatic attraction, resulting in metal nanoparticles. As MNPs are synthesized, phytochemical compounds will stabilize them to prevent agglomeration (Shafey, 2020) (Fig. 18.2).

18.3.1 Reduction mechanism

Although the plant-mediated synthesis of MNPs is still complicated in terms of their precise mechanism and the components involved, it has been suggested that there is a bioreduction stage during the synthesis of MNPs in which metal ions/salts are reduced from their monovalent or divalent forms to zerovalent oxidations by bioactive compounds contained in the extracts of plants (Aboyewa et al., 2021a,b). This means that the metals can accept electron pairs from reducing agents when the reduction takes place on their surface (Bindhu et al., 2020). Physical observation of a change in color in the reaction medium determines that the reduced metal atoms are nucleating. These stages of bioreduction and nucleation proceed to produce larger particles from smaller particles due to their mechanical interaction, leading to a more thermodynamically stable form of MNPs. The metallic nanoparticles' shape and morphology are determined by the stability potentials of bioactive plant compounds during the termination phase of the nanoparticle synthesis (Aboyewa et al., 2021a,b). Many natural biomolecules in the extract of plants such as proteins, enzymes, vitamins, polysaccharides, amino acids, and organic acids like citrates possibly have the ability to reduce metal ions (Mohammadlou et al., 2016). Also, flavonoids, phenolics, terpenoids, sugars, ketones, aldehydes, carboxylic acids, and amides are the most important phytochemicals for the bioreduction of metallic nanoparticles (Singh et al., 2018; Yazdanian et al., 2022). Ag NPs and Au NPs, for example, were reduced and stabilized by phenols, sugars, proteins, and other phytochemical constituents found in the flower extract of *Aglaia elaeagnoidea* (Manjari et al., 2017).

Flavonoids play an important role as a reducing agent for the synthesis of MNPs and also bind with metal ions. A variety of fruits and vegetables contain flavonoids, which are a class of natural compounds including anthocyanins,

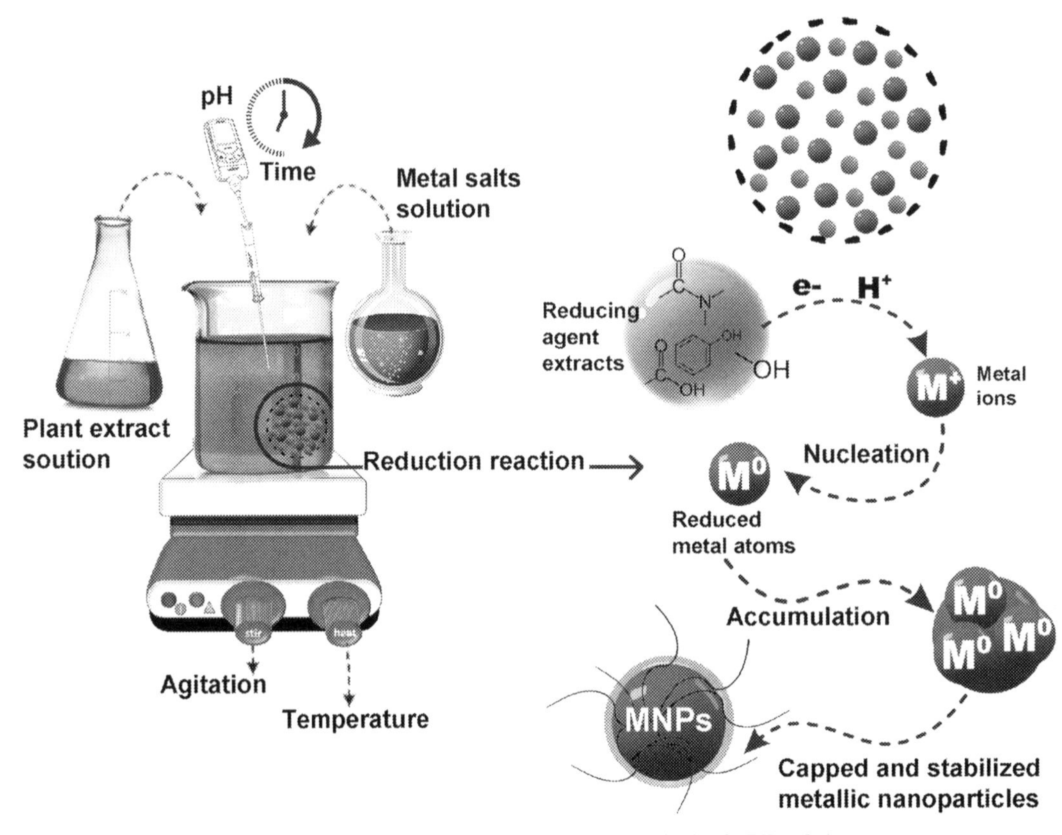

FIGURE 18.2 Schematic representation of biochemical mechanism suggested to the synthesis of MNPs of plant extracts.

isoflavonoids, flavonols, chalcones, flavones, and flavanones. These natural phenolic compounds are able to reduce metal ions more effectively. The reduction process of metal ions into nanoparticles occurs as a result of tautomeric transformation into flavonoids because the conversion of the enol-form to the keto-form results in the release of reactive atoms of hydrogen. As a result, flavonoids are involved in the reduction and chelation to synthesize MNPs through nucleation, stabilization, and growth (Shafey, 2020). This means that flavones and phenols as reduction and stabilization agents are potentially able to reduce and stabilize MNPs by hydration, electrostatic, van der Waals forces, and steric interactions (Yazdanian et al., 2022). Recent studies have been conducted on the impact of *Ocimum basilicum* (Family *Lamiaceae*) on the synthesis of nanoparticles such as silver, titanium dioxide, and zinc oxide (Michael and Moses, 2021). They reveal that the plant extract of *Ocimum* contains high amounts of phytochemicals such as phenolics (rosmarinic acid) and flavonoids (luteolin), playing an important role in the metal reduction processes. In this bioreduction, by forming luteolin enol forms, reactive hydrogen can be liberated, which is involved in converting metal ions to metallic nanoparticles. In addition, it should be mentioned that although the phenolic compounds of rosemarinic acid (hydroxyl and carboxyl groups) are converted into the form of keto, resulting in releasing reactive hydrogen to make an intermediate and highly reactive form of enol, having two hydroxyl groups on a single carbon atom makes this compound unstable, and therefore it tends to return to its form of keto. As a result, both cases lead to the liberation of reactive hydrogen, which contributes to the synthesis of metallic nanoparticles (Elumalai et al., 2019).

Phytochemicals called polyphenols (commonly found in fruits, vegetables, and seeds) are derived from secondary metabolites of plants, and they are also composed of two or more phenolic groups (Hano and Tungmunnithum, 2020). These compounds such as gallic acid, ellagic acid, ferulic acid, caffeic acid, quercetin, and p-coumaric acid chelate metal ions with their nearby hydroxyl groups of phenolic, so bioreduction involves oxidizing these compounds to produce quinones and transferring free protons and electrons to reduce metal ions to zerovalent nanoparticles. According to the literature reports, polyphenolic compounds can readily be oxidized with hydroxyl radicals to produce metal-phenolate compounds (Md Ishak et al., 2019). In another study, authors reported that *Solanum nigrum* plant leaf extract contains polyphenols that give the Ag NPs a negative charge, resulting in generating electrostatic repulsive forces, thereby inhibiting particle agglomeration (Vijilvani et al., 2020).

18.3.2 Potential capping and stabilizing agent

In MNPs' synthesis, a crucial step is the application of capping agents, known as "binding molecules," for controlling the surface functionalization and stabilization of metallic nanoparticles (Sharma et al., 2021; Sidhu et al., 2022). In general, these molecules, interacting strongly with the metal nanoparticle surface, modify metallic nanoparticles by modulating their surface chemistry, morphology, and size distribution (Sharma et al., 2019). For example, due to adding a negative charge to the surface of reduced metals, polyphenols have a capping effect (Bindhu et al., 2020). In addition, as a result of the complex structures formed between the metallic ions in the precursor salts, they prevent agglomeration and increase the reduction kinetics (Sharma et al., 2019). Due to the fact that metallic nanoparticle stability in aquatic environments or biological fluids is critical, various functional groups derived from plant extracts can be used to alter the surface of metallic nanoparticles in order to make them more effective in targeted applications, especially biomedicine, which requires both stability and biocompatibility (Ocsoy et al., 2018). Therefore these capping agents are well-dispersed, biosoluble, biodegradable, nontoxic, and biocompatible and, under mild reaction conditions, provide desired characteristics by regulating the growth, size, morphology, aggregation, and chemical and physical characteristics of synthesized metallic nanoparticles (Gnanasangeetha and Suresh, 2020; Chugh et al., 2021). These changes in physicochemical and biological properties result from the steric effects of stabilizing agents that can adsorb at the surface of a metallic core (Javed et al., 2020). As a result, plant biological extracts acting as biocompatible capping agents or stabilizers inhibit agglomeration (known as steric stabilization) and promote stability by modifying the biological activities, surface chemistry, morphology, and magnetic properties of MNPs. It also controls the metallic nanoparticle interaction in the medium where they were prepared (Javed et al., 2020; Khan, 2022; Sidhu et al., 2022). Furthermore, they have shown great potential for the production of nontoxic, surface-functionalized, and monodispersed MNPs for use in various applications such as medical utilizations (Sidhu et al., 2022). As mentioned earlier, proteins, vitamins, polysaccharides, as well as amino acids can all apply as both reducing and stabilizing agents (Shafey, 2020; Vanlalveni et al., 2021). For instance, it has been reported that alkaloids, amino acids, proteins, and flavonoids found in *Mangifera indica* flower extract can reduce Ag ions to Ag NPs and stabilize the synthesized silver nanoparticles (Ameen et al., 2019).

As one of the most important components of plant extracts, proteins are efficient capping agents in the green synthesis of metal nanoparticles, providing biocompatible functional groups that are effective at modifying metallic nanoparticle surfaces because of their inherently biodegradable and sustainable properties (Shafey, 2020; Khan, 2022). The functional groups of amino acids found in proteins have been studied for their efficacy to stabilize MNPs during the synthesis of metal nanoparticles (Li et al., 2007). The study of Ag nanoparticle synthesis by *Commiphora gileadensis* stem extract confirmed the possible role of proteins as stabilizing agents and capping agents (Al-Zahrani et al., 2022).

18.4 Various sources of plant extract employed in the synthesis of metallic nanoparticles

Due to the various natural phytochemical compounds present in plants, which act as reduction and stabilization agents in bioreduction reactions, several metallic nanoparticles (including cobalt, copper, silver, gold, palladium, platinum, zinc oxide, and magnetite) have been successfully synthesized using extracts of plants or different plant parts such as stems, roots, leaves, fruits, flowers, peels, and seeds (Singh et al., 2018) (Fig. 18.3).

18.4.1 Metallic nanoparticles' synthesis using leaf extracts of plants

As compared with other plant parts, leaf extracts are considered the primary source for metal nanoparticle synthesis due to their abundance of metabolites, rejuvenation, and nondevastating properties. Plant leaf extracts can be easily obtained, and moreover, they contain a wide variety of phytochemicals (such as alkaloids, terpenoids, phenolic acids, sugars, polyphenols, and proteins), playing an important role as reducing and stabilizing agents in the bioreduction process of metal nanoparticles and their formation from their precursors. The use of different leaf plants to synthesize MNPs has been studied extensively in recent decades (Shafey, 2020). Recent research was conducted by Lomelí-Rosales et al. using leaf extract of *Capsicum chinense*, taking account of agro-industrial waste, to synthesize Au NPs and Ag NPs. Despite the leaf extract containing polyphenols, reducing sugars, and amino acids, the result illustrated that only reducing sugars acted as reducing and stabilizing agents (Lomelí-Rosales et al., 2022). Likewise, FTIR analysis of *Azadirachta indica* leaf extract revealed that alkaloids and flavonoids were responsible for reducing and capping synthesized Ag NPs (Pawar et al., 2022). Table 18.1 shows additional MNP synthesized by plant leaf extracts.

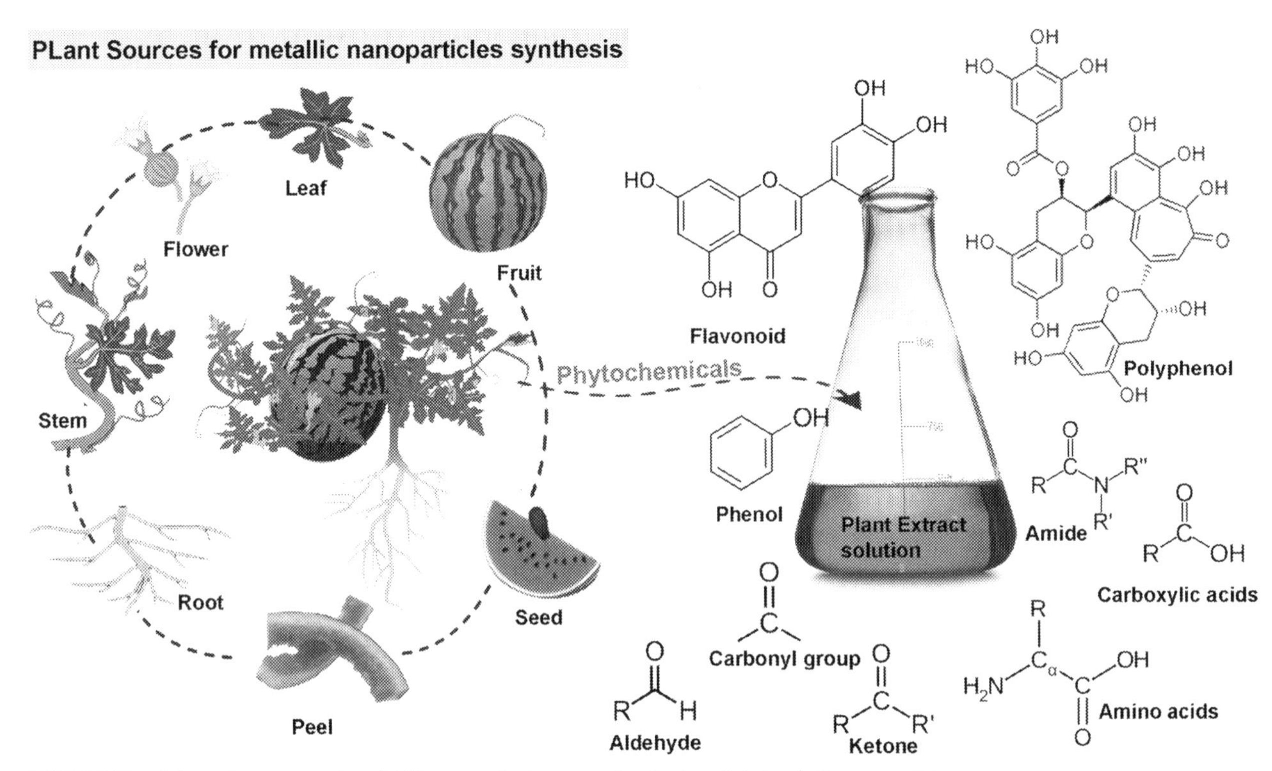

FIGURE 18.3 Schematic representation of different synthesis sources for plant-mediated metallic nanoparticles.

18.4.2 Metallic nanoparticles' synthesis using stem extract

A great deal of effort has also been put into biosynthesizing various types of MNPs from plant stem bark extract, which comes from the stem of woody plants, due to phytochemical compounds in various extracts of stem bark such as tannins, flavonoids, saponins, phenols, and alkaloids present in the aqueous stem bark extract of the plant (Wan Mat Khalir et al., 2020). Al-Zahrani et al. found that chemical functional groups of proteins, amines, and polyphenols found in the stem extract of *C. gileadensis* played a major role as both reductants and stabilizers in the Ag NPs synthesis because of their chemical interactions with silver ions (Al-Zahrani et al., 2022). Likewise, *Entada spiralis* stem extract was utilized as reducing and stabilizing agents for the biosynthesis of Ag NPs. Based on the FTIR analysis, the hydroxyl groups of the saponin compound in *E. spiralis* extract donated hydrogen atoms and electrons in the reduction of Ag ions to metallic Ag. After that, as a result of van der Walls forces between negatively charged functional groups and oxygen surrounding the Ag NP surface, Ag NPs were stabilized by saponin functional groups. In fact, aponin compounds contain negative charge functional groups that stabilize synthesized Ag NPs (Wan Mat Khalir et al., 2020). More additional MNP synthesized by stem extract are illustrated in Table 18.2.

18.4.3 Metallic nanoparticles' synthesis using flower extract

Since flowers are readily accessible and are rich in phytochemicals (flavonoids, tannins, saponins, terpenoids, coumarins, alkaloids, coumarins, saponins, sterol, and flavones), they have been considered as a significant plant part for reducing and stabilizing MNPs. Furthermore, flowers that are usually used for synthesizing MNPs are usually popular medicinal plants (such as *Clitoria ternatea*, *Malva Sylvestris*, etc.), which are medicinally valuable (Kumar Bachheti et al., 2020). The recent study conducted by Alahmdi et al. indicated that phenols are abundant in the flower extract of *C. ternatea*, suggesting they were used to cap ZnO NPs and reduce them as well (Alahmdi et al., 2022). In another study, the results of phytochemical analysis of the aqueous extract of *Jatropha integerrima Jacq* flower indicated the presence of anthocyanin, coumarin, carbohydrate, glycoside, protein, tannin, phenol, and saponin in the related extract. However, the bands of O–H in the FTIR clearly demonstrated the reduction and capping of Ag nanoparticles by phenols and amino acids in the flower extract (Suriyakala et al., 2022). More additional MNP synthesized by flower extract are illustrated in Table 18.2.

TABLE 18.1 Metallic nanoparticles synthesized from leaf extracts.

Plant name	Size (nm) (SEM/TEM)	Morphology	Bioapplications	References
Ag NPs				
Calotropis gigantea	8–30	Spherical	- Larvicidal - Anticancer	Mol et al. (2022)
Pimenta dioica	32	Spherical	- Larvicidal	Kumar et al. (2022)
Aristolochia bracteolata Lam	29.30–46.32	Spherical	- Larvicidal	Narayanan et al. (2022)
Capsicum chinense	20.24 ± 0.24	Spherical	- Antioxidant	Lomelí-Rosales et al. (2022)
Urena lobate	20	Spherical	- Adsorption (MB dye)	Gowda et al. (2022)
Aloe vera	30–80	Spherical	- Antimicrobial	Arshad et al. (2022)
Guettarda Speciosa	30	Spherical	- Antimicrobial	Deivanathan and Prakash (2022)
Muntingia calabura	15.52	Spherical	- Antimicrobial	Vankudoth et al. (2022)
Cinnamomum tamala	25–30	Spherical	- Antimicrobial	Al Mashud et al. (2022)
Rumex nervosus	20–70	Spherical	- Antimicrobial	Alshameri et al. (2022)
Prunella vulgaris	2–20	Spherical	- Antimicrobial	Ganaie et al. (2022)
Origanum majorana	26.63	Spherical	- Antimicrobial	Yassin et al. (2022a)
Azadirachta indica	20–100	Spherical	- Antimicrobial	Pawar et al. (2022)
Oxalis griffithii	30	Spherical	- Antimicrobial	Singla et al. (2022)
Naringi crenulata	32–78	Spherical	- Antimicrobial	Chinnathambi et al. (2023)
Cleome gynandra	ND	Spherical	- Antibacterial - Antifungal	Kannan et al. (2022)
Stachys schtschegleevii	31.43	Spherical	- Antibacterial - Antifungal - Antioxidant	Shahabadi and Mahdavi (2022)
Leucas biflora (Vahl)Sm.	40–98	Hexagonal	- Anticancer	Chitra et al. (2022)
Ocimum sanctum L. A	20	Face-centered cubic	- Antioxidant	Melkamu and Jeyaramraja (2022)
Green tea (*C. sinensis*)	39.12 ± 5.33 43.11 ± 6.42	Face-centered cubic	- Anticancer	Liu et al. (2022)

(*Continued*)

TABLE 18.1 (Continued)

Plant name	Size (nm) (SEM/ TEM)	Morphology	Bioapplications	References
		Au NPs		
C. chinense	15.05 ± 0.27	Spherical	- Antioxidant	Lomelí-Rosales et al. (2022)
Spinacia oleracea	11–23	Spherical	- Antioxidant	Zhu et al. (2022)
Alhagi maurorum	18.36	Spherical	- Antioxidant - Antiprostate cancer	Zhao et al. (2022)
Angelica keiskei	10–40	Spherical	- Antimicrobial	Krishnaraj et al. (2022)
Limonia acidissima	100	Spherical	- Antimicrobial - Wound healing properties	Kabeerdass et al. (2022)
		ZnO NPs		
Phoenix roebelenii	15.6	Spherical	-Photocatalytic (MB dye)	Aldeen et al. (2022)
Annona reticulata	10–28	Spherical, Cubical, and Hexagonal	- Photocatalytic (MB dye)	Selvam et al. (2022a)
Euphorbia milii	16.11	Hexagonal wurtzite	- Photocatalytic (MB dye)	Venkatesan, Suresh, Ramu, Kandasamy, et al. (2022a)
Solanum trilobatum	25	Hexagonal wurtzite	- Photocatalytic (MB dye)	Venkatesan, Suresh, Ramu, Arumugam, et al. (2022b)
Trigonella foenum-graecum L	24, 33, 44	Hexagonal wurtzite	- Photocatalytic (MB and EBT dyes)	Kermani et al. (2022)
Elaeagnus indica	20–30	ND	- Photocatalytic (MB dye)	Indhira et al. (2022)
M. calabura	16.02	Irregular-shaped & clusters	- Photocatalytic (2,4-D)	Vinayagam et al. (2022)
Delonix elata	19.6	Hexagonal	- Photocatalytic (CV dye)	Karthik et al. (2022)
Syzygium cumini	20–30	Floral	- Antibacterial - Photocatalytic (MB dye)	Padmavathi et al. (2022)
O. majorana	12.467 ± 1.36	Multiplicity (Spherical, Hexagonal)	- Antimicrobial	Yassin, et al. (2022b)
Rivina humilis	14.4	Irregular & Circular	- Antioxidant - Antimicrobial	Annapoorani et al. (2022)
A. indica	50–120 30–70	Rod-shaped	- Anticancer	Rani et al. (2022)

O. majorana	32	Spherical	- Antioxidant	Khaleghi et al. (2022)
Dysphania ambrosioides	7–130	Spherical	- Antibacterial	Álvarez-Chimal et al. (2022)
Manilkara zapota	50–86	Spherical	- Antibacterial	Maseera et al. (2022)
		CoO NPs		
M. calabura	27.59	Irregular-shaped and clusters	- Photocatalytic (MB dye)	Vinayagam et al. (2023)
		CuO NPs		
Canthium coromandelicum	33	Spherical	- Photocatalytic (MB and MO dyes)	Selvam et al. (2022b)
Eucalyptus globulus	85.80	Spherical	- Adsorption (MO dye)	Alhalili (2022)
Phragmanthera austroarabica	44.6 ± 2.7	Spherical	- Antibacterial	Alahdal et al. (2023)
Centella asiatica aqueous	24–34	Diverse (flower like cubical)	- Photocatalytic (MB dye)	Joy Prabu et al. (2022)
Elaeagnus indica	30–40	ND	- Photocatalytic (MB dye)	Indhira et al. (2022)
		TiO_2 NPs		
Acorus calamus	15–40	Semispherical	- Photocatalytic (RhB dye)	Ansari et al. (2022)
Ficus religiosa (Peepal tree)	83.22 ± 1.50	Spherical	- Larvicidal - Antibacterial	Soni and Dhiman (2020)
Phyllanthus niruri	23	Spherical	- Photocatalytic (MB dye)	Shimi et al. (2022)
Lippia adoensis (Kusaayee)	Various sizes	Spherical	- Antibacterial	Gudata et al. (2022)
Luffa acutangula	10–49	Hexagonal	- Antibacterial	Anbumani et al. (2022)
Coleus aromaticus	12	Hexagonal	- Antibacterial - Antifungal - Antidiabetic - Antioxidant	Anupong et al. (2023a)

MB dye, Methylene blue dye; *2,4-D*, 2,4-dichlorophenoxy acetic acid; *CV dye*, crystal violet dye; *EBT dye*, eriochrome black T dye; *ND*, not defined.

18.4.4 Metallic nanoparticles' synthesis using plant peel extract

Almost every part of plants even plant peel, which can be considered as plant waste materials, is extremely valuable. Thus green nanotechnology has attracted a great deal of attention for using these parts as promising sources of biomolecules for the synthesis and stabilization of metallic NPs, in which their extracts serve as reducing and capping agents. Furthermore, the effective recycling and use of plant peel wastes would be both an environment-friendly and a cost-effective approach as well as would assist in resolving environmental issues. Therefore the use of fruit/vegetable peels as renewable precursors can be widely expanded in the future, since they are relatively easy to replenish (Kumar et al., 2020) (Table 18.2).

18.5 Effect of the plant extract on the synthesis and characteristics of metallic nanoparticles

The different methods used to prepare MNPs can result in significant variations in the type, concentration, and composition of the bioactive molecules among compounds within different plants, particularly medicinal plants (Dube et al., 2020). In fact, plant extracts contain numerous agents for reduction and stabilization, which have a prominent role in the synthesis of nanoparticles. In other words, nanoparticles' type, shape, size, and morphology are greatly influenced by the nature of plant extracts (Sharma et al., 2019; Aboyewa et al., 2021a,b). Moreover, there are many differences in the type of plants and the interaction of aqueous metal ions with plant extracts, and also there are many parameters such as temperature, pressure, pH, and reaction time that lead to MNPs displaying various physical, chemical, and biological characteristics (Pal et al., 2019; Siakavella et al., 2020). Generally, the most significant determinant of the morphology of synthesized MNPs is the phytochemical compounds present in plant extract because biochemical-reducing agents are present in various concentrations in plant extracts. There are regional and seasonal differences in biochemical reductant concentrations in most extracts of plants, resulting in variation in synthesized nanoparticles from batch to batch (Kulkarni and Muddapur, 2014). Additionally, reports suggest that phytochemical properties of the plant extract have an important part in the MNPs' bioactivities and plant extract-derived MNPs may have bioactivities similar to the plant extract (Aboyewa et al., 2021a,b; Tyavambiza et al., 2021).

18.6 Applications of plant extract-mediated synthesized metallic nanoparticles

Metallic nanoparticles have distinctive and special characteristics such as conductivity, catalytic activity, biocompatibility, and large surface-to-volume ratio, leading to various applications in biomedical sciences, water treatment, agriculture, etc. As mentioned above, phytochemicals of plant extracts can influence the material type, size, shape, composition, and charge of the surface, which are important properties of MNPs for their use in different fields (Bandala, 2022).

18.6.1 Biomedical applications

In biomedicine, metallic nanoparticles are used for a number of medical applications such as drug delivery, vaccination, imaging, and biosensors (Abd Elkodous et al., 2021) (Fig. 18.4). Recently, a number of concerns have been expressed about the safety of nanoparticle-based synthetic antibiotics regarding their effect on humans and the environment, resulting in using natural products such as plant extracts as an effective alternative for antimicrobial surfaces since they offer nontoxic and ecologically friendly properties (Adeyemi et al., 2022). For example, catechin, a polyphenol compound and strong natural antibiotic, is very abundant in a lot of plants (like gooseberries, blueberries, strawberries, apples, kiwi, grape seeds, green tea, cocoa, etc.) (Bae et al., 2020). It is a well-known biological component responsible for the synthesis of metallic nanoparticles from plant extracts, such as catechin-Cu-nanoparticles, which have shown an effective antimicrobial activity and great antimicrobial potential against multidrug resistant microbes (Adeyemi et al., 2022).

Two significant parameters determine whether metallic nanoparticles are effective antimicrobials: (1) the material used for synthesis and (2) the particle size. According to recent studies, MNP size has a significant impact on both antimicrobial activity and traveling around the human body without being affected by the reticuloendothelial system (RES) and the body's defense cells (Singh et al., 2018).

Plant-mediated Ag NPs, the most popular inorganic nanoparticles in medicine, have been extensively explored for treating fungal and viral infections, as well as inflammation, as shown in Table 18.1 (Vanlalveni et al., 2021). The Au

TABLE 18.2 Metallic nanoparticles synthesized from stem, flower, seed, fruit, root, and peel (or shell) extracts.

Plant name	Plant part used	Size (nm) (SEM/TEM)	Shape	Bioapplications	References
			Ag NPs		
Commiphora gileadensis	Stem	13	Spherical	- Anticancer	Al-Zahrani et al. (2022)
Securidaca inappendiculata	Stem	10–15	Spherical	- Antibacterial - Antioxidant	Jayeoye et al. (2021)
Entada spiralis	Stem	18.49 ± 4.23	Spherical	- Biomedical - Antibacterial	Wan Mat Khalir et al. (2020)
Jatropha integerrima	Flower	17–45	Spherical	- Antibacterial	Suriyakala et al. (2022)
Malva sylvestris	Flower	20–30	Spherical	- Antibacterial	Mehdizadeh et al. (2021)
Syzygium cumini	Flower	37.21–46.48	Oval	- Antibacterial	de Carvalho Bernardo et al. (2021)
	Seed	36.25–77.01	Oval	- Antibacterial	
Phoenix dactylifera	Seed	14	Spherical	- Anticancer	Farshori et al. (2022)
Raphanus sativus L.	Seed	5–20	Spherical	- Anticancer	Khare et al. (2022)
Panicum miliaceum	Seed (grain)	10–25	Rectangular	- Antioxidant - Antidiabetic - Antiinflammatory - Larvicidal - Insecticidal	Velsankar et al. (2022)
Ribes rubrum L. (red currants)	Fruit	8–59	Spherical	- Antifungal - Antibacterial	Rizwana et al. (2022)
Tomato	Fruit	73	Spherical	- Catalyst	Weng et al. (2022)
Orange	Fruit	24	Spherical		
Grapefruit	Fruit	31	Spherical		
Conocarpus lancifolius	Fruit	5–30	Spherical	- Antibacterial	Oves et al. (2022)
P. dioica	Fruit	94 ± 29	Spherical	- Antibacterial	Kaithal et al. (2022)
Sambucus ebulus	Fruit	35–50	Spherical	- Antibacterial	Hashemi et al. (2022)
Glycosmis pentaphylla	Fruit	17	Spherical	- Antifungal - Antibacterial	Dutta et al. (2022)
Baccaurea ramiflora (Latka)	Fruit	30–90	Spherical and cubical	- Antibacterial	Banerjee et al. (2022)

(*Continued*)

TABLE 18.2 (Continued)

Plant name	Plant part used	Size (nm) (SEM/TEM)	Shape	Bioapplications	References
Dragon fruit (*Hylocereus undatus*)	Peel	10–50	Spherical	- Anticancer	Shyamalagowri et al. (2022)
Passiflora edulis	Peel	25	Spherical	- Antibacterial - Catalyst (nitrophenols)	My-Thao Nguyen et al. (2021)
Black chickpea (*Cicer arietinum*)	Peel	47.46 ± 0.42	Spherical	- Antibacterial	Mouriya et al. (2023)
Pomelo	Peel	35–40	Spherical	- Antibacterial	Barbhuiya et al. (2022)
Rubus ellipticus	Root	13.85–34.30	Spherical	- Antibacterial	Khanal et al. (2022)
Acacia nilotica	Bark	20–50	Variable shapes	- Anticancer - Antidiabetic - Antioxidant	Zubair et al. (2022)
Au NPs					
Euphorbia neriifolia L.	Stem	23–25	Spherical	- Antibacterial - Antifungal	Ul Haq, Ullah (2022)
Carica papaya	Fruit	12 ± 2.31	Spherical	- Antioxidant - Anticancer	Anadozie et al. (2022)
Red Dragon Pulp	Fruit	17	Spherical	- Antifungal	Dutta et al. (2022)
Hylocereus polyrhizus	Fruit	8.65–36.2	Spherical	- Antioxidant	Al-Radadi (2022)
Rosa canina	Fruit	20–30	Globular	Heterogeneous catalyst (MB, RhB, and 4-NP)	Alikhani et al. (2022)
P. edulis	Peel	25	Spherical	- Antibacterial - Catalyst (4-NP)	My-Thao Nguyen et al. (2021)
Spondias dulcis (Anacardiaceae)	Peel	36.75 ± 11.36	Spherical	- Cancer treatment	Pechyen et al. (2022)
Citrus		Au NPs-U: 13.65 ± 3.90 Au NPs-NU: 16.80 ± 4.41		- Antiinflammatory	Gao et al. (2022)
ZnO NPs					
Clitoria ternatea	Flower	84	Rob	- Anticancer - Antibacterial	Alahmdi et al. (2022)
Cassia auriculata	Flower	41.25	Hexagonal	- Antimicrobial	Seshadri (2021)
Myrica esculenta	Fruit	31.67	Thin pellets-like	- Photocatalytic (MB dye)	Lal et al. (2022)

				- Antimicrobial	
Aegle marmelos (L.)	Fruit	22.5	Hexagonal wurtzite	- Antibacterial - Antibiofilm	Senthamarai and Malaikozhundan (2022)
Sechium edule	Fruit	30–100	Spherical and triangular	- Photocatalytic (MB dye)	Bharathi et al. (2022)
Physalis minima	Fruit	50–150	Spherical	- Photocatalytic (MB dye) - Antioxidant	Yazhiniprabha et al. (2022)
Plantain	Peel	20	Spherical	- Antimicrobial	Imade et al. (2022)
Onion	Peel	20–80	Spherical	- Agricultural	Modi et al. (2022)
Watermelon	Peel	40–88	Spherical	- Photocatalytic (metronidazole)	Al-Gheethi et al. (2022)
Punica granatum	Peel	20–40	Spherical and hexagonal	- Antimicrobial	Shaban et al. (2022)
			CuO NPs		
Solanum macrocarpon	Fruit	Heating Method: 35.60 ± 6.24Microwave Irradiation Method: 47.14 ± 6.18	Spherical	- Electrochemical	Okpara et al. (2021)
Prickly pear	Peel	40	Spherical	- Catalytic (4-NP)	Badri et al. (2021)
			Cu_2O NPs		
Kiwi (*Actinidia deliciosa*)	Fruit	93.1	Spherical	- Antibacterial	Kodasi et al. (2023)
			TiO_2 NPs		
Limon-citrus	Fruit	ND	Spherical	- Antibacterial	Ouerghi et al. (2022)
Citrus limon	Fruit	17 ± 1	Spherical	- Dye-sensitized solar cells	Singh et al. (2022)
Caricaceae (Papaya)	Shell	15	Spherical	- Antifungal	Saka et al. (2022)
			CoO NPs		
Orange	Peel	14.2–22.7	Octahedral	- Antibacterial - Antioxidant	Anupong et al. (2023b)
			NiO NPs		
Orange	Fruit	21–130	Cubic and triangular	- Antibacterial - Antioxidant	Narayanan et al. (2023)
			Pt NPs		
Nymphaea tetragona	Flower	4.04 ± 1.31 2.01 ± 0.80	ND	- Antioxidant	Zhang et al. (2022)

MB dye, methylene blue dye; 2,4-D, 2,4-dichlorophenoxy acetic acid; CV dye, crystal violet dye; EBT dye, eriochrome black T dye; 4-NP, 4-nitrophenol; RhB, rhodamine B; ND, not defined.

NPs synthesized using plant extract can also serve as powerful tools in the development of antibacterial agents owing to their nontoxic nature, capacity to be functionalized, and photothermal effects (Bharadwaj et al., 2021; Santhosh et al., 2022). Various plan-mediated ZnO NPs with different morphologies exhibit strong antibacterial activity against a wide variety of bacteria. This is due to their unique antimicrobial properties resulting from their interaction with the surface of bacteria cells and/or entering their interior (Akintelu and Folorunso, 2020; Xu et al., 2021). The high surface area, outstanding surface morphology, crystal structure, and photocatalytic properties of TiO_2 NPs in nature play major roles in determining their physicochemical characteristics, and consequently their antimicrobial activity. Recently, plan-mediated TiO_2 NPs with various sizes, shapes, morphologies, and crystal structures have demonstrated excellent antibacterial and biodegradation properties (Zhu et al., 2019; Verma et al., 2022). A wide range of microorganisms has been tested with plant-mediated CuO NPs due to their distinctive biological and antibacterial properties, which are highly effective and less expensive than gold, palladium, and silver nanoparticles with similar characteristics. Furthermore, the plant-mediated synthesis of CuO NPs has gained popularity as a next-generation antibiotic because of its ease, environmental friendliness, practicality, and profitability (Bouafia et al., 2020). More details are illustrated in Table 18.1.

18.6.2 Wastewater treatment applications

As time passes, the demand for safe and hygienic drinking water increases. As a result, metals and metal oxides as semiconductor nanoparticles have recently gained great attention in recent materials' research because they have the potential to oxidize toxic pollutants (Singh et al., 2018). Furthermore, the environmental friendliness, sustainability, and affordability of metal nanoparticles prepared via green synthesis approaches have made them effective for water purification, specifically in removing a wide variety of pollutants in the water, both organic and inorganic (M^{n+} and R) (Fig. 18.4), such as heavy metals, toxic chemicals, pesticides, and radioactive materials from aquatic ecosystems. Thus the toxic organic compounds can be completely mineralized as a result of their outstanding adsorption and catalytic abilities (Ugwu et al., 2022).

Metallic NPs have high surface adsorption rates and high volume-to-surface ratios, which make them excellent catalysts. Plant-mediated green synthesis of metallic nanoparticles as catalysts, including Au NPs (Hosny et al., 2022), Ag NPs (Hashemi et al., 2022), CuO NPs (Shelke et al., 2022), ZnO NPs (Belén Perez Adassus et al., 2022), and Pd NPs (Olajire and Mohammed, 2019), has shown a remarkable catalytic ability as a reducing agent for the reduction of various cationic and anionic organic dyes such as methylene blue (MB) (Bandala, 2022; Parmar and Sanyal, 2022). Furthermore, researchers have shown that TiO_2 nanoparticles synthesized from leaf extracts from *Phyllanthus niruri* (Shimi et al., 2022) and *Acorus calamus* (Ansari et al., 2022) were highly photocatalytically active in removing MB and RhB dyes from water by oxidizing and degrading them.

The outstanding disinfectant properties of Ag NPs without producing any disinfection byproducts have led to their increasing use in various water treatment systems, especially in point-of-use devices for household drinking water treatment. Based on recent research, in addition to acting as a sensor material for detecting various pollutants (such as heavy metals, pathogenic microbes, and pesticides), Au nanoparticles can interact with water-soluble organomercury compounds, and are therefore capable of acting as adsorbents for removing organic pollutants, including heavy metals. Similarly, Au NPs are excellent catalysts for water remediation. Zerovalent iron (nZVI) nanoparticles display admirable adsorption properties and a strong reducing tendency for the removal of dissolved polychlorinated contaminants, inorganic anions, and heavy metals due to their extraordinarily small size and exceptionally high specific surface area. nZVI nanoparticles are the most commonly applied nanoparticles for the effective removal of dissolved heavy metals in aqueous solutions and wastewater owing to their high chemical reactivity, cost efficiency, and environmental friendliness (Patanjali et al., 2019). It was demonstrated that nZVI synthesized from flaxseed gum extract could act as an effective in situ remediation material for both acidic and alkaline soil and water contaminated with Cr (VI) (Izadi et al., 2022).

18.6.3 Agricultural applications

Applied nanotechnology in agriculture is called nanoagriculture, which applies new technologies to increase crop yields and performance with a lower environmental impact (less waste and runoff) (Nayantara and Kaur, 2018). In agriculture, nanofertilizers, nanoherbicides, nanocoating, and nanopesticides are different forms of using nanoparticles (Nair et al., 2022) (Fig. 18.4). The nanofertilizer—an effective, cost-effective, and environmentally friendly option to synthetic fertilizers—provides nutrient release slowly and steadily that not only improves the growth of plants but also preserves the

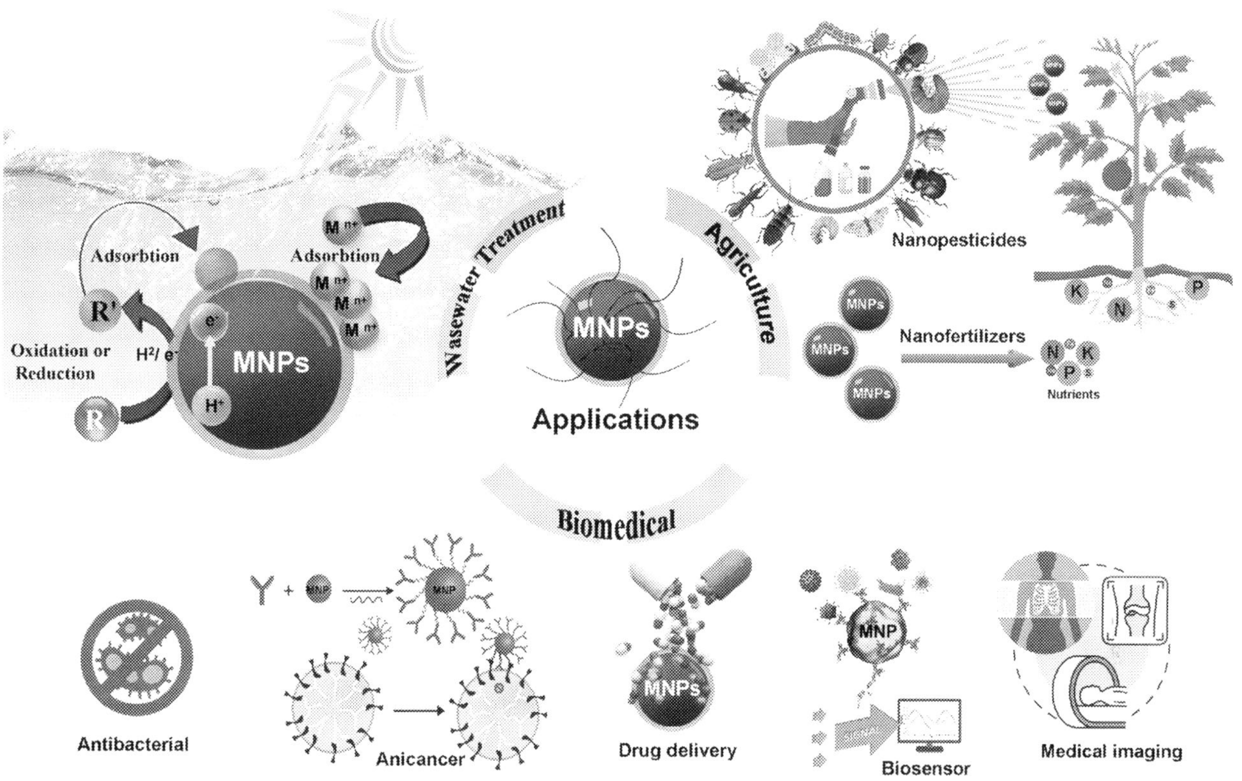

FIGURE 18.4 Schematic representation of possible applications of metallic nanoparticles in wastewater treatment, agriculture, and biomedical.

diversity of useful microbes (Kalwani et al., 2022). According to recent studies, metal and metal-oxide nanoparticles are known to offer a variety of merits to plants via their application in the micronutrient delivery, stimulation, and activation of defense mechanisms of plants, as well as the prevention and control of plant pathogens (Cartwright et al., 2020; Nair et al., 2022). Biosynthesized MNPs from various biological sources can be used in agriculture. In fact, plants can uptake metal nanoparticles as micronutrients from soil and consume them. However, they are not considered among the main sources of soil pollution due to their small concentration in soil (Nayantara and Kaur, 2018). ZnO NPs synthesized by plants, which were used as a nutrient for plants, demonstrated that the germination and growth of mung and wheat seeds were improved over control seeds (Modi et al., 2022). The results obtained by Walid et al. also indicated that the vegetative growth parameters, flowering and fruit set percentage, and yield of ten-year-old lemon trees increased significantly as a result of spraying Ag NPs on the foliage (Mosa et al., 2022).

18.7 Future prospects

Over the last decades, despite conducting wide research in further exploration for green synthesis of metal nanomaterials using many plants, there is room for improvement on the principal issues of biosynthesis, especially the selection of plants containing the main source of phytoconstituents, which play a key role in reducing, stabilizing, and capping, synthesis process, as well as MNP quality. Currently, in many research groups, metal nanoparticles have been synthesized by using local and readily available plants or plant extracts. In spite of their promising performance in synthesizing metal nanoparticles, these plants may be constrained by their lack of supply for industrial and large-scale production of metal nanoparticles. Hence, the potential of producing metal nanoparticle products at a large scale from local plants should be further investigated. The ability to accurately identify particular phytoconstituent molecules responsible for reducing, stabilizing, and capping nanoparticles in the mechanisms of biosynthesis is another issue to consider. In other words, the conducted studies have been able to confirm the role of plant extract in the biosynthesis of MNPs, however the precise reactions responsible for the synthesis remain unclear. It is therefore suggested that using some techniques that could help us to identify the molecules responsible for the reduction and stabilization in the biosynthesis process could help us to better understand the mechanisms of biosynthesis and the chemical reactions involved. In addition, there is a great concern about the large variation in size and shape of the plant-mediated synthesized metal nanoparticles, making a green synthesis of MNPs unsuitable for large-scale production or difficult to control particle size during synthesis that results in their tremendous limitations in practical applications. Thus there is a need for a comprehensive investigation of biosynthesized metallic nanoparticles that incorporates techniques from biological engineering, nanotechnology, bioprocessing, chemical engineering, and plant biology.

18.8 Conclusion

In recent years, plant extracts with beneficial phytochemicals have been demonstrated to be an effective greener method for the synthesis of metallic nanoparticles due to their simplicity, environmental friendliness, and biocompatibility. A further benefit of these phytochemicals is their ability to modify metallic nanoparticle properties, such as their size and shape, so they can be applied to a variety of uses. A comprehensive overview of bioreduction mechanisms and recent research into the role of different plant parts in the synthesis of MNPs is presented in this chapter.

Acknowledgment

The authors would like to express their gratitude to the School of Environment, College of Engineering, University of Tehran.

References

Abd Elkodous, M., et al., 2021. Recent advances in waste-recycled nanomaterials for biomedical applications: waste-to-wealth. Nanotechnology Reviews 1662–1739. Available from: https://doi.org/10.1515/ntrev-2021-0099.

Aboyewa, J.A., Sibuyi, Nicole, R.S., et al., 2021a. Gold nanoparticles synthesized using extracts of cyclopia intermedia, commonly known as honeybush, amplify the cytotoxic effects of doxorubicin. Nanomaterials 11 (1), 1–16. Available from: https://doi.org/10.3390/nano11010132.

Aboyewa, J.A., Sibuyi, Nicole, R.S., et al., 2021b. Green synthesis of metallic nanoparticles using some selected medicinal plants from Southern Africa and their biological applications. Plants 10 (9), 1929. Available from: https://doi.org/10.3390/plants10091929.

Adeyemi, J.O., et al., 2022. Plant extracts mediated metal-based nanoparticles: synthesis and biological applications. Biomolecules. Biomolecules. Available from: https://doi.org/10.3390/biom12050627.

Akintelu, S.A., Folorunso, A.S., 2020. A review on green synthesis of zinc oxide nanoparticles using plant extracts and its biomedical applications. BioNanoScience. Springer, pp. 848–863. Available from: 10.1007/s12668-020-00774-6.

Al Mashud, M.A., et al., 2022. Green synthesis of silver nanoparticles using *Cinnamomum tamala* (Tejpata) leaf and their potential application to control multidrug resistant Pseudomonas aeruginosa isolated from hospital drainage water. Heliyon 8 (7). Available from: https://doi.org/10.1016/j.heliyon.2022.e09920.

Al-Gheethi, A., et al., 2022. Metronidazole photocatalytic degradation by zinc oxide nanoparticles synthesized in watermelon peel extract; advanced optimization, simulation and numerical models using machine learning applications. Environmental Research 212, 113537. Available from: https://doi.org/10.1016/j.envres.2022.113537.

Al-Radadi, N.S., 2022. Biogenic proficient synthesis of (Au-NPs) via aqueous extract of Red Dragon Pulp and seed oil: characterization, antioxidant, cytotoxic properties, anti-diabetic anti-inflammatory, anti-Alzheimer and their anti-proliferative potential against cancer cell. Saudi Journal of Biological Sciences 29 (4), 2836–2855. Available from: https://doi.org/10.1016/j.sjbs.2022.01.001.

Al-Zahrani, S.A., et al., 2022. Anticancer potential of biogenic silver nanoparticles using the stem extract of *Commiphora gileadensis* against human colon cancer cells. Green Processing and Synthesis 11 (1), 435–444. Available from: https://doi.org/10.1515/gps-2022-0042.

Alahdal, F.A.M., et al., 2023. Green synthesis and characterization of copper nanoparticles using *Phragmanthera austroarabica* extract and their biological/environmental applications. Sustainable Materials and Technologies 35, e00540. Available from: https://doi.org/10.1016/j.susmat.2022.e00540. December 2022.

Alahmdi, M.I., et al., 2022. Green nanoarchitectonics of ZnO nanoparticles from Clitoria ternatea flower extract for in vitro anticancer and antibacterial activity: inhibits MCF-7 cell proliferation via intrinsic apoptotic pathway. Journal of Inorganic and Organometallic Polymers and Materials 32 (6), 2146–2159. Available from: https://doi.org/10.1007/s10904-022-02263-7.

Aldeen, T.S., Ahmed Mohamed, H.E., Maaza, M., 2022. ZnO nanoparticles prepared via a green synthesis approach: physical properties, photocatalytic and antibacterial activity. Journal of Physics and Chemistry of Solids 160, 110313. Available from: https://doi.org/10.1016/j.jpcs.2021.110313.

Alhalili, Z., 2022. Green synthesis of copper oxide nanoparticles CuO NPs from *Eucalyptus globulus* leaf extract: adsorption and design of experiments. Arabian Journal of Chemistry 15 (5). Available from: https://doi.org/10.1016/j.arabjc.2022.103739.

Alikhani, N., et al., 2022. Green synthesis of gold nanoparticles (Au NPs) using *Rosa canina* fruit extractand evaluation of its catalytic activity in the degradation of organic dye pollutants of water. Inorganic Chemistry Communications 139, 109351. Available from: https://doi.org/10.1016/j.inoche.2022.109351.

Alshameri, A.W., et al., 2022. Rumex nervosus mediated green synthesis of silver nanoparticles and evaluation of its in vitro antibacterial, and cytotoxic activity. OpenNano 8, 100084. Available from: https://doi.org/10.1016/j.onano.2022.100084. September.

Álvarez-Chimal, R., et al., 2022. Influence of the particle size on the antibacterial activity of green synthesized zinc oxide nanoparticles using *Dysphania ambrosioides* extract, supported by molecular docking analysis. Arabian Journal of Chemistry 15 (6), 103804. Available from: https://doi.org/10.1016/j.arabjc.2022.103804.

Alyamani, A.A., et al., 2021. Green fabrication of zinc oxide nanoparticles using phlomis leaf extract: characterization and in vitro evaluation of cytotoxicity and antibacterial properties. Molecules (Basel, Switzerland) 26 (20), 6140. Available from: https://doi.org/10.3390/molecules26206140.

Ameen, F., et al., 2019. Phytosynthesis of silver nanoparticles using *Mangifera indica* flower extract as bioreductant and their broad-spectrum antibacterial activity. Bioorganic Chemistry 88, 102970. Available from: https://doi.org/10.1016/j.bioorg.2019.102970. May.

Anadozie, S.O., et al., 2022. Synthesis of gold nanoparticles using extract of *Carica papaya* fruit: evaluation of its antioxidant properties and effect on colorectal and breast cancer cells. Biocatalysis and Agricultural Biotechnology 42, 102348. Available from: https://doi.org/10.1016/j.bcab.2022.102348.

Anbumani, D., et al., 2022. Green synthesis and antimicrobial efficacy of titanium dioxide nanoparticles using *Luffa acutangula* leaf extract. Journal of King Saud University - Science 34 (3), 101896. Available from: https://doi.org/10.1016/j.jksus.2022.101896.

Annapoorani, A., et al., 2022. Eco-friendly synthesis of zinc oxide nanoparticles using Rivina humilis leaf extract and their biomedical applications. Process Biochemistry 112, 192–202. Available from: https://doi.org/10.1016/j.procbio.2021.11.022.

Ansari, A., et al., 2022. Green synthesis of TiO_2 nanoparticles using acorus calamus leaf extract and evaluating its photocatalytic and in vitro antimicrobial activity. Catalysts 12 (2), 181. Available from: https://doi.org/10.3390/catal12020181.

Anupong, W., On-uma, R., Jutamas, K., Salmen, S.H., et al., 2023a. Antibacterial, antifungal, antidiabetic, and antioxidant activities potential of *Coleus aromaticus* synthesized titanium dioxide nanoparticles. Environmental Research 216, 114714. Available from: https://doi.org/10.1016/j.envres.2022.114714. August 2022.

Anupong, W., On-uma, R., Jutamas, K., Joshi, D., et al., 2023b. Cobalt nanoparticles synthesizing potential of orange peel aqueous extract and their antimicrobial and antioxidant activity. Environmental Research 216, 114594. Available from: https://doi.org/10.1016/j.envres.2022.114594. October 2022.

Arshad, H., et al., 2022. Synthesis of Aloe vera-conjugated silver nanoparticles for use against multidrug-resistant microorganisms. Electronic Journal of Biotechnology 55, 55–64. Available from: https://doi.org/10.1016/j.ejbt.2021.11.003.

Badri, A., et al., 2021. Green synthesis of copper oxide nanoparticles using Prickly Pear peel fruit extract: characterization and catalytic activity. Inorganic Chemistry Communications. Elsevier, p. 109027. Available from: https://doi.org/10.1016/j.inoche.2021.109027.

Bae, J., et al., 2020. Activity of catechins and their applications. Biomedical Dermatology 4 (1), 1–10. Available from: https://doi.org/10.1186/s41702-020-0057-8.

Bandala, E.R., 2022. Photocatalytic applications of biogenic nanomaterials. Sustainable nanotechnology for environmental remediation. Elsevier, pp. 383–396. Available from: https://doi.org/10.1016/B978-0-12-824547-7.00023-0.

Banerjee, S., et al., 2022. Synthesis of silver nanoparticles using underutilized fruit *Baccaurea ramiflora* (Latka) juice and its biological and cytotoxic efficacy against MCF-7 and MDA-MB 231 cancer cell lines. South African Journal of Botany 145, 228–235. Available from: https://doi.org/10.1016/j.sajb.2021.09.016.

Barbhuiya, R.I., et al., 2022. Ultrasound-assisted rapid biological synthesis and characterization of silver nanoparticles using pomelo peel waste. Food Chemistry 385, 132602. Available from: https://doi.org/10.1016/j.foodchem.2022.132602.

Belén Perez Adassus, M., Spetter, C.V., Lassalle, V.L., 2022. Biofabrication of ZnO nanoparticles from *Sarcocornia ambigua* as novel natural source: a comparative analysis regarding traditional chemical preparation and insights on their photocatalytic activity. Journal of Molecular Structure 1256, 132460. Available from: https://doi.org/10.1016/j.molstruc.2022.132460.

Bharadwaj, K.K., et al., 2021. Green synthesis of gold nanoparticles using plant extracts as beneficial prospect for cancer theranostics. Molecules. Multidisciplinary Digital Publishing Institute (MDPI). Available from: https://doi.org/10.3390/molecules26216389.

Bharathi, D., et al., 2022. Photocatalytic degradation of Rhodamine B using green-synthesized ZnO nanoparticles from *Sechium edule* polysaccharides. Applied Nanoscience (Switzerland) 12 (8), 2477–2487. Available from: https://doi.org/10.1007/s13204-022-02502-w.

Bindhu, M.R., et al., 2020. Green synthesis and characterization of silver nanoparticles from *Moringa oleifera* flower and assessment of antimicrobial and sensing properties. Journal of Photochemistry and Photobiology B: Biology 205, 111836. Available from: https://doi.org/10.1016/j.jphotobiol.2020.111836.

Bouafia, A., Laouini, S.E., Ouahrani, M.R., 2020. A review on green synthesis of CuO nanoparticles using plant extract and evaluation of antimicrobial activity. Asian Journal of Research in Chemistry 13 (1), 65. Available from: https://doi.org/10.5958/0974-4150.2020.00014.0.

Cartwright, A., et al., 2020. A review of metal and metal-oxide nanoparticle coating technologies to inhibit agglomeration and increase bioactivity for agricultural applications. Agronomy 10 (7), 1018. Available from: https://doi.org/10.3390/agronomy10071018.

Chinnathambi, A., et al., 2023. Synthesis of AgNPs from leaf extract of *Naringi crenulata* and evaluation of its antibacterial activity against multidrug resistant bacteria. Environmental Research 216, 114455. Available from: https://doi.org/10.1016/j.envres.2022.114455. August 2022.

Chitra, K., et al., 2022. Synthesis of silver nanoparticles using *Leucas biflora* (Vahl) Sm. Leaf extracts and their activity on breast cancer (MDA-MB-231) cells. Materials Letters 312, 131706. Available from: https://doi.org/10.1016/j.matlet.2022.131706. November 2021.

Chugh, D., Viswamalya, V.S., Das, B., 2021. Green synthesis of silver nanoparticles with algae and the importance of capping agents in the process. Journal of Genetic Engineering and Biotechnology 19 (1), 126. Available from: https://doi.org/10.1186/s43141-021-00228-w.

de Carvalho Bernardo, W.L., et al., 2021. Antimicrobial effects of silver nanoparticles and extracts of *Syzygium cumini* flowers and seeds: periodontal, cariogenic and opportunistic pathogens. Archives of Oral Biology 125, 105101. Available from: https://doi.org/10.1016/j.archoralbio.2021.105101. December 2020.

Deivanathan, S.K., Prakash, J.T.J., 2022. Green synthesis of silver nanoparticles using aqueous leaf extract of Guettarda Speciosa and its antimicrobial and anti-oxidative properties. Chemical Data Collections 38, 100831. Available from: https://doi.org/10.1016/j.cdc.2022.100831.

Dube, P., et al., 2020. Antibacterial activity of biogenic silver and gold nanoparticles synthesized from *Salvia africana*-lutea and *Sutherlandia frutescens*. Nanotechnology 31 (50). Available from: https://doi.org/10.1088/1361-6528/abb6a8.

Dutta, T., et al., 2022. Green synthesis of antimicrobial silver nanoparticles using fruit extract of *Glycosmis pentaphylla* and its theoretical explanations. Journal of Molecular Structure 1247, 131361. Available from: https://doi.org/10.1016/j.molstruc.2021.131361.

Elumalai, D., et al., 2019. Bio fabricated of silver nanoparticles using *Ocimum basilicum* and its efficacy of antimicrobial and antioxidant activity. Iranjournals. Nlai. Ir 3, 103–124. Available from: https://www.researchgate.net/publication/327110614_Bio_fabricated_of_silver_nanoparticles_using_Ocimum_basilicum_and_its_efficacy_of_antimicrobial_and_antioxidant_activity.

Ettadili, F.E., et al., 2022. Recent advances in the nanoparticles synthesis using plant extract: applications and future recommendations. Journal of Molecular Structure 1248, 131538. Available from: https://doi.org/10.1016/j.molstruc.2021.131538.

Farshori, N.N., et al., 2022. Green synthesis of silver nanoparticles using *Phoenix dactylifera* seed extract and its anticancer effect against human lung adenocarcinoma cells. Journal of Drug Delivery Science and Technology 70, 103260. Available from: https://doi.org/10.1016/j.jddst.2022.103260. March.

Ganaie, S.A., Zahoor, I., Singh, R., 2022. Prunella vulgaris leaf extract assisted green synthesis of silver nanoparticles: antimicrobial activity. Materials Today: Proceedings, (xxxx). Available from: https://doi.org/10.1016/j.matpr.2022.09.520.

Gao, L., et al., 2022. Ultrasound-assisted green synthesis of gold nanoparticles using citrus peel extract and their enhanced anti-inflammatory activity. Ultrasonics Sonochemistry 83, 105940. Available from: https://doi.org/10.1016/j.ultsonch.2022.105940.

Gnanasangeetha, D., Suresh, M., 2020. A review on green synthesis of metal and metal oxide nanoparticles. Nature Environment and Pollution Technology 19 (5), 1789–1800. Available from: https://doi.org/10.46488/NEPT.2020.v19i05.002.

Gowda, S.A.M., Goveas, L.C., Dakshayini, K., 2022. Adsorption of methylene blue by silver nanoparticles synthesized from *Urena lobata* leaf extract: kinetics and equilibrium analysis. Materials Chemistry and Physics 288, 126431. Available from: https://doi.org/10.1016/j.matchemphys.2022.126431. June.

Gudata, L., et al., 2022. Investigation of TiO_2 nanoparticles using leaf extracts of *Lippia adoensis* (Kusaayee) for antibacterial activity. Journal of Nanomaterials 1–8. Available from: https://doi.org/10.1155/2022/3881763. Edited by A. Roy, 2022.

Hano, C., Tungmunnithum, D., 2020. Plant polyphenols, more than just simple natural antioxidants: oxidative stress, aging and age-related diseases. Medicines 7 (5), 26. Available from: https://doi.org/10.3390/medicines7050026.

Hashemi, Z., et al., 2022. Sustainable green synthesis of silver nanoparticles using *Sambucus ebulus* phenolic extract (AgNPs@SEE): optimization and assessment of photocatalytic degradation of methyl orange and their in vitro antibacterial and anticancer activity: sustainable green'. Arabian Journal of Chemistry 15 (1), 103525. Available from: https://doi.org/10.1016/j.arabjc.2021.103525.

Hosny, M., et al., 2022. Facile synthesis of gold nanoparticles for anticancer, antioxidant applications, and photocatalytic degradation of toxic organic pollutants. ACS Omega 7 (3), 3121–3133. Available from: https://doi.org/10.1021/acsomega.1c06714.

Imade, E.E., et al., 2022. Green synthesis of zinc oxide nanoparticles using plantain peel extracts and the evaluation of their antibacterial activity. Scientific African 16, e01152. Available from: https://doi.org/10.1016/j.sciaf.2022.e01152.

Indhira, D., et al., 2022. Biomimetic facile synthesis of zinc oxide and copper oxide nanoparticles from *Elaeagnus indica* for enhanced photocatalytic activity. Environmental Research 212, 113323. Available from: https://doi.org/10.1016/j.envres.2022.113323.

Ishtiaq, M., et al., 2019. Ionic liquid-based colloidal nanoparticles: applications in organic synthesis. Metal Nanoparticles for Drug Delivery and Diagnostic Applications. Elsevier, pp. 279–299. Available from: 10.1016/B978-0-12-816960-5.00015-X.

Izadi, N., et al., 2022. The removal of Cr(VI) from aqueous and saturated porous media by nanoscale zero-valent iron stabilized with flaxseed gum extract: synthesis by continuous flow injection method. Korean Journal of Chemical Engineering 39 (2), 1–12. Available from: https://doi.org/10.1007/s11814-022-1069-4.

Jadoun, S., et al., 2021. Green synthesis of nanoparticles using plant extracts: a review. Environmental Chemistry Letters. Springer, pp. 355–374. Available from: https://doi.org/10.1007/s10311-020-01074-x.

Javed, R., et al., 2020. Role of capping agents in the application of nanoparticles in biomedicine and environmental remediation: recent trends and future prospects. Journal of Nanobiotechnology 18 (1), 172. Available from: https://doi.org/10.1186/s12951-020-00704-4.

Jayeoye, T.J., et al., 2021. Multifarious biological applications and toxic Hg^{2+} sensing potentiality of biogenic silver nanoparticles based on *Securidaca inappendiculata* hassk stem extract. International Journal of Nanomedicine 16, 7557–7574. Available from: https://doi.org/10.2147/IJN.S325996. September.

Joy Prabu, H., et al., 2022. Laser induced plant leaf extract mediated synthesis of CuO nanoparticles and its photocatalytic activity. Environmental Research 212, 113295. Available from: https://doi.org/10.1016/j.envres.2022.113295.

Kabeerdass, N., et al., 2022. *Limonia acidissima* leaf mediated gold nanoparticles synthesis and their antimicrobial and wound healing properties. Materials Letters 314, 10–13. Available from: https://doi.org/10.1016/j.matlet.2022.131893. February.

Kaithal, P., Rohit Lall, R.K., Preetam Verma, A., 2022. Green synthesis of silver nanoparticles from *Pimenta dioica* berries and its antimicrobial potential. International Journal of Current Microbiology and Applied Sciences 11 (1), 308–314. Available from: https://doi.org/10.20546/ijcmas.2022.1101.037.

Kalwani, M., et al., 2022. Effects of nanofertilizers on soil and plant-associated microbial communities: emerging trends and perspectives. Chemosphere 287, 132107. Available from: https://doi.org/10.1016/j.chemosphere.2021.132107. May 2021.

Kannan, K., et al., 2022. Green synthesis silver nanoparticles using medicinal plant and antimicrobial activity against human pathogens. Materials Today: Proceedings 69, 1346–1350. Available from: https://doi.org/10.1016/j.matpr.2022.08.506. 2022.

Karthik, P., et al., 2022. Plant-mediated biosynthesis of zinc oxide nanoparticles from *Delonix elata*: a promising photocatalyst for crystal violet degradation. Inorganic Chemistry Communications 146, 110122. Available from: https://doi.org/10.1016/j.inoche.2022.110122. October.

Kermani, M., et al., 2022. The photocatalytic, cytotoxicity, and antibacterial properties of zinc oxide nanoparticles synthesized using *Trigonella foenum-graecum* L extract. Environmental Science and Pollution Research 1–13. Available from: https://doi.org/10.1007/s11356-022-23518-3.

Khaleghi, S., Khayatzadeh, J., Neamati, A., 2022. Biosynthesis of zinc oxide nanoparticles using *Origanum majorana* L. leaf extract, its antioxidant and cytotoxic activities. Materials Technology 37 (13), 2522–2531. Available from: https://doi.org/10.1080/10667857.2022.2044218.

Khan, M., 2022. Alternative capping agent for green metallic nanoparticle synthesis, frontiers in Nanotechnology. https://www.azonano.com/news.aspx?newsID = 38632 (accessed 08.07.22).

Khanal, L.N., et al., 2022. Green synthesis of silver nanoparticles from root extracts of *Rubus ellipticus* Sm. and comparison of antioxidant and antibacterial activity. Journal of Nanomaterials 1–11. Available from: https://doi.org/10.1155/2022/1832587. Edited by T. Premkumar, 2022.

Khare, S., Singh, R.K., Prakash, O., 2022. Green synthesis, characterization and biocompatibility evaluation of silver nanoparticles using radish seeds. Results in Chemistry 4, 100447. Available from: https://doi.org/10.1016/j.rechem.2022.100447. May.

Kodasi, B., et al., 2023. Adept green synthesis of Cu_2O nanoparticles using Kiwi fruit *(Actinidia deliciosa)* juice and Studies on their cytotoxic activity and antimicrobial evaluation. Journal of Trace Elements and Minerals 3, 100044. Available from: https://doi.org/10.1016/j.jtemin.2022.100044. December 2022.

Krishnaraj, C., Young, G.M., Yun II, S., 2022. In vitro embryotoxicity and mode of antibacterial mechanistic study of gold and copper nanoparticles synthesized from *Angelica keiskei* (Miq.) Koidz. leaves extract. Saudi Journal of Biological Sciences 29 (4), 2552–2563. Available from: https://doi.org/10.1016/j.sjbs.2021.12.039.

Kulkarni, A.G., De Britto, S., Jogaiah, S., 2021. Economic considerations and limitations of green synthesis vs chemical synthesis of nanomaterials. Advances in Nano-Fertilizers and Nano-Pesticides in Agriculture. Woodhead Publishing, pp. 459–468. Available from: https://doi.org/10.1016/b978-0-12-820092-6.00018-5.

Kulkarni, N., Muddapur, U., 2014. Biosynthesis of metal nanoparticles: a review. Journal of Nanotechnology. Hindawi Publishing Corporation. Available from: https://doi.org/10.1155/2014/510246.

Kumar, H., et al., 2020. Fruit extract mediated green synthesis of metallic nanoparticles: a new avenue in pomology applications. International Journal of Molecular Sciences 21 (22), 8458. Available from: https://doi.org/10.3390/ijms21228458.

Kumar, D., et al., 2022. Fabrication and characterization of noble crystalline silver nanoparticles from *Pimenta dioica* leave extract and analysis of chemical constituents for larvicidal applications. Saudi Journal of Biological Sciences 29 (2), 1134–1146. Available from: https://doi.org/10.1016/j.sjbs.2021.09.052.

Kumar Bachheti, R., et al., 2020. Biogenic fabrication of nanomaterials from flower-based chemical compounds, characterization and their various applications: a review. Saudi Journal of Biological Sciences 27 (10), 2551–2562. Available from: https://doi.org/10.1016/j.sjbs.2020.05.012.

Lal, S., et al., 2022. Antioxidant, antimicrobial, and photocatalytic activity of green synthesized ZnO-NPs from *Myrica esculenta* fruits extract. Inorganic Chemistry Communications 141, 109518. Available from: https://doi.org/10.1016/j.inoche.2022.109518.

Li, S., et al., 2007. Green synthesis of silver nanoparticles using *Capsicum annuum* L. Extract. Green Chemistry 9 (8), 852–885. Available from: https://doi.org/10.1039/b615357g.

Liu, Y., et al., 2022. Ecofriendly and enhanced biogenic synthesis of silver nanoparticles using deep eutectic solvent-based green tea extracts. Journal of Cleaner Production 379, 134655. Available from: https://doi.org/10.1016/j.jclepro.2022.134655. October.

Lomelí-Rosales, D.A., et al., 2022. Green synthesis of gold and silver nanoparticles using leaf extract of *Capsicum chinense* plant. Molecules (Basel, Switzerland) 27 (5), 1692. Available from: https://doi.org/10.3390/molecules27051692.

Manjari, G., et al., 2017. Facile *Aglaia elaeagnoidea* mediated synthesis of silver and gold nanoparticles: antioxidant and catalysis properties. Journal of Cluster Science 28 (4), 2041–2056. Available from: https://doi.org/10.1007/s10876-017-1199-8.

Maseera, R., et al., 2022. Green synthesis and evaluation of zinc oxide nanoparticles from *Manilkara zapota* leaf extract. International Journal of Pharmaceutical Sciences and Nanotechnology 15 (1), 5822–5830. Available from: https://doi.org/10.37285/ijpsn.2022.15.1.10.

Md Ishak, N.A.I., Kamarudin, S.K., Timmiati, S.N., 2019. Green synthesis of metal and metal oxide nanoparticles via plant extracts: an overview. Materials Research Express 6 (11), 112004. Available from: https://doi.org/10.1088/2053-1591/ab4458.

Mehdizadeh, S., et al., 2021. Biosynthesis of silver nanoparticles using *Malva sylvestris* flower extract and its antibacterial and catalytic activity. Chemical Methodologies 5 (4), 356–366. Available from: https://doi.org/10.22034/CHEMM.2021.132490.

Melkamu, Z., Jeyaramraja, P.R., 2022. Optimization of the synthesis of silver nanoparticles using the leaf extract of *Ocimum sanctum* and evaluation of their antioxidant potential. Nano Express 3 (3). Available from: https://doi.org/10.1088/2632-959X/ac8fac.

Michael, M.L., Moses, J.A., 2021. Characterization of silver nanoparticles synthesized using *Ocimum basilicum* seed extract. Letters in Applied NanoBioScience 11 (2), 3411–3420. Available from: https://doi.org/10.33263/LIANBS112.34113420.

Miu, B.A., Dinischiotu, A., 2022. New green approaches in nanoparticles synthesis: an overview. Molecules (Basel, Switzerland) 27 (19). Available from: https://doi.org/10.3390/molecules27196472.

Modi, S., et al., 2022. Onion peel waste mediated-green synthesis of zinc oxide nanoparticles and their phytotoxicity on mung bean and wheat plant growth. Materials 15 (7), 2393. Available from: https://doi.org/10.3390/ma15072393.

Mohammadlou, M., Maghsoudi, H., Jafarizadeh-Malmiri, H., 2016. A review on green silver nanoparticles based on plants: synthesis, potential applications and eco-friendly approach. International Food Research Journal 23 (2), 446–463.

Mol, R.L.D., et al., 2022. Biomimetic green approach on the synthesis of silver nanoparticles using *Calotropis gigantea* leaf extract and its biological applications. Applied Nanoscience 12 (8), 2489–2495. Available from: https://doi.org/10.1007/s13204-022-02513-7.

Mosa, W.F.A., et al., 2022. Impact of silver nanoparticles on lemon growth performance: insecticidal and antifungal activities of essential oils from peels and leaves. Frontiers in Plant Science 13, 1–15. Available from: https://doi.org/10.3389/fpls.2022.898846. May.

Mouriya, G.K., et al., 2023. Green synthesis of *Cicer arietinum* waste derived silver nanoparticle for antimicrobial and cytotoxicity properties. Biocatalysis and Agricultural Biotechnology 47, 102573. Available from: https://doi.org/10.1016/j.bcab.2022.102573. December 2022.

My-Thao Nguyen, T., et al., 2021. Biosynthesis of metallic nanoparticles from waste *Passiflora edulis* peels for their antibacterial effect and catalytic activity. Arabian Journal of Chemistry 14 (4), 103096. Available from: https://doi.org/10.1016/j.arabjc.2021.103096.

Nabi, G., et al., 2018. A review on novel eco-friendly green approach to synthesis TiO_2 nanoparticles using different extracts. Journal of Inorganic and Organometallic Polymers and Materials 28 (4), 1552–1564. Available from: https://doi.org/10.1007/s10904-018-0812-0.

Nair, G.M., Sajini, T., Mathew, B., 2022. Advanced green approaches for metal and metal oxide nanoparticles synthesis and their environmental applications. Talanta Open. Elsevier, p. 100080. Available from: https://doi.org/10.1016/j.talo.2021.100080.

Narayanan, M., et al., 2022. Phyto-fabrication of silver nanoparticle using leaf extracts of *Aristolochia bracteolata* Lam and their mosquito larvicidal potential. Process Biochemistry 121, 163–169. Available from: https://doi.org/10.1016/j.procbio.2022.06.022. June.

Narayanan, M., et al., 2023. Characterization of NiONPs synthesized by aqueous extract of orange fruit waste and assessed their antimicrobial and antioxidant potential. Environmental Research 216, 114734. Available from: https://doi.org/10.1016/j.envres.2022.114734. October 2022.

Nayantara, Kaur, P., 2018. Biosynthesis of nanoparticles using eco-friendly factories and their role in plant pathogenicity: a review. Biotechnology Research and Innovation 2 (1), 63–73. Available from: https://doi.org/10.1016/j.biori.2018.09.003.

Ocsoy, I., et al., 2018. Biomolecules incorporated metallic nanoparticles synthesis and their biomedical applications. Materials Letters 212, 45–50. Available from: https://doi.org/10.1016/j.matlet.2017.10.068.

Okpara, E.C., et al., 2021. Green synthesis of copper oxide nanoparticles using extracts of *Solanum macrocarpon* fruit and their redox responses on SPAu electrode. Heliyon 7 (12), e08571. Available from: https://doi.org/10.1016/j.heliyon.2021.e08571.

Olajire, A.A., Mohammed, A.A., 2019. Green synthesis of palladium nanoparticles using *Ananas comosus* leaf extract for solid-phase photocatalytic degradation of low density polyethylene film. Journal of Environmental Chemical Engineering 7 (4), 103270. Available from: https://doi.org/10.1016/j.jece.2019.103270.

Ouerghi, O., et al., 2022. Limon-citrus extract as a capping/reducing agent for the synthesis of titanium dioxide nanoparticles: characterization and antibacterial activity. Green Chemistry Letters and Reviews 15 (3), 483–490. Available from: https://doi.org/10.1080/17518253.2022.2094205.

Oves, M., et al., 2022. Green synthesis of silver nanoparticles by *Conocarpus lancifolius* plant extract and their antimicrobial and anticancer activities. Saudi Journal of Biological Sciences 29 (1), 460–471. Available from: https://doi.org/10.1016/j.sjbs.2021.09.007.

Padmavathi, R., et al., 2022. *Syzygium cumini* leaf extract exploited in the green synthesis of zinc oxide nanoparticles for dye degradation and antimicrobial studies. Materials Today: Proceedings 69, 1200–1205. Available from: https://doi.org/10.1016/j.matpr.2022.08.257. 2022.

Pal, G., Rai, P., Pandey, A., 2019. Green synthesis of nanoparticles: a greener approach for a cleaner future. Green Synthesis, Characterization and Applications of Nanoparticles. Elsevier, pp. 1–26. Available from: https://doi.org/10.1016/b978-0-08-102579-6.00001-0.

Parmar, M., Sanyal, M., 2022. Extensive study on plant mediated green synthesis of metal nanoparticles and their application for degradation of cationic and anionic dyes. Environmental Nanotechnology, Monitoring and Management. Elsevier, p. 100624. Available from: https://doi.org/10.1016/j.enmm.2021.100624.

Patanjali, P., et al., 2019. Nanotechnology for water treatment: a green approach. Green Synthesis, Characterization and Applications of Nanoparticles. Elsevier, pp. 485–512. Available from: https://doi.org/10.1016/B978-0-08-102579-6.00021-6.

Patel, P., et al., 2020. Cellular and molecular impact of green synthesized silver nanoparticles. Engineered Nanomaterials - Health and Safety. IntechOpen. Available from: https://doi.org/10.5772/intechopen.90717.

Pawar, A.A., et al., 2022. *Azadirachta indica*-derived silver nanoparticle synthesis and its antimicrobial applications. Journal of Nanomaterials 1–15. Available from: https://doi.org/10.1155/2022/4251229. Edited by H. M. Ahmed, 2022.

Pechyen, C., et al., 2022. Biogenic synthesis of gold nanoparticles mediated by *Spondias dulcis* (Anacardiaceae) peel extract and its cytotoxic activity in human breast cancer cell. Toxicology Reports 9, 1092–1098. Available from: https://doi.org/10.1016/j.toxrep.2022.04.031. March.

Rani, N., et al., 2022. *Azadirachta indica* leaf extract mediated biosynthesized rod-shaped zinc oxide nanoparticles for in vitro lung cancer treatment. Materials Science and Engineering: B 284, 115851. Available from: https://doi.org/10.1016/j.mseb.2022.115851. June.

Rizwana, H., et al., 2022. Sunlight-mediated green synthesis of silver nanoparticles using the berries of *Ribes rubrum* (red currants): characterisation and evaluation of their antifungal and antibacterial activities. Molecules (Basel, Switzerland) 27 (7), 2186. Available from: https://doi.org/10.3390/molecules27072186.

Saka, A., et al., 2022. Biosynthesis of TiO_2 nanoparticles by Caricaceae (Papaya) shell extracts for antifungal application. Scientific Reports 12 (1), 15960. Available from: https://doi.org/10.1038/s41598-022-19440-w.

Santhosh, P.B., Genova, J., Chamati, H., 2022. Green synthesis of gold nanoparticles: an eco-friendly approach. Chemistry (Weinheim an der Bergstrasse, Germany) 4 (2), 345–369. Available from: https://doi.org/10.3390/chemistry4020026.

Selvam, K., et al., 2022a. *Annona reticulata* leaves-assisted synthesis of zinc oxide nanoparticles and assessment of cytotoxicity and photocatalytic impact. Materials Letters 309, 131379. Available from: https://doi.org/10.1016/j.matlet.2021.131379.

Selvam, K., et al., 2022b. Enhanced photocatalytic activity of novel *Canthium coromandelicum* leaves based copper oxide nanoparticles for the degradation of textile dyes. Environmental Research 211, 113046. Available from: https://doi.org/10.1016/j.envres.2022.113046.

Senthamarai, M.D., Malaikozhundan, B., 2022. Synergistic action of zinc oxide nanoparticle using the unripe fruit extract of *Aegle marmelos* (L.) - antibacterial, antibiofilm, radical scavenging and ecotoxicological effects. Materials Today Communications 30, 103228. Available from: https://doi.org/10.1016/j.mtcomm.2022.103228.

Seshadri, V.D., 2021. Zinc oxide nanoparticles from *Cassia auriculata* flowers showed the potent antimicrobial and in vitro anticancer activity against the osteosarcoma MG-63 cells. Saudi Journal of Biological Sciences 28 (7), 4046–4054. Available from: https://doi.org/10.1016/j.sjbs.2021.04.001.

Shaban, A.S., et al., 2022. *Punica granatum* peel extract mediated green synthesis of zinc oxide nanoparticles: structure and evaluation of their biological applications. Biomass Conversion and Biorefinery. Available from: https://doi.org/10.1007/s13399-022-03185-7.

Shafey, A.M.E., 2020. Green synthesis of metal and metal oxide nanoparticles from plant leaf extracts and their applications: a review. Green Processing and Synthesis 9 (1), 304–339. Available from: https://doi.org/10.1515/gps-2020-0031.

Shahabadi, N., Mahdavi, M., 2022. Green synthesized silver nanoparticles obtained from *Stachys schtschegleevii* extract: ct-DNA interaction and in silico and in vitro investigation of antimicrobial activity. Journal of Biomolecular Structure and Dynamics. Available from: https://doi.org/10.1080/07391102.2022.2028680.

Sharma, D., Kanchi, S., Bisetty, K., 2019. Biogenic synthesis of nanoparticles: a review. Arabian Journal of Chemistry 12 (8), 3576–3600. Available from: https://doi.org/10.1016/j.arabjc.2015.11.002.

Sharma, P., et al., 2021. Capping agent-induced variation of physicochemical and biological properties of α-Fe_2O_3 nanoparticles. Materials Chemistry and Physics 258, 123899. Available from: https://doi.org/10.1016/j.matchemphys.2020.123899.

Shelke, H.D., et al., 2022. Multifunctional Cu2SnS3 nanoparticles with enhanced photocatalytic dye degradation and antibacterial activity. Materials 15 (9), 1–15. Available from: https://doi.org/10.3390/ma15093126.

Shimi, A.K., et al., 2022. Photocatalytic activity of green construction TiO_2 nanoparticles from *Phyllanthus niruri* leaf extract. Journal of Nanomaterials 1–11. Available from: https://doi.org/10.1155/2022/7011539. Edited by A. Ahmed, 2022.

Shyamalagowri, S., et al., 2022. In vitro anticancer activity of silver nanoparticles phyto-fabricated by *Hylocereus undatus* peel extracts on human liver carcinoma (HepG2) cell lines. Process Biochemistry 116, 17–25. Available from: https://doi.org/10.1016/j.procbio.2022.02.022.

Siakavella, I.K., et al., 2020. Effect of plant extracts on the characteristics of silver nanoparticles for topical application. Pharmaceutics 12 (12), 1–17. Available from: https://doi.org/10.3390/pharmaceutics12121244.

Sidhu, A.K., Verma, N., Kaushal, P., 2022. Role of biogenic capping agents in the synthesis of metallic nanoparticles and evaluation of their therapeutic potential. Frontiers in Nanotechnology. Frontiers Media S.A 105. Available from: https://doi.org/10.3389/fnano.2021.801620.

Singh, J., et al., 2018. "Green" synthesis of metals and their oxide nanoparticles: applications for environmental remediation. Journal of Nanobiotechnology 16 (1), 84. Available from: https://doi.org/10.1186/s12951-018-0408-4.

Singh, S., et al., 2022. Green synthesis of TiO_2 nanoparticles using *Citrus limon* juice extract as a bio-capping agent for enhanced performance of dye-sensitized solar cells. Surfaces and Interfaces 28, 101652. Available from: https://doi.org/10.1016/j.surfin.2021.101652. November 2021.

Singla, S., et al., 2022. Green synthesis of silver nanoparticles using *Oxalis griffithii* extract and assessing their antimicrobial activity. OpenNano 7, 100047. Available from: https://doi.org/10.1016/j.onano.2022.100047. February.

Solgi, M., Taghizadeh, M., 2020. Biogenic synthesis of metal nanoparticles by plants. Biogenic Nano-Particles and their Use in Agro-ecosystems. Springer, Singapore, pp. 593–606. Available from: https://doi.org/10.1007/978-981-15-2985-6_27.

Soni, N., Dhiman, R.C., 2020. Larvicidal and antibacterial activity of aqueous leaf extract of Peepal (*Ficus religiosa*) synthesized nanoparticles. Parasite Epidemiology and Control 11, e00166. Available from: https://doi.org/10.1016/j.parepi.2020.e00166.

Suriyakala, G., et al., 2022. Phytosynthesis of silver nanoparticles from *Jatropha integerrima* Jacq. flower extract and their possible applications as antibacterial and antioxidant agent. Saudi Journal of Biological Sciences 29 (2), 680–688. Available from: https://doi.org/10.1016/j.sjbs.2021.12.007.

Tyavambiza, C., et al., 2021. The antimicrobial and anti-inflammatory effects of silver nanoparticles synthesised from *Cotyledon orbiculata* aqueous extract. Nanomaterials 11 (5), 1343. Available from: https://doi.org/10.3390/nano11051343.

Ugwu, E.I., et al., 2022. Application of green nanocomposites in removal of toxic chemicals, heavy metals, radioactive materials, and pesticides from aquatic water bodies. Sustainable Nanotechnology for Environmental Remediation. Elsevier, pp. 321–346. Available from: https://doi.org/10.1016/B978-0-12-824547-7.00018-7.

Ul Haq, T., Ullah, R., 2022. Green synthesis and characterization of gold nanoparticles (Au-NPs) using stem extract of *Euphorbia neriifolia* L. and evaluation of their antibacterial and antifungal potential. International Journal of Nanoscience. Available from: https://doi.org/10.1142/S0219581X22500089.

Vanitha, G., 2021. Eco-friendly synthesis of some novel metal nanoparticles mediated by ocimum Basilicum-Lamiaceae (Thiru Neetru Pathilai) leaves extract. International Journal for Research in Applied Science and Engineering Technology 9 (1), 548–561. Available from: https://doi.org/10.22214/ijraset.2021.32881.

Vankudoth, S., et al., 2022. Green synthesis, characterization, photoluminescence and biological studies of silver nanoparticles from the leaf extract of *Muntingia calabura*. Biochemical and Biophysical Research Communications 630, 143–150. Available from: https://doi.org/10.1016/j.bbrc.2022.09.054.

Vanlalveni, C., et al., 2021. Green synthesis of silver nanoparticles using plant extracts and their antimicrobial activities: a review of recent literature. RSC Advances 11 (5), 2804–2837. Available from: https://doi.org/10.1039/D0RA09941D.

Velsankar, K., et al., 2022. Green synthesis of silver oxide nanoparticles using *Panicum miliaceum* grains extract for biological applications. Advanced Powder Technology 33 (7), 103645. Available from: https://doi.org/10.1016/j.apt.2022.103645.

Venkatesan, S., Suresh, S., Ramu, P., Kandasamy, M., et al., 2022a. Biosynthesis of zinc oxide nanoparticles using *Euphorbia milii* leaf constituents: characterization and improved photocatalytic degradation of methylene blue dye under natural sunlight. Journal of the Indian Chemical Society 99 (5), 100436. Available from: https://doi.org/10.1016/j.jics.2022.100436.

Venkatesan, S., Suresh, S., Ramu, P., Arumugam, J., et al., 2022b. Methylene blue dye degradation potential of zinc oxide nanoparticles bioreduced using *Solanum trilobatum* leaf extract. Results in Chemistry 4, 100637. Available from: https://doi.org/10.1016/j.rechem.2022.100637. November.

Verma, V., et al., 2022. A review on green synthesis of TiO_2 NPs: synthesis and applications in photocatalysis and antimicrobial. Polymers 14 (7). Available from: https://doi.org/10.3390/polym14071444.

Vijilvani, C., et al., 2020. Antimicrobial and catalytic activities of biosynthesized gold, silver and palladium nanoparticles from *Solanum nigurum* leaves. Journal of Photochemistry and Photobiology B: Biology 202, 111713. Available from: https://doi.org/10.1016/j.jphotobiol.2019.111713.

Vinayagam, R., et al., 2022. Rapid photocatalytic degradation of 2, 4-dichlorophenoxy acetic acid by ZnO nanoparticles synthesized using the leaf extract of *Muntingia calabura*. Journal of Molecular Structure 1263, 133127. Available from: https://doi.org/10.1016/j.molstruc.2022.133127.

Vinayagam, R., et al., 2023. Green synthesized cobalt oxide nanoparticles with photocatalytic activity towards dye removal. Environmental Research 216 (P4), 114766. Available from: https://doi.org/10.1016/j.envres.2022.114766.

Wan Mat Khalir, W.K.A., et al., 2020. Biosynthesized silver nanoparticles by aqueous stem extract of *Entada spiralis* and screening of their biomedical activity. Frontiers in Chemistry 8, 620. Available from: https://doi.org/10.3389/fchem.2020.00620.

Weng, X., et al., 2022. Biosynthesis of silver nanoparticles using three different fruit extracts: characterization, formation mechanism and estrogen removal. Journal of Environmental Management 316, 115224. Available from: https://doi.org/10.1016/j.jenvman.2022.115224.

Xu, J., et al., 2021. A review of the green synthesis of ZnO nanoparticles using plant extracts and their prospects for application in antibacterial textiles. Journal of Engineered Fibers and Fabrics. Available from: https://doi.org/10.1177/15589250211046242. SAGE PublicationsSage UK: London, England.

Yassin, M.T., Mostafa, A.A.-F., et al., 2022a. Facile green synthesis of silver nanoparticles using aqueous leaf extract of *Origanum majorana* with potential bioactivity against multidrug resistant bacterial strains. Crystals 12 (5), 603. Available from: https://doi.org/10.3390/cryst12050603.

Yassin, M.T., Al-Askar, A.A., et al., 2022b. Green synthesis of zinc oxide nanocrystals utilizing *Origanum majorana* leaf extract and their synergistic patterns with colistin against multidrug-resistant bacterial strains. Crystals 12 (11), 1513. Available from: https://doi.org/10.3390/cryst12111513.

Yazdanian, M., et al., 2022. The potential application of green-synthesized metal nanoparticles in dentistry: a comprehensive review. Bioinorganic Chemistry and Applications 1–27. Available from: https://doi.org/10.1155/2022/2311910. Edited by V. De Matteis, 2022.

Yazhiniprabha, M., et al., 2022. Biomimetically synthesized *Physalis minima* fruit extract-based zinc oxide nanoparticles as eco-friendly biomaterials for biological applications. Journal of Drug Delivery Science and Technology 73, 103475. Available from: https://doi.org/10.1016/j.jddst.2022.103475.

Ying, S., et al., 2022. Green synthesis of nanoparticles: current developments and limitations. Environmental Technology and Innovation 102336. Available from: https://doi.org/10.1016/j.eti.2022.102336. Elsevier.

Zhang, Y., et al., 2022. Green synthesis of platinum nanoparticles by *Nymphaea tetragona* flower extract and their skin lightening, antiaging effects. Arabian Journal of Chemistry 104391. Available from: https://doi.org/10.1016/j.arabjc.2022.104391.

Zhao, W., et al., 2022. Green synthesis, characterization and determination of anti-prostate cancer, cytotoxicity and antioxidant effects of gold nanoparticles synthesized using *Alhagi maurorum*. Inorganic Chemistry Communications 141, 109525. Available from: https://doi.org/10.1016/j.inoche.2022.109525.

Zhu, B., et al., 2022. Formulation and characterization of a novel anti-human endometrial cancer supplement by gold nanoparticles green-synthesized using *Spinacia oleracea* L. Leaf aqueous extract. Arabian Journal of Chemistry 15 (3), 103576. Available from: https://doi.org/10.1016/j.arabjc.2021.103576.

Zhu, X., Pathakoti, K., Hwang, H.-M., 2019. Green synthesis of titanium dioxide and zinc oxide nanoparticles and their usage for antimicrobial applications and environmental remediation. Green Synthesis, Characterization and Applications of Nanoparticles. Elsevier, pp. 223–263. Available from: 10.1016/b978-0-08-102579-6.00010-1.

Zubair, M., et al., 2022. Green synthesis and characterization of silver nanoparticles from Acacia nilotica and their anticancer, antidiabetic and antioxidant efficacy. Environmental Pollution 304, 119249. Available from: https://doi.org/10.1016/j.envpol.2022.119249. April.

Chapter 19

The intra- and extracellular mechanisms of microbially synthesized nanomaterials and their purification

Nathania Puspitasari[1,2], Ery Susiany Retnoningtyas[1], Chintya Gunarto[1,2] and Felycia Edi Soetaredjo[1,2]
[1]*Department of Chemical Engineering, Widya Mandala Surabaya Catholic University, Surabaya, East Java, Indonesia,* [2]*Collaborative Research Center for Zero Waste and Sustainability, Widya Mandala Surabaya Catholic University, Surabaya, East Java, Indonesia*

19.1 Introduction

Nanotechnology, which involves creating functional systems at the molecular level, is one of the scientific and technology fields that is growing the fastest. The word "nanotechnology" has gained enormous traction in recent years due to its numerous uses in agriculture, health, food, textiles, cosmetics, and electronics industries. Nanotechnology is linked to the production of nanomaterials (NMs) with improved properties that distinguish them from bulk materials. NMs consist of one or more components having at least one dimension between 1 and 100 nm, for example, nanoparticles, composite materials, nanofibers, and nanostructured surfaces (Borm et al., 2006; Verma et al., 2019, 2018). NMs have become more prominent in technological breakthroughs due to their superior performance compared with their bulk counterparts in terms of mechanical, electrical, and magnetic behavior, as well as chemical characteristics (Jeevanandam et al., 2018; Lloyd et al., 2011). These NMs can be classified into the following types based on their size and characteristics, that is, carbon-based NMs, composite-based NMs, organic-based NMs, and inorganic-based NMs (Kolahalam et al., 2019; Zhang et al., 2012). Currently, metal-based NMs such as silver (Ag), zinc (Zn), lead (Pb), gold (Au), iron (Fe), carbon (C), and copper (Cu) have attracted great interest among researchers (Khan et al., 2021; Zhang et al., 2023).

The synthesis of NMs can be prepared by various techniques, including a top-down approach and a bottom-up approach (self-assembly). These techniques are further divided into subclasses based on the operation and reaction conditions. The bottom-up approach also known as a building-up process involves constructing a structure atom by atom, molecule by molecule, or by self-arrangements. Techniques such as sedimentation and reduction through green synthesis, spinning, and biochemical synthesis serve as examples of this method. In the top-down approach, physical and chemical techniques are used to reduce the size of the appropriate starting components. NMs have been synthesized using conventional physical techniques such as electrospinning, radiolysis, spray pyrolysis, ultrasonication, and photo-irradiation (Bhardwaj et al., 2019, 2018, 2017; Khan et al., 2019) However, chemical techniques have attracted more interest than physical techniques due to their greater ability to control the size and structure of NMs. Sol–gel, solvothermal, coprecipitation, and template-based approaches are the major chemical techniques. The accessible and widely used physical and chemical methods for producing NMs are energy-intensive, contain hazardous chemicals, and require a high temperature for reaction (Abid et al., 2022; Nasaruddin et al., 2021). Although there are many physicochemical ways to synthesize NMs, it is still necessary to develop nontoxic, low-cost, high-yield, low-energy, and eco-friendly methods particularly for applications in the fields of human health and medicine. Therefore numerous strategies for the bio-based synthesis of NMs have been explored to establish sustainable and cost-effective bioproduction alternatives. For instance, various flavonoids found in biomass waste produced from fruit residues can chelate metal ions and reduce them into nanoparticles (Aswathi et al., 2022; Putro et al., 2022). Several researchers have reported the production of graphene utilizing pulp waste and biodegradable waste from paper cups (Shukla et al., 2020.; Singh et al., 2021).

Other biosynthesis pathways of NMs using microbes involving bacteria, fungi, yeast, and algae have been widely reported due to their reducing characteristics, which are often responsible for reducing metal compounds in particular

Green and Sustainable Approaches Using Wastes for the Production of Multifunctional Nanomaterials. DOI: https://doi.org/10.1016/B978-0-443-19183-1.00004-0

NMs. Microorganisms can be used in nanotechnology as a green technology for sustainable development strategies due to the use of cleaner production as well as the preservation of natural resources. For instance, fungus-mediated methods include simple procedures for the nanosynthesis of inorganic substances such as $CuAlO_2$, which requires low-temperature conditions (Ahmad et al., 2007). Moreover, fungal biomass was also essential for chemically synthesized BiOCl nanoplates with sizes between 150 and 200 nm to break down into extremely tiny particles (<10 nm) without affecting their crystalline structure (Chung et al., 2016). Researchers have recently exploited a variety of biological extracts to synthesize metallic NMs by following direct techniques and employing microbial extracts as a source of reductants. With the use of biological resources, it is feasible to get the specific size, shape, and monodispersity of NMs either extracellularly or intracellularly (de Jesus et al., 2021). This chapter reviews the current works in green synthesis of NMs by microbes that focused on their intra- and extracellular mechanisms, purification techniques, characterizations, and applications. The difficulties of elaborating this technology at a large-scale level and the prospects of biological synthesis approaches are also highlighted in the last section.

19.2 Microbially synthesized nanomaterials

19.2.1 Intracellular and extracellular mechanisms

Since the formation of the Earth, biological organisms and inorganic materials have been in continual touch with each other. The interactions between inorganic substances and living things have drawn more attention from scientists in recent years. Numerous microorganisms produce various inorganic compounds either extracellularly or intracellularly, and the mechanisms vary from one organism to another (Fariq et al., 2017; Hulkoti and Taranath, 2014). By using several synthesis components, including microorganisms, plant extracts, and other biological components, NMs are synthesized through biological processes (Saravanan et al., 2021). Due to their ease of cultivation, rapid growth, and potential to thrive under ambient conditions, microbes such as bacteria, algae, yeast, and fungi are typically selected for synthesis in NMs. Interestingly, microbes can detoxify and accumulate heavy metals in the presence of reductase enzymes, which play a crucial role in reducing metal salts into NMs (Ovais et al., 2018). Different biological agents and various metal solutions have varying effects on the production of NMs.

There are two categories for microbial production of NMs. The first category is biosorption, which does not require energy use and involves the attachment of metal ions found in aqueous solutions to the cell wall. Stable NMs are formed as a result of interactions with the cell wall or peptides (Egan-Morriss et al., 2022; Pantidos, 2014). The prospective processes for the biosorption of the metal on microbes consist of physical processes including ion exchange, complexation, precipitation, and physisorption. Microbes typically secrete lipopolysaccharide, glycoprotein, and other exopolysaccharide compounds that have anionic structural groups for positive metal adhering to negative charges of the cell wall. Chitin was shown to be the primary component of the fungal cell wall, and it is associated with the complex formation of heavy metals, which leads to the synthesis of NMs (L. Wang et al., 2018). Few researchers have reported the biosynthesis of copper NMs via the biosorption method from *Rhodotorula mucilaginosa* biomass. The spherical form of the produced NMs made them accessible for simultaneous pollution removal and NMs synthesis. The formation of metallic molybdenum NMs by *Clostridium pasteurianum* has also been the subject of another investigation (Nordmeier et al., 2018; Salvadori et al., 2014).

Meanwhile, bioreduction occurs when metal ions are chemically reduced by living organisms into more stable forms. Numerous species can utilize metabolism metal reduction, in which the reduction of a metal ion is linked to the oxidation of an enzyme. As a consequence, stable and inert metallic NMs are formed, which may be removed safely from a polluted material. The synthesis of NMs may be triggered by several substances found in microbial cells, notably amides, amines, alkaloids, carbonyl groups, proteins, pigments, and other reducing agents (Quintero-Quiroz et al., 2019; Sable et al., 2020). Some microbes usually release chemicals with a high capacity for oxidation or reduction of metal ions to produce zero-valent or magnetic NMs. Additionally, these organisms are easy to handle and susceptible to genetic manipulation (Puspitasari et al., 2021; Puspitasari and Lee, 2021).

It is well known that both intracellular and extracellular proteins, enzymes, lipids, and chelating activity of DNA subunits are actively involved as reducing agents throughout the biosynthesis process. These bioactive substances have high reduction potential and can donate H^+ ions to reduce metal ions from a higher oxidation form to a lower oxidation form (Dauthal and Mukhopadhyay, 2016; Srivastava et al., 2021). According to the site where NMs are generated, extracellular and intracellular syntheses become the most common processes of biosynthesis (Fig. 19.1). NMs can be accumulated in the periplasm, cytoplasmic membrane, and cell wall when observed under a microscope.

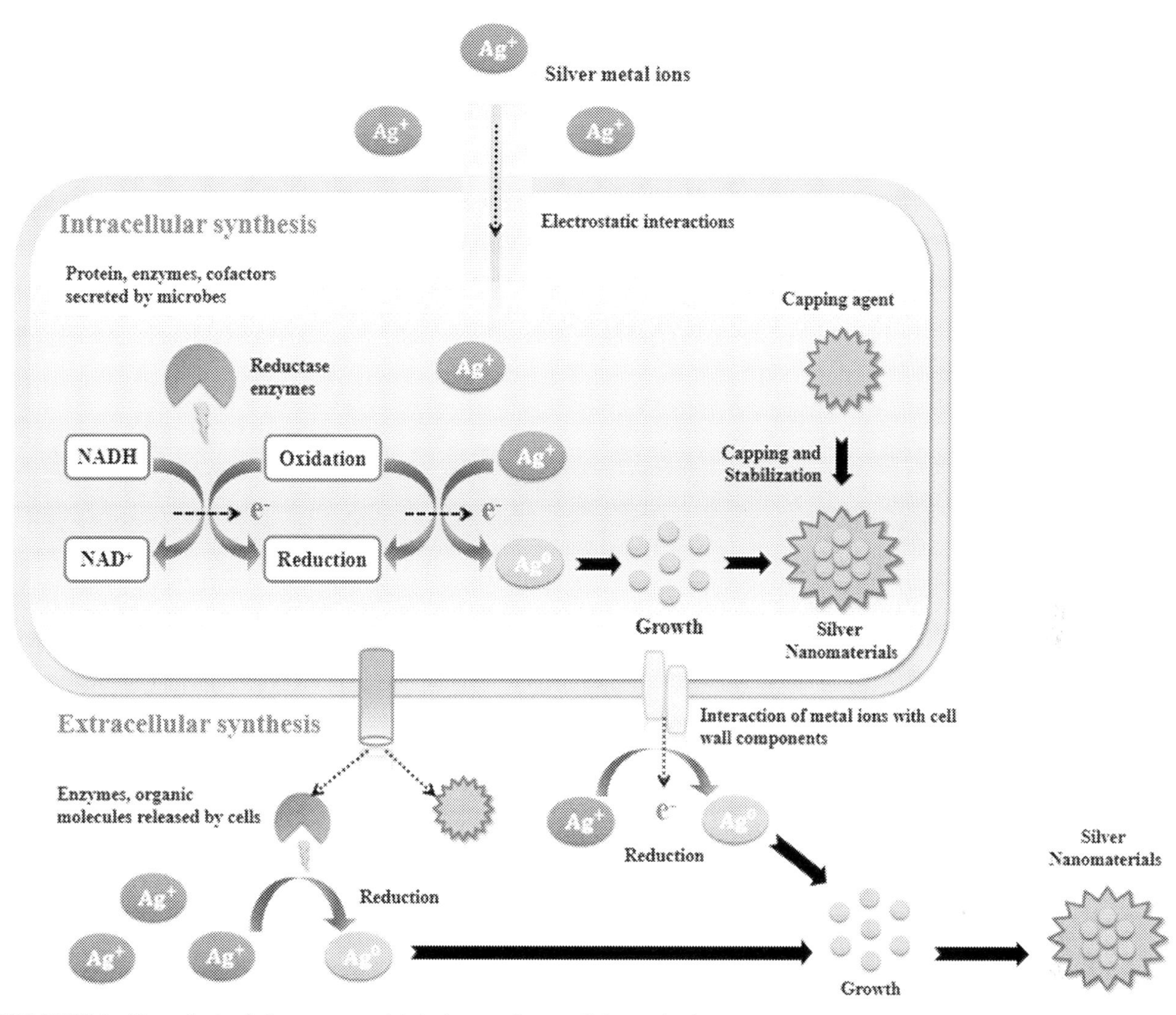

FIGURE 19.1 Biosynthesis of silver nanomaterials by intra- and extracellular mechanisms.

In the extracellular approach, NMs are produced outside cells by capturing metal ions on their surfaces and reducing ions in the presence of microbe-secreted enzymes (Li et al., 2011). Cofactors such as reduced nicotinamide adenine dinucleotide (NADH) and reduced nicotinamide adenine dinucleotide phosphate-reliant enzymes both have crucial roles as reductants through electron transfer from NADH through NADH-reliant enzymes. For example, the release of NADH and NADH-reliant enzymes is an important process in the extracellular biosynthesis of silver nanomaterials (AgNMs) by microbes. The bioreduction of silver is initiated by NADH-reliant reductase enzymes found in microbes by electron transfer from NADH (He et al., 2007). As a result, silver ions (Ag^{+}) receive electrons and are reduced (Ag^{0}), resulting in the generation of enlarged metal nuclei and the formation of stable AgNMs within cell-free supernatant. Precursor concentration, pH, temperature, and reaction time are some limiting factors affecting the size and properties of NMs.

The intracellular approach includes transporting ions into the inner space of microbial cells to produce NMs when the enzymes are present. Microbial cells and sugar molecules are primarily involved in the intracellular process of metal bioreduction. The interactions between intracellular enzymes and positively charged groups are the main mechanism for the trapping of metal ions from the media and their subsequent reduction within the cell. This resulted from NMs being produced as a result of enzymatic reduction and metal ion transport across membranes (Dauthal and Mukhopadhyay, 2016). In order to release the biosynthesized NMs from intracellular production, additional processes are needed such as ultrasonic treatment or interactions with the appropriate detergents. In contrast, extracellular biosynthesis is inexpensive, requires less complex downstream processing, and supports large-scale production of NMs to investigate its possible uses. Therefore, the extracellular method for biosynthesis of NMs has been the main subject of several studies compared with the intracellular method (Das et al., 2014). An extensive list of the microbes used in synthesizing NMs is provided in Table 19.1.

19.2.2 Synthesis of nanomaterials using bacteria

Bacteria have become one of the most useful research subjects due to their abundance in the environment and their ability to endure harsh circumstances. Additionally, they can grow rapidly, and their cultivation is easy to control, such as temperature, pH, oxygenation, and incubation time. Optimizing these conditions is crucial since different sizes of NMs are needed for various applications including optics, catalysts, and antimicrobials (He et al., 2007). Bacteria typically produce intracellular or extracellular inorganic substances, which can be employed for the biosynthesis of NMs. *Bacillus marisflavi* was shown to produce AuNMs with a particle size of 14 nm. AuNMs synthesis from bacterial cell-free extract occurred extracellularly, and the color changed from light yellow to bluish-purple. The production of AuNMs was indicated by the presence of bluish-purple color caused by surface plasmon resonance (Nadaf and Kanase, 2019).

19.2.3 Synthesis of nanomaterials using fungi

Researchers across the world frequently utilize fungi for NMs synthesis using both intracellular and extracellular processes. It is well known that using fungi to produce metal oxide or NMs is an effective technique with clear morphology (Ijaz et al., 2020). Fungi produce more NMs than bacteria because their intracellular enzymes function as biological substances that increase the bioaccumulation capacity and metal resistance (Kalpana and Devi Rajeswari, 2018). Significant advantages include the ease of scaling up and downstream processing, economic feasibility, and the presence of mycelia, which supplies a high surface area (Mohanpuria et al., 2008). The most well-known fungi for synthesizing silver and gold nanomaterials are *Fusarium* sp., *Penicillium* sp., and *Aspergillus* sp. (Shah et al., 2015). The extracellular production of AgNMs was carried out using *Penicillium* sp. The enzyme induction was facilitated by the existence of silver nitrate in the cell culture broth, and optimal synthesis was shown at pH 6 with a substrate concentration of about 1.5 mM (Shareef et al., 2017; Spagnoletti et al., 2019).

19.2.4 Synthesis of nanomaterials using yeast

Due to their improved function and stability, yeasts have been considered a highly efficient source of NMs synthesis. Additionally, they can capture large amounts of potentially toxic metals. The present study on yeast focuses mostly on the production of nanocrystalline quantum semiconductors, notably cadmium sulfide (CdS) and zinc sulfide (ZnS) nanomaterials. The biosynthesis of silver and gold NMs was mainly carried out by *S. cerevisiae* and other silver-resistant yeast strains (Korbekandi et al., 2016). The production of silica NMs is another use of *S. cerevisiae* in the nanomaterial generation process. The NMs were produced when yeast extract and sodium silicate (precursor solution) were added. One potential mechanism involves the interaction of yeast extract and sodium silicate in an aqueous medium to generate sodium hydroxide and silica oxide NMs (Zamani et al., 2020).

19.2.5 Synthesis of nanomaterials using algae

It has been reported that algae play a significant part in the biological synthesis of NMs and the buildup of certain toxic metals. Large-scale algae production is mostly utilized to synthesize gold, silver, and possibly zinc oxide NMs. Algae are recognized for their capacity to transform toxic metals into their harmless equivalents (Ong et al., 2021). For example, *Sargassum muticum* was employed in the production of ZnO NMs and was found to have antiapoptotic and antiangiogenesis properties in $HepG_2$ cells (Yang and Cui, 2008). Furthermore, *Staphylococcus aureus* and *Pseudomonas aeruginosa* were effectively inhibited by the NMs, with inhibition zones of 13.33 and 15.17 mm, respectively (Bhuyar et al., 2020).

19.3 Purification methods of biosynthesized nanomaterials

The biosynthesized NMs can be purified by several methods including chromatography, magnetic fields, density gradient centrifugation, and electrophoresis (Table 19.2).

TABLE 19.1 Biosynthesis of various nanomaterials using microbes and their applications.

No.	Microbe	Type of nanomaterial	Synthesis location	Physicochemical parameters			Size (nm)	Shape	Application	References
				Temperature	pH	Incubation time				
Bacteria										
1.	*Geobacillus* spp.	Silver (Ag)	Extracellular	55°C	7.5	48 h	<100	Spherical	–	Cekuolyte et al. (2023)
2.	*Vibrio alginolyticus*	Gold (Au)	Extracellular	40°C	7	14 h	100–150	Irregular	Anticancer and antioxidant	Shunmugam et al. (2021)
3.	*Marinomonas* sp. ef1	Cooper (Cu)	Extracellular	22°C	–	48 h	10–70	Spherical/ ovoidal	Antimicrobial	John et al. (2021)
4.	*Shewanella loihica* PV-4	Palladium (Pd)	Extracellular	30°C	7	72 h	4–10	Spherical	Catalyst for Cr (VI) reduction	W. Wang et al. (2018)
5.	*Nocardiopsis flavascens* RD30	Silver (Ag)	Extracellular	30°C	–	72 h	5–50	Spherical	Cytotoxicity	Ranjani et al. (2018)
6.	*Pseudoalteromonas lipolytica*	Silver (Ag)	Extracellular	28°C	6.5–7	72 h	5–15	Spherical	Dye decolorization	Kulkarni et al. (2018)
7.	*Shewanella loihica* PV-4	Platinum (Pt)	Extracellular	30°C	7	48 h	2–6	–	Dye decomposition	Ahmed et al. (2018)
8.	*Desulfovibrio* sp. LS4	Maghemite (Fe_2O_3)	Extracellular	30°C	7.8	35 days	18	Round	Iron nanoparticle formation in saltpan sediment	Das et al. (2018)
9.	*Enterococcus faecalis*	Selenium (Se)	Extracellular	37°C	7	24 h	29–195	Spherical	Antibacterial	Shoeibi and Mashreghi (2017)
10.	*Pseudomonas aeruginosa* JP-11	Cadmium sulfide (CdS)	Extracellular	50°C	–	20 h	20–40	Spherical	Cadmium removal from aqueous solution	Raj et al. (2016)
Fungi										
1.	*Penicillium oxalicum*	Silver (Ag)	Extracellular	28°C	–	24 h	10–50	Spherical	Antimicrobial, anticancer, antioxidant	Gupta et al. (2022)
2.	*Trichoderma longibrachiatum*	Silver (Ag)	Extracellular	55°C	7	24 h	5–50	Spherical	Biosafety assessment	Cui et al. (2022)

(*Continued*)

TABLE 19.1 (Continued)

No.	Microbe	Type of nanomaterial	Synthesis location	Physicochemical parameters			Size (nm)	Shape	Application	References
				Temperature	pH	Incubation time				
3.	*Periconium sp.*	Zinc oxide (ZnO)	Extracellular	45°C	5	24 h	16–78	Quasi-spherical	Antioxidant, antibacterial	Ganesan et al. (2020)
4.	*Lignosus rhinocerotis*	Gold (Au)	Extracellular	65°C	4.5	2.5 h	49.5–82.4	Spherical	Antibacterial	Katas et al. (2019)
5.	*Trichoderma asperellum*	Copper oxide (CuO)	Extracellular	40°C	–	24 h	110	Spherical	Photothermolisis on human lung carcinoma	Saravanakumar et al. (2019)
6.	*Rhodotorula mucilaginosa*	Silver (Ag)	Extracellular	25°C	7	168 h	13.7	Spherical	Antifungal, catalyst, cytotoxicity	Cunha et al. (2018)
7.	*Aspergillus niger*	Zinc oxide (ZnO)	Extracellular	32°C	6.2	48 h	53–69	Spherical	Antibacterial, dye degradation	Kalpana et al. (2018)
8.	*Penicillium chrysogenum*	Platinum (Pt)	Extracellular	100°C	–	12 h	5–40	Spherical	Cytotoxicity	Subramaniyan et al. (2017)
9.	*Cladosporium cladosporioides*	Gold (Au)	Extracellular	30°C	7	48 h	60	Round	Antioxidant, antibacterial	Joshi et al. (2017)
10.	*Rhizopus stolonifer*	Silver (Ag)	Extracellular	40°C	–	48 h	2.86	Spherical	–	AbdelRahim et al. (2017)
Yeast										
1.	*Saccharomyces cerevisiae*	Iron oxide (Fe_2O_3)	Extracellular	30°C	–	2–3 days	70–100	Spherical	Antimicrobial	Asha Ranjani et al. (2022)
2.	*Pichia kudriavzevii* HA	Silver (Ag)	Extracellular	30°C	–	72 h	29.6–30.14	Round /cubic	Anticancer	Ammar et al. (2021)
3.	*Saccharomyces cerevisiae*	Silica	Intracellular	29°C	6–11	1 h	40–70	Spherical	Oil recovery	Zamani et al. (2020)
4.	*Saccharomyces cerevisiae*	Silver (Ag)	Intracellular	25°C	7	24 h	2–20	Spherical	Biocatalyst	Korbekandi et al. (2016)
5.	*Magnusiomyces ingens* LH-F1	Gold (Au)	Extracellular	30°C	–	24 h	80.1	Sphere/ triangle/ hexagon	Catalytic reduction of nitrophenols	Zhang et al. (2016)

Algae										
1.	*Spirogyra hyalina*	Silver (Ag)	Extracellular	60°C	–	24 h	52.7	Spherical	Antimicrobial	Abdullah et al. (2022)
2.	*Coelastrella aeroterrestrica*	Silver (Ag)	Extracellular	30°C	–	24 h	14.5	Hexagon	Antimicrobial, anticancer, antioxidant	Hamida et al. (2022)
3.	*Padina* sp.	Silver (Ag)	Extracellular	60°C	–	48 h	25–60	Spherical/oval	Antibacterial	Bhuyar et al. (2020)
4.	*Colpomenia sinuosa*	Iron oxide (Fe_3O_4)	Extracellular	30°C	2	1 h	11.24–33.71	Nanospheres	Antibacterial, antifungal	Salem et al. (2019)
5.	*Spirulina platensis*	Palladium (Pd)	Extracellular	70°C	–	20 min	10–20	Spherical	Adsorbent	Sayadi et al. (2018)

TABLE 19.2 Purification methods of biosynthesized nanomaterials by various microbes

Type	Microbe	NMs	Synthetic location	Purification method	Application	References
Chromatography						
Fungi	*Talaromyces purpurogenus* (pigment)	Ag	Extracellular	Two steps: – Centrifugation (6,700 × g, 4°C, 20 min) – Thin Layer Chromatography	Biomedical	Bhatnagar et al. (2022)
Bacteria	*Acinetobacter* sp. (lignin peroxidase)	Au, Se	Extracellular	Two steps sequentially: – Anion exchange chromatography – Gel filtration chromatography (lignin peroxidase)	Biocatalyst	Wadhwani et al. (2018)
Bacteria	*Escherichia coli* (sulfite reductase)	Au	Extracellular	Two steps: – Column chromatography (sulfite reductase) – Centrifugation (80,000 × g, 20 min) (mixed sulfite reductase AuNMs)	Biocatalyst	Gholami-Shabani et al. (2015)
Bacteria	*Pseudomonas aeruginosa* (rhamnolipids)	Ag	Extracellular	Two steps: – Gel column chromatography (rhamnolipids) – Centrifugation (mixed rhamnolipids - AgNMs)	Biosurfactant	Ganesh et al. (2010)
Magnetic fields						
Bacteria	*Magnetospirillum magneticum*	Mag Mn	Intracellular	Two steps: – Centrifugation (8,000 × g, 10°C, 20 min) – Neodymium magnets	Magnetic tumor targeting	Designed Research; K (2022)
Bacteria	*Magnetospirillum gryphiswaldense*	Mag Mn	Intracellular	Two steps: – Column-based magnetic – Ultracentrifugation	Biomedical and Biotechnology	Rosenfeldt et al. (2021)
Fungi	*Mixed fungi*	Fe_3O_4	Intracellular	Two steps: – Centrifugation (500 rpm, 10°C, 20 min) – Permanent magnets	Cleaning agent	Sayed et al. (2021)
Fungi	*Aspergillus niger*	FeS and Fe_3O_4	Intracellular	Permanent magnets	Biomedical	Abdeen et al. (2016)
Density gradient centrifugation						
Fungi	*Aspergillus flavus*	Fe	Extracellular	Centrifugation (5000 rpm, 5 min)	Extraction and Clarification	Hassan et al. (2022)
Bacteria	*Bacillus subtilis*	Ag	Extracellular	Centrifugation twice (10,000 rpm, 5 min)	Antibacterial	Alsamhary (2020)
Bacteria	*Actinomycetes* sp.	Ag	Extracellular	Centrifugation (15,000 rpm, 15 min)	Antimicrobial	Al-Dhabi et al. (2018)
Fungi	*Pleurotus ostreatus* (Laccase)	Au	Extracellular	Centrifugation (2415 × g, 15 min, 4°C)	Decolorization	El-Batal et al. (2015)

(*Continued*)

TABLE 19.2 (Continued)

Type	Microbe	NMs	Synthetic location	Purification method	Application	References
Electrophoresis						
Bacteria	*Streptomyces spiralis; Streptomyces rochei*	Ag	Extracellular	Agarose gel electrophoresis 1%	Antibacterial	Mabrouk et al. (2021)
Fungi	*Aspergillus tubingensis; Bionectria ochroleuca*	Ag	Extracellular	Electrophoresis (sodium dodecyl sulfate-polyacrylamide gel)	Antimicrobial	Rodríguez-González et al. (2020)
Bacteria	*Staphylococcus aureus*	Ag	Intracellular and Extracellular	Agarose gel electrophoresis 0.7%	Biosensors	Amin et al. (2019)

19.3.1 Chromatography

Chromatography is a method for separating mixtures of substances based on variations in how fast the different components spread through a given media. These media are the stationary phase and mobile phase. The stationary phase can be solid or liquid while the mobile phase can be liquid or gas. This chromatography can be used for purification and separation in the biosynthesis of NMs. Several uses of chromatographic methods in the purification of NMs synthesis are described. Current researchers widely use intracellular enzymes in producing AuNM for various applications (Gholami-Shabani et al., 2015). The enzyme is an agent in reducing the metal NMs to be stable material. Enzymes produced by microbes (e.g., *Acinetobacter* sp.) extracellularly and intracellularly after purification by anion exchange and gel filtration chromatography were used to produce Au and Se nanomaterials (Wadhwani et al., 2018).

19.3.2 Magnetic fields

Magnetic fields are purification methods that use magnetic properties to separate and purify NMs, particularly iron (Fe) NMs. One magnetotactic bacterium is *Magnetospirillum gryphiswaldense*, which can move along magnetic field lines due to magnetosomes (Mag Mn). Magnetosomes produced by intracellular bacteria are membrane-enclosed single-domain ferromagnetic NMs (Rosenfeldt et al., 2021). The purification of synthetic materials containing Fe by bacteria consists of two stages: (1) cell wall breakdown and (2) separation-purification. For the breakdown of cell walls, sonification and ultracentrifugation methods can be used, while column-based magnetic (neodymium magnet) can be used for the separation-purification method (Hamdous et al., 2017; Raschdorf et al., 2018; Rosenfeldt et al., 2021).

19.3.3 Density gradient centrifugation

Density gradient centrifugation is the simple purification method of NMs extracellular synthesis. The process of centrifugation is used to separate particles from a solution based on their size, shape, density, medium viscosity, and rotor speed. The density gradient centrifugation method may be required more than once in some cases. For example, *Nocardiopsis* sp. cultures were centrifuged at $10,000 \times g$, 4°C for 10 minutes up to three times after incubation, and 5 mL of each strain's cell-free supernatant was then subjected to 50 mL of an aqueous solution containing 1×10^{-3} M $HAuCl_40.3H_2O$. Subsequently, the samples were centrifuged again at high speed after the reaction for a certain time to separate the produced AuNMs (Manivasagan et al., 2015). Extracellular purification of AgNMs synthesized using *Bacillus subtilis* can be performed by centrifugation method at 10,000 rpm for 5 minutes twice (Alsamhary, 2020).

19.3.4 Electrophoresis

Electrophoresis is the term used to describe the movement and separation of charged particles (ions) caused by electric fields. Two electrodes (anode, cathode) with opposing charges are joined by a conducting liquid known as an electrolyte to form an electrophoretic system. Agarose gel electrophoresis is usually used to purify and separate NMs based on size and shape. For example, 1% agarose gel electrophoresis (Bio-Rad) was used to purify AgNMs generated by fungi isolated from mangroves (Rodrigues et al., 2013). Another work on AgNMs that utilized amplified DNA fragments from *Streptomyces* sp. was separated using TBE buffer containing ethidium bromide (1 g/mL) on 1% agarose gel electrophoresis (Mabrouk et al., 2021). The synthesis of AgNMs by *S. aureus* can be carried out intracellularly and extracellularly so that the purification process requires cell wall lysis (Triton-X100), as well as separation using centrifugation and gel electrophoresis (Amin et al., 2019).

19.4 Characterization of biosynthesized nanomaterials

Biosynthesized nanomaterials characterizations were determined by various techniques, such as spectroscopic technique, microscopic technique, and diffraction technique. Nanomaterials characterization plays a huge role in various applications of nanomaterials. Each technique has a different purpose, methods, and instruments, which will be discovered below.

19.4.1 Spectroscopic techniques

The spectroscopic technique is a measurement to examine the content of the materials, specifically nanomaterials and the surface properties in a mixture solution. It uses various types of instruments, such as UV-Vis Spectroscopy, Fourier Transform Infrared (FTIR), and Raman Scattering, which have distinctive methods. UV-Vis Spectroscopy aims to detect and monitor the size and shape of metal ions of NMs with particle sizes between 2 and 100 nm (Begum et al., 2018; Kumar et al., 2020). Another spectroscopy technique commonly used in NMs is FTIR, to observe the functional group, composition, and interinteraction of molecules (Alessio et al., 2017; Kamnev et al., 2021). In addition, FTIR could identify and classify several microorganisms, such as *Bacillus* (Procacci et al., 2021), *Escherichia coli* (Farouk et al., 2022), Pseudomonas (Lee et al., 2019), and *S. aureus* (Hong et al., 2022).

19.4.2 Microscopic techniques

The microscopic technique is used to determine the physical morphology, texture, and size of the NMs. Several instruments included microscopic techniques, such as the optical microscope, scanning electron microscope (SEM), and transmission electron microscope (TEM). SEM performs morphology, size, and shape of nanoparticles between 0.001 and 5 μm (Maheshwari et al., 2018). In addition, compositional information could be collected by energy dispersive X-ray (EDX) and mapping analysis with an SEM instrument. TEM could observe material with a particle size of up to 1 nm due to high image resolutions, thus real size and structures are detected (Sierra, 2019). The NMs microbially synthesized keep developing with various raw materials, microorganisms, and methods to acquire wider and better applications of NMs. Moreover, High-Resolution TEM (HR-TEM) can provide the morphology of the samples and identify the crystal structure from the atomic scale to thin layer of samples (Javed et al., 2018). All SEM, TEM, and HR-TEM perform best in solid samples, usually powder, fiber, and membrane.

19.4.3 Diffraction techniques

One of the diffraction techniques well known in NMs characterization is XRD, which provides data on the crystallography and structure of the material, also the lattice parameter of samples (Mourdikoudis et al., 2018). Various peaks in the 2θ range show different molecules, for example, Ag nanoparticles appear at 27.81, 32.16, 38.12, 44.3, 46.21, 54.83, 57.39, 64.42, and 77.45 degrees (Meng, 2015), while TiO_2 nanoparticles show peaks at 25.23, 37.71, 47.72, and 62.54 degrees (Toro et al., 2020). XRD performs well in solid, dry, and homogeneous materials. However, for suspension of NMs, measurement of hydrodynamic diameter could be conducted by dynamic light scattering. Liquid NMs with high viscosity, such as liposomes (Zong et al., 2022), polymeric micelles (Ghezzi et al., 2021), nano gels (Ahmed et al., 2020; Pourjavadi et al., 2020), and microemulsion (Gunarto et al., 2020), are required for dilution to have an accurate measurement.

19.5 Challenges and limitations

The NMs are produced from various sources of microbes and have been developed rapidly since the 21st century. Over the years, different methods, sources, and analyses have been carried out and resulted in different types of NMs based on their structure and sizes. However, obtaining homogeneous NMs with the same methods and type of microbe is still challenging due to the unpredictable growth and ability of the microbes. Therefore, more experiments are essential in determining and observing the microorganism in NMs systems. Purification steps of NMs by either intra- or extracellular are considered expensive on an industrial scale as the process requires advanced equipment such as nanofiltration to enhance the purity of NMs. Another limitation in NMs microbially synthesized is an insufficient yield. However, the discovery of a cost-effective NMs biosynthesis alternative can be carried out by utilizing waste materials.

19.6 Conclusions and future outlook

In this chapter, green and sustainable approaches of microbially synthesized nanomaterials were summarized, as well as the intra–extracellular mechanisms and purification methods of NMs. Nanomaterials are synthesized by several types of microbes, such as bacteria, fungi, yeast, and algae. Several researchers are manipulating the DNA of microbes to improve the yield of NMs. In addition, the combination of synthesis mechanism, intra–extracellular in a system is likely to produce a higher amount of nanomaterial. However, it required an established and complete process of purification for industrial production. On the other hand, utilization of NMs specifically in medical applications is possibly overabsorbed due to their tiny size and excellent efficient absorption toward the human body.

References

Abdeen, M., Sabry, S., Ghozlan, H., El-Gendy, A.A., Carpenter, E.E., 2016. Microbial-physical synthesis of Fe and Fe_3O_4 magnetic nanoparticles using *Aspergillus niger* YESM1 and supercritical condition of ethanol. Journal of Nanomaterials 2016. Available from: https://doi.org/10.1155/2016/9174891.

AbdelRahim, K., Mahmoud, S.Y., Ali, A.M., Almaary, K.S., Mustafa, A.E.Z.M.A., Husseiny, S.M., 2017. Extracellular biosynthesis of silver nanoparticles using *Rhizopus stolonifer*. Saudi Journal of Biological Sciences 24, 208–216. Available from: https://doi.org/10.1016/j.sjbs.2016.02.025.

Abdullah, Al-Radadi, N.S., Hussain, T., Faisal, S., Ali Raza Shah, S., 2022. Novel biosynthesis, characterization and bio-catalytic potential of green algae (*Spirogyra hyalina*) mediated silver nanomaterials. Saudi Journal of Biological Sciences 29, 411–419. Available from: https://doi.org/10.1016/j.sjbs.2021.09.013.

Abid, N., Khan, A.M., Shujait, S., Chaudhary, K., Ikram, M., Imran, M., et al., 2022. Synthesis of nanomaterials using various top-down and bottom-up approaches, influencing factors, advantages, and disadvantages: a review. Advances in Colloid and Interface Science. Available from: https://doi.org/10.1016/j.cis.2021.102597.

Ahmad, A., Jagadale, T., Dhas, V., Khan, S., Patil, S., Pasricha, R., et al., 2007. Fungus-based synthesis of chemically difficult-to-synthesize multifunctional nanoparticles of CUAIO2. Advanced Materials 19, 3295–3299. Available from: https://doi.org/10.1002/adma.200602605.

Ahmed, S., Alhareth, K., Mignet, N., 2020. Advancement in nanogel formulations provides controlled drug release. International Journal of Pharmaceutics 584, 119435. Available from: https://doi.org/10.1016/j.ijpharm.2020.119435.

Ahmed, E., Kalathil, S., Shi, L., Alharbi, O., Wang, P., 2018. Synthesis of ultra-small platinum, palladium and gold nanoparticles by *Shewanella loihica* PV-4 electrochemically active biofilms and their enhanced catalytic activities. Journal of Saudi Chemical Society 22, 919–929. Available from: https://doi.org/10.1016/j.jscs.2018.02.002.

Al-Dhabi, N.A., Mohammed Ghilan, A.K., Arasu, M.V., 2018. Characterization of silver nanomaterials derived from marine *Streptomyces* sp. Al-Dhabi-87 and its in vitro application against multidrug resistant and extended-spectrum beta-lactamase clinical pathogens. Nanomaterials 8. Available from: https://doi.org/10.3390/nano8050279.

Alessio, P., Aoki, P.H.B., Furini, L.N., Aliaga, A.E., Leopoldo Constantino, C.J., 2017. Spectroscopic Techniques for Characterization of Nanomaterials, Nanocharacterization Techniques. Elsevier Inc. Available from: https://doi.org/10.1016/B978-0-32349778-7.00003-5.

Alsamhary, K.I., 2020. Eco-friendly synthesis of silver nanoparticles by *Bacillus subtilis* and their antibacterial activity. Saudi Journal of Biological Sciences 27, 2185–2191. Available from: https://doi.org/10.1016/j.sjbs.2020.04.026.

Amin, Z.R., Khashyarmanesh, Z., Fazly Bazzaz, B.S., Noghabi, Z.S., 2019. Does biosynthetic silver nanoparticles are more stable with lower toxicity than their synthetic counterparts? Iranian Journal of Pharmaceutical Research.

Ammar, H.A., el Aty, A.A.A., el Awdan, S.A., 2021. Extracellular myco-synthesis of nano-silver using the fermentable yeasts *Pichia kudriavzevii*HA-NY2 and *Saccharomyces uvarum*HA-NY3, and their effective biomedical applications. Bioprocess and Biosystems Engineering 44, 841–854. Available from: https://doi.org/10.1007/s00449-020-02494-3.

Asha Ranjani, V., Tulja Rani, G., Sowjanya, M., Preethi, M., Srinivas, M., Nikhil, M., 2022. Yeast mediated synthesis of iron oxide nano particles: its characterization and evaluation of antibacterial activity. International Research Journal of Pharmacy and Medical Sciences (IRJPMS) 5, 12–16.

Aswathi, V.P., Meera, S., Maria, C.G.A., Nidhin, M., 2022. Green synthesis of nanoparticles from biodegradable waste extracts and their applications: a critical review. Nanotechnology for Environmental Engineering. Available from: https://doi.org/10.1007/s41204-022-00276-8.

Begum, R., Farooqi, Z.H., Naseem, K., Ali, F., Batool, M., Xiao, J., et al., 2018. Applications of UV/Vis spectroscopy in characterization and catalytic activity of noble metal nanoparticles fabricated in responsive polymer microgels: a review. Critical Reviews in Analytical Chemistry 48, 503–516. Available from: https://doi.org/10.1080/10408347.2018.1451299.

Bhardwaj, A.K., Kumar, V., Pandey, V., Naraian, R., Gopal, R., 2019. Bacterial killing efficacy of synthesized rod shaped cuprous oxide nanoparticles using laser ablation technique. SN Applied Sciences 1. Available from: https://doi.org/10.1007/s42452-019-1283-9.

Bhardwaj, A.K., Shukla, A., Maurya, S., Singh, S.C., Uttam, K.N., Sundaram, S., et al., 2018. Direct sunlight enabled photo-biochemical synthesis of silver nanoparticles and their bactericidal efficacy: photon energy as key for size and distribution control. Journal of Photochemistry and Photobiology. B, Biology 188, 42–49. Available from: https://doi.org/10.1016/j.jphotobiol.2018.08.019.

Bhardwaj, A.K., Shukla, A., Mishra, R.K., Singh, S.C., Mishra, V., Uttam, K.N., et al., 2017. Power and time dependent microwave assisted fabrication of silver nanoparticles decorated cotton (SNDC) fibers for bacterial decontamination. Frontiers in Microbiology 8. Available from: https://doi.org/10.3389/fmicb.2017.00330.

Bhatnagar, S., Ogbonna, C.N., Ogbonna, J.C., Aoyagi, H., 2022. Effect of physicochemical factors on extracellular fungal pigment-mediated biofabrication of silver nanoparticles. Green Chemistry Letters and Reviews. Available from: https://doi.org/10.1080/17518253.2022.2036376.

Bhuyar, P., Rahim, M.H.A., Sundararaju, S., Ramaraj, R., Maniam, G.P., Govindan, N., 2020. Synthesis of silver nanoparticles using marine macroalgae *Padina* sp. and its antibacterial activity towards pathogenic bacteria. Beni-Suef University Journal of Basic and Applied Sciences 9. Available from: https://doi.org/10.1186/s43088-019-0031-y.

Borm, P.J.A., Robbins, D., Haubold, S., Kuhlbusch, T., Fissan, H., Donaldson, K., et al., 2006. The potential risks of nanomaterials: a review carried out for ECETOC. Particle and fibre toxicology. Available from: https://doi.org/10.1186/1743-8977-3-11.

Cekuolyte, K., Gudiukaite, R., Klimkevicius, V., Mazrimaite, V., Maneikis, A., Lastauskiene, E., 2023. Biosynthesis of silver nanoparticles produced using *Geobacillus* spp. bacteria. Nanomaterials 13, 702. Available from: https://doi.org/10.3390/nano13040702.

Chung, I.M., Park, I., Seung-Hyun, K., Thiruvengadam, M., Rajakumar, G., 2016. Plant-mediated synthesis of silver nanoparticles: their characteristic properties and therapeutic applications. Nanoscale Research Letters. Available from: https://doi.org/10.1186/s11671-016-1257-4.

Cui, X., Zhong, Z., Xia, R., Liu, X., Qin, L., 2022. Biosynthesis optimization of silver nanoparticles (AgNPs) using *Trichoderma longibranchiatum* and biosafety assessment with silkworm (*Bombyx mori*. Arabian Journal of Chemistry 15. Available from: https://doi.org/10.1016/j.arabjc.2022.104142.

Cunha, F.A., Cunha, M., da, C.S.O., da Frota, S.M., Mallmann, E.J.J., Freire, T.M., et al., 2018. Biogenic synthesis of multifunctional silver nanoparticles from *Rhodotorula glutinis* and *Rhodotorula mucilaginosa*: antifungal, catalytic and cytotoxicity activities. World Journal of Microbiology and Biotechnology 34. Available from: https://doi.org/10.1007/s11274-018-2514-8.

Das, K.R., Kowshik, M., Praveen Kumar, M.K., Kerkar, S., Shyama, S.K., Mishra, S., 2018. Native hypersaline sulphate reducing bacteria contributes to iron nanoparticle formation in saltpan sediment: a concern for aquaculture. Journal of Environmental Management 206, 556–564. Available from: https://doi.org/10.1016/j.jenvman.2017.10.078.

Das, V.L., Thomas, R., Varghese, R.T., Soniya, Ev, Mathew, J., Radhakrishnan, E.K., 2014. Extracellular synthesis of silver nanoparticles by the Bacillus strain CS 11 isolated from industrialized area. 3 Biotech 4, 121–126. Available from: https://doi.org/10.1007/s13205-013-0130-8.

Dauthal, P., Mukhopadhyay, M., 2016. Noble metal nanoparticles: plant-mediated synthesis, mechanistic aspects of synthesis, and applications. Industrial & Engineering Chemistry Research 55, 9557–9577. Available from: https://doi.org/10.1021/acs.iecr.6b00861.

de Jesus, R.A., de Assis, G.C., de Oliveira, R.J., Costa, J.A.S., da Silva, C.M.P., Bilal, M., et al., 2021. Environmental remediation potentialities of metal and metal oxide nanoparticles: mechanistic biosynthesis, influencing factors, and application standpoint. Environmental Technology & Innovation. Available from: https://doi.org/10.1016/j.eti.2021.101851.

Designed Research, K, J.W.M., 2022. Magnetosome-inspired synthesis of soft ferrimagnetic nanoparticles for magnetic tumor targeting. Proceedings of the National Academy of Sciences of the United States of America 119. Available from: https://doi.org/10.1073/pnas.

Egan-Morriss, C., Kimber, R.L., Powell, N.A., Lloyd, J.R., 2022. Biotechnological synthesis of Pd-based nanoparticle catalysts. Nanoscale Advances. Available from: https://doi.org/10.1039/d1na00686j.

El-Batal, A.I., Elkenawy, N.M., Yassin, A.S., Amin, M.A., 2015. Laccase production by *Pleurotus ostreatus* and its application in synthesis of gold nanoparticles. Biotechnology Reports 5, 31–39. Available from: https://doi.org/10.1016/j.btre.2014.11.001.

Fariq, A., Khan, T., Yasmin, A., 2017. Microbial synthesis of nanoparticles and their potential applications in biomedicine. Journal of Applied Biomedicine. Available from: https://doi.org/10.1016/j.jab.2017.03.004.

Farouk, F., Essam, S., Abdel-Motaleb, A., El-Shimy, R., Fritzsche, W., Azzazy, H.M.E.S., 2022. Fast detection of bacterial contamination in fresh produce using FTIR and spectral classification. Spectrochimica Acta Part A, Molecular and Biomolecular Spectroscopy 277, 121248. Available from: https://doi.org/10.1016/j.saa.2022.121248.

Ganesan, V., Hariram, M., Vivekanandhan, S., Muthuramkumar, S., 2020. *Periconium* sp. (endophytic fungi) extract mediated sol-gel synthesis of ZnO nanoparticles for antimicrobial and antioxidant applications. Materials Science in Semiconductor Processing 105. Available from: https://doi.org/10.1016/j.mssp.2019.104739.

Ganesh, C.K., Mamidyala, S.K., Das, B., Sridhar, B., Sarala Devi, G., Karuna, M.S.L., 2010. Synthesis of biosurfactant-based silver nanoparticles with purified rhamnolipids isolated from *Pseudomonas aeruginosa* BS-161R. Journal of Microbiology and Biotechnology 20, 1061–1068. Available from: https://doi.org/10.4014/jmb.1001.01018.

Ghezzi, M., Pescina, S., Padula, C., Santi, P., Del Favero, E., Cantù, L., et al., 2021. Polymeric micelles in drug delivery: an insight of the techniques for their characterization and assessment in biorelevant conditions. Journal of Controlled Release 332, 312–336. Available from: https://doi.org/10.1016/j.jconrel.2021.02.031.

Gholami-Shabani, M., Shams-Ghahfarokhi, M., Gholami-Shabani, Z., Akbarzadeh, A., Riazi, G., Ajdari, S., et al., 2015. Enzymatic synthesis of gold nanoparticles using sulfite reductase purified from *Escherichia coli*: a green eco-friendly approach. Process Biochemistry 50, 1076–1085. Available from: https://doi.org/10.1016/j.procbio.2015.04.004.

Gunarto, C., Ju, Y.H., Putro, J.N., Tran-Nguyen, P.L., Soetaredjo, F.E., Santoso, S.P., et al., 2020. Effect of a nonionic surfactant on the pseudoternary phase diagram and stability of microemulsion. Journal of Chemical & Engineering Data 65, 4024–4033. Available from: https://doi.org/10.1021/acs.jced.0c00341.

Gupta, P., Rai, N., Verma, A., Saikia, D., Singh, S.P., Kumar, R., et al., 2022. Green-based approach to synthesize silver nanoparticles using the fungal endophyte *Penicillium oxalicum* and their antimicrobial, antioxidant, and in vitro anticancer potential. ACS Omega 7, 46653–46673. Available from: https://doi.org/10.1021/acsomega.2c05605.

Hamdous, Y., Chebbi, I., Mandawala, C., le Fèvre, R., Guyot, F., Seksek, O., et al., 2017. Biocompatible coated magnetosome minerals with various organization and cellular interaction properties induce cytotoxicity towards RG-2 and GL-261 glioma cells in the presence of an alternating magnetic field. Journal of Nanobiotechnology 15. Available from: https://doi.org/10.1186/s12951-017-0293-2.

Hamida, R.S., Ali, M.A., Almohawes, Z.N., Alahdal, H., Momenah, M.A., Bin-Meferij, M.M., 2022. Green synthesis of hexagonal silver nanoparticles using a novel microalgae coelastrella aeroterrestrica strain BA_Chlo4 and resulting anticancer, antibacterial, and antioxidant activities. Pharmaceutics 14. Available from: https://doi.org/10.3390/pharmaceutics14102002.

Hassan, S.S., Duffy, B., Williams, G.A., Jaiswal, A.K., 2022. Biofabrication of magnetic nanoparticles and their use as carriers for pectinase and xylanase. OpenNano 6. Available from: https://doi.org/10.1016/j.onano.2021.100034.

He, S., Guo, Z., Zhang, Y., Zhang, S., Wang, J., Gu, N., 2007. Biosynthesis of gold nanoparticles using the bacteria *Rhodopseudomonas capsulata*. Materials Letters 61, 3984–3987. Available from: https://doi.org/10.1016/j.matlet.2007.01.018.

Hong, J.S., Kim, D., Jeong, S.H., 2022. Performance evaluation of the IR biotyper ® system for clinical microbiology : application for detection of *Staphylococcus aureus* sequence type 8 strains. Antibiotics 11.

Hulkoti, N.I., Taranath, T.C., 2014. Biosynthesis of nanoparticles using microbes-a review. Colloids and Surfaces B, Biointerfaces 121. Available from: https://doi.org/10.1016/j.colsurfb.2014.05.027.

Ijaz, I., Gilani, E., Nazir, A., Bukhari, A., 2020. Detail review on chemical, physical and green synthesis, classification, characterizations and applications of nanoparticles. Green Chemistry Letters and Reviews 13. Available from: https://doi.org/10.1080/17518253.2020.1802517.

Javed, Y., Ali, K., Akhtar, K., Jawaria, Hussain, M.I., Ahmad, G., et al., 2018. Chapter 5 TEM for atomic-scale study: fundamental, instrumentation, and applications in nanotechnology. Handbook of Materials Characterization. Available from: https://doi.org/10.1007/978-3-319-92955-2.

Jeevanandam, J., Barhoum, A., Chan, Y.S., Dufresne, A., Danquah, M.K., 2018. Review on nanoparticles and nanostructured materials: history, sources, toxicity and regulations. Beilstein Journal of Nanotechnology. Available from: https://doi.org/10.3762/bjnano.9.98.

John, M.S., Nagoth, J.A., Zannotti, M., Giovannetti, R., Mancini, A., Ramasamy, K.P., et al., 2021. Biogenic synthesis of copper nanoparticles using bacterial strains isolated from an Antarctic consortium associated to a psychrophilic marine ciliate: characterization and potential application as antimicrobial agents. Marine drugs 19. Available from: https://doi.org/10.3390/md19050263.

Joshi, C.G., Danagoudar, A., Poyya, J., Kudva, A.K., BL, D., 2017. Biogenic synthesis of gold nanoparticles by marine endophytic fungus-*Cladosporium cladosporioides* isolated from seaweed and evaluation of their antioxidant and antimicrobial properties. Process Biochemistry 63, 137–144. Available from: https://doi.org/10.1016/j.procbio.2017.09.008.

Kalpana, V.N., Devi Rajeswari, V., 2018. A review on green synthesis, biomedical applications, and toxicity studies of ZnO NPs. Bioinorganic Chemistry and Applications. Available from: https://doi.org/10.1155/2018/3569758.

Kalpana, V.N., Kataru, B.A.S., Sravani, N., Vigneshwari, T., Panneerselvam, A., Devi Rajeswari, V., 2018. Biosynthesis of zinc oxide nanoparticles using culture filtrates of *Aspergillus niger*: antimicrobial textiles and dye degradation studies. OpenNano 3, 48–55. Available from: https://doi.org/10.1016/j.onano.2018.06.001.

Kamnev, A.A., Dyatlova, Y.A., Kenzhegulov, O.A., Vladimirova, A.A., Mamchenkova, P.V., Tugarova, A.V., 2021. Fourier transform infrared (FTIR) spectroscopic analyses of microbiological samples and biogenic selenium nanoparticles of microbial origin: sample preparation effects. Molecules 26. Available from: https://doi.org/10.3390/molecules26041146.

Katas, H., Lim, C.S., Nor Azlan, A.Y.H., Buang, F., Mh Busra, M.F., 2019. Antibacterial activity of biosynthesized gold nanoparticles using biomolecules from *Lignosus rhinocerotis* and chitosan. Saudi Pharmaceutical Journal 27, 283–292. Available from: https://doi.org/10.1016/j.jsps.2018.11.010.

Khan, M., Khan, M.S.A., Borah, K.K., Goswami, Y., Hakeem, K.R., Chakrabartty, I., 2021. The potential exposure and hazards of metal-based nanoparticles on plants and environment, with special emphasis on ZnO NPs, TiO_2 NPs, and AgNPs: a review. Environmental Advances. Available from: https://doi.org/10.1016/j.envadv.2021.100128.

Khan, I., Saeed, K., Khan, I., 2019. Nanoparticles: properties, applications and toxicities. Arabian Journal of Chemistry. Available from: https://doi.org/10.1016/j.arabjc.2017.05.011.

Kolahalam, L.A., Kasi Viswanath, I.v, Diwakar, B.S., Govindh, B., Reddy, V., Murthy, Y.L.N., 2019. Review on nanomaterials: synthesis and applications. Materials Today: Proceedings. Elsevier Ltd., pp. 2182–2190. Available from: https://doi.org/10.1016/j.matpr.2019.07.371.

Korbekandi, H., Mohseni, S., Jouneghani, R.M., Pourhossein, M., Iravani, S., 2016. Biosynthesis of silver nanoparticles using Saccharomyces cerevisiae. Artificial Cells, Nanomedicine, and Biotechnology 44, 235–239. Available from: https://doi.org/10.3109/21691401.2014.937870.

Kulkarni, R., Harip, S., Kumar, A.R., Deobagkar, D., Zinjarde, S., 2018. Peptide stabilized gold and silver nanoparticles derived from the mangrove isolate *Pseudoalteromonas lipolytica* mediate dye decolorization. Colloids and surfaces. A, Physicochemical and engineering aspects 555, 180–190. Available from: https://doi.org/10.1016/j.colsurfa.2018.06.083.

Kumar, H., Bhardwaj, K., Kuča, K., Kalia, A., Nepovimova, E., Verma, R., et al., 2020. Flower-based green synthesis of metallic nanoparticles: applications beyond fragrance. Nanomaterials 10. Available from: https://doi.org/10.3390/nano10040766.

Lee, J., Ahn, M.S., Lee, Y.L., Jie, E.Y., Kim, S.G., Kim, S.W., 2019. Rapid tool for identification of bacterial strains using Fourier transform infrared spectroscopy on genomic DNA. Journal of Applied Microbiology 126, 864–871. Available from: https://doi.org/10.1111/jam.14171.

Li, X., Xu, H., Chen, Z.S., Chen, G., 2011. Biosynthesis of nanoparticles by microorganisms and their applications. Journal of Nanomaterials. Available from: https://doi.org/10.1155/2011/270974.

Lloyd, J.R., Byrne, J.M., Coker, V.S., 2011. Biotechnological synthesis of functional nanomaterials. Current Opinion in Biotechnology. Available from: https://doi.org/10.1016/j.copbio.2011.06.008.

Mabrouk, M., Elkhooly, T.A., Amer, S.K., 2021. Actinomycete strain type determines the monodispersity and antibacterial properties of biogenically synthesized silver nanoparticles. Journal of Genetic Engineering and Biotechnology 19. Available from: https://doi.org/10.1186/s43141-021-00153-y.

Maheshwari, R., Todke, P., Kuche, K., Raval, N., Tekade, R.K., 2018. Micromeritics in pharmaceutical product development. Dosage Form Design Considerations (I), 599–635. Available from: https://doi.org/10.1016/B978-0-12-814423-7.00017-4. Volume.

Manivasagan, P., Alam, M.S., Kang, K.H., Kwak, M., Kim, S.K., 2015. Extracellular synthesis of gold bionanoparticles by *Nocardiopsis* sp. and evaluation of its antimicrobial, antioxidant and cytotoxic activities. Bioprocess and biosystems engineering 38. Available from: https://doi.org/10.1007/s00449-015-1358-y.

Meng, Y., 2015. A sustainable approach to fabricating ag nanoparticles/PVA hybrid nanofiber and its catalytic activity. Nanomaterials 5, 1124–1135. Available from: https://doi.org/10.3390/nano5021124.

Mohanpuria, P., Rana, N.K., Yadav, S.K., 2008. Biosynthesis of nanoparticles: technological concepts and future applications. Journal of Nanoparticle Research. Available from: https://doi.org/10.1007/s11051-007-9275-x.

Mourdikoudis, S., Pallares, R.M., Thanh, N.T.K., 2018. Characterization techniques for nanoparticles: comparison and complementarity upon studying nanoparticle properties. Nanoscale 10, 12871–12934. Available from: https://doi.org/10.1039/c8nr02278j.

Nadaf, N.Y., Kanase, S.S., 2019. Biosynthesis of gold nanoparticles by Bacillus marisflavi and its potential in catalytic dye degradation. Arabian Journal of Chemistry 12, 4806–4814. Available from: https://doi.org/10.1016/j.arabjc.2016.09.020.

Nasaruddin, R.R., Chen, T., Yao, Q., Zang, S., Xie, J., 2021. Toward greener synthesis of gold nanomaterials: from biological to biomimetic synthesis. Coordination chemistry reviews. Available from: https://doi.org/10.1016/j.ccr.2020.213540.

Nordmeier, A., Merwin, A., Roeper, D.F., Chidambaram, D., 2018. Microbial synthesis of metallic molybdenum nanoparticles. Chemosphere 203, 521–525. Available from: https://doi.org/10.1016/j.chemosphere.2018.02.079.

Ong, H.C., Tiong, Y.W., Goh, B.H.H., Gan, Y.Y., Mofijur, M., Fattah, I.M.R., et al., 2021. Recent advances in biodiesel production from agricultural products and microalgae using ionic liquids: opportunities and challenges. Energy Conversion and Management. Available from: https://doi.org/10.1016/j.enconman.2020.113647.

Ovais, M., Khalil, A.T., Ayaz, M., Ahmad, I., Nethi, S.K., Mukherjee, S., 2018. Biosynthesis of metal nanoparticles via microbial enzymes: a mechanistic approach. International Journal of Molecular Sciences. Available from: https://doi.org/10.3390/ijms19124100.

Pantidos, N., 2014. Biological synthesis of metallic nanoparticles by bacteria, fungi and plants. Journal of Nanomedicine & Nanotechnology 05. Available from: https://doi.org/10.4172/2157-7439.1000233.

Pourjavadi, A., Doroudian, M., Bagherifard, M., Bahmanpour, M., 2020. Magnetic and light-responsive nanogels based on chitosan functionalized with Au nanoparticles and poly(: N-isopropylacrylamide) as a remotely triggered drug carrier. New Journal of Chemistry 44, 17302–17312. Available from: https://doi.org/10.1039/d0nj02345k.

Procacci, B., Rutherford, S.H., Greetham, G.M., Towrie, M., Parker, A.W., Robinson, C.V., et al., 2021. Differentiation of bacterial spores via 2D-IR spectroscopy. Spectrochimica Acta Part A, Molecular and Biomolecular Spectroscopy 249, 119319. Available from: https://doi.org/10.1016/j.saa.2020.119319.

Puspitasari, N., Lee, C.K., 2021. Class I hydrophobin fusion with cellulose binding domain for its soluble expression and facile purification. International Journal of Biological Macromolecules 193, 38–43. Available from: https://doi.org/10.1016/j.ijbiomac.2021.10.089.

Puspitasari, N., Tsai, S.L., Lee, C.K., 2021. Class I hydrophobins pretreatment stimulates PETase for monomers recycling of waste PETs. International Journal of Biological Macromolecules 176, 157–164. Available from: https://doi.org/10.1016/j.ijbiomac.2021.02.026.

Putro, J.N., Edi Soetaredjo, F., Irawaty, W., Budi Hartono, S., Santoso, S.P., Lie, J., et al., 2022. Cellulose nanocrystals (CNCs) and its modified form from durian rind as dexamethasone carrier. Polymers (Basel) 14. Available from: https://doi.org/10.3390/polym14235197.

Quintero-Quiroz, C., Acevedo, N., Zapata-Giraldo, J., Botero, L.E., Quintero, J., Zárate-Trivinõ, D., et al., 2019. Optimization of silver nanoparticle synthesis by chemical reduction and evaluation of its antimicrobial and toxic activity. Biomaterials Research 23. Available from: https://doi.org/10.1186/s40824-019-0173-y.

Raj, R., Dalei, K., Chakraborty, J., Das, S., 2016. Extracellular polymeric substances of a marine bacterium mediated synthesis of CdS nanoparticles for removal of cadmium from aqueous solution. Journal of Colloid and Interface Science 462, 166–175. Available from: https://doi.org/10.1016/j.jcis.2015.10.004.

Ranjani, A., Gopinath, P.M., Ananth, S., Narchonai, G., Santhanam, P., Thajuddin, N., et al., 2018. Multidimensional dose–response toxicity exploration of silver nanoparticles from *Nocardiopsis flavascens* RD30. Applied Nanoscience (Switzerland) 8, 699–713. Available from: https://doi.org/10.1007/s13204-018-0824-7.

Raschdorf, O., Bonn, F., Zeytuni, N., Zarivach, R., Becher, D., Schüler, D., 2018. A quantitative assessment of the membrane-integral sub-proteome of a bacterial magnetic organelle. Journal of Proteomics 172, 89–99. Available from: https://doi.org/10.1016/j.jprot.2017.10.007.

Rodrigues, A.G., Ping, L.Y., Marcato, P.D., Alves, O.L., Silva, M.C.P., Ruiz, R.C., et al., 2013. Biogenic antimicrobial silver nanoparticles produced by fungi. Applied Microbiology and Biotechnology 97, 775–782. Available from: https://doi.org/10.1007/s00253-012-4209-7.

Rodríguez-González, V., Obregón, S., Patrón-Soberano, O.A., Terashima, C., Fujishima, A., 2020. An approach to the photocatalytic mechanism in the TiO_2-nanomaterials microorganism interface for the control of infectious processes. Applied Catalysis B. Available from: https://doi.org/10.1016/j.apcatb.2020.118853.

Rosenfeldt, S., Mickoleit, F., Jörke, C., Clement, J.H., Markert, S., Jérôme, V., et al., 2021. Towards standardized purification of bacterial magnetic nanoparticles for future in vivo applications. Acta biomaterialia 120, 293–303. Available from: https://doi.org/10.1016/j.actbio.2020.07.042.

Sable, S.V., Kawade, S., Ranade, S., Joshi, S., 2020. Bioreduction mechanism of silver nanoparticles. Materials Science and Engineering C 107. Available from: https://doi.org/10.1016/j.msec.2019.110299.

Salem, D.M.S.A., Ismail, M.M., Aly-Eldeen, M.A., 2019. Biogenic synthesis and antimicrobial potency of iron oxide (Fe_3O_4) nanoparticles using algae harvested from the Mediterranean Sea, Egypt. Egyptian Journal of Aquatic Research. Available from: https://doi.org/10.1016/j.ejar.2019.07.002.

Salvadori, M.R., Ando, R.A., Oller Do Nascimento, C.A., Corrêa, B., 2014. Intracellular biosynthesis and removal of copper nanoparticles by dead biomass of yeast isolated from the wastewater of a mine in the Brazilian Amazonia. PLoS ONE 9. Available from: https://doi.org/10.1371/journal.pone.0087968.

Saravanakumar, K., Shanmugam, S., Varukattu, N.B., MubarakAli, D., Kathiresan, K., Wang, M.H., 2019. Biosynthesis and characterization of copper oxide nanoparticles from indigenous fungi and its effect of photothermolysis on human lung carcinoma. Journal of Photochemistry and Photobiology. B, Biology 190, 103–109. Available from: https://doi.org/10.1016/j.jphotobiol.2018.11.017.

Saravanan, A., Kumar, P.S., Karishma, S., Vo, D.V.N., Jeevanantham, S., Yaashikaa, P.R., et al., 2021. A review on biosynthesis of metal nanoparticles and its environmental applications. Chemosphere 264. Available from: https://doi.org/10.1016/j.chemosphere.2020.128580.

Sayadi, M.H., Salmani, N., Heidari, A., Rezaei, M.R., 2018. Bio-synthesis of palladium nanoparticle using Spirulina platensis alga extract and its application as adsorbent. Surfaces and INTERFACES 10, 136–143. Available from: https://doi.org/10.1016/j.surfin.2018.01.002.

Sayed, H., Sadek, H., Abdel-Aziz, M., Mahmoud, N., Sabry, W., Genidy, G., et al., 2021. Biosynthesis of iron oxide nanoparticles from fungi isolated from deteriorated historical gilded cartonnage and its application in cleaning. Egyptian Journal of Archaeological and Restoration Studies 11, 129–145. Available from: https://doi.org/10.21608/ejars.2021.210365.

Shah, M., Fawcett, D., Sharma, S., Tripathy, S.K., Poinern, G.E.J., 2015. Green synthesis of metallic nanoparticles via biological entities. Materials. Available from: https://doi.org/10.3390/ma8115377.

Shareef, J.U., Navya Rani, M., Anand, S., Rangappa, D., 2017. Synthesis and characterization of silver nanoparticles from Penicillium sps. Materials Today: Proceedings. Elsevier Ltd., pp. 11923–11932. Available from: https://doi.org/10.1016/j.matpr.2017.09.113.

Shoeibi, S., Mashreghi, M., 2017. Biosynthesis of selenium nanoparticles using Enterococcus faecalis and evaluation of their antibacterial activities. Journal of Trace Elements in Medicine and Biology 39, 135–139. Available from: https://doi.org/10.1016/j.jtemb.2016.09.003.

Shukla, K., Verma, A., Verma, L., Rawat, S., Singh, J., 2020. A novel approach to utilize used disposable paper cups for the development of adsorbent and its application for the malachite green and rhodamine-B dyes removal from aqueous solutions.

Shunmugam, R., Renukadevi Balusamy, S., Kumar, V., Menon, S., Lakshmi, T., Perumalsamy, H., 2021. Biosynthesis of gold nanoparticles using marine microbe (Vibrio alginolyticus) and its anticancer and antioxidant analysis. Journal of King Saud University - Science 33. Available from: https://doi.org/10.1016/j.jksus.2020.101260.

Sierra, C.F.E., 2019. Fundamentals of transmission electron microscopy, the technique with the best resolution in the world. Bogota 0–6.

Singh, M.P., Bhardwaj, A.K., Bharati, K., Singh, R.P., Chaurasia, S.K., Kumar, S., et al., 2021. Biogenic and non-biogenic waste utilization in the synthesis of 2D materials (graphene, h-BN, g-C2N) and their applications. Frontiers in Nanotechnology. Available from: https://doi.org/10.3389/fnano.2021.685427.

Spagnoletti, F.N., Spedalieri, C., Kronberg, F., Giacometti, R., 2019. Extracellular biosynthesis of bactericidal Ag/AgCl nanoparticles for crop protection using the fungus Macrophomina phaseolina. Journal of Environmental Management 231, 457–466. Available from: https://doi.org/10.1016/j.jenvman.2018.10.081.

Srivastava, M., Srivastava, N., Saeed, M., Mishra, P.K., Saeed, A., Gupta, V.K., et al., 2021. Bioinspired synthesis of iron-based nanomaterials for application in biofuels production: a new in-sight. Renewable and Sustainable Energy Reviews. Available from: https://doi.org/10.1016/j.rser.2021.111206.

Subramaniyan, S.A., Sheet, S., Vinothkannan, M., Yoo, D.J., Lee, Y.S., Belal, S.A., et al., 2017. One-pot facile synthesis of Pt nanoparticles using cultural filtrate of microgravity simulated grown *P. chrysogenum* and their activity on bacteria and cancer cells. Journal of nanoscience and nanotechnology 18, 3110–3125. Available from: https://doi.org/10.1166/jnn.2018.14661.

Toro, R.G., Diab, M., de Caro, T., Al-Shemy, M., Adel, A., Caschera, D., 2020. Study of the effect of titanium dioxide hydrosol on the photocatalytic and mechanical properties of paper sheets. Materials 13. Available from: https://doi.org/10.3390/ma13061326.

Verma, S.K., Das, A.K., Gantait, S., Kumar, V., Gurel, E., 2019. Applications of carbon nanomaterials in the plant system: a perspective view on the pros and cons. Science of the Total Environment. Available from: https://doi.org/10.1016/j.scitotenv.2019.02.409.

Verma, S.K., Das, A.K., Patel, M.K., Shah, A., Kumar, V., Gantait, S., 2018. Engineered nanomaterials for plant growth and development: a perspective analysis. Science of the Total Environment. Available from: https://doi.org/10.1016/j.scitotenv.2018.02.313.

Wadhwani, S.A., Shedbalkar, U.U., Singh, R., Chopade, B.A., 2018. Biosynthesis of gold and selenium nanoparticles by purified protein from *Acinetobacter* sp. SW 30. Enzyme and Microbial Technology 111, 81–86. Available from: https://doi.org/10.1016/j.enzmictec.2017.10.007.

Wang, L., Liu, X., Lee, D.J., Tay, J.H., Zhang, Y., Wan, C.L., et al., 2018. Recent advances on biosorption by aerobic granular sludge. Journal of Hazardous Materials. Available from: https://doi.org/10.1016/j.jhazmat.2018.06.010.

Wang, W., Zhang, B., Liu, Q., Du, P., Liu, W., He, Z., 2018. Biosynthesis of palladium nanoparticles using: *Shewanella loihica* PV-4 for excellent catalytic reduction of chromium(VI). Environmental Science: Nano 5, 730–739. Available from: https://doi.org/10.1039/c7en01167a.

Yang, D.P., Cui, D.X., 2008. Advances and prospects of gold nanorods. Chemistry, an Asian journal. Available from: https://doi.org/10.1002/asia.200800195.

Zamani, H., Jafari, A., Mousavi, S.M., Darezereshki, E., 2020. Biosynthesis of silica nanoparticle using *Saccharomyces cervisiae* and its application on enhanced oil recovery. Journal of Petroleum Science and Engineering 190. Available from: https://doi.org/10.1016/j.petrol.2020.107002.

Zhang, X., Qu, Y., Shen, W., Wang, J., Li, H., Zhang, Z., et al., 2016. Biogenic synthesis of gold nanoparticles by yeast *Magnusiomyces ingens* LH-F1 for catalytic reduction of nitrophenols. Colloids and surfaces. A, Physicochemical and engineering aspects 497, 280–285. Available from: https://doi.org/10.1016/j.colsurfa.2016.02.033.

Zhang, L., Wang, L., Jiang, Z., Xie, Z., 2012. Synthesis of size-controlled monodisperse Pd nanoparticles via a non-aqueous seed-mediated growth. Nanoscale Research Letters 7.

Zhang, Q., Zhang, H., Hui, A., Lu, Y., Wang, A., 2023. Incorporation of Ag NPs/palygorskite into chitosan/glycyrrhizic acid films as a potential anti-bacterial wound dressing. Results in Materials 18. Available from: https://doi.org/10.1016/j.rinma.2023.100396.

Zong, T.-X., Silveira, A.P., Morais, J.A.V., Sampaio, M.C., Muehlmann, L.A., Zhang, J., et al., 2022. Recent advances in antimicrobial nano-drug delivery systems. Nanomaterials 12, 1855. Available from: https://doi.org/10.3390/nano12111855.

Chapter 20

Fundamental scope of nanomaterial synthesis from wastes

Naveed Qasim Abro[1], Najma Memon[1], Muhammad Siddique Kalhoro[2], Sakib Hussain Laghari[1] and Zafar Ali[3]

[1]*National Centre of Excellence in Analytical Chemistry, University of Sindh, Jamshoro, Pakistan,* [2]*Institute of Physics, University of Sindh, Jamshoro, Pakistan,* [3]*Chemistry Department, University of Turbat, Balochistan, Pakistan*

Abbreviations

Ag	silver
Ag NPs	silver nanoparticle
Al	aluminum
AMD	acid mine drainage
ASR	automobile shredders residue
As	arsenic
Au	gold
Bt	billion tons
CNOs	carbon nano-onions
CNTs	carbon nanotubes
Cd	cadmium
CDs	carbon dots
CH_4	methane
CH_4N_2S	thiourea
CO_2	carbon dioxide
$CoFe_2O_4$	cobalt ferrite
CO_3O_4 NPs	cobalt oxide nanoparticles
CRT	cathode ray tube
Cr	chromium
Cu_2O	copper oxide
Cu-Sn	copper-tin
Cu	copper
E-waste	electronic waste
EDS	energy dispersive spectroscopy
FsHAp	fish-scale hydroxyapatite
FsCol	fish-scale collagen
Fe	iron
F_2O_3	iron oxide
Fe_3O_4 NPs	iron oxide nanoparticles
Hg	mercury
$H_2S_2O_3$	thiosulfate
LED	light emitting diodes
Li	lithium
Li-ion	lithium-ion

Green and Sustainable Approaches Using Wastes for the Production of Multifunctional Nanomaterials. DOI: https://doi.org/10.1016/B978-0-443-19183-1.00007-6

$InBO_3$ NPs indium borate
Li-S lithium-sulfur
MnO manganese oxide
Mn manganese
Mt million tons
MWCNT multiwalled carbon nanotube
Ni-Cd nickel-cadmium
NMs nanomaterials
nm nanometer
Ni-MH nickel-metal hydride
Ni–Mn–Zn nickel–manganese–zinc
NiO nickel oxide
NPs nanoparticles
P-CQDs phosphorus-Doped carbon quantum dots
Pb lead
PbO_2 lead oxide
PCB printed circuit board
Pd palladium
Pt platinum
PVC polyvinyl chloride
$PbSO_4$ lead sulfate
QDs quantum dots
RGO reduced graphene oxide
Rh rhodium
SiC silicon carbide
Sn tin
SiO_2 silicon dioxide
SnO tin-oxide
Si_3N_4-NWs silicon-nitride nanowires
TiO_2 titanium oxide
V_2O_5 vanadium pentoxide
Zn-C zinc-carbon
$ZnMn_2O_4$ zinc manganite
Zn zinc
ZnS zinc sulfide
ZnO zinc oxide

20.1 Introduction

In the current era, science and engineering have been revolutionized by the impact of nanoscience. It may include the analysis, progress, synthesis, and processing of material and their structure on a nanoscale basis. A particle size equal to 100 or <100 nm is known as a nanomaterial (NM). The key dissimilarity between non-NMs and NMs is their size. Particle sizes <100 nm in all dimensions are referred to as a NMs, while >100 nm are non-NMs. Likewise, naked eyes can't see NMs while non-NMs can be seen through a simple microscope (Samaddar et al., 2018). NMs are considered as versatile, because they may have shorter dimension, quantum effect, and higher area to volume ratio, and these types of properties allow them to perform well in various type of applications such as solar energy, medical sciences, drug delivery, water treatment, detection of various analytes, biomedical research, nano-sensors, nano-sorbents, fuel cells, etc. (De, 2022, Bratovcic, 2019). Besides this, NM has demonstrated higher adsorption volume in gas as well as in liquid phase. Properties like chemical, physical, morphological, thermal, mechanical, optical, magnetic, and conductivity may come from the doping of various types of materials with the NMs. The NMs may be classified as inorganic (Wang et al., 2021b), organic (Zhang et al., 2021a), carbon (Maiti et al., 2018), and composite based (Ali and Andriyana, 2020). The manufacturing or synthesis of NMs may require traditional automation along with an eco-friendly mechanism. They may also require the additional energy along with natural resources. Two types of wastes are available namely biodegradable or organic waste (Kou et al., 2021), and nonbiodegradable or inorganic waste (Bisht et al., 2022).

The organic waste may include the agriculture, forestry, fruits, and vegetables. The source of organic waste may be classified accordingly, such as peels, leaves, sugarcane bagasse, husk of rise, corn, coconut, eggshells, wheat straw etc, that can break into carbon dioxide (CO_2), methane (CH_4), and other organic molecules (Abu Yazid et al., 2017). The number of NMs can be synthesized by using organic waste such as carbon nanotubes (CNTs) (Chen et al., 2021a), graphene (Raghavan et al., 2017), graphene oxide (Liou and Wang, 2020), fullerenes (Baskar et al., 2022), quantum dots (QDs) (Dhandapani et al., 2020), etc. The inorganic waste includes industrial, electronic, and mining waste. The printed circuit board (PCB), spent batteries, all electronic appliances, automobile shredder's residue, and aluminum cans are included in electronic and industrial waste, whereas mine tailing, metallurgical slag, and acid mine drainage (AMD) are included in mining waste (Das et al., 2018). The number of NMs can be synthesized from inorganic waste such as copper oxide (Cu_2O), lead (Pb), copper-tin (Cu-Sn), palladium (pd), zinc (Zn), zinc oxide (ZnO), manganese oxide (MnO), cobalt ferrite ($CoFe_2O_4$), iron oxide (F_2O_3), silver (Ag), and gold (Au) NMs (Ulman et al., 2018; Zhan et al., 2017; Hossain et al., 2018; Singh et al., 2017). Scientists from developing countries are searching for suitable solutions in order to collect, treat, and use the waste into valuable products (NMs). Conversion of waste into NMs is considered as an attractive aspect because the waste is easily accessible and low cost, while the synthesis of NMs is a creative method to recycle or treat the waste. The sustainable progress of waste-derived NMs is evolving debate in current annums (Oliveira et al., 2017; Mihai, 2020).

This chapter demonstrated fundamental scope of synthesis of NMs using cost-effective, green, and sustainable methods from agriculture and electronic waste. Various applications of NMs derived from waste that are reported in literature are also added herein.

20.1.1 Electronic and electrical waste or electronic waste

Electronic and electrical waste or electronic waste includes all types of electronic appliances along with their components such as batteries and cords, temperature exchange machines (heaters, air conditions, fans), mini machines (roaster, digital camcorders, calculators, dry razor, and coffee pot), heavy machines (dry cleaning, dish washer, solar panels), screen equipment (monitors, television, tablets, phones, laptops), and standard lamps (low pressure, fluorescent lamps) (Kang et al., 2020). The composition of electronic waste may vary from device to device. IT and telecommunication appliances may have a more significant number of precious and rare earth metals (Rodríguez-Padrón et al., 2020; Brewer et al., 2022). The rapid interest in electronic devices is growing day by day and it may be a dangerous situation in order of increasing electronic waste. However, the countless brand-new high-tech equipment are coming into the market every single day with short life span and unfortunately the waste bins are full of million tons (Mt) of old gadgets (Ganesh and Sharma, 2022). According to the statement of electronic waste global monitoring system, around 54 Mt of electronic waste was produced in the year of 2019 and is predicted to increase up to 75 Mt by 2030, 3%–5% increase per year (Fares and Radaydeh, 2022). Another research elaborates that around 2 billion tons (Bt) of electronic waste was generated and only 30% of that was recycled. It is predicted that around 3.5 Bt of electronic waste will be produced till 2050 globally (Jain et al., 2022). Nearly 25 Mt of electronic waste was produced by Asia and most of it was thrown away at junkyard and moved to the well settled states, where it may be transformed through informal way of recycling which is also known as "backyard recycling or transformation" that may discharge the toxic emissions (Ohajinwa et al., 2018). There are few major components of all electronic waste appliances such as PCB, which is one of the most common components used in all electronic appliances, that may consist of metals like palladium, cobalt, lithium, zinc, manganese, lead, nickel, gold, Sn, Ag, Cd, Cr, Al, Fe, Pt, etc. (Annamalai and Gurumurthy, 2021). Cathode ray tube (CRT) is another component of electronic waste, mainly used in glass of electronics that may consist of 20%–30% lead oxide and phosphorus (Yao et al., 2018). Polyvinyl chloride (PVC) is also a main component of all the electronic appliances that may consist of polymers (Turner and Filella, 2017). Besides this, the production of battery electronic waste that is around 15 billion per annum worldwide is also under debate, the batteries may be composed of Li-ion, Li-S, Zn-C, Ni-MH, Ni-Cd, etc (Roy et al., 2021). All electronic devices are composed of hazardous materials (Kumar et al., 2017). The dumping of electronic waste at open landfills may contaminate the environment, it can leach out in soil and water bodies, and it may disturb the human health (Saha et al., 2021). The toxic metal such as Cd present in electronic waste may disturb the neurodevelopment in young children because it is redox inactive that may not produce free radicals, hence it can join with sulfhydryl protein group and may cause no oxidative stress (Suhani et al., 2021). Mercury is used in lighting flat screen devices that may cause brain damage with central nervous system (Basu et al., 2022). Cr may cause disorders in neurodevelopment. As gallium arsenide may be present in light emitting diodes, it may cause disorder in heart rhythm, breathing, or bladder cancer. Cu is considered as a highly poisonous metal and it may cause disorder in the ecological system (Prasad et al., 2021; Yadav et al., 2021; Covre et al., 2022). This data shows that electronic waste contains hazardous materials that should not be dumped in landfills and affects the environment. The precise metals and other useful components can be recovered by using proper recycling methods. Informal recycling is the rehabilitation of metals by using nontechnological and immoral strategy with minimum or no safety, uncontrolled, and unsystematized way. It may release the toxic material to the environment. While formal recovery methods consist of pretreatment process by using one or more physical separation methods, as long as to retrieve metals via electronic waste. Informal recycling may have less than 80% efficiency as compared to formal recycling. Higher efficiency values may be achieved as more than 90% via advanced recycling from electronic waste (Chen and Gao, 2021). The present objective of a circular economy is the capability to alter electronic waste into precious products such as NMs by using green, cost-effective, more efficient, and eco-friendly methods, so that our environment remains safe from hazardous or toxic release of chemicals.

20.1.2 Batteries

The word battery can be defined as "A device that can store the chemical energy and may convert it into electrical energy." Due to handy electric source, battery or storage cells may be considered attractive and have significant value. They may be classified within binary groups such as disposable and nondisposable storage cells. The disposable storage cells may be used only once, although nondisposable storage cells can be used for long time. Essential classes of storage cells may be classified as follows: Li-ion, Li-S, Ni-Cd storage cells, etc. In the current era, popular storage cell is lithium-ion (Li-ion), which is used more commonly (Gao et al., 2021). Li may have lower molar mass and density because of which it is extremely applicable for set down in overlaid materials (Chombo and Laoonual, 2020). The types of storage cells are broadly suggested for recycling various types of metals such as Co, Li, Zn, Mn, Pb, Ni, Au, Pd, rare earth elements from them. At least 60% $PbSO_4$, 20%−50% PbO_2, and 4% Pb can be obtained from arid Pb paste from waste Pb storage cells. Later, the dumping of storage cells at inappropriate place may contaminate the soil along with water bodies. It may cause harsh human health issues. According to research, waste storage cells may simply be reprocessed in small as well as industrial scale in order to extract useful metals and convert them into useful NMs such as ZnO, $ZnMn_2O_4$, Ni−Mn−Zn, graphene nanocomposites or MnO_2, perovskite quantum dots, and ZnS NPs (Das et al., 2021; Assefi et al., 2019b). Assefi et al. (Assefi et al., 2018) synthesized NiO nanocuboid from spent Ni-Cd batteries by using hydrothermal method. Zhang et al. (He et al., 2021) synthesized graphene by utilizing chemical reduction method from waste Li-ion storage cell.

20.1.3 Instruments and machines

The metals such as mercury, lead, iron, silver, gold, platinum, copper, palladium, and rhodium may be retrieved against PCB of various electronics appliances such as computers, mobile phones, electrical gadgets, supercapacitor, laptops, automobiles, radios, stereos, televisions, etc. (Díaz-Martínez et al., 2019). In addition, nano-zero-valent Cu particles, CNTs, Cu_2O NPs, and Cu_2O/TiO_2 were also recovered from electronic waste. However, electronic waste other than batteries has comparably attracted less curiosity for the synthesis or recovery of NMs or metals. PCBs may contain Fe 4%−8%, Cu 11%−13%, Sn 2%−3%, Al 1%−3%, Ni 1%−2%, Zn 0.5%−2%, Pb 1%−2%, Ag 0.008%, 0.005% Ag, 0.002% Pd, and 0.004% Pt. In the current era, PCB from electronic waste is under debate and may have outstanding future because of the number of metals that can be recovered from them as mentioned above Zhou et al. (2021) synthesized Au NPs by using microemulsion method from spent PCBs of mobile phone. Martins et al. (2021) synthesized Cu NPs from spent PCBs by using chemical reduction method.

20.2 Recycling strategies of retrieved metals against the electronic waste

20.2.1 Formal recycling

The type of recycling which consists of pretreatment process involves one or more physical separation methods for metal recovery against electronic waste. The distribution of formal recycling may be as follows; pyrometallurgy, hydrometallurgy, and biohydrometallurgy (Priya et al., 2017). Gold, silver, aluminum, copper, cobalt, lead, nickel, palladium, manganese, and zinc may be recovered from the electronic waste by using the abovementioned methods. In the pyrometallurgical (thermal) method, smelting of electronic waste is done at the high temperature in the furnace, then the refractories are ejected as slag which will be proceeded further for the recovery of metals (Makuza et al., 2021). The benefits of WPCBs pretreatment along with pyrometallurgical method were also discussed to improve the efficiency

(Faraji et al., 2022). A variety of metals was retrieved by using *Chromobacterium violaceum* and *Pseudomonas fluorescens* in biohydrometallurgy process (Magoda and Mekuto, 2022). Similarly, the hydrometallurgical (chemical) method uses a particular class of lixiviants, such as thiourea, halide, thiosulfate, and acids that may extract away from metals by farming lasting metal complexes (Larouche et al., 2020). The green approach to recover silver from leaching solutions replaces hydrometallurgical routes and these approaches are often used for silver recovery and may need strong acidic solutions (Muscetta et al., 2023). Besides this, biohydrometallurgy is the process in which particular class of microorganisms was used to recover metals either in the existence of cell free extract or cell media. The use of microbial agents made this process more efficient, eco-friendly or green method, less use of energy, and may need minimum number of labors in order to recover metals (Desmarais et al., 2020).

20.2.2 Biohydrometallurgy or biometallurgy

Biohydrometallurgy or biometallurgy is considered as innovative, eco-friendly, green, and energy conservative concept (Nigam et al., 2021). This method may be established upon extract capability against the microbes defined as the biomining in order to transform the metallic components into dissolved resultant which may be recovered later from the solution. However, the biohydrometallurgy uses minimal amount of energy and chemicals, due to this reason it is the most favorable and economic-friendly method. According to most of the industrialists, it is green and clean method which is sustainable among all methods for the recovery of metals (Heydarian et al., 2018). This process may be handled into two different approaches via single stage or dual stage. In the single stage, both metal and microbes are added together with the nutritional medium, later on the microbial growth takes place and consequently, disintegration of metallic components may occur within the resultant. In majority of the cases, the existence of metal with high concentration may active the cellular toxicity and it can affect the growth of microbial. Hence, it may reduce the recovery of metals. Besides this, the dual stage process involves the microbial cultivation which may take place in nourishing media unless and until it achieves the stationary phase position for their period of growth and that may be achieved by using three different mechanisms, *acidolysis*, *complexolysis*, and *redoxolysis* (Lyu et al., 2021). Later on, the electronic waste material will be recycled and then it is allowed for recovery. Researchers believe utilizing a free cell supernatant in order to extract the metals. For such a process, cells are gathered once it achieved the stationary phase conditions. Used up media include all the active metabolites needed against disunion of metallic components from electronic waste and will prevent physical association between microbe and metals, thus it may avoid their connection in order of inhibitory effects and recovery may be relatively high as compared to other methods. *Acidobacillus thiooxidans, Acidobacillus ferrooxidans, sulfolobus* sp., *acidithiobacillus* sp., *Leptospirillum* sp., etc. are the broadly utilized microorganisms in the leaching process (Priya et al., 2018). The bioleaching process was used to recover the metals by using *Pleurotus florida, Trametes versicolor*, and *Aspergillus niger* (Kaur et al., 2022). The process of bioleaching was also found affective in order to retrieve the metals from electronic waste (Brindhadevi et al., 2023). The enzymatic bioleaching process was used to recover metals from discarded cellphone PCBs (Van Yken et al., 2023).

20.3 Synthetic approaches for the nanomaterials from electronic waste

NMs can be synthesized by using the number of methods available such as physical, chemical, and biomethods. The most frequently used physical methods used for the synthesis of NMs are chemical vapor deposition, laser ablation, ball mining, etc. However, there may be few limits in order to synthesize NMs by using physical methods. The chemical synthetic approaches may involve wet chemical method, hydrothermal treatment, coprecipitation, microemulsion, etc. as shown in Table 20.1. Similarly, biological methods are also available based on microorganisms and plants extracts. Hybrid methods are also available such as electrochemical and chemical vapor deposition (Kolahalam et al., 2019). All the abovementioned approaches may require a precursor that may be salt, metal, or compounds rich with carbon or may have different composition other than carbon and that may be utilized against the NMs synthetic strategy as shown in Fig. 20.1. Electronic waste is a nonbiodegradable waste and the best option as precursor that may contain rich metals, metal oxides, and polymeric materials. However, this scheme will provide highly efficient, dual method strategy in order of valorization of electronic waste and to synthesis an NM from them as shown in Table 20.2.

20.3.1 Synthesis of nanomaterials from fish wastes

Fish wastes are important sources of chemicals, materials, and fuel. Different nanomaterials are synthesized from fish wastes. Fish scales are a solid waste that primarily contain collagen and chitin, two of the most common natural useful

TABLE 20.1 Benefits or drawbacks of various approaches involved in nanomaterial's synthesis from electronic waste.

Approaches	Benefits	Drawbacks	References
Hydrometallurgical	– It may recover the zinc metal that may exist into the composite form on the reduced cluster. – From electronics, thrown away zinc may be recovered by using this method.	– Consume higher quantity of acid and water. – Less pure quality of ZnO may be obtained from this method.	Tunsu et al. (2015)
Biometallurgy	– High efficiency – Low cost and needs lower equipment.	– Harder to cultivate bacteria – Long leaching time required	Zheng et al. (2018)
Coprecipitation	– It may be cost-effective because it doesn't involve in expensive equipment. – Higher yield at low temperature	– Particle may be agglomerated. – Unlimited use of chemicals.	Bader et al. (2014)
Sol–gel	– Repeatable and simple – High purity product can be achieved. – By adjusting synthesis, morphology can be altered	– Agglomerated particles may be obtained. – Unsuitable for large scale. – Higher use of chemicals.	Dimesso and Technology (2016)
Hydrothermal	– Higher grade pure and crystalline product may be obtained. – Cost-effective and can be managed in lab scale. – By controlling the experimental variables, properties of a product can be altered.	– Higher pressure and temperature required. – Problems may occur in reproducibility and reliability.	Jamkhande et al. (2019)
Mechanochemical	– Reliable at industrial-scale production. – Low-cost production may be obtained.	– Impurities in milling atmosphere can affect the process. – Greater milling time required.	Kumar et al. (2020)
Thermal decomposition	– Time-efficient process. – Large-scale production can easily be handled. – Doesn't use expensive raw material or instruments.	– High chemical usage. – More time is required.	Odularu (2018)
Green synthesis	– Greater usage of biosafe process. – No discharge of toxic waste. – Greater stability and reproducibility. – Doesn't use expensive raw material or instruments.	– No uniform distribution	Iravani et al. (2017)

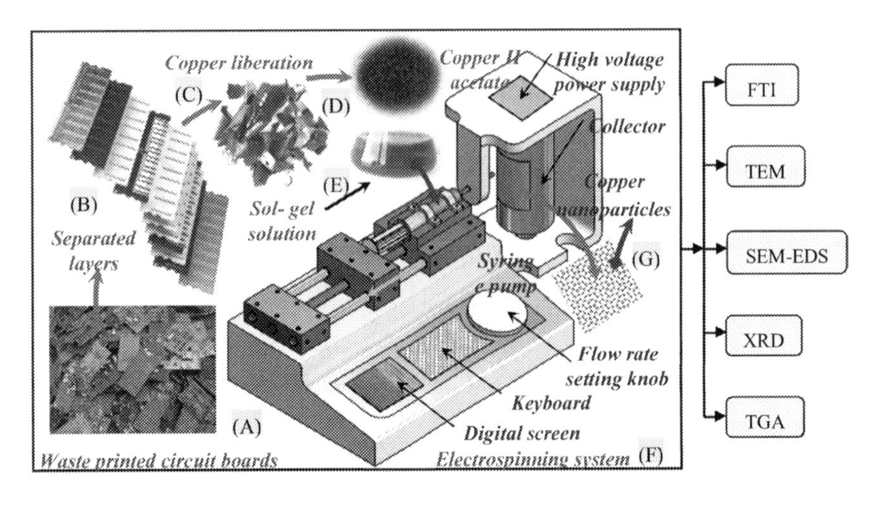

FIGURE 20.1 Synthesis of copper nanoparticles from electronic waste: (A) waste printed circuit board, (B) separated layers, (C) copper liberation, (D) copper II acetate, (E) sol–gel solution, (F) electrospinning system, and (G) copper nanoparticles (Yousef et al., 2018a). *Adapted with permission Ref from Yousef, S., Tatariants, M., Makarevičius, V., Lukošiūtė, S.-I., Bendikiene, R., Denafas, G., 2018a. A strategy for synthesis of copper nanoparticles from recovered metal of waste printed circuit boards. Journal of Cleaner Production, 185, 653–664.*

TABLE 20.2 Electronic waste-derived nanomaterial (NMs) including their synthesis, size, yield, and applications.

Waste source	NMs	Synthesis method	Size (nm)	Yield %	Morphology (shape)	Application	References
PCB	Ag NPs	Chemical reduction	15	75	Spherical	Medicine	Najafi et al. (2021)
PCB	SnO, Ag NPs	Microwave irradiation	100	90	Spherical	Optical device	Cerchier et al. (2017)
PCB	Cu NPs	Sol–gel/ electrospinning	5–50	84	Spherical	Sensor	El-Nasr et al. (2020)
PCB	Cu NPs	Electrospinning technique	30–70	84	Spherical	Sensor	Yousef et al. (2018b)
PCB	Cu NPs	Ultrasonic treatment	7	84	Spherical	Biomedical	Tatariants et al. (2018a)
PCB	Cu NPs	Microemulsion method	20–50	84	Spherical	Heat conduction	Mdlovu et al. (2018)
PCB	SnO_2, PbO_2 NPs	Ultrasonic treatment	1–7	99	Spherical	Gas sensors	Tatariants et al. (2018b)
Electronic waste	SiC NPs	Sol–gel	40–90	89	Spherical	Catalyst	Rajarao et al. (2014)
LTCC	Ag NPs	Wet chemical reduction	5–100	90	Spherical	Automotive industry	Swain et al. (2018)
PCB	Cu_2O NPs	Electrochemical technique	43	70	Spherical	flexible printed electronics	Kwon et al. (2019)
Spent battery	Mn–Zn ferrite composite	Hydrothermal	25	500	Spherical or cubic	Pb2 + removal	Niu et al. (2020)
Spent battery	Poly-metallic NPs	Chemical reduction	50	67	Spherical	Blue 4 dye removal	Nascimento et al. (2018b)
Plastic	Si_3N_4-NWs	Carbo-thermal reduction	500	80	Rod like	Chemical reactors	Maroufi et al. (2018)
PCB	CuO NPs	Electrochemical	11	90	Spherical	Supercapacitor	Rajkumar et al. (2022)
Spent battery	CO_3O_4 NPs	Electrochemical	47	91	Spherical	Electro-chromic devices	Dhiman et al. (2021)
PCB	Cu_2O	Chemical reduction	100	33	Spherical	Dye removal	Nascimento et al. (2018a)
LCD	$InBO_3$ NPs	Precipitating method	110	90	Spherical	Luminescence material	Assefi et al. (2019a)
PCB	Cu-Sn NPs	Chemical reduction	<500	80	Spherical	Sensor	(Shokri et al., 2017)

(Continued)

TABLE 20.2 (Continued)

Waste source	NMs	Synthesis method	Size (nm)	Yield %	Morphology (shape)	Application	References
Electronic waste	Fe_3O_4 NPs	Hydrothermal treatment	100	90	Spherical	Sensor	Akilarasan et al. (2018)
Spent battery	ZnO NPs	Sol–gel	5	95	Spherical	Bioimaging	Deep et al. (2016)
Sim card	CuO NPs	Wet chemical	100	90	Spherical	Electrochemical capacitor	Rajkumar et al. (2020)
Spent battery	MnO, ZnO NPs	Thermal decomposition	50	80	Spherical	Catalysis	Farzana et al. (2018)
Spent battery	ZnO NPs	Sol–gel	100	80	Spherical	Sensor	Hassan et al. (2021a)
PCB	AgNPs	Chemical reduction	67	90	Spherical	Biomedical	Caldas et al. (2021)
TBBPA	MWCNTs/Fe_3O_4	Chemical Reduction	23	80	Cubic	Pb^{+2} removal	Ji et al. (2012)
Electronic waste	V_2O_5/RGO/pt nanocomposites	Sol–gel	50	90	Thin stacked flakes	Photocatalyst	Mohan et al. (2019)
Spent battery	ZnO NPs	Solvothermal	40	90	Cotton ball	Sensors	Hassan et al. (2021b)
Jatropha podagrica Leaf	Fe_2O_3 NPs	Hydrothermal	60	91	Spherical	Dye removal	Haseena et al. (2022)
Neem leaves	ZnO–NiO–NPs	Hydrothermal	15	70	Spherical	Sensor	Hessien et al. (2021)
Bamboo	CDs	Hydrothermal	5	54	Spherical	Biosensors	Chen et al. (2021b)
Lichen	Ag NPs	Mechanochemical	23	90	Spherical	Biosensors	Goga et al. (2021)
Salvia extract	ZnO NPs	Hydrothermal	27	80	Spherical	Optoelectronics	Alrajhi et al. (2021)
Okra	Cu-doped NiO-NPs	Sol–gel	16	78	Cubic	Photocatalytic	Ghazal et al. (2021)
Penicillium fungi	ZrO_2 NPs	Anodization	50	75	Spherical	Biomedical	Tran et al. (2022)
Aspergillus sydowii fungus	Ag NPs	Biological	1–24	70	Spherical	Medical diagnosis	Wang et al. (2021a)
Lemanea manipurensis algae	MnO_2 NPs	Chemical method	24	90	Spherical	Catalysis	Borah et al. (2022)
PCB	CNTs	Chemical vapor deposition	30	90	Cylindrical	Energy storage	Iqbal et al. (2020)
CD/DVD	CDs	Hydrothermal	5	80	Spherical	Sensor	Ng (2014)
Spent PV	SiO_2	Chemical method	10–40	90	Spherical	Ink printing	Sapra et al. (2021)
Spent battery	Cdo NPs	Precipitation method	27	90	Spherical	Phototransistor	Golev et al. (2019)

mialaters. Various parts of fish waste are rich in carbon, hydrogen, oxygen, and nitrogen. The rapid microwave pyrolysis of fish scales (Xin et al., 2022) synthesized carbon nano-onions (CNOs). One of the most useful carbons is CNOs. Many industries, including electronics, photovoltaic, catalyst, and biomedical diagnostics, have used nanoforms extensively. Similarly, another research group (Campalani et al., 2021) synthesized a new class of carbon dots using fish scales as a source material. In the absence of any other reagent, they used the hydrothermal technique. They mentioned that the production of luminous carbon nanoparticles with a high capacity for photoelectron transfer uses fisheries wastes, which provides a carbon source. The elemental examination, morphology, and surface characteristics revealed that the nitrogen concentration was 3.21 C/N. Since CDs naturally contain nitrogen doping, the nanoparticles' photocatalytic activity was a result. Mudhafar et al. (2022) also investigated composites of natural fish-scale hydroxyapatite (FsHAp), fish-scale collagen (FsCol), and nano-silver (nAg) with different content ratios (Ag). Collagen is referred to be a natural polymer, whereas hydroxyapatite is a natural ceramic. The FsHA/FsCol/AgNPs composite, which is made using a simple mixing method, is used as a bone filler. The AgNPs were added to the FsCol/FsHA to provide the composite its antibacterial properties. In terms of degradation rate and swelling ratio, the composite with an 80:20 (FsHA: FsCol) ratio showed the greatest stability. Sinha et al. (2014) presented a low-cost and environment-friendly approach for the synthesis of self-assembled silver nanoparticles. They used an easy irradiation technique with an aqueous extract of *Labeo rohita* fish scales. Gelatin is thought to be the main component responsible for both the stabilization and reduction of the self-assembled silver nanoparticle. They did not use any reducing or stabilizing agents because gelatin serves as a reducer and stabilizer for the self-assembled silver nanoparticles. The various factors, such as temperature, concentration, and pH, can be used to control silver nanoparticle's size. Further research revealed that while Ag NP's size grows with rising temperature, it decreases with rising concentration and pH. Many reduced aromatic nitrocompounds into their respective amino derivatives are using Ag NPs as a catalyst. Large-scale synthesis can be accomplished using this method for synthesis of Ag NPs. Nowadays, a huge number of attempts has been made to convert the waste materials into the useful NMs (Zhang et al., 2021b). The cyanobacteria were used in the synthesis of silver, gold, copper, and zinc (Bhardwaj & Naraian, 2021). Mobarak et al. (2022) used fish waste for CuO nanomaterial's synthesis. After being heated, the fish scale's collagen content changed into gelatin, which in turn changed the precursor material into CuO NPs. The mono-clinic lattice structure of the CuO NPs was identified by X-ray diffraction (XRD) analysis, which also confirmed their production.

In another study, Campalani et al. (2022) used collagen which was extracted from the scales of *Mugil cephalus* and carbon dots (CDs) synthesized from the scales of *Dicentrarchus labrax*. These materials were combined to make hybrid films with UV-blocking properties, by casting a mixture of gelatin, glycerol (15%), and CDs (0%, 1%, 3%, and 5%). So, fish wastes contain special materials for innovative product's synthesis.

20.3.2 Synthesis of nanomaterials from expired medicine (pharmaceutical wastes)

Expired medicine disposing is an environment and an economical issue as there are no guidelines and other usage for expired medicine. They are threat to the environment and also an economical loss. The ideal method for reusing them, according to Jha and Prasad (2019), is to develop a way to turn waste into wealth by turning them into nanomaterial. Devi et al. (2021) presented a method for turning pharmaceutical waste (i.e., expired medications) into fluorescent carbon quantum dots with surface functionalities. The surface functionalities of the synthesized P-CQDs were found due to $-OH$, $C=O$ and $C=C$ group. By using an expired medicine of flouroquinolone (norfloxacin) and tinidazole combination, Jha and Prasad (2012) synthesized metal (Au), oxide (ZrO_2), and chalcognide (CdS) nanoparticles. Both of the ingredients of norfloxacin and tinidazole are able to carry out redox reactions due to the fundamental chemical structures of their molecules. The quinolone's nucleus contains a ketonic group, and an extra fluorine atom, both of which serve for a redox reaction. In the same way, the $S=O$ group may function in the tinidazole complex as one that withdraws protons. Therefore, it is confirmed that these nanosynthesizers are due to the generation of a zweitter-ion. However, drug's chemical characteristics, biological action, and disposition can all be affected by fluorine atom alternation. Jha and Prasad (2019) again synthesized silver and zinc oxide nanoparticles from chicken slaughtered wastes and expired drug of Norflox-TZ. The presence of metallothioneine and the synthesis of zweitter-ions aids expired medicine into nanoparticles transformation. From this approach, they produced metal as well as oxide nanoparticles. These steps proved to be fruitful to the environmental cleanup and reuse of expired medicines. Valian et al. (2023) used expired B12 vitamin vials as a capping agent and reductant in the sol–gel technique for the synthesis of adolinium titanate ($GT = Gd_2Ti_2O_7$) nanoparticles. These GT/N-GQD (nitrogen-doped graphene quantum dots) nanocomposites were prepared by the synthesis of N-GQDs in the presence of the GT nanoparticles. Then, using visible light and photocatalytic processes, these nanocomposites were employed for photodegradation of hydroxychloroquine sulfate, a multipurpose medication with a long half-life. This nanocomposite was used as an electrochemical sensor to measure the amount of the drug present in human blood serum, tablets, and wastewater. In light of this, it is possible to reuse expired drugs to synthesize nanomaterials that can be used in a variety of fields.

20.3.3 Synthesis of nanomaterials from hair waste

Barber hair causes an environmental issue. At barbershops and beauty parlors, a lot of human hair waste is produced. But economically and practically, the hair waste is useful a material. Im et al. (2020) synthesized hair-silver nanoparticle (Ag NPs) composites using a one- or two-step process that combined heat treatment with thermal or UV reduction. After Ag ion adsorption in the one-step procedure, untreated hair was thermally reduced between 250°C and 450°C, which partially oxidized the silver ion. During the synthesis of hair-silver nanoparticle composite, the silver ions thermally reduced. Here, heat-treated hair is used for the silver-ion adsorption. Alternatively, a two-step procedure was used to synthesize composites of hair and Ag NP. Pramanick et al. (2016) describe a novel technique for synthesis of glassy carbon microfibers from human hair. Long hollow carbon fibers may be produced using a simple pyrolysis process because to the distinctive coaxial structure of human hair. The Raman scattering and EDS studies revealed that glassy carbon constituted the majority of the chemical composition of pyrolyzed hair. Energy storage devices and electrochemical sensors are two examples of applications for glassy carbon. The electrochemical sensors' performance was compared to that of unmodified sensors, and the results for the carbons made from hair were noticeably better. Pretreatment of human hair samples allowed for the fabrication of many glassy carbon fibers. For the electrochemical detection of dopamine and ascorbic acid, glassy carbon produced from human hair has been successfully used. Nevertheless, the carbon microfibers made from hair have a lower surface-to-volume ratio than CNTs. The key benefits of employing carbon microfibers made from human hair over CNTs are their extremely low cost and simplicity of production. The properties of the human hair samples both before and after pyrolysis would improve the electrochemical sensing performance. Singh et al. (2020) demonstrated that human hair is the best source material for producing extremely fluorescent carbon dots with a variety of surface properties. They observed that the synthetic CDs' fluorescence and peaks intensity may be used for qualitative and quantitative analyses of chloroform with a 3 ppb limit of

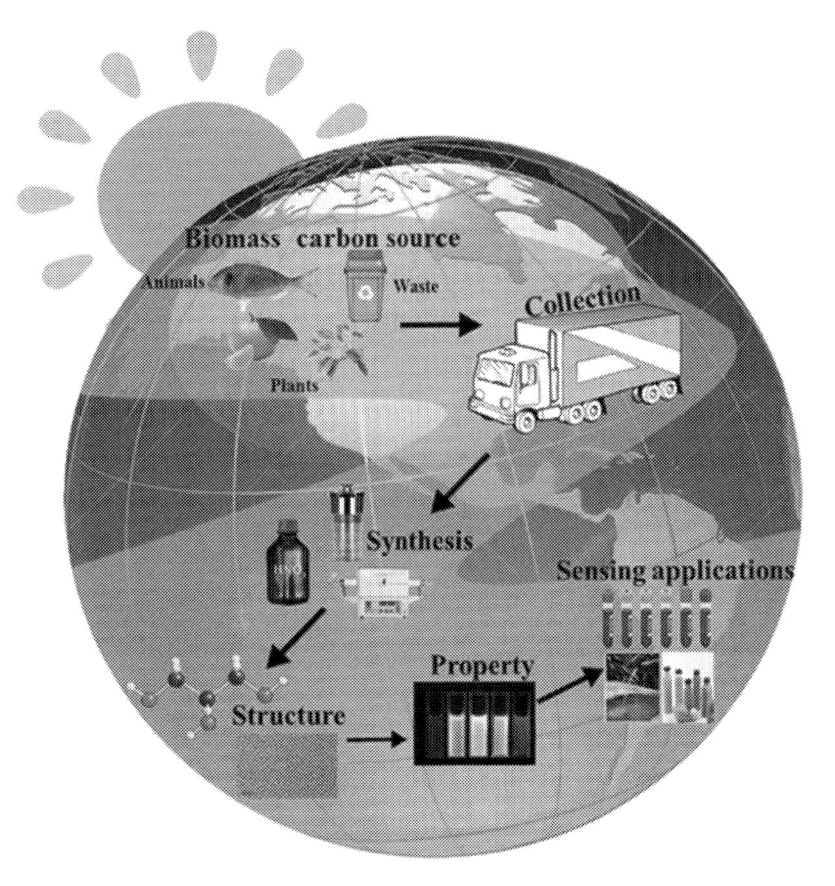

FIGURE 20.2 Synthesis of nanomaterials from Biowaste Adapted with Permission license under Attribution 4.0 International (CC BY 4.0) (Lou et al., 2021). *From Lou, Y., Hao, X., Liao, L., Zhang, K., Chen, S., Li, Z., et al., 2021. Recent advances of biomass carbon dots on syntheses, characterization, luminescence mechanism, and sensing applications. Nano Select, 2, 1117–1145.*

detection. They observed that pyridinic N oxide, which is unique to CDs (OCDs) made in an autoclave and their microwave-produced counterpart, imparts sensitive and particular sensing capacities toward chloroform. Chloroform is found in water produced from chlorination of water, which is cancer causing, as shown in Fig. 20.2. Therefore, fluorescent carbon dots are suitable for chloroform determination in chlorinated water reservoir. Animal's hair fiber with hollow structure is also a sustainable protein biomass material, which could be an excellent carrier after carbonization. These hair wastes are animal protein fibers, which are quite rich in carbon and nitrogen elements to provide carbon and nitrogen for the catalysts, which means nitrogen can be introduced into the composites based on the introduction of carbon doping. Chen et al. (2022) worked on rabbit's hair waste and designed a novel TiO_2/CRFs nanocomposites photocatalyst by combining carbonization, impregnation, and calcination processes with improved photocatalytic performance and excellent stability.

TiO_2/CRFs are naturally doped with carbon and nitrogen elements from waste rabbit hair, which effectively reduce the compounding of photo-generated electron-hole pairs and the band gap of TiO_2, and the absorption of visible light is enhanced, thereby greatly improving the photocatalytic activity of TiO_2.

Akintayo et al. (2020) suggested in his studies that goat's fur-based nanomaterial of silver nanoparticles was synthesized of spherical with size ranging from 11 to 32 nm. They showed larvicidal activities of AF-AgNPs against larvae of the vector of Plasmodium parasites (60%–100% after 12 hours of exposure), including their anticoagulant and thrombolytic activities. This indicates their relevance in biomedical applications. So, hair/fur wastes due to their unique chemical composition and structure are valuable biomaterials for nanomaterial's synthesis.

20.4 Conclusion

Organic or electronic wastes are useful sources for production of nanomaterials. A variety of methods are reported for synthesizing nanomaterials, which includes chemical, biological, and thermal methods. All the methods reported so far have certain advantages and disadvantages, however none of the methods has found applications in production of nanoparticles on a larger scale. Therefore, it is highly recommended that new researches should focus on the development of scalable and eco-friendly processes to produce commercially viable nanoparticles.

Fish wastes contain chitin and collagen as a natural biopolymer. Carbon dots, as one of the promising luminescent materials, are synthesized from fish scale which contains naturally, nitrogen-doped carbonaceous nanostructured photocatalysts. Similarly, medicines pose as one of the environmental threatening and economical losses after date expiry. But due to different chemical composition, functional group and atom which have different nanomaterials can be synthesized for such value-added fluorescent carbon quantum dots, metals NPs, metal oxide, and composite. The hair waste also is beneficial from economic and practical aspects. It was found that hair/Ag NP and composites which are capable in degradation reaction of methylene blue (MB) have potential for the treatment and purification of dye's pollutant. Glassy carbon microfiber synthesized from human hairs has potential of sensor and energy storage capacity. It means that waste is a green source of NMs. Different NMs can be synthesized following different routes of synthesis and have potential applications. Besides this, electronic waste is also promising in order to synthesize NMs. Various types of NMs such as silver, copper gold, cadmium, chromium, tin, and Mn–Zn ferrite, were successfully synthesized from batteries, PCBs, and other parts of electronic appliance waste, not last but least with wide range of applications such as gas sensors, catalyst, dye removal, bioimaging, etc.

The NMs synthesized from wastes have different potential and functionalities. Similarly, different waste source has different functionality. The NMs synthesized from biowaste and electronic wastes are different from chemical sources. Henceforth, applicability of NPs obtained from waste sources should be investigated distinctly and waste source correlations with application areas may be developed.

It is also recommended that feasible methods using circular economy approach for recycling of e-waste should be developed because enormous dispose-off of electronic components is going to be the next global challenge.

References

Abu Yazid, N., Barrena, R., Komilis, D., Sánchez, A.J.S., 2017. Solid-state fermentation as a novel paradigm for organic waste valorization: a review. Sustainability 9, 224.

Akilarasan, M., Kogularasu, S., Chen, S.-M., Chen, T.-W., Lou, B.-S.J.R.A., 2018. A novel approach to iron oxide separation from e-waste and bisphenol A detection in thermal paper receipts using recovered nanocomposites. RSC Advances 8, 39870–39878.

Akintayo, G., Lateef, A., Azeez, M., Asafa, T., Oladipo, I., Badmus, J., et al., 2020. IOP conference series: materials science and engineering Synthesis, Bioactivities and Cytogenotoxicity of Animal Fur-mediated Silver Nanoparticles. IOP Publishing, p. 012041.

Ali, A., Andriyana, A., 2020. Properties of multifunctional composite materials based on nanomaterials: a review. RSC Advances 10, 16390–16403.

Alrajhi, A., Ahmed, N.M., Al Shafouri, M., Almessiere, M.A.J.M.S.I.S.P., 2021. Green synthesis of zinc oxide nanoparticles using salvia officials extract. Materials Science in Semiconductor Processing 125, 105641.

Annamalai, M., Gurumurthy, K., 2021. Characterization of end-of-life mobile phone printed circuit boards for its elemental composition and beneficiation analysis. Journal of the Air & Waste Management Association 71, 315–327.

Assefi, M., Maroufi, S., Mayyas, M., Sahajwalla, V.J.J.O.E.C.E., 2018. Recycling of Ni-Cd batteries by selective isolation and hydrothermal synthesis of porous NiO nanocuboid. Journal of Environmental Chemical Engineering 6, 4671–4675.

Assefi, M., Maroufi, S., Sahajwalla, V.J.E.S., Research, P., 2019a. Recycling of the scrap LCD panels by converting into the InBO3 nanostructure product. Environmental Science and Pollution Research 26, 36287–36295.

Assefi, M., Maroufi, S., Yamauchi, Y., Sahajwalla, V.J.A.S.C., Engineering, 2019b. Core–shell nanocatalysts of Co_3O_4 and NiO shells from new (discarded) resources: sustainable recovery of Cobalt and Nickel from spent lithium-ion batteries, Ni–Cd batteries, and LCD panel. ACS Sustainable Chemistry Engineering 7, 19005–19014.

Bader, N., Benkhayal, A.A., Zimmermann, B.J.I.J.C.S., 2014. Co-precipitation as a sample preparation technique for trace element analysis: an overview. International Journal of Chemical Sciences 12, 519–525.

Baskar, A.V., Benzigar, M.R., Talapaneni, S.N., Singh, G., Karakoti, A.S., Yi, J., et al., 2022. Self-assembled fullerene nanostructures: synthesis and applications. Advanced Functional Materials 32, 2106924.

Basu, N., Abass, K., Dietz, R., Krummel, E., Rautio, A., Weihe, P., 2022. The impact of mercury contamination on human health in the Arctic: a state of the science review. Science of the Total Environment 831, 154793.

Bhardwaj, A.K., Naraian, R.J.B., 2021. Cyanobacteria as biochemical energy source for the synthesis of inorganic nanoparticles, mechanism and potential applications: a review. Biotech 11, 1–16.

Bisht, B., Lohani, U., Kumar, V., Gururani, P., Sinhmar, R.J.C.R.I.F.S., Nutrition, 2022. Edible hydrocolloids as sustainable substitute for non-biodegradable materials. Critical Reviews in Food Science and Nutrition 62, 693–725.

Borah, D., Rout, J., Gogoi, D., Ghosh, N.N., Bhattacharjee, C.R.J.I.C.C., 2022. Composition controllable green synthesis of manganese dioxide nanoparticles using an edible freshwater red alga and its photocatalytic activity towards water soluble toxic dyes. Inorganic Chemistry Communications 138, 109312.

Bratovcic, A.J.I.J.O.M.S.A.E., 2019. Different applications of nanomaterials and their impact on the environment. International Journal of Material Science and Engineering 5, 1–7.

Brewer, A., Dror, I., Berkowitz, B.J.C., 2022. Electronic waste as a source of rare earth element pollution: leaching, transport in porous media, and the effects of nanoparticles. Chemosphere 287, 132217.

Brindhadevi, K., Barcelo, D., Lan Chi, N.T., Rene, E.R., 2023. E-waste management, treatment options and the impact of heavy metal extraction from e-waste on human health: scenario in Vietnam and other countries. Environmental Research 217, 114926.

Caldas, M.P.K., Martins, T.A.G., De Moraes, V.T., Tenório, J.A.S., Espinosa, D.C.R.J.J.O.E.C.E., 2021. Synthesis of Ag nanoparticles from waste printed circuit board. Journal of Environmental Chemical Engineering 9, 106845.

Campalani, C., Cattaruzza, E., Zorzi, S., Vomiero, A., You, S., Matthews, L., et al., 2021. Biobased carbon dots: from fish scales to photocatalysis. Nanomaterials 11, 524.

Campalani, C., Causin, V., Selva, M., Perosa, A., 2022. Fish-waste-derived gelatin and carbon dots for biobased UV-blocking films. ACS Applied Materials & Interfaces 14, 35148–35156.

Cerchier, P., Dabalà, M., Brunelli, K.J.J., 2017. Synthesis of SnO_2 and Ag nanoparticles from electronic wastes with the assistance of ultrasound and microwaves. JOM 69, 1583–1588.

Chen, L., Gao, M., 2021. Formal or informal recycling sectors? Household solid waste recycling behavior based on multi-agent simulation. Journal of Environmental Management 294, 113006.

Chen, Y.-Y., Jiang, W.-P., Chen, H.-L., Huang, H.-C., Huang, G.-J., Chiang, H.-M., et al., 2021b. Cytotoxicity and cell imaging of six types of carbon nanodots prepared through carbonization and hydrothermal processing of natural plant materials. RSC Advances 11, 16661–16674.

Chen, Y., Wang, C., Chen, J., Wang, S., Ju, J., Kang, W., 2022. Preparing biomass carbon fiber derived from waste rabbit hair as a carrier of TiO_2 for photocatalytic degradation of methylene blue. Polymers 14, 1593.

Chen, Q., Wang, H., Tang, X., Ba, Z., Zhao, X., Wang, Y., et al., 2021a. One-step synthesis of carbon quantum dot-carbon nanotube composites on waste eggshell-derived catalysts for enhanced adsorption of methylene blue. Journal of Environmental Chemical Engineering 9, 106222.

Chombo, P.V., Laoonual, Y.J.J.O.P.S., 2020. A review of safety strategies of a Li-ion battery. Journal of Power Sources 478, 228649.

Covre, W.P., Ramos, S.J., Pereira, W., Souza, E.S., Martins, G.C., Teixeira, O.M.M., et al., 2022. Impact of copper mining wastes in the Amazon: properties and risks to environment and human health. Journal of Hazardous Materials 421, 126688.

Das, S., Chakraborty, J., Chatterjee, S., Kumar, H.J.E.S.N., 2018. Prospects of biosynthesized nanomaterials for the remediation of organic and inorganic environmental contaminants. Environmental Science: Nano 5, 2784–2808.

Das, H.T., Elango Balaji, T., Mahendraprabhu, K., Vinoth, S., 2021. Cost-effective nanomaterials fabricated by recycling spent batteries. Waste Recycling Technologies for Nanomaterials Manufacturing. Springer.

De, A.J.C.R., 2022. Nanomaterial synthesis from end-of-cycle products: a sustainable way of waste valorisation. ChemBioEng Reviews .

Deep, A., Sharma, A.L., Mohanta, G.C., Kumar, P., Kim, K.H., 2016. A facile chemical route for recovery of high quality zinc oxide nanoparticles from spent alkaline batteries. Waste Management 51, 190–195.

Desmarais, M., Pirade, F., Zhang, J., Rene, E.R.J.B.T.R., 2020. Biohydrometallurgical processes for the recovery of precious and base metals from waste electrical and electronic equipments: current trends and perspectives. Bioresource Technology Reports 11, 100526.

Devi, P., Jindal, N., Kim, K.-H., Thakur, A., 2021. Nanostructures derived from expired drugs and their applications toward sensing, security ink, and bactericidal material. Science of the Total Environment 764, 144260.

Dhandapani, E., Duraisamy, N., Raj, R.M.J.M.T.P., 2020. Green synthesis of carbon quantum dots from food waste. Materials Today: Proceedings .

Dhiman, S., Gupta, B.J.E.T., Innovation, 2021. Co_3O_4 nanoparticles synthesized from waste Li-ion batteries as photocatalyst for degradation of methyl blue dye. Environmental Technology & Innovation 23, 101765.

Dimesso, L.J.H.O.S.-G.S., Technology, 2016. Pechini processes: an alternate approach of the sol–gel method, preparation, properties, and applications. Handbook of Sol-Gel Science 2, 1–22.

Díaz-Martínez, M.E., Argumedo-Delira, R., Sánchez-Viveros, G., Alarcón, A., Mendoza-López, M.J.C.M., 2019. Microbial bioleaching of Ag, Au and Cu from printed circuit boards of mobile phones. Current Microbiology 76, 536–544.

El-Nasr, R.S., Abdelbasir, S., Kamel, A.H., Hassan, S.S.J.S., Technology, P., 2020. Environmentally friendly synthesis of copper nanoparticles from waste printed circuit boards. Separation and Purification Technology 230, 115860.

Faraji, F., Golmohammadzadeh, R., Pickles, C.A.J.J.O.E.M., 2022. Potential and current practices of recycling waste printed circuit boards: a review of the recent progress in pyrometallurgy. Journal of Environmental Management 316, 115242.

Fares, M.M., Radaydeh, S.K.J.P.C., 2022. Multifunctional sustainable quercetin polyphenol/functionalized carbon nanotubes; semi-transparent conductive films and 3D printing inks. Polymer Composites 43, 2318–2328.

Farzana, R., Rajarao, R., Hassan, K., Behera, P.R., Sahajwalla, V.J.J.O.C.P., 2018. Thermal nanosizing: novel route to synthesize manganese oxide and zinc oxide nanoparticles simultaneously from spent Zn–C battery. Journal of Cleaner Production 196, 478–488.

Ganesh, S., Sharma, P., 2022. The prerequisite of E-waste recycling-a review study. Journal of Physics: Conference Series. IOP Publishing, 012027.

Gao, Y., Guo, Q., Zhang, Q., Cui, Y., Zheng, Z.J.A.E.M., 2021. Fibrous materials for flexible Li–S battery. Advanced Energy Materials 11, 2002580.

Ghazal, S., Khandannasab, N., Hosseini, H.A., Sabouri, Z., Rangrazi, A., Darroudi, M.J.C.I., 2021. Green synthesis of copper-doped nickel oxide nanoparticles using okra plant extract for the evaluation of their cytotoxicity and photocatalytic properties. Ceramics International 47, 27165–27176.

Goga, M., Balaz, M., Daneu, N., Elecko, J., Tkacikova, L., Marcincinova, M., et al., 2021. Biological activity of selected lichens and lichen-based Ag nanoparticles prepared by a green solid-state mechanochemical approach. Materials Science & Engineering C-Materials for Biological Applications 119, 111640.

Golev, A., Corder, G.D., Rhamdhani, M.A.J.M.E., 2019. Estimating flows and metal recovery values of waste printed circuit boards in Australian e-waste. Minerals Engineering 137, 171–176.

Haseena, S., Jayamani, N., Shanavas, S., Duraimurugan, J., Haija, M.A., Kumar, G.S., et al., 2022. Bio-synthesize of photocatalytic Fe_2O_3 nanoparticles using *Leucas aspera* and *Jatropha podagrica* leaf extract for an effective removal of textile dye pollutants. Optik 249, 168275.

Hassan, K., Hossain, R., Farzana, R., Sahajwalla, V.J.A.C.A., 2021a. Microrecycled zinc oxide nanoparticles (ZnO NP) recovered from spent Zn–C batteries for VOC detection using ZnO sensor. Anal. Chim. Acta 1165, 338563.

Hassan, K., Hossain, R., Sahajwalla, V.J.S., Chemical, A.B., 2021b. Novel microrecycled ZnO nanoparticles decorated macroporous 3D graphene hybrid aerogel for efficient detection of NO2 at room temperature. Sensors and Actuators B: Chemical 330, 129278.

Hessien, M., Da'na, E., Taha, A.J.C.I., 2021. Phytoextract assisted hydrothermal synthesis of ZnO–NiO nanocomposites using neem leaves extract. Ceramics International 47, 811–816.

Heydarian, A., Mousavi, S.M., Vakilchap, F., Baniasadi, M.J.J.O.P.S., 2018. Application of a mixed culture of adapted acidophilic bacteria in two-step bioleaching of spent lithium-ion laptop batteries. Journal of Power Sources 378, 19–30.

He, K., Zhang, Z.Y., Zhang, F.S., 2021. Synthesis of graphene and recovery of lithium from lithiated graphite of spent Li-ion battery. Waste Management 124, 283–292.

Hossain, R., Nekouei, R.K., Mansuri, I., Sahajwalla, V.J.A.S.C., Engineering, 2018. Sustainable recovery of Cu and Sn from problematic global waste: exploring value from waste printed circuit boards. ACS Sustainable Chemistry & Engineering 7, 1006–1017.

Im, D.S., Hong, B.M., Kim, M.H., Park, W.H., 2020. Formation of human hair-Ag nanoparticle composites via thermal and photo-reduction: a comparison study. Colloids and Surfaces A: Physicochemical and Engineering Aspects 600, 124995.

Iqbal, A., Jan, M.R., Shah, J., Rashid, B.J.M.E., 2020. Dispersive solid phase extraction of precious metal ions from electronic wastes using magnetic multiwalled carbon nanotubes composite. Minerals Engineering 154, 106414.

Iravani, A., Akbari, M., Zohoori, M.J.I.J.S.E.A., 2017. Advantages and disadvantages of green technology; goals, challenges and strengths. International Journal of Applied Science and Engineering 6, 272–284.

Jain, S., Sharma, T., Gupta, A.K.J.R., Reviews, S.E., 2022. End-of-life management of solar PV waste in India: situation analysis and proposed policy framework. Renewable and Sustainable Energy Reviews 153, 111774.

Jamkhande, P.G., Ghule, N.W., Bamer, A.H., Kalaskar, M.G.J.J.O.D.D.S., Technology, 2019. Metal nanoparticles synthesis: an overview on methods of preparation, advantages and disadvantages, and applications. Journal of Drug Delivery Science 53, 101174.

Jha, A.K., Prasad, K., 2012. Synthesis of nanomaterials using expired medicines: an eco-friendly option. Nanotechnology Development 2, e7. -e7.

Jha, A.K., Prasad, K., 2019. Nanomaterials from biological and pharmaceutical wastes–a step towards environmental protection. Materials Today: Proceedings 18, 1465–1471.

Ji, L., Zhou, L., Bai, X., Shao, Y., Zhao, G., Qu, Y., et al., 2012. Facile synthesis of multiwall carbon nanotubes/iron oxides for removal of tetrabromobisphenol A and Pb (II). Journal of Materials Chemistry 22, 15853–15862.

Kang, K.D., Kang, H., Ilankoon, I., Chong, C.Y.J.J.O.C.P., 2020. Electronic waste collection systems using Internet of Things (IoT): household electronic waste management in Malaysia. Journal of Cleaner Production 252, 119801.

Kaur, P., Sharma, S., Albarakaty, F.M., Kalia, A., Hassan, M.M., Abd-Elsalam, K.A.J.S., 2022. Biosorption and bioleaching of heavy metals from electronic waste varied with microbial genera. Sustainability 14, 935.

Kolahalam, L.A., Viswanath, I.K., Diwakar, B.S., Govindh, B., Reddy, V., Murthy, Y.J.M.T.P., 2019. Review on nanomaterials: synthesis and applications. Materials Today: Proceedings 18, 2182–2190.

Kou, X., Zhao, Q., Xu, W., Xiao, Z., Niu, Y., Wang, K.J.J.O.R.M., 2021. Biodegradable materials as nanocarriers for drugs and nutrients. Journal of Renewable Materials 9, 1189.

Kumar, A., Holuszko, M., Espinosa, D.C.R.J.R., Conservation, Recycling, 2017. E-waste: an overview on generation, collection, legislation and recycling practices. Resources, Conservation and Recycling 122, 32–42.

Kumar, M., Xiong, X., Wan, Z., Sun, Y., Tsang, D.C., Gupta, J., et al., 2020. Ball milling as a mechanochemical technology for fabrication of novel biochar nanomaterials. Bioresource Technology. 312, 123613.

Kwon, Y.-T., Yune, S.-J., Song, Y., Yeo, W.-H., Choa, Y.-H.J.A.A.E.M., 2019. Green manufacturing of highly conductive Cu_2O and Cu nanoparticles for photonic-sintered printed electronics. ACS Applied Electronic Materials 1, 2069–2075.

Larouche, F., Tedjar, F., Amouzegar, K., Houlachi, G., Bouchard, P., Demopoulos, G.P., et al., 2020. Progress and status of hydrometallurgical and direct recycling of Li-ion batteries and beyond. Materials (Basel) 13, 801.

Liou, T.H., Wang, P.Y., 2020. Utilization of rice husk wastes in synthesis of graphene oxide-based carbonaceous nanocomposites. Waste Management 108, 51–61.

Lou, Y., Hao, X., Liao, L., Zhang, K., Chen, S., Li, Z., et al., 2021. Recent advances of biomass carbon dots on syntheses, characterization, luminescence mechanism, and sensing applications. Nano Select 2, 1117–1145.

Lyu, J., Liu, Y., Lyu, X., Ma, Z., Zhou, J.J.J.O.C.P., 2021. Efficient bromine removal and metal recovery from waste printed circuit boards smelting flue dust by a two-stage leaching process. Journal of Cleaner Production 322, 129054.

Magoda, K., Mekuto, L.J.R., 2022. Biohydrometallurgical recovery of metals from waste electronic equipment: current status and proposed process. Recycling, 2022 - mdpi 7, 67.

Maiti, D., Tong, X., Mou, X., Yang, K., 2018. Carbon-based nanomaterials for biomedical applications: a recent study. Front. Pharmacol. 9, 1401.

Makuza, B., Tian, Q., Guo, X., Chattopadhyay, K., Yu, D.J.J.O.P.S., 2021. Pyrometallurgical options for recycling spent lithium-ion batteries: a comprehensive review. Journal of Power Sources 491, 229622.

Maroufi, S., Mayyas, M., Nekouei, R.K., Assefi, M., Sahajwalla, V.J.A.S.C., Engineering, 2018. Thermal nanowiring of e-waste: a sustainable route for synthesizing green Si_3N_4 nanowires. ACS Sustainable Chemistry & Engineering 6, 3765–3772.

Martins, T.A.G., Falconi, I.B.A., Pavoski, G., De Moraes, V.T., Baltazar, M.D.P.G., Espinosa, D.C.R.J.J.O.E.C.E., 2021. Green synthesis, characterization, and application of copper nanoparticles obtained from printed circuit boards to degrade mining surfactant by Fenton process. Journal of Environmental Chemical Engineering 9, 106576.

Mdlovu, N.V., Chiang, C.-L., Lin, K.-S., Jeng, R.-C.J.J.O.C.P., 2018. Recycling copper nanoparticles from printed circuit board waste etchants via a microemulsion process. Journal of Cleaner Production 185, 781–796.

Mihai, F.-C., 2020. Electronic waste management in Romania: pathways for sustainable practices. Handbook of Electronic Waste Management. Elsevier.

Mobarak, M.B., Hossain, M.S., Chowdhury, F., Ahmed, S., 2022. Synthesis and characterization of CuO nanoparticles utilizing waste fish scale and exploitation of XRD peak profile analysis for approximating the structural parameters. Arabian Journal of Chemistry 15, 104117.

Mohan, H., Selvaraj, D., Kuppusamy, S., Venkatachalam, J., Park, Y.J., Seralathan, K.K., et al., 2019. E-waste based V_2O_5/RGO/Pt nanocomposite for photocatalytic degradation of oxytetracycline. Environmental Progress & Sustainable Energy 38, 13123.

Mudhafar, M., Zainol, I., Alsailawi, H., Aiza Jaafar, C., 2022. Synthesis and characterization of fish scales of hydroxyapatite/collagen–silver nanoparticles composites for the applications of bone filler. Journal of the Korean Ceramic Society 59, 229–239.

Muscetta, M., Clarizia, L., Race, M., Pirozzi, F., Marotta, R., Andreozzi, R., et al., 2023. A novel green approach for silver recovery from chloride leaching solutions through photodeposition on zinc oxide. Journal of Environmental Management 330, 117075.

Najafi, A., Khoeini, M., Khalaj, G., Sahebgharan, A.J.M.R.E., 2021. Synthesis of silver nanoparticles from electronic scrap by chemical reduction. Materials Research Expressyou 8, 125009.

Nascimento, M.A., Cruz, J.C., Dos Reis, M.F., Carvalho Damasceno, D.E., Reis, O.I., Reis, E.L., et al., 2018a. Synthesis of polymetallic nanoparticles from printed circuit board waste and application in textile dye remediation. Journal of Environmental Chemical Engineering 6, 5580–5586.

Nascimento, M.A., Cruz, J.C., Rodrigues, G.D., De Oliveira, A.F., Lopes, R.P.J.J.O.C.P., 2018b. Synthesis of polymetallic nanoparticles from spent lithium-ion batteries and application in the removal of reactive blue 4 dye. Journal of Cleaner Production 202, 264–272.

Ng, S.M.J.T.I.E.A.C., 2014. Sustainable alternative in environmental monitoring using carbon nanoparticles as optical probes. Trends in Environmental Analytical Chemistry 3, 36–42.

Nigam, S., Jha, R., Singh, R.P.J.M.T.P., 2021. A different approach to the electronic waste handling–a review. Materials Today: Proceedings 46, 1519–1525.

Niu, Z., Feng, W., Huang, H., Wang, B., Chen, L., Miao, Y., et al., 2020. Green synthesis of a novel Mn–Zn ferrite/biochar composite from waste batteries and pine sawdust for Pb2 + removal. Chemosphere 252, 126529.

Odularu, A.T., 2018. Metal nanoparticles: thermal decomposition, biomedicinal applications to cancer treatment, and future perspectives. Bioinorganic Chemistry and Applications 2018, 9354708.

Ohajinwa, C.M., Van Bodegom, P.M., Vijver, M.G., Peijnenburg, W., 2018. Impact of informal electronic waste recycling on metal concentrations in soils and dusts. Environmental Research 164, 385–394.

Oliveira, L.S., Oliveira, D.S., Bezerra, B.S., Pereira, B.S., Battistelle, R.A.G.J.J.O.C.P., 2017. Environmental analysis of organic waste treatment focusing on composting scenarios. Journal of Cleaner Production 155, 229–237.

Pramanick, B., Cadenas, L.B., Kim, D.-M., Lee, W., Shim, Y.-B., Martinez-Chapa, S.O., et al., 2016. Human hair-derived hollow carbon microfibers for electrochemical sensing. Carbon. N. Y. 107, 872–877.

Prasad, S., Yadav, K.K., Kumar, S., Gupta, N., Cabral-Pinto, M.M.S., Rezania, S., et al., 2021. Chromium contamination and effect on environmental health and its remediation: a sustainable approaches. Journal of Environmental Management 285, 112174.

Priya, A., Hait, S.J.E.S., Research, P., 2017. Comparative assessment of metallurgical recovery of metals from electronic waste with special emphasis on bioleaching. Environmental Science and Pollution Research 24, 6989–7008.

Priya, A., Hait, S.J.W., Valorization, B., 2018. Feasibility of bioleaching of selected metals from electronic waste by *Acidiphilium acidophilum*. Waste and Biomass Valorization 9, 871–877.

Raghavan, N., Thangavel, S., Venugopal, G.J.A.M.T., 2017. A short review on preparation of graphene from waste and bioprecursors. Applied Materials Today 7, 246–254.

Rajarao, R., Ferreira, R., Sadi, S.H.F., Khanna, R., Sahajwalla, V.J.M.L., 2014. Synthesis of silicon carbide nanoparticles by using electronic waste as a carbon source. Materials Letters 120, 65–68.

Rajkumar, S., Elanthamilan, E., Balaji, T.E., Sathiyan, A., Jafneel, N.E., Merlin, J.P.J.J.O.A., et al., 2020. Recovery of copper oxide nanoparticles from waste SIM cards for supercapacitor electrode material. Journal of Alloys 849, 156582.

Rajkumar, S., Elanthamilan, E., Wang, S.-F., Chryso, H., Balan, P.V.D., Merlin, J.P.J.R., et al., 2022. One-pot green recovery of copper oxide nanoparticles from discarded printed circuit boards for electrode material in supercapacitor application. Resources, Conservation and Recycling 180, 106180.

Rodríguez-Padrón, D., Alothman, Z.A., Osman, S.M., Luque, R.J.C.O.I.G., Chemistry, S., 2020. Recycling electronic waste: prospects in green catalysts design. Current Opinion in Green and Sustainable Chemistry 25, 100357.

Roy, J.J., Rarotra, S., Krikstolaityte, V., Zhuoran, K.W., Cindy, Y.D., Tan, X.Y., et al., 2021. Green recycling methods to treat lithium-ion batteries E-waste: a circular approach to sustainability. Advanced Materials e2103346.

Saha, L., Kumar, V., Tiwari, J., Rawat, S., Singh, J., Bauddh, K.J.E.T., et al., 2021. Electronic waste and their leachates impact on human health and environment: global ecological threat and management. Environmental Technology & Innovation 24, 102049.

Samaddar, P., Ok, Y.S., Kim, K.-H., Kwon, E.E., Tsang, D.C.J.J.O.C.P., 2018. Synthesis of nanomaterials from various wastes and their new age applications. Journal of Cleaner Production 197, 1190–1209.

Sapra, G., Chaudhary, V., Kumar, P., Sharma, P., Saini, A.J.M.T.P., 2021. Recovery of silica nanoparticles from waste PV modules. Materials Today: Proceedings 45, 3863–3868.

Shokri, A., Pahlevani, F., Levick, K., Cole, I., Sahajwalla, V.J.J.O.C.P., 2017. Synthesis of copper-tin nanoparticles from old computer printed circuit boards. Journal of Cleaner Production 142, 2586–2592.

Singh, A., Eftekhari, E., Scott, J., Kaur, J., Yambem, S., Leusch, F., et al., 2020. Carbon dots derived from human hair for ppb level chloroform sensing in water. Sustainable Materials and Technologies 25, e00159.

Singh, R., Mahandra, H., Gupta, B.J.W.M., 2017. Recovery of zinc and cadmium from spent batteries using Cyphos IL 102 via solvent extraction route and synthesis of Zn and Cd oxide nanoparticles. Waste Management 67, 240–252.

Sinha, T., Ahmaruzzaman, M., Sil, A., Bhattacharjee, A., 2014. Biomimetic synthesis of silver nanoparticles using the fish scales of *Labeo rohita* and their application as catalysts for the reduction of aromatic nitro compounds. Spectrochimica Acta Part A: Molecular and Biomolecular Spectroscopy 131, 413–423.

Suhani, I., Sahab, S., Srivastava, V., Singh, R.P.J.C.O.I.T., 2021. Impact of cadmium pollution on food safety and human health. Current Opinion in Toxicology 27, 1–7.

Swain, B., Shin, D., Joo, S.Y., Ahn, N.K., Lee, C.G., Yoon, J.H., 2018. Synthesis of submicron silver powder from scrap low-temperature co-fired ceramic an e-waste: understanding the leaching kinetics and wet chemistry. Chemosphere 194, 793–802.

Tatariants, M., Yousef, S., Sakalauskaitė, S., Daugelavičius, R., Denafas, G., Bendikiene, R.J.W.M., 2018a. Antimicrobial copper nanoparticles synthesized from waste printed circuit boards using advanced chemical technology. Waste Management 78, 521–531.

Tatariants, M., Yousef, S., Skapas, M., Juskenas, R., Makarevicius, V., Lukošiūtė, S.-I., et al., 2018b. Industrial technology for mass production of SnO_2 nanoparticles and PbO_2 microcube/microcross structures from electronic waste. Journal of Cleaner Production 203, 498–510.

Tran, T.V., Nguyen, D.T.C., Kumar, P.S., Din, A.T.M., Jalil, A.A., Vo, D.N., 2022. Green synthesis of ZrO_2 nanoparticles and nanocomposites for biomedical and environmental applications: a review. Environmental Chemistry Letters 20, 1309–1331.

Tunsu, C., Petranikova, M., Gergorić, M., Ekberg, C., Retegan, T.J.H., 2015. Reclaiming rare earth elements from end-of-life products: a review of the perspectives for urban mining using hydrometallurgical unit operations. Hydrometallurgy 156, 239–258.

Turner, A., Filella, M.J.S.O.T.T.E., 2017. Bromine in plastic consumer products–Evidence for the widespread recycling of electronic waste. Science of the Total Environment 601, 374–379.

Ulman, K., Maroufi, S., Bhattacharyya, S., Sahajwalla, V.J.J.O.C.P., 2018. Thermal transformation of printed circuit boards at 500° C for synthesis of a copper-based product. Journal of Cleaner Production 198, 1485–1493.

Valian, M., Soofivand, F., Khoobi, A., Yousif, Q.A., Salavati-Niasari, M., 2023. A green approach: eco-friendly synthesis of $Gd_2Ti_2O_7$/N-GQD nanocomposite and photo-degradation and electrochemical measurement of hydroxychloroquine as a perdurable drug. Arabian Journal of Chemistry 16, 104401.

Wang, D., Xue, B., Wang, L., Zhang, Y., Liu, L., Zhou, Y.J.S.R., 2021a. Fungus-mediated green synthesis of nano-silver using *Aspergillus sydowii* and its antifungal/antiproliferative activities. Scientific Reports 11, 1–9.

Wang, X., Zhong, X., Li, J., Liu, Z., Cheng, L., 2021b. Inorganic nanomaterials with rapid clearance for biomedical applications. Chemical Society Reviews 50, 8669–8742.

Xin, Y., Odachi, K., Shirai, T., 2022. Fabrication of ultra-bright carbon nano-onions via a one-step microwave pyrolysis of fish scale waste in seconds. Green Chemistry 24, 3969–3976.

Yadav, M.K., Saidulu, D., Gupta, A.K., Ghosal, P.S., Mukherjee, A.J.J.O.E.C.E., 2021. Status and management of arsenic pollution in groundwater: a comprehensive appraisal of recent global scenario, human health impacts, sustainable field-scale treatment technologies. Journal of Environmental Chemical Engineering 9, 105203.

Yao, Z., Ling, T.-C., Sarker, P., Su, W., Liu, J., Wu, W., et al., 2018. Recycling difficult-to-treat e-waste cathode-ray-tube glass as construction and building materials: a critical review. Renewable and Sustainable Energy Reviews 81, 595–604.

Van Yken, J., Cheng, K.Y., Boxall, N.J., Nikoloski, A.N., Moheimani, N., Valix, M., et al., 2023. An integrated biohydrometallurgical approach for the extraction of base metals from printed circuit boards. Hydrometallurgy 216, 105998.

Yousef, S., Tatariants, M., Makarevičius, V., Lukošiūtė, S.-I., Bendikiene, R., Denafas, G., 2018a. A strategy for synthesis of copper nanoparticles from recovered metal of waste printed circuit boards. Journal of Cleaner Production 185, 653–664.

Yousef, S., Tatariants, M., Makarevičius, V., Lukošiūtė, S.-I., Bendikiene, R., Denafas, G.J.J.O.C.P., 2018b. A strategy for synthesis of copper nanoparticles from recovered metal of waste printed circuit boards. Journal of Cleaner Production 185, 653–664.

Zhang, N., Song, X., Jiang, H., Tang, C.Y.J.S., Technology, P., 2021a. Advanced thin-film nanocomposite membranes embedded with organic-based nanomaterials for water and organic solvent purification: a review. Separation and Purification Technology 269, 118719.

Zhang, N., Song, X., Jiang, H., Tang, C.Y.J.S., Technology, P., 2021b. Advanced thin-film nanocomposite membranes embedded with organic-based nanomaterials for water and organic solvent purification: A review 269, 118719.

Zhan, L., Xiang, X., Xie, B., Gao, B.J.P.T., 2017. Preparing lead oxide nanoparticles from waste electric and electronic equipment by high temperature oxidation-evaporation and condensation. Powder Technology 308, 30–36.

Zheng, X., Zhu, Z., Lin, X., Zhang, Y., He, Y., Cao, H., et al., 2018. A mini-review on metal recycling from spent lithium ion batteries. Engineering 4, 361–370.

Zhou, W., Liang, H., Xu, H.J.J.O.H.M., 2021. Recovery of gold from waste mobile phone circuit boards and synthesis of nanomaterials using emulsion liquid membrane. Journal of Hazardous Materials 411, 125011.

Chapter 21

Application of nanomaterials synthesized using agriculture waste for wastewater treatment

Pubali Mandal[1], Manoj Kumar Yadav[2], Abhradeep Majumder[3] and Partha Sarathi Ghosal[4]

[1]Department of Civil Engineering, Birla Institute of Technology and Science Pilani, Pilani, Rajasthan, India, [2]Department of Civil and Environmental Engineering, Indian Institute of Technology Patna, Patna, Bihar, India, [3]School of Environmental Science and Engineering, Indian Institute of Technology Kharagpur, Kharagpur, West Bengal, India, [4]School of Water Resources, Indian Institute of Technology Kharagpur, Kharagpur, West Bengal, India

21.1 Introduction

One of the critical issues is the security of the world's water supply because of excessive consumption of water resources and increasing water bodies' pollution. Discharge of effluent without appropriate treatment and different geological and anthropogenic activities result in poor water quality and pose a potential threat to human health (Deshpande et al., 2020). Several organic and inorganic pollutants could exist in wastewater, some of which are persistent in nature. Furthermore, various chemical compounds generated from daily human activities and water-borne pathogens may create serious concerns for living organisms. Therefore, wastewater collection and treatment are needed to manage water scarcity problems.

Several techniques can be used, such as biological methods, chemical methods (oxidation, adsorption, coagulation, flocculation, and disinfection), and physical methods (filtration, floatation, etc.) for wastewater treatment (Lu and Astruc, 2020). Owing to reactivity and high ratio of surface to volume, nanomaterials (NMs) have gained immense recognition recently by depicting high efficiency. The developments and modifications of NMs into nanosorbents, nanocatalysts, and nanodisinfectants led to a promising avenue in wastewater treatment (Singh et al., 2021). Several chemical and physical methods can be employed for metal and metal oxide (M & MO) nanomaterials synthesis (Bhardwaj et al., 2019). However, the methods are expensive and require high energy. Studies have been conducted to innovate green methods for the preparation of NMs to overcome the drawbacks of chemical and physical methods. The M & MO NMs synthesis using plant extract has gained popularity as the method is simple, cost-effective, and eco-friendly. Agricultural waste may create adverse ecological impacts when they are not properly handled. The common management techniques for agricultural waste include disposal in landfills, composting, animal feed, and open burning. Millions of tons of agricultural waste are landfilled every year. The disposal is associated with negative effects, for example, landfill gas/leachate production and groundwater, as well as air pollution (Jovanov et al., 2018). Burning of agricultural waste in open lands is another conventional method, especially for developing countries. Open burning is associated with extreme deterioration of air quality due to the emission of various greenhouse gases (CO_2, CH_4, and N_2O), particulate matter, and several other air contaminants (CO, NH_3, SO_2, and volatile organics) (Tyagi et al., 2016). Although composting and animal feeding are alternative options, a very limited portion of agricultural waste goes for the purpose (Kee et al., 2021). The utilization of zero-cost and biodegradable agricultural waste to synthesize nanosorbents, nanocatalysts, and nanodisinfectants has made NMs more sustainable (Omran and Baek, 2022). The agricultural sector contributes a significant portion of biowaste emissions. The reusing of agricultural waste in the preparation of NMs will definitely help in attaining a circular economy.

Several industries, such as textile, carpet, food, plastic, leather, and print, are the major users of dyes. A large amount of dyes is discharged every year from the textile industry, presenting serious consequences to the environment.

Green and Sustainable Approaches Using Wastes for the Production of Multifunctional Nanomaterials. DOI: https://doi.org/10.1016/B978-0-443-19183-1.00019-2

Dyes are difficult to degrade by biological and photodegradation due to their high stability-promoting resistance (Akpomie and Conradie, 2020a, 2020b). Phenolic compounds are also used in several industries, such as leather, print, textile, food, paper, dye production, pulp mill, and plastics. The release of these compounds may cause serious damage to human health due to their carcinogenic nature (Adebayo and Areo, 2021). Antibiotics play a major role in medical sciences due to their ability to inhibit microbial growth. As the usage of antibiotics is continuously increasing, different types of antibiotics are seldom a constituent of wastewater (Gopal et al., 2020). Among the most prevalent contaminants in wastewater are toxic heavy metals. Production industries such as metal, cement, pigment, battery, and petrochemical are contributing different types of toxic heavy metals in wastewater (Sharififard et al., 2018). Microbial contamination is the most usual and widespread pollution of water. The formation of disinfection by-products is the major limitation associated with conventional disinfection methods.

This chapter concentrates on the utilization of agricultural waste to synthesize nanosorbents, nanocatalysts, and nanodisinfectants through a green approach to apply in wastewater treatment. The chapter documented several agricultural wastes that can be used for the preparation of NMs. Types of NMs and their modifications into nanosorbents, nanocatalysts, and nanodisinfectants have also been described. The different types of contaminant removal, which are covered in this chapter are dyes, phenolic compounds, pharmaceuticals, inorganics, and pathogenic bacteria (Fig. 21.1). The current challenges and future perspectives are also discussed.

21.2 Wastewater treatment using nanomaterials synthesized using agricultural waste

The removal of several contaminants from wastewater using NMs prepared from agricultural waste is documented in Tables 21.1 and 21.2.

21.2.1 Removal of dyes and phenolic compounds

Abarna et al. (2019) prepared photocatalytically active ZnO nanoparticles using lemon peel and through sol–gel method. The lemon peel provided excellent control on the aggregation of nanoparticles (NPs) and thereby complete decolorization of crystal violet dye happened in 2.5 hours under visible light irradiation. *Punica granatum* peel extract was applied as a stabilizing and reducing agent to prepare $Ag/Fe_3O_4/RGO$ nanocomposite (Adyani and Soleimani, 2019). The degradation ability of the NPs as catalyst was tested to reduce 4-nitrophenol, methyl green, methylene blue, and methyl orange using $NaBH_4$, and complete reduction was obtained for the pollutants within 53, 84, 98, 76 seconds, respectively. Banana peel was used to prepare cellulose/SiO_2 nanocomposite (Ali, 2018). The developed biosorbent was

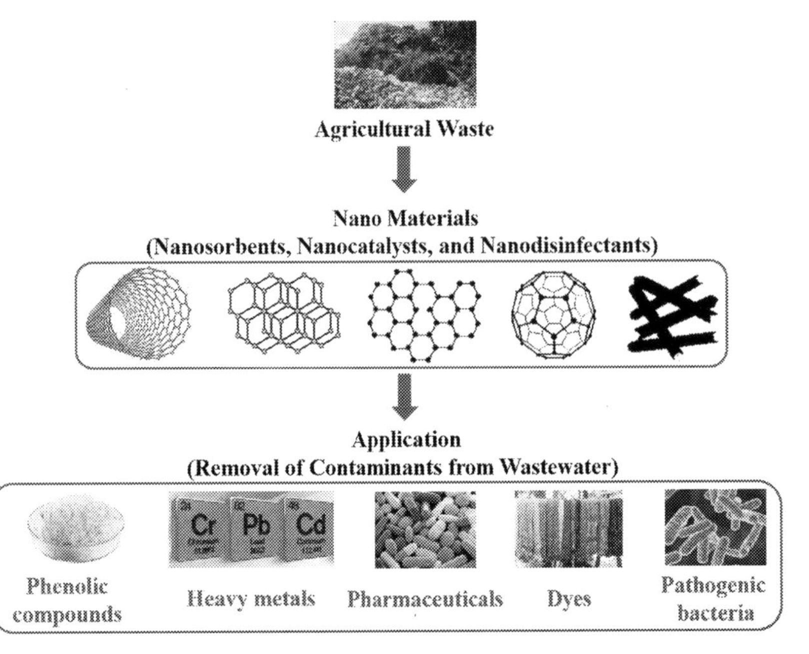

FIGURE 21.1 Application of agricultural waste for the treatment of wastewater.

TABLE 21.1 Application of nanosorbents and nanocatalysts prepared using agricultural waste.

Type of nanoparticles	Agricultural waste used	Surface area (m^2/g)	Pollutant	Removal mechanism	Removal potential	Reusability	References
Nano zinc oxide	Lemon peel	30	Crystal violet	Photocatalytic degradation	~100% (2.5 h)	5 cycles	Abarna et al. (2019)
$Ag/Fe_3O_4/RGO$ nanocomposites	*Punica granatum* peel	–	Methylene blue	Photocatalytic degradation	~100% (98 s)	5 cycles	Adyani and Soleimani (2019)
			Methyl green	Photocatalytic degradation	~100% (84 s)	5 cycles	
			Methyl orange	Photocatalytic degradation	~100% (76 s)	5 cycles	
Zn_2SnO_4	*Garcinia mangostana* fruit peel	5531	Methylene blue	Photodegradation	85% (90 min)	-	Angasa et al. (2020)
Fe_3O_4 nanoparticles	*Musa acuminata*	3.8	Bromophenol blue	H-bonding, hydrophobic, and electrostatic interactions	95.2% and 8.12 mg/g	3 cycles	Akpomie and Conradie (2020b)
Ag nanoparticles	*Terminalia bellerica* kernel	–	Eosin yellow	Catalytic reduction	71% (60 min)	–	Sherin et al. (2020)
			Methylene blue		77% (60 min)		
			Methyl orange		82.5% (60 min)		
Nitrogenous silane nanoparticles	*Helianthus annuus* husk	15.26	Methyl blue, Remazol, Acid red 88, Acid green 25, Brilliant blue R	Monolayer adsorption	416.67–714.29 mg/g	–	Truskewycz et al. (2020)
$ZnAl_2O_4@ZnO$ nanocomposite	Tragacanth gum	44.6	Acid orange 7 dye	Photocatalytic degradation	96% (15 min)	–	Azar et al. (2020)
Silver nanoparticles	Hyperpigmented tomato (*Solanum lycopersicum* L.) skins		Methylene blue	Photocatalytic degradation	~100% (90 min)	–	Carbone et al. (2020)

(Continued)

TABLE 21.1 (Continued)

Type of nanoparticles	Agricultural waste used	Surface area (m^2/g)	Pollutant	Removal mechanism	Removal potential	Reusability	References
Ag nanoparticles	*Derris trifoliata*	–	2-Nitrophenols	Catalytic reduction	~100% (9 min)	–	Cyril et al. (2020)
			4-Nitrophenols		~100% (6 min)		
			Azo violet		~100% (8 min)		
Au nanoparticles	*Derris trifoliata*	–	2-Nitrophenols	Catalytic reduction	~100% (9 min)	–	
			4-Nitrophenols		~100% (8 min)		
			Azo violet		~100% (32 min)		
Nitrogen doped carbon dots	*Azadirachta indica* seeds	–	Safranin-O	Langmuir- Hinshelwood mechanism	95% (6 min)	–	Dhanush and Sethuraman (2020)
Silica nanoparticle	*Bambusa vulgaris*	60.40	Congo red	Multilayer adsorption onto heterogeneous active sites	172 mg/g	–	Durairaj et al. (2019)
N-doped carbon supported silver nanoparticles	*Prunus mume*	–	Methylene blue	Catalytic reductive degradation	~100%, (15 min)	10 cycles	Edison et al. (2020)
			Methyl orange		~100%, (30 min)		
Ceria nanoparticles	*Cydonia oblonga miller*	–	Rhodamine B	Photocatalytic degradation	92% (180 min)	–	Elahi et al. (2020)
Silver nanoparticles	*Saussurea costus* root	–	Safranin dye	Catalytic degradation	~85% (4320 min)	–	Abd El-Aziz et al. (2021)
Silver nanoparticle	Banana waste peduncles	–	Methylene blue	Photodegradation	72.4% (12 min)	8 cycles	El-Desouky et al. (2021)
$NiFe_2O_4$ nanoparticles	Pine needles	634.51	Remazol black 5	Photocatalytic oxidation	98%	3 cycles	Gupta et al. (2020)
Iron nanoparticles	*Artocarpus heterophyllus* (Jackfruit) peel	–	Basic fuchsin	Oxidative degradation	~88% (20 min)	–	Jain et al. (2021)

ZnO nano-flowers	Panax plants	–	Methylene blue	Photocatalyst degradation	>99% (80 min)	5 cycle	Kaliraj et al. (2019)
			Eosin Y		>99% (90 min)		
			Malachite green		>99% (110 min)		
Manganese nanoparticles	*Cinnamomum verum* bark	–	Congo red	Photocatalytic degradation	78.5% (60 min)	–	Kamran et al. (2019)
ZnO nanoparticles	Rambutan (*Nephelium lappaceum* L.) peel	69.01	Methyl orange	Photodegradation	84% (120 min)	–	Karnan and Selvakumar (2016)
ZnO nanoparticles	Loquat (*Eriobotrya japonica*) seed	–	Methylene blue	Photocatalytic degradation	82.1%	–	Shabaani et al. (2020)
ZnO nanoparticles	*Passiflora foetida* fruit peels	0.83	Methylene blue	Photocatalytic degradation	93.25% (70 min)	5 cycles	Khan et al. (2021)
			Rhodamine B		91.06% (70 min)		
Iron oxide nanoparticles	*Citrus paradisi* peel	–	Methyl rose	Adsorption	97% (6 h)	–	Kumar et al. (2020)
			Methylene blue		81% (6 h)		
			Methyl orange		90% (6 h)		
Tin dioxide nanoparticles	*Citrus paradisi*	7.62	Methyl orange	Photocatalytic degradation	~100% (20 min)	–	Luque et al. (2021)
			Methylene blue		~100% (1 h)		
			Rhodamine B		~100% (1 h)		
$MgFeCrO_4$ spinel nanoparticles	Tragacanth gum	34.4	Direct black 122	Photodegradation	96% (60 s)	4 cycle	Moradnia et al. (2020)
TiO_2 nanoparticles	Lemon peel	–	Rhodamine B	Photocatalytic degradation	70%	–	Nabi et al. (2022)
Zinc oxide nanoparticles	*Abelmoschus esculentus* mucilage	–	Methylene blue	Photocatalytic degradation	~100% (60 min)	–	Prasad et al. (2019)
			Rhodamine B,		~100% (50 min)		
Ferric vanadate nanoparticles	Pomegranate, apple, grapes, orange and banana peel waste	–	Methylene blue	Photocatalytic degradation	99% (80 min)	–	Priya et al. (2019)
Zinc oxide nanoparticles	*Phoenix dactylifera*	31.32	Methylene blue	Photocatalytic degradation	90% (3 h)	–	Rambabu et al. (2021)
			Eosin yellow		90% (3 h)		
Nickel oxide nanoparticles	*Calotropis gigantea* latex	11.25	Congo red	Photocatalytic degradation	95% (35 min)	10 cycle	Roopan et al. (2019)
			4-Nitro phenol		94% (50 min)		
NiFe nanoparticles	Pomegranate (*P. granatum*) peel	1.53	Tetracycline	Adsorption and degradation	93% (90 min)	–	Ravikumar et al. (2020)
Au/Fe_3O_4 nanoparticles	*Piper auritum*	–	Methyl orange	Sonocatalytic degradation	91% (15 min)	–	Ruíz-Baltazar and Jesús (2021)

(Continued)

TABLE 21.1 (Continued)

Type of nanoparticles	Agricultural waste used	Surface area (m^2/g)	Pollutant	Removal mechanism	Removal potential	Reusability	References
nZVI-Cu nanoparticles	Pomegranate peel	60	Tetracycline	Dehydration, functional groups loss, and epimerization	95%	3 cycles	Gopal et al. (2020)
$Ag/Fe_3O_4/RGO$ nanocomposites	*P. granatum* peel	–	4-Nitrophenol	Photocatalytic degradation	~100% (53 s)	5 cycles	Adyani and Soleimani (2019)
$NiFe_2O_4$ nanoparticles	Pine needles	634.51	4-Nitrophenols	Catalytic reduction	>95% (5 min)	3 cycles	Gupta et al. (2020)
Multiwalled carbon nanotubes	Pecan (*Carya illinoinensis*, leaves)	94.2	Glyphosate	Adsorption	21.17 mg/g and 84.08%	–	Diel et al. (2021)
MSW nanoparticles	*Moringa oleifera* seed waste	4.6	Chlorpyrifos	Hydrogen bonding, hydrophobicity, electrostatic interactions, and $\pi-\pi$ interactions	25 mg/g and 81% (30 min)	5 cycles	Hamadeen et al. (2021)
Silica nanoparticle	*B. vulgaris*	60.40	Cadmium	Multilayer adsorption onto heterogeneous active sites	133 mg/g	–	Durairaj et al. (2019)
Silver nanoparticle	Banana waste peduncles	–	Cr(VI)	Photocatalytic reduction	97% (12 min)	–	El-Desouky et al. (2021)
Titanium dioxide nanoparticles	Lemon peel biomass	–	Nickel	Adsorption	~15 mg/g and 90%	–	Herrera-Barros et al. (2020)
Nano-silica	*Saccharum officinarum* leaves	75.44	Pb(II)	Adsorption	148 mg/g	–	Kaliannan et al. (2019)
			Zn(II)		137 mg/g		
Nano-scale sugar beet	Sugar beet processing	33.38	Cd	Monolayer adsorption on homogeneous surface	99% and 33.3 mg/g	–	Lashen et al. (2022)
			Cu		91% and 1111 mg/g		
Iron nanoparticles	Korla fragrant pear (*Pyrus sinkiangensis Yu*) peel	–	Cr(VI)	Monolayer adsorption on homogeneous surface	99.1% (120 min) and 46.62 mg/g	–	Rong et al. (2020)

TABLE 21.2 Application of nanodisinfectants prepared using agricultural waste.

Type of nanoparticles	Agricultural waste used	Size of NPs	Tested Pathogens	Method	Antibacterial activity	References
Zinc oxide nanoparticles	Loquat (*Eriobotrya japonica*) seed	18–27 nm	*S. typhimurium, B. cereus, S. aureus*, and *E. faecalis*	Agar disk diffusion assay	Successful only for gram-positive bacteria	Shabaani et al. (2020)
Zinc oxide nanoparticles	*Phoenix dactylifera*	30 nm	*Staphylococcus aureus, Pseudomonas aeruginosa, Streptococcus pyogenes*, and *Proteus mirabilis*	In vitro disk diffusion technique	Strong antibacterial property for strain of gram-positive and negative bacteria	Rambabu et al. (2021)
Copper oxide nanoparticles	*Zea mays* L. Dry husk	36–73 nm	*Escherichia coli, Staphylococcus aureus, P. aeruginosa*, and *Bacillus licheniformis*	Well diffusion method	More effective in inhibiting the growth of *E. coli* and *S. aureus*	Nwanya et al. (2019)
Silver nanoparticles and gold nanoparticles	*Passiflora edulis* peel	7–25 nm	*E. coli, Bacillus subtilis*, and *Staphylococcus, aureus*	Disk diffusion method	Strong antibacterial activity against all three strains	Nguyen et al. (2021)
Manganese nanoparticles	*Cinnamomum verum* bark	<100 nm	*S. aureus* and *E. coli*	Disk diffusion method	Antimicrobial activity order: *E. coli* > *S. aureus*	Kamran et al. (2019)
Silver nanoparticle	Banana waste peduncles	~14.1 nm	*Candida albicans, S. aureus, Aspergillus niger*, and *E. coli*	Disk diffusion method	Enhanced antimicrobial activity	El-Desouky et al. (2021)
Titanium dioxide nanoparticles	Pomegranate peel extract	~92 nm	*P. aeruginosa, E. coli*, and *S. aureus*	Well diffusion method, microbial inhibition concentration, minimum bactericidal concentration and live/dead cell assay	Highest inhibition property against *S. aureus* than *E. coli* and *P. aeruginosa*	Abu-Dalo et al. (2019)

able to remove 99.5% of methylene blue dye in 2 hours, and adsorption capacity was 78.75 mg/g. The impregnated magnetite nanoparticle onto *Musa acuminata* peel was used for the adsorption of bromophenol blue dye by Akpomie and Conradie (2020a). The uptake capacity of the prepared composite was 8.12 mg/g. In a study (Angasa et al., 2020), Zn_2SnO_4 nanoparticles were synthesized using *Garcinia mangostana* fruit peel extract. The Zn_2SnO_4 prepared without extract exhibited 78.93% ± 2.87% degradation efficiency in 90 minutes in comparison with 85.94% ± 2.24% when green synthesis was adopted. $ZnAl_2O_4$@ZnO nanocomposite was prepared through sol–gel method using tragacanth gel (Azar et al., 2020). The developed nanocomposite exhibited excellent photocatalytic property to degrade acid orange 7 under visible light (96% removal in 15 minutes when initial dye concentration was 40 mg/L and catalyst dose was 0.02 g, 75% reduction in total organic carbon). The SnO_2 quantum dots were prepared using extract of *Parkia speciosa Hassk* pods and employed for the photocatalytic removal of acid yellow 23 dye using UV_{254} light (Begum and

Ahmaruzzaman, 2018). The degradation efficiency of 98% in 24 minutes was obtained under UV_{254} light when initial dye concentration was 5 mg/L and catalyst dose was 20 mg. Carbone et al. (2020) synthesized Ag NPs using the extract of hyperpigmented tomato (*Solanum lycopersicum* L.) skins, and the prepared catalyst was able to significantly decrease the concentration of methylene blue within 90 minutes. Ag NPs and Au NPs were prepared utilizing extract of *Derris trifoliata* seed (Cyril et al., 2020). The biosynthesized Ag nanoparticles have demonstrated better catalytic activity in comparison with Au NPs in degrading 4-nitrophenol (6 minutes for Ag and 8 minutes for Au) and azo violet (8 minutes for Ag and 32 minutes for Au) by $NaBH_4$. However, equal removal time was required for 2-nitrophenol (9 minutes for both catalysts). CuO nanoparticles were prepared from flowers and seed extract of *Madhuca longifolia* (Das et al., 2018). Photodegradation of methylene blue was able to remove 77% in 150 minutes. The preparation of highly fluorescent phyto-derived nitrogen-doped carbon dots (PDNCDs) was accomplished from seeds of neem (Dhanush and Sethuraman, 2020). The PDNCDs have been employed as catalyst when reduction of Safranin-O dye was tried using $NaBH_4$ as reducing agent. 95% reduction efficiency was obtained within a short period of 6 minutes. Silica nanoparticles were synthesized using *Bambusa vulgaris*. The developed material exhibited multilayer adsorption of Congo red (adsorption capacity of 172 mg/g) onto heterogeneous active sites (Durairaj et al., 2019). Unripe *Prunus mume* fruit extract has been used as reducing agent and carbon inception for the development of N-doped carbon supported Ag NPs (Edison et al., 2020). The prepared NPs were used for the catalytic reductive degradation of methyl orange and methylene blue using $NaBH_4$, and obtained degradation efficiencies were 99.96% and 99.9% after 30 minutes and 15 minutes, respectively. Ceria nanoparticles were prepared using seed extract of *Cydonia oblonga miller* as capping and stabilizing material (Elahi et al., 2020). The prepared material exhibited 92% of rhodamine B degradation in 180 minutes. Ag NP synthesized by *Saussurea costus* root aqueous extract was also proven efficient for degrading 84.6% of safranin dye at 72 hours (Abd El-Aziz et al., 2021). Instead of plant extract, Ag NPs were synthesized using banana waste product by El-Desouky et al. (2021), and the reduction efficiency of methylene blue was 72.4% in 12 minutes. Magnetically retrievable nanocomposite, $NiFe_2O_4$ nanoparticles were prepared utilizing pine needles (Gupta et al., 2020). The developed catalyst was utilized for photocatalytic oxidation of Remazol Black 5 and 4-nitrophenols. and 98% and >95% of degradation efficiencies were achieved. Ismail et al. (2018) synthesized Ag NPs dispersed on taro (*Colocasia esculenta*) plant using the rhizome as supporting agent. The prepared catalyst was applied to reduce nitroarenes such as picric acid (90% in 12 minutes), 4-nitroaniline (96.3% in 13 minutes) and organic azo dyes such as Congo red (96.9% in 12 minutes), Methyl red (96.29% in 14 minutes), and Rhodamine B (97.78% in 6 minutes) using $NaBH_4$. The catalyst was also efficient in complete reduction of dye mixture (Methyl orange + Methyl red in 9 minutes and Methyl orange + Methyl red + Congo red in 10 minutes) and mixture of nitroarenes (4-nitrophenol + 2-nitrophenol in 11 minutes, 4-nitroaniline + 4-nitrophenol + 2-nitrophenol in 9 minutes). Iron nanoparticles were also synthesized using *Artocarpus heterophyllus*, commonly known as Jackfruit peel extract as capping and stabilizing agents (Jain et al., 2021). The ability of the prepared catalyst was reported as Fenton-like, which has been justified by the removal of Fuchsin Basic dye (87.5% in 20 minutes). ZnO resembling Quaker ladies flower was developed utilizing panos extract (Kaliraj et al., 2019). The photocatalyst demonstrated >99% removal efficiency within 80 minutes, 90 minutes, and 110 minutes when the initial concentration of methylene blue, eosin Y, and methyl green was 15 mg/L. The catalyst was also able to degrade low concentration (5 mg/L) of the dyes within 30 minutes–40 minutes. Manganese NPs were prepared using bark extracts of *Cinnamomum verum* (Kamran et al., 2019). The photocatalytic activity was proven by 78.5% of Congo red dye degradation in 60 minutes when 0.06 g/L of Mn NPs dose was used. As a new and low-cost method of developing ZnO NPs from peel extract of rambutan (*Nephelium lappaceum* L.), Karnan and Selvakumar (2016) prepared ZnO NPs for photodegradation of methyl orange dye. The decolorization efficiency of 83.99% was obtained within 120 minutes of illumination. Kaur et al. (2018) prepared three different types of nanoparticles [iron-silver (FeO/Ag NPs), iron gold (FeO/Au NPs), and iron nanoparticles (FeONPs)] utilizing pomegranate fruit peel extract. 90% of the aniline dye was degraded after 120 minutes by FeO/Au NPs, whereas FeONPs and FeO/Ag NPs exhibited 90% degradation achieved in 150 minutes. Sonochemical activation can also be adopted for the synthesis of NPs. Ag NPs were prepared using phytoconstituents of peel of *Longan* (*Euphorbia longana* Lam.) fruit by reducing Ag^+ (Khan et al., 2018), which was able to degrade 99% of methylene blue photocatalytically within 7 minutes of reaction time. In another study, the reduction of metal salts to prepare ZnO nanoparticles was done using *Passiflora foetida* fruit peels extract (Khan et al., 2021), and the NPs depicted good photocatalytic property for methylene blue degradation (93.25% in 70 minutes) and rhodamine B (91.06% in 70 minutes). Khodadadi et al. (2017a) used *Achillea millefolium* L. extract to reduce Ag^+ ions for Ag NPs synthesis. The reduction process generates highly dispersed Ag NPs on the surface of peach kernel. The catalytic property of the prepared NPs was studied for the removal of 4-nitrophenol, methylene blue, and methyl orange. Removal of methyl orange, methylene blue, and 4-nitrophenol using Ag NPs/peach kernel shell and $NaBH_4$ is completed in 1.45 minutes, 50 seconds, and 48 seconds, respectively.

The high efficiency was obtained due to lower agglomeration of Ag NPs and higher surface area of the peach kernel shell. The research group (Khodadadi et al., 2017b) also investigated synthesis of Pd NPs at apricot kernel shell substrate utilizing extract of *Salvia hydrangea* and its application in the removal of 4-nitrophenol, Congo red, methylene blue, methyl orange, and rhodamine B at room temperature. The application of $NaBH_4$ and 3 mg catalyst reduced methyl orange and methylene blue at no time, whereas 4-nitrophenol, rhodamine B, and Congo red took 36, 16, and 15 seconds, respectively. Iron oxide nanoparticles were prepared using extracts of *Citrus paradisi* peel by Kumar et al. (2020). The synthesized NPs were used for the removal of synthetic dyes due to its adsorptive capacity. The decolorization efficiencies were obtained as 96.65% for 50 mg/L of Methyl rose, 80.76% for 10 mg/L of methylene blue, and 89.64% for 10 mg/L of methyl orange in 6 hours. *C. paradisi* extract has also been used for the synthesis of tin dioxide nanoparticles (Luque et al., 2021). The NPs were used for the degradation of methylene blue, methyl orange, rhodamine B under solar and UV radiation. The NPs (1 g/L) were able to degrade 100% of methyl orange in 20 minutes, whereas methylene blue and rhodamine B took a little longer, that is, 60 minutes (when UV radiation was used). The $MgFeCrO_4$ NP developed by Moradnia et al. (2020) by the green (Tragacanth gum) sol–gel procedure was photocatalytically active and was able to reduce 96% of DB122 dye within 60 seconds. Nabi et al. (2022) used lemon peel as capping and reducing material in the preparation of TiO_2 nanoparticles. Photocatalytic degradation of the rhodamine B dye has been attempted using the modified NPs, and more than 70% degradation efficiency was obtained, which is better than that of regular TiO_2 NPs. Fe_3O_4@polythiophen-Ag NP was prepared using grapefruit peel extract for the reduction process (Najafineja et al., 2018). The developed catalyst showed excellent catalytic activity by degrading methyl orange and methylene blue in 60 seconds. Ag NPs and Au NPs were prepared utilizing extract of waste *Passiflora edulis* peel for reduction and stabilization (Nguyen et al., 2021). The biosynthesized Ag nanoparticles have demonstrated excellent catalytic activity in comparison with Au NPs in degrading 2- and 3-nitrophenol ($>$20 minutes for both the catalysts), 4-nitrophenol (8 minutes for Ag and 12 minutes for Au), rhodamine B (6 minutes for Ag and 12 minutes for Au), methyl orange (8 minutes for Ag and 12 minutes for Au), rhodamine 6 G (12 minutes for Ag and 18 minutes for Au), eosin Y (12 minutes for Ag and 18 minutes for Au) by $NaBH_4$. Nakkala et al. (2018) synthesized Ag NPs using aqueous rhizome extract of *Acorus calamus* and applied it for the reduction of dyes and constituent compounds of dye. 4-nitrophenol, 3-nitrophenol, picric acid, Congo red, coomassie brilliant blue, rhodamine B, eosin Y, eriochrome black T, methylene blue, methyl orange, phenol red cresol red, acridine orange, and Methyl red were reduced using the developed silver NPs and $NaBH_4$, and the reduction time was 8, 9, 16, 10, 5, 3, 8, 12, 15, 3, 6, 60, 16, and 18 minutes, respectively. Prasad et al. (2019) used *Abelmoschus esculentus* (okra) mucilage for green synthesis of ZnO NPs. The photocatalyst was tested using several dyes such as methylene blue and rhodamine B. 100% degradation of methylene blue solution within 60 minutes was achieved when 125 mg of the catalyst was used, and 100 mg of the same was necessary for 100% removal of rhodamine B within 50 minutes. Ferric vanadate nanoparticles was prepared by Priya et al. (2019) using peel waste of pomegranate, apple, grapes, orange, and banana. The photocatalyst degraded 99% of methylene blue in 80 minutes. In another study (Radini et al., 2018), *Trigonella foenum-graecum* seed extract was used to prepare Fe NPs, which were able to degrade methyl orange (95% in 90 minutes) photocatalytically. ZnO nanoparticles has also been prepared using pulp waste as an effective bioreductant (Rambabu et al., 2021). Photocatalytic degradation of eosin yellow and methylene blue dyes was attempted, and 90% degradation efficiency was obtained for both the pollutants in 3 hours. Roopan et al. (2019) utilized *Calotropis gigantea* latex as capping agent to prepare nickel oxide nanoparticles through green synthesis method. The developed NiO NPs degraded 95% of diazo dye Congo red in 35 minutes and 94% of 4-nitrophenol in 50 minutes under solar light irradiation. Rovani et al. (2018) prepared silica nanoparticles utilizing sugarcane waste ash. The multilayer adsorption sites showed adsorption capacity of 230 mg/g for acid orange 8 dye. Ruíz-Baltazar and Jesús (2021) used *Piper auritum* as reducing agent and ultrasonication helped in the creation of plasmonic nanoparticles (Au/Fe_3O_4). Methyl orange was sonocatalytically degraded, and 91.2% removal was achieved at 15 minutes. Rupa et al. (2018) used the extract of *Rubus coreanus* immature fruit as a reducing and capping material in the preparation of Rc-ZnO NPs. The developed NPs was used to degrade malachite green dye in dark and light (90% at 4 hours) conditions. Shabaani et al. (2020) investigated nanoparticle synthesis using seed extract of loquat (*Eriobotrya japonica*). The developed ZnO NPs showed highest degradation of methylene blue when NP concentration was 12 mg/mL under UV light irradiation. In another study (Sherin et al., 2020), Ag NPs were prepared using novel *Terminalia bellerica* kernel extract. The catalyst showed excellent reduction ability of 4-nitrophenol (87% in 60 minutes), methylene blue (76.9% in 60 minutes), eosin yellow (71% in 60 minutes), and methyl orange (82.5% in 60 minutes) by $NaBH_4$. Truskewycz et al. (2020) prepared nitrogenous silane nanoparticles as an efficient adsorbent to remove mixed dye solutions. Utilization of *Helianthus annuus* husk extract in the NPs preparation helped in increasing surface area and pore volume, which resulted in an increase of adsorption efficiency from 416.67 to 714.29 mg/L when treatment of mixture of acid red 88, remazol brilliant blue R, methyl blue, and acid Green 25 was attempted.

21.2.2 Removal of antibiotics and pesticides

NiFe nanoparticles were synthesized using pomegranate (*P. granatum*) peel. The developed nanoadsorbent was used to remove tetracycline antibiotic from aqueous medium, and 93% of removal efficiency was obtained in 90 minutes (Ravikumar et al., 2020). Gopal et al. (2020) attempted preparation of bentonite-supported bimetallic nZVI-Cu NPs using pomegranate rind. The NMs were able to remove 95% of tetracycline antibiotic. Multiwalled carbon nanotubes were synthesized using extract of pecan (*Carya illinoinensis*) leaves (Diel et al., 2021). The developed biosorbent was able to remove 84.08% of glyphosate, an herbicide, and the adsorption capacity was 21.17 mg/g. *Moringa oleifera* seed nanoparticles were developed using *M. oleifera* seed waste extract by Hamadeen et al. (2021). The developed material was able to remove Chlorpyrifos, an organophosphate insecticide. The obtained efficiency of the NMs was 81% in 30 minutes, and the adsorption capacity was 25 mg/g.

21.2.3 Removal of inorganics

Silica nanoparticles were prepared from leaves ash of *B. vulgaris* through sol–gel technique, and the material was used for the removal of Cd (Durairaj et al., 2019). Multilayer adsorption onto heterogeneous active sites of the biosorbent resulted in an adsorption capacity of 133 mg/g. Silver nanoparticles were prepared from banana waste peduncles and used for photocatalytic reduction of Cr(VI) (El-Desouky et al., 2021). The obtained removal efficiency was 97% in 12 minutes. Titanium dioxide nanoparticles were prepared using lemon peel as biowaste source (Herrera-Barros et al., 2020). The developed NMs were utilized for Nickel adsorption. The obtained adsorption capacity was 15.36 mg/g. The modification of the NMs helped in increasing Ni removal efficiency from 78% to 90%. Jabasingh et al. (2018) synthesized iron oxide NPs by extract of bagasse from sugarcane. Adsorption percentage of 96.12% in 180 minutes indicated that the developed NMs could potentially be applied for Cr(VI) adsorption. Agricultural waste *Saccharum officinarum* leaves were utilized for nano-silica synthesis by Kaliannan et al. (2019). The adsorbent was applied for the removal of Pb^{2+} and Zn^{2+}, and the adsorption rates were obtained as 148 and 137 mg/g, respectively. Lashen et al. (2022) prepared two novel NMs using sugar beet processing residuals. The applicability of the developed biosorbent was investigated for Cd and Cu sorption in water. The obtained removal efficiencies for Cd and Cu were 99% and 91%, respectively. The NMs acted on monolayer adsorption and obtained adsorption rate of 33.3 and 1111 mg/g for Cd and Cu. Peel extract of Korla fragrant pear (*Pyrus sinkiangensis* Yu) was utilized for the preparation of iron NPs and used to remove Cr^{6+} (Rong et al., 2020). The monolayer adsorption on homogeneous surface resulted in removal efficiency of 99.1% in 120 minutes and adsorption capacity of 46.62 mg/g. Sebastian et al. (2018) used coconut husk extract to prepare iron oxide NPs. The adsorption capacity of the prepared NMs was 9.6 mg/g for Cd.

21.2.4 Removal of pathogenic bacteria

Green synthesis of titanium dioxide NPs was accomplished by Abu-Dalo et al. (2019) utilizing peel of pomegranate. The pomegranate-modified TiO_2 showed antibacterial properties against Gram-negative and positive bacteria. Maximum inhibition was obtained against *S. aureus* than *P. aeruginosa* and *E. coli*. Plant extract of *M. longifolia* was used as reducing material to develop CuO nanoparticles (Das et al., 2018). The nanodisinfectant showed good antibacterial activity against *S. aureus*, B. *subtilis*, and *E. coli*. Banana waste peduncles were applied to prepare silver NPs by El-Desouky et al. (2021). The synergistic effect of Ag NPs promoted good antimicrobial properties of Ag NPs/Degussa nanocomposite for *S. aureus*, *Aspergillus niger*, *Candida albicans*, and *E. coli*. Kamran et al. (2019) synthesized Mn nanoparticles utilizing bark extracts of *C. verum*. The potential of the NPs was proved by the good microbial property against *S. aureus* and *E. coli*. Kaur et al. (2018) investigated three nanoparticles, that is, iron-silver NPs (FeO/Ag NPs), iron NPs (FeONPs), and iron-gold NPs (FeO/Au NPs). Peel extract of pomegranate was utilized for the preparation of all three types of nanoparticles. The developed NPs showed good antibacterial and antifungal properties against *A. alternate*, *P. aeruginosa*, *R. solani*, *M. luteus*, *A. flavus*, and *S. enterica*. However, among all the three NPs, FeO/Ag NPs exhibited higher antimicrobial properties. Nguyen et al. (2021) used *P. edulis* peel to prepare silver and gold nanoparticles. The NPs were tested for antimicrobial properties, and it shows strong activity for several strains, that is, *S. aureus*, *Bacillus subtilis*, and *Escherichia coli*. Dry husk extract of *Zea mays* L. was applied to prepare CuO NPs (Nwanya et al., 2019). The antimicrobial activity was tested for four strains, that is, *E. coli*, *P. aeruginosa*, *B. licheniformis*, and *S. aureus*. However, the NPs inhibited the growth of *E. coli* and *S. aureus* the most. Radini et al. (2018) attempted preparation of zero-valent iron nanoparticles and stabilization of it using *T. foenum-graecum*. The developed NPs were checked for antibacterial activities, and they showed proficient activity for *S. aureus* and *E. coli*. Rambabu et al. (2021) attempted preparation of ZnO NPs using date pulp waste, which acted as bioreductant. The modified ZnO NPs exhibited good antimicrobial properties against *S. aureus*, *P. mirabilis*, *P. aeruginosa*, and *S. pyogenes*. Zinc oxide nanoparticles were synthesized using Loquat seed (*E. japonica*) extract (Shabaani et al., 2020). The NPs demonstrated antibacterial activities but only forGgram-positive bacteria (*Listeria monocytogenes*, *Bacillus cereus*, *Enterococcus faecalis*, *Salmonella typhimurium*, and *S. aureus*). Fe_3O_4-Ag NPs prepared using stem of grape (*Vitis vinifera)* showed potential antibacterial properties for *E. coli, S. aureus, B. subtilis*, and *P. aeruginosa* (Venkateswarlu et al., 2015). Zaheer (2018) synthesized Ag NPs using palm date fruit pericarp extract. *S. aureus*, *C. albicans*, and *E. coli* were tested for inhibition, and the results indicated that the developed Ag NPs have good potential for antimicrobial activities.

21.3 Mechanism and functions of nanocatalysts, nanoadsorbent, and nanodisinfectant

Nanocatalysts, which are used to degrade organic pollutants by photocatalytic degradation, decompose the organic materials into several products, and conversion into CO_2, H_2O, and inorganic ions happens during the mineralization process. The degradation and mineralization occur through the generation of active species with high oxidation potential. The nanoadsorbent used for the removal of inorganics acts on the separation process by electrostatic/chemical interaction between the target metal ions and active sites (Anirudhan and Suchithra, 2010; Ehrampoush et al., 2015). The occurrence of multilayer adsorption has also been reported (Ehrampoush et al., 2015). With the replacement of OH^-, ligands bound to the solid surface, which eventually increases cation binding by reducing repulsion between adsorbent and adsorbate (Jain et al., 2015). The nanodisinfectants show high antimicrobial potential by destroying the cell membrane. The destruction process may be by direct contact or the release of toxic metal ions.

21.4 Summary, present status, conclusion, and future outlook

Nanomaterials have gained widespread application in water and wastewater treatment in recent times. In this regard, synthesis of NMs from biodegradable, low or zero-cost, and readily accessible agricultural waste and their application for wastewater treatment are highlighted in this chapter. The chapter demonstrates nanoparticles developed through green synthesis and their utilization for the removal of several organics (dyes, phenols, pesticides, antibiotics etc.), inorganics (Cd, Cr, Ni, etc.), and pathogenic microorganisms from aqueous medium. The nanomaterials discussed in this chapter broadly are of three types, nanosorbent, nanocatalyst, and nanodisinfectant.

The biomodified NMs showed satisfactory treatment efficiency and sometimes even better than the conventional ones. Furthermore, the recyclability of the NPs for wastewater treatment is promising. The waste extract materials are employed as reducing, capping, and stabilizing materials in the synthesis of NMs. The chapter emphasizes on the useful resources of agricultural waste. Several NMs prepared using agricultural waste showed significant antibacterial activity, which makes them excellent nanodisinfectants.

However, the applicability of different types of NMs for wastewater treatment is still limited. Most of the studies are focusing on specific pollutants in synthetic solutions. However, investigations on diverse types of pollutants and studies on real wastewater incorporating synergistic effects of several ions and other compounds commonly present in wastewater are the need of the hour for large-scale implementation. In addition, the cytotoxic properties of the biomodified nanosorbent, nanocatalyst, and nanodisinfectant need to be understood. The cost-effectiveness studies addressing the differences in green-synthesized and conventionally synthesized nanomaterials will encourage the adaptability of biomodified nanomaterials.

References

Abarna, B., Preethi, T., Rajarajeswari, G.R., 2019. Lemon peel guided sol-gel synthesis of visible light active nano zinc oxide. Journal of Environmental Chemical Engineering 7, 102742.

Abd El-Aziz, A.R.M., Gurusamy, A., Alothman, M.R., Shehata, S.M., Hisham, S.M., Alobathani, A.A., 2021. Silver nanoparticles biosynthesis using Saussurea costus root aqueous extract and catalytic degradation efficacy of safranin dye. Saudi Journal of Biological Sciences 28, 1093–1099.

Abu-Dalo, M., Jaradat, A., Albiss, B.A., Al-Rawashdeh, N.A.F., 2019. Green synthesis of TiO_2 NPs/pristine pomegranate peel extract nanocomposite and its antimicrobial activity for water disinfection. Journal of Environmental Chemical Engineering 7, 103370.

Adebayo, M.A., Areo, F.I., 2021. Removal of phenol and 4-nitrophenol from wastewater using a composite prepared from clay and *Cocos nucifera* shell: kinetic, equilibrium and thermodynamic studies. Resources, Environment and Sustainability 3, 100020.

Adyani, S.H., Soleimani, E., 2019. Green synthesis of Ag/Fe_3O_4/RGO nanocomposites by Punica granatum peel extract: catalytic activity for reduction of organic pollutants. International Journal of Hydrogen Energy 44, 2711–2730.

Akpomie, K.G., Conradie, J., 2020a. Biogenic and chemically synthesized *Solanum tuberosum* peel–silver nanoparticle hybrid for the ultrasonic aided adsorption of bromophenol blue dye. Scientific Reports 10, 17094.

Akpomie, K.G., Conradie, J., 2020b. Efficient synthesis of magnetic nanoparticle-*Musa acuminata* peel composite for the adsorption of anionic dye. Arabian Journal of Chemistry 13, 7115–7131.

Ali, S.M., 2018. Fabrication of a nanocomposite from an agricultural waste and its application as a biosorbent for organic pollutants. International Journal of Environmental Science and Technology 15, 1169–1178.

Angasa, E., Eka, Y., Jamarun, N., Arief, S., 2020. Improving the morphological, optical, and photocatalytic properties of octahedral Zn_2SnO_4 using *Garcinia mangostana* fruit peel extract. Vacuum 182, 109719.

Anirudhan, T.S., Suchithra, P.S., 2010. Equilibrium, kinetic and thermodynamic modeling for the adsorption of heavy metals onto chemically modified hydrotalcite. Indian Journal of Chemical Technology 17, 247–259.

Azar, B.E., Ramazani, A., Fardood, S.T., Morsali, A., 2020. Green synthesis and characterization of $ZnAl_2O_4$@ZnO nanocomposite and its environmental applications in rapid dye degradation. Optik (Stuttg) 208, 164129.

Begum, S., Ahmaruzzaman, M., 2018. Green synthesis of SnO_2 quantum dots using *Parkia speciosa* Hassk pods extract for the evaluation of anti-oxidant and photocatalytic properties. Journal of Photochemistry and Photobiology. B, Biology 184, 44–53.

Bhardwaj, A.K., Kumar, V., Pandey, V., Naraian, R., Gopal, R., 2019. Bacterial killing efficacy of synthesized rod shaped cuprous oxide nanoparticles using laser ablation technique. SN Applied Sciences 1, 1426.

Carbone, K., Angelis, A.D., Mazzuca, C., Santangelo, E., Macchioni, V., et al., 2020. Microwave-assisted synthesis of catalytic silver nanoparticles by hyperpigmented tomato skins: a green approach. LWT 133, 110088.

Cyril, N., George, J.B., Nair, P.V., Joseph, L., Sunila, C.T., Smitha, V.K., et al., 2020. Catalytic activity of Derris trifoliata stabilized gold and silver nanoparticles in the reduction of isomers of nitrophenol and azo violet. Nano-Structures and Nano-Objects 22, 100430.

Das, P., Ghosh, S., Ghosh, R., Dam, S., Baskey, M., 2018. *Madhuca longifolia* plant mediated green synthesis of cupric oxide nanoparticles: a promising environmentally sustainable material for waste water treatment and efficient antibacterial agent. Journal of Photochemistry and Photobiology. B, Biology 189, 66–73.

Deshpande, B.D., Agrawal, P.S., Yenkie, M.K.N., Dhoble, S.J., 2020. Prospective of nanotechnology in degradation of waste water: a new challenges. Nano-Structures & Nano-Objects 22, 100442.

Dhanush, C., Sethuraman, M.G., 2020. Influence of phyto-derived nitrogen doped carbon dots from the seeds of *Azadirachta indica* on the NaBH4 reduction of Safranin-O dye. Diamond and Related Materials 108, 107984.

Diel, J.C., Franco, D.S.P., Igansi, A.V., Cadaval, T.R.S., Pereira, H.A., Nunes, I., et al., 2021. Green synthesis of carbon nanotubes impregnated with metallic nanoparticles: characterization and application in glyphosate adsorption. Chemosphere 283, 131193.

Durairaj, K., Senthilkumar, P., Velmurugan, P., Dhamodaran, K., Kadirvelu, K., Kumaran, S., 2019. Sol-gel mediated synthesis of silica nanoparticle from *Bambusa vulgaris* leaves and its environmental applications: kinetics and isotherms studies. Journal of Sol-Gel Science and Technology 90, 653–664.

Edison, T.N.J.I., Atchudan, R., Karthik, N., Balaji, J., Xiong, D., Lee, Y.R., 2020. Catalytic degradation of organic dyes using green synthesized N-doped carbon supported silver nanoparticles. Fuel 280, 118682.

Ehrampoush, M.H., Miria, M., Salmani, M.H., Mahvi, A.H., 2015. Cadmium removal from aqueous solution by green synthesis iron oxide nanoparticles with tangerine peel extract. Journal of Environmental Health Science and Engineering 13, 84.

Elahi, B., Mirzaee, M., Darroudi, M., Kazemi Oskuee, R., Sadri, K., Gholami, L., 2020. Role of oxygen vacancies on photocatalytic activities of green synthesized ceria nanoparticles in Cydonia oblonga miller seeds extract and evaluation of its cytotoxicity effects. Journal of Alloys and Compounds 816, 152553.

El-Desouky, N., Shoueir, K.R., El-Mehasseb, I., El-Kemary, M., 2021. Bio-inspired green manufacturing of plasmonic silver nanoparticles/Degussa using banana waste peduncles: photocatalytic, antimicrobial, and cytotoxicity evaluation. Journal of Materials Research and Technology 10, 671–686.

Gopal, G., Sankar, H., Natarajan, C., Mukherjee, A., 2020. Tetracycline removal using green synthesized bimetallic nZVI-Cu and bentonite supported green nZVI-Cu nanocomposite: a comparative study. Journal of Environmental Management 254, 109812.

Gupta, K., Kaushik, A., Tikoo, K.B., Kumar, V., Singhal, S., 2020. Enhanced catalytic activity of composites of $NiFe_2O_4$ and nano cellulose derived from waste biomass for the mitigation of organic pollutants. Arabian Journal of Chemistry 13, 783–798.

Hamadeen, H.M., Elkhatib, E.A., Badawy, M.E.I., Abdelgaleil, S.A.M., 2021. Green low cost nanomaterial produced from *Moringa oleifera* seed waste for enhanced removal of Chlorpyrifos from wastewater: mechanism and sorption studies. Journal of Environmental Chemical Engineering 9, 105376.

Herrera-Barros, A., Bitar-Castro, N., Villabona-Ortíz, Á., Tejada-Tovar, C., González-Delgado, Á.D., 2020. Nickel adsorption from aqueous solution using lemon peel biomass chemically modified with TiO_2 nanoparticles. Sustainable Chemistry and Pharmacy 17, 100299.

Ismail, M., Khan, M.I., Bahadar, S., Akhtar, K., Ali, M., Asiri, A.M., 2018. Catalytic reduction of picric acid, nitrophenols and organic azo dyes via green synthesized plant supported Ag nanoparticles. Journal of Molecular Liquids 268, 87–101.

Jabasingh, S.A., Belachew, H., Yimam, A., 2018. Iron oxide induced bagasse nanoparticles for the sequestration of Cr^{6+} ions from tannery effluent using a modified batch reactor. Journal of Applied Polymer Science 135, 46683.

Jain, R., Jordan, N., Schild, D., van Hullebusch, E.D., Weiss, S., Franzen, C., et al., 2015. Adsorption of zinc by biogenic elemental selenium nanoparticles. Chemical Engineering Journal 260, 855–863.

Jain, R., Mendiratta, S., Kumar, L., Srivastava, A., 2021. Green synthesis of iron nanoparticles using Artocarpus heterophyllus peel extract and their application as a heterogeneous Fenton-like catalyst for the degradation of Fuchsin Basic dye. Current Research in Green and Sustainable Chemistry 4, 100086.

Jovanov, D., Vujić, B., Vujić, G., 2018. Optimization of the monitoring of landfill gas and leachate in closed methanogenic landfills. Journal of Environmental Management 216, 32–40.

Kaliannan, D., Palaninaicker, S., Palanivel, V., Mahadeo, M.A., Ravindra, B.N., Jae-Jin, S., 2019. A novel approach to preparation of nano-adsorbent from agricultural wastes (*Saccharum officinarum* leaves) and its environmental application. Environmental Science and Pollution Research 26, 5305–5314.

Kaliraj, L., Chan, J., Jahan, E., Abid, S., Lu, J., Yang, D.C., 2019. Synthesis of panos extract mediated ZnO nano-flowers as photocatalyst for industrial dye degradation by UV illumination. Journal of Photochemistry and Photobiology. B, Biology 199, 111588.

Kamran, U., Bhatti, H.N., Iqbal, M., Jamil, S., Zahid, M., 2019. Biogenic synthesis, characterization and investigation of photocatalytic and antimicrobial activity of manganese nanoparticles synthesized from *Cinnamomum verum* bark extract. Journal of Molecular Structure 1179, 532–539.

Karnan, T., Selvakumar, S.A.S., 2016. Biosynthesis of ZnO nanoparticles using rambutan (Nephelium lappaceum L.) peel extract and their photocatalytic activity on methyl orange dye. Journal of Molecular Structure 1125, 358–365.

Kaur, P., Thakur, R., Malwal, H., Manuja, A., Chaudhury, A., 2018. Biosynthesis of biocompatible and recyclable silver/iron and gold/iron core-shell nanoparticles for water purification technology. Biocatalysis and Agricultural Biotechnology 14, 189–197.

Kee, S.H., Chiongson, J.B.V., Saludes, J.P., Vigneswari, S., Ramakrishna, S., Bhubalan, K., 2021. Bioconversion of agro-industry sourced biowaste into biomaterials via microbial factories – a viable domain of circular economy. Environmental Pollution 271, 116311.

Khan, A.U., Yuan, Q., Khan, Z.U.H., Ahmad, A., Khan, F.U., Tahir, K., et al., 2018. An eco-benign synthesis of AgNPs using aqueous extract of Longan fruit peel: antiproliferative response against human breast cancer cell line MCF-7, antioxidant and photocatalytic deprivation of methylene blue. Journal of Photochemistry and Photobiology. B, Biology 183, 367–373.

Khan, M., Ware, P., Shimpi, N., 2021. Synthesis of ZnO nanoparticles using peels of Passiflora foetida and study of its activity as an efficient catalyst for the degradation of hazardous organic dye. SN Applied Sciences 3 (528), 1–17.

Khodadadi, B., Bordbar, M., Nasrollahzadeh, M., 2017a. *Achillea millefolium* L. extract mediated green synthesis of waste peach kernel shell supported silver nanoparticles: application of the nanoparticles for catalytic reduction of a variety of dyes in water. Journal of Colloid and Interface Science 493, 85–93.

Khodadadi, B., Bordbar, M., Nasrollahzadeh, M., 2017b. Green synthesis of Pd nanoparticles at Apricot kernel shell substrate using Salvia hydrangea extract: catalytic activity for reduction of organic dyes. Journal of Colloid and Interface Science 490, 1–10.

Kumar, B., Smita, K., Galeas, S., Sharma, V., Guerrero, V.H., Debut, A., et al., 2020. Characterization and application of biosynthesized iron oxide nanoparticles using Citrus paradisi peel: a sustainable approach. Inorganic Chemistry Communications 119, 108116.

Lashen, Z.M., Shams, M.S., El-Sheshtawy, H.S., Slaný, M., Antoniadis, V., Yang, X., et al., 2022. Remediation of Cd and Cu contaminated water and soil using novel nanomaterials derived from sugar beet processing- and clay brick factory-solid wastes. Journal of Hazardous Materials 428, 128205.

Lu, F., Astruc, D., 2020. Nanocatalysts and other nanomaterials for water remediation from organic pollutants. Coordination Chemistry Reviews 408, 213180.

Luque, P.A., Garrafa-Gálvez, H.E., Nava, O., Olivas, A., Martínez-Rosas, M.E., Vilchis-Nestor, A.R., et al., 2021. Efficient sunlight and UV photocatalytic degradation of methyl orange, methylene blue and rhodamine B, using citrus × paradisi synthesized SnO_2 semiconductor nanoparticles. Ceramics International 47, 23861–23874.

Moradnia, F., Fardood, S.T., Ramazani, A., Gupta, V.K., 2020. Green synthesis of recyclable $MgFeCrO_4$ spinel nanoparticles for rapid photodegradation of direct black 122 dye. Journal of Photochemistry and Photobiology A: Chemistry 392, 112433.

Nabi, G., Ain, Q., Tahir, M.B., Riaz, K.N., Rafique, M., Hussain, S., et al., 2022. Green synthesis of TiO_2 nanoparticles using lemon peel extract: their optical and photocatalytic properties. International Journal of Environmental Analytical Chemistry 102, 434–442.

Najafineja, M.S., Mohammadi, P., Afsahi, M.M., Sheibani, H., 2018. Green synthesis of the Fe_3O_4 @ polythiophen-Ag magnetic nanocatalyst using grapefruit peel extract: application of the catalyst for reduction of organic dyes in water. Journal of Molecular Liquids 262, 248–254.

Nakkala, J.R., Mata, R., Raja, K., Chandra, V.K., Sadras, S.R., 2018. Green synthesized silver nanoparticles: catalytic dye degradation, in vitro anticancer activity and in vivo toxicity in rats. Materials Science & Engineering C 91, 372–381.

Nguyen, T.M.-T., Nguyen, T.A.-T., Pham, N.T.-V., Ly, Q.V., Tran, T.T.-Q., Thach, T.D., et al., 2021. Biosynthesis of metallic nanoparticles from waste Passiflora edulis peels for their antibacterial effect and catalytic activity. Arabian Journal of Chemistry 14, 103096.

Nwanya, A.C., Razanamahandry, L.C., Bashir, A.K.H., Ikpo, C.O., Nwanya, S.C., Botha, S., et al., 2019. Industrial textile effluent treatment and antibacterial effectiveness of *Zea mays* L. Dry husk mediated bio-synthesized copper oxide nanoparticles. Journal of Hazardous Materials 375, 281–289.

Omran, B.A., Baek, K.-H., 2022. Valorization of agro-industrial biowaste to green nanomaterials for wastewater treatment: approaching green chemistry and circular economy principles. Journal of Environmental Management 311, 114806.

Prasad, A.R., Garvasis, J., Oruvil, S.K., Joseph, A., 2019. Bio-inspired green synthesis of zinc oxide nanoparticles using *Abelmoschus esculentus* mucilage and selective degradation of cationic dye pollutants. Journal of Physical and Chemistry of Solids 127, 265–274.

Priya, R., Stanly, S., Anuradha, R., Sagadevan, S., Muthukrishnan, L., 2019. A waste to worth approach in preparing Ferric vanadate nanoparticles using peel extract for photocatalytic dye degradation induced by UV light. Optik - International Journal for Light and Electron Optics 194, 163085.

Radini, I.A., Hasan, N., Malik, M.A., Khan, Z., 2018. Biosynthesis of iron nanoparticles using *Trigonella foenum*-graecum seed extract for photocatalytic methyl orange dye degradation and antibacterial applications. Journal of Photochemistry and Photobiology. B, Biology 183, 154–163.

Rambabu, K., Bharath, G., Banat, F., Show, P.L., 2021. Green synthesis of zinc oxide nanoparticles using *Phoenix dactylifera* waste as bioreductant for effective dye degradation and antibacterial performance in wastewater treatment. Journal of Hazardous Materials 402, 123560.

Ravikumar, K.V.G., Kubendiran, H., Ramesh, K., Rani, S., Mandal, T.K., Pulimi, M., et al., 2020. Batch and column study on tetracycline removal using green synthesized NiFe nanoparticles immobilized alginate beads. Environmental Technology & Innovation 17, 100520.

Rong, K., Wang, J., Zhang, Z., Zhang, J., 2020. Green synthesis of iron nanoparticles using Korla fragrant pear peel extracts for the removal of aqueous Cr(VI). Ecological Engineering 149, 105793.

Roopan, S.M., Elango, G., Priya, D.D., Asharani, I.V., Kishore, B., Vinayprabhakar, S., et al., 2019. Sunlight mediated photocatalytic degradation of organic pollutants by statistical optimization of green synthesized NiO NPs as catalyst. Journal of Molecular Liquids 293, 111509.

Rovani, S., Santos, J.J., Corio, P., Fungaro, D.A., 2018. Highly pure silica nanoparticles with high adsorption capacity obtained from sugarcane waste ash. ACS Omega 3, 2618–2627.

Ruíz-Baltazar, A.D., Jesús, 2021. Sonochemical activation-assisted biosynthesis of Au/Fe_3O_4 nanoparticles and sonocatalytic degradation of methyl orange. Ultrasonics Sonochemistry 73, 105521.

Rupa, E.J., Anandapadmanabana, G., Mathiyalagan, R., Yang, D.-C., 2018. Synthesis of zinc oxide nanoparticles from immature fruits of *Rubus coreanus* and its catalytic activity for degradation of industrial dye. Optik - International Journal for Light and Electron Optics 172, 1179–1186.

Sebastian, A., Nangia, A., Prasad, M.N.V., 2018. A green synthetic route to phenolics fabricated magnetite nanoparticles from coconut husk extract: implications to treat metal contaminated water and heavy metal stress in *Oryza sativa* L. Journal of Cleaner Production 174, 355–366.

Shabaani, M., Rahaiee, S., Zare, M., Jafari, S.M., 2020. Green synthesis of ZnO nanoparticles using loquat seed extract; biological functions and photocatalytic degradation properties. LWT 134, 110133.

Sharififard, H., Shahraki, Z.H., Rezvanpanah, E., Rad, S.H., 2018. A novel natural chitosan/activated carbon/iron bio-nanocomposite: sonochemical synthesis, characterization, and application for cadmium removal in batch and continuous adsorption process. Bioresource Technology 270, 562–569.

Sherin, L., Sohail, A., Amjad, U., Mustafa, M., Jabeen, R., Ul-Hamid, A., 2020. Facile green synthesis of silver nanoparticles using *Terminalia bellerica* kernel extract for catalytic reduction of anthropogenic water pollutants. Colloid Interface Sci Commun 37, 100276.

Singh, M.P., Bhardwaj, A.K., Bharati, K., Singh, R.P., Chaurasia, S.K., Kumar, S., et al., 2021. Biogenic and non-biogenic waste utilization in the synthesis of 2D materials (graphene, h-BN, g-C_2N) and their applications. Frontiers in Nanotechnology 3.

Truskewycz, A., Taha, M., Jampaiah, D., Shukla, R., Ball, A.S., Cole, I., 2020. Interfacial separation of concentrated dye mixtures from solution with environmentally compatible nitrogenous-silane nanoparticles modified with *Helianthus annuus* husk extract. Journal of Colloid and Interface Science 560, 825–837.

Tyagi, P., Kawamura, K., Fu, P., Bikkina, S., Kanaya, Y., Wang, Z., 2016. Impact of biomass burning on soil microorganisms and plant metabolites: a view from molecular distributions of atmospheric hydroxy fatty acids over Mount Tai. Journal of Geophysical Research: Biogeosciences 121, 2684–2699.

Venkateswarlu, S., Kumar, N., Prathima, B., Anitha, B., Jyothi, N.V.V., K., 2015. A novel green synthesis of Fe_3O_4-Ag core shell recyclable nanoparticles using *Vitis vinifera* stem extract and its enhanced antibacterial performance. Physica B 457, 30–35.

Zaheer, Z., 2018. Biogenic synthesis, optical, catalytic, and in vitro antimicrobial potential of Ag-nanoparticles prepared using Palm date fruit extract. Journal of Photochemistry and Photobiology. B, Biology 178, 584–592.

Chapter 22

Nanomaterial synthesis from the plant extract and tree part

Thi Thao Truong and Minh Quy Bui
TNU-University of Sciences, Thai Nguyen City, Thai Nguyen, Vietnam

22.1 Introduction

Nanomaterials (NMs) are comprised of objects less than 100 nm in at least one dimension, including nanorods, nanofilms, nanofibers, nanowires, and nanoparticles (NPs). It means that matter exists in the state of individual atoms, molecules, or clusters of atoms/molecules. Since the fourth century AD, NMs were used by the Romans in the dichroic Lycurgus cup from the British Museum collection. It appeared slate green when illuminated from the outside but glowed red when illuminated from the inside because of the presence of silver-gold NPs 50–100 nm; or the presence of Ag or Cu NPs in glittering "luster" ceramic glazes which were used in the Islamic world, and then in Europe between the 9th and the 17th centuries; or "Damascus steel" in South Asia and the Middle East before the 18th century. It is legendary for its strength durability and ability to hold a keen edge, due to the use of carbon nanotubes (Bayda et al., 2020). Nanotechnology NMs have become one of the fastest-growing fields of research in the past few decades. The applicability of NMs also expands, becoming multifunctional NMs. NMs are currently being studied and applied in all areas of production and life: optical, electronic, magnetic, agriculture, pharmaceutical, cosmetic, food, transportation, medical, environment, industrial products (printing, coating, textile, glass, ceramics), and chemical industry; for example, cobalt ferrite NPs in photocatalytic, electrochemical sensing, and antibacterial applications (Singh, 2022). ZnO NPs are consumed for the production of various cross-linked rubber; used in medicines for epilepsy, diarrhea, suppositories, temporary fillings, and sunscreen; in the textile industry: self-cleaning, waterproof, and UV-blocking textile fabrics; photoelectronic, electronic equipment; surface acoustic wave devices; UV lasers; sensors, solar cells; in paints, inks; in treating the environment as a photocatalysis and an absorbent; an essential nutrient, an animal feed additive, and an artificial fertilizer (Kolodziejczak-Radzimska et al., 2014).

Physical and chemical pathways or combinations usually carry out nanoscale synthesis processes. The most prominent physical methods today are evaporation-condensation (for metals, alloys, and ceramics), combustion synthesis (for ceramic oxides), arc discharge/plasma (for fullerenes and other related materials), laser/electron beam heating (for carbon nanotubes, carbon nanocapsules, and carbon nanoparticles), and laser ablation technique (metal oxides, carbides, and nitrides). Chemical methods are often simple, allow synthesis in large quantities, and are feasible with most nanomaterials, such as chemical reduction, microemulsion, sol–gel process, precipitation/co-precipitation, and oxidation process. Nowadays, it is quite common to combine chemical and physical methods such as UV-initiated photoreduction, microwave-assisted synthesis, electric dispersion reaction, and irradiation method. The most significant disadvantage of the chemical methods is that they use many chemicals and often have byproducts. With global awareness of sustainable development, green, and environmentally friendly production, the current priority research and development direction are biological methods (biosynthesis), such as the synthesis of nanoparticles using plant extracts, fungi, yeast, actinomycetes, or synthesized from biological material as lignin, cellulose, chitosan, etc. Plants are the most abundant, readily available, and inexpensive source of raw materials, carrying within themselves large quantities of minerals and organic compounds that play a variety of roles in chemical reactions such as reducing agents, oxidants, reagents, surfactants, solvents, and additives (Bhardwaj and Naraian 2021). So plant-mediated synthesis of NPs is attracting significant attention nowadays. Many NMs were synthesized using plant extracts like Ag, Au, Cu, Fe, Co, ZnO, Fe_xO_y, TiO_2, and organic-based NMs such as lignin, cellulose, and chitosan. This chapter focuses on the synthesis and application of NMs from plant extract and other tree parts.

Green and Sustainable Approaches Using Wastes for the Production of Multifunctional Nanomaterials. DOI: https://doi.org/10.1016/B978-0-443-19183-1.00006-4

22.2 Methods of synthesizing multifunctional nanomaterials from plant extract and other parts

According to the component, NMs were classified into three main categories: inorganic-based NMs, organic-based NMs, and carbon-based NMs.

22.2.1 Inorganic nanomaterials

Multifunctional inorganic NMs are mainly metal NPs such as Ag, Au, Cu, Fe, Co, Ti, Pt, and metal oxide or a composite of metal-metal, oxide-oxide, or metal-oxide metal.

22.2.1.1 Metal nanoparticles

Commonly, metal NPs were synthesized by the chemical reduction of metal ions ($HAuCl_4$, $AgNO_3$) to metal by a reducing agent as $NaBH_4$ or trisodium citrate ($Na_3C_6H_5O_70.2H_2O$), and then using a cationic surfactant to stabilization. In green synthesis, plant extract acted as a reducing agent, a surfactant, and a stabilizing agent without using an organic solvent.

Metal NPs were synthesized by adding the extract drop by drop to the metal ion solution under stirring at a temperature for a time: Au NPs were obtained from *olive leaf* extract in the size of 17−21 nm and had good dispersibility (Khalil et al., 2012). Ag NPs were prepared from *Cupressus sempervirens* L. pollen extract, ranging from 11.70 to 24.36 nm (Turunc et al., 2021). Ag NPs synthesized by *Emblica officinalis* fruit extract were around 15 nm (Ramesh et al., 2015).

The process of adding the metal precursor to the extract can be done simply by mixing two solutions: a solution of Au precursors and *Dracaena Draco* L extracts put on a hotplate without stirring to form Au NPs 0.9−2.8 nm (Luna et al., 2019); the solution of $AgNO_3$ and *Handroanthus heptaphyllus* (Vell.) Mattos aqueous extract was put in a water bath at 50°C for a time and obtained the Ag NPs size of 8.6 nm−29.0 nm (Pereira et al., 2020); or the mixture of $AgNO_3$ and *Tamarindus indica* natural fruit extract in a microwave oven for a time and Ag NPs was 6−8 nm (Jayaprakash et al., 2017). The NaOH solution was added as an accelerator to enhance the reactions by some researchers, to form Ag NPs (Li et al., 2021), Pt NPs (Nellore et al., 2013), Fe NPs (Puthukkara et al., 2021), Cu−Ni hybrid (Abdullah et al., 2022), and Au−Ag bimetallic (Elemike et al., 2019b).

22.2.1.2 Metal oxide

The metal oxide was synthesized by some methods: co-precipitation, sol−gel, hydrothermal, and sintering. The essence is to carry out a chemical reaction between a metal precursor (salt) and a reactant (an alkaline solution, a salt, or an organic compound) to form insoluble and thermally unstable products of that metal, such as hydroxides, salt, or complexes, which are decomposed under heat to form oxides. In principle, to obtain a nanoscale product, the reaction could perform on the organic solvent, use a surfactant, and use in combination with mechanical action such as stirring and grinding. However, using organic solvents or surfactant is unfriendly to the environment. The plant extract includes functional groups and can act as a reactant, a surfactant, a complexant, or all of the above.

The co-precipitation was performed in several different ways: (**PM1**) the metal ions solution was added to the extract at a certain temperature for a specific time, and then the $NaOH/NH_3$ solution was added to the above mixture solution to certain pH, then the suspension was centrifuged, washed, and dried (or then calcined): Xiu et al. (2022) prepared Fe_2O_3 below 100 nm by *Centaurea alba* extract; ZnO about 20 nm in size was prepared by plantain peel extracts (Imade et al., 2022). Yang et al. (2018) prepared γ-Fe_2O_3 20−40 nm in size by *Tridax* leaf extracts. The process was also performed with slight modification (**PM2**): the alkali solution was added to the metal ion solution first, and then the extract was added to this mixture: ZnO NPs were synthesized by extract of *Banana peel* crude (Ruangtong et al., 2020) which were in nanorod: length 272.86÷526.52 nm, width 81.22÷423.78 nm; ZnO NPs were about 15.58÷18.09 nm, and were synthesized by *Thymbra spicata extract* (Şahin et al., 2021), or (Shabaani et al., 2020) synthesized ZnO by using extract of Loquat *(Eriobutria japonica)* seed. Alternatively, some extracts can precipitate metal ions without adding an alkaline solution (**PM3**): CuO NPs were prepared by *Simarouba glauca* leaf extract with the average length and breadth of 10−30 nm (Kalaiyan et al., 2020). Velsankar et al. (2020) prepared CuO in size range of 20−40 nm by *Allium sativum* extract; ZnO 25 nm was prepared by *Moringa oleifera* leaf extract (Letsholathebe et al., 2019) and ZnO 5−30 nm by *Dysphania ambrosioides* extract (Álvarez-Chimal et al., 2021). Haseena et al. (2021) prepared α-Fe_2O_3 19.3 and 20.4 nm with *Phyllanthus niruri* and *Moringa stenopetala* leaf extract, respectively.

The sol–gel method (**SM**) was performed only by dropwise extract to the metal ion solution at a temperature under continuous stirring for several hours until a gel was formed. The gel was dried and then calcined to a specific temperature. For example, ZnO was prepared in a size range of 8 to 30 nm, E_g of 2.77 eV by *Hibiscus sabdariffa* flower (Jamaica) extracts (Soto-Robles et al., 2019), and ZnO 49 nm by *Solanum nigrum* leaf extract (Muthuvel et al., 2020).

The hydrothermal method (**HM**): a homogenous mixture of extract and a metal ion solution (and Alkali) were conducted in an autoclave at certain temperatures (from 180°C to 300°C) for several hours: ZnO 31 nm was prepared in the presence of *Prosopis juliflora* leaf extract (Soto-Robles et al., 2021).

22.2.1.3 Nanocomposite of metal and metal oxide

The composite of oxides, for example, CuO-modified $CoTiO_3$ by SM with *Catharanthus roseus* leaf extract (Yulizar et al., 2020).; CuO-modified ZnO by SM with *Piper chaudocanum* L. leave extract (Truong et al., 2022); ZnO-SnO_2 by coprecipitation method with *Solanum macrocarpon* fruit extract (Onwudiwe et al., 2021); ZnO-TiO_2 by coprecipitation with *Hibiscus subdariffa* leaf extract (Suganthi et al., 2020).

These composites were synthesized in two ways: (1) 1-stage reaction according to one of the above methods, for example, a solution of $HAuCl_4$ and $CuSO_4$ in *Stigmaphyllon ovatum* leaf extract was mixed with stirring, then centrifuged, washed, and dried. The obtained Au–CuO NPs were about 6.40 nm as compared with CuO 4.42 nm (Elemike et al., 2019a). (2) The oxide was synthesized first, and then metal was dispersed on the oxide surface in the second stage. For example, Ag/Fe_3O_4 nanocomposite (Huang et al., 2022), Cu-Hibiscus@/Fe_3O_4 nanoparticles (Wang et al., 2022), or Fe_3O_4/Thyme-Cu nanoparticles (Tang et al., 2022).

22.2.2 Organic-based nanomaterials

Organic-based nanomaterials are growing in popularity, such as lignin, cellulose, starch, gum, alginate, pectin, chitin, and chitosan-based NMs. They were prepared from renewable biomass resources. Billions of tons of excess biomass are generated around the world every year.

22.2.2.1 Lignin-based nanomaterials

Lignin is one of two major constituents in the plant cell wall (with cellulose), a complex organic biopolymer (noncarbohydrate) that contains numerous polyphenols with the monomer being phenyl propane united, as a binding component between hemicellulose and cellulose. Lignin NPs (LNPs) were synthesized from plant or block lignin by physical treatment (mechanical method such as ball milling, low-temperature milling, high shear homogenizer (Nair et al., 2014), ultrasonication (Gilca et al., 2015; Garcia Gonzalez et al., 2017), microwave-assisted methods), or chemical treatment (acid precipitation, alkali precipitation, flash precipitation, antisolvent precipitation, polymerization, solvent-exchange, water-in-oil microemulsion, and interfacial cross-linking).

Some chemical treatment processes are as: (1) Lignin was isolated from the biomass by acid-alkali treatment (Trevisan and Rezende, 2020) or only alkali treatment and then precipitated by H_2SO_4 (Bertolo et al., 2019). The lignin separation from the black liqueur was also precipitated by calcium oxide (Behboudi et al., 2021). (2) Lignin was isolated by extraction exchange solvent: Lignin NPs were prepared from hardwood (*aspen* and *eucalyptus*) and softwood (*pine* and *herbaceous* corn stover) by exchange of the DMSO solvent and water, and the lignin NPs separation efficiency reached 17.5% to 29.4%, particle size 20 to 100 nm, and was relatively stable at pH 4–10 (Chen et al., 2020). The raw sugarcane bagasse was pretreated with 1:1 ethanol/water solution (w/w), then exchanged by water (Bertolo et al., 2019); or hammer-milled poplar wood was extracted by A fresh p-toluenesulfonic acid solution in a polypropylene column, then the collected liquid phase was diluted by distilled water, centrifuged, further filtrated using a sand core funnel to obtain supernatant colloidal lignin NPs (Wang et al., 2021).

With many functional groups in the structure, lignin can easily enhance surface functionalization, and some desirable properties of lignin. For example, carboxylation of lignin (Figueiredo et al., 2017), functionalized tyrosinase-lignin NPs (Capecchi et al., 2018), hydrophobic lignin NPs (Ding et al., 2021), and lignin hydroxymethylation (Căpraru et al., 2012). Besides, the functionalization of lignin by metals/metal oxides creates new materials with many outstanding advantages which have received great attention in recent years: Lignin nonmagnetic iron nanoparticles (do Nascimento Junior et al., 2021), lignin/zinc oxide hybrid NPs (Del Buono et al., 2022), and organosolv lignin/Fe_3O_4 NPs (de Araújo Padilha et al., 2020). Lignin is also utilized to prepare nanocomposite with organic components for various applications: Biocomposites composed of lignin and natural rubber (Ikeda et al., 2017), lignin was added to the PVA solution to form the electrospinning solution (Lee et al., 2019), PVA/lignin nanomicelle NCs films (Zhang et al., 2020), PVA-lignin nanocomposites (Luo et al., 2021), poly(methyl methacrylate) (Yang et al., 2018), polyethylene (Wang et al., 2016), poly(lactic acid) (Yang et al., 2015), and polystyrene co-butyl acrylate (Jairam et al., 2013).

22.2.2.2 Cellulose nanocrystal and cellulose-based nanocomposite

The preparation of cellulose nanocrystals from plants consists of 2 stages: pretreatment and extraction.

There are four pretreatment methods: chemical, physical, physicochemical, and biological. In chemical pretreatment, lignocellulosic was treated using inorganic acid, alkali, organic solvents, or ionic liquids. The most common is acid: HCl, H_2SO_4 (Rajan et al., 2020; Wang et al., 2020), H_3PO_4, or organic acid as formic acid and p-toluenesulfonic acid. Physical pretreatments commonly used are chipping and milling or new physical treatment methods as gamma rays, ultrasonication, microwaving, and electron beams. Pretreatments by temperature elevation, irradiation, steam explosion (superheated steam, hydrothermal, and steam explosion), or auto-hydrolysis are called physicochemical pretreatments methods. The high temperatures and stream led to the broken lignocellulosic structure, reducing the biomass size but at a low cost and saving energy consumption. However, to increase the yield of nanocellulose (NC), physicochemical pretreatment is often used in combination with chemical or biological pretreatments. For example, Formosan alder biomass was pretreated by a steam explosion followed by bleaching, with an α-cellulose yield of 91.32% compared with 51.66% as only pretreatments by the steam explosion (Teo and Wahab, 2020). Biological pretreatments are biological reactions with microorganisms (bacteria or fungi) or cellulosic enzymes (ligninase, xylanase, and cellulase) to break the structure of lignocellulosic biomass. Biological pretreatment is more efficient than steam explosions (Teo and Wahab, 2020). However, biological pretreatment is often combined with chemical pretreatment to reduce pretreatment times, minimizing chemical usage and the risk of secondary pollution (Norrahim et al., 2021).

The extraction of NC was done after the pretreatment, mainly by chemical methods (usually hydrolysis, oxidation, organosolv), and physical or biological methods. This process is often done in combination with modified cellulose. The chemical method remains the simplest and the most common method for NC extraction. The essence is that the hydrolysis of amorphous cellulose is the mainly used acid (sulfuric acid (Maiti et al., 2013) or hydrochloric acid (Hastuti et al., 2018), formic acid, acetic acid, phosphoric acid (Yogi et al., 2017), and chlorine oxide) or alkali. There are many factors such as temperature, time, precursor (biomass), acid type, and concentration which affected significantly the production yield and properties of NC. Generally, the concentration of acid, reaction time, and temperature need to be increased to a particular value. But, the chemical method leaves acid and base residues in wastewater, so the current trend in NC extraction is reducing chemicals by ball milling (solid-state synthesis) and enzymatic hydrolysis. Although ball milling is quite simple, it consumes energy intensive and is small in capacity and inappropriate for large-scale production. Enzymatic hydrolysis is friendlier in nature, but low production yields and much slower rates are issues that need to be considered (Teo and Wahab, 2020).

Modified cellulose nanomaterials expanded the applicability of cellulose. The modified process was done in combination with pretreatment, isolated cellulose, or synthesized directly from the NC and active ingredients. For example, low copper NPs (<1 mol.%) stabilized on NC were synthesized by dispersing NC with $CuSO_4$ in water in the presence of hydrazine hydrate stirred for 2 hours at room temperature (Dutta et al., 2019). Acetylation, carbamation, esterification, etherification, silanization, peptide linkage/amidation, sulfated NCs surfaces, nucleophilic substitution reactions, hydrogen bonding, and ion pair formation graft copolymerization to surface functionalization of NCs.

In addition, many nanomaterials based on organic are being studied at present, such as lignin, alginate-derived nanomaterials, chitin and chitosan-derived nanomaterials, gum-derived nanomaterials, pectin-derived nanomaterials, and starch-derived nanomaterials (Calvino et al., 2020; Nasrollahzadeh et al., 2020, 2021b; Zhang et al., 2021).

22.2.3 Carbon-based nanomaterials

Carbon nanomaterials consist of graphene, carbon nanotubes, crystalline diamond, diamond-like carbon, and recently biochar, all display exceptional physical, chemical, and electrochemical properties that lead to their widespread applications. But, most of them are prepared from bulk carbon derived from coal. A new type of material is rich in carbon and was prepared from biomass with nanostructure and a variety of applications, and it is biochar/hydrochar and modified biochar/hydrochar, collectively known as carbonized material. Carbonized material is a porous structure (pore diameter in nanoscale), amorphous, stable, low-density carbon, multifunctional properties, biodegradable, and derived from many different types of biological residues. They are prepared by following one or two stages: (1) carbonization of the biomass in an inert atmosphere—pyrolyzed (biochar) or by hydrothermal (hydrochar) and (2) activation of the material. The carbonization is carried out mainly by pyrolysis (clean, chopped, the dry feedstock is heated to a specific temperature, usually around 300°C–700°C under inert gas or no oxygen) and hydrothermal (feedstock was distributed in DW conducted in an autoclave at temperatures from 180°C to 260°C). Although the hydrothermal is more energy efficient than pyrolysis and raw materials do not need to be dried, the hydrothermal takes a longer time, and the properties of carbonized materials are not as good as the pyrolysis method. Therefore, pyrolysis has been carried out more. In recent years, microwave-assisted pyrolysis, which shortens the time and improves the quality of biochar, may be a preeminent solution for preparing biochar. The activation of carbonized material is conducted by physical activation, chemical activation, or a combination of them. Physical activation is performed after carbonization by suitable oxidizing gases such as carbon dioxide and/or steam at high temperatures (Shim et al., 2015; Gao et al., 2020). In chemical activation, the carbonization and activation can be performed simultaneously: the raw materials or biochars are first impregnated with some chemical activators (acid, alkali, or metal compound), and then the impregnated materials are carbonized (Cataldo et al., 2018; Zhang and Leiviskä, 2020; Ahmed et al., 2021). Sometimes, chemical activation and biochar were dispersed together to the microparticle level, such as biochar mixed with sublimed sulfur homogeneously before pyrolyzing (Huang et al., 2019), or biochar and ZnO soaked in ethanol under sonication and stirred (Hu et al., 2019). Physical activation is a simple process that does not require chemicals, but the ability to improve porosity is not as good as chemical activation. Therefore, many researchers have studied the combination of physical-chemical activation. For example, hemp stem biochar was activated by phosphoric acid and steam in a nitrogen atmosphere (Wu et al., 2018), and petroleum coke was activated by KOH or NaOH and steam (Lupul et al., 2015).

22.3 Application of nanomaterials from plants

NMs are studied and applied in many fields, such as biomedical, environmental treatment, sensors, magnetic, and optical materials, but most are applications in biomedicine and water treatment.

22.3.1 Water treatments

Currently, many conventional methods are used to treat water, such as chemical oxidation, coagulation, membrane separation, ion exchange, reverse osmosis, adsorption, and photocatalytic. Among these methods, the most popular are adsorption and photocatalysis, which have low production costs, high selectivity, and applicability in low concentrations (Queiroz et al., 2022). They can be applied in both batches and continuous processes. In addition, the adsorbent and photocatalytic materials can be reused in other cycles. The utilization of NMs derived from plant extracts and tree parts (bio-NMs) has emerged as a promising alternative to mainstream methods. They include metal NPs, metal oxide NPs, lignin NPs, modified lignin Nps, NCs, modified NCs, graphene, and biochar. The pollutants that have been usually investigated involve heavy metals, dyes, organic pollutants, and antibiotics.

Some adsorbents are synthesized from plant extracts, such as nano zero-valent iron (nZVI NPs), synthesized from Nettle and Thyme extracts, that were used to adsorb the antibiotic with an adsorption capacity of 1667 and 1428 mg/g, respectively (Leili et al., 2018). nZVI NPs produced using *Ricinus communis* Linn leaf extract had a 98% effectiveness in adsorbing tetracycline (Abdelfatah et al., 2021). Fe_3O_4 nanoparticles synthesized from Eucalyptus leaf extract adsorbed 95.13% of the phosphate (Gan et al., 2018). NC and their nanocomposites (ZnO/NC, Ag/NC) synthesized from bagasse pulp can adsorb Pb (II) over 94% (Badawy et al., 2021). Biochar and cobalt-gadolinium-modified biochar synthesized from *Camellia oleifera* shells adsorbed Ciprofloxacin and tetracycline (Hu et al., 2021). These nanobiomaterials have also been investigated for the recycling process. The materials have a high rate of reuse and can be recycled three to five times. Even after five regeneration cycles, the ZIF-8/biochar nanocomposite maintained its ability to adsorb ketamine ($>90\%$ of the initial adsorption capacity) (Liu et al., 2021). Cellulose nanocrystal and cellulose-based nanocomposite were used for water treatment (Nasrollahzadeh et al., 2021a,b). It was suggested that the synergistic effects of pore-filling, $\pi-\pi$ interaction, hydrogen bonding, chelation, and electrostatic interaction between the surface of the adsorbent and the pollutant work in concert explain how the adsorption process works (Badawy et al., 2021; Liu et al., 2021).

Photocatalyst is a method that is widely used in water treatment. According to Samadi et al. (2021), Barberry leaf extracts' nZVI NPs function as photocatalysts to eliminate the Cr (VI). Ag NPs and modified Ag NPs are frequently used as photocatalysts and catalysts to remove hazardous organic compounds. Vanaamudan et al. (2016) investigated the removal of Rhodamine-6G, Reactive Blue-21, and Reactive Red-141 dyes by Ag NPs produced from palm shell extract. ZnO NPs are metal oxides that have potential uses as photocatalysts and are of interest to many scientists, such as Congo red, methylene blue, malachite green, and methyl orange more than 91% (Chakraborty et al., 2020; Alharthi et al., 2021; Sadiq et al., 2021b; Vasantharaj et al., 2021). ZnO NPs prepared from Cyanometra ram flora leaves extract removed Rhodamine B 98% within 200 minutes under sunlight (Varadavenkatesan et al., 2019). The dispersing NC with $CuSO_4$ (Dutta et al., 2019) was used for biodegradable support, catalysts in the oxidation reaction of sulfides, and primary alcohols. The process mechanism is also discussed in the investigation. An electron is encouraged to go from the valence band to the conduction band by the energy this lamp delivers, which results in an electronic vacancy or hole in the valence band. That is, electron-hole pairs support redox reactions on the catalyst surface and result in the production of superoxide ions (O_2^-) and hydroxyl-free radicals (OH^-). These species produce CO_2 and H_2O as products while also acting as potent oxidants to mineralize the organic pollutants (Sheik Mydeen et al., 2020; Abdelfatah et al., 2021; Nasrollahzadeh et al., 2021a; Sadiq et al., 2021).

22.3.2 Biomedical field

Plant extracts and tree parts are frequently used in biomedicine due to their low cytotoxic potential, particularly in the green synthesis of metal and oxide nanoparticles. Although these substances have a wide range of medicinal applications, this research concentrated on applications in antibacterial, antioxidant, and anticancer nanobiomaterials.

Metal NPs, such as Ag NPs, Au NPs, and Cu NPs, are frequently used as antibacterial, anticancer, and antioxidant agents. *Conocarpus Lancifolius* fruit-derived Ag NPs exhibit antibacterial efficacy against *S. pneumonia* and *S. aureus* (Oves et al., 2022). Au NPs produced from *Elaeocarpus ganitrus* bark extract exhibit effective antibacterial *P. desmolyticum* and *S. aureus* (Vinay et al., 2021). Cu NPs are made from *Couroupita guianensis Aubl*'s leaves stem, petal, and bark extracts effectively against *B. subtilis* and *E. coli.* (Logambal et al., 2022). *Spinacia oleracea* extract can be used to create Ag NPs that can destroy leukemia (HL60) cells and antioxidants (Zangeneh, 2020). On AgNPs, leukemia-causing cells (32D-FLT3- ITD Human, HL60) were also investigated with a solution made from the leaf extract of *Melissa officinalis* (Ahmeda et al., 2019). Au-NPs-doped Fig extract enhanced the antioxidant activities in kidney-damaged tissue and reduced the oxidative toxicity induced by cisplatin-induced acute renal injury (El-Sayed et al., 2019). Further, Au NPs have been researched regarding leukemia (K562, C1498) (Datkhile et al., 2017) and prostate cancer's PC3 cell line (Vinay et al., 2021).

In the area of biomedicine, ZnO NPs have been produced with various plant extracts and extensively researched as an antioxidant, anticancer, and antibacterial agent: ZnO NPs produced from crude banana peel extract can inhibit the growth of HepG2 liver cancer cells, A431 lung cancer cells, SW620 cervical cancer cells, and bacterias (*B. subtilis, S. epidermidis*, and *E. coli*) (Ruangtong et al., 2020). ZnO NPs from the leaves of the Java plum (*Syzygium cumini*) were investigated as an antioxidant and antileukemic agent (Arumugam et al., 2021). Fe_3O_4 NPs have been investigated in targeted anticancer drug delivery, photo-thermal therapy, and bioimaging (Barabadi et al., 2017; Yew et al., 2020).

Not only do metal nanoparticles and metal oxide nanoparticles display excellent biological applications but so do their composites made from plant extracts. Under the influence of *Ocimum basilicum* leaf extract, combining ZnO NPs with RGO confirmed the material's ability to be antibacterial (*Cocci, E. coli*) and antioxidant (Malik et al., 2022). The antibacterial (*S. aureus, B. subtilis, K. pneumonia*, and *E. coli*) and anticancer (Hela) properties of synthetic Chitosan-CuO NPs composite materials fabricated with *Psidium guajava* leaf extract were examined by Karthikeyan et al. (2021).

Plant-based nanocellulose, carbon nanomaterial (NC, CNM), and their biodegradable composites are often applied as a multifunction material in the medical industry for adhesion barriers, artificial skins, and tissue engineering. Additionally, polyurethane/NC composites are suitable natural resources for the production of medical protective equipment, including nonlatex condoms, medical disposables, surgical gloves, and gowns (Gaddam et al., 2015; Bhat et al., 2019). The graphene-2D produced from *Drepanostachyum falcatum* extract displayed excellent fluorescence and photoluminescent capabilities, low cytotoxicity, and widespread intracellular localization, and medical and biological applications were studied (Tewari et al., 2022). The actual mechanism of biofunctionalized nanoparticles' antibacterial action is unknown. Some putative antimicrobial activity pathways, including mechanical damage, gene toxicity, oxidative stress injury, and chemisorption, were proposed by the researchers (Bhavyasree and Xavier, 2022; Mostafavi et al., 2022; Noah and Ndangili, 2022): The NPs enhance the production of reactive oxygen species, which leads to protein breakdown and DNA damage, eventually leading to cancer cell death; or they may promote apoptosis by stimulating various intracellular pathways.

Although nanomaterials are promising in the biomedical field, their use in clinical trials remains challenging. First, although their toxicity has decreased compared to chemically made nanomaterials, a careful review of their effectiveness, potential for bioaccumulation, and possible human immune reactions are still needed. Other obstacles also prevent the use of nanobiomaterials in cancer therapy, particularly problems related to the absorption, diffusion, and penetration of NPs into cells to assure their high efficacy. Unfortunately, not much information about the majority of these mechanisms has been published in the literature on the aforementioned nanostructures. Additionally, research on antivirals, mainly those responsible for the pandemic COVID-19, has not been addressed.

22.3.3 Other applications

Al-Radadi (2022a,b) studied organic acids obtained from citrus wastes to develop biodegradable polymers and functional materials for food processing, chemical and other industries, and pharmaceuticals. Shell residues can be processed to produce fibers and fabrics, 3D printing materials, carbon nanoparticles for bioimaging, energy storage materials, and nanostructured materials for various applications, leaving no residue waste. For example, the fluorescent carbon quantum dots were synthesized by *Eleusine coracana*, which was used as a sensor probe for selective detection of Cu^{2+} (Murugan et al., 2019). Fe_3O_4 carbon dots from lemon and grapefruit extracts acted as photoluminescence sensor for detecting *E. coli* bacteria (Ahmadian-Fard-Fini et al., 2018). Fluorescent carbon nanodots were investigated to use in optical applications (Wang et al., 2015). ZnO NP-embedded nitrogen-doped carbon sheet was synthesized from peach fruit and was used as a high-performance glucose biosensor (Muthuchamy et al., 2018). The electrochemical sensor also employed ZnO NPs using orange (*Citrus sinensis*) peel aqueous extract (Wicaksono et al., 2022). SnO_2 NPs about 7.1 nm were synthesized by papaya leaves extract to use as a liquid petroleum gas sensor (Jadhav and Kokate, 2019). ZnPor/rGO/CuO composite was synthesized by mechanical dispersion method using *Calotropis procera* extract and used for solar cells application (Arjmand et al., 2022). Additionally, it has been discovered that greenly produced TiO_2 NPs work well as larvicidal agents against a variety of parasitic insect species (Nadeem et al., 2018). Chitosan-ZnO NPs were synthesized by Alamdari et al. (2022) using *Mentha pulegium* extract for packaging, applications, and preservation of the fruits for up to eight days at 23°C.

22.4 Conclusion and future prospective

The science of NMs and nanotechnology has developed in recent years but has ushered in outstanding achievements. Applications of NMs are met in almost all areas of life and production. Therefore, the synthesis of NMs has become the focus of researchers worldwide. Along with that is the awareness of the importance of green chemistry and sustainable development, which requires the synthesis processes as well as the materials themselves, to be greener and more environmentally friendly, safer, and more energy-efficient processes, resulting in materials having highly reusable, biocompatible, or biodegradable, reducing the using chemicals and disposing of byproducts and toxic chemicals, gradually replacing of chemicals method. Therefore, researchers are finding an innovative alternative with new techniques such as ultrasonication, microwaves, enzymes, organosolv process, etc., using combined methods such as physicochemical and biochemical methods, moving toward sustainable and commercially viable large-scale production of NMs. Plant extract and tree parts rich in organic matter and minerals have become reactants, oxidizing agents, reducing agents, solvents, surfactants, and complexing agents to replace traditional chemicals, satisfying greener requirements, and more environmentally friendly. Moreover, this feedstock is available, inexpensive, and easy to regenerate, meeting sustainable development needs. Research results so far show that inorganic NPs, organic-based NMs, and carbon-based NMs can all be synthesized in the presence of plant extract or from tree parts with nanoscale and many outstanding features and high applicability.

However, there are still many issues that need to be further studied, such as:

- The biochemical composition of plant extract and other tree parts needs to be clarified further, thereby clarifying the effects of the amount of feedstock, time, and temperature of the synthesis process on the size NPs, aggregation state, morphology, as well as other physical and chemical properties of prepared NMs.
- The stability of the extract is a limitation, but the stability and degradation of the prepared NMs have not been elucidated and need to be studied more closely.
- Optimize the green route synthesis NMs according to the trend of green chemistry and sustainable development for large-scale production.
- Currently, applied research is mainly focused on biomedical and environmental remediation. The number of applied research in other fields is few and needs to be studied.
- Current research also focuses on synthesizing some NMs such as Ag, Au, Cu, ZnO, Fe_2O_3, SnO_2, lignin, cellulose, and biochar. Many other NMs need to be further studied, especially nanocomposites of organic and inorganic materials.

References

Abdelfatah, A.M., et al., 2021. Efficient adsorptive removal of tetracycline from aqueous solution using phytosynthesized nano-zero valent iron. Journal of Saudi Chemical Society 25 (12), 101365. Available from: https://doi.org/10.1016/J.JSCS.2021.101365.

Abdullah, et al., 2022. Green synthesis and characterization of copper and nickel hybrid nanomaterials: investigation of their biological and photocatalytic potential for the removal of organic crystal violet dye. Journal of Saudi Chemical Society 26 (4), 101486. Available from: https://doi.org/10.1016/j.jscs.2022.101486.

Ahmadian-Fard-Fini, S., Salavati-Niasari, M., Ghanbari, D., 2018. Hydrothermal green synthesis of magnetic Fe_3O_4-carbon dots by lemon and grape fruit extracts and as a photoluminescence sensor for detecting of *E. coli* bacteria. Spectrochimica Acta Part A: Molecular and Biomolecular Spectroscopy 203, 481–493. Available from: https://doi.org/10.1016/J.SAA.2018.06.021.

Ahmed, W., et al., 2021. Enhanced adsorption of aqueous Pb(II) by modified biochar produced through pyrolysis of watermelon seeds. Science of the Total Environment 784, 147136. Available from: https://doi.org/10.1016/j.scitotenv.2021.147136.

Ahmeda A., et al., 2019. Preparation, formulation, and chemical characterization of silver nanoparticles using Melissa officinalis leaf aqueous extract for the treatment of acute myeloid leukemia in vitro and in vivo conditions. Appl Organometal Chem. 2019, e5378. https://doi.org/10.1002/aoc.5378.

Al-Radadi, N.S., 2022a. Laboratory scale medicinal plants mediated green synthesis of biocompatible nanomaterials and their versatile biomedical applications. Saudi Journal of Biological Sciences 29 (5), 3848–3870. Available from: https://doi.org/10.1016/j.sjbs.2022.02.042.

Al-Radadi, N.S., 2022b. Single-step green synthesis of gold conjugated polyphenol nanoparticle using extracts of Saudi's myrrh: their characterization, molecular docking and essential biological applications. Saudi Pharmaceutical Journal 30 (9), 1215–1242. Available from: https://doi.org/10.1016/j.jsps.2022.06.028.

Alamdari, S., et al., 2022. Green synthesis of multifunctional ZnO/chitosan nanocomposite film using wild Mentha pulegium extract for packaging applications. Surfaces and Interfaces 34, 102349. Available from: https://doi.org/10.1016/j.surfin.2022.102349. June.

Alharthi, M.N., et al., 2021. Green synthesis of zinc oxide nanoparticles by *Ziziphus jujuba* leaves extract: environmental application, kinetic and thermodynamic studies. Journal of Physics and Chemistry of Solids 158, 110237. Available from: https://doi.org/10.1016/j.jpcs.2021.110237.

Álvarez-Chimal, R., et al., 2021. Green synthesis of ZnO nanoparticles using a *Dysphania ambrosioides* extract. Structural characterization and antibacterial properties. Materials Science and Engineering: C 118, 111540. Available from: https://doi.org/10.1016/j.msec.2020.111540.

Arjmand, F., et al., 2022. The first and cost effective nano-biocomposite, zinc porphyrin/CuO/reduced graphene oxide, based on *Calotropis procera* plant for perovskite solar cell as hole-transport layer under ambient conditions. Journal of Materials Research and Technology 16, 1008–1020. Available from: https://doi.org/10.1016/j.jmrt.2021.12.012.

Arumugam, M., et al., 2021. Green synthesis of zinc oxide nanoparticles (ZnO NPs) using *Syzygium cumini*: potential multifaceted applications on antioxidants, cytotoxic and as nanonutrient for the growth of *Sesamum indicum*. Environmental Technology & Innovation 23, 101653. Available from: https://doi.org/10.1016/j.eti.2021.101653.

Badawy, A.A., et al., 2021. Utilization and characterization of cellulose nanocrystals decorated with silver and zinc oxide nanoparticles for removal of lead ion from wastewater. Environmental Nanotechnology, Monitoring & Management 16, 100501. Available from: https://doi.org/10.1016/J.ENMM.2021.100501.

Barabadi, H., et al., 2017. Anti-cancer green bionanomaterials: present status and future prospects. Green Chemistry Letters and Reviews 10 (4), 285–314. Available from: https://doi.org/10.1080/17518253.2017.1385856.

Bayda, S., et al., 2020. The history of nanoscience and nanotechnology: from chemical-physical applications to nanomedicine. Molecules (Basel, Switzerland) 25 (1), 1–15. Available from: https://doi.org/10.3390/molecules25010112.

Behboudi, G., et al., 2021. Optimized synthesis of lignin sulfonate nanoparticles by solvent shifting method and their application for adsorptive removal of dye pollutant. Chemosphere 285, 131576. Available from: https://doi.org/10.1016/j.chemosphere.2021.131576. June.

Bertolo, M.R.V., et al., 2019. Lignins from sugarcane bagasse: renewable source of nanoparticles as Pickering emulsions stabilizers for bioactive compounds encapsulation. Industrial Crops and Products 140, 111591. Available from: https://doi.org/10.1016/j.indcrop.2019.111591. February.

Bhardwaj, A.K., Naraian, R., 2021. Cyanobacteria as biochemical energy source for the synthesis of inorganic nanoparticles, mechanism and potential applications: a review'. 3 Biotech 11 (10), 445. Available from: https://doi.org/10.1007/s13205-021-02992-5.

Bhat, A.H., et al., 2019. Cellulose an ageless renewable green nanomaterial for medical applications: an overview of ionic liquids in extraction, separation and dissolution of cellulose. International Journal of Biological Macromolecules 129, 750–777. Available from: https://doi.org/10.1016/J.IJBIOMAC.2018.12.190.

Bhavyasree, P.G., Xavier, T.S., 2022. Green synthesised copper and copper oxide based nanomaterials using plant extracts and their application in antimicrobial activity: review. Current Research in Green and Sustainable Chemistry 5, 100249. Available from: https://doi.org/10.1016/j.crgsc.2021.100249.

Calvino, C., et al., 2020. Development, processing and applications of bio-sourced cellulose nanocrystal composites. Progress in Polymer Science 103. Available from: https://doi.org/10.1016/j.progpolymsci.2020.101221.

Capecchi, E., et al., 2018. Functionalized tyrosinase-lignin nanoparticles as sustainable catalysts for the oxidation of phenols. Nanomaterials 8 (6). Available from: https://doi.org/10.3390/nano8060438.

Căpraru, A.M., et al., 2012. Chemical and spectral characteristics of annual plant lignins modified by hydroxymethylation reaction. Cellulose Chemistry and Technology 46 (9–10), 589–597.

Cataldo, S., et al., 2018. Biochar from byproduct to high value added material – a new adsorbent for toxic metal ions removal from aqueous solutions'. Journal of Molecular Liquids 271, 481–489. Available from: https://doi.org/10.1016/j.molliq.2018.09.009.

Chakraborty, S., et al., 2020. Averrhoe carrambola fruit extract assisted green synthesis of zno nanoparticles for the photodegradation of Congo red dye. Surfaces and Interfaces 19, 100488. Available from: https://doi.org/10.1016/j.surfin.2020.100488.

Chen, Y., et al., 2020. Fabrication of spherical lignin nanoparticles using acid-catalyzed condensed lignins. International Journal of Biological Macromolecules 164, 3038–3047. Available from: https://doi.org/10.1016/j.ijbiomac.2020.08.167.

Datkhile, K.D., Durgawale, P.D., Patil, M.N., 2017. Biogenic silver nanoparticles are equally cytotoxic so as chemically synthesized silver nanoparticles. Biomedical and Pharmacology Journal 10 (1), 337–344. Available from: https://doi.org/10.13005/bpj/1114.

de Araújo Padilha, C.E., et al., 2020. Organosolv lignin/Fe_3O_4 nanoparticles applied as a β-glucosidase immobilization support and adsorbent for textile dye removal. Industrial Crops and Products 146, 112167. Available from: https://doi.org/10.1016/j.indcrop.2020.112167. January.

Del Buono, D., et al., 2022. Synthesis of a lignin/zinc oxide hybrid nanoparticles system and its application by nano-priming in maize. Nanomaterials 12 (3), 1–17. Available from: https://doi.org/10.3390/nano12030568.

Ding, H., et al., 2021. High hydrophobic poly(lactic acid) foams impregnating one-step Si–F modified lignin nanoparticles for oil/organic solvents absorption. Composites Communications 25, 100730. Available from: https://doi.org/10.1016/j.coco.2021.100730. February.

do Nascimento Junior, J.R., et al., 2021. Enhancement of biohydrogen production in industrial wastewaters with vinasse pond consortium using lignin-mediated iron nanoparticles. International Journal of Hydrogen Energy 46 (54), 27431–27443. Available from: https://doi.org/10.1016/j.ijhydene.2021.06.009.

Dutta, A., et al., 2019. Copper nanoparticles immobilized on nanocellulose: a novel and efficient heterogeneous catalyst for controlled and selective oxidation of sulfides and alcohols. Catalysis Letters 149 (1), 141–150. Available from: https://doi.org/10.1007/s10562-018-2615-x.

Elemike, E.E., Onwudiwe, D.C., Nundkumar, N., Singh, M., 2019a. CuO and Au-CuO nanoparticles mediated by *Stigmaphyllon ovatum* leaf extract and their anticancer potential. Inorganic Chemistry Communications 104, 93–97. Available from: https://doi.org/10.1016/j.inoche.2019.03.039. January.

Elemike, E.E., Onwudiwe, D.C., Nundkumar, N., Singh, M., et al., 2019b. Green synthesis of Ag, Au and Ag-Au bimetallic nanoparticles using *Stigmaphyllon ovatum* leaf extract and their in vitro anticancer potential. Materials Letters 243, 148–152. Available from: https://doi.org/10.1016/j.matlet.2019.02.049.

El-Sayed, M.S., et al., 2019. Effect of Ficus carica L. leaves extract loaded gold nanoparticles against cisplatin-induced acute kidney injury. Colloids Surf B Biointerfaces 184, 110465. Available from: http://doi.org/10.1016/j.colsurfb.2019.110465.

Figueiredo, P., et al., 2017. Functionalization of carboxylated lignin nanoparticles for targeted and pH-responsive delivery of anticancer drugs. Nanomedicine: Nanotechnology, Biology, and Medicine 12 (21), 2581–2596. Available from: https://doi.org/10.2217/nnm-2017-0219.

Gaddam R.R., et al., 2015. Bulk synthesis of green carbon nanomaterials from Desmostachya bipinnata for the development of functional polyurethane hybrid coatings. Progress in Organic Coatings 79, 37–42. https://doi.org/10.1016/j.porgcoat.2014.11.001.

Gan, L., et al., 2018. Effects of cetyltrimethylammonium bromide on the morphology of green synthesized Fe_3O_4 nanoparticles used to remove phosphate. Materials Science and Engineering: C 82, 41–45. Available from: https://doi.org/10.1016/j.msec.2017.08.073.

Gao, Y., et al., 2020. Porous bamboo-like CNTs prepared by a simple and low-cost steam activation for supercapacitors. International Journal of Energy Research 44 (13), 10946–10952. Available from: https://doi.org/10.1002/er.5672.

Garcia Gonzalez, M.N., et al., 2017. Lignin nanoparticles by ultrasonication and their incorporation in waterborne polymer nanocomposites. Journal of Applied Polymer Science 134 (38), 1–10. Available from: https://doi.org/10.1002/app.45318.

Gilca, I.A., et al., 2015. Obtaining lignin nanoparticles by sonication. Ultrasonics Sonochemistry 23, 369–375. Available from: https://doi.org/10.1016/j.ultsonch.2014.08.021.

Haseena, S., et al., 2021. Investigation on photocatalytic activity of bio-treated α-Fe_2O_3 nanoparticles using Phyllanthus niruri and Moringa stenopetala leaf extractagainst methylene blue and phenol molecules: kinetics, mechanismand stability. Journal of Environmental Chemical Engineering 9, 104996. Available from: https://doi.org/10.1016/j.jece.2020.104996.

Hastuti, N., et al., 2018. Hydrochloric acid hydrolysis of pulps from oil palm empty fruit bunches to produce cellulose nanocrystals. Journal of Polymers and the Environment 26 (9), 3698–3709. Available from: https://doi.org/10.1007/s10924-018-1248-x.

Hu, Y., et al., 2019. An efficient adsorbent: simultaneous activated and magnetic ZnO doped biochar derived from camphor leaves for ciprofloxacin adsorption. Bioresource Technology 288, 1–8. Available from: https://doi.org/10.1016/j.biortech.2019.121511. April.

Hu, B., et al., 2021. Cobalt-gadolinium modified biochar as an adsorbent for antibiotics in single and binary systems. Microchemical Journal 166, 106235. Available from: https://doi.org/10.1016/j.microc.2021.106235.

Huang, S., et al., 2019. Sulfurized biochar prepared by simplified technic with superior adsorption property towards aqueous Hg(II) and adsorption mechanisms. Materials Chemistry and Physics 238. Available from: https://doi.org/10.1016/j.matchemphys.2019.121919. April.

Huang, T., et al., 2022. Green synthesis of Ag/Fe_3O_4 nanoparticles using Mentha extract: preparation, characterization and investigation of its anti-human lung cancer application. Journal of Saudi Chemical Society 26 (4), 101505. Available from: https://doi.org/10.1016/j.jscs.2022.101505.

Ikeda, Y., et al., 2017. Reinforcing biofiller "lignin" for high performance green natural rubber nanocomposites. RSC Advances 7 (9), 5222–5231. Available from: https://doi.org/10.1039/c6ra26359c.

Imade, E.E., et al., 2022. Green synthesis of zinc oxide nanoparticles using plantain peel extracts and the evaluation of their antibacterial activity. Scientific African 16, e01152. Available from: https://doi.org/10.1016/j.sciaf.2022.e01152.

Jadhav, D.B., Kokate, R.D., 2019. Green synthesis of SnO_2 using green papaya leaves for nanoelectronics (LPG sensing) application. Materials Today: Proceedings 26, 998–1004. Available from: https://doi.org/10.1016/j.matpr.2020.01.180.

Jairam, S., et al., 2013. Encapsulation of a biobased lignin-saponite nanohybrid into polystyrene Co-butyl acrylate (PSBA) latex via miniemulsion polymerization. ACS Sustainable Chemistry & Engineering 1, 1630–1637. Available from: https://doi.org/10.1021/sc4003196.

Jayaprakash, N., et al., 2017. Green synthesis of Ag nanoparticles using Tamarind fruit extract for the antibacterial studies. Journal of Photochemistry and Photobiology B: Biology 169, 178–185. Available from: https://doi.org/10.1016/j.jphotobiol.2017.03.013.

Kalaiyan, G., et al., 2020. Green synthesis of CuO nanostructures with bactericidal activities using Simarouba glauca leaf extract. Chemical Physics Letters 761 (3), 138062. Available from: https://doi.org/10.1016/j.cplett.2020.138062.

Karthikeyan, C., et al., 2021. Biocidal (bacterial and cancer cells) activities of chitosan/CuO nanomaterial, synthesized via a green process. Carbohydrate Polymers 259, 117762. Available from: https://doi.org/10.1016/j.carbpol.2021.117762.

Khalil, M.M.H., et al., 2012. Biosynthesis of Au nanoparticles using olive leaf extract. 1st Nano updates. Arabian Journal of Chemistry 5 (4), 431–437. Available from: https://doi.org/10.1016/j.arabjc.2010.11.011.

Kolodziejczak-Radzimska, A., et al., 2014. Zinc oxide-from synthesis to application: a review. Materials 7 (4), 2833–2881. Available from: https://doi.org/10.3390/ma7042833.

Lee, E., et al., 2019. Crosslinking of lignin/poly(vinyl alcohol) nanocomposite fiber webs and their antimicrobial and ultraviolet-protective properties. Textile Research Journal 89 (1), 3–12. Available from: https://doi.org/10.1177/0040517517736468.

Leili, M., et al., 2018. Green synthesis of nano-zero-valent iron from Nettle and Thyme leaf extracts and their application for the removal of cephalexin antibiotic from aqueous solutions. Environmental Technology 39 (9), 1158–1172. Available from: https://doi.org/10.1080/09593330.2017.1323956.

Letsholathebe, D., et al., 2019. Green synthesis of ZnO doped *Moringa oleifera* leaf extract using Titon yellow dye for photocatalytic applications. Materials Today: Proceedings 475–479. Available from: https://doi.org/10.1016/j.matpr.2020.05.119.

Li, W., et al., 2021. Antimicrobial activity of sliver nanoparticles synthesized by the leaf extract of *Cinnamomum camphora*. Biochemical Engineering Journal 172, 108050. Available from: https://doi.org/10.1016/j.bej.2021.108050.

Liu, Y., et al., 2021. A new nanocomposite assembled with metal organic framework and magnetic biochar derived from pomelo peels: a highly efficient adsorbent for ketamine in wastewater. Journal of Environmental Chemical Engineering 9 (5), 106207. Available from: https://doi.org/10.1016/j.jece.2021.106207.

Logambal, S., et al., 2022. Synthesis and characterizations of CuO nanoparticles using Couroupita guianensis extract for and antimicrobial applications. Journal of King Saud University - Science 34 (3), 101910. Available from: https://doi.org/10.1016/j.jksus.2022.101910.

Luna, M., et al., 2019. Biosynthesis of uniform ultra-small gold nanoparticles by aged Dracaena Draco L extracts. Colloids and Surfaces A: Physicochemical and Engineering Aspects 581, 123744. Available from: https://doi.org/10.1016/j.colsurfa.2019.123744.

Luo, T., et al., 2021. Innovative production of lignin nanoparticles using deep eutectic solvents for multifunctional nanocomposites. International Journal of Biological Macromolecules 183, 781–789. Available from: https://doi.org/10.1016/j.ijbiomac.2021.05.005.

Lupul, I., et al., 2015. Tailoring of porous texture of hemp stem-based activated carbon produced by phosphoric acid activation in steam atmosphere. Journal of Porous Materials 22 (1), 283–289. Available from: https://doi.org/10.1007/s10934-014-9894-4.

Maiti, S., et al., 2013. Preparation and characterization of nano-cellulose with new shape from different precursor. Carbohydrate Polymers 98 (1), 562–567. Available from: https://doi.org/10.1016/j.carbpol.2013.06.029.

Malik, A.R., et al., 2022. Green synthesis of RGO-ZnO mediated *Ocimum basilicum* leaves extract nanocomposite for antioxidant, antibacterial, antidiabetic and photocatalytic activity. Journal of Saudi Chemical Society 26 (2), 101438. Available from: https://doi.org/10.1016/j.jscs.2022.101438.

Mostafavi, E., et al., 2022. Antineoplastic activity of biogenic silver and gold nanoparticles to combat leukemia: beginning a new era in cancer theragnostic. Biotechnology Reports 34, e00714. Available from: https://doi.org/10.1016/j.btre.2022.e00714.

Murugan, N., et al., 2019. Green synthesis of fluorescent carbon quantum dots from Eleusine coracana and their application as a fluorescence "turn-off" sensor probe for selective detection of Cu2. Applied Surface Science 476, 468–480. Available from: https://doi.org/10.1016/J.APSUSC.2019.01.090.

Muthuchamy, N., et al., 2018. High-performance glucose biosensor based on green synthesized zinc oxide nanoparticle embedded nitrogen-doped carbon sheet. Journal of Electroanalytical Chemistry 816, 195–204. Available from: https://doi.org/10.1016/j.jelechem.2018.03.059.

Muthuvel, A., et al., 2020. Effect of chemically synthesis compared to biosynthesized ZnO-NPs using *Solanum nigrum* leaf extract and their photocatalytic, antibacterial and in-vitro antioxidant activity. Journal of Environmental Chemical Engineering 8 (2), 103705. Available from: https://doi.org/10.1016/j.jece.2020.103705.

Nadeem, M., et al., 2018. The current trends in the green syntheses of titanium oxide nanoparticles and their applications. Green Chemistry Letters and Reviews 11 (4), 492–502. Available from: https://doi.org/10.1080/17518253.2018.1538430.

Nair, S.S., et al., 2014. High shear homogenization of lignin to nanolignin and thermal stability of nanolignin-polyvinyl alcohol blends. ChemSusChem 7 (12), 3513–3520. Available from: https://doi.org/10.1002/cssc.201402314.

Nasrollahzadeh, M., et al., 2020. Recent progresses in the application of cellulose, starch, alginate, gum, pectin, chitin and chitosan based (nano)catalysts in sustainable and selective oxidation reactions: a review. Carbohydrate Polymers 241, 116353. Available from: https://doi.org/10.1016/j.carbpol.2020.116353. April.

Nasrollahzadeh, M., et al., 2021a. Green-synthesized nanocatalysts and nanomaterials for water treatment: current challenges and future perspectives. Journal of Hazardous Materials 401, 123401. Available from: https://doi.org/10.1016/j.jhazmat.2020.123401.

Nasrollahzadeh, M., et al., 2021b. Starch, cellulose, pectin, gum, alginate, chitin and chitosan derived (nano)materials for sustainable water treatment: a review. Carbohydrate Polymers 251, 116986. Available from: https://doi.org/10.1016/j.carbpol.2020.116986. September 2020.

Nellore, J., et al., 2013. Bacopa monnieri phytochemicals mediated synthesis of platinum nanoparticles and its neurorescue effect on 1-methyl 4-phenyl 1,2,3,6 tetrahydropyridine-induced experimental parkinsonism in zebrafish. Journal of Neurodegenerative Diseases 2013, 1–8. Available from: https://doi.org/10.1155/2013/972391.

Noah, N.M., Ndangili, P.M., 2022. Green synthesis of nanomaterials from sustainable materials for biosensors and drug delivery. Sensors International 3, 100166. Available from: https://doi.org/10.1016/j.sintl.2022.100166.

Norrrahim, M.N.F., et al., 2021. Greener pretreatment approaches for the valorisation of natural fibre biomass into bioproducts. Polymers 13 (17). Available from: https://doi.org/10.3390/polym13172971.

Onwudiwe, D.C., et al., 2021. Hexavalent chromium reduction by ZnO, SnO_2 and ZnO-SnO_2 synthesized using biosurfactants from extract of *Solanum macrocarpon*. South African Journal of Chemical Engineering 38, 21–33. Available from: https://doi.org/10.1016/j.sajce.2021.07.002. June.

Oves, M., et al., 2022. Green synthesis of silver nanoparticles by Conocarpus Lancifolius plant extract and their antimicrobial and anticancer activities. Saudi Journal of Biological Sciences 29 (1), 460–471. Available from: https://doi.org/10.1016/j.sjbs.2021.09.007.

Pereira, T.M., et al., 2020. Modulating physical, chemical, and biological properties of silver nanoparticles obtained by green synthesis using different parts of the tree *Handroanthus heptaphyllus* (Vell.) Mattos. Colloid and Interface Science Communications 34, 100224. Available from: https://doi.org/10.1016/j.colcom.2019.100224.

Puthukkara, A.R.P., et al., 2021. Plant mediated synthesis of zero valent iron nanoparticles and its application in water treatment. Journal of Environmental Chemical Engineering 9 (1), 104569. Available from: https://doi.org/10.1016/j.jece.2020.104569.

Queiroz, R.N., et al., 2022. Adsorption of polycyclic aromatic hydrocarbons from wastewater using graphene-based nanomaterials synthesized by conventional chemistry and green synthesis: a critical review. Journal of Hazardous Materials 422, 126904. Available from: https://doi.org/10.1016/j.jhazmat.2021.126904.

Rajan, K., et al., 2020. Investigating the effects of hemicellulose pre-extraction on the production and characterization of loblolly pine nanocellulose. Cellulose 27 (7), 3693–3706. Available from: https://doi.org/10.1007/s10570-020-03018-8.

Ramesh, P.S., et al., 2015. Plant mediated green synthesis and antibacterial activity of silver nanoparticles using Emblica officinalis fruit extract. Spectrochimica Acta - Part A: Molecular and Biomolecular Spectroscopy 142, 339–343. Available from: https://doi.org/10.1016/j.saa.2015.01.062.

Ruangtong, J., et al., 2020. Green synthesized ZnO nanosheets from banana peel extract possess anti-bacterial activity and anti-cancer activity. Materials Today Communications 24, 101224. Available from: https://doi.org/10.1016/j.mtcomm.2020.101224.

Sadiq, H., et al., 2021. Green synthesis of ZnO nanoparticles from *Syzygium cumini* leaves extract with robust photocatalysis applications. Journal of Molecular Liquids 335, 116567. Available from: https://doi.org/10.1016/j.molliq.2021.116567.

Şahin, B., et al., 2021. Superior antibacterial activity against seed-borne plant bacterial disease agents and enhanced physical properties of novel green synthesized nanostructured ZnO using Thymbra spicata plant extract. Ceramics International 47 (1), 341–350. Available from: https://doi.org/10.1016/j.ceramint.2020.08.139.

Samadi, Z., et al., 2021. Facile green synthesis of zero-valent iron nanoparticles using barberry leaf extract (GnZVI@BLE) for photocatalytic reduction of hexavalent chromium. Bioorganic Chemistry 114, 105051. Available from: https://doi.org/10.1016/j.bioorg.2021.105051.

Shabaani, M., et al., 2020. Green synthesis of ZnO nanoparticles using loquat seed extract; Biological functions and photocatalytic degradation properties. LWT 134, 110133. Available from: https://doi.org/10.1016/j.lwt.2020.110133.

Sheik Mydeen, S., et al., 2020. Biosynthesis of ZnO nanoparticles through extract from Prosopis juliflora plant leaf: antibacterial activities and a new approach by rust-induced photocatalysis. Journal of Saudi Chemical Society 24 (5), 393–406. Available from: https://doi.org/10.1016/j.jscs.2020.03.003.

Shim, T., et al., 2015. Effect of steam activation of biochar produced from a giant Miscanthus on copper sorption and toxicity. Bioresource Technology 197, 85–90. Available from: https://doi.org/10.1016/j.biortech.2015.08.055.

Singh, A.K., 2022. A review on plant extract-based route for synthesis of cobalt nanoparticles: photocatalytic, electrochemical sensing and antibacterial applications. Current Research in Green and Sustainable Chemistry 5, 100270. Available from: https://doi.org/10.1016/j.crgsc.2022.100270. January.

Soto-Robles, C.A., et al., 2019. Study on the effect of the concentration of Hibiscus sabdariffa extract on the green synthesis of ZnO nanoparticles. Results in Physics 15, 102807. Available from: https://doi.org/10.1016/j.rinp.2019.102807.

Soto-Robles, C.A., et al., 2021. Biosynthesis, characterization and photocatalytic activity of ZnO nanoparticles using extracts of *Justicia spicigera* for the degradation of methylene blue. Journal of Molecular Structure 1225. Available from: https://doi.org/10.1016/j.molstruc.2020.129101.

Suganthi, N., et al., 2020. *Hibiscus sabdariffa* leaf extract mediated 2-D fern-like ZnO/TiO_2 hierarchical nanoleaf for photocatalytic degradation. FlatChem 24, 100197. Available from: https://doi.org/10.1016/j.flatc.2020.100197. September.

Tang, X., Li, X., Sun, Z., 2022. Green supported Cu nanoparticles on modified Fe_3O_4 nanoparticles using thymbra spicata flower extract: investigation of its antioxidant and the anti-human lung cancer properties. Arabian Journal of Chemistry 15 (6), 103816. Available from: https://doi.org/10.1016/j.arabjc.2022.103816.

Teo, H.L., Wahab, R.A., 2020. Towards an eco-friendly deconstruction of agro-industrial biomass and preparation of renewable cellulose nanomaterials: a review. International Journal of Biological Macromolecules 161, 1414–1430. Available from: https://doi.org/10.1016/j.ijbiomac.2020.08.076.

Tewari, C., et al., 2022. Green and cost-effective synthesis of 2D and 3D graphene-based nanomaterials from *Drepanostachyum falcatum* for bioimaging and water purification applications. Chemical Engineering Journal Advances 10, 100265. Available from: https://doi.org/10.1016/j.ceja.2022.100265.

Trevisan, H., Rezende, C.A., 2020. Pure, stable and highly antioxidant lignin nanoparticles from elephant grass. Industrial Crops and Products 145, 112105. Available from: https://doi.org/10.1016/j.indcrop.2020.112105. January.

Truong, T.T. et al., 2022. Adsorption characteristics of Copper (II) ion on Cu-doped ZnO nanomaterials based on green synthesis from Piper Chaudocanm L. Leaves extract. *Colloid and Polymer Science* [Preprint], (Ii). Available from: https://doi.org/10.1007/s00396-022-05028-3.

Turunc, E., et al., 2021. Green synthesis of silver nanoparticles using pollen extract: characterization, assessment of their electrochemical and antioxidant activities. Analytical Biochemistry 621, 114123. Available from: https://doi.org/10.1016/j.ab.2021.114123.

Vanaamudan, A., et al., 2016. Palm shell extract capped silver nanoparticles—As efficient catalysts for degradation of dyes and as SERS substrates. Journal of Molecular Liquids 215, 787–794. Available from: https://doi.org/10.1016/J.MOLLIQ.2016.01.027.

Varadavenkatesan, T., et al., 2019. Photocatalytic degradation of Rhodamine B by zinc oxide nanoparticles synthesized using the leaf extract of *Cyanometra ramiflora*. Journal of Photochemistry and Photobiology B: Biology 199, 111621. Available from: https://doi.org/10.1016/J.JPHOTOBIOL.2019.111621.

Vasantharaj, S., et al., 2021. Enhanced photocatalytic degradation of water pollutants using bio-green synthesis of zinc oxide nanoparticles (ZnO NPs. Journal of Environmental Chemical Engineering 9 (4), 105772. Available from: https://doi.org/10.1016/j.jece.2021.105772.

Velsankar, K., et al., 2020. Green synthesis of CuO nanoparticles via *Allium sativum* extract and its characterizations on antimicrobial, antioxidant, antilarvicidal activities. Journal of Environmental Chemical Engineering 8 (5), 104123. Available from: https://doi.org/10.1016/j.jece.2020.104123.

Vinay, S.P., et al., 2021. In-vitro antibacterial, antioxidant and cytotoxic potential of gold nanoparticles synthesized using novel Elaeocarpus ganitrus seeds extract. Journal of Science: Advanced Materials and Devices 6 (1), 127–133. Available from: https://doi.org/10.1016/j.jsamd.2020.09.008.

Wang, J., et al., 2015. Large-scale green synthesis of fluorescent carbon nanodots and their use in optics applications. Advanced Optical Materials 3 (1), 103–111. Available from: https://doi.org/10.1002/adom.201400307.

Wang, S., et al., 2016. Preparation of polyethylene/lignin nanocomposites from hollow spherical lignin-supported vanadium-based Ziegler–Natta catalyst. Polymers for Advanced Technologies 27 (10), 1351–1354. Available from: https://doi.org/10.1002/pat.3803.

Wang, J., et al., 2020. Adsorption and desorption of cellulase on/from lignin pretreated by dilute acid with different severities. Industrial Crops and Products 148, 112309. Available from: https://doi.org/10.1016/j.indcrop.2020.112309. October 2019.

Wang, H., et al., 2021. Colloidal lignin nanoparticles from acid hydrotropic fractionation for producing tough, biodegradable, and UV blocking PVA nanocomposite. Industrial Crops and Products 168, 113584. Available from: https://doi.org/10.1016/j.indcrop.2021.113584. May.

Wang, C., et al., 2022. Bio-supported of Cu nanoparticles on the surface of Fe_3O_4 magnetic nanoparticles mediated by Hibiscus sabdariffa extract: evaluation of its catalytic activity for synthesis of pyrano[3,2-c]chromenes and study of its anti-colon cancer properties. Arabian Journal of Chemistry 15 (6), 103809. Available from: https://doi.org/10.1016/j.arabjc.2022.103809.

Wicaksono, W.P., et al., 2022. Formaldehyde electrochemical sensor using graphite paste-modified green synthesized zinc oxide nanoparticles. Inorganic Chemistry Communications 143, 109729. Available from: https://doi.org/10.1016/j.inoche.2022.109729.

Wu, J., et al., 2018. Impacts of amount of chemical agent and addition of steam for activation of petroleum coke with KOH or NaOH. Fuel Processing Technology 181, 53–60. Available from: https://doi.org/10.1016/j.fuproc.2018.09.018. September.

Xiu, J., et al., 2022. Facile preparation of Fe_2O_3 nanoparticles mediated by *Centaurea alba* extract and assessment of the anti-atherosclerotic properties. Arabian Journal of Chemistry 15 (1), 103493. Available from: https://doi.org/10.1016/j.arabjc.2021.103493.

Yang, W., et al., 2015. Effect of processing conditions and lignin content on thermal, mechanical and degradative behavior of lignin nanoparticles/polylactic (acid) bionanocomposites prepared by melt extrusion and solvent casting. European Polymer Journal 71, 126–139. Available from: https://doi.org/10.1016/j.eurpolymj.2015.07.051.

Yang, W., et al., 2018. Role of lignin nanoparticles in UV resistance, thermal and mechanical performance of PMMA nanocomposites prepared by a combined free-radical graft polymerization/masterbatch procedure. Composites Part A: Applied Science and Manufacturing 107, 61–69. Available from: https://doi.org/10.1016/j.compositesa.2017.12.030.

Yew, Y.P., et al., 2020. Green biosynthesis of superparamagnetic magnetite Fe_3O_4 nanoparticles and biomedical applications in targeted anticancer drug delivery system: a review. Arabian Journal of Chemistry 13 (1), 2287–2308. Available from: https://doi.org/10.1016/J.ARABJC.2018.04.013.

Yogi, A., et al., 2017. Synthesis of cellulose nanocrystals from oil palm empty fruit bunch by using phosphotungstic acid. ICNSE 2017: 19th International Conference on Nanomaterials Science and Engineering 4 (10), 74549.

Yulizar, Y., et al., 2020. CuO-modified $CoTiO_3$ via *Catharanthus roseus* extract: a novel nanocomposite with high photocatalytic activity. Materials Letters 277, 128349. Available from: https://doi.org/10.1016/j.matlet.2020.128349.

Zangeneh, M.M., 2020. Green synthesis and formulation a modern chemotherapeutic drug of *Spinacia oleracea* leaf aqueous extract conjugated silver nanoparticles; chemical characterization and analysis of their cytotoxicity, antioxidant, and anti-acute myel. Applied Organometallic Chemistry 34 (1). Available from: https://doi.org/10.1002/aoc.5295.

Zhang, X., et al., 2020. High performance PVA/lignin nanocomposite films with excellent water vapor barrier and UV-shielding properties. International Journal of Biological Macromolecules 142, 551–558. Available from: https://doi.org/10.1016/j.ijbiomac.2019.09.129.

Zhang, N., et al., 2021. Advanced thin-film nanocomposite membranes embedded with organic-based nanomaterials for water and organic solvent purification: a review. Separation and Purification Technology 269, 118719. Available from: https://doi.org/10.1016/j.seppur.2021.118719.

Zhang, R., Leiviskä, T., 2020. Surface modi fi cation of pine bark with quaternary ammonium groups and its use for vanadium removal. Chemical Engineering Journal 385, 123967. Available from: https://doi.org/10.1016/j.cej.2019.123967. November 2019.

Chapter 23

Recent advances in agriculture waste for nanomaterial production

Manish Gaur[1], Charu Misra[2], Anand Kumar Bajpayee[3] and Abhishek Kumar Bhardwaj[4]

[1]Centre of Biotechnology, Institute of Inter-Disciplinary Sciences, University of Allahabad, Prayagraj, Uttar Pradesh, India, [2]Department of Pharmacy, School of Chemical Sciences and Pharmacy, Central University of Rajasthan, Ajmer, Rajasthan, India, [3]Department of Zoology, MLK [PG] College, Balrampur, Uttar Pradesh, India, [4]Department of Environmental Science, Amity School of Life Sciences, Amity University, Gwalior, Madhya Pradesh, India

23.1 Introduction

According to the United Nations, the global annual agricultural biomass production is around 30 billion tons (Feng et al., 2020). Press Information Bureau estimated that 62 million agricultural waste is produced yearly, which has an annual growth rate of 4% in India (Maji et al., 2020). Modern technologies need us to minimize costs while doing thorough risk analyses of the operations and maximizing productivity. Advances in nanomaterials including fullerenes, nanofibers, quantum dots, and CNTs (carbon nanotubes) have been made recently. These materials may be used in various agricultural fields due to their distinctive physical, optical, and mechanical characteristics. These may be included in nanosensors for important uses such as soil analysis, pesticide release, water supply, and the control of other biochemicals. This chapter discusses the green synthesis of NPs, their characteristics, and their significant biological applications. This chapter provides a thorough, insightful, and comprehensible review of this creative and sustainable trending subject.

Advancement of technologies increased, which improved agricultural production and quality, but at the same time worldwide population has grown enormously. Therefore there is one way to utilize this waste, which will be the key factor of a growing economy. Avoiding the reuse of residues and waste, a significant amount of agricultural-based biomass production is wasted. There are enormous opportunities to fill this gap by utilizing waste through numerous innovations in agriculture production. Nanotechnology is expected to make a significant impact in this area as well. Significant amounts of food and agricultural products are wasted each year on a massive level. According to one estimate, 1/3 of the food produced each year is wasted for human consumption (almost 1.6 G tons). However, according to a different estimate, agricultural products, including food and nonfood items, are lost annually at a rate of 6 G tons (Javad, Akhtar and Naz, 2020). Initially, in the process, we need to reduce this extra loss of agricultural products, which unwillingly increases the burden on our resources. It will minimize the demand for food supply even with an increasing population that will naturally lower the usage of chemical fertilizers and pesticides. There are significant issues associated with crop harvesting, such as crop waste, which makes up roughly 80% of all agricultural biomass. This waste is typically burned by farmers, which results in significant harmful gaseous emissions. Thus it is necessary to handle waste in a systematic manner to recover, recycle, and repurpose agricultural waste. Agricultural waste is not waste; it will be wealth when it is used properly and becomes a part of income and economy. Numerous applications have been proposed, including the fabrication of biotechnological goods, the synthesis of nanoparticles (NPs), and the manufacturing of manure and biofuel production, among others. This chapter discusses recent research studies of NMs synthesis using citrus and orange waste, banana peel, lignocellulose fibers, barley grain, corn hub, rice husk, wheat straw, rice residues, wood fibers, sawdust, and different types of agricultural wastes.

Green and Sustainable Approaches Using Wastes for the Production of Multifunctional Nanomaterials. DOI: https://doi.org/10.1016/B978-0-443-19183-1.00008-8

23.2 Various forms of agricultural waste

Agricultural waste is mainly categorized into two parts: agricultural residues and agricultural industry residues. In the category of agricultural residues, it is further classified into two parts: field residues and process residues. Field residues include stems, shells. Stacks and process residues contain husk, pulp, bagasse, and molasses. Agricultural industry residues are the waste that remains after making food packaging products such as potato peel, orange feel, soybean oil cake, groundnut oil cake, coconut oil cake, and many others (Sadh, Duhan and Duhan, 2018).

The majority of waste comes from products such as fruits, cotton, tea, vegetables, coffee, and cereals. These waste materials can be utilized to make organic solvents such as ethanol, acetone, and butanol as well as biofuels (Kannah et al., 2020). Moreover, pineapple leaf fiber is being researched as a fiberglass replacement. It is suggested to be utilized in construction, especially (Lopattananon, Payae and Seadan, 2008). Another illustration of lignocellulosic waste that is a significant contributor is waste from the wine-making process. These leftover wastes can be utilized as cultivation media, to alter the soil, to produce lactic acid, phenolic acid, antioxidant chemicals, and biosurfactants, as well as a low-cost heavy metal adsorbent (Nanni, Parisi and Colonna, 2021).

Orange juice is among the most popular juices consumed globally. Following extraction, almost half of the fruit's weight is converted into citrus peel waste of orange (CPWO), a cheap substance that works best for the creation of first-generation bioethanol. The researchers also demonstrated that orange peel waste could be fermented by the strain *Saccharomyces cerevisiae* that can also yield substantial amounts of glucose and ethanol by using an innovative direct steam injection method. CPWO is a useful bioresource for the manufacture of nanocellulose, d-limonene, and bioethanol (20% g/g of dry CPWO) (Tsukamoto, Durán and Tasic, 2013).

Due to its ecological attributes and renewable nature, pineapple biomass has gained interest in tropical areas. Currently, the production of pineapple leaves is a waste. These fibers can be obtained for industrial sectors without spending additional charges. The composition consists of a system of vascular bundles constructed of bundles of fibrous cells that were recovered after all epidermal tissues had been physically removed. On the other hand, the fibers are quite hygroscopic and reasonably priced. Because of its high cellulose content and 14 degrees microfibrilliary angle, pineapple leaf fibers have been shown to have multicellular and lignocellulosic top mechanical qualities. On the behalf of above application of waste, it can be utilized in synthesis of various nanoparticles discussed Fig. 23.1.

23.3 Types of nanomaterial synthesized from agricultural wastes

23.3.1 Synthesis of carbon nanomaterials from agricultural wastes

A range of agricultural wastes with a high carbon content, such as cellulose, hemicellulose, and lignin, are employed as carbon precursors. Numerous agricultural by-products have been studied for a long time and are still introducing new interest as precursors for activated carbon (Fathy, Basta and Lotfy, 2020). Due to their exceptional mechanical, thermal,

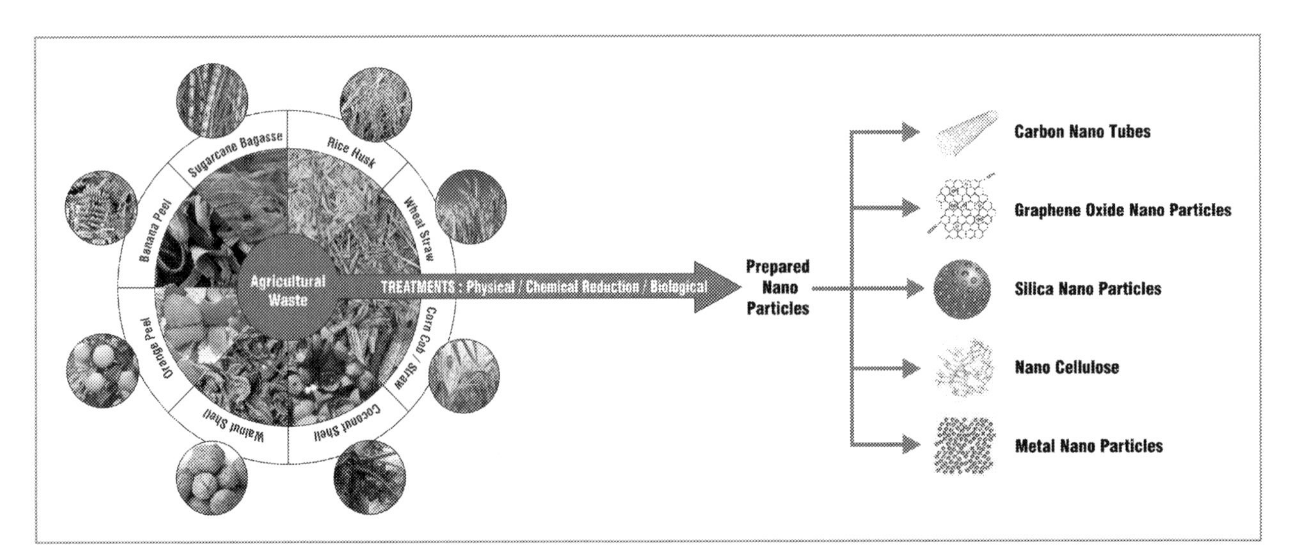

FIGURE 23.1 Numerous agricultural wastes were used for the synthesis of various types of potential nanomaterials through physical, chemical, and biological methods.

and electrical qualities, CNTs have recently gained more recognition. As a result, they offer enormous promise in applications for the environment and energy. According to Goodell et al., wood fiber can be used to construct CNTs with outer diameters in range of 10 and 20 nm through a continuous oxidation process at 240°C led by several cyclic oxidations at 400°C with a restricted amount of oxygen gas, even though pure cellulose and lignin did not produce any CNTs (Goodell et al., 2008). They suggested that the spatial and molecular architecture of the cell wall is the cause of the acquired morphology of tubular carbon atoms, which is closely related to the hexagonal arrangement of carbon atoms in the graphite sheets. The synthesis of carbon-based nanostructures also utilized the chemical vapour deposition (CVD)-assisted heat treatment approach. Sevilla and Fuertes used hydrothermal treatment of cellulose at 250°C proceeded by heat processing of impregnated hydrothermal char with nickel salt at 900°C to construct graphitic carbon nanostructures with coil-like patterns. Alvesa et al. reported the catalytic reduction of bioethanol by-products in horizontal stainless steel at temperatures ranging from 600°C to 1000°C to generate CNMs from the pyrolysis of both sugarcane bagasse and corn grains (Alves et al., 2011). They determined that the pyrolysis of sugarcane bagasse produced multiwalled CNTs (MWCNTs) with longer and straight tubes that ranged in length around 50 um and changed in diameter from 20 to 50 nm. The numerous types of CNMs produced by the pyrolysis of corn grains at 1000°C have diameters between 100 and 300 nm and long, entangled, cord-like structures with rugged walls. Zhu et al. reported that to create CNTs using CVD at temperatures between 1200°C and 1400°C while employed ethanol vapor as a carbon source and bamboo charcoal as a catalyst. They noticed that the bamboo charcoal, which contains calcium silicate, served as a catalyst to enhance the nucleation of MWCNTs (Zhu et al., 2012). Fathy recently developed CNT bundles using two different methods: (1). the CVD technique employing camphor as a carbonaceous gas, and (2) hydrothermal treatment of pretreated rice straw impregnated with iron and/or nickel metals at 250°C for 2 hours to form hydrochar utilized as a substrate (Fathy, El-Shafey and Khalil, 2013). The hydrochar made from rice straw was carbonized at 800°C while the carbon gas from the cracked camphor was evaporated at 250°C for two hours to generate CNTs growth with outer diameters of 22–66 nm. The findings demonstrated that the use of Fe and Ni oxides as catalysts improved the graphitization of CNTs onto carbonized hydrochar, which was produced hydrothermally from pretreated rice straw. CNT yield was achieved at 44% with enhanced surface area per gram.

Based on its ability to generate a high-energy-density plasma, microwave irradiation has recently been used to manufacture CNMs, such as carbon nanofibers, carbon nanowalls, and CNTs. Recently, Wang et al. reported that microwave plasma irradiation was used to create graphenated-CNT (g-CNTs) materials from rice husk waste. The resulting nanocarbons were confirmed to be g-CNTs, in which the graphene sheets primarily constituted 2–6 layers, and the CNTs had a length among many 10μm and a diameter of 50–200 nm. Table 23.1 shows synthesized CNTs from various carbon sources from different methods (Wang et al., 2015). Alves et al. developed a new pyrolysis reactor to produce MWCNTs with approximately 50 μm length and 20–50 nm widths from leftover sugarcane bagasse. In this study, CNTs were produced on stainless steel meshes made of AISI 304 by catalytically converting the carbon content of gases produced during the pyrolysis of sugarcane bagasse (Alves et al., 2011). Similar to this, during the pyrolysis of wood sawdust, Bernd et al. (2017) developed multiple carbon nanostructures of CNTs and carbon nanofibers (CNFs). This procedure involved heating a tubular reactor filled with a reducing agent (commercial zinc), calcite (used as a bed material), and a catalyst (ferrocene or Fe/Mo/MgO) to 750°C for 3 hours devoid the use of any other gases, such as nitrogen, air, or external carbon gas. The obtained samples of CNT and CNF were grown in a variety of sizes and forms (Bernd et al., 2017).

23.3.2 Synthesis of graphene oxide from agricultural wastes

Worldwide, 120 million tons of rice husks are generated each year. In most cases, rice residues have been burned due to their low commercial value either dumped without a purpose. By utilizing it in the best way, rice waste and sugarcane bagasse are used to prepare graphene. Graphene holds excellent electrical conductivity, high strength, and good thermal conductivity, which contribute in many fields and research such as medicine, energy, biosensing, drug delivery, desalination, and diagnostic (Tuck et al., 2012; Gaur et al., 2021).

Sugar is heavily consumed in sweets and made from sugarcane. Sugarcane is largely produced in agriculture. Sugar is prepared from the juice of sugarcane by crushing so remaining material is dry, fiber known as sugarcane bagasse. The fibers were processed with crushing and grinding to make powder form through several repeated processes. It is estimated that 0.5 g of powder sugarcane mixed with 0.1 g of ferrocene was kept in a crucible and set into a muffle furnace at 300°C as shown in Fig. 23.2. Then a black solid product was formed and collected at room temperature (Somanathan et al., 2015).

TABLE 23.1 Carbon nanostructures were shown in the following; synthesized from different agricultural waste, their size, used catalyst, temperature, and growth time.

Carbon nanostructure	Methods of preparation	Using agricultural waste	Size (nm)	Used catalyst/temperature/ growth time	References
Carbon nanotubes	Microwave assisted	Tea-tree extract	10–40 nm	Fast reaction time, eco-friendly	Mostafavi et al. (2022)
Carbon quantum dots	Hydrothermal approach	Corn stalk shell	<10 nm	Without use of chemical reagent	Li et al. (2021)
Carbon nanotubes	pyrolysis	Walnut shells	<15 nm	Iron/700°C/ 4 h	Roquia et al. (2021)
Multiwalled Carbon nanotubes	pyrolysis	Chickpea peel	7 nm	400°C in aqueous medium	Singh et al. (2021)
Carbon quantum dots and carbon nanofibers	Ball milling assisted hydrothermal	Pine fruit	< 10 nm	Heated at 180°C for 4.5 h	Shahba and Sabet (2020)
Mixture of SWNTs and MWNTs	Catalytic hydrothermal pyrolysis	Rice straw	4–47	Fe-Ni/Al_2O_3, 830°C/60 min	Lotfy, Fathy and Basta (2018)
MWCNTs	CVD	Camphor	22–66	Fe-Ni/rice straw hydrochar as substrate /800°C/ 120 min	Fathy (2017)
MWCNTs	CVD	Camphor	20–40	Fe-Ni/rice straw or organic xerogel as substrate/ 750°C/ 60 min	Fathy, Lotfy and Basta (2017)
Carbon nanosheets	Conventional pyrolysis	Sugarcane bagasse	10–150 nm in thickness	KOH/ 600°C–900°C	Niu et al. (2017)
Carbon dots	Hydrothermal treatment	Wheat straw	<2 nm	250°C/10 h	Yuan et al. (2015)

Rice wastes are used in the production of graphene together with a range of chemical activators. Furthermore, potassium hydroxide (KOH) is the chemical agent most usually used in the production of CNMs because it promotes and increases the porosity of the resulting CNMs. A continuous 3D network is produced by activation with KOH, leading to high volumetric capacitance. By effectively removing contaminants such as silica and disordered carbon, it also improves the quality and integrity of the graphene nanomaterial (Fatihah et al., 2020; Raghavan, Thangavel and Venugopal, 2017). Muramatsu reported a new way for preparation of graphene from rice husk by using KOH for activation and basic calcination at a high temperature 850°C, which creates graphene sheets with high crystallinity, clean and nano-sized range. Corrugated graphene and crystalline graphene nanostructures were formed in single-layer and multi-layer manner. Monolayer and bilayer graphene sheets synthesized with stable edges and corrugated graphene were found in nanometer size with clear boundaries and nonhexagonal carbon rings, which have been characterized by XRD and Raman techniques. However, carbon black was used here to prevent oxidation while it was substituted with rice husk itself (Muramatsu et al., 2014).

Graphene nanosheets are synthesized by the carbonization of brown-rice husks followed by a one-stage KOH-activation process for the design of a sustainable electrochemical energy-storage electrode as shown in Fig. 23.3.

Uda et al. demonstrated that graphene was synthesized from rice straw ash by the using same protocol with a defined ratio and KOH at 700°C. Graphene was characterized by scanning electron microscope (SEM) with energy-dispersive X-rays, transmission electron microscope (TEM), and atomic force microscope (AFM) (Uda et al., 2020). In comparison to graphene, porous graphene shows distinctive physical properties, which increases the potential application of graphene. Othman synthesized the porous graphene by using rice husk ashes in different temperature conditions. Solid residues had formed at defined stabilization temperatures of 100°C, 200°C, 300°C, and 400°C and were denoted with their temperature variation as GRHA100, GRHA200, GRHA300, and GRHA400. GRHA shows graphene from

FIGURE 23.2 Graphical illustration of preparation of graphene oxide from sugarcane agro waste (Somanathan et al., 2015).

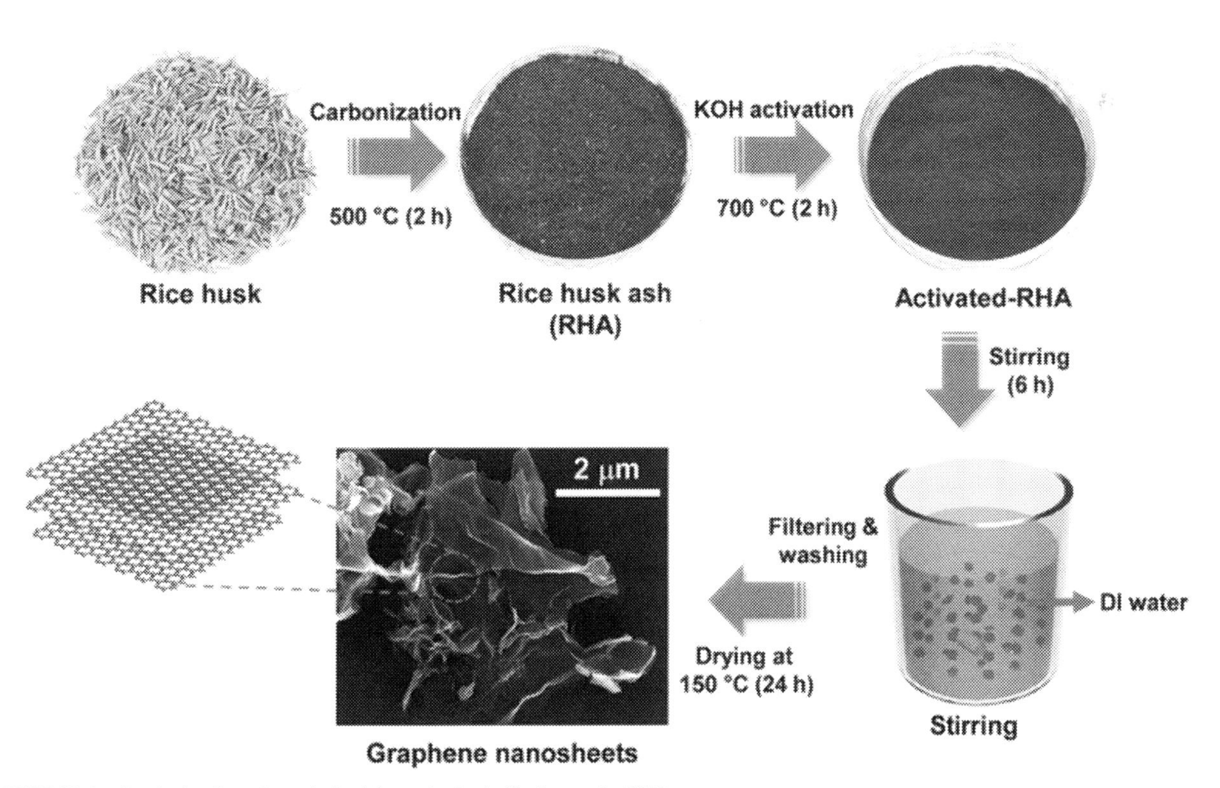

FIGURE 23.3 Synthesis of graphene derived from rice husk (Sankar et al., 2017).

Rice Husk Ashes. KOH was used for chemical activation at a 1:5 impregnated ratio. This porous graphene was characterized with TEM, Raman spectroscopy, and XRD (Che Othman et al., 2020).

The preparation of graphene oxide nanoparticles was also reported by agricultural waste and employed in various applications. Seitzhanova et al. prepared graphene oxide by using the rice husk. Synthesis process includes four steps: preliminary carbonization, desilication by using NaOH and mixing with KOH at definite ratio, then keeping in air for 2 hours at 850°C; after that the product was washed with water upto found pH 7 and then dried at 105°C. The resulting graphene layers had a large amount of carbon and rich in inorganic matter. The product was functionalized by H_2SO_4 with continuous stirring upto a day at regulated temperature, and then the mixture was centrifuged, washing upto getting neutral pH, and finally filtered with PVDF membrane. Graphene oxide was characterized by Fourier-transform infrared spectroscopy (FTIR), Raman spectroscopy, SEM, and TEM (Seitzhanova et al., 2019).

Jacob MV et al. reported the synthesis of high-quality graphene film from a tea plant's natural extract (*Melaleuca alternifolia*) without use of any catalyst. Prepared graphene film has shown excellent properties for coating in electronic devices, good contact angle and memristive behavior made it better. Graphene coating has been prepared by using plasma-enhanced CVD method, except use of catalyst. A specially designed quartz tube reactor with a heater coiled onto it and under the control of a variable voltage controller made up the experimental setup. The system was designed to operate at an 800°C temperature and 500 W of input RF energy. As substrates, 100 nm thermal oxide–coated Si wafers that had been cleaned were employed. Before deposition, Si materials were pretreated with plasma for 1 minute. As a carbon source, 99.90% pure *M. alternifolia* oil was utilized, which is volatile at ambient temperature, so no need of heating to transfer the vapors from the flask into the chamber. It has been characterized by UV-Vis spectroscopy, X-ray photoelectron spectroscopy, SEM, AFM, and confocal Raman spectroscopy (Jacob et al., 2015).

In a different work, graphene was synthesized using rice bran as a source of carbon. Hashmi et al. demonstrated the successful preparation of GO from agricultural waste under ideal circumstances and at a relatively low temperature, in both individual and tricomposite form (sugarcane bagasse, rice bran, and orange peel). The carbonaceous precursor, powdered rice bran, was combined with ferrocene and fired for 15 minutes at 300°C in a muffle kiln, producing solid products that are black in color and kept for 30 minutes at room temperature. Graphene oxide after preparation has been characterized by SEM, TEM, FTIR, XRD, TGA, and DTA (Differential Thermal Analysis). This straightforward, fast, one-step, and environmentally friendly method for synthesizing graphene oxide has demonstrated that a variety of solid carbonaceous agricultural wastes may be used as a feedstock to create high-quality, added-value GO without the need for sanitization (Hashmi et al., 2020).

Tohamy et al. used a similar approach to produce GO by independently oxidizing rice straw, sugarcane bagasse, lignin, and mature beech pinewood sawdust with ferrocene as a catalyst in a muffled atmosphere. SEM, TEM, FTIR, EDS (energy-dispersive spectroscopy), and Raman spectroscopy were used for characterization. Then, an adsorption technique for wastewater treatment was developed using the GO produced from rice straw. An effective and affordable adsorbent for removing Ni (II) from contaminated water was developed (Tohamy et al., 2020). The term "pyrolysis" refers to the thermal disintegration of organic compounds at high temperatures in either a vacuum or an inert atmosphere; this results in a change in chemical composition. Carbonization is a type of extreme pyrolysis that mostly produces carbon residue. The process of pyrolyzing biomass is incredibly efficient for creating a variety of products with additional value. It entails heating the organic material found in the biomass precursor, followed by dehydration, degradation, and carbonization (Zhou, 2017).

By adjusting the thermal heating conditions, any carbonaceous waste material can be converted to graphene oxide without the need for graphite or other hazardous compounds. Without the use of an inert atmosphere, Naik et al. showed the pyrolytic synthesis of graphene oxide from a nongraphitic carbonaceous source. The waste products from the agricultural industry, such as waste newspaper, rice husk, and sugarcane bagasse, were utilized as a starting point for the synthesis of graphene oxide nanoflakes. The precursors underwent pyrolysis (at the ideal temperature) with a very limited amount of air, which caused the breakdown of cellulosic structures and the formation of monomeric glucose. This caused the monomeric glucose to simultaneously undergo oxidation, condensation, and aromatization. The findings revealed GO in the form of brownish-blackish powders or flakes that had been crushed, cleaned, and dried for a day in an oven. The nanoflakes had a graphitic composition. XRD, EDS, SEM, and HR-TEM were used to describe graphene oxide nanoflakes (Naik et al., 2020) (Table 23.2).

23.3.3 Production of silica nanoparticles from agricultural waste

The hard protective pods of rice seeds known as rice husks (or rice hulls) make up 20% of the total amount of rice produced. Rice mill industries currently prepare large quantities of rice husk as one of the most abundant agricultural wastes because it contains high levels of cellulose (28%–35%), hemicellulose (12%–29%), and lignin (15%–20%).

TABLE 23.2 Synthesized graphene oxide nanostructures from various agricultural wastes as carbon sources, their size and synthesis process condition.

GO nanostructures	Methods of preparation	Using agricultural waste	Size (nm)	Used catalyst/ temperature/ growth time	References
GO	Hummer's method	Mixture of rice bran, orange peel and sugarcane bagasse	23.99 nm	Ferrocene /300°C/ 15 min	Hashmi et al. (2020)
GO based carbonaceous nanocomposite	Modified hummer's method	Rice husk	4.35 nm	H_3PO_4, $ZnCl_2$/ 30°C–40°C/2 h	Liou and Wang (2020)
GO	Hummer's method with modification	Sugarcane and rice straw mixture	250.9 nm (pore size)	$ZnCl_2$/300°C/ 10 min	Tohamy et al. (2020)
Potassium-doped GO nanosheet	Solvothermal process	Oak fruit	2.4	–/120°C/4 h	Tewari et al. (2019)
Reduced GO nanosheet	Modified hummer's method	Sugarcane bagasse	–	Without any oxidizing agent/ 450°C/	Debbarma, Naik and Saha (2019)
GO	Modified hummer's method	Coconut husk ash	0.14 nm	KMnO4/60°C/1 h	Sahila Grace, Prabha and Malar (2018)
GO nanoplatelets	Hummer's method	Rice straw biomass	60–100 nm	$KMnO_4$ as (oxidizer) /450°C/ 45 min	Goswami et al. (2017)

It's interesting to note that the husk of this plant has a high silica concentration up to 12%, primarily in the form of amorphous, hydrated SiO_2, which is similar to the form of the majority of other biosphere items (Zamani, Marjani and Mousavi, 2019). Nanoparticle's size was managed by surfactant concentration, ageing time, and temperature. For the manufacturing of silica NPs, Thuc et al. used acid-washed Vietnamese rice husk. When rice husk is calcined, small amounts of cations are removed, which increases the amount of SiO_2 that is extracted. Next, sodium hydroxide is added to make a sodium silicate solution. The cetyltrimethylammonium bromide (CTAB) solution in the mixture of water and butanol was then gradually added to this sodium silicate solution. By adding H_2SO_4, silica nanoparticles (SiO_2NPs) were precipitated at pH 4 (Le, Thuc and Thuc, 2013).

In another study, Wang et al. reported that the extraction of lignocellulose from rice husk by the treatment of ionic liquid and further calcination was done to separate rice husk residues to form 70 nm size of SiO_2NPs having a surface area of 241.1 m^2/g at 700°C in 2 hours duration. The surface area has been increased by using HCl at the place of ionic liquid with 283.3 m^2/g. Another amorphous form functionalized SiO_2 nanoparticle has been prepared from rice husk by dissolving in ionic liquids to extract lignocellulosic biomass and then thermally treated to collect from the solution (Chen et al., 2013). Rice husk ash was used for the synthesis of mesoporous silica NPs by using sodium silicate solution, NaOH solution, and CTAB employed as structure-directing agent, and in this reaction fusion temperature is shown as a critical factor to determine the size of mesopore and pore volume. It has been established that the silica materials have mesoporous structures with average pore diameters of 11.8–25.8 nm, specific surface areas of 211–518 m^2/g, and pore volumes ranging from 1.36 to 1.92 cm^3/g. Additionally, tetraethylenepentamine could be dispersed on these silica materials to act as support for CO_2 adsorption. In order to prepare, Andas et al. synthesized sodium silicate solution stirred rice husk that had been acid-treated and cleaned it in NaOH solution to produce a dark brown solution. Mesoporous cobalt catalysts (Co@SiO_2) with a large surface area were created by adding a cobalt (II) acid solution to a sodium silicate solution until the pH reached 3 (Scheme 23.1).

Andas et al. demonstrated that rice husk–derived silica bed was used in synthesis and optimization of immobilized silver NPs (Andas and Adam, 2016). Lin et al. optimized the development of layered sodium silicates by using of rice husk. As the temperature and synthesis time were increased, phase transitions were to be seen. Produced silica with a

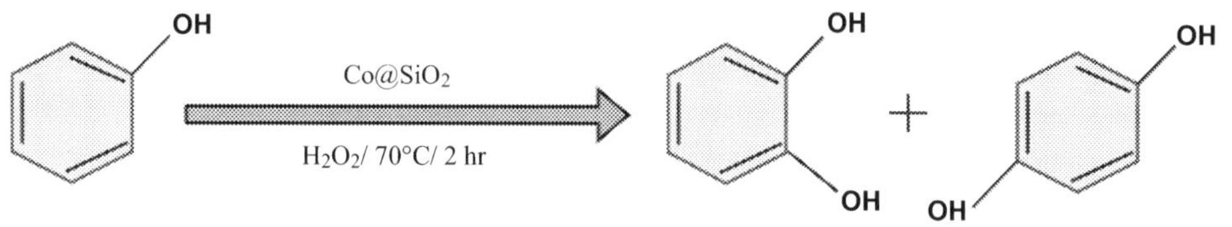

OH
Co@SiO2
H2O2/ 70°C/ 2 hr
OH
OH
+
OH
OH

SCHEME 23.1 Oxidation of phenol catalyzed by $Co@SiO_2$.

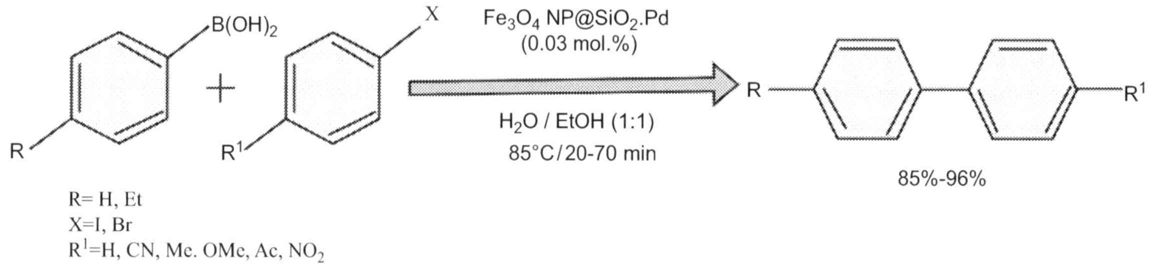

B(OH)2
+
X
R
R1
Fe3O4 NP@SiO2.Pd
(0.03 mol.%)
H2O / EtOH (1:1)
85°C/20-70 min
R
R1
85%-96%

R= H, Et
X=I, Br
R^1=H, CN, Me. OMe, Ac, NO_2

SCHEME 23.2 $Fe_3O_4NP@SiO_2$-Pd catalyzed Suzuki reaction.

larger percentage of the phase displayed stronger Mg^{2+} and Ca^{2+} binding capabilities, making it a suitable replacement for phosphorus-based detergents that are harmful to the environment (Deng et al., 2016). Mukti et al. developed a method in which synthesis of hierarchical ZSM-5 at less than 100°C temperature used sodium aluminate and silica from rice husks, with tetrapropylammonium bromide acting as a structure-directing agent (Kadja et al., 2017). Zeolite has shown porous crystalline silicates having spherical morphology. Zeolite synthesis is the most studied field due to various pore sizes, structures, and chemical stability, so numerous applications in the area of industrial processes in catalysis, absorbance, separation, chemical sensor, electronics, and pharmaceutical. The synthesized zeolite has shown microporosity and intercrystallite mesoporosity from their inherit nanocrystallites structure. Mor et al. synthesized 15 nm porous silicate particles from rice husk agricultural residues by using of acidification of sodium silicate solution (Mor et al., 2017). Hsu et al. described preparation of carbon–silica composite nanostructure from rice residues by washing with acid in unreactive environment, then hydrothermal treatment to minimize the impurities (Hsieh et al., 2017). Rangari et al. described the pyrolysis of rice husk particles in a high-pressure and -temperature reactor to form silica–carbon hybrid NPs with a size of 14 nm. Using a 3D printing technology, this nanomaterial was employed to create a nanocomposite from Ecoflex, a biodegradable compostable polymer (Biswas, Jeelani and Rangari, 2017). Additional testing was done on this 3D printed biocomposite to determine its mechanical and thermal characteristics. The inclusion of silica-based nanomaterial demonstrated a significant improvement in thermal stability in TGA and tensile testing. Fe_3O_4 NPs and silica produced from rice husks were mixed together to produce silica-coated magnetic NPs ($Fe_3O_4NP@SiO_2$), which were then utilized to support palladium ($Fe_3O_4NP@SiO_2$-Pd) (Khazaei, Khazaei and Nasrollahzadeh, 2017) (Scheme 23.2).

The sources such as rice husk, groundnut shell, bamboo leaves, and sugarcane bagasse are considered as a waste material. In recent years, sugarcane bagasse ash has been extensively used for silica NPs production because bagasse is easily available in large scale and is low-cost raw source. The bagasse constitutes have almost 80%–95% of SiO_2, which was analyzed by different countries. For the production of silica, bagasse was given pretreatment with acid (H_2SO_4, HNO_3) and base (NaOH, KOH). The pure crystalline form of nanosilica was reported by using three different methods: electrochemical, ball milling, and sol–gel procedures. Electrochemical method yields highly porous confined colloidal silica NPs. Although ball milling procedure is easy, operational inexpensive method synthesizing a range of size of nanosilica NPs, sol–gel is used for surface modification and controlled size distribution of nanosilica NPs (Chen et al., 2013).

Akhayere and coworker synthesized magnetic nanosilica by using barley husk waste. This nanosilica was used to eliminate the contamination of petroleum and its by-products from polluted water. These were synthesized by environment-friendly chemicals in two steps: acid reflux followed by heat treatment; after that particles were added into magnetic solution of Fe_3O_4 NPs. Nanosilica particles were characterized by SEM, FTIR, and XRD. Magnetic nanosilica particles eliminated petroleum contaminations, which depended on pH, contact point, reusability, and sorbent's weight (Akhayere, Vaseashta and Kavaz, 2020) (Table 23.3)

TABLE 23.3 Synthesized silica nanostructures from various agricultural wastes as carbon source from different synthesis methods and their attributes.

Silica nanostructures	Methods of preparation	Using agricultural waste	Size (nm)	Used to improve preparation	References
Nanostructured silica powder	Precipitation method	Rice hull ash	–	Calcined at 900°C/ 3 h	Mirasol and Shimomura (2023)
Amorphous silica oxide NPs	Environmentally friendly short method	Rice husks	<100 nm	calcination	Zarei et al. (2021)
Mesoporous silica NPs	Sol–gel technique	Rice husk	50–60 nm	Calcined at 550°C/ 4 h	Usgodaarachchi et al. (2021)
Silica NPs	Precipitation method and post-biotransformation with *Fusarium culmorum*	Corn cobs husks	40–70 nm	Lack of chemicals	Pieła et al. (2020)
Silica NPs	Modified sol–gel treatment	Maize stalk wastes	Below 30 nm	Sodium silicate reduces agglomeration	Adebisi et al. (2019)
Silica NPs	Acid treatment and combustion method	Barley grain wastes	~150 nm	700°C	Akhayere, Kavaz and Vaseashta (2019)
Amorphous silica nanodisks	Slow gelation and freeze-drying process	Rice straw ash	172 nm	Base dissolution and acid precipitation/ 575°C for gasification	Lu and Hsieh (2012)

23.3.4 Preparation of nanocellulose

However, at the moment, the majority of the usage of biomass waste, following a recycling or recovery operation, is mostly limited to a few minimal value-added products. For instance, the sugar industry frequently uses bagasse, one of the major agricultural wastes, to generate steam and electric power by direct combustion. The primary component of biomass wastes is cellulose, and its effective use in a high-value manner may open a new path to a promising future for biomass waste management. Nanocellulose is a cellulosic substance with one dimension in the nanoscale range, and its primary sources include bacteria, plants, and microbes. The extraction of nanocellulose for use from various biomass sources has drawn more and more interest (Yu et al., 2021). In addition to having a lot of hydroxyl groups, nanocellulose has some aldehyde and carboxyl groups for further chemical treatment and modification. Due to the material's nanostructure, which exhibits a very high specific surface area, active functionalization groups, mechanical strength, and crystallinity, nanocellulose has a diverse range of uses in various fields (Thakur et al., 2021). In fact, lignocellulosic agricultural waste can be employed to develop several types of nanocellulose that can be extensively used in the production of various materials in the domains of cosmetics, drugs, biomedicine, and many other industries. Several techniques, including chemical treatment, mechanical treatment, and chemo-mechanical treatments, are used to separate fibers from cellulose. Chemicals such as sodium hydroxide and KOH are frequently used in cellulose extraction processes. Excessive use of chemical can cause environmental issues. It has been established that procedures for pretreatment that are friendly to the environment, such as the use of superheated steam, hot compressed water, and steam explosion, are beneficial (Nor Faiz Norrrahim et al., 2021).

Pea is diversely used vegetable in which pea seeds are eatable, and covering part is a vegetable waste. Chen et al. used pea hull fibers for the synthesis of nanowhiskers, a type of crystal nanofibers having less than 100 nm in diameter, by using acid hydrolysis, and reported nanowhiskers were measured at 400–240 nm and 12–7 nm in length and diameter, respectively. The nanowhiskers from pea hull fibers and pea starch can be used together to produce a novel nanocomposite material as a possible expansion for composite materials. This material would exhibit some distinctive or enhanced properties, including excellent UV absorption, increased transparency, increased tensile strength, and passable water resistance (Chen et al., 2009). Some other researcher used apple stem to synthesize nanocellulose by the treatments of chemicals. Phantong et al. extracted lignin from raw apple stem for the synthesis of nanocellulose. Lignin was extracted by using a particular buffer solution made up of acetic acid, sodium chlorite and diluted with water. Then alkali treatment was given to find out the cellulose. Acid treatments were used to extracted nanocellulose, which has ~20 nm diameter, good degree of crystallinity, high thermal stability (Phanthong et al., 2015).

Abraham et al. reported synthesis of nanocellulose from agricultural residues such as pineapple leaves and the jute fibers. Cellulose removal and formation of nanocellulose were achieved through steam explosion. The produced nanocellulose, which can be employed as a polymer reinforcing agent, has extensive surface area and reported diameter of 5–40 nm. As a result, a physical tool is frequently employed or suggested in conjunction with other techniques to extract high-quality nanocellulose (Yu et al., 2021). Furthermore some researchers used mechanochemical esterification method for synthesis of nanocellulose. Kang et al. reported preparation of nanocellulose by extracting cellulose from corn cob residues and obtained diameter ~3 nm nanocellulose particles, which were used in nanopaper formation. Nanopaper is a type of film-making material with higher strength and mainly produced from nanocellulose by deposition of crystals and fibers. It has excellent optical properties and higher transparency rate. These enhanced characteristics of nanopaper may be used to improve a developed material's electronics and other unique qualities (Kang et al., 2017).

23.3.5 Preparation of metal nanoparticles

Green nanotechnology is tremendously growing in the synthesis of metal nanoparticles due to minimizing the use of harmful chemicals, more expenses on product processing, and encouraging biosynthesis (Bhardwaj and Naraian 2020; Bhardwaj et al., 2018). Agricultural waste has the potential to become a part of sustainable economy by using it as a raw material for nanomaterials synthesis.

Metal and metal oxide NPs can be prepared by using different types of agricultural waste and their properties improved. AgNPs have been prepared from metal salt by the using reducing power of agricultural wastes, which have phenolic compounds, flavanols, flavones, gallic acid, enzyme, anthocyanidins, and other organic compounds (Bhardwaj and Naraian 2021). In the past few years, many other weeds such as *Cyperus rotundus*, *Medicago sativa*, *Gloriosa superba*, and *Tinospora cordiofolia* were used for the synthesis of nanostructures and also from crop residues reported such as orange peel, banana peel, pomegranate peel, coconut shell, wheat straw, and rice husk (Thangadurai et al., 2021). Orange peel water extract was reported for the synthesis of AgNPs from silver nitrate salt at constant stirring and heating upto 75°C (Skiba and Vorobyova, 2019). Coconut shell extract was used for synthesis of AgNPs and evaluated their antibacterial activity against human pathogens. Spherical-shaped AgNPs were in range of 14.2–22.96 nm, and characterization was done by UV–visible spectroscopy, FTIR, XRD, and electron microscopy (Sinsinwar et al., 2018). Banana peel extract was used as reducing and capping agent for the preparation of Mn_3O_4 from the reduction of $KMnO_4$, which has extensive electrochemical performance (Yan et al., 2014). Banana peel extract was also used to synthesize bioinspired palladium NPs (PdNPs) by boiling of peel, crushing, and then acetone precipitation. PdNPs were shown spherical morphology, characterized and exhibited catalytic activity (Bankar et al., 2010). Baiocco D. et al. utilized agricultural waste and produced AgNPs by using phenolic extract of bilberry wastes and spent coffee grounds. Aqueous ethanol was used as extraction solvent. AgNPs were successfully achieved with spherical shape and 10–20 nm in size range in 3 hours–5 hours at ambient temperature (Baiocco et al., 2016). AgNPs were reported to be synthesized by using xylan of wheat bran, which acted like both reducing and stabilizing agent. AgNPs were better in scavenging activity with 20 nm size range (Harish, Uppuluri and Anbazhagan, 2015). Gold NPs were prepared from agricultural waste rice bran. Rice bran extracted was found accountable for reduction of Au^{3+} to Au^0 (Malhotra et al., 2014). Metal NPs have become useful in nanoform and easy to prepare, so widely used and explored by researchers. Zin oxide (ZnO) NPs were prepared from *Brassica oleracea botrytis* leaves, which shall remain in market and be considered as agricultural waste. These leaves also reported for the preparation of copper, iron, and lead NPs. ZnONPs have shown antibacterial activity, also used in medical, cosmetic, food industries, and textiles (Gupta and Chandra, 2020).

23.4 Conclusion and future perspectives

Agricultural wastes are the remains in the fields and low nutritional value and may include antinutritional substances, which limit their use in animal feeds. It has been discovered that using agricultural waste as a feedstock for the production of nanostructured materials is an effective waste management method. Due to their effective usage in the creation of several nanoparticles with distinctive features, chemicals produced from plants and microorganisms have demonstrated the promise of nanoscience activities. Carbon nanotube is one of the most trustworthy carbon nanomaterials owing to many new characteristics, including its large surface area and important functional groups. There are several scientific domains where graphene oxide was used. It is important to effectively stress on contemporary methods for the synthesis of affordable and environmentally acceptable nanoparticles based on agricultural waste. Synthesis of graphene oxide nanoparticles, silica nanoparticles, nanocellulose, and metal nanoparticles has been discussed from easily available agricultural wastes. Agricultural waste has been better utilized to create useful nanomaterials and minimize environmental pollution by efficient use of waste. Therefore through the nanotechnology, the agricultural waste revolutionized in the economy development.

References

Adebisi, J.A., et al., 2019. Green production of silica nanoparticles from maize stalk. An International Journal Taylor & Francis 38 (6), 667–675. Available from: https://doi.org/10.1080/02726351.2019.1578845.

Akhayere, E., Kavaz, D., Vaseashta, A., 2019. Synthesizing nano silica nanoparticles from barley grain waste: effect of temperature on mechanical properties. Polish Journal of Environmental Studies 28 (4), 1–9. Available from: https://doi.org/10.15244/pjoes/91078.

Akhayere, E., Vaseashta, A., Kavaz, D., 2020. Novel magnetic nano silica synthesis using barley husk waste for removing petroleum from polluted water for environmental sustainability'*Sustainability 2020, Vol. 12, Page 10646*Multidisciplinary Digital Publishing Institute 12 (24), 10646. Available from: https://doi.org/10.3390/SU122410646.

Alves, J.O., et al., 2011. Catalytic conversion of wastes from the bioethanol production into carbon nanomaterials. Applied Catalysis B: Environmental 106 (3–4), 433–444. Available from: https://doi.org/10.1016/J.APCATB.2011.06.001. Elsevier.

Andas, J., Adam, F., 2016. One-pot synthesis of nanoscale silver supported biomass-derived silica. Materials Today: Proceedings 3 (6), 1345–1350. Available from: https://doi.org/10.1016/J.MATPR.2016.04.013. Elsevier.

Baiocco, D., et al., 2016. Production of metal nanoparticles by agro-industrial wastes: a green opportunity for nanotechnology', *Chemical engineering transactions*. Italian Association of Chemical Engineering - AIDIC 47, 67–72. Available from: https://doi.org/10.3303/CET1647012.

Bankar, A., et al., 2010. Banana peel extract mediated novel route for the synthesis of palladium nanoparticles. Materials Letters 64 (18), 1951–1953. Available from: https://doi.org/10.1016/J.MATLET.2010.06.021. North-Holland.

Bernd, M.G.S., et al., 2017. Synthesis of carbon nanostructures by the pyrolysis of wood sawdust in a tubular reactor. Journal of Materials Research and Technology 6 (2), 171–177. Available from: https://doi.org/10.1016/J.JMRT.2016.11.003. Elsevier.

Bhardwaj, A.K., Naraian, R., 2020. Green synthesis and characterization of silver NPs using oyster mushroom extract for antibacterial efficacy. Journal of Chemistry, Environmental Sciences and its Applications 7 (1), 13–18.

Bhardwaj, A.K., Naraian, R., 2021. Cyanobacteria as biochemical energy source for the synthesis of inorganic nanoparticles, mechanism and potential applications: a review. 3 Biotech 11 (10), 445.

Bhardwaj, A.K., Shukla, A., Maurya, S., Singh, S.C., Uttam, K.N., Sundaram, S., et al., 2018. Direct sunlight enabled photo-biochemical synthesis of silver nanoparticles and their bactericidal efficacy: photon energy as key for size and distribution control. Journal of Photochemistry and Photobiology B: Biology 188, 42–49.

Biswas, M.C., Jeelani, S., Rangari, V., 2017. Influence of biobased silica/carbon hybrid nanoparticles on thermal and mechanical properties of biodegradable polymer films. Composites Communications 4, 43–53. Available from: https://doi.org/10.1016/J.COCO.2017.04.005. Elsevier.

Che Othman, F.E., et al., 2020. Methane adsorption by porous graphene derived from rice husk ashes under various stabilization temperatures. Carbon Letters 30 (5), 535–543. Available from: https://doi.org/10.1007/S42823-020-00123-3. *2020 30:5*. Springer.

Chen, H., et al., 2013. Extraction of lignocellulose and synthesis of porous silica nanoparticles from rice husks: a comprehensive utilization of rice husk biomass. ACS Sustainable Chemistry and Engineering 1 (2), 254–259. Available from: https://doi.org/10.1021/SC300115R/ASSET/IMAGES/MEDIUM/SC-2012-00115R_0005.GIF. American Chemical Society.

Chen, Y., et al., 2009. Bionanocomposites based on pea starch and cellulose nanowhiskers hydrolyzed from pea hull fibre: effect of hydrolysis time. Carbohydrate Polymers 76 (4), 607–615. Available from: https://doi.org/10.1016/J.CARBPOL.2008.11.030. Elsevier.

Debbarma, J., Naik, M.J.P., Saha, M., 2019. From agrowaste to graphene nanosheets: chemistry and synthesis. Taylor & Francis, 27 (6), 482–485. Available from: https://doi.org/10.1080/1536383X.2019.1601086.

Deng, M., et al., 2016. Simple process for synthesis of layered sodium silicates using rice husk ash as silica source. Journal of Alloys and Compounds 683, 412–417. Available from: https://doi.org/10.1016/J.JALLCOM.2016.05.115. Elsevier.

Fathy, N.A., 2017. Carbon nanotubes synthesis using carbonization of pretreated rice straw through chemical vapor deposition of camphor. RSC Advances 7 (45), 28535–28541. Available from: https://doi.org/10.1039/C7RA04882C. Royal Society of Chemistry.

Fathy, N.A., Basta, A.H., Lotfy, V.F., 2020. Novel trends for synthesis of carbon nanostructures from agricultural wastes. Carbon Nanomaterials for Agri-food and Environmental Applications. Elsevier, pp. 59–74. Available from: https://doi.org/10.1016/B978-0-12-819786-8.00004-9.

Fathy, N.A., El-Shafey, O.I., Khalil, L.B., 2013. Effectiveness of alkali-acid treatment in enhancement the adsorption capacity for rice straw: the removal of methylene blue dye, ISRN Physical Chemistry, 2013. Hindawi Publishing Corporation, p. 15. Available from: https://doi.org/10.1155/2013/208087.

Fathy, N.A., Lotfy, V.F., Basta, A.H., 2017. Comparative study on the performance of carbon nanotubes prepared from agro- and xerogels as carbon supports, Journal of Analytical and Applied Pyrolysis, 128. Elsevier, pp. 114–120. Available from: https://doi.org/10.1016/J.JAAP.2017.10.019.

Fatihah, N., et al., 2020. Graphene from waste and bioprecursors synthesis method and its application: a review. Malaysian Journal of Fundamental and Applied Sciences 16 (3), 342–350.

Feng, Y., et al., 2020. Upgrading agricultural biomass for sustainable energy storage: bioprocessing, electrochemistry, mechanism, Energy Storage Materials, 31. Elsevier, pp. 274–309. Available from: https://doi.org/10.1016/J.ENSM.2020.06.017.

Gaur, M., et al., 2021. Biomedical applications of carbon nanomaterials: fullerenes, quantum dots, nanotubes, nanofibers, and graphene. Materials 14 (20), 5978. Available from: https://doi.org/10.3390/MA14205978. *2021, Vol. 14, Page 5978*. Multidisciplinary Digital Publishing Institute.

Goodell, B., et al., 2008. Carbon nanotubes produced from natural cellulosic materials. Journal of Nanoscience and Nanotechnology 8 (5), 2472–2474. Available from: https://doi.org/10.1166/JNN.2008.235.

Goswami, S., et al., 2017. Graphene oxide nanoplatelets synthesized with carbonized agro-waste biomass as green precursor and its application for the treatment of dye rich wastewater, Process Safety and Environmental Protection, 106. Elsevier, pp. 163–172. Available from: https://doi.org/10.1016/J.PSEP.2017.01.003.

Gupta, V., Chandra, N., 2020. Biosynthesis and antibacterial activity of metal oxide nanoparticles using Brassica oleracea subsp. botrytis (L.) leaves, an agricultural waste. Proceedings of the National Academy of Sciences India Section B - Biological Sciences 90 (5), 1093–1100. Available from: https://doi.org/10.1007/S40011-020-01184-0/FIGURES/7. Springer.

Harish, B.S., Uppuluri, K.B., Anbazhagan, V., 2015. Synthesis of fibrinolytic active silver nanoparticle using wheat bran xylan as a reducing and stabilizing agent, Carbohydrate Polymers, 132. Elsevier, pp. 104–110. Available from: https://doi.org/10.1016/J.CARBPOL.2015.06.069.

Hashmi, A., et al., 2020. Muffle atmosphere promoted fabrication of graphene oxide nanoparticle by agricultural waste. Fullerenes Nanotubes and Carbon Nanostructures 28 (8), 627–636. Available from: https://doi.org/10.1080/1536383X.2020.1728744/SUPPL_FILE/LFNN_A_1728744_SM8237.DOCX. Taylor and Francis Inc.

Hsieh, Y.Y., et al., 2017. Rice husk agricultural waste-derived low ionic content carbon–silica nanocomposite for green reinforced epoxy resin electronic packaging material, Journal of the Taiwan Institute of Chemical Engineers, 78. Elsevier, pp. 493–499. Available from: https://doi.org/10.1016/J.JTICE.2017.06.010.

Jacob, M.V., et al., 2015. Catalyst-free plasma enhanced growth of graphene from sustainable sources. Nano Letters 15 (9), 5702–5708. Available from: https://doi.org/10.1021/ACS.NANOLETT.5B01363/SUPPL_FILE/NL5B01363_SI_001.PDF. American Chemical Society.

Javad, S., Akhtar, I., Naz, S., 2020. Nanomaterials and agrowaste. Nanoagronomy. Springer, Cham, pp. 197–207. Available from: https://doi.org/10.1007/978-3-030-41275-3_11.

Kadja, G.T.M., et al., 2017. The effect of structural properties of natural silica precursors in the mesoporogen-free synthesis of hierarchical ZSM-5 below 100 °C. Advanced Powder Technology 28 (2), 443–452. Available from: https://doi.org/10.1016/J.APT.2016.10.017. Elsevier.

Kang, X., et al., 2017. Thin cellulose nanofiber from corncob cellulose and its performance in transparent nanopaper. ACS Sustainable Chemistry and Engineering 5 (3), 2529–2534. Available from: https://doi.org/10.1021/ACSSUSCHEMENG.6B02867/ASSET/IMAGES/MEDIUM/SC-2016-028676_0014.GIF. American Chemical Society.

Kannah, R.Y., et al., 2020. Valorization of food waste for bioethanol and biobutanol production. Food Waste to Valuable Resources: Applications and Management. Academic Press, pp. 39–73. Available from: https://doi.org/10.1016/B978-0-12-818353-3.00003-1.

Khazaei, A., Khazaei, M., Nasrollahzadeh, M., 2017. Nano-$Fe_3O_4@SiO_2$ supported Pd(0) as a magnetically recoverable nanocatalyst for Suzuki coupling reaction in the presence of waste eggshell as low-cost natural base. Tetrahedron 73 (38), 5624–5633. Available from: https://doi.org/10.1016/J.TET.2017.05.054. Pergamon.

Le, V.H., Thuc, C.N.H., Thuc, H.H., 2013. Synthesis of silica nanoparticles from Vietnamese rice husk by sol–gel method. Nanoscale Research Letters 8 (1), 1–10. Available from: https://doi.org/10.1186/1556-276X-8-58. *2013 8:1*. SpringerOpen.

Li, Z., et al., 2021. Green synthesis of carbon quantum dots from corn stalk shell by hydrothermal approach in near-critical water and applications in detecting and bioimaging, Microchemical Journal, 166. Elsevier, p. 106250. Available from: https://doi.org/10.1016/J.MICROC.2021.106250.

Liou, T.H., Wang, P.Y., 2020. Utilization of rice husk wastes in synthesis of graphene oxide-based carbonaceous nanocomposites, Waste Management, 108. Pergamon, pp. 51–61. Available from: https://doi.org/10.1016/J.WASMAN.2020.04.029.

Lopattananon, N., Payae, Y., Seadan, M., 2008. Influence of fiber modification on interfacial adhesion and mechanical properties of pineapple leaf fiber-epoxy composites. Journal of Applied Polymer Science 110 (1), 433–443. Available from: https://doi.org/10.1002/APP.28496. John Wiley & Sons, Ltd.

Lotfy, V.F., Fathy, N.A., Basta, A.H., 2018. Novel approach for synthesizing different shapes of carbon nanotubes from rice straw residue. Journal of Environmental Chemical Engineering 6 (5), 6263–6274. Available from: https://doi.org/10.1016/J.JECE.2018.09.055. Elsevier.

Lu, P., Hsieh, Y.Lo, 2012. Highly pure amorphous silica nano-disks from rice straw, Powder Technology, 225. Elsevier, pp. 149–155. Available from: https://doi.org/10.1016/J.POWTEC.2012.04.002.

Maji, S., et al., 2020. Agricultural Waste: Its Impact on Environment and Management Approaches'. Springer, Singapore, pp. 329–351. Available from: https://doi.org/10.1007/978-981-15-1390-9_15.

Malhotra, A., et al., 2014. Multi-analytical approach to understand biomineralization of gold using rice bran: a novel and economical route. RSC Advances 4 (74), 39484–39490. Available from: https://doi.org/10.1039/C4RA05404K. The Royal Society of Chemistry.

Mirasol, E.S., Shimomura, M., 2023. Synthesis and characterization of nanostructured silica powders using rice hull ash-based sodium silicate solution by precipitation and calcination, Materials Letters: X, 17. Elsevier, p. 100175. Available from: https://doi.org/10.1016/J.MLBLUX.2022.100175.

Mor, S., et al., 2017. Nanosilica extraction from processed agricultural residue using green technology, Journal of Cleaner Production, 143. Elsevier, pp. 1284–1290. Available from: https://doi.org/10.1016/J.JCLEPRO.2016.11.142.

Mostafavi, E., et al., 2022. Eco-friendly synthesis of carbon nanotubes and their cancer theranostic applications. Materials Advances 3 (12), 4765–4782. Available from: https://doi.org/10.1039/D2MA00341D. Royal Society of Chemistry.

Muramatsu, H., et al., 2014. Rice husk-derived graphene with nano-sized domains and clean edges. Small (Weinheim an der Bergstrasse, Germany) 10 (14), 2766–2770. Available from: https://doi.org/10.1002/SMLL.201400017. John Wiley & Sons, Ltd.

Naik, M.J.P., et al., 2020. Graphene oxide nanoflakes from various agrowastes. Materialwissenschaft und Werkstofftechnik 51 (3), 368–374. Available from: https://doi.org/10.1002/MAWE.201900061. John Wiley & Sons, Ltd.

Nanni, A., Parisi, M., Colonna, M., 2021. Wine by-products as raw materials for the production of biopolymers and of natural reinforcing fillers: a critical review. Polymers 13 (3), 381. Available from: https://doi.org/10.3390/POLYM13030381. *2021, Vol. 13, Page 381*. Multidisciplinary Digital Publishing Institute.

Niu, Q., et al., 2017. Large-size graphene-like porous carbon nanosheets with controllable N-doped surface derived from sugarcane bagasse pith/chitosan for high performance supercapacitors, Carbon, 123. Pergamon, pp. 290–298. Available from: https://doi.org/10.1016/J.CARBON.2017.07.078.

Nor Faiz Norrahim, M. et al., 2021. Polymers greener pretreatment approaches for the valorisation of natural fibre biomass into bioproducts. Available from: https://doi.org/10.3390/polym13172971.

Phanthong, P., et al., 2015. Extraction of nanocellulose from raw apple stem. Journal of the Japan Institute of Energy 94 (8), 787–793. Available from: https://doi.org/10.3775/JIE.94.787. The Japan Institute of Energy.

Pieła, A., et al., 2020. Biogenic synthesis of silica nanoparticles from corn cobs husks. Dependence of the productivity on the method of raw material processing, Bioorganic Chemistry, 99. Academic Press, p. 103773. Available from: https://doi.org/10.1016/J.BIOORG.2020.103773.

Raghavan, N., Thangavel, S., Venugopal, G., 2017. A short review on preparation of graphene from waste and bioprecursors, Applied Materials Today, 7. Elsevier, pp. 246–254. Available from: https://doi.org/10.1016/J.APMT.2017.04.005.

Roquia, A. et al., 2021. Synthesis and characterisation of carbon nanotubes from waste of Juglans regia (walnut) shells, https://doi.org/10.1080/1536383X.2021.1900123. Taylor & Francis, 29 (11) 860–867. Available from: https://doi.org/10.1080/1536383X.2021.1900123.

Sadh, P.K., Duhan, S., Duhan, J.S., 2018. Agro-industrial wastes and their utilization using solid state fermentation: a review. Bioresources and Bioprocessing 5 (1), 1–15. Available from: https://doi.org/10.1186/S40643-017-0187-Z/FIGURES/4. Springer.

Sahila Grace, A., Prabha, G.S., Malar, L., 2018. Synthesis and characterization of graphene oxide from coconut husk ash. Oriental Journal of Chemistry. Available from: https://doi.org/10.13005/ojc/360220.

Sankar, S., et al., 2017. Ultrathin graphene nanosheets derived from rice husks for sustainable supercapacitor electrodes. New Journal of Chemistry 41 (22), 13792–13797. Available from: https://doi.org/10.1039/C7NJ03136J. The Royal Society of Chemistry.

Seitzhanova, M.A., et al., 2019. The characteristics of graphene obtained from rice husk and graphite. Eurasian Chemico-Technological Journal 21 (2), 149–156. Available from: https://doi.org/10.18321/ECTJ825. al-Farabi Kazakh State National University.

Shahba, H., Sabet, M., 2020. Two-step and green synthesis of highly fluorescent carbon quantum dots and carbon nanofibers from pine fruit. Journal of fluorescence 30 (4), 927–938. Available from: https://doi.org/10.1007/S10895-020-02562-7/FIGURES/20. Springer.

Singh, V., et al., 2021. Chickpea peel waste as sustainable precursor for synthesis of fluorescent carbon nanotubes for bioimaging application. Carbon Letters 31 (1), 117–123. Available from: https://doi.org/10.1007/S42823-020-00156-8/METRICS. Springer.

Sinsinwar, S., et al., 2018. Use of agricultural waste (coconut shell) for the synthesis of silver nanoparticles and evaluation of their antibacterial activity against selected human pathogens. Microbial Pathogenesis 124, 30–37. Available from: https://doi.org/10.1016/J.MICPATH.2018.08.025. Academic Press.

Skiba, M.I., Vorobyova, V.I., 2019. Synthesis of silver nanoparticles using orange peel extract prepared by plasmochemical extraction method and degradation of methylene blue under solar irradiation, Advances in Materials Science and Engineering, 2019. Hindawi Limited. Available from: https://doi.org/10.1155/2019/8306015.

Somanathan, T., et al., 2015. Graphene oxide synthesis from agro waste. Nanomaterials 5, 826–834. Available from: https://doi.org/10.3390/nano5020826.

Tewari, C., et al., 2019. A simple, eco-friendly and green approach to synthesis of blue photoluminescent potassium-doped graphene oxide from agriculture waste for bio-imaging applications. Materials Science and Engineering: C 104, 109970. Available from: https://doi.org/10.1016/J.MSEC.2019.109970. Elsevier.

Thakur, V., et al., 2021. Recent advances in nanocellulose processing, functionalization and applications: a review. Materials Advances 2 (6), 1872–1895Royal Society of Chemistry. Available from: https://doi.org/10.1039/D1MA00049G.

Thangadurai, D., et al., 2021. Nanomaterials from agrowastes: past, present, and the future. Handbook of Nanomaterials and Nanocomposites for Energy and Environmental Applications. Springer, Cham, pp. 471–487. Available from: https://doi.org/10.1007/978-3-030-36268-3_43.

Tohamy, H.-A.S. et al., 2020. Preparation of eco-friendly graphene oxide from agricultural wastes for water treatment. Available from: https://doi.org/10.5004/dwt.2020.25652.

Tsukamoto, J., Durán, N., Tasic, L., 2013. Nanocellulose and bioethanol production from orange waste using isolated microorganisms. Journal of the Brazilian Chemical Society 24 (9), 1537–1543. Available from: https://doi.org/10.5935/0103-5053.20130195. Sociedade Brasileira de Química.

Tuck, C.O., et al., 2012. Valorization of biomass: deriving more value from waste. Science 337 (6095), 695–699. Available from: https://doi.org/10.1126/SCIENCE.1218930. American Association for the Advancement of Science.

Uda, M.N.A., et al., 2020. Simple and green approach strategy to synthesis graphene using rice straw ash. IOP Conference Series: Materials Science and Engineering 864 (1), 012181. Available from: https://doi.org/10.1088/1757-899X/864/1/012181. IOP Publishing.

Usgodaarachchi, L., et al., 2021. Synthesis of mesoporous silica nanoparticles derived from rice husk and surface-controlled amine functionalization for efficient adsorption of methylene blue from aqueous solution. Current Research in Green and Sustainable Chemistry 4, 100116. Available from: https://doi.org/10.1016/J.CRGSC.2021.100116. Elsevier.

Wang, Z., et al., 2015. Nanocarbons from rice husk by microwave plasma irradiation: from graphene and carbon nanotubes to graphenated carbon nanotube hybrids. Carbon 94, 479–484. Available from: https://doi.org/10.1016/J.CARBON.2015.07.037. Pergamon.

Yan, D., et al., 2014. Supercapacitive properties of Mn_3O_4 nanoparticles bio-synthesized from banana peel extract. RSC Advances 4 (45), 23649–23652. Available from: https://doi.org/10.1039/C4RA02603A. The Royal Society of Chemistry.

Yu, S., et al., 2021. Nanocellulose from various biomass wastes: its preparation and potential usages towards the high value-added products. Environmental Science and Ecotechnology 5, 100077. Available from: https://doi.org/10.1016/J.ESE.2020.100077. Elsevier.

Yuan, M., et al., 2015. One-step, green, and economic synthesis of water-soluble photoluminescent carbon dots by hydrothermal treatment of wheat straw, and their bio-applications in labeling, imaging, and sensing. Applied surface science 355, 1136–1144. Available from: https://doi.org/10.1016/J.APSUSC.2015.07.095. North-Holland.

Zamani, A., Marjani, A.P., Mousavi, Z., 2019. Agricultural waste biomass-assisted nanostructures: synthesis and application. Green Processing and Synthesis 8 (1), 421–429. Available from: https://doi.org/10.1515/GPS-2019-0010/ASSET/GRAPHIC/J_GPS-2019-0010_FIG_012.JPG. De Gruyter.

Zarei, V., et al., 2021. Environmental method for synthesizing amorphous silica oxide nanoparticles from a natural material. Processes 9 (2), 334. Available from: https://doi.org/10.3390/PR9020334. *2021, Vol. 9, Page 334*. Multidisciplinary Digital Publishing Institute.

Zhou, H., 2017. Combustible solid waste thermochemical conversion: a study of interactions and influence factors. https://books.google.com/books?hl = en&lr = &id = PjYtDgAAQBAJ&oi = fnd&pg = PR7&ots = sxdeVTX6Rp&sig = ieoeTizdOOOiEO5j8z4WHLcHunA (accessed 01.10.22).

Zhu, J., et al., 2012. Synthesis of multiwalled carbon nanotubes from bamboo charcoal and the roles of minerals on their growth. Biomass and Bioenergy 36, 12–19. Available from: https://doi.org/10.1016/J.BIOMBIOE.2011.08.023. Pergamon.

Chapter 24

Nanomaterials′ synthesis from the fruit wastes

Swati Rose Toppo
Department of Microbiology and Bioinformatics, Atal Bihari Vajpayee Vishwavidyalaya, Bilaspur, Chhattisgarh, India

24.1 Nanotechnology in pomology science

India stands second for the world's largest producer of fruits and vegetables. It is due to diverse agroclimatic conditions, the cultivation of numerous horticultural crops and aromatic plants is possible in India. Every year during cultivation, harvesting, processing, and consumption of these foodstuffs left behind massive quantities of lignocellulosic biomass. Leftovers, from the fruit industries such as fruit peel and pulp after consumption of edible part of fruits, are separated and dumped into municipal landfills that cause serious pollution and disposal problems. These wastes can be exploited as raw material for various biotechnological products like low-cost biosorbent material, for production of biochemical and biofuels, as substrate for producing enzymes, enriched proteins and metabolites, microbial pigments, antibiotics bioactive compounds, organic acids, and nanomaterials. Diversification of these residues into value-added products will help reduce pollution load, waste handling, and management issues from the environment. Thus, employment of fruit wastes for biotechnological applications serves duple purpose: **(i) engendering capital from waste and (ii) as an effectual solid-waste diminution**.

Since everything that is, biotic and abiotic components in nature is composed of atoms and nanoscience is the technological power to perceive and regulate atoms and molecules discretely. Nanotechnology is an innovative field of science that deals with designing, production, and manipulation of atoms and molecules at nanoscale. The name nano is derived from Greek word "nano" meaning dwarf. It is one billionth part of a meter or 10^{-9} m (Holdren, 2011). Nanomaterials are very minute makeups ranging from 0.1 to 100 nm. Microscopic size and properties different from macromaterials make them significantly important. Nanomaterials possess properties such as electrical conductance, magnetism, chemical reactivity, optical effects, and physical strength that vary from macromaterials due to their smaller size. These nanomaterials form a linkage between macromaterials and their respective nanoparticles (Boisseau and Loubaton, 2011). The development of nanotechnology has recently placed a strong emphasis on green procedures that make use of moderate reaction conditions and nontoxic precursors to support environmental sustainability.

The goal of green nanoparticle synthesis is to reduce waste production and use environmentally friendly methods. In recent years, green production of metallic nanoparticles has emerged as a fresh and exciting area of study. The creation of green nanoparticles has grown in significance over the past few years due to its many advantages, including ease of scaling up for large-scale synthesis, low cost, good stability of the nanoparticles produced, and nontoxic byproducts. Green synthesis has drawn attention for the synthesis of different metal and metal oxide nanoparticles since chemical synthesis methods result in the presence of harmful chemical species adsorbed on the surface of nanoparticles. Green synthesis methods using various fruit wastes are dependable, long-lasting, and environmentally acceptable building blocks for the synthesis of metal and metal oxide nanoparticles.

24.2 Types of nanomaterials (Madeeha Ansari et al., 2020)

There are three groups of nanomaterials based on dimensions (Hett, 2004).

Green and Sustainable Approaches Using Wastes for the Production of Multifunctional Nanomaterials. DOI: https://doi.org/10.1016/B978-0-443-19183-1.00018-0

24.2.1 One-dimensional nanoparticles

Nanomaterials less than 100 nm size with one dimension. Examples include: nanowires and nanorods used in manufacturing of various chemical and biological sensors, solar cells, IT systems, and optical devices.

24.2.2 Two-dimensional nanoparticles

Nanomaterials with size less than 100 nm along two dimensions at least. Examples include: carbon nanotube fibers and platelets.

24.2.3 Three-dimensional nanoparticles

Metallic nanomaterials size less than 100 nm in all dimensions. Examples include: Quantum dots, dendrimers, and hollow spheres are three-dimensional nanoparticles.

24.2.4 Classification of nanomaterials on the basis of structural configuration

Metallic nanoparticles, nanocrystals quantum dots, carbon nanotubes, liposomes, dendrimer, polymeric micelles, and polymeric nanoparticles.

24.2.5 Classification on the basis of chemical form

Organic nanoparticles (carbon-based nanoparticles) and inorganic nanoparticles can be further categorized into *magnetic nanoparticles* (iron), *noble metal nanoparticles* (silver, gold, and platinum), and *semiconductor nanoparticles* (silica, zinc oxide, and so on) (Liu et al., 2020).

24.3 Synthesis of nanomaterials

Nanoparticles show entirely novel and enhanced properties due to specific characteristics' dissimilarities with macromaterials, in size, distribution, and structures of the particles. With reduction in the particle size, nanoparticles represent more surface area to volume portion (Vinay et al., 2018). Nanoparticles can be synthesized via chemical, physical, and biological methods by mainly two approaches:

- Top-to-bottom synthesis
- Bottom-to-up synthesis

24.3.1 Top-to-bottom approach

It refers to the breakdown of suitable given macromaterial into particles in size range of nanometer due to reduction in size by various methods. This includes physical methods (Fig. 24.1). Various nanoparticles have been synthesized using this technique such as silver nanoparticles, gold nanoparticles, lead, and fullerene nanoparticles.

24.3.1.1 Physical method

In this method, the particular macromaterial is placed on a boat, in the center of the tube furnace (used to generate high atmospheric pressure), which is then allowed to vaporize and carried through a gas (Samberg et al., 2009; Prathna et al., 2011; Sintubin et al., 2011; Kumar et al., 2014a).

24.3.2 Bottom-to-up approach

It refers to the synthesis of nanoparticles using either chemical or biological methods. In bottom-to-top approach, atoms are self-assembled into new nuclei forming the particles of nano size.

Chemical Method: Numerous chemical techniques have been employed for the synthesis of nanomaterials, as mentioned in Fig. 24.1 (Yuvakkumar et al., 2015) but the most common is chemical reduction. Both organic and inorganic compounds are used as reducing agents. Example: sodium borohydride (NaBH4), ascorbate, elemental hydrogen,

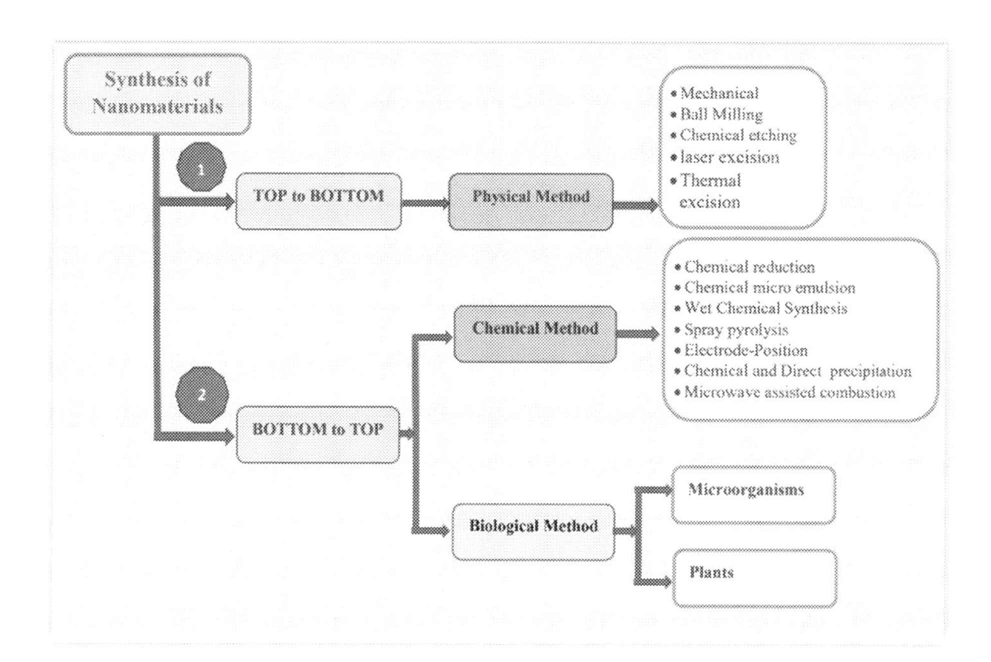

FIGURE 24.1 Methods for synthesis of nanomaterials.

Tollen's reagent, N, N-dimethyl formamide, sodium citrate, Tollen's reagent, and copolymers of polyethylene glycol (Tran and Le, 2013; Iravani et al., 2014).

Chemical technique requires capping agents to control and stabilize the size of the nanoparticles and to avoid their aggregation which are toxic and lead to noneco-friendly byproducts. Thus, there is an increasing demand for green nanotechnology. Plants provide a better platform for nanoparticles' synthesis as they are free from toxic chemicals as well as provide natural capping agents (Vinay et al., 2018).

24.3.2.1 Biological method

Here, biotic agents such as microorganisms (viruses, bacteria, and fungi) as well as plant materials are used for production of nanoparticles (Reddy et al.,2012).

24.4 Biosynthesis of nanomaterial

Microorganisms play important role in nanoparticles' synthesis and help in the production of reducing agents (reductase enzymes) that reduce metals into metallic nanoparticles. This approach is an eco-friendly and cost-effective tool, as natural metabolic pathway of microorganisms has been encountered for nanoparticles' synthesis both extra and intracellularly.

Nanomaterials' biosynthesis, called as green synthesis, is also very simple, eco-friendly, cost-effective, and easily scaled up for the production of nanoparticles at large scale with the use of plant materials. Nanomaterials' biosynthesis is more beneficial than using microorganisms as keeping cultures of microbes involve complex procedures. Various plant parts including their extract rich in phytochemicals have been used for the synthesis of nanoparticles, which act as both reducing and capping agents (Ramesh et al., 2015; Xiao et al., 2016).

24.5 Some of the fruit wastes known for synthesis of nanomaterials

24.5.1 Banana

Banana is the second largest produced fruit, accounting for 16% of the total fruit production worldwide. India is the largest producer of banana, accounting for 27% of the world's total banana production. Banana (Musa sp.,) belongs to the Family: *Musaceae*. The average weight of fruit is about 125 g that contains 25% dry matter and 75% water. Banana peel (BP) consists of about 30%–40% (w/w) of fresh banana. The composition of ripped BP is as follows: crude protein (8%), ether extract (6.2%), soluble sugars (13.8%), and total phenolic compounds (4.8%). Main components of BP are cellulose, hemicellulose, chlorophyll, pectin, and other low-molecular weight compounds as starch, glucose, and fructose (Pathak et al., 2017). Flavonoids and other bioactive substances with some anticancer properties are considered for antioxidant activity and antitumor function (Liu et al., 2020).

24.5.2 Orange

*Citrus sinesis*is genus of flowering plants in the family Rutaceae (Rao et al., 2015a,b). Orange (*Citrus sinensis* (L.)) is the most common among fruit trees grown worldwide (Pandharipande and Makode, 2012). It is the best source of vitamin C and good for health and skin. Major constituent of OP is cellulose, hemicellulose, lignin, pectin (galacturonic acid), chlorophyll pigments, and other low-molecular weight compounds (e.g., limonene). Traditionally, OP is treated to obtain volatile and nonvolatile fractions of essential oils and flavoring compounds. In addition, OP has been reported to have germicidal, antioxidant, and anticarcinogenic properties, and thus may be effective against breast and colon cancers, skin inflammation, muscle pain, stomach upset, and ringworm (Vinay et al., 2018). Orange peels are boon for skin, as they possess antimicrobial and antiinflammatory properties. The dried powdered OP can be used to scrub and exfoliate skin. OP can be used in bath oil, room freshen air, face creams, mosquito repellent, and weight loss. It is a great cleanser, helps to cure acne and pus-filled pimples, and removes blackheads, dark spots, and pigmentation. Orange peel acts as a reducing agent for the synthesis of TiO_2 because it contains citric acid as a main source (Torrado et al., 2011).

24.5.3 Citrus

Citrus limetta, Family: Rutaceae is the world's largest produced fruit, accounting for 23% of the world's total fruit production. Citrus peel is a potential source of certain essential oils and yields about 0.5–3 kg oil/ton of fruit, which are used for various purposes such as pharmaceuticals, confectioneries, cosmetics, alcoholic beverages, and also for improving the shelf-life and safety of various foodstuff. Besides, CP is also rich in pectin (Pathak et al., 2017).

24.5.4 Orange

Citrus aurantium L., Eco-enzymes derived from their peels have been shown to have antimicrobial and antiinflammatory properties due to high content of polyphenolic compounds (Mavani et al., 2020).

24.5.5 Lemon

Citrus limon, Family: Rutaceae peel's outer layer is called flavedo, a rich source of essential oils, that has been used since early times in flavoring and fragrance industries. The major component of lemon peel is albedo, which is a spongy and cellulosic layer under the flavedo and has high dietary fiber content (Pathak et al., 2017).

24.5.6 Pomegranate

Punica granatum is one of the most leading crops grown in Saudi Arabia (Kaur et al., 2016). The pomegranate peel is an agro-industrial waste, which contributes about 60% of the weight of pomegranate fruit. Pomegranate peel is a rich source of tannins, flavonoids, polyphenols, and some anthocyanins as delphinidins, cyanidins, etc. Antioxidant properties are due to the presence of these polyphenolic phytochemicals (Vinay et al., 2018), therefore extensively used in the food industry and in traditional medicines to treat various ailments. The pomegranate peel extract (PE) is also identified to have elevated nutrient value, consisting of minerals (potassium), vitamins, phenolics, flavonoids, antioxidants, and anticarcinogenic properties and shows antibacterial property against bacterial strains of *E. coli, P. aeruginosa*, and *S. aureus*.

24.5.7 Avocado

Persea americana belongs to the family *Lauraceae* and genus *Persea*. The edible fruit is native to America but adaptable to the tropics. Avocado is traditionally cultivated for food and medicinal purposes because of its high nutritive contents and therapeutic properties, thus used in traditional and complementary medicine to treat different diseases in subtropic and tropic zones (Adebayo et al., 2019).

24.5.8 Dragon fruit

Hylocereus undatus or red pitaya is a tropical fruit which belongs to the cactus family *Cactaceae*. The pulp of dragon fruit is rich in vitamin C and water-soluble fiber. Usually, dragon fruit is consumed directly or being processed into juice for functional drink. The key chemical ingredients in dragon fruit peel are betanin, phyllocactin, hylocerenin, beta-cyanin, pectin, triterpenoids, and steroids which have many pharmacological activities such as antitumor, antiinflammatory, and antioxidant properties. Organic compounds in the peel could have the ability for the reduction of metal salt ion in the synthesis of metal nanoparticle (Phongtongpasuk et al., 2016).

24.5.9 Papaya (Carica papaya)

Papaya peel is rich in papain that exhibits significant antibacterial efficacy against *Enterococcus faecalis* (Mavani et al., 2020). It is found that 0.8% of papain is equally effective as 1.0% NaOCl in inhibiting the growth of *E. faecalis*. In comparison to NaOCl, it has less harmful effects on vital tissues as its proteolytic activities selectively target unhealthy tissues where α1-antitrypsin plasmatic antiprotease is absent. Phytochemicals found in the papaya peel eco-enzyme demonstrate a potential antiinflammatory effect as well, which minimizes the chronic inflammatory process and tissue destruction, particularly in cases of apical periodontitis.

24.5.10 Pineapple (Ananas comosus)

Eco-enzyme derived from pineapple peels have been shown to have antimicrobial as well as antiinflammatory properties. The high content of polyphenolic compounds and flavonoids in pineapple peel extracts are responsible for their excellent antimicrobial and antioxidant activities. Bromelain from pineapple extracts is shown to be effective in killing *E. faecalis* by disrupting the peptidoglycan and polysaccharide components of bacterial cell membranes (Mavani et al., 2020).

24.5.11 Grape

Vitis vinifera is among the world's major fruits and is extensively utilized in juice and wine industries. During processing, a significant amount of solid byproduct pomace (GP) is generated. The GP consists of grape skins and seeds which are rich in phytochemicals, namely phenolic acids, flavonoids, anthocyanins, resveratrol, and proanthocyanidins. The presence of these valuable phytochemicals makes GP a vital source for the green synthesis of AgNPs.

24.6 Types of nanomaterials produced using various fruit waste

Nanoparticles size ranges from 1 to 100 nm. They are synthesized in different sizes and shapes, such as triangular, spherical, irregular, etc. (Vinay et al., 2018). Because of high specific surface area, surface defects, and quantum effects, nanomaterial displays optical, electronic, magnetic, and catalytic properties that differ from their macroscopic counterpart materials. These unique properties enhance their usage in many fields such as biological imaging, nanosensors, biomedicine, photoelectric conversion, and so on (Liu et al., 2020). Nanoparticles can be categorized into organic nanoparticles (carbon-based nanoparticles) and inorganic nanoparticles. There are three categories of inorganic nanoparticles - magnetic nanoparticles (iron), noble metal nanoparticles (silver, gold, and platinum), and semiconductor nanoparticles (silica, zinc oxide, and so on).

The biological synthesis technique follows bottom-to-top approach for nanomaterial's synthesis. Fruit wastes can be efficiently used for the production of nanomaterials of wide range of particle sizes and shapes using green, chemistry-based techniques (Ghosh et al., 2019). The diverse types of nanoparticles prepared using different fruits' extracts have been discussed below:

24.6.1 Noble metal nanoparticles

24.6.1.1 Silver nanoparticles

Silver nanoparticles (AgNPs) have gained significant attention due to their biochemical and catalytic activity owing to their large surface area in contrast to other particles with equivalent chemical structures (Kumar et al., 2020) and extraordinary antimicrobial properties (Ghosh et al., 2019). Due to this unique physiochemical and superior antimicrobial properties, nowadays Ag have been amalgamated into a multiple of biomedical procedures and pharmaceuticals (Sotiriou and Pratsinis, 2011). Reduction of $Ag+$ ions is an important step in AgNPs' synthesis and the reduction of $Ag+$ ions in fruit waste extract liquid has been revealed in various studies. For example, Ahmad and Sharma (2012) showed the formation of spherical Ag nanoparticles ranging in size from 5 to 35 nm, using pineapple peel extract. Extracts of *C. sinensis* (Orange) peel were used to synthesize spherical-shaped crystalline Ag nanoparticles ranging in size from 3 to 12 (Konwarha et al., 2011). Basavegowda et al. (2013) produced spherical Ag nanoparticles ranging in size from 5 to 20 nm using an extract of *Citrus unshiu* (Mandarin) peel. *Tanacetum vulgare*, often known as tansy fruit, is good source of Ag and Au nanoparticles. Ag nanoparticles usually are spherical with an average size of 16 nm, whereas Au nanoparticles with a triangular plate shape has an average size of about 11 nm. AgNPs are synthesised in two steps:

Step 1: $Ag+$ ions are reduced to Ag°
Step 2: Clustering of colloidal AgNPs to form oligomeric clusters which finally get stabilized (Kumar et al., 2020a,b).

The reduction of $Ag+$ ions requires biological catalysts, that is, enzymes, obtained from different biological sources like microbes, fruits extracts, plants, etc. Moreover, a variety of fruit extracts have already been known for synthesizing AgNPs with diverse biological potential (Kumar et al., 2020a,b).

24.6.1.2 Gold nanoparticles

In recent times, gold nanoparticles (AuNPs) have gained significant consideration due to their biocompatibility, optical, and physical (shape and size) properties. AuNPs of diverse morphology and size are extensively employed in medicine for different purposes such as for the detection of tumors, as a drug carrier, etc. (Kumar et al., 2020a,b).

Gold (Au) nanoparticles has wide range of applications in fields of catalysis, biomedicine, biosensors, pharmaceuticals, imaging, and diagnostics (Ghosh et al., 2019). Krishnaswamy et al. (2014) stated the creation of Au nanoparticles with size ranging from 20 to 25 nm having spherical shape from grape wastes. Ghodake et al. (2010) synthesized triangular and hexagonal crystalline gold nanoparticles with size range from 200 to 500 nm using *Pyrus* sp. (pear) extract. The Au nanoparticles of size ranging from 6.03 ± 2.77 nm to 18.01 ± 3.67 nm were also reported to biosynthesis from *Mangifera indica* (mango) peel extract (Yang et al., 2014). Several studies reveal that the category and number of biomolecules existing in fruit peel extract can influence nanoparticle formation and their subsequent stability.

24.6.2 Semiconductor nanoparticles

24.6.2.1 Copper oxide nanoparticles

Copper oxide (Cu_2O) is one of the best transition metal oxides having narrow bandgap, that is, ~ 2.0 eV and shows distinct features like significant electrochemical activity, improved redox potential, high specific surface area, and incomparable stability in solutions. Owing to its promising application in the field of antifouling coatings, biocidal agents, catalysis, sensors/biosensors, electrochemistry, and energy storage, researchers working in the field of nanotechnology choose these nanomaterials after the noble metal nanoparticles (Bhardwaj et al., 2017). These nanoparticles are broadly used for nonenzymatic sensing of clinical analytes due to their ability to promote electron transfer reaction even at low potential (Verma and Kumar, 2019; Kumar et al., 2020a,b).

24.6.2.2 Zinc oxide nanoparticles

ZnONPs have fascinated scientists working in the field of nanotechnology due to their varied application in areas such as biomedical, electronics, and optical sector. Synthesis of ZnONPs is considered to be easy, cost-effective, and benign. ZnONPs have been primarily known for antiinflammatory properties and wound healing in the medical sector. Nowadays, ZnONPs are mainly used in cosmetic products like sunscreen lotions, as they reveal intrinsic UV filtering potential. Other than this, ZnONPs are also used in drug delivery system as they are known for their anticancerous, antifungal, antimicrobial, and antidiabetic properties (Agarwal, et al., 2017; Kumar et al., 2020a,b).

24.6.2.3 Titanium dioxide nanoparticles

Titanium dioxide (TiO_2) is a white color metal oxide, which has advanced properties like hydrophobic, nonwet ability, high surface-to-volume ratio, and large band gap. Hence, it has various applications such as in self-cleaning devices, dye sensitized solar cell, photocatalysis, electrochemistry, antibacterial products, and textiles (Ibrahim et al., 2013;

Senic et al., 2011; Saowaluk et al., 2011). Rao et al. (2015a,b) synthesized TiO_2 nanoparticles by using orange (*C. sinensis)* peel extract via green synthesis method. Orange peel acts as a reducing agent for synthesis of TiO_2 because it contains citric acid as a main source (Torrado et al., 2011).

24.6.3 Other types of nanoparticles

Recently, a few studies have revealed that fruit waste is used for the formation of several other types of nanoparticles (Ghosh et al., 2019). For example, palladium (Pd) nanoparticles were produced using liquid extract of banana peel. The nanoparticles formed were crystalline and irregular in shape with average size of 50 nm (Bankar et al., 2010). Lakshmipathy et al. (2015) also synthesized Pd nanoparticles with a mean particle size of 96 nm by using watermelon rind extract. Other types of metal oxide nanoparticles produced are magnesium oxide (MgO) from *C. sinensis* (orange) peels and manganese (II, III) oxide (Mn_3O_4) from banana peels extract.

Rao et al. (2015a,b) efficiently produced spherical MgO nanoparticles with average particle size of 29 nm from *C. sinensis* (orange) peel extract. Banana peel extract was explored by Yan et al. (2014) to biofabricate spherical-shaped Mn_3O_4 nanoparticles ranging in size from 20 to 50 nm. Their study also reported that synthesized nanoparticles have highly capacitive properties and their potential usage in highly stable electrodes based on Mn_3O_4. Orange residues generally have no marketable worth but the residue consists of soluble, insoluble carbohydrates, pectin, essential oils, cellulose, and hemicellulose which are the basis of various industries (Awan et al., 2013). The orange nanocellulose fiber's production from these residues thus not only eases costs but also plays a remarkable role in the development of qualitative and sustainable textiles (Sumera et al., 2020).

24.7 Biosynthesis methods

Green synthesis of nanoparticles mediated by fruit extracts has been comprehended to contain a high amount of reducing agents (Kumar et al., 2020a,b). For example, fruits like blueberries, blackberries, *Cornus mas* L., *Citrullus lanatus*, grape, *Terminalia arjuna*, and *Punica granatum* L. comprise a high number of anthocyanins, ascorbic acid, phenolic compounds, flavonoids, saccharides, and other vitamins (Timoszyk, 2018). The synthesis of NPs from fruits has an overbiogenic methodology of synthesis, because the biogenic procedure for NP synthesis is facilitated by microbes, and microbial culture should be axenic and must be kept in a germ-free environment. Moreover, the isolation of nanoparticles from microbial culture during the processing of extraction process is relatively complicated. Furthermore, transformation of metallic salts (soluble) into the elemental oxide/elemental NPs is time consuming (Kumar et al., 2020a,b).

A general scheme of biosynthesis of varied nanoparticles using fruit extracts has been exemplified in Fig. 24.2.

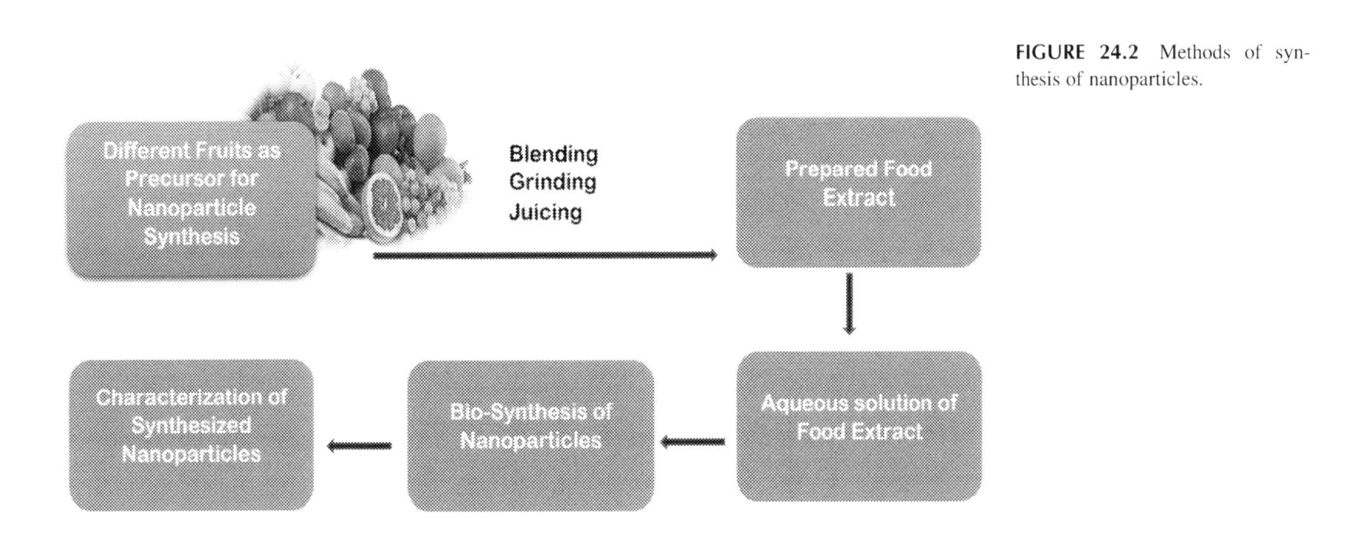

FIGURE 24.2 Methods of synthesis of nanoparticles.

24.7.1 Collection of fruit wastes

The peels of fruits like banana, orange, citrus, lemon, dragon fruit, and pomegranate and other fruits can be collected or purchased from a local fruit market. Their leaves, sticks, and other unwanted fragments were then removed. The peels should thoroughly wash with tap water followed by washing with double-distilled water to remove any physical impurity. All chemicals should be of analytical grade and all solutions should be prepared with deionized water.

24.7.2 Preparation of fruit peel extract and synthesis of nanoparticles

Fruit peel extract can be obtained at various stages of preparation.

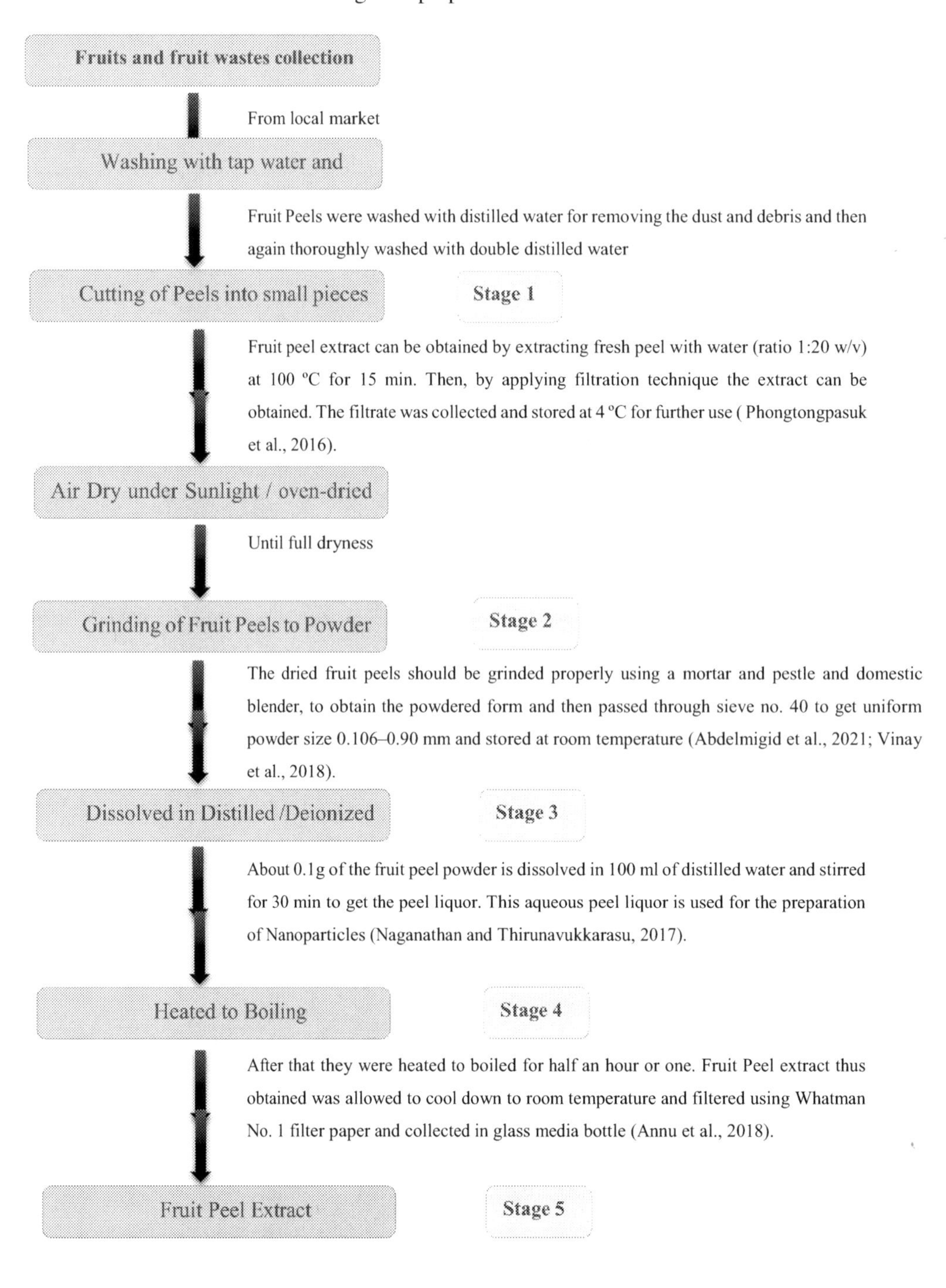

24.7.3 There are various protocols adopted by different researchers for nanoparticles' synthesis

1. A mixture of silver nitrate aqueous solution ($AgNO_3$, 10−3 M, Alfa Aeser, UK) mixed with fruit extract (stage 2) is prepared in the ratio 9:1. Mixture is then kept at 80°C with continuous stirring until there is change in color of extract from pale yellow (PPE) to dark brown, indicating formation of Ag nanoparticles (AgNP). The reaction mixtures are centrifuged at 15,000 rpm for10 minutes to get Ag nanoparticles. The supernatant is discarded, the pellet is washed with sterile distilled water (3x), and centrifuged for removal of any adsorbed substances on nanoparticles' surfaces. Biofabricated silver nanoparticles (AgNPs) are then oven dried (60°C) overnight in glass petri dishes and scraped out (Abdelmigid et al., 2021).
2. A reaction mixture contained 220 mL of dragon fruit extract (stage 1) in 110 mL of 10 mM silver nitrate ($AgNO_3$) solution. The solution was stirred for 24 hours on a magnetic stirrer at room temperature. Silver nanopaticles were separated from the reaction mixture by centrifugation at 12,000 rpm for 15 minutes at 4°C. The pellet was redispersed in deionized water and centrifuged again at the same condition as mentioned above to gain the purity of nanosilver. Later, lyophilization was performed to obtain silver nanoparticle powder (Stage 5) (Phongtongpasuk et al., 2016).
3. Powdered sample (100 gm) of fruit peel (stage 2) was extracted with 800 mL ethyl alcohol by using Soxhlet extraction method that takes about 6 hours. The mixture is then filtered through a Whatman filter paper (No. 2) to remove peel particles. The extracts were filtered and evaporated to dryness under reduced pressure at 60°C in a rotary evaporator (Buchi, Singapore). The extracts were kept in dark bottles and stored in cold conditions at 4°C (Vinay et al., 2018). To a 50 mL of freshly prepared 0.001 M silver nitrate solution, 5 mL of 0.002 M sodium borohydride solution was added with continuous stirring and kept aside for 15 minutes in a clean 250 mL beaker till a clear and slightly dark solution is obtained. The clear solution is then heated and maintained in water bath at 45°C for 30 minutes (solution A). Pomegranate peel extract in different concentrations that is, 200 mg (P20) and 500 mg (P50) was dissolved in 4 mL of Milli Q water separately in test tube and heated slightly to get yellow brownish-colored solution. This solution was added dropwise into the solution A with continuous stirring using glass rod for 30 minutes−45 minutes till clear tea brown solution is obtained. The obtained solution is cooled to room temperature and 0.5 mL of 0.5 mcg/mL PVP solution (polyvinylpyrrolidone) was added as a stabilizer and filtered to get clear tea brown-colored solution of silver nanoparticles. This solution was stored in a dark place in an air tight container until further use. The same procedure is repeated for the orange peel extract to get clear orange color solution of silver nanoparticles.
4. Bioreduction of AgNPs-silver nitrate is used as a source of silver and its aqueous solution is prepared by taking 250 mL 4 mM silver nitrate in an Erlenmeyer flask. 10 mL of this prepared silver nitrate was added to different concentrations of FWE (Stage 4) in separate beakers further placed in dark chamber in order to minimize the photoreduction of silver nitrate solution at room temperature. In place of varying concentration of silver nitrate, the concentration of FWEs was varied and change in color of solution was observed from colorless to yellowish to reddish brown, which is used as an advantageous candidate for bioreduction of $Ag+$ to $Ag0$ (Annu et al., 2018).
5. Aqueous peel liquor (Stage 3) is used for the preparation of SNP for which 3 mL of fruit peel liquor is supplemented to 47 mL of (0.1 mM) silver nitrate and mixed well to observe the change in color of the solution which implies that the reduction was completed within a short period (2 minutes) at room temperature. The appearance of yellowish-brown color confirms the formation of SNP (Naganathan and Thirunavukkarasu, 2017).
6. AgNPs, AuNPs, and Ag−AuNPs were synthesized using (avocado) fruit peel extract (Stage 3) which was kept in air tight container and placed in a dark cupboard for 24 hours. The extract was filtered by using Whatman No. 1 filter paper and was centrifuged at 4000 rpm for 20 minutes. The supernatant collected was used without further purification. 1 mL of the extract was added to a reaction vessel containing 24 mL of 1 mM silver nitrate ($AgNO_3$) and chloroauric acid ($HAuCl_4$) solution to reduce the silver and gold ions, while that of alloy was prepared by the mixture of 1 mM $AgNO_3$ and 1 mM $HAuCl_4$ in ratio 4:1 (v/v), respectively. The reaction was carried out under static condition at room temperature (30°C ± 2°C) for 2 hours. All the materials were held at ambient condition (30°C ± 2°C) to observe for change in color as a result of formation of nanoparticles (Adebayo et al., 2019).
7. Grape pomace (GP) obtained was oven dried at 550°C till the constant weight was achieved and pulverized into powdered form. For the organic solvent extraction of phytochemicals, the GP powder was immersed in pure ethanol in 1:20 proportion. The reaction mixture was kept for 24 hours under dark condition by continuous stirring (200 rpm). The reaction suspension was then centrifuged at 5000 × g for 15 minutes, and the supernatant was concentrated in the rotary evaporator. The lyophilized violet-brown hygroscopic powder was stored in dark. Silver nitrate (1 mM) solution and aqueous GPE extract were mixed together in a ratio of 20:1 in 500 mL of Erlenmeyer

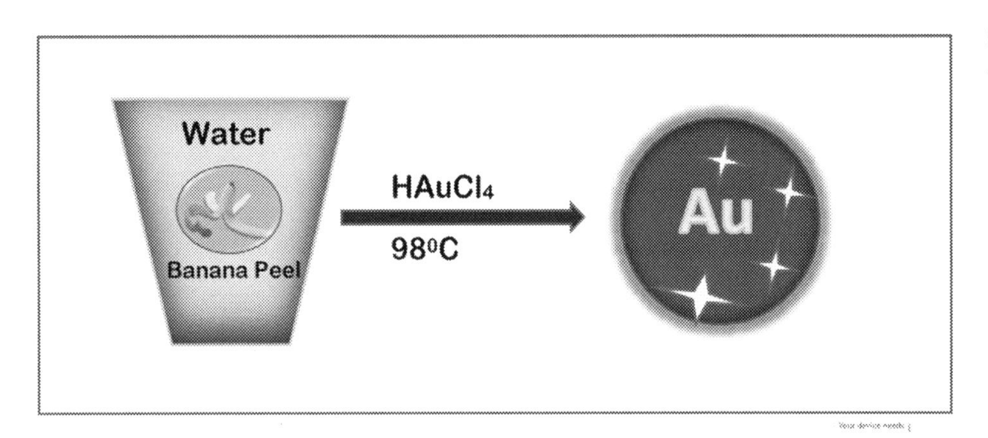

FIGURE 24.3 Synthesis route of Au-dendrite.

conical flask. The reaction suspension was homogeneously mixed and incubated at 500°C for 20 minutes for the biosynthesis of AgNPs. The successful synthesis of GPE-AgNPs was indicated by a change in color from light yellow to dark brown in 20 minutes. The colored reaction mixture obtained after the process was centrifuged for 20 minutes at 12,000 rpm to remove the unreacted reductants (Labogene, 1736 R, Lillerød, Denmark). The AgNP was washed twice with Milli-Q water and the obtained pellet of the AgNPs was dried in an oven at 600°C and stored in sealed vial at 4°C.

8. Synthesis of Au-dendrite (Fig. 24.3): One hundred grams of (banana) peel was taken and pulverized. 100 mL of deionized water was added to it and heated at 98°C for 30 minutes with continuous stirring. The heated mixture was then filtered. After filtration, acetone of the same volume was added to the filtrate and centrifuged to obtain concentrate. Supernatant was discarded and pellet was then dried at 70°C for 12 hours. Successively, 120 mg of the dried powder was mixed with 1 mL chloroauric acid (0.02 M) and 23 mL deionized water. The mixed liquid was then heated for 3 minutes at 98°C to obtain Au-dendrite (Liu et al., 2020).
9. Synthesis of Cu nanoparticles—CuNPs were synthesized using a biological agent, that is, the peel extract of pomegranate as a catalyst. The pomegranate peel extract was prepared in methanol using Soxhlet apparatus according to the method as described by Hegazy and Ibrahium (2012). For this, peels were air-dried in a vacuum oven at 40°C for 48 hours and grounded to fine powder; 100 g powdered sample was extracted with 1000 mL methanol at room temperature by the Soxhlet extraction method for 8 hours. 0.05 M $CuSO_4$ (50 mL) aqueous solution was added to 50 mL peel extract dropwise with continuous stirring. After heating at 80°C for 10 minutes, it was continuously stirred for 4 hours at 40°C. The solution was centrifuged at 10,000 rpm, at 4°C for 30 minutes and the pellet was dissolved in distilled water for periodic probe sonication for 5 seconds for 5 minutes at 30°C ± 0.5°C. Nanosuspension thus obtained was dried in oven at 70°C for 24 hours to obtain nanoparticles in powder form for further experiments (Kaur et al., 2016).
10. Synthesis of semiconductor nanoparticles -Orange peel was collected from the left over material of eaten orange fruit in the lab, making it into small pieces. A 50 g of orange peel was directly taken into the beaker and extracted with 150 mL of water for 2 hours at 90°C. The extract was filtered using Whatman Filter Paper. The filtrate is stored for further synthesis of nanoparticles (Balashanmugam, et al., 2013). Dissolve 1.5 N of titanium tetra isopropoxide in 100 mL of distilled water for synthesis of the TiO_2 nanoparticles. To the reaction mixture with constant stirring, extract is added dropwise up to achieve pH 7. The mixture is then subjected to continuous constant stirring for 3 hours at room temperature. In this process, formation of nanoparticles occurs which is separated using Whatman Filter Paper and washed with distilled water repeatedly to remove the byproducts. The obtained wet nanoparticles were dried at 80°C for overnight (Rao et al., 2015a,b).

24.8 Characterization of nanomaterials synthesized from fruit wastes

The nanoparticles of silver, copper, gold, titanium oxide, and zinc oxide can be characterized by using the techniques like UV-visible spectroscopy, transmission electron microscopy (TEM), X-ray diffraction (XRD) analysis, scanning electron microscopy (SEM), etc.

24.8.1 UV-Vis spectroscopy

Nanomaterial product samples can be subjected to UV–Visible spectroscopic studies for the detection of surface plasmon resonance (SPR) property (Kaur et al.,2016). The measurement of absorption spectra of the synthesized nanoparticles (Annu et al., 2018) is done in aqueous sample, before drying, using a UV-vis-NIR spectrophotometer (UV-1601, Shimadzu, UV-2450, Shimadzu, Japan), at room temperature at wavelength range from 200 to 800 nm. Autoclaved distilled water is taken for reference in measurement (Abdelmigid et al., 2021). The particular liquid fruit peel extracts were taken as blank in the same cuvette throughout the experiment with 1 nm resolution from 350 to 700 nm. For example, for the surface plasmon resonance property of the grape peel extract-AgNPs, the absorbance spectrum is measured in the range of 300–650 nm at fixed time intervals by UV–Visible spectrophotometer, Optizen, Model-2120 (Saratale et al., 2020).

8.1.2.Reduction of nanoparticles is an important step in synthesis process. For the green synthesis NPs, a 20 mg of the respective lyophilized extract (extract from 5th stage of Synthesis) was added to 10 mL aqueous solution of silver nitrate (0.1 M). The reaction mixture was incubated at 90°C for 20 minutes and the reduction of $Ag+$ ions was monitored (Sotoa et al., 2019) by UV–Vis spectrophotometry (Spectra Max Tunable Microplate Reader, Molecular Devices Co., Sunnyvale, CA, USA). Samples of 250 μL were taken at 5 minutes, 10 minutes, 15 minutes, and 20 minutes of reaction and the UV–Vis spectra were recorded in the wavelength range of 200–800 nm.

24.8.2 Spectroscatter

It is used for characterising the size distribution of particles suspended in the solvent. The size of synthesized NPs can be analyzed by DLS with the help of instrument DLS Spectroscatter 201 (Annu et al., 2018).

24.8.3 The X-ray diffractometer

The X-ray pattern of the NPs is recorded by an X-ray diffractomer (Sotoa et al., 2019) (Dmax 2100 Rigaku Americas, Texas, USA) that has a K αCu radiation generator ($k = 1.5418$ A°). Pattern of XRD is plotted with help of software Origin Lab and compared with JCPDS card no. 040783 (Abdelmigid et al., 2021).

The X-ray diffractometer (e.g., XRD, Pan Analytical, X-pert pro, Netherland) study is also used to check the surface crystalline shape of the biosynthesized nanoparticles. XRD was used at 30 kV and 100 mA, and the spectrum was recorded by CuKα radiation with a wavelength of 1.5406 Å in the 2θ (from the range of 20 degrees–80 degrees). Crystallinity and phase of the NPs were characterized by X-ray diffractometer (Rigaku, Tokyo, Japan) (Saratale et al., 2020).

24.8.4 Fourier-transform infrared spectroscopy

Fourier-transform infrared spectroscopy (FTIR) spectroscopy (Agilent Technologies, Santa Clara, CA, USA) is used to find out the possible functional groups present in fruit peel extract and sodium citrate (at wavelength range between 450 and 4000 cm^{-1}) that are accountable to nanoparticle synthesis and stabilization (Abdelmigid et al., 2021). With FTIR spectroscopy, the interactions of peel extract and NPs were analyzed. For FTIR measurements, sample is finely grounded with KBr and the pellet is prepared which is further placed in the analyzer-FTIR spectrophotometer (Affinity-FTIR, Shimadzu, Japan) in the range of 400–4000 cm^{-1}. Fourier-transform infrared (FTIR) spectra of the GPE and synthesized NPs is recorded by taking spectra in the range of 400–4000 cm^{-1} on Fourier-transform infrared spectrophotometer (Perkin-Elmer, Norwalk, CT, USA) to identify the organic functional groups present on the synthesized NPs (Saratale et al., 2020).

24.8.5 Microscope

For SEM, dried and powdered nanoparticles are coated with gold using a Chessington Sputter Coater (108 Auto, thickness controller MTM-10, UK) for 10 minutes before scanning. To check surface morphology of NPs, scanning electron microscope at 20 kV can be used with different magnifications of ×500, ×2000, and ×3000 with scale bars of 50, 10, and 5 μm, respectively (Abdelmigid et al., 2021).

Morphological studies of the synthesized NPs are discovered by transmission electron microscope (Hitachi FEI Philips Morgagni 268D TEM, Japan) at an accelerating voltage of 100 kV. For TEM, the specimen should be suspended in distilled water, dispersed ultrasonically to separate individual particles, and one or two drops of the suspension should be deposited onto carbon-coated copper grids and air-dried well, before mounting and observing under TEM. The surface morphology, size, imaging, and exterior arrangement of synthesized NPs is interpreted by high-resolution TEM (HRTEM, Tecnai G2 20 S-TWIN, FEI Company, Loughborough, UK). The particle size and shape of NPs were evaluated by measuring the median size of about 100 diverse NPs acquired in TEM image (Annu et al., 2018; Saratale et al., 2020). All measurements should be performed at 25°C.

24.8.6 Size distribution and zeta potential

The size and zeta potential of NPs can be measured using a particle size analyzer (Zeta sizer ZS-90, Malvern, UK) at 25°C with a detection angle 90 degrees. In the procedure, to prevent accumulations of nanoparticles, the powdered samples were freshly resuspended in 0.9% saline solution and sonicated for 30 minutes at high speed before assessment (Abdelmigid et al., 2021). The diameters of the NPs are obtained by measuring at least 50 particles with Image J software. Infrared absorption spectra of dried NPs and lyophilized extracts are recorded by DRIFT (Spectrum GX spectrophotometer, Perkin Elmer, Massachusetts, USA with a diffuse reflectance accessory, Pike Technology model) (Sotoa et al., 2019).

24.8.7 Biomedical assays of nanomaterials

24.8.7.1 Antimicrobial activities

24.8.7.1.1 Antibacterial activities

1. The antibacterial properties of the synthesized nanoparticles are investigated using the liquid culture method. Clinical isolates of *Escherichia coli* (stool), *E. coli* (ATCC 25922), *Bacillus subtilis* (ATCC 6633), *Klebsiella pneumoniae* (Urine), *Listeria monocytogenes* (ATCC 19111), *Proteus vulgaris* (Abdomen), *Pseudomonas aeruginosa* (ATCC 27853), *Staphylococcus aureus* (ATCC 25923), *S. aureus* (Ear), and *Streptococcus pyogenes* (Sputum) can be used as test organisms. Peptone water is used for the growth of bacterial isolates by inoculating with 1 mL of 1 × 106 CFU/mL of test organism. The NPs (1 mL each of gradient concentrations from 20 to 80 μg/mL) is added into test tubes containing 8 mL of peptone water. An aliquot of 1 mL from the test bacterial suspensions is separately inoculated into each tube and incubated at 37°C for 24 hours. Three replicates were maintained for each test and the control experiment contained 1 mL of each bacterial suspension without adding the nanoparticles. The proliferation of the bacterial isolates was measured as optical density (OD) at 600 nm using UV–visible spectrophotometer (Cecil, USA) (Adebayo et al.,2019).
2. The antimicrobial activity of fruit waste aqueous extract (100 μg/mL) as well as chloramphenicol (500 μg/mL) was included as a reference. Growth inhibition studies were conducted by measuring the optical density (OD600) with a Bioscreen C Microbiological Growth Analyzer (Labsystems). In a 100-well microtitre plate containing 30 μL of bacterial cultures (≈ 109 CFU/mL) in TSB, different volumes (20, 40, 60, 80, and 100 μL) of the freshly prepared NPs were added and completed with TSB to a total volume of 250 μL. Then the OD600 was recorded every 30 minutes over 48 hours at 35°C. Growth data was entered to the Baranyi model using DMFit v3.0 (ComBase; http://www.combase.cc/tools/) to estimate the lag phase (h), maximum population density (ODmax), and growth rate (μ = OD/h) (Sotoa et al., 2019).

   The percentage of growth inhibition was estimated by using the formula:

   % Bacterial Growth Inhibition = Control OD-Test OD/Control OD x 100%
3. The antibacterial property of the biosynthesized nanoparticles is also evaluated by well diffusion method. Plates of the Mueller-Hinton agar and Mackonkey agar medium (HiMedia) (Annu et al., 2018) are used to inoculate different pathogenic bacterial strains, for example, *Enterobacter aerogenes*, *K. pneumoniae (K. pneumoniae), P. aeruginosa*, and methicillin-resistant *S. aureus* (MRSA) and in tryptic soy broth *E. coli* (ATCC 25922) and *L. monocytogenes* (ATCC 35152) strains were cultivated (Sotoa et al., 2019). A total of 5–7 mm diameter wells are made by cut agar surface by sterile blue tips. For each plate, five wells were cut, one for each -Fruit peel extract, antibiotic, and three different concentrations of AgNPs. A total of 100 μL of fruit peel extract and 100 μL of the antibiotic ciprofloxacin 750 (1.6 mg/mL) can be added as a reference and positive control, respectively. Different concentrations viz. 2, 4, and 8 mg/mL of nanoparticles should be suspended in distilled water and properly sonicated. A total of 100 μL each concentration should be added per well. Plates were incubated at 37°C overnight. After incubation, diameters of zone of inhibition can be measured using a standard metric ruler (Abdelmigid et al., 2021). Inhibition zones were recorded in millimeters (mm). The antimicrobial activity of CuNPs was tested by the agar well diffusion method

against *M. luteus* MTCC 1809, *P. aeruginosa* MTCC 424, *Salmonella enteric* MTCC 1253, and *E. aerogenes* MTCC 2823 in vitro (Kaur et al., 2016).

4. Plates of nutrient agar plates are inoculated with bacterial cultures for example, Gram-negative (*E. coli*) and Gram-positive (*S. aureus*). The antibacterial activity of NPs was accomplished by determining the zone of inhibition. Sterile Whatman-1 filter paper discs (5 mm diameter) are used to load NP's solution (10, 15, and 20 μL) from stock solution (2 mg/mL). The plates were allowed to stand for 1 hour for the perfusion of the samples and were kept at 37°C for 24 hours for incubation. The zone of inhibition, in millimeter, surrounding the discs was measured. Deionized water is used as a negative control and fruit peel extract as a positive control. Antibacterial potency of NPs is compared with the standard check antibiotics such as ampicillin, tetracycline, kanamycin, vancomycin, and streptomycin (Saratale et al., 2020).

24.8.7.1.2 Antifungal activities

The antifungal activities were determined according to mycelial growth inhibition test. The prepared nanoparticles are dispersed in potato dextrose agar (PDA). Fungal culture discs of 6 mm of 48-h-old cultures of various fungus for example, *Aspergillus niger, Aspergillus fumigatus, Fusarium solani, Aspergillus favus*, and *Candida albicans* can be inoculated onto the surface of the PDA plates at the middle. The control plates lacked the nanoparticles. All the plates were incubated at 28°C ± 2°C for 72 hours. The radial fungal growths in all the plates are measured to calculate percentage growth inhibitions (Adebayo et al.,2019).

Percentage growth inhibition = D of F in control − D of F in test / D of F in control × 100%

where *D* is the diameter of fungal growth on the PDA plates.

24.8.7.2 Antioxidant activity

The antioxidant activity is determined using 2,2-diphenyl-1-picrylhydrazyl, DPPH (Sigma-Aldrich, USA). The reaction mixture consisted of 1 mL of graded concentrations, each NPs reacted with 4 mL of methanolic DPPH (0.1 mM) and kept in the dark at room temperature for 30 minutes, while 1 mL of methanol and 4 mL of methanolic DPPH are used as blank and the absorbance was read at the end of the experiment at 517 nm. The effective concentration (EC50) of NPs that decreases the initial concentration of the DPPH radical by 50% was obtained by interpolation from linear regression analysis. The procedure was repeated for standard antioxidant compounds, namely BHA and ascorbic acid (Annu et al., 2018; Adebayo et al., 2019).

24.9 Multifunctional application of nanoparticles produced by green methods

24.9.1 Applications in medicine

Nanoparticles have very useful applications in medicine (Ijaz et al., 2020 and OV Salata, 2004), namely:

- Used in fluorescent biological labeling
- In drugs and gene delivery
- In biological detection of pathogens
- Useful in finding/detection of proteins
- In probing of the DNA structure
- For tissue and cell engineering
- Used for tumor destruction through heating (hyperthermia)
- Helpful for separation and purification of biological cells and molecules
- In MRI improvement by contrast enhancement
- By using many kinds of biosensors that are based on nanoparticle's diagnosis of different diseases.
- In phagokinetic studies
- Cellular imaging

24.9.2 Tumor detection and treatment

In a study, it was mentioned that the CuO NPs nanoparticles that were created using black soya beans damaged the mitochondrial membrane in Hela cells. The effects of iron oxide nanoparticles produced by *Psoralea corylifolia* seeds aqueous extract on cancer cells were demonstrated using the sulforhodamine assay. Their study demonstrated the dose-dependent suppression of human breast cancer cells by iron oxide nanoparticles. Low amounts of substances can also suppress the growth of renal tumor cells in vitro condition.

24.9.3 Antimicrobial and antibiofilm agents

Recent findings of Ajlouni Abdul-Wali et al. (2022) revealed that green synthesized AP-AgNPs were efficient antimicrobial and antibiofilm agents against selected bacteria such as *E. coli, S. aureus, methicillin-resistant S. aureus*, multidrug-resistant *P. aeruginosa*, multidrug-resistant *A. baumannii*, and fungus *C. albicans*. Also, their results of the biofilm inhibition experiments showed that AP-AgNPs more potently prevent the development of biofilms in Gram-negative bacteria compared to Gram-positive bacteria and the fungus *C. albicans*. The plant extract from *A. pseudocotula Boiss*. hence has a great potential for producing various metallic nanoparticles as well as other valuable nanomaterials.

24.9.4 Use as a catalyst in chemistry

Nickel, lead, silver, and platinum NPs have been employed as unique metal catalysts in chemical processes. The dissociative adsorption of hydrogen and oxygen molecules on gold's surface below 200°C is still not possible, as gold nanoparticles are not reactive during these reactions. Gold material is utilized as a catalyst in hydrogenation and oxidation reactions. Haruta (2003) however, found that gold nanoparticles function well as catalysts. This condition is seen in cluster formations and Au-NPs particles can be employed as catalysts because gold clusters are extremely stable. The catalytic activity of catalyzed oxidation reactions increases as the size of the gold nanoparticle decreases.

24.9.5 Use of nanoparticles in agriculture and food industries

Nanotechnology has found its way in the field of food sector. New functional materials and devices are designed employing nanotechnology for food preservation and biosecurity. Bayer Corporation created impermeable plastic packaging using nanotechnology wherein food can be preserved. Genetic modifications to the agricultural plant's structure are also possible.

24.9.6 Bioremediation and waste water treatment

Ying et al. (2022) reported the use of metal nanoparticles to remove contaminants because metal NPs have a large surface area, strong adsorption, and high reducibility. The bactericidal efficacy of silver NPs prepared using Oyster Mushroom Extract and tested against *E. coli* confirmed the lowest 50 μg/L concentration of silver NPs as effectively bactericidal for the purposes of potential water disinfection, killing of bacteria, disinfection of medical equipment, wound washings, preservation of food stuffs, and in hand sanitization. The approach of green synthesis of silver NPs is fast, simple, environmentally friendly, and economically viable as suggested by Bhardwaj and Naraian (2020). The effectiveness of NZVI produced by *Spinacia oleracea* (spinach) to remove COD and BOD was 73.82% and 60.31%, respectively, after 15 days (Turakhia et al., 2018). The chromium was broken down by the NZVIs created by mandarin, lemon, lime, and orange, with degradation efficiencies ranging from 12% to 37%. (Machado et al., 2014). Weng et al., 2016 produced Fe-NPs from eucalyptus leaf extracts and utilized them to remove mixed Cu and Cr (VI) with respective efficiencies of 33.0% and 58.9%. The removal efficiency was 45.2% and 74.2%, respectively, when Cu and Cr (VI) were treated separately. Obviously, combined metal ions have worse removal efficiency. Not all metal nanoparticles created by green synthesis, meanwhile, can remove environmental toxins. Concerns are raised by these poor removal rates for NZVIs, particularly regarding the wasteful use or waste of iron. Nanoscale metals produced by green synthesis are ineffective in removing hazardous metal mixtures. Practically, heavy metals and other pollutants are combined with other contaminants in sewage, and rarely does a waste solely contain one heavy metal. Thus, efficiency must be increased for this kind of practical application to use greenly manufactured nanoscale materials. However, there is currently little knowledge regarding the uses of nanoscale materials produced through green synthesis.

24.9.7 Applications for generating energy

Due to nonrenewable nature of fossil fuels, they have certain limitations and will not be sufficient in the years to come. As a result, researchers are attempting to look for the materials that are conveniently accessible and inexpensive, and to produce green energy sources. The following qualities made NPs the ideal candidate, according to scientists, for these reasons: big expanse of surface, applications of photocatalysis, catalytic process, and optical performance. NPs are commonly employed to produce energy from the electrochemical water and photoelectrochemical splitting processes. Some more advanced options for producing energy include reduction, water splitting, electrochemical CO_2, and from fuels. Solar cells and piezoelectric generators are also among them. NPs can also store energy at the nanoscale level and can be utilized in energy storage applications. Nanogenerators have developed the capacity to transform mechanical energy into electrical energy by utilizing piezoelectricity, however it is an uncommon method for energy production.

24.9.8 Use of nanoparticles in microwiring

The ideal choice in the electronics sector for producing a printed wiring board is a paste of metal nanoparticles. The melting point of metal nanoparticles is lower than that of bulk metals. On a polymer base material, circuit creation is achievable using traditional electric conduction paste. The thickness of wire is reduced to a nano level if nanoscale particles are employed. The ink-jet technique is crucial for the creation of wiring at the nanoscale. The ink-jet process is less expensive and less time consuming than commonly used conventional procedures including photolithographic technologies and vacuum evaporation. Gold can be used to create a paste of metal nanoparticles. Nanoparticles in electronics applications: Since a few years ago, there has been an increase in interest in printed electronics because of their potential to be more affordable than current silicon printing methods. Electronics printed with various inks are anticipated to flow quickly. Furthermore, CNTs, organic, and ceramic nanoparticles may be present in these inks. One-dimensional semiconductors and metals offer unique structural, electrical, and optical capabilities that serve as the fundamental building blocks for the creation of electronic, photonic, and sensor technologies. Today, new semiconducting materials are being found gradually in the electrical sector. But vacuum tubes have been replaced by diodes, transistors, and even tiny chips.

24.10 Future aspects

The following suppositions and assumptions are presented to get beyond the restrictions and flaws in existing research, and ideally put environmentally friendly nanoscale metals to use practically.

24.10.1 Ideal raw materials

Researchers should first think about the seasonal and regional availability for resource collection before thinking about how straightforward the extraction method would be for nanoparticles' synthesis that could avoid seasonal time constraints and is potentially a good alternative material (Pictures 24.1 and 24.2). *Filicium decipiens, Mediterranean cypress* (Cupressus sempervirens), *Murraya koenigii*, and *Stachys lavandulifolia, Urtica dioica*, Eucalyptus, and grape leaves are some plants; collections of these plant materials was not limited by time or season, so they have high practicability for green synthesis in terms of availability (Ying et al., 2022).

PICTURE 24.1 Fruits resources for nanoscience.

PICTURE 24.2 Fruit waste resources for nanomaterial synthesis.

24.10.2 Use of wastes

Agricultural waste, which complies with the waste usage concept for environmental preservation, is another appropriate material. For instance, *Eichhornia crassipes*, also known as water hyacinth, is an invasive weed species with an amazing rate of growth and reproduction that has led to significant ecological issues. Grape seed is a frequent byproduct of the wine industry that is utilized in green synthesis (Gao et al., 2016). Mandarin, orange, lime, and lemon peelings, as well as waste from pistachios, were all employed in the synthesis of nanoscale zero-valent iron (NZVI). These materials provide a viable solution to the demand for resource recycling (Machado et al., 2014; Soliemanzadeh et al., 2016). By using garbage, it will not only assist decrease waste but also prevent major ecological harm (Wei et al., 2017). Moreover, acquiring citron juice is less complicated than getting plant extracts (Shende et al., 2015). In order to make the material extraction process as easy as feasible, future research should also take the feasibility and simplicity of the procedure into account.

24.10.3 Product enhancement

Future studies should concentrate on creating consistently small-sized particles with a lot of surface area. For instance, *Rose canina* fruit extract's Pd NPs can be recycled seven times without significantly losing their activity (Veisi et al., 2016). The catalytic activity to produce 2-aryl benzimidazoles under room temperature did not substantially decrease for the Fe_3O_4 NPs produced by *Passiflora tripartita var. mollissima* fruit extract, making them recyclable for four times (Kumar et al., 2014). Another important aspect of nanoparticle quality is product stability. It is particularly noteworthy that the freeze-dried NZVI produced by grape seed extract may be kept directly under ambient settings without losing effectiveness, making it advantageous for practical manufacturing (Gao et al., 2016).

24.10.4 Limit energy consumption

In order to limit the use of energy, scientists need to pay attention to nonheating procedures to address the issue of excessive energy usage in various synthesis processes. Based on the current energy-saving trials, for instance, using hazardous chemical reagents, high reaction energy, or high reaction temperature could be avoided to manufacture Fe NPs when employing leaf extracts from *Hippophae rhamnoides*, Gardenia, and Henna (Naseem and Farrukh, 2015; Nasrollahzadeh and Mohammad Sajadi, 2015). Ping et al. (2018) synthesized Ag nanoparticles from grape seed extract at reaction temperatures below 100°C, and *Genipa americana* fruit extract-produced Au nanoparticles at temperatures between 22°C and 25°C are examples of low-energy extraction techniques. There are already several low-energy extraction techniques in use, such as grape seed extract used to synthesize Ag nanoparticles at reaction temperatures below 100°C or *G. americana* fruit extract to synthesize Au nanoparticles at temperatures between 22°C and 25°C (Kumar et al., 2016a). The needed temperature was 80°C when the nettle and thyme leaf extracts were employed to create Fe NPs (Leili et al., 2018).

24.10.5 Preservation

Metals at the nanoscale need to be stored as well. The cost of storage or preservation will significantly drop if ambient conditions can be maintained for nanoscale metals. The stability of metals at the nanoscale is correlated with their storage. The more inexpensively they can be kept, the more stable would be the NPs. For example, the aqueous guava extract-based Cu NPs were stable at room temperature for 15 days. PEG 6000 was utilized as a capping agent to stabilize the metal colloid in storage for 15 days since Cu NPs can rapidly oxidize copper oxide (Caroling et al., 2015).

24.11 Conclusions

The concept of "green synthesis or biosynthesis" of nanomaterials is on demand for sustainable environment. Biosynthesis of nanomaterial is finding its way by exploiting wastes generated from fruit industries as well. The most important thing that should be taken in consideration is that the enormously huge extents of fruit wastes are generated worldwide. This offers striking and renewable source of biomolecules and bioactive compounds' production. Fruit peels contain valuable compounds such as bioactive chemicals, phenolics, antioxidants, enzymes, carboxylic acid, etc. Nanobiomaterial from fruit wastes may be considered to be an attractive product having biodegradable, removable, or biocompatible properties. Biologically made nanoparticles may have a varied range of biomedical applications as they are free from relatively harmful chemicals and solvents. Ag, Au, and Cu nanoparticles are known have a broad spectrum of antimicrobial properties, also nanomaterial synthesized from fruit waste such as Au-dendrites have other biomedical properties including anticancer and antitumor proliferation properties, therefore could be compatibly applied in several therapeutic procedures. TiO_2 nanomaterial finds its applications in self-cleaning devices, dye sensitized solar cell, photocatalysis, electrochemistry, antibacterial products, and textiles. Other types of nanoparticles produced from fruit peel extract such as Pd and magnesium oxide (MgO) nanoparticles could be successfully used for degrading several azo dyes and to remove arsenic metal ions from aqueous solutions respectively, thus aiding in biodegradation and bioremediation. At present, there are serious alarms about antibiotic resistance, complex synthesis methods, environmental pollution, and the high production costs associated with conventional method of nanomaterials preparations, thus there should be positive aspects for the biosynthesis of these very rich bionanomaterials.

In this chapter, the green synthesis of Au, Ag, Fe, Cu, and Pd NPs has been discussed. Research on various plant materials specially fruit wastes shows that it is possible to produce nanoscale metals in an ecologically benign manner. Recently, a few research on environmentally friendly metal nanoscale manufacturing have been published. The manufacturing and usage of green synthesized nanomaterials, however, are hampered by several challenges and limitations, such as low yield, irregular particle sizes, challenging extraction procedures, seasonal and regional raw material availability, and other problems that must be handled.

So, the future of research must concentrate on boosting the production of nanoscale metal particles, utilizing cheap raw materials, and putting simple energy-saving technologies into practise.

References

Abdelmigid, H.M., Morsi, M.M., Hussien, N.A., Alyamani, A.A., Al Sufyani, N.M., 2021. Comparative analysis of nanosilver particles synthesized by different approaches and their antimicrobial efficacy. Hindawi Journal of Nanomaterials Article Id 2204776, 12 Pages Https://Doi.Org/10.1155/2021/2204776.

Adebayo, A.E., Okel, A.M., Lateef, A., Oyatokun, A.A., Abisoye, O.D., Itunu, P.A., et al., 2019. Biosynthesis of silver, gold and silver–gold alloy nanoparticles using persea Americana fruit peel aqueous extract for their biomedical properties. Nanotechnology For Environmental Engineering 4, 13.

Agarwal, H., Kumar, S.V., Rajeshkumar, S., 2017. A review on green synthesis of zinc oxide nanoparticles -an eco-friendly approach. Resource-Efficient Technologies 3, 406–413.

Ahmad, N., Sharma, S., 2012. Green synthesis of silver nanoparticles using extracts of ananas comosus. Green and Sustainable Chemistry 2, 141–147.

Ajlouni A.-W., E.H. Hamdan, R.A.E. Alshalawi, M.R. Shaik, M. Khan, M. Kuniyil, et al., 2022. Green synthesis of silver nanoparticles using aerial part extract of the anthemis pseudocotula boiss. Plant and their biological activity. Molecules 2023, 28, 246. Available from: https://doi.org/10.3390/molecules28010246, https://www.mdpi.com/journal/molecules.

Annu, Shakeel, A., Gurpreet, K., Praveen, S., Sandeep, S., Saiqa, I., 2018. Fruit waste (peel) as bio-reductant to synthesize silver nanoparticles with antimicrobial, antioxidant and cytotoxic activities. Journal of Applied Biomedicine 16, 221–231.

Awan, A.R., Manfredo, A., Pleiss, J.A., 2013. Lariat sequencing in a unicellular yeast identifies regulated alternative splicing of exons that are evolutionarily conserved with humans. Proceedings of the National Academy of Sciences 110, 12762–12767.

Balashanmugam, P., Nandhini, R., Vijayapriyadharshini, V., Kalaichelvan, P.T., 2013. Biosynthesis of silver nanoparticles from orange peel extract and its antibacterial activity against fruit and vegetable pathogens. International Journal of Innovative Research in Science & Engineering 1 (2), 6.

Bankar, A., Joshi, B., Kumar, A.R., Zinjarde, S., 2010. Banana peeled extract mediated novel route for the synthesis of palladium nanoparticles. Materials Letters 64, 1951–1953.

Basavegowda, N., Rok Lee, Y., 2013. Synthesis of silver nanoparticles using Satsuma mandarin (Citrus unshiu) peel extract: a novel approach towards waste utilization. Materials Letters 109, 31–33.

Bhardwaj, A.K., Naraian, R., 2020. Green synthesis and characterization of silver NPs using oyster mushroom extract for antibacterial efficacy. Journal of Chemistry. Environmental Sciences and its Applications 7 (1), 13–18.

Bhardwaj, A., Shukla, A., Singh, S.C., Uttam, K.N., Nath, G., Gopal, R., 2017. Green synthesis of Cu_2O hollow microspheres. Advanced Materials Proceedings 2 (2), 132–138.

Boisseau, P., Loubaton, B., 2011. Nanomedicine, nanotechnology in medicine. Comptes Rendus Physique 12 (7), 620–636.

Caroling, G., P., M.N., Vinodhini, E., Mercy Ranjitham, A., Shanthi, P., 2015. Biosynthesis of copper nanoparticles using aqueous guava extract-characterization and study of antibacterial effects. International Journal of Pharmacy and Biological Sciences 5 (2), 25–43.

Gao, J.-F., Li, H.-Y., Pan, K.-L., Si, C.-Y., 2016. Green synthesis of nanoscale zero-valent iron using a grape seed extract as a stabilizing agent and the application for quick decolorization of azo and anthraquinone dyes. RSC Advances 6 (27), 22526–22537.

Ghodake, G., Deshpande, N., Lee, Y., Jin, E., 2010. Pear fruit extract-assisted room-temperature biosynthesis of gold nanoplates. Colloids and Surfaces B, Biointerfaces 75, 584–589.

Ghosh, P.R., Fawcett, D., Sharma, S.B., Poinern, G.E.J., 2019. Production of high-value nanoparticles via biogenic processes using aquacultural and horticultural food waste. Materials 10, 852. Available from: https://doi.org/10.3390/ma10080852.

Haruta, M., 2003. When gold is not noble: catalysis by nanoparticles. Chemical Record (New York, N.Y.) 3 (2), 75–87.

Hegazy, A.E., Ibrahium, M.I., 2012. Antioxidant activities of orange peel extract. World Applied Sciences Journal 18, 684–688.

Hett, A., 2004. Nanotechnology: small matter, many unknowns. Swiss Reinsurance Company, Zurich.

Holdren, Jp, 2011. The National Nanotechnology Initiative Strategic Plan Report at Subcommittee on Nanoscale Science, Engineering And Technology Of Committee On Technology. National Science Technology Council (Nstc), Arlington.

Ibrahim, D.M., Chandiran, A.K., Gratzel, M., Nazeeruddin, M.K., Shivashankar, S.A., 2013. Controlled Synthesis of TiO_2 Nanoparticles and Nanospheres Using a Microwave Assisted Approach for Their Application in Dye-sensitized Solar Cells, Electronic Supplementary Material. *The Royal Society of Chemistry*, p. 7.

Ijaz, I., Gilani, E., Nazir, A., Bukhari, A., 2020. Detail review on chemical, physical and green synthesis, classification, characterizations, and applications of nanoparticles. Green Chemistry Letters and Reviews 13 (3), 223–245.

Iravani, S., Korbekandi, H., Mirmohammadi, S.V., Zolfaghari, B., 2014. Synthesis of silver nanoparticles: chemical, physical and biological methods. Research in Pharmaceutical Sciences 9 (6), 385–406.

Kaur, P., Thakur, R., Chaudhury, A., 2016. Biogenesis of copper nanoparticles using peel extract of punica granatum and their antimicrobial activity against opportunistic pathogens. Green Chemistry Letters and Reviews 9 (1), 33–38. Available from: https://doi.org/10.1080/17518253.2016.1141238.

Konwarha, R., Gogoia, B., Philip, R., Laskarb, M.A., Karaka, N., 2011. Biomimetic preparation of polymersupported free radical scavenging, cytocompatible and antimicrobial ‘green’ silver nanoparticles using aqueous extract of Citrus sinensis peel. Colloids and Surfaces B, Biointerfaces 84, 338–345.

Krishnaswamy, K., Vali, H., Orsat, V., 2014. Value-adding to grape waste: green synthesis of gold nanoparticles. Journal of Food Engineering 142, 210–220.

Kumar, H., Bhardwaj, K., Dhanjal, D.S., Nepovimova, E., Sen, F., Regassa, H., et al., 2020. Review fruit extract mediated green synthesis of metallic nanoparticles: a new avenue in pomology applications. International Journal of Molecular Science 21, 8458. Available from: https://doi.org/10.3390/Ijms21228458.

Kumar, H., Bhardwaj, K., Kuča, K., Kalia, A., Nepovimova, E., Verma, R., et al., 2020a. Flower-based green synthesis of metallic nanoparticles: applications beyond fragrance. Nanomaterials 10, 766.

Kumar, H., Bhardwaj, K., Nepovimova, E., Kuča, K., Dhanjal, D.S., Bhardwaj, S., et al., 2020b. Antioxidant functionalized nanoparticles: a combat against oxidative stress. Nanomaterials 10, 1334.

Kumar, D.A., Palanichamy, V., Roopan, S.M., 2014a. Green synthesis of silver nanoparticles using Alternanthera dentata leaf extract at room temperature and their antimicrobial activity. Spectrochimica Acta Part A, Molecular and Biomolecular Spectroscopy 127, 168–171.

Kumar, B., Smita, K., Cumbal, L., Camacho, J., Hernández-Gallegos, E., Chávez-López, M.de G., et al., 2016a. One pot phytosynthesis of gold nanoparticles using Genipa americana fruit extract and its biological applications. Materials Science & Engineering C-Materials for Biological Applications 62, 725–731.

Kumar, B., Smita, K., Cumbal, L., Debut, A., 2014. Biogenic synthesis of iron oxide nanoparticles for 2-arylbenzimidazole fabrication. Journal of Saudi Chemical Society 18 (4), 364–369.

Lakshmipathy, R., Palakshi Reddy, B., Sarada, N.C., Chidambaram, K., Khadeer Pasha, S., 2015. Watermelon rind-mediated green synthesis of noble palladium nanoparticles: catalytic application. Applied Nanoscience 5, 223–228.

Leili, M., Fazlzadeh, M., Bhatnagar, A., 2018. Green synthesis of nano-zero-valent iron from Nettle and Thyme leaf extracts and their application for the removal of cephalexin antibiotic from aqueous solutions. Environmental Technology 39 (9), 1158–1172.

Liu, Y., Song, X., Cao, F., Li, F., Wang, M., Yang, Y., et al., 2020. Peel-derived dendrite-shaped Au nanomaterials with dual inhibition toward tumor growth and migration. International Journal of Nanomedicine 15, 2315–2322.

Machado, S., Pinto, S.L., Grosso, J.P., Nouws, H.P.A., Albergaria, J.T., Delerue-Matos, C., 2014. Utilization of food industry wastes for the production of zero-valent iron nanoparticles. The Science of the Total Environment 496, 233–240.

Madeeha, A., Kiran, S., Shakil, A., 2020. Nanotechnology: A Breakthrough in Agronomy. Nanotechnology And Plant Disease Diagnosis and Management. Springer Nature Switzerland AG ISBN 978-3-030–41275-3 (eBook), 1: 1–22 Doi: 10.1007/978-3-030–41275-3_7.

Mavani, H.A.K., Tew, I.M., Wong, L., Yew, H.Z., Mahyuddin, A., Ahmad Ghazali, R., et al., 2020. Antimicrobial efficacy of fruit peels eco-enzyme against enterococcus faecalis: an in vitro study. International Journal of Environmental Research and Public Health 17 (5107), 12. pages.

Naganathan, K., Thirunavukkarasu, S., 2017. Green way genesis of silver nanoparticles using multiple fruit peels waste and its antimicrobial, antioxidant and anti-tumor cell line studies. IOP Conf. Series: Materials Science and Engineering 191, 012009. Available from: https://doi.org/10.1088/1757-899X/191/1/012009.

Naseem, T., Farrukh, M.A., 2015. Antibacterial activity of green synthesis of iron nanoparticles UsingLawsonia inermisandgardenia jasminoidesleaves extract. Journal of Chemistry 1–7.

Nasrollahzadeh, M., Mohammad Sajadi, S., 2015. Green synthesis of copper nanoparticles using Ginkgo biloba L. leaf extract and their catalytic activity for the Huisgen [3 + 2] cycloaddition of azides and alkynes at room temperature. Journal of Colloid and Interface Science 457, 141–147.

Pandharipande, S., Makode, H., 2012. Separation of oil and pectin from orange peel and study of effect of Ph of extracting medium on the yield of pectin. Journal of Engineering Research and Studies 3 (2), 6–9.

Pathak, P.D., Mandavgane, S.A., Kulkarni, B.D., 2017. Fruit peel waste: characterization and its potential uses. Current Science 113, 1–11. Available from: https://doi.org/10.18520/Cs/V113/I03/444-454.

Phongtongpasuk, S., Poadang, S., Yongvanich, N., 2016. Environmental-friendly method for synthesis of silver nanoparticles from dragon fruit peel extract and their antibacterial activities. Energy Procedia 89, 239–247.

Ping, Y., Zhang, J., Xing, T., Chen, G., Tao, R., Choo, K.-H., 2018. Green synthesis of silver nanoparticles using grape seed extract and their application for reductive catalysis of direct orange 26. Journal of Industrial and Engineering Chemistry 58, 74–79.

Prathna, T.C., Chandrasekaran, N., Raichur, A.M., Mukherjee, A., 2011. Kinetic evolution studies of silver nanoparticles in a bio-based green synthesis process. Colloids and Surfaces. A, Physicochemical and Engineering Aspects 377 (1–3), 212–216.

Ramesh, M., Anbuvannan, M., Viruthagiri, G., 2015. Green synthesis of ZnO nanoparticles using Solanum nigrum leaf extract and their antibacterial activity. Spectrochimica Acta, Part A: Molecular and Biomolecular Spectroscopy 136, 864–870.

Rao, G.K., Ashok, C.H., Venkateswara Rao, K., Shilpa Chakra, C.H., Akshaykranth, A., 2015b. Eco- friendly synthesis of MgO nanoparticles from orange fruit waste. International Journal of Advanced Research in Physical Science 2, 1–6.

Rao, K.G., Ashok, Ch, Venkateswara Rao, K., Shilpa Chakra, Ch, Rajendar, V., 2015a. Synthesis of TiO_2 nanoparticles from orange fruit waste. International Journal of Multidisciplinary Advanced Research Trends 2 (1), 82–90.

Reddy, G.A.K., Joy, J.M., Mitra, T., Shabnam, S., Shilpa, T., 2012. Nano silver–a review. International Journal of Advances in Pharmacy 2 (1), 09–15.

Salata, O.V., 2004. Applications of nanoparticles in biology and medicine. Journal of Nanobiotechnology . Available from: http://www.jnanobiotechnology.com/content/2/1/3:1-6.

Samberg, M.E., Oldenburg, S.J., Monteiro-Riviere, N.A., 2009. Evaluation of silver nanoparticle toxicity in skin in vivo and keratinocytes in vitro. Environmental Health Perspectives 118 (3), 407–413.

Saowaluk, B., Sutthisripok, W., Sikong, L., 2011. Antibacterial activity of TiO_2 and Fe^{3+} doped TiO_2 nanoparticles synthesized at low temperature. Advanced Materials Research 214, 197–201.

Saratale, G.D., Saratale, Ri. J.G., Kim, D.-S., Kim, D.-Y., Shin, H.-S., 2020. Exploiting fruit waste grape pomace for silver nanoparticles synthesis, assessing their antioxidant, antidiabetic potential and antibacterial activity against human pathogens: *a novel approach*. Nanomaterials 10 (1457), 17. Available from: https://doi.org/10.3390/nano10081457.

Senic, Z., Bauk, S., Todorovic, M.V., Pajic, N., Samolov, A., Rajic, D., 2011. Application of TiO_2 nanoparticles for obtaining self-decontaminating smart textiles. Scientific Technical Review 61 (3–4), 63–72.

Shende, S., Ingle, A.P., Gade, A., Rai, M., 2015. Green synthesis of copper nanoparticles by Citrus medica Linn. (Idilimbu) juice and its antimicrobial activity. World Journal of Microbiology and Biotechnology 31 (6), 865–873.

Sintubin, L., De Gusseme, B., Van der Meeren, P., Pycke, B.F., Verstraete, W., Boon, N., 2011. The antibacterial activity of biogenic silver and its mode of action. Applied Microbiology and Biotechnology 91 (1), 153–162.

Soliemanzadeh, A., Fekri, M., Bakhtiary, S., Mehrizi, M.H., 2016. Biosynthesis of iron nanoparticles and their application in removing phosphorus from aqueous solutions. Chemical Ecology 32 (3), 286–300.

Sotiriou, G.A., Pratsinis, S.E., 2011. Engineering nanosilver as an antibacterial, biosensor and bioimaging material. Current Opinion in Chemical Engineering 1, 3–10.

Sotoa, K.M., Quezada-Cervantesa, C.T., Hernández-Iturriagaa, M., Luna-Bárcenasb, G., Vazquez-Duhaltc, R., Mendozaa, S., 2019. Fruit peels waste for the green synthesis of silver nanoparticles with antimicrobial activity against foodborne pathogens. LWT - Food Science and Technology 103, 293–300.

Sumera, J., Iqra, A., Shagufta N., 2020. Nanomaterials and agrowaste. Nanotechnology and plant disease diagnosis and management. *Springer Nature Switzerland* AG ISBN 978-3-030–41275-3 (eBook), 11: 197–206. Doi: 10.1007/978-3-030–41275-3_7.

Timoszyk, A., 2018. A review of the biological synthesis of gold nanoparticles using fruit extracts: scientific potential and application. Bulletin of Materials Science 41, 154.

Torrado, A.M., Cortés, S., Salgado, J.M., Max, B., Rodríguez, N., Bibbins, B.P., et al., 2011. Citric acid production from orange peel wastes by solid-state fermentation. Brazilian Journal of Microbiology 42, 394–409.

Tran, Q.H., Le, A.T., 2013. Silver nanoparticles: synthesis, properties, toxicology, applications and perspectives. Advances in Natural Sciences: Nanoscience and Nanotechnology 4 (3), 033001.

Turakhia, B., Turakhia, P., Shah, S., 2018. Green synthesis of zero valent iron nanoparticles from Spinacia oleracea (spinach) and its application in waste water treatment. Journal of Advanced Research in Applied Sciences 5 (1), 46–51.

Veisi, H., Rashtiani, A., Barjasteh, V., 2016. Biosynthesis of palladium nanoparticles using Rosa caninafruit extract and their use as a heterogeneous and recyclable catalyst for Suzuki-Miyaura coupling reactions in water. Applied Organometallic Chemistry 30 (4), 231–235.

Verma, N., Kumar, N., 2019. Synthesis and biomedical applications of copper oxide nanoparticles: an expanding horizon. ACS Biomaterials Science & Engineering 5, 1170–1188.

Vinay, C.H., Goudanavar, P., Acharya, A., 2018. Development and characterization of pomegranate and orange fruit peel extract-based silver nanoparticles. Journal of Manmohan Memorial Institute of Health Science 4 (1), 72–85. Available from: https://doi.org/10.3126/Jmmihs.V4i1.21146.

Wei, Y., Fang, Z., Zheng, L., Tsang, E.P., 2017. Biosynthesized iron nanoparticles in aqueous extracts of Eichhornia crassipes and its mechanism in the hexavalent chromium removal. Applied surface science 399, 322–329.

Weng, X., Jin, X., Lin, J., Naidu, R., Chen, Z., 2016. Removal of mixed contaminants Cr (VI) and Cu (II) by green synthesized iron-based nanoparticles. Ecological Engineering 97, 32–39.

Xiao, L., Liu, C., Chen, X., Yang, Z., 2016. Zinc oxide nanoparticles induce renal toxicity through reactive oxygen species. Food and Chemical Toxicology: An International Journal Published for the British Industrial Biological Research Association 90, 76–83.

Yan, D., Zhang, H., Chen, L., Zhu, G., Wang, Z., Xu, H., et al., 2014. Super-capacitive properties of Mn_3O_4 nanoparticles biosynthesized from banana peel extract. RSC Advances. 4, 23649–23652.

Ying, S., Guan, Z., Ofoegbu, P.C., Clubb, P., Rico, C., He, F., et al., 2022. Green synthesis of nanoparticles: current developments and limitations. Environmental Technology & Innovation 26, 102336.

Yuvakkumar, R., Suresh, J., Saravanakumar, B., Nathanael, A.J., Hong, S.I., Rajendran, V., 2015. Rambutan peels promoted biomimetic synthesis of bioinspired zinc oxide nanochains for biomedical applications. Spectrochimica Acta Part A, Molecular and Biomolecular Spectroscopy 137, 250–258.

Index

Note: Page numbers followed by "*f*" and "*t*" refer to figures and tables, respectively.

D

G

H

I

N

O

P

X

Y

Z

Printed in the United States
by Baker & Taylor Publisher Services